WEED BIOLOGY AND MANAGEMENT

Weed Biology and Management

Edited by

INDERJIT

Department of Botany,
University of Delhi

KLUWER ACADEMIC PUBLISHERS
DORDRECHT / BOSTON / LONDON

A C.I.P. Catalogue record for this book is available from the Library of Congress.

ISBN 1-4020-1761-8

Published by Kluwer Academic Publishers,
P.O. Box 17, 3300 AA Dordrecht, The Netherlands.

Sold and distributed in North, Central and South America
by Kluwer Academic Publishers,
101 Philip Drive, Norwell, MA 02061, U.S.A.

In all other countries, sold and distributed
by Kluwer Academic Publishers,
P.O. Box 322, 3300 AH Dordrecht, The Netherlands.

Printed on acid-free paper

Printed in the Netherlands.

Dedication

I feel honored to dedicate the book, "*Weed Biology and Management*" to Dr. Jamuna Sharan Singh, Professor of Botany at the Banaras Hindu University, Varanasi, India. Much of the present research on Himalayan and Dry Tropical Environments and Ecosystems can be attributed to Dr. J. S. Singh. Born on December 26, 1941, Dr. Singh earned his Ph.D. degree from the Banaras Hindu University. In 1984, Dr Singh joined Banaras Hindu University as Professor of Botany. He is the Fellow of Third World Academy of Sciences, Indian National Science Academy, Indian Academy of Sciences and National Academy of Sciences, India. He has received several prestigious awards such as Shanti Swarup Bhatnagar Award in 1980, Pitamber Pant National Environment Fellowship in 1984, Pranavanand Saraswati Award in 1985, Dr Birbal Sahni Gold Medal in 1999 and Professor S.B. Seksena Memorial Medal in 1999. His international membership includes: Interdisciplinary Committee of the World Cultural, Mexico; Board of Governers, Board of Directors and Programme (Advisory) Committee of International Centre for Integrated Mountain Development, International Association for Vegetation Science.

Dr. Singh has done pioneering work on grassland ecosystem, and his research findings added to the better understanding of structure and function of grassland ecosystem. Dr. Singh has supervised 42 doctoral students, and several M.S. students. He has published over 388 research publications. It is my privilege to dedicate this volume to a great ecologist, Professor J. S. Singh.

Contents

Preface

Weeds hold an enigmatic and sometimes-controversial place in agriculture, where they are generally reviled, grudgingly tolerated, and occasionally admired. In most cases, growers make considerable effort to reduce the negative economic impact of weeds because they compete with crops for resources and hinder field operations, thereby affecting crop productivity and quality, and ultimately the sustainability of agriculture.

Weed control in production agriculture is commonly achieved through the integration of chemical, biological, and mechanical management methods. Chemicals (herbicides) usually inhibit the growth and establishment of weed plants by interfering with various physiological and biochemical pathways. Biological methods include crop competition, smother crops, rotation crops, and allelopathy, as well as specific insect predators and plant pathogens. Mechanical methods encompass an array of tools from short handled hoes to sophisticated video-guided robotic machines. Integrating these technologies, in order to relieve the negative impacts of weeds on crop production in a way that allows growers to optimize profits and preserve human health and the environment, is the science of weed management.

The aim of the book is to provide insight and recent progress in the science of weed research. Articles presenting the novel and critical appraisals of specific topics are included. The volume starts with an overview on ecology and management of invasive plants, followed by invasive ecology of weeds. Chapter 3 describes adoption strategies of crop-associated weeds. Evolutionary biology of foxtail species has been discussed in the Chapter 4. Chapter 5 discusses important aspects of aquatic weed management. Although attention is commonly directed to the negative aspects of weeds, the benefits of weeds are discussed in Chapter 6 from an agroecological standpoint. Following chapters (Chapter 7-11) discuss ecological and physiological aspects of weed/crop competition and chemical control. Chapters 12-21 discuss characteristics of weed management systems using information obtained in recent research. The final four chapters (Chapter 22, 23, 24 and 25) discuss the weed management with specific examples. Thus, in totality, book covers the physiology, ecology, and management of weeds.

I would like to express my gratitude to the people who made this volume possible. First, I am indebted to authors for willingness to contribute their work to this volume, and for their cooperation in revising the manuscripts. I am grateful to referees that helped the authors to improve their original work. People who helped in reviewing the manuscript are: Krishna Reddy, L.H. Ziska, Robert Kremer, Douglas Buhler, Eric Galllandt, Calvin Messersmith, Anne Legere, Matt Liebman, James Hazen, Robyn Gaskin, John Nalewaja, Maria Olosfdotter, Steve Duke, Donald Penner, Steve Radosevich, John Cardina, Nick Jordan, James Kells, Kassim Al-Khatib, Joseph M. DiTomaso, Reid Smeda, Martin Kroff, M.P. Greaves, David Shaw, Leslie Weston, Chester Foy, James Westwood, Jim Pratley, Steve Murphy, Prasanta Bhowmik, Jodie Holt, Barabara Booth, Bill Dyer, J.J. Smit, Christopher Preston, Rod Heisey, Barry Brecke, James Bounce, John Lindquist, John Teasdale, Kelvin Murphy, R. Charudattan, Caelone Chase, Baruch Rubin, David Gealy, Barney Caton, Albert Fischer, Robert Norris, Shane Frisen and Sidnei Thomaz. I sincerely thank Noeline Gibson and Sandra Oomkes of Kluwer Academic Publishers for their assistance throughout the preparation of the book. I will end the preface with acknowledging some people who always helped and encouraged me in various ways – Professor C. R. Babu, Professor S. R. Yadav and Dr. B. Hari Gopal. I trust that the information presented in this volume will truly provide an ecological perspective on weed management, and will help the scientific community toward a better understanding of the subject. Equally I hope that it will stimulate young scientists to take up studies on the enduring problems – and perhaps opportunities – that weeds present in the production of food and fiber crops on which humans depend.

Inderjit
Department of Botany
University of Delhi

August, 2003

CHAPTER 1

INVASIVE PLANTS

HAMID R. MASHHADI[1] & STEVEN R. RADOSEVICH[2]

[1]*Department of Crop and Weed Science, Ferdowsi University of Mashhad, IRAN;*
[2]*Department of Forest Science, Oregon State University, Corvallis, Oregon USA*

Invasive plant species, unlike agriculture weeds, can successfully occupy and spread to new habitats without further assistance from humans. These plants can apparently invade into new areas already fully occupied by native vegetation, displacing many species. An increasing trade-based global economy, world-wide transport of biological commodities, and opportunities for trans-world travel during the last century have increased the introduction and subsequent colonization of exotic species in many parts of the world (Table 1). Thus, plant invasions from the intentional or unintentional transport of plants to new areas now seriously threaten the biodiversity, structure and function of many of the world's ecosystems (Crosby, 1986; Ewel et al., 1999; Enserink, 1999).

Recent estimates of the number of species of exotic vascular plants in different countries reveal the extensiveness of the invasive plant problem. For example, Atkinson and Cameron (1993) reported that over fifty percent of the existing vascular plants in New Zealand are exotic in origin, while forty percent of the total flora in the British Isles are introduced from other areas of the world (Ellis, 1994). In Australia, 1500 to 2000 plant species have been introduced since European settlement of which over 200 species are now considered to be noxious weeds (Humphries et al., 1991; Parsons and Cuthbertson, 1992). Lonsdale (1994) reported that of over 2000 introductions of exotic pasture species to northern Australia, only five percent are recommended for use and thirteen percent are now listed as weeds. Stein (1996) found that nurseries in the USA are currently selling twenty of the worst weeds in that country, including *Lythrum salicaria* and *Lonicera japonica.* The National Association of Exotic Pest Plants in the USA note that nurseries in Florida have sold 80 percent of that state's exotic pest plants (Tenenbaum, 1996). Rejmanek and Randall (1994) also report a dramatic and exponential increase in alien plant species in California over the last century (Figure 1). The rate of invasions in the San Francisco Bay area, for example, has increased from an average of one new species every 55 weeks during 1851-1960 to one new species every 14 weeks during 1961-1995 (Cohen and Carlton, 1998).

Inderjit (ed.), Weed Biology and Management, 1-28.

Table1. Pathways of harmful introductions into the United States, 1980-1993. Note that 159 other known introductions in the United States during this time period are not included in the table because either their pathway of introduction had not been identified or they had not been characterized by experts as economically, environmentally, or otherwise harmful, as of 1993. Adapted from table 3-1, pp. 101-105 in OTA (1993).

Type of pathway	Number of introductions	Percent of total introductions
Importation via international trade	38	82.6
Introduced from elsewhere, then spread	4	8.7
Escaped ornamental plant	1	2.2
Escaped from aquaculture confinement	1	2.2
Containment of smuggled product	2	4.3
Total	46	100

The magnitude of the plant invasion problem suggests that, if it continues at its current pace, the earth's flora could eventually homogenize to only a few highly successful plant species (McNeely, 1999; Luken and Thieret, 1996). However, invasive species are not uniformly distributed across the world. Warm and mesic regions are more prone to be invaded than cold dry ones. Thus, high elevations and the arctic climate do not generally favor invasive plants (Woodward, 1987). These observations further suggest that native flora of tropical and temperate regions are most at risk of exotic introduction of invasive plants.

The problem of exotic plant invasion is by itself a challenging arena, but it becomes an even more difficult problem when placed in the context of a changing world. In the following sections of this chapter, the impacts of many elements of global environmental change that relate to invasive plants are examined. Next, the ecology of invasive plants is explored, especially in relation to the invasive process. The roles of intrinsic biological factors and the extrinsic environment are also discussed. Finally, several generalized approaches used to manage invasive plants are described.

PLANT INVASIONS AND GLOBAL CHANGE

To many people, the concept of global change is perceived as only being climate change. However, there are other elements of global change as well - all of which concern biologists and relate to invasive plants.

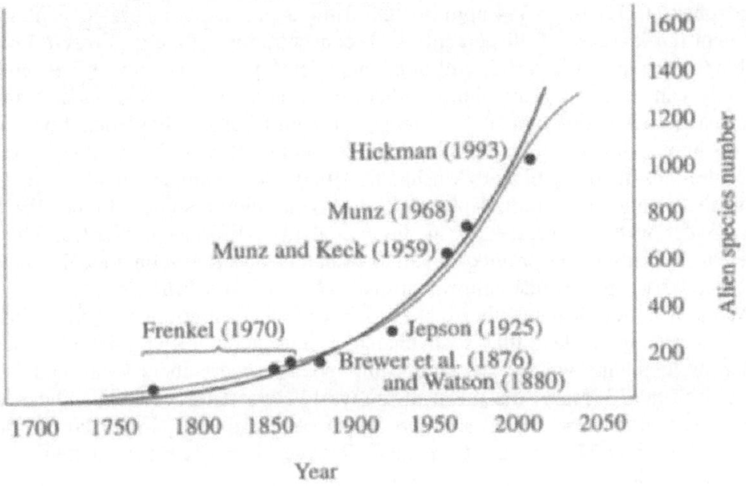

Figure 1. Number of alien species of vascular plants in California flora, 1700-1993. From Rejmanek and Randall, 1994. (References in this figure are cited in Rejmanek and Randall, 1994)

Global Warming

Although meteorologists do not agree about the rate at which the earth's climate is warming, they all generally accept that global warming is taking place - mainly because of accumulation of greenhouse gases (Watson et al., 1995). An inevitable outcome of global warming is the expansion and contraction of potential ranges of plant species, and the migration of plant species into new territory. Global warming may allow existing species to migrate to new locations by making formerly unsuitable habitats more suitable. Warming trends also may permit longer breeding seasons and thus allow species that were biologically unable to mate previously to hybridize (Crooks and Soule, 1999). Hybridization of native species with closely related exotic species after introduction also could give rise to new phenotypes capable of invading new habitats.

Increase in CO_2 Concentration

Atmospheric CO_2 rose to 364 ppm in 1997 from its pre-industrial era concentration of 280 ppm (an increase of 30 percent). CO_2 concentration is likely to reach 560 ppm, double the pre-industrial level, within the next century. A raise in CO_2 concentration not only can increase plant photosynthetic rate but also can raise plant water use efficiency (WUE), although such a change is not uniform among plants. Species with the C_3 photosynthetic pathways tend to be more affected by a change in CO_2 concentrations than C_4 plants (Mashhadi, 1989). For example, Field et al. (1996) report that an increase in atmospheric CO_2 concentration results in higher WUE and thus greater water savings early in the season in California grasslands. The extra water storage in the soil provides a suitable habitat for late season annuals, including certain invasive species like yellow starthistle (*Centaurea solstitialis*).

While it is not clear whether increases in CO_2 concentrations will favor invasive plant species, it is clear that such change brings about differential physiological responses in plants which may favor some species over others (Mashhadi, 1998). Holm et al. (1977) report that most of the world's worst weeds are C_4 plants while, with exception of corn, millet, sorghum, sugar cane and a few minor crops, most of the world's principle crops are C_3 plants. However, it is not clear whether C_3 or C_4 plants will be losers or winners in a future world that will be both higher in CO_2 and warmer (Patterson, 1995; Mooney and Hofgaard, 1999). *Bromus tectorum* is an invasive plant that has established extensively throughout the Great Basin of the United States. Smith et al. (1987) report that one cause of this plant's success could be the rise in atmospheric CO_2 concentration over the last century. Dukes (2000) also speculates that leguminous shrubs might become especially invasive as rising CO_2 stimulates nitrogen fixation.

Nitrogen Deposition

The alteration of the global biogeochemical nitrogen cycle is also an important element of global change. Human induced nitrogen fixation through the production of synthetic fertilizers and fossil fuel burning is now greater than natural fixation. This has caused unprecedented eutrophication of the biosphere. Annual airborne nitrogen deposition of 30-60 Kg N ha^{-1} has been reported in Germany (Umweltbundesamt, 1997). Such continuous nitrogen deposition may cause irreversible loss of biodiversity due to the competitive exclusion of organisms not adapted to nutrient rich sites. Both alien and native species may expand to such nutrient rich locations, however. Scherer et al. (2000) report that most successful invasive species in central Europe are favored by high atmospheric nitrogen deposition.

Land-Use Change

With increasing world population, a dramatic increase in human-induced disturbances of land is taking place. As the human need for food and shelter increases, more land has been devoted to agriculture, urban areas, roads, and industrial uses. Hobbs and Hopkins (1990) propose four common land-use categories: (i) conservation, where no deliberate modification or disturbance of land takes place, (ii) utilization where natural plants are harvested for wood and forage by humans and animals, (iii) replacement where intensified agricultural and horticultural systems are in place and introduced species replace natural species on the land, and (iv) removal, where urban and industrial development take place. There also are many examples where spread of invasive species has been facilitated by roads and transportation.

Changes in land use, degradation, and human disturbance all provide opportunities for invasion, and invading species in turn can force changes in land use or modifications in management. For example, Mooney and Hobbs (2000) express a two-way relationship between fire and plant invasion. They report that the use of fire in management generally increases the invasibility of ecosystems, but invasive plants also can alter fire regimes - as has been noted for *Bromus tectorum* in the sagebrush steppe of the Great Basin.

Trade Globalization

In contrast to what many believe, neither international releases nor contraband are responsible for most introduction of harmful invasive species. According to a 1993 U.S. Office of Technology Assessment (OTA) report, unintentional importation through international trade is responsible for over 80 percent of the harmful species introduced into the United States (Table1). Furthermore, the global economy has greatly increased in recent years. Imports of agricultural products and industrial raw materials - those with the greatest potential to contribute to the problem of invasive species - amounted to $ 482 billion in 1990, up from $55 billion in 1965 (McNeely, 2000).

International laws regulating the unintentional introduction of harmful species through trade are generally weak, since more value is usually placed on free trade than on prevention of invasive species. Under many international trade agreements like GATT (General Agreement on Tariffs and Trade) and NAFTA (North American Free Trade Agreement) bans and restrictions may not discriminate needlessly against imported goods, that is, protectionist trade barriers cannot be disguised as environmental protection measures (Jenkins, 1999). In other words, it is only through a science-based risk assessment that a restriction on trade or import of a commodity can be made. When addressing the necessity to regulate invasive species, the Convention of Biological Diversity (CBD) article 8 asks both trade parties - by using phrases like "as far as possible and as appropriate" - to prevent the introduction, to control or to eradicate alien species that threaten ecosystems (Jenkins, 1996 a, b).

Increased Travel

As world travel becomes easier, more people tend to travel either for business or pleasure. People around the world visit metropolitan cities that are the centers for business. Many parts of the world also are popular for tourist visits and are thus exposed to inadvertent introduction of invasive species. State and federal inspections in Hawaii, a state visited by nearly seven million tourists annually and which imports eighty percent of its consumable goods, intercepted 2,275 individual non-indigenous invertebrates in a single year, including 259 species not known to already occur in Hawaii (Holt, 1999). MacDonald et al. (1988) report that in Hawaiian parks there is a linear relationship between the number of invading species and the number of park visitors.

ECOLOGY OF INVASIVE PLANTS

Weed is an anthropocentric term applied to plants when they are undesirable in a particular location. There is substantial literature on the origin and biological characteristics of present day weed species (Salisbury, 1961; Baker, 1974; Holzner and Numata, 1982; Radosevich et al., 1997). Weeds are plants that are most often associated with agricultural production systems. These plants, agrestals, require human management to maintain their populations (Hilbig, 1982). Other weeds are less specific than agrestals, adapting more generally to human disturbed environments.

Rejmanek (2000) describes much of the flora of California as weeds, colonizers (taxa appearing early in vegetation succession) and naturalized species (non-native species that form sustainable populations without direct human assistance). As shown in Figure 2, not all naturalized taxa are invasive, nor are all colonizers considered to be weeds. However, many weeds, either native or non-native, are colonizers. Although these three categories of vegetation represent different points of view, it seems that successful invaders are plants that colonize and successfully spread in areas where they are not indigenous.

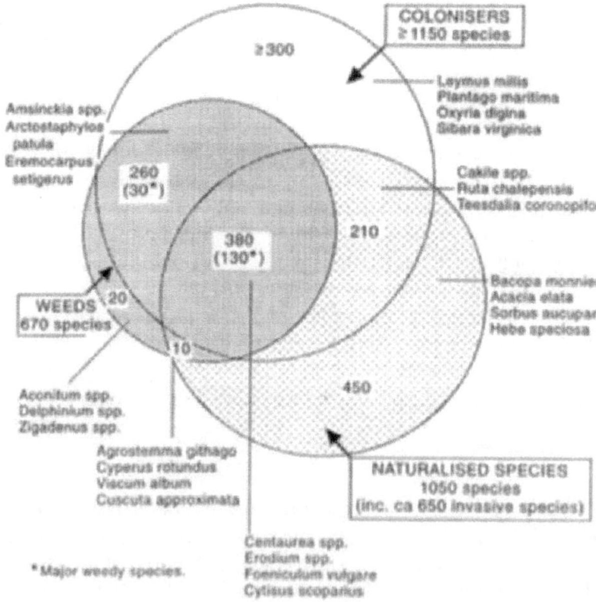

Figure 2. Weeds, colonizers and naturalized species (including invaders) are three overlapping but not identical concepts reflecting three different viewpoint: anthropocentric (weeds are plants growing where they are not desired), ecological (colonizers appear early in succesional series) and biogeographical (naturalized species are species locally establish or spreading in areas where they are not native). A subset of naturalized species are invaders, namely those non-native species that are spreading. Estimated species numbers and examples of species representing seven resulting categories of Californian vascular flora, which consists of 4840 species (Rejmanek and Randall,1994). Native weeds that are not colonizers (about 20 species) are mostly poisonous plants growing in successionally advanced communities. From Rejmanek, 2000.

Invasion Process

Groves (1986) and Cousens and Mortimer (1995) divided the process of invasion by plants into three phases: *introduction*, *colonization*, and *naturalization*. Although such phases are arbitrary, they suggest that plant invasions occur at different geographic and biological scales and can be organized according to stages of population development (Table 2). Sauer (1988) and Forman (1995) also indicate that processes determining the stability of plant populations during migration are scale-dependent, and range from individual plants to metapopulations as indicated in Table 2.

Table 2. Ecological processes, patterns and scales at different phases of plant invasion. From Radosevich et al. (2003).

Phase of invasion	Ecological process	Ecological pattern	Scale
Introduction	Dispersal Immigration Survival	Species recruitment	Individual
Colonization	Birth Death Immigration Emigration	Patch expansion	Population
Naturalization	Birth Death Immigration Emigration	Range expansion	Metapopulation

Habitat Suitability

Disturbed ecosystems (either by human or natural causes) often have a higher susceptibility to invasions than those that spend long periods in late successional phases. For example, semi-arid grasslands may be more vulnerable to invasion by exotic plant species because these areas have spatially open niches; sites that are devoid of vascular plants for some or most of the year (Baker, 1986). Plants with life forms dissimilar to the native vegetation also have invaded some ecosystems. The conversion to annual grasslands from tussock grasslands in California and the invasion of *Opuntia stricta* into Australia (where no members of Cactaceae existed previously) are examples where biological characteristics play an important role in the invasion process, thus the intrinsic biology of the plant species and the extrinsic nature of the ecosystem are both important factors in determining the success of invasive plant species.

Extrinsic Factors

The extrinsic factors most responsible for floristic growth and persistence of invasive species are climate, soil and land use. Climate drives both abiotic and biotic thresholds for growth in ecosystems. For example, the retreat of forests to mountainous areas over the last 10,000 years is a direct result of changing climate (Betancourt et al., 1990). Climate also can change habitat suitability over short timescales through drought, seasonal frost and flooding (Nilson and Muller, 1980). Invasive species tend to be adapted to a variety of habitats, but usually invade areas with similar climates to their native range first, then adapt to other climates later (Panetta and Mitchell, 1991); however notable exceptions also occur (Mack, 1995). Topography and elevation also influence climate and the species that can grow at a given location.

Many studies have shown that both plant growth and plant invasions occur within certain ranges of soil types (Huenneke et al., 1990). Although soil-mapping units often have indistinct boundaries, species usually are most productive within certain soil types. One criterion that could be used to study or develop predictive models of invasions is the isolation of soil mapping units suitable for each target species.

Land use and the condition of the environment is a third general driving component in habitat suitability and invasion. Changes in land use are thought to be the single most important factor in species extinction (Cousens and Mortimer,1995) and to have strong influences on invasable sites (Elton, 1958 as cited by Hodkinson and Thompson, 1997). Humans to some degree have modified most of the world's ecosystems and this has direct affects on invasion suitability. Some ecosystems are altered by weed presence (D'Antonio and Vitousek, 1992) through an increase in fire frequency, nitrogen depletion or addition, or allelopathic chemicals.

Intrinsic Factors

There is no clear-cut description or list of attributes that make a plant species more invasive than another. However, some useful generalizations have been made. For example Rejmanek (2000) lists the following biological characteristics as being responsible for invasiveness:

- Fitness homeostasis or the ability of an individual or population to maintain relatively constant fitness over a range of environments. This is equivalent to Baker's (1974, 1995) "general purpose genotype."
- Small genome size - usually associated with short minimum generation time, short juvenile period, small seed size, high leaf area ratio, high relative growth rate.
- Dispersed easily by humans and animals.
- Ability to vegetatively propagate. This is an especially important characteristics in aquatic environments (Auld et al., 1983; Henderson, 1991) and high latitude (Pysek, 1997).
- Alien plants belonging to exotic genera are more invasive than are alien species with native congeners. This may be partly because of an absence or limited number of resident natural enemies for that species (Darwin, 1859; Rejmanek, 1999).
- Plant species without dependence on specific mutualisms (root symbiosis, pollinators, seed dispersers, etc.) (Baker, 1974; Richardson et al., 2000).
- Tall plants tend to invade mesic plant communities.

- Persistent seed banks—seeds with different inherent dormancies that provide a random appearance through time and guarantees their survival and persistence.

Phases of the Invasion Process

Plant species invasions can generally be depicted by a logistic growth curve that encompass the three phases of the invasion process (Groves, 1986; Cousens and Mortimer, 1995; Figures 3 and 4).

Introduction Phase

Theoretical population growth curves can be generated for species given assumptions of their environment, initial population size, intrinsic growth rate, and time. The earliest phase of such population curves must result from introduction of the plant species. However small populations are often undetected during the introduction phase, and plant invasions are most likely to fail at this point due to stochasticity (Mack, 1995), or to a lack of a minimum critical patch size (Latore et al., 1998). Seedlings from newly introduced seeds must compete with the established flora that is well adapted to the site and occupies space, so chance of establishment is low. Thus few invasions proceed to the colonization phase. Most factors, aside from suitability to the ecosystem, that influence the rate of the introduction are extrinsic to the disseminating seed. Prediction of the introduction phase is strongly related to these extrinsic factors (Figure 4).

Role of Disturbance on Introduction. Disturbance is believed to be a major factor favoring plant invasions. Grime (1979) defines disturbance as the removal or damage of plant biomass. Pickett and White (1985) view disturbance as 'any relatively discrete event in time that disrupts ecosystem, community or population structure and changes resources or substrate availability or the physical environment'. Plant invasion may result from direct destruction of vegetation or indirectly by changing resource levels (Davis et al., 2000; Booth et al., 2003) or other conditions that subsequently affect the development of plants. Disturbance may be caused by large-scale fire, floods and storms or smaller scale events like soil turnover or vegetation removal by animals or humans (Hobbs, 1991).

Disturbance does not always lead to invasion, but it may provide a temporary location or "safe site" for a potential invasive species to establish a founding population. Some form of disturbance usually accompanies the success of many invasive species. Humphries et al. (1991) lists a number of major "environmental" weeds in Australia along with their association to at least one type of disturbance (Table 3). Multiple disturbances usually increase the chance of successful invasion. For example, Hobbs and Atkins (1988) conducted an experiment in two woodland and three shrubland sites. They planted *Avena fatua* seeds into four types of disturbance treatments at each site: (1) undisturbed, (2) disturbed, (3) fertilized, and (4) disturbed and fertilized. They observed that disturbance in general, and multiple disturbance in particular, increased the density and biomass of the introduced species (Hobbs, 1991; Figure 5).

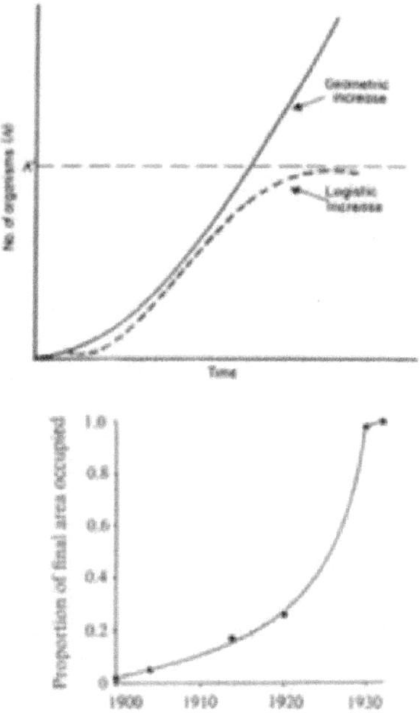

Figure 3. (Top) Geometric (solid line) versus logistic (dashed line) population growth over time. K is the carrying capacity of the environment for a population showing density-dependent logistic growth (dashed line). (Bottom) Increase in the area occupied by *Bromus tectorum* in western North America. Modified from Radosevich et al. 1997.

The Tens Rule. The inflow of exotic plant species into new regions is a continuous process and many believe it is happening at an ever-faster rate. However, research suggests that only a few alien species that make it to a new region actually become established there (Cousens and Mortimer, 1995), and then only a small percentage of the established species become invasive. Williamson (1993, 1996) estimated that only ten percent of all introduced species into the British flora actually became established and then only ten percent of those were invasive enough to be considered as pests. He presented this finding as the "tens" rule for plant invasions. Kowarik (1995) similarly reported that of 3150 woody species introduced into Brandenburg and Berlin, Germany, ten percent spread beyond the initial site of introduction, two percent became established, and half of those naturalized. Kowarik also noted this ratio (10:2:1) for the Central European flora of vascular plants (12,000 species). Weeda (1987), examining a number of "neophytes" and alien species in the Netherlands, estimated that approximately 1%, penetrated the natural vegetation - supporting the later estimates of Williamson and Kowarik. Kornas (1990) found that of the 799 plant species introduced around Montpelier, France, 692 species failed to become established.

Colonization Phase
Colonization of invasive plant species is best characterized by an often prolonged lagtime after introduction followed by exponential population growth (Figures 3 and 4). During colonization, the species often becomes noticeable and control efforts to halt its spread begin.

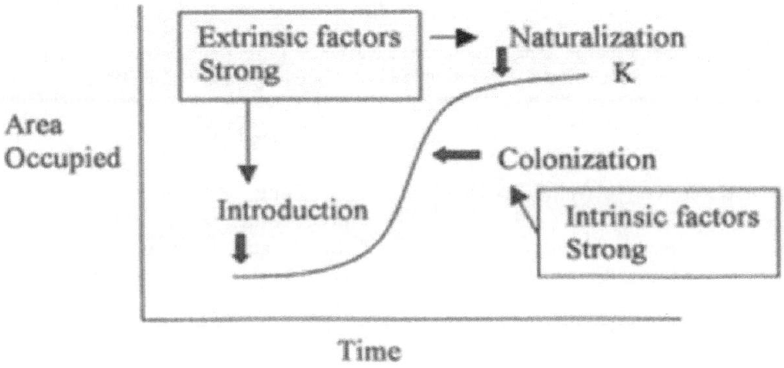

Figure 4. Growth curve depicting the phases of expanding populations. From Radosevich et al. (2003).

Lag Phase of Population Growth. Although some invasive species may experience rapid population growth after introduction (usually insects, like African bee), most invasive species have a long lag phase between initial introduction and subsequent rapid population growth. According to Crooks and Soule (1999), three categories of population lag are recognized:

1. Inherent lag times caused by the nature of population growth and range expansion,
2. Prolonged lag times caused by environmental factors related to improving ecological conditions caused by nature or humans like soil nutrient enrichment, climate change, dispersal vectors, and intra-specific interactions.
3. Genetic factors that usually concern fitness.

Some species have a so-called 'general purpose' genotype (Baker, 1995) that enables them to grow over a wide range of environments. If an introduced species lacks such characteristics, it will be confined to a restricted area until a genetic change occurs through recombination, introgression, or to a lesser degree mutation, and adapt to the new environment. The likelihood of overcoming genetic lag or fitness deficit is proportional to the population size and the rate of genetic adaptation (Crooks and Soule, 1999).

Past performance of an exotic species is a poor predictor of its lag time in a new habitat; however, some plant characteristics like length of time to reproduce have a profound effect on lag time. Kowarik (1995) in a historical reconstruction of the invasion dynamics of 184 alien woody species near Bradenburg, Germany found that only six percent of the species spread within 50 years after their first introduction to the area. Twenty-five percent lagged up to 100 years, 51% lagged for 200 years, 14% for up to 300 years and 4% for more than 300 years (Figure 6).

Table 3. Species identified as the most serious environmental weeds in Australia, and types of disturbance favouring spread and/or establishment (derived from Humphries et al., 1991).

Species	Disturbance type
Acacia nilotica (L.) Willd. Ex Del.	Cattle dung, flooding
Cenchrus ciliaris L.	Floods, cattle
Chrysanthemoides monilifera ssp. *rotundata* (DC.) Norlindh	Rabbits, cattle, roading, sandmining, natural disturbances
Cryptostegia grandiflora R. Br.	Drought followed by floods, fire, grazing
Eichhornia crassipes (C. Martius) Solms	Human interference, nutrient enrichment
Mimosa pigra L.	Clearing, water level fluctuations
Myrsiphyllum asparagoides (L.) Willd.	?
Parkinsonia aculeata L.	Flooding
Pennisetum polystachyon (L.) Schultes	Vehicular spread along roads, small scale disturbance by pigs, bandicoots
Prosopis spp.	Flooding
Salvinia molesta D. Mitch.	Human transport, nutrient enrichment
Tamarix aphylla (L.) Karst	Flooding, high rainfall years
Thunbergia grandiflora (Rottler) Roxb.	Edge or gap formation

Life form had a significant affect on lag phase. On average, trees had a longer lag phase (170 years) than shrubs (131 years).

Exponential Phase of Population Growth. The speed at which colonization proceeds is closely related to the intrinsic rate of increase for the plant species. Hence, predictions of colonization rates and management options during this phase of the invasion process should focus on the intrinsic biology of the species with minor influences from extrinsic factors.

For example Maxwell et al. (1988) used a population demographic approach to examine population growth and management options of leafy spurge (*Euphorbia esula*). Following Watson (1985), they divided the life history of leafy spurge into five stages: seeds, buds, seedling, vegetative shoots, and flowering shoots (Figure 7, top). By identifying these stages, the process of population development was determined. It was found that three important transitions - basal buds to vegetative shoots (G2), the number of basal buds that flowering shoots produced (V5), and the number of basal buds that vegetative shoots produced (V4) - were sensitive to their own density. When these three density dependent functions were simultaneously included in their model, initial exponential growth followed by decline and eventual stabilization of the simulated population was predicted (Figure 7, middle). They then subjected the simulated population to several management tactics: a single application of picloram (Figure 7, middle) and several levels of a foliage-feeding herbivore (Figure 7, bottom) on leafy spurge. The accuracy of the population dynamics model was striking in predicting both population growth and outcomes of actual management treatments.

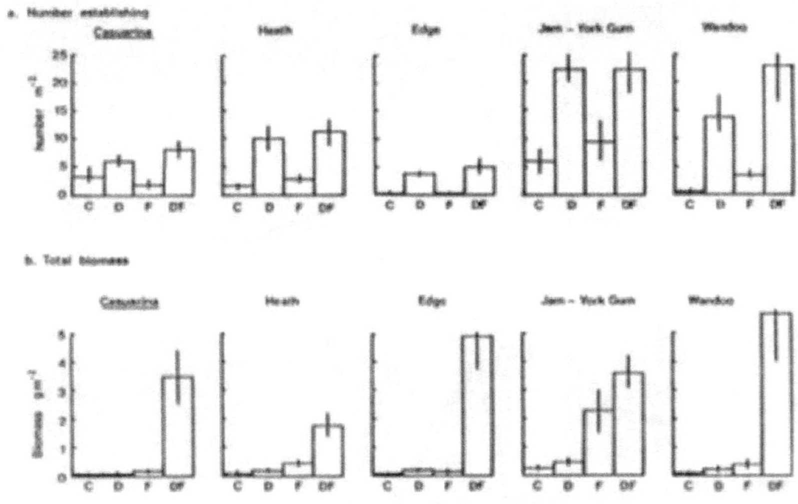

Figure 5. Responses of *Avena fatua* L. sown into experimental plots (each 1 m^2) in five different vegetation communities in the Western Australian wheatbelt. Treatments are: C, control; D, disturbed; F, fertilized; DF, disturbed and fertilized. Plant communities are: *Casuarina*, shrubland dominated by *Allocasuarina campestris*; heath, species rich sandplain heath; edge, regenerating community on previously-cleared sandplain heath; jam-york gum, *Eucalyptus loxophlebal, Acacia acuminata* woodland; wandoo, *Eucalyptus wandoo* woodland. Histograms give mean values ± 1 S.E. (n=4, treatments with different letters are significantly different at $p < 0.05$. (a) Number of seedlings established in each plot, (b) total dry weight per plot. From Hobbs, 1991.

Sources and Satellites. Colonization occurs when plants in a founding population reproduce and increase sufficiently to become self-perpetuating. Cousens and Mortimer (1995) described the radial expansion of such self-perpetuating patches as an advancing front on all sides with the rate of proportional increase conforming to the equation:

$$\text{(Equation 1)} \qquad (dA/dt)/A = 2\pi r^2 t$$

where A equals the area occupied, r is the radius of the population, and t is time in years or generations. In other words, a patch should grow by half the distance of each previous generation (Cousens and Mortimer, 1995), assuming successful recruitment continues within the founding population and the rate of expansion is constant. Although this model of range expansion seems intuitively reasonable, it has received only limited examination in field studies (Auld and Coote, 1980; Moody and Mack, 1988).

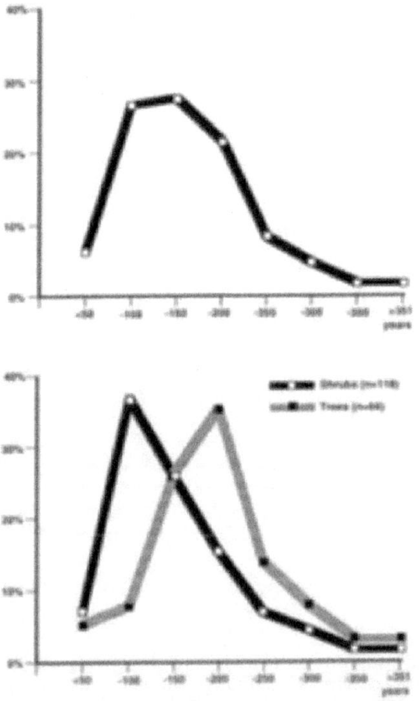

Figure 6. Time lags between the introduction and beginning of invasion of woody species in Brandenburg. (a) Calculation for 184 woody species grouped in 50-year-classes, (b) differentiation between trees (n = 66) and shrubs (including woody climbers, n = 118). From Kowarik, 1995.

It is possible for some individuals from a founding *source* population to disperse widely (Figure 8) and to produce new *satellite*, populations of the species. In this case, dispersing individuals recruited from an already established *source* population should behave much the same and have the same requirements as a new introduction.

A common method to study movement of dispersing invasive species is through reaction-diffusion models (Skellman, 1951; Holmes et al., 1994). These models combine a parameter of diffusion with deterministic population growth to describe the spread of an organism over an area or through a plant community. Accordingly,

$$\text{(Equation 2)} \qquad dA/dt = \pi 4rD$$

where A, t, and r are area, time, and radius of the population, respectively and D is a diffusion parameter somewhat like "spread" or "environmental porosity." The diffusion parameter, D, in this equation assumes a normal distribution and has been used to model animal populations with reasonable accuracy (Andow et al., 1990; Holmes et al., 1994). Unfortunately, estimates of spread by plants using reaction-diffusion models have largely underestimated the area being invaded, often by orders of magnitude. The model outcomes also have proven difficult to extrapolate to real invasions because D is difficult to quantify experimentally.

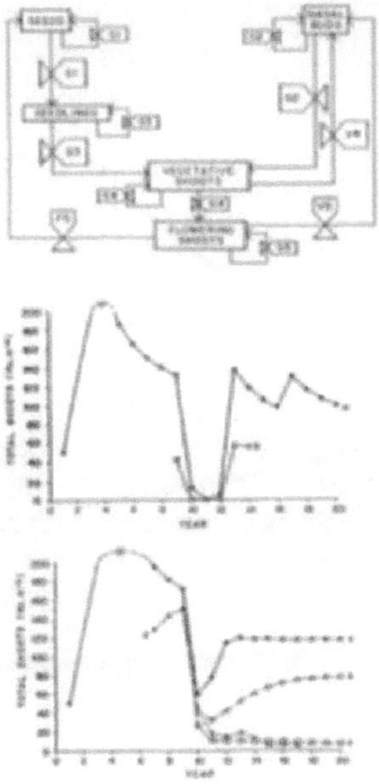

Figure 7. (top) Diagrammatic model of a leafy spurge population. (middle) Leafy spurge population simulation with density dependent functions and a single application of picloram simulated at year 10 (a) and (b) observed effects of picloram application. (bottom) Leafy spurge population simulation with density dependent functions simulating the introduction of a foliage-feeding herbivore at year 10 that removed (a) 40, (b) 50 and (c) 60 percent of the stems. Also shown are (d) observed effects of sheep feeding on leafy spurge. (Modified from Radosevich et al., 1997).

A relevant research and management question emerges from such analyses which is whether to contain "source" or "satellite" populations of an invading species after successful introduction has occurred. Cousens and Mortimer (1995) and Moody and Mack (1988) suggest that the preferred containment strategy would be to remove satellite populations as they occur over space and through time, since these populations expand more rapidly and potentially cover greater area than the front of a source population. For example, Ghersa et al. (1999, 2000) determined that satellite populations of *Sorghum halepense* that were uniformly distributed over a previously vacant area occupied that area more quickly than the advancing front of an adjacent source population. None-the-less, this strategy of expansion has received relatively little experimental attention by field scientists. Land managers continue to control source rather than satellite populations of invading species (Moody and Mack, 1988; Cronk and Fuller, 1995; Zamora and Thill, 1999), perhaps because such populations are much more obvious than the smaller satellite populations or control tactics seem more effective.

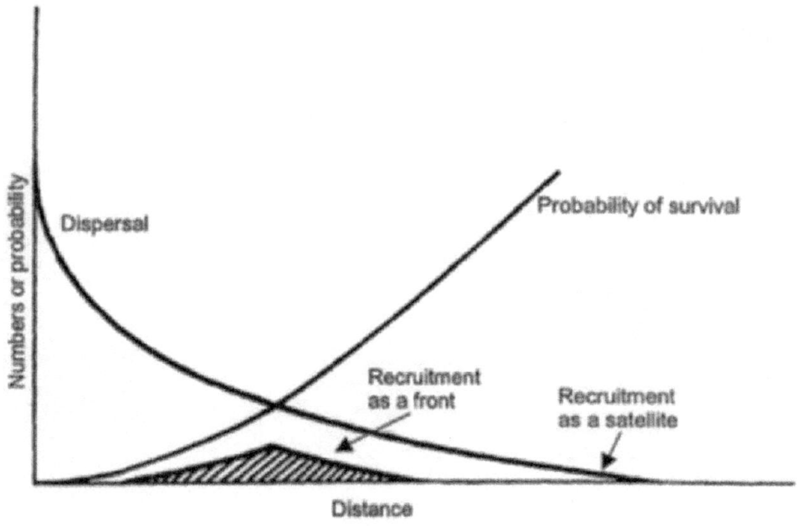

Figure 8. Recruitment of new genotypes as a function of the number of dispersed seeds and the probability of juvenile survival. (From Cook, 1980 in O.T. Solbrig, ed. Demography and Evolution in Plant Popualtions; modified from Radosevich et al., 1997).

Naturalization Phase

A species becomes naturalized in its new environment when it successfully establishes new self-perpetuating populations, is dispersed widely throughout a region, and is incorporated into the resident flora. At some carrying capacity, K, the population approaches a quasi-threshold density where its population growth may remain near one, i.e., stabilize and not expand very quickly (naturalization) (Figures 3 and 4). The K density occurs when niche occupancy and available resources limit the rate of spread. This phase is controlled by extrinsic factors so predictions of risk for populations approaching K should also be focused on extrinsic parameters. Most agencies remove weeds from their target list (e.g., noxious weed lists) in this phase as they are too difficult or expensive to eradicate.

MANAGEMENT APPROACHES AND STRATEGIES FOR INVASIVE PLANTS

Management Approaches

Rejmanek (2000) notes that five general approaches have been used to study invasive plant species for management purposes. These approaches are: stochastic, empirical-taxon specific, evaluation of biological characters, evaluation of habitat compatibility, and experimentation.

Stochastic

This approach is the most robust but least helpful generalization about invasive plant ecology, i.e. the probability of invasion success increases with initial population size and the number of introduction attempts, residence time and spatial distribution. Mulvancy (unpublished data cited by Rejmanek, 2000) observed a strong correlation between the amount of planting and the probability of a woody plant becoming naturalized in southeastern Australia. In Venezuela, a significant correlation was observed between the total number of known localities and the minimum residence time (the years since the first record in the country) (Rejmanek, 2000). Owen et al. (1997) found a similar correlation in New Zealand conservation land (Table 4), and localities and the minimum residence time (the years since the first record in the country) also was found (Rejmanek, 2000).

Table 4. Ecological weeds on conservation land in New Zealand (data in Owen, 1997)

Distribution type	Mean minimum residence time (years)	Number of localities (n)
Established with an isolated distribution	26	8
Established with a limited distribution, but spreading	65	52
Widely distributed and extending their ranges	83	88
Widespread, having reached almost all suitable habitats	118	11

The shortcoming of this approach is that it tells relatively little about the potential impact of the species or the environmental and biological factors necessary to manage it.

Empirical-Taxon Specific
This management approach is based on information about the invasiveness of a species elsewhere. Information about the population growth of a particular species in a specific habitat can help land managers when confronted with a new introduction. Knowing the experiences of others also can help land managers make wise decisions to control and/or eradicate the invasive species. Williams et al. (2000) report that eighty percent of the exotic weed species in New Zealand are also described as invasive outside the country.

Evaluation of Biological Characters
Rapoport (1991) reports that ten percent of the 200,000 vascular plant species are potential invaders, but fifteen percent of these have actually invaded an area somewhere. There are certainly some characters that make some plants more invasive than others, and understanding how and why certain biological characters promote invasiveness in a species is an important management tool. Statistical tools like discriminate analysis, multiple logistic regression, path analysis and classification, and regression trees can be used to assess biological characters responsible for invasiveness.

Evaluation of Habitat Compatibility
Invasive characteristics of plant species occcur only when habitat conditions are favorable for them (see extrinsic factor, above). Habitat compatibility, which is also referred to as ecosystem vulnerability can be determined if data on species and ecosystem characteristics exist (Rejmanek, 2000).

Experimentation
Experiments could be conducted to test how effectively the first four management approaches prevented or reduced the invasive potential of a species in a particular environment. However, the time lag that is inherent in most invasive episodes (Kowarik, 1995) usually make such experiments unappealing to land managers.

Management Priorities

Hobbs and Humphries (1995) suggest a framework for setting priorities for management of invasive plants based on land value and the degree of site disturbance (risk of invasion). Their diagram is modified here to show four distinct regions of management (Figure 9). In the bottom right corner are sites of high value that are relatively undisturbed, i.e. the risk of invasion is low. Such sites could be free of the invasive species and, according to Hobbs and Humphries, the management objective should be to keep them that way. Wilderness areas are good examples of such locations where monitoring and localized containment around trailheads should keep such areas free of invading species. Hobbs and Humphries suggest that such locations should be treated as "fortresses," and management resources should be directed at minimizing human induced disturbance and the dispersal and establishment of invasive species.

In the top right corner of the diagram (Figure 9) are locations of high value that are subject to greater levels of disturbance (risk) and, hence, are more susceptible to invasion. Hobbs and Humphries believes the management objective here should be to manage these sites by reducing or removing any disturbance factors, controlling current populations, and preventing further introductions. Such areas also could be likely locations for restoration. In the bottom left corner of the diagram (Figure 9) are sites of low value that are subject to low levels of disturbance. These sties should require little or no management input, but require constant monitoring so that local colonies of invasive plants do not spread. At the top left are sites of low value that are subject to high levels of disturbance (risk). Grasslands of California are good examples of such locations where invasive exotic plants have spread so extensively that they have naturalized, replacing the native bunchgrass species. Although such locations may be subject to rapid change and extensive invasion, they should be regarded as low priority for management because attempts to restore the native vegetation to its pristine condition are unlikely to succeed.

The four regions depicted in Figure 9 establish clear management priorities. However, many locations represent areas where the determinations of management priorities are more difficult. Unfortunately, the prevailing trend is one of transition from the bottom right to the top left of Figure 9, as environmental degradation continues and plant communities come into higher risk of invasion.

Management Strategies

Land managers routinely use several general strategies to manage invasive plant species. Various tools and tactics are also employed to reduce the prevalence of exotic invasive species once they are detected. The most common strategies used against invasive plants are prevention, early detection and containment.

Prevention

Like many other pest management problems or disease epidemics, prevention is a key tool used against invasive plants. Hobbs and Humphries (1995) believe that much stronger quarantine laws should be developed and rigorously implemented. For example North American seed and nursery catalogues offer over 59,000 plant species and varieties for sale to national and international markets (Issacson, 1996), most of them exotic to their proposed marketplace. Besides such deliberate introduction of exotic plant species, inadvertent introduction of non-indigenous plants also will continue. Improved global transportation, increased free trade and the continued quest of economic growth will all result in an expanding exchange of organisms among biogeographic regions of the world (Jenkins, 1996 a, b).

However as Ewel et al. (1999) points out, "introductions of non-indigenous organisms can be both a boon and a bane to society." These authors indicate that over ninety percent of the world's food comes from nine crops (Sattaur, 1989; Prescott-Allen and Prescott-Allen, 1990), each of which is cultivated in habitats far beyond their natural range. Three genera, *Eucalyptus*, *Pinus* and *Tectona* comprise eighty-five percent of the industrial forestry plantations of the world, all considered exotic in several regions of the world (Evans, 1992). According to the 1996 New Zealand

Department of Statistics, ninety-five percent of that nation's export earnings are derived from alien species. Therefore, policy makers must consider the beneficial gains of deliberate introduction of exotic species a well as their harmful effects to the environment. The question is would strong quarantine laws deprive humans of the benefits of non-indigenous species? Further, would such laws be effective and practical in the present world?

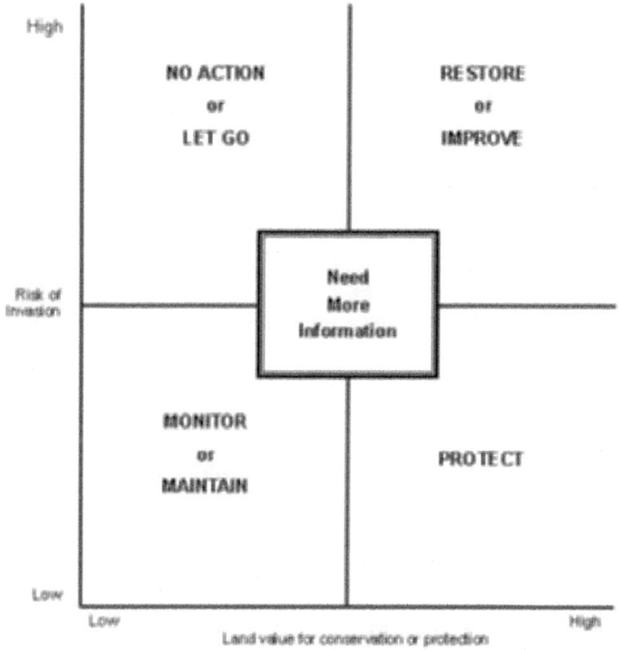

Figure 9. Assessment of management priorities for a region based on the relative value of different sites for conservation and/or production, and their relative degree of risk to invasion. Modified from Hobbs and Humphries, 1995.

Perring (2000) conducted an economic analysis to address the effectiveness of prevention on the management of invasive plant species. He considered the value of the cost of screening to be the product of three factors: (a) the accuracy of the screening process, (b) the probability that a species will be invasive and (c) the cost of screening error as noted by Smith et al. (1999). The accuracy of screening is the proportion of species that have been appropriately dealt with, i.e. the number of invaders rejected by the process versus the number of non-invaders accepted. From this proportion Perring derived a likelihood ratio that depends on the probabilities of whether the introduced species escapes to become a casual species, that a casual species will naturalize, and that a naturalized species will become a pest. In general, this likelihood ratio is very low as Williamson's (1996) tens rule suggests. The effectiveness of the screening system is then determined by the product of the likelihood ratio, the probability of an introduced species becoming a pest, and the cost if it does.

Early Detection

Since invasive plants usually demonstrate a long lag period before exponential growth, they could be recognized and eradicated at that time. However, early detection usually requires an informed and vigilant public. Therefore education and information transfer about the risks of an invasive species is usually a priority for early detection. Land managers who can act quickly to eradicate an invasive plant usually avoid the long-term cost of containing an invasive plant species, and the damage it causes to the land (Chippendale, 1991; Figure 10).

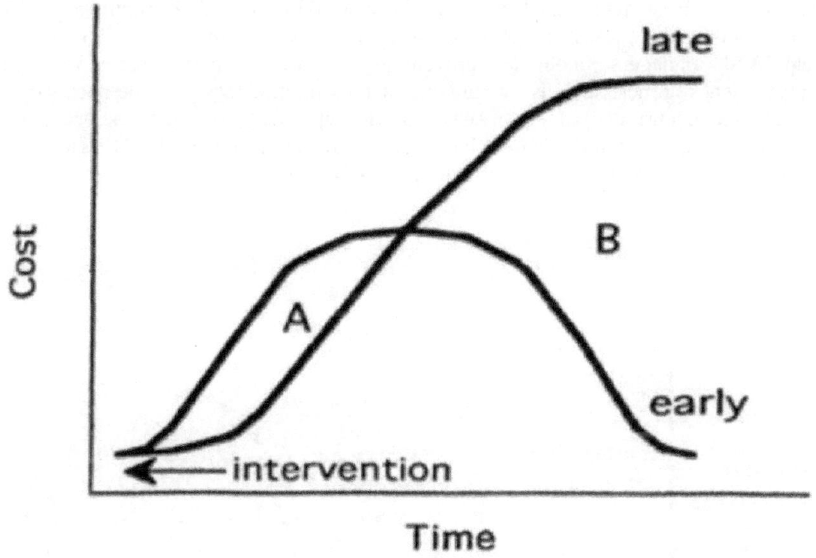

Figure 10. Total social cost of a plant invasion (incorporates the costs of damage due to the invasion and the costs of control) in relation to timing of intervention (early versus late). Costs of early expenditure (area A) and the resulting benefit (area B), after Chippendale, 1991.

Containment or Control

Chippendale (1991) overlaid possible management strategies onto a diagram of invasive species population growth (Figures 11 and 4). He suggests that quarantine is only an effective strategy until a species introduction takes place. After introduction and throughout the lag phase, which may take many years, eradication of the species should be the primary objective of land managers. After the lag phase and during the rapid growth or colonization of a species, land managers should abandon efforts to eradicate and try to control or contain the population from building. If no action is taken to contain the invasive species from spreading, it will eventually become naturalized, occupying the land close to its carrying capacity and replacing many local species. At naturalization, it is difficult to devise strategies to manage such a widespread problem unless substantial amount of time and resources are spent. The ease of treatment of an invasive plant problem decrease as time proceeds after successful introduction (Figure 11).

Management in the Social Context

Land managers and management agencies tend to direct resources toward control of already major weed species, and much less to the prevention, early detection, or containment of new exotic plants. Most major weed-control programs get underway only after a particular species is an obvious problem and most biocontrol research is directed at these recognized problem species (Hobbs and Humphries, 1995). Understanding the causes for the appearance of an invasive species in a habitat may help land managers undertake preventive actions. Changing prevailing land management regimes, such as overgrazing or logging practices, may be necessary to prevent the occurrence of an invasive species episode. However, socioeconomic factors of the area often prevent such changes from occurring readily. Therefore only small or incremental changes in current socioeconomic causal factors may be possible.

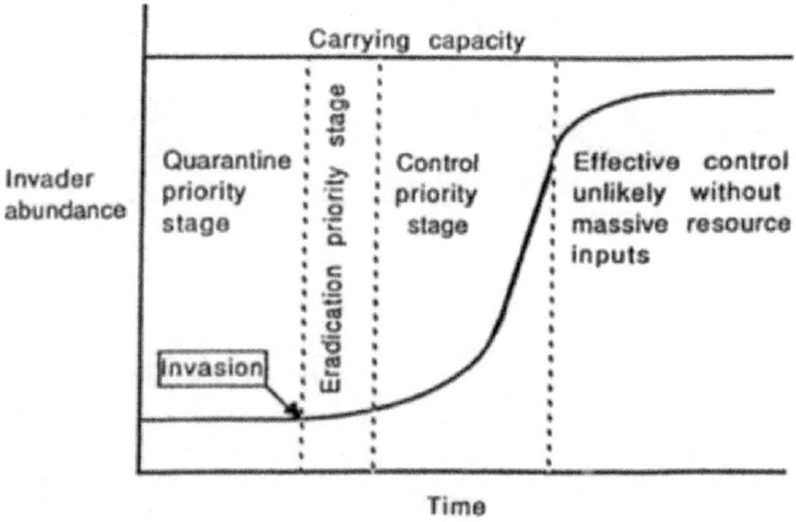

Figure 11. Phases of weed invasion and priorities for action at each phase. Ease of treatment of an invasion problem declines from left to right (after Chippendale, 1991).

Figure 12 demonstrates the course of a plant invasion from early introduction to its development into a major weed problem. At each stage of the invasion, human activities may act to encourage the spread, and changes in human behavior are usually required to reduce the impact of current plant invasions or to minimize future problems (Hobbs and Humphries, 1995).

Risk Assessment

Modeling weed invasions has been attempted in numerous studies over the last decade with varying success. Unfortunately, unquantifiable parameters and overgeneralizations have led to disagreement about the value of invasion models. Predictive models have most often used analytical diffusion equations to examine plant spread over a generalized area (Skellman, 1951; Holmes et al., 1994). However, the future of invasion modeling probably lies in the interplay between descriptions of species biology (intrinsic factors), and potential habitats and landscape scale connections (extrinsic factors) (Figure 4). Intrinsic factors include physiological growth parameters, growth rates, and demography, while extrinsic drivers, e.g. soil, climate, and land use, are the environmental baseline for assessing invasion risk. However, much of the above information is only now becoming available for invasive species and their pertinent habitats. A need for local and subsequently global databases is recognized (Wade, 1997), but the infrastructure for such a database is currently not present locally or globally.

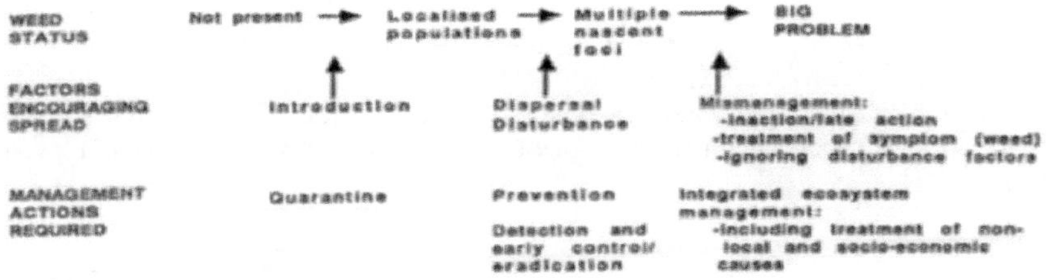

Figure 12. Phases of weed invasion, showing factors encouraging spread and management actions required at each stage. From Hobbs and Humphries, 1995.

A GIS approach

Present advanced information technology software, like Geographical Information Service (GIS), makes it possible to store and collate local data of invasive plants, and their pertinent ecosystem characteristics. The use of such software could enable managers to use local databases and then predict the vulnerability of certain ecosystems that might be invaded by particular species. (Figure 13).

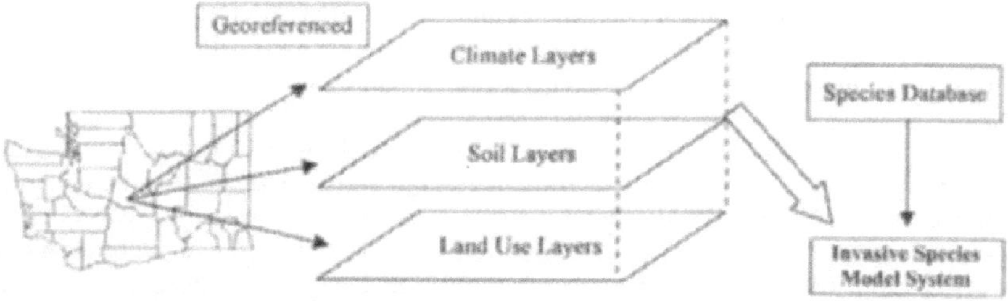

Figure 13. Schematic for the process faced by policy-makers and managers who must decide which nonindigenous species to either eradicate or control, once already present in the system, or keep out, if not already present in the system. From Radosevich et al. (2003).

Land managers also would be able to identify sites with the environmental characteristics necessary for invasive plant establishment and take means to alleviate them.

Global Risk Assessment

Some countries, like Great Britain, have made progress in creating the necessary framework for information surveillance of invasive species. For example, the Botanical Society of British Isles (BSBI) established an Alien Study Group in 1993 with the primary aim to study alien plants established in Britain and Ireland that are increasing and pose a threat to native plant species. The group compiles a register of alien plants and seeks volunteers to adopt a species and monitor its spread. The group gathers this information into a national mapping scheme for alien plants of the British Isles. The aim of this effort is to further increase awareness of alien plant species, including their accurate identification (Ewel et al., 1999). However, the efforts of various organizations, like the BSBI, remain poorly integrated and there is great need for such efforts to be recognized and implemented by government institutions.

SUMMARY

Plant invasions from the accidental or intentional introduction of new species to a region or area have altered many of the world's ecosystems. This problem is magnified by many factors that contribute to global change. Invasive plants, unlike other weed species, can displace wild plants from wild habitats without further human disturbance and therefore threatens the biodiversity of many habitats.

The invasion process is an effective way to study invasive plants. This process consists of three phases; introduction, colonization, and naturalization. Introduction is the initial successful entry of an invasive plant into a new area. It is usually aided by disturbance. Colonization often involves a prolonged lag time before the next phase of the process begins. Rapid exponential growth and spread of the new population also occurs during colonization. Naturalization exists when the new population inhabits all available niches, and carrying capacity is reached. Both intrinsic factors of biology and extrinsic factors of environment are necessary for successful invasions.

There are several approaches and strategies used by land managers to prevent, detect, monitor, contain and/or control invasive plant species. However, managers need tools that directly aid in the detection and monitoring of invasive plants while they remain in the small satellite population phase when early detection and control are most successful. Models general enough to predict risk for many species, but also adaptable enough to changing criteria for different species provide the greatest hope to achieve some predictable success against invasive species at an affordable cost.

REFERENCES

Andow D.A., Kareiva P.M., Levin S.A., Okubo A. Spread of invading organisms. Landscape Ecol 1990; 4:177-188.

Atkinson E.I., Cameron E.K. Human influence on the terrestrial biota and biotic communities of New Zealand. Trends Ecol Evol 1993; 8:447-451.

Auld B.A., Coote B.G. A model of a spreading plant population. Oikos 1980; 34:287-292.

Auld B.A., Hosking J., McFadyen R.E. Analysis of the spread of tiger pear and parthenium weed. Aust Weeds 1983; 2:56-60.

Baker H.G. The evolution of weeds. Ann Rev Ecol Syst 1974; 5:1-24.

Baker H.G. "Patterns of plant invasion in North America." In *Ecology of Biological Invasions of North America and Hawaii*, H.A. Mooney and J.A. Drake, eds, New York, NY: Springer-Verlag, 1986.

Baker H.G. "Aspects of the genecology of weeds." In *Genecology and Ecogeographic Races*, A.R Kruckeberg, R.B. Walker, A.E. Leviton, eds. San Francisco, CA: Pacific Division American Association for the Advancement of Science, 1995; pp. 189-224.

Betancourt J.L., VanDevender T.R., Martin P.S. *Fossil Packrat Middens: the Last 40,000 Years of Biotic Change in the Arid West.* Tucson: University of Arizona Press, 1990.

Booth B.D., Murphy S.D., Swanton C.S. "Invasive ecology of weeds." In *Weed Biology and Management*, Inderjit, ed. The Netherlands: Kluwer Academic Publishers, 2003; pp. 29-45.

Chippendale J.F. Potential Returns to Research on Rubber Vine (*Cryptostegia grandiflora*). M.S. Thesis. University of Queensland, Brisbane, 1991.

Cohen A.N., Carlton J.T. Accelerating invasion rate in a highly invaded estuary. Science 1998; 279:555-558.

Cousens R., Mortimer M. *Dynamics of Weed Populations.* New York, NY: Cambridge University Press, 1995.

Crooks J., Soulé M.E. "Lag times in population explosions of invasive species: causes and implications." In *Invasive Species and Biodiversity Management*, O.T. Sandlund, S.J. Schei, A. Vikens, eds. The Netherlands: Kluwer Academic Publishers, 1999; pp. 103-125.

Cronk Q.C.B., Fuller J.L. *Plant Invaders, a 'People and Plants' Conservation Manual. World Wide Fund for Nature.* London, UK: Chapman and Hall, 1995.

Crosby A.W. *Ecological Imperialism, The Biological Expansion of Europe, 900-1900.* New York, NY: Cambridge University Press, 1986.

D'Antonio C.M., Vitousek P.M. Biological invasions by exotic grasses, the grass/fire cycle, and global change. Ann Rev Ecol Syst 1992; 23:63-87.

Darwin C. *The Origin of Species by Means of Natural Selection.* Murray, London, 1859.

Davis M.A., Grime J.P., Thompson K. Fluctuating resources in plant communities: a general theory of invasability. J Ecol 2000; 88:528-534.

Dukes S.J. "Will the increase atmospheric CO_2 concentration affect the success of invasive species?" In *Invasive Species in a Changing World*, H.A. Mooney, R.J. Hobbs, eds. Washington, D.C.: Island Press, 2000; pp. 95-113.

Ellis R.G. Alien study group. BSBI News 1994; 65:44-45.

Enserink M. Biological invaders sweep in. Science 1999; 285:1834-1836.

Evans J. *Plantation Forestry in the Tropics.* 2nd ed. Oxford: Clarendon Press, 1992.

Ewel J. et al. Deliberate introductions of species: research needs. BioScience 1999; 49:619-630.

Field C.B., Chapin III F.S., Chiariello N.R., Holland E.A., Mooney H.A. "The Jasper Ridge CO_2 experiment: design and motivation." In *Carbon Dioxide and Terrestrial Ecosystems*, G.W. Koch, H.A. Mooney, eds. San Diego, CA: Academic Press. 1996; pp. 121-145.

Forman R.T.T. *Land Mosaics, The Ecology of Landscapes and Regions.* NewYork: Cambridge University Press, 1995.

Ghersa C.M., León R.J.C. "Successional changes in the agroecosystems of the Rolling Pampas. In *Ecosystems of Disturbed Ground,* L. R. Walker, ed. The Netherlands: Elsevier, 1999.

Ghersa C.M., Satorre E.H., Benech Arnold R.L., Martínez-Ghersa M.A. Advances in weed management strategies. Field Crops Res 2000; 67:95-105.

Grime J.P. *Plant Strategies and Vegetation Processes.* New York: John Wiley, 1979.

Groves R.H. "Invasion of Mediterranean ecosystems by weeds." In *Resilience in Mediterranean-type Ecosystems,* B. Dell, A.J.M. Hopkins, B.B. Lamont, eds. Junk: Dordrecht, 1986; pp. 129-45.

Henderson L. Alien invasive *Salix* spp. (willows) in the grassland biome of South Africa. S Afr For J 1991; 157:91-95.

Hilbig W. "Preservation of agrestal weeds." In *Biology and Ecology of Weeds*, W. Holzner, N. Numata. eds. The Hague: Dr W. Junk Publishers, 1982; pp. 57-59.

Hobbs R.J. Disturbance a precursor to weed invasion in native vegetation. Plant Protect Quart 1991; 6:99-104.

Hobbs R.J., Atkins L. The effect of disturbance and nutrient addition on native and introduced annuals in the Western Ausralian wheat belt. Aust J Ecol 1988; 13:171-179.

Hobbs R.J., Hopkins. From frontier to fragments: European impact on Australian vegetation. Proc Ecol Soc Aust 1990; 16:93-114.

Hobbs R.J., Humphries, S.E. An integrated approach to the ecology and management of plant invasions. Conser Biol 1995; 9:761-770.

Hodkinson D.J., Thompson K. Plant dispersal: the role of man. J Appl Ecol 1997; 34:1484-1496.

Holm L.G., Plucknett D.L., Pancho, Herberger J.P. *The World's Worst Weeds, Distribution and Biology.* Honolulu, HI: University of Hawaii Press, 1977.

Holmes E.E., Lewis M.A., Banks J. E, Veit R.R. Partial differential equations in ecology: spatial interactions and population dynamics. Ecology 1994; 75:17-29.

Holzner W., Numata M. *Biology and Ecology of Weeds.* The Hague: Dr. W. Junk Publishers, 1982.

Huenneke L.F., Hamburg S.P., Koide R.., Mooney H.A., Vitousek P.M. Effects of soil resources on plant invasion and community structure in Californian serpentine grassland. Ecology 1990; 71:478-491.

Humphries S.E., Groves R.H., Mitchell D.S. "Plant invasions and Australian ecosystems: a status review and management directions." In *Plant Invasions: The Incidence of Environmental Weeds in Australia.* Canberra, Australia: Australian National Parks and Wildlife Service, 1991; pp. 1-127.

Isaacson R.T. *The Andersen Horticultural Library's Source List of Plants and Seeds: A completely Revised Listing of 1993-1996 Catalogues.* 4th ed. Chanhassen (MN): Andersen Horticultural Library, 1996.

Jenkins P.T. "Free trade and exotic species introductions." In *Invasive Species Biodiversity and Management*, O.T. Sandlund, P.J. Schei, A. Viken, eds. Dordrecht, The Netherlands: Kluwer Academic Publishers. 1996; pp. 145-147.

Jenkins P.T. Free trade and exotic species introduction. Conser Biol 1996a; 10:300-302.

Jenkins T.J. "Trade and exotic species introductions." In *Invasive Species and Biodiversity Management*, O.T. Sandlund, P.J. Schei, A. Viken, eds. Dordrecht, The Netherlands: Kluwer Academic Publishers, 1999b; pp. 229-235.

Kornas J. "Plant invasions in Central Europe: historical and ecological perspectives." In *Biological invasions in Europe and the Mediterranean Basin*, F. di Castri, A.J. Hansen, M. Debussche, eds. Dordrecht, The Netherlands: Kluwer Academic Publishers, 1990; pp. 19-36.

Kowarik I. "Time lags in biological invasions with regard to the success and failure of alien species." In *Plant Invasions, General Aspects and Special Problems*. P. Pysek, K. Prach, M. Rejmánek, M. Wade, eds. SPB Academic Publishers, 1995; pp. 15-38.

Latore J., Gould P., Mortimer A.M. Spatial dynamics and critical patch size of annual plant populations. J Theort Biol 1998; 190: 277-285.

Lonsdale W.M. Inviting trouble: introduced pasture species in northern Australia. Aust J Ecol 1994; 19:345-354.

Luken J.O., Thieret J.W. Amore honeysuckle, its fall from grace. Bioscience 1996; 46:18-24.

MacDonald I.A.W., Loope L.L., Usher M.B., Hamann O. "Wildlife conservation and the invasion of nature Reserves by introduced species: a global perspective." In *Biological Invasions, a Global Perspective*. J.A. Drake, H.A. Mooney, F. di Casteri, R.H. Groves, F.J. Kruger, M. Rejmánek, M. Williamson, eds. Chichester, UK: John Wiley, 1988; pp. 215-255.

Mack R.N. "Invading plants: their potential contribution to population biology." In *Studies in Plant Demography*, J. White, ed. London, U.K.: Academic Press, 1995; pp. 127-142

Mashhadi HR. Ecophysiological characteristics of C_3, C_4 and CAM plants and their implication in agriculture. Agric Sci Technol 1989; 5: 41-58.

Mashhadi HR. Weeds and changing climate. Proceedings of Fourth Iranian Crop Production and Breeding Congress. Isfahan, Iran, Isfahan University of Technology Press, 1998; pp. 44-62.

Maxwell B.D., Wilson M.V., Radosevich S.R. Population modeling approach for evaluating leafy spurge (*Euphorbia esula*) development and control. Weed Technol 1988; 2:132-138.

McNeely J.A. "The future of alien species: changing social views." In *Invasive Species in a Changing World*, H.A. Mooney, R.J. Hobbs, eds. Washington, D.C.: Island Press, 2000; pp. 171-190.

McNeely J.A. "The great reshuffling: how species help feed the global economy." In *Invasive Species Biodiversy and Management*, O.T. Sandlund, P.J. Schei, A. Viken, eds. Dordrecht, The Netherlands: Kulwer Academic Purblishers, 1999; pp. 11-33.

Moody M.E., Mack R. N. Controlling the spread of plant invasions: the importance of nascent foci. J Appl Ecol 1988; 25:1009-1021.

Mooney H.A., Hobbs R.J. "Global change and invasive species: where do we go from here." In *Invasive Species in a Changing World*, H.A. Mooney, R.J. Hobbs eds. Washington, D.C.: Island Press. 2000; pp. 425-434.

Mooney H.A., Hofgaard. "Biological Invasions and Global Change." In *Invasive Species and Biodiversity Management*, O.T. Sandlund, P.J. Sahei, A. Viken, eds. Dordrecht, The Netherlands: Kluwer Academic Publishers, 1999; pp. 139-148.

Nilson E.T., Muller C. H. A comparison of the relative naturalization ability of two *Schinus* species in Southern California. Bull Torrey Bot Club 1980; 107:51-56.

Office of Technology Assessment. *Harmful Non-indigenous Species in the United States*. Washington, DC: US Government Printing Office. Report OTA-F-565, 1993.

Owen S.J. *Ecological Weeds on Conservation Land in New Zealand: a Database*. Wellington: Department of Conservation, Wellington, 1997.

Panetta F.D., Mitchell N. D. Homoclime analysis and the prediction of weediness. Weed Res 1991; 31: 273-284.

Parson W.T., Cuthbertson E.G. *Noxious weeds of Australia*. Melbourne: Inkata Press, 1992.

Patterson D.T. Weeds in a changing climate. Weed Sci 1995; 43:685-701.

Perring C. The economics of biological invasions. Proceedings of the workshop on best management practices for preventing and controlling invasive alien species. South Africa, 2000.

Pickett S.T.A., White, P.S. *The Ecology of Natural Disturbance and Patch Dynamics*. Orlando, Florida: Academic Press, 1985

Prescott-Allen R., Prescott-Allen E. How many plants feed the world? Conser Biol 1990; 4:365-374.

Pysek P. "Clonality and plant invasions: can a trait make a difference?" In *The Ecology and Evolution of Clonal Plants*, H. de Kroon, H. van Groenendael, eds. Leiden: Backhuys 1997; pp. 405-427.

Radosevich S.R., Holt J., Ghersa C.M. *Weed Ecology, Implications for Management*. New York: John Wiley & Sons, 1997.

Radosevich S.R., Stubbs M.M., Ghersa C.M. Plant invasions: mechanisms, processes and patterns. Weed Sci 2003; (in press).

Rapoport E.H. "Tropical versus temperate weeds: a glance into the present and future." In *Ecology of Biological Invasion in the Tropics*, P.S. Ramakrishnan, ed. New Delhi: International Scientific Publications, 1991; pp. 441-451.

Rejmánek M. "Invasive plant species and invasible ecosystems." In *Invasive Species and Biodiversity Management*, O.T. Sandlund, P.J. Schei, A. Vilken, eds. Dordrecht, The Netherlands: Kluwer, 1999; pp. 79-102.

Rejmánek M. Invasive plants: approaches and predictions. Aust Ecol 2000; 25:497-506.

Rejmánek M., Randall J.J. Invasive alien plants in California: 1993 summary and comparison with other areas in North America. Madroño 1994; 41:161-177.

Richardson D.M., Allsopp N., D'Antonio C., Milton S.J., Rejmánek, M. Plant invasions – the role of mutualisms. Biol Rev 2000; 75: 65-93.

Salisbury E.J . *Weeds and Aliens.* London: Collins, 1961.

Sauer J.D. Plant Migration: The Dynamics of Geographic Patterning in seed Plant Species. *University of California Press, 1988.*

Sattaur O. The shrinking gene pool. New Scient 1989; 1675:37-41.

Scherer-Lorenzen Elend A., Nöllert S., Schulze E.D. "Plant invasions in Germany: general aspects and impact of nitrogen deposition." In *Invasive Species in a Changing World*, H.A. Mooney, R.J. Hobbs, eds. Washington, D.C.: Island Press, 2000; pp. 351-368.

Skellman J.G. Random dispersal in theoretical populations. Biomatrika 1951; 38:196-218.

Smith C.S., Lonsdale W.M., Fortune J. When to ignore advice: invasion predictions and decision theory. Biol Invasion 1999; 1:89-96.

Smith S.D., Strain B.R., Sharkey T.D. Effects of CO_2 enrichment on four Great Basin grasses. Funct Ecol 1987; 1:139-143.

Stein S. *Planting Noah's Garden.* New York: Houghton, 1996.

Tenenbaum D. Weeds from hell. Technol Rev 1996; 99:32-40.

Umweltbundesamt. Daten zur umwelt. Berlin: Erich Schmidt, 1997.

Wade M. "Predicting plant invasions: making a start." In *Plant Invasions: Studies from North American and Europe*, J.H. Brock, M. Wade, P. Pysek, D. Green, eds. Leiden, Backhuys Publishers, 1997.

Watson A.K. "Integrated management of leafy spurge." In *Leafy Spurge,* A.K. Watson, ed, Weed Sci. Soc. Am. Champaign, IL. 1985; pp. 93-104.

Watson R.T., Zinyowera M.C., Moss R.H. *Climate Change: Impacts, Adaptations, and Mitigation of Climate Change.* The IPCC Second Assessment Report. New York: Cambridge University Press, 1995.

Weeda E.J. "Invasions of vascular plants and mosses into the Netherlands." In Proceedings of the Koninklijke Nederlandse Akademie van Wetenschapen, Amsterdam. Series C: Biological and Medical Sciences, W. Joenje, K. Bakker, L. Vlijm, eds, 1987; pp. 19-29.

Williams P.A., Nicol E., Newfield M. "Assessing the risk to indigenous biota of new plant taxa not yet in New Zealand." In *First International Workshop on Weed Risk Assessment*, J. G. Virture, R. H. Groves eds. Adelaide, Australia: Commonwealth Scientific and Industrial Research Organization, 2000.

Williamson M. Invaders, weeds and the risk from genetically manipulated organisms. Experimentia 1993; 49:219-224.

Williamson M. *Biological Invasions. Population and Community* Series 15. Chapman and Hall. London, 1996.

Woodward F.I. *Climate and Plant Distribution.* Cambridge: Cambridge University Press, 1987.

Zamora D.L., Thill D.C. "Early detection and eradication of new weed infestations." In *Biology and Management of Noxious Rangeland Weeds*, R.L. Sheley, J.K. Petroff, eds. Corvallis, OR: Oregon State University Press, 1999.

CHAPTER 2

INVASIVE ECOLOGY OF WEEDS IN AGRICULTURAL SYSTEMS

BARBARA D. BOOTH[1], STEPHEN D. MURPHY[2] & CLARENCE J. SWANTON

[1] *Department of Plant Agriculture, University of Guelph, Guelph, Ontario, Canada.* [2] *Department of Environmental and Resource Studies, University of Waterloo, Waterloo, Ontario, Canada*

No community is truly safe from plant invaders. Nor should it be. Succession is the normal process of species invading and replacing other species (Huston, 1994; Johnstone, 1986; Rejmánek, 1989), although this connection between invasion and succession is often ignored (Davis et al., 2001). The problem is that some plant invasions have high ecological and economic costs associated with them (Parker et al., 1999; Pimentel et al., 2000, 2001). Some economic costs of invasion are easily quantifiable (e.g. the cost of weed control, yield loss) whereas other are not (e.g., damage to ecosystems, loss of recreational land, aesthetics). The cost of invasive plants to crop and pasture land in the United States is calculated as well over 34 billion dollars annually (Pimentel et al., 2000), while in India, the cost stands even higher at 38 billion per year (Pimentel et al., 2001). Any ability to predict what species could invade will therefore have economic and ecological benefits.

Agricultural weeds are essentially species that are successful invaders in disturbed habitats (Swanton et al., 1993). Agricultural weed ecology should therefore be put in the broader context of plant invasions. There is a large body of ecological theory considering how, why, and when a species will invade a particular habitat. In this chapter, we give an overview of what makes a community invasible, what traits are typical of invasive species, and how trait-environment interactions determine whether an invasion succeeds. Furthermore, we discuss how this knowledge can then be used to begin to predict invasions. Currently, we understand why an invasion *has* occurred, but we are not very good at predicting whether an invasion *will* occur. While many agricultural weeds have persisted for decades or centuries, we have to be prepared for new weed problems to emerge because new non-native species are continually introduced (National Research Council, 2002). In addition, changes related to crop selection, management system, and the environment will also lead to shifts in weed species composition. Our ultimate goal should be to develop ecologically-based predictive management systems, rather than purely reactive systems.

Inderjit (ed.), Weed Biology and Management, 29-45.

WHAT MAKES A COMMUNITY INVASIBLE?

Not all plant communities are equally invasible. Some communities are more invasible than others while some seem to be more resistant to invasions. Being able to characterize or predict which communities or habitat types are easier to invade would have obvious practical benefits. For example, we could work on protecting vulnerable habitats and not worrying about ones resistant to exotic invasions. Agricultural systems are generally thought of as invasible, because they are species poor, disturbed, nutrient rich environments with high light penetration below the crop canopy. Thus, there seems to be plenty of opportunities for the establishment of newly introduced weeds. But if this were true, we really should be asking why there are so *few* agricultural weeds. The point is that management creates opportunities for only certain types of species to invade. So, how do we identify habitats likely to be invaded by weeds? Three habitat features that are generally touted as encouraging invasions are disturbance, low species richness, and resource availability.

Disturbance

Disturbance is often cited as a precursor to invasions (Baker, 1965; Crawley, 1987; Fox and Fox, 1986; Hobbs and Huenneke, 1992; Orians, 1986) and is said to provide a window of opportunity for invasion (Hobbs, 2000). In fact, disturbance is an integral part of any ecosystem (natural or managed) and specific types of disturbances may be required to maintain a community. Fire, for example, is a natural part of some ecosystems (e.g., boreal forest) and is sometimes used to control invasions by exotic species (e.g., *Centaurea solstitialis* - yellow starthistle in California grassland) (Hastings and DiTomaso, 1996; Kyser and DiTomaso, 2002).

While disturbance is a natural part of an ecosystem, human activities can alter the disturbance regime and intensity (Hobbs, 2000). A change in the type, frequency and extent of a disturbance in a community will create different invasion opportunities (Crawley et al., 1999; Lonsdale, 1999; Stohlgren et al., 1999). That is why weed communities differ with differing management systems (Swanton et al., 1999). For example, the expansion of *Hordeum jubatum* (foxtail barley) and *Taraxacum officinale* (dandelion) were facilitated by an increase in reduced tillage (Derksen et al., 1993; Shrestha et al., 2002). Seemingly innocuous species may become weeds after a new type of disturbance occurs. Therefore, while disturbance does not necessarily cause invasions, certain types of disturbance may encourage the invasion by some weeds (Dukes and Mooney, 1999; Hobbs, 1989; Vitousek et al., 1996) if it changes or removes the abiotic or biotic constraints acting on the community.

Management practices such as tillage, nutrient addition, and herbicide use are the primary agents of disturbance in agricultural landscapes. Tillage is a general disturbance, removing all vegetation in its path. It selects for weed species that are able to recover through a persistent seedbank, or vegetative sprouting from rootstock. High levels of tillage select for ruderals which are fast growing annuals with rapid seed production. Ruderals are typical of most annual weeds in a conventional cropping system. Weeds are able to take advantage of changing nutrient status after the addition of fertilizers. For example, weeds that rapidly take up nitrogen and respond with increased growth will likely have an advantage over species that respond slowly. Most herbicides are selective for specific weed spectra, thereby lending themselves to weed escapes of resistant or tolerant species, or to species with germination patterns that avoid control from the residual herbicides. A change in herbicide usage results in a different spectra of weeds. For example, the move to non-residual herbicides within cropping year may lead to the establishment of winter annual weeds in corn and soybean residue (Chicoine and Hahn, 2001).

Disturbance, however, is not always necessary for an invasion to occur, and even intact natural ecosystems can be invaded. For example, King and Grace (2000) found that *Imperata cylindrica* (cogongrass) did not require disturbance gaps to invade wet pine savanna. In some cases, disturbance may even inhibit invasion (Lodge, 1993). In general, species will invade when there is a change in the disturbance regime that removes the abiotic or biotic constraints acting on a community.

Species Richness

Elton (1958) hypothesized that there was a negative relationship between native species richness (diversity) and community invasibility, although similar observations date back to Darwin (1859). Elton suggested that communities with many species would be invasion resistant, while species-poor communities will be highly invasible. Although some lines of evidence used by Elton are flawed (Goodman, 1975), there are other reasons to expect such a pattern. MacArthur (1955) suggested that a more diverse community will have more links among species and therefore if a link is broken the system will still be relatively intact. Also, niches in species rich communities will be more saturated and therefore there will be few niches for invading species (Moore et al., 2001). Species-rich communities are thought to sequester limiting resources more than species poor communities, therefore, invasive species will be less able to establish in species poor communities (Knops et al., 1999; Tilman, 1987, 1988).

There is empirical evidence to support the hypothesis that species poor communities are more invasible (Case, 1990; Doak et al., 1998; Knops et al., 1999; Rejmánek, 1989; Tilman, 1997, 1999). Fox and Fox (1986), for example, reviewed three data sets from two shrubland and one heathland site in Australia and noted a negative relationship between native species richness and the percent of species that were non-native. Tilman (1997, 1999) also observed a negative relationship between species richness and invasibility showing that high species richness was related to a decrease in the invasion success in prairie grasslands. Van der Putten et al. (2000) tested the effect of late-successional species richness on the invasion success of early successional species (arable weeds) at five European sites. They found that sites with 15 species were less invasible than sites with four species, although it was unclear whether this effect was due to the inhibition of weeds by higher biomass, or by the effects of richness itself. Although there is some evidence to support the hypothesis that species rich communities are less invasible, Elton's idea does remain controversial (e.g., Crawley et al., 1999; Levine and D'Antonio, 1999; Lonsdale, 1999; Mack et al., 2000).

There is also evidence to the contrary, that areas with high species richness are more vulnerable to invasion than species poor communities (Levine, 2000; Palmer and Maurer, 1997; Planty-Tabacchi et al., 1996; Robinson et al., 1995; Stohlgren et al., 1998, 1999). May (1973) proposed that diverse communities are continually gaining and losing species and therefore invasion is part of habitat function. Huston and D'Angelis (1994) highlighted that since species rich communities have high levels of habitat heterogeneity, introduced species are likely to find suitable habitats to invade. The evidence supporting this relationship is typically observational or correlation studies in natural habitats (Levine, 2000). Results typically show that species rich communities have the highest number of non-native invasive species (Planty-Tabacchi et al., 1996; Robinson et al., 1995; Stohlgren et al., 1998, 1999, 2001).

There may, in fact, be no general relationship between richness and invasibility, or it may be that we are asking too general a question. The lack of an obvious relationship between richness and invasibility does not mean that richness is not important; it means that richness is too wide a measure and has little explanatory power (Crawley et al., 1999). Species richness may simply be too broad a factor to explain the relative invasibility of communities (Levine and D'Antonio, 1999) because it is more likely an aggregation of other variables that are determinants of invasibility. As richness increases, for example, so may propagule pressure and spatial heterogeneity, both of which increase invasibility. Heterogeneity will increase because species will differ in height, canopy architecture, and rooting depth (Palmer and Maurer, 1997). This will allow more species to invade an area because there will be more types of microenvironments available. In addition, species rich communities may occur in habitats that have more resources available which could allow more introduced species to invade. Simply put, diverse communities may have more invasive species because the environment is favourable to many species (Levine, 2000; Levine and D'Antonio, 1999; Lonsdale, 1999). Factors that allow more species to co-exist may promote invasion (Levine, 2000) giving the appearance of cause-and-effect.

A community's composition can complicate our interpretation of richness effects. Observed effects of richness on invasibility can be caused by either the overriding effect of one or several keystone species or the effect of richness itself. Having more species increases the chance that one capable of excluding the introduced species will be present (Grime, 1997; Hooper and Vitousek, 1998; Lepš et al., 2001). In linking to agricultural systems, crop monocultures (low species richness) may vary in their susceptibility to invasion. Forage and cereal crops, for example, may be less invasible than row crops because they have high seeding rates and early canopy development and therefore less opportunity for invasion.

A final reason why it may be impossible to detect species richness and invasibility relationships is that different patterns will emerge depending on the scale of observation. Stohlgren et al. (1999) observed a negative relationship between species richness and invasibility at the 1-m^2 scale but a positive relationship at the landscape scale (1000-m^2) and biome scale in central and western USA. The relationship between species richness and habitat invasibility is not simple, and therefore we cannot conclude that agricultural systems are invasible because they are species poor.

Resource Availability

Davis et al. (2000) recently proposed the hypothesis of fluctuating resource availability to explain why the invasibility of a community will change over time and space. This theory incorporates ideas of both species richness and disturbance to explain invasibility. Davis et al. (2000) stressed the importance of unused available resources (Figure 1). Therefore, a nutrient rich community will not be invasible as long as the current community is using all the available nutrients. Conversely, a nutrient poor community will be invasible if the community is not sequestering all of the nutrients. This hypothesis is linked to other hypotheses of invasibility. For example, disturbance has often been associated with invasion because it reduces or eliminates competitors or because it increases the availability of resources (D'Antonio, 1993; Hobbs, 1989). According to Davis et al. (2000) disturbance is only one mechanism that alters resource availability. When the community is disturbed, or when a species is removed, then released resources become available to other colonists.

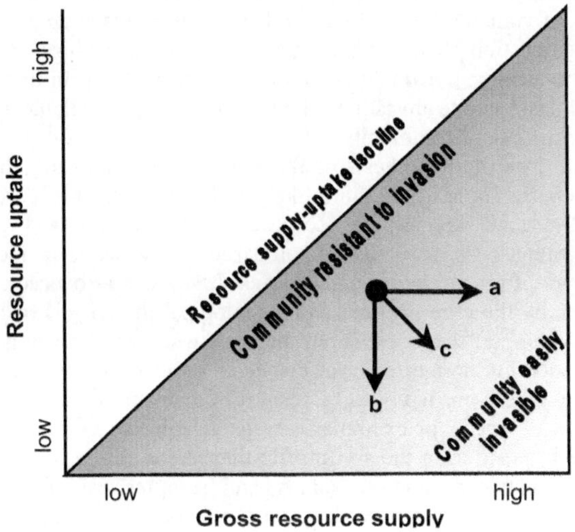

Figure 1. Theory of fluctuating resource availability. The invasibility of a community will depend on the ratio of gross resource supply and resource uptake. A community (•) becomes more invasible if gross resource supply increases (vector a), if resource uptake decreases (vector b), or if both happen simultaneously (vector c) (Redrawn from Davis et al., 2000).

Even the removal of a rare species can increase the invasibility of a community by changing the nutrient dynamics within a community. Lyon and Schwartz (2001), for example, experimentally removed rare species and found that that the exotic *Lolium multiflorum* (Italian ryegrass) established more in these plots than in either undisturbed control plots or when an equal biomass of more common species was removed. The authors speculated that when rare species were removed, resources were released that the dominant species could not use, and this allowed *L. multiflorum* to invade.

Phenomena other than disturbance can also release nutrients. For example, herbivory by the leaf beetle *Trirhabda canadensis* in experimental monocultures of *Solidago missouriensis* (Missouri goldenrod) increased the availability of water, nitrate and light (Brown, 1994). Even though the relative growth rate of *S. missouriensis* increased in response to herbivory, more species were able to invade compared to monocultures without the beetle.

Changing the abiotic environment can also alter the nutrient status of a community. The environment changes in regular predictable ways such as the seasonal fluctuations in temperature and precipitation or in unpredictable ways; for example, when we experience a particularly dry summer. Such natural fluctuations in the abiotic environment could affect whether a species can invade and/or persist in a community. For example, a drought may create a window of opportunity for invasion if some species are killed and more light becomes available. Alternatively, a flood may bring in an influx of nutrients encouraging the invasion of species able to take advantage of high nutrient and moisture availability.

The hypothesis of fluctuating resource availability helps to explain why a community's invasibility changes over time. A community may have regular or intermittent periods of vulnerability to invasion dependent on resource availability and use. This is why the question '*Which* communities are more invasible?' is misleading. Rather, we should be asking '*When* is a community invasible?' According to Davis et al. (2000) communities become invasible when resources are available in excess of what can be used by the extant community.

What application does the hypothesis of fluctuating resource availability have for agricultural ecosystems? These systems are likely to have repeated resource additions due to inputs of nutrients and water. Resources are also released following a disturbance such as tillage (e.g., oxidation of soil carbon) or when herbicide use causes the release of nutrients from decaying plant tissue. Resources also become available when nutrient uptake by the crop declines as physiological maturity is reached and as the crop residue remaining after harvest is returned to the soil and broken down into its constituent components. Furthermore, because species richness in agricultural systems is managed to be low, there are likely to be resources available in excess of what the crop species can use. A weed, for example, may gain better access to nutrients if its roots are in a different position in the soil profile than the crop's roots. The key is to consider the balance between resource supply and resource use. Even short-term increases in resource availability can have long-term effects on community composition (Davis and Pelsor, 2001).

Clearly, we cannot predict whether a community will be invasible based on simple habitat characteristics. A community's susceptibility to invasion is dependent on many factors and will change over time. If communities are not inherently invasible or invasion resistant, what other factors should we consider in our attempt to understand invasions? The other side of the story is the weed species that do the invading. For example, only certain types of weeds will invade a corn field. Why is this? Only certain weeds have the set of traits that allow it to invade the habitat created by the corn management system.

WHAT TRAITS ARE TYPICAL OF INVASIVE WEEDS?

Early research into invasions focused on identifying traits typical of invasive weeds. Baker (1965) listed the traits of an ideal weed and said that a plant possessing all of the traits would be "a formidable weed indeed" (Table 1). Not all 'formidable' weeds have many of these traits, and species with only few of these traits can be weeds. *Impatiens glandulifera* (Himalayan balsam), for example, is an important weed in Britain and yet possesses only two of Baker's characteristics. Other less pernicious weeds such as *Veronica persica* (Persian speedwell) and *Stellaria media* (common chickweed) have seven of Baker's traits (Perrins et al., 1993). The difficulty with creating lists of typical weed traits is that there are always exceptions (Mack, 1996), and most non-weeds will posses at least some of the traits.

Table 1. Baker's (1965) list of the ideal characteristics of weeds.

1. Germinates in a wide range of environmental conditions.
2. Long-lived seeds that are internally controlled so that germination is discontinuous.
3. Rapid growth from vegetative through flowering stage.
4. Self-compatible, but not completely autogamous or apomictic.
5. Cross-pollination (when present) by wind or generalist insects.
6. Seeds produced continuously throughout the growth period.
7. Seed production occurs under a wide range of environmental conditions.
8. High seed output when environmental conditions are favourable.
9. Propagules (seeds) adapted to short- and long-distance dispersal.
10. If perennial, has a high rate of vegetative reproduction or regeneration from fragments.
11. If perennial, ramet attachments fragment easily, so it is difficult to pull from the ground.
12. Strong potential to compete interspecifically via allelopathy, rosettes, rapid growth, and other means.

Grime (1977, 1979) expanded the use of traits and developed the concept of life history strategies as groups of species with the same sets of traits. He defined three main strategies: ruderals (R), stress-tolerators (S) and competitors (C), as well as intermediate groups (e.g., competitive ruderals, CR). Each strategy is associated with a set of traits based on the established stages of a plant's life cycle. R-strategists are adapted to disturbed but potentially productive environments and are able to maximize growth and reproduction. S-strategists are adapted to harsh and stressful environments where competition and disturbance are rare. C-strategists are able to maximize resource use in undisturbed but productive habitats by increasing their vegetative growth and reducing allocation to reproduction. Weeds in cultivated annual crops tend to be R-strategists while C and S strategists are more likely in no-till systems.

Baker's and Grime's schemes are based on fairly simple traits and in Baker's case, are only vaguely defined (Perrins et al., 1993). We are beginning to understand that simple trait lists do not adequately describe or explain weed invasions. Lodge (1993) summarized traits associated with weediness but noted that when tested, many of these were statistically rejected, or else there were too many exceptions to make them useful. For example, the idea that a high '*r*' (intrinsic rate of population increase) increases invasiveness is not supported (Williamson, 1989). Are there, then, general features typical of invasive plants?

General Features of Invasive Species

Life History Traits

A number of attempts have been made at characterizing weeds by their life history traits. Pyšek et al. (1995), for example, compared the native flora of the Czech Republic with 132 exotic species introduced since 1492. They compared trait lists of native species with those of successful invaders and found that compared to native species, exotic species tended to be seed producing annuals, and there were more competitive (C) and competitive-ruderal (CR) species, and fewer stress tolerant species (S) according to Grimes' strategies. The characteristics of successful invaders differed depending on whether they invaded semi-natural or human-made habitats. Species that invaded human-made habitats tended to prefer drier sites and were either seed producing annuals or survived by rhizomes, tubers or bulbs. This approach shows that no single trait can be used to explain the weediness of a species, but that general sets of traits may be characteristic of plant groups.

Perrins et al. (1992) identified three traits that were consistently associated with annual plants classified as weeds in Britain. These were the abilities to flower late in the season, disperse in the gut of animals, and germinate in the spring. These roughly relate to some of Baker's traits but excludes most of them. Others researchers, however, have not been able to detect such generalities (Smith and Knapp, 2001). Thompson et al. (1995), for example, found that species of exotic and native plants in England, Scotland, the Republic of Ireland and the Netherlands whose ranges were expanding were almost indistinguishable as far as trait lists.

Large Native Geographical Range

Goodwin et al. (1999) tested whether simple biological attributes (life form, height, length of flowering period, and size of native range) could be used to characterize the invasiveness of a species. They compared pairs of congenerics where one species had successfully invaded New Brunswick, Canada and the other had not. They found that the size of a species's native range was the single best predictor of invasiveness and could predict correctly whether a species would be invasive or non-invasive for about 70% of the species tested. Range seems to be important for two reasons. First, widespread species may be more widely dispersed simply because they are in more locations and therefore are more likely to picked up dispersal agents. Second, widespread species are usually adapted to a wide range of environmental conditions and therefore are more likely to survive in a new habitat following dispersal. Native habitat range has been found to be important by others (Rejmánek, 1995; Roy et al., 1991). For example, species that are widespread in southern Africa will tend to become weeds if introduced into Australia (Scott and Panetta, 1993).

Taxonomic Patterns

Detecting patterns of plant families or other taxa that are more likely to posses invasive species can be difficult because the data are often anecdotal, there may be problems with proper identification and because taxonomy is not consistent world wide (Heywood, 1989; Pyšek, 1998). Such an endeavor would require a detailed census in every country where all plant species were identified correctly and classified as native or exotic.

Nevertheless, we can make some generalities about what taxa are more likely to have invasive species. For example, Pyšek (1998) examined data from exotic floras in 26 regions of the world. He found that invasive families were spread throughout the phylogenetic tree but that some families with many invasive species were concentrated together. For example, the subclasses Caryophyllidae and Asteridae contained many families with high numbers of invasive species. These results were in terms of sheer number of invasive species; however, if we take the family size into account then a different pattern emerges. Here, the families with the highest proportion of the species that are exotic somewhere in the world are Papaveraceae, Chenopodiaceae, Amaranthaceae, Brassicaceae, Polygonaceae and Poaceae. When we divide plant families into groups that have an over and under-representation of invasive species, we can see some patterns. Daehler (1998) looked at the taxonomic distribution of agricultural weeds and classified them as over- or under-represented depending on how many weedy species you would expect based on family size. He concluded that Amaranthaceae, Convolvulaceae, Cyperaceae, Papaveraceae, Poaceae, and Polygonaceae have more terrestrial agricultural weeds than expected and that Araceae, Asclepiadaceae, Bromeliaceae, Gesneriaceae, Melastomataceae and Orchidaceae had fewer than expected.

INTERACTION OF SPECIES TRAITS WITH HABITAT CHARACTERISTICS

Much of weed ecology is based on in-depth studies of specific weed populations in the habitats that they invade. This garners much species-specific data that are useful in building weed population models under specific management systems. An alternative approach is to identify species traits that confer weediness in the context of habitat characteristics (Booth and Swanton, 2002). If we follow a trait-based approach, then we must think in terms of 'invasive traits' rather than 'invasive species' because the abiotic and biotic environment removes (filters) traits rather than species from the community (Booth and Swanton, 2002; Weiher and Keddy, 1999). The composition of a plant community is determined by the interaction of species traits and the environment. A weed will fail to invade when it does not possess the suite of traits necessary to survive and reproduce in that new environment. A trait is only advantageous under certain circumstances. For example, herbicide resistance is an advantageous trait only if the plant is growing in a habitat where the specific herbicide is used. In habitats where the herbicide is not applied, the herbicide resistant trait may, in fact, be detrimental if it is associated with a loss of fitness (Parks et al., 1996).

Both the species-based and trait-based approaches have useful applications. In many ways, a trait-based approach is simpler than a species-based approach because it does not require an in-depth knowledge of how every species behaves in every habitat type. Rather it requires an understanding of how traits interact with the environment. A trait-based approach has the added advantage of being more likely to lead to general principles because the results are not species specific.

Of course, a trait-based approach necessitates being able to determine what traits are biologically relevant. Any scheme to predict invasiveness should include both vegetative and reproductive traits; the former influences the acquisition and storage of resources and the latter influences recolonization ability (Díaz et al., 1999a). Several authors (Díaz et al., 1999b; Gitay et al., 1999; Lavorel et al., 1997; McIntyre et al., 1999; Weiher et al., 1999) have presented trait lists designed for natural communities, while Booth and Swanton (2002) have summarized these and put them in the context of agricultural ecosystems (Table 2).

Once important traits are identified, they can be related to environmental characteristics using a number of multivariate statistical techniques. Often groups of species with similar traits can be identified; these form functional groups. Species within a functional group serve similar ecological functions within a community and respond to ecological filters in a similar manner. Smith et al. (1997) review a number of ways to construct functional groups. Nobel and Slatyer (1980) proposed that the functional group approach "could be applied to man-modified communities, and to such phenomena as weed invasion or suppression".

Díaz et al. (1992, 1999b), for example, related important plant traits to climatic variables using Detrended Correspondence Analysis (DCA). By multiplying a species-by-site matrix with a trait-by-species matrix, they obtained a trait-by-site matrix that could then be subjected to DCA. This procedure identified what traits were important along a climatic gradient and under various land-use (disturbance) regimes. In addition, a Two Way Indicator Species Analysis (TWINSPAN) was used to separate species into functional groups.

Similarly, Kleyer (1999) used clustering techniques to identify functional groups of plants in the agricultural landscape of Germany. He sampled the vegetation at 129 sites including fields, mown and abandoned grassland, roadsides, and wet and dry meadows. Species were characterized by a number of traits, and using a cluster analysis were placed in functional groups. At each site, he also assessed disturbance (tillage, herbicide use, trampling and cutting) and resource supply (soil moisture and nutrient availability). Using these data, the author could place functional groups along disturbance and resource supply gradients and was able to draw conclusions about what functional groups were more likely to occur under specific combinations of disturbance regime and resource supply. Kleyer also examined how response planes overlapped to predict which of the individual traits might be more likely to occur in specific disturbance resource combinations. For example, functional groups with extensive lateral extension tended to occur at low disturbance intensity at all levels of resource supply. Also, as disturbance intensity and/or resource supply increased so did the presence of functional groups with early reproduction.

Advantageous Traits will vary with the Stage of Invasion

One issue that can complicate our ability to detect which plant traits are advantageous is that traits conferring a benefit will change depending on the species's stage of invasion. A successful invasion requires that a species arrive, become established and reproduce (i.e. colonize), and then spread (naturalize) (Richardson et al., 2000b; Wade, 1997). Therefore, traits useful at all stages of invasion should be used when we try to identify invasive weedy species (Kolar and Lodge, 2001).

The first step of an invasion is for a speciess' propagules to disperse to a recipient site. Propagule pressure (the number and rate at which propagules are released) is an important factor in the success of an exotic species (Scott and Panetta, 1993). Most introductions fail because species die in transit or shortly after they arrive (Lodge, 1993). Traits that are important to dispersal are related to seed and fruit morphology (e.g. size, weight presence of wings or awns) and nutritional content.

Table 2. List of traits that could be used as a starting point for a trait-based approach to weed invasions. Some or all of the traits could be selected to record, and of these only some traits would be ecologically important. Based on lists in Díaz et al. (1999b), Díaz Barradas et al. (1999), Kleyer (1999) and McIntyre et al. (1999) and modified from Booth and Swanton (2002).

Vegetative Traits	**Reproductive Traits**
Size	Seed size (max. length)
Height:width ratio	Seed shape (length:width:depth)
Specific leaf area	Seed number (per plant)
Life cycle	Weight of dispersal unit (fruit or seed)
Life span	Seed dispersal
General form (erect, rosette, vine)	Fruit type
Leaf form (evergreen, deciduous)	Season of germination
Leaf angle	Age of first reproduction
Leaf size	Peak period of flower and fruit production
Leaf shape (length:width)	Pollination mode
Photosynthetic metabolism	
Potential relative growth rate	**Agricultural Specific Traits**
Drought avoidance	Herbicide tolerance and resistance
Palatability	Weed size relative to crop
Leaf texture	Seedling emergence timing relative to crop
Root morphology	
Maximum rooting depth	
Extent of clonal expansion	
Resprouting ability	
Mycorrhizal associations	
Storage organ	

After a species arrives, individuals must be able to germinate and grow in the new habitat. The important traits here relate to the speciess' abiotic requirements for germination and growth. A weed with good dispersal ability may not invade a new habitat if its seed or seedling requirements are not met. Following the establishment of individuals, a species must reproduce to maintain a self-sustaining population. This depends on its ability to reproduce sexually or asexually. Asexual reproduction has the obvious benefit of not requiring pollinators. If a non-native species cannot self-pollinate or if specialist pollinators are absent the population will not persist. While many species of *Ficus* (figs) were introduced into Florida, the three species that became invasive did so only after their specialist pollinator wasps were also introduced (Richardson et al., 2000a). When a species is first established, populations will be small and isolated. Therefore, for a population to succeed, it must be able to persist at low population numbers. Genetic effects (such as inbreeding depression) and ecological effects (low pollination rate) can cause small populations to become extinct.

As the weed expands in number of individuals and in the area covered, it will be present in many subpopulations that may interact following metapopulation dynamics. Even established populations may become extinct due to low density effects. Species that can grow but not reproduce will be dependent on continual reintroduction. For example, *Agrostemma githago* (corn cockle) was a weed in Britain only as long as it was reintroduced during sowing of corn (Firbank, 1988). With the advent of corn seed cleaning, *A. githago* virtually disappeared.

Some species are self-persistent, but are not weedy because they lack the traits necessary to compete with the crops or other weeds. A weed's impact on its community will depend on whether it has a similar or different function to other species already in the community (e.g. a nitrogen fixer in a nitrogen-poor environment) or whether it can outcompete other species with a similar function. Therefore, even a widely dispersed exotic species will not be a 'weed' unless it posses traits that allow it to interfere with native species.

A species must have the right set of traits and be present at the right place and time for it to invade. As we begin to understand how traits and the environment interact, we can change the way weed invasions are examined and move from explaining why a weed has successfully invaded a community to predicting whether it will invade. Weed ecology would be a stronger science if we were able to predict rather than just explain invasions.

MOVING FROM EXPLAINING INVASIONS TO PREDICTING INVASIONS

At this point, we should distinguish between *a posteriori* explanations of why a weed has invaded, and *a priori* predictions of what species are likely to invade. A *posteriori* explanations of invasive species are done after the invasion has occurred, and therefore are not necessarily useful in predicting whether another species will invade. For example, after a weed has invaded, we can scan Baker's list of ideal weed characteristics (Table 1) and select the ones applicable to the species of interest. Wild-proso millet (*Panicum miliaceum*), for example, is a successful invader of corn and soybean in Ontario, Canada because it has dormant, hard-coated seeds with a prolonged emergence pattern and it is tolerant to several herbicides. When other weeds that were more competitive than proso millet were removed with herbicides, the distribution of proso millet increased. When we look back, it is easy to tell why proso millet became a weed, but its weediness was not predicted before hand. While adding to our knowledge of invasives, such *a posteriori* explanations do not necessarily increase our power of prediction. They do, however, provide the basis for informed hypotheses.

Correct prediction has two parts (Daehler and Carino, 2000; Pheloung et al., 1999). First, to correctly identify the suite of traits that a weed will require to be a successful invaders, and second, to correctly identify which traits will make a species an unsuccessful invader. The first is useful to predict what species we want to prevent from invading or to minimize its spread. The second is useful to decide what species we do not need to worry about. Identifying both invaders and non-invaders will save money because it would allow us to focus management efforts.

What Predictions can be made?

Is it possible to make predictions about what types of weeds will invade a habitat? One way to simplify the approach of prediction is to identify traits associated with specific groups of weeds. For example, one could look at traits of weeds in temperate habitats, herbaceous weeds or weeds in one taxa (Pyšek, 2001). A prediction made for a subgroup of species lowers its general applicability but strengthens its power of prediction. In the examples below, we present some predictions made about groups of species.

Scott and Panetta (1993) used multiple regression analysis to predict what southern African species were likely to become agricultural weeds in Australia. They concluded that species likely to become weeds were i) already weedy in southern Africa, ii) occurred in a wide range of climate types in their native range, and iii) had congeneric species that were weeds. Using this information, the authors looked at 55 species introduced since 1950 and identified seven species likely to become agricultural weeds in Australia. They also identified 20 species in southern Africa that would likely become weeds if introduced into Australia.

Rejmánek and Richardson (1996) used Discriminant Function Analysis (DFA) to predict the invasive ability of pines in the southern hemisphere. Of the ten traits included in the analysis, only three (seed mass, minimum juvenile period, and interval between large seed crops) were required to separate invasive from non-invasive pines. When the pine discriminant function was applied to 40 highly invasive woody angiosperms, 95% were correctly identified as invasive species.

Reichard and Hamilton (1997) also used DFA as well as Classification and Regression Trees (CART) analysis to predict the invasion of woody plants into North America. Their goal was to decide which species should be admitted into North America and which should be banned. The DFA incorporated 11 biological and geographical traits and explained 67% of the data variance. Validation studies correctly classified 97% of invasive species. The DFA was less successful at predicting what species would *not* invade as it correctly classified only 71% of non-invaders. The CART analysis created a 'pruned tree' that used only four variables. It correctly classified 73% of invaders and 82% of non-invaders in validation studies. The variables that best detected invasiveness were the ability to reproduce vegetatively and to invade elsewhere, whereas being native to North America, an interspecific hybrid (of nursery origin) and not invasive elsewhere were used to distinguish non-invasive species.

CONCLUSIONS

Agricultural weeds are successful invaders of disturbed habitats. In order to improve on weed management, weed science must adjust from being a descriptive and reactive science to being a more ecologically based predictive science. If we are to make advances in predicting the invasibility of weeds, we must understand what makes a plant community invasible, what traits are typical of invasive weeds, and how advantageous traits change with the environment and with the stage of invasion.

REFERENCES

Baker, H.G. "Characteristics and modes of origin of weeds." In *The Genetics of Colonizing Species*, H.G. Baker, G.L. Stebbins, eds. New York: Academic Press, 1965; pp. 147-168.

Booth B.D., Swanton C.J. Assembly theory applied to weed communities. Weed Sci 2002; 50:2-13.

Brown D.G. Beetle folivory increases resource availability and alters plant invasion in monocultures of goldenrod. Ecology 1994; 75:1673-1683.

Case T.K. Invasion resistance arises in strongly interacting species-rich model competition communities. Phil Nat Acad Sci 1990; 87:9610-9614.

Chicoine T., Hahn K. Managing winter and spring annuals. Crop Insights 2001; 11:1-4.

Crawley M.J. "What makes a community invasible?" In *Colonization, Succession and Stability*, A.J. Gray, M.J. Crawley, P.J. Edwards, eds. Oxford: Blackwell Science, 1987; pp. 429-543.

Crawley M.J., Brown S.L., Heard M.S., Edwards G.R. Invasion-resistance in experimental grassland communities: species richness or species identity? Ecol Letters 1999; 2:140-148.

Daehler C.C. The taxonomic distribution of invasive angiosperm plants: ecological insights and comparison to agricultural weeds. Biol Cons 1998; 84:167-180.

Daehler C.C., Carino D.A. Predicting invasive plants: prospects for a general screening system based on current regional models. Biol Inv 2000; 2:93-102.

D'Antonio C.M. Mechanisms controlling invasion of coastal plant communities by the alien succulent *Carpobrotus edulis*. Ecology 1993; 74:83-95.

Darwin C. *The Origin of Species*. London: Reprinted by Penguin Books, 1859

Davis M.A, Pelsor M. Experimental support for a resource-based mechanistic model of invasibility. Ecol Letters 2001; 4:421-428.

Davis M.A., Grime J.P., Thompson K. Fluctuating resources in plant communities: a general theory of invasibility. J Ecol 2000; 88:528-534.

Davis M.A., Thompson K., Grime J.P. Charles S. Elton and the dissociation of invasion ecology from the rest of ecology. Divers Distr 2001; 7:97-102.

Derksen D.A., Lafond G.P., Thomas A G., Loeppky H.A., Swanton C.J. Impact of agronomic practices on weed communities: tillage systems. Weed Sci 1993; 41:409-417.

Díaz S, Acosta A., Cabido M. Morphological analysis of herbaceous communities under different grazing regimes. J Veg Sci 1992; 3:689-696.

Díaz S., Cabido M., Casanoves F. "Functional implications of trait-environment linkages in plant communities." In *Ecological Assembly Rules: Perspectives, Advances, Retreats*, E. Weir, P. A. Keddy eds. Cambridge: Cambridge University Press, 1999a; pp. 339-362.

Díaz S., Cabido M., Zak M., Martínez Carretero E., Araníbar J. Plant functional traits, ecosystem structure and land-use history along a climatic gradient in central-western Argentina. J Veg Sci 1999b; 10:651-660.

Díaz Barradas M.C., Zunzunegui M., Tirado R., Ain-Lhout F., García Novo F. Plant functional types and ecosystem function in Mediterranean shrubland. J Veg Sci 1999; 10:709-716.

Doak D.F., Bigger D., Harding E.K., Marvier M.A., O'Malley R.E., Thomson D. The statistical inevitability of stability-diversity relationships in community ecology. Am Nat 1998; 151:264-276.

Dukes J.S., Mooney H.A. Does global change increase the success of biological invaders. Trends Ecol Evol 1999; 14:135-139.

Elton C.S. *The Ecology of Invasions by Animals and Plants*, London: Methuen and Co, 1958.

Firbank L.G. Biological flora of the British Isles, No 165: *Agrostemma githago* L. J Ecol 1988; 76:1232-1246.

Fox M.D., Fox B.J. "The susceptibility of natural communities to invasion". In *Ecology of Biological Invasions*, R.H. Groves, J.J. Burdon, eds. Cambridge, UK: Cambridge University Press, 1986; pp. 57-66.

Gitay H., I. R. Nobel, J. H. Connel. Deriving functional types for rain-forest trees. J Veg Sci 1999; 10:641-650.

Goodman D. The theory of diversity-stability relationships in ecology. Quart Rev Biol 1975; 50:237-266.

Goodwin B.J, McAllister A.J., Fahrig L. Predicting invasiveness of plant species based on biological information. Cons Biol 1999; 13:422-426.

Grime J.P. Evidence for the existence of three primary strategies in plants and its relevance to ecological and evolutionary theory. Am Nat 1977; 111:1169-1194.

Grime J.P. *Plant Strategies and Vegetational Processes*, Chichester, UK: John Wiley and Sons, 1979.

Grime J.P. Biodiversity and ecosystem function: the debate deepens. Science 1997; 277:1260-1261.

Hastings M.S., DiTomaso J.M. Fire controls yellow starthistle in California grasslands. Rest Manage Notes 1996; 14:124-128.

Heywood V.H. "Patterns, extents and modes of invasion by terrestrial plants." In *Biological Invasions: A Global Perspective*, J.A. Drake, H.A. Mooney, F. di Castri, R.H. Groves, F.J Kruger, M. Rejmánek, M. Williamson, eds. Chichester, UK: John Wiley & Sons, 1989; pp. 31-60.

Hobbs R.J. "The nature and effects of disturbance relative to invasions." In *Biological Invasions: A Global Perspective*, J.A. Drake, H.A. Mooney, F. di Castri, R.H. Groves, F.J Kruger, M. Rejmánek, M. Williamson, eds. Chichester, UK: John Wiley & Sons, 1989; 389-405.

Hobbs R.J. "Land-use changes and invasions." In *Invasive Species in a Changing World*, H.A. Mooney, R.J. Hobbs eds. Washington DC: Island Press, 2000; pp. 55-64.

Hobbs R.K., Huenneke L.F. Disturbance, diversity and invasion: implications for conservation. Cons Biol 1992; 6:324-337.

Hooper D.U., Vitousek P.M. Effects of plant composition and diversity on nutrient cycling. Ecol Monogr 1998; 68:121-149.

Huston M.A. *Biological Diversity: The Coexistence of Species on Changing Landscapes*. Cambridge, UK: Cambridge University Press, 1994.

Huston M.A., D'Angelis D.L. Competition and coexistence: the effects of resource transport and supply rates. Am Nat 1994; 144:954-977.

Johnstone I.M. Plant invasion windows: a time-based classification of invasion potential. Biol Rev 1986; 61:369-394.

King S.E., Grace J.B. The effects of gap size and disturbance type on invasion of wet pine savanna by cogongrass, *Imperata cylindrica* (Poaceae). Am J Bot 2000; 87:1279-1286.

Kleyer M. Distribution of plant functional types along gradients of disturbance intensity and resource supply in an agricultural landscape. J Veg Sci 1999; 10:697-708.

Knops J.M.H., Tilman D., Haddad N.M., Naeem S., Mitchell C.E., Haarstad J., Ritchie M.E., Howe K.M., Reich P.B., Siemann E., Groth J. Effects of plant species richness on invasion dynamics, disease outbreaks, insect abundances and diversity. Ecol Letters 1999; 2:286-293.

Kolar C.S., Lodge D.M. Progress in invasion biology: predicting invaders. Trends Ecol Evol 2001; 16:199-204.

Kyser G.B., DiTomaso J.M. Instability in a grassland community after the control of yellow starthistle (*Centaurea solstitialis*) with prescribed burning. Weed Sci 2002; 50: 648-657.

Lavorel S., McIntyre S., Lansberg J., Forbes T.D.A. Plant functional classifications: from general groups to specific groups based on response to disturbance. Trends Ecol Evol 1997; 12:474-478.

Lepš J., Brown V.K., Diaz Len T.A., Gormsen D., Hedlund K., Kailová J., Korthals G.W., Mortimer S.R., Rodriguez-Barrueco C., Roy J., Santa Regina I., van Dijk C., van der Putten W.H. Separating the chance effect from other diversity effects in the functioning of plant communities. Oikos 2001; 92:123-134.

Levine J.M., D'Antonio C.M. Elton revisited: a review of evidence linking diversity and invasibility. Oikos 1999; 87:15-26.

Levine J.M. Species diversity and biological invasions: relating local process to community pattern. Science 2000; 288:852-854.

Lodge D.M. Biological invasions: lessons for ecology. Trends Ecol Evol 1993; 8:133-137.

Lonsdale W.M. Global patterns of plant invasions and the concept of invasibility. Ecology 1999; 80:1522-1536.

Lyon K.G., Schwartz M.W. Rare species loss alters ecosystem function – invasion resistance. Ecol Letters 2001; 4:358-365.

MacArthur R.H. Fluctuations of animal populations and a measure of community stability. Ecology 1955; 36:533-536.

Mack R.N. Predicting the identity and fate of plant invaders: emergent and emerging approaches. Biol Cons 1996; 78:107-121.

Mack R.N., Simberloff D., Lonsdale W.M., Evans H., Clout M., Bazzaz F.A. Biotic invasions: causes, epidemiology, global consequences, and control. Ecol Appl 2000; 10:689-710.

May R. *Stability and Complexity in Model Ecosystems*. Princeton, NJ: Princeton Univ. Press, 1973.

McIntyre S., Lavorel S., Landsberg J., Forbes T.D.A. Disturbance response in vegetation - towards a global perspective on functional traits. J Veg Sci 1999; 10:621-630.

Moore J.L., Mouquet N., Lawton J.H., Loreau M. Coexistence, saturation and invasion resistance in simulated plant assemblages. Oikos 2001; 94:303-314.

National Research Council. Committee on the Scientific Basis for Predicting the Invasion Potential of Nonindigenous Plants and Plant Pests. *Predicting Invasions of Nonindigenous Plants and Plant Pests*. Washington, DC: National Academy of Science, 2002.

Nobel I.R., Slatyer R.O. The use of vital attributes to predict successional changes in plant communities subject to recurrent disturbances. Vegetatio 1980; 43:5-21

Orians G.H. "Site characteristics favoring invasions." In *Ecology of Biological Invasions of North America and Hawaii*, New York: Springer-Verlag, 1986; pp. 133-148.

Palmer M.W., Maurer T.A. Does diversity beget diversity? A case study of crops and weeds. J Veg Sci 1997; 8:235-240.

Parker I.M., Simberloff D., Lonsdale W.M., Goodell K., Wonham M., Kareiva P.M., Williamson M.H., Von Holle B., Moyle P.B., Byers J.E., Goldwasserr L. Impact: toward a framework for understanding the ecological effects of invaders. Biol Invasions 1999; 1:3-19.

Parks R.J., Curran W.S., Roth G.W., Hartwig N.L., Calvin D.D. Herbicide susceptibility and biological fitness of triazine-resistant and susceptible common lambsquarters (*Chenopodium album*). Weed Sci 1996; 44:517-523.

Perrins J., Fitter A., Williamson M. Population biology and rates of invasion of three introduced *Impatiens* species in the British Isles. J Biogeog 1993; 20:33-44.

Perrins J., Williamson M., Fitter A. A survey of differing views of weed classification: implications for regulation of introductions. Biol Cons 1992; 60:47-56.

Pheloung P.C., Williams P.A., Halloy S.R. A weed risk assessment model for use as a biosecurity tool evaluating plant introductions. J Env Manag 1999; 57:239-251.

Pimentel D.A., McNair S., Janecka J., Wightman J., Simmonds C., O'Connell C., Wong E., Russel L., Zern J., Aquino T., Tsomondo T. Economic and environmental threats of alien plant, animal and microbe invasions. Agric Ecosys Env 2001; 84:1-20.

Pimentel D.A., Lach L., Zuniga R., Morrison D. Environmental and economic costs of non-indigenous species in the United States. BioScience 2000; 50:53-65.

Planty-Tabacchi A., Tabacchi E., Naiman R.J., DeFerrari C., Décamps H. Invasibility of species-rich communities in riparian zones. Cons Biol 1996; 10:598-607.

Pyšek P. Is there a taxonomic pattern to plant invasions? Oikos 1998; 92:282-294.

Pyšek P. Past and future predictions in plant invasions: a field test by time. Diver Dist 2001; 7:145-151.

Pyšek P., Prach K., Smilauer P. "Relating invasion success to plant traits: an analysis of the Czech alien flora." In *Plant Invasions: General Aspects and Special Problems*. P. Pyšek, K. Prach, M. Rejmánek, M. Wade, eds. Amsterdam: SPA Academic Publishing, 1995; pp. 39-60.

Reichard S.H., Hamilton C.W. Predicting invasions of woody plants introduced into North America. Cons Biol 1997; 11:193-203.

Rejmánek M. "Invasibility of plant communities." In *Biological Invasions: A Global Perspective*, J. Drake, H.A. Mooney, F. DiCastri, R.H. Groves, F.J. Kruger, M. Rejmánek, M. Williamson. Chichester, UK: John Wiley & Sons, 1989; pp. 369-388.

Rejmánek M. "What makes a species invasive?" In *Plant Invasions: General Aspects and Special Problems*, P. Pyšek, K. Prach, M. Rejmánek, M. Wade, eds. Amsterdam: SPA Academic Publishing, 1995; pp. 3-13.

Rejmánek M., Richardson D.M. What attributes make some plant species more invasive? Ecology 1996; 77:1655-1661.

Richardson D.M., Allsopp N., D'Antonio C.M., Milton S.J., Rejmánek M. Plant invasion – the role of mutualisms. Biol Rev 2000a; 75:65-93.

Richardson D.M., Pyšek P., Rejmánek M., Barbour M.G., Panetta F.D., West C.J. Naturalization and invasion of alien plants: concepts and definitions. Div Distri 2000b; 6:93-107.

Robinson G.R., Quinn J.F., Stanton M.L. Invasibility of experimental habitat in California winter annual grassland. Ecology 1995; 76:786-794.

Roy J., Navas M.L., Sonie L. "Invasions by annual brome grasses: a case study challenging the homocline approach to invasions." In *Biogeography of Mediterranean Invasions*, R.H. Groves, F. di Castri, eds. Cambridge, UK: Cambridge University Press, 1991; pp. 207-224.

Scott J.K., Panetta F.D. Predicting the Australian weed status of southern African plants. J Biogeog 1993; 20:87-93.

Shrestha A., Knezevic S.Z., Roy R.C., Ball-Coelh B.R., Swanton C.J. Effect of tillage, cover crop and crop rotation on the composition of weed flora in a sandy soil. Weed Res 2002; 42:76-87

Smith M.D., Knapp A.K. Physiological and morphological traits of exotic, invasive exotic, and native plant species in tallgrass prairie. Int J Plant Sci 2001; 162:782-792.

Smith T.M. Shugart H.H., Woodward F.I. (eds). *Plant Functional Types*. Cambridge, UK: Cambridge University Press, 1997.

Stohlgren T.J., Binkley D., Chong G.W., Kalkham M.A., Schell L.D., Bull K.A., Otsuki Y. Newman G., Bashkin M., Son Y. Exotic plants invade hot spots of native plant diversity. Ecol Mono 1999; 69:25-46.

Stohlgren T.J., Bull K.A., Otsuki Y., Villa C., Lee M. Riparian zones as havens for exotic plant species in the central grasslands. Pl Ecol 1998; 138:113-125.

Stohlgren T.J., Otsuki Y., Villa C., Lee M., Belnap J. Patterns of invasions: a case example of native species hotspots and rare habitats. Biol Inv 2001; 3:37-50.

Swanton C.J., Shrestha A., Roy R.C., Ball-Coelho B.R., Knezevic S.Z. Effect of tillage systems, N, and cover crop on the composition of weed flora. Weed Sci 1999; 47:454-461.

Swanton C.J., Clements D.R., Derksen D.A. Weed succession under conservation tillage: a hierarchical framework for research and management. Weed Tech 1993; 7:286-297.

Thompson K., Hodgson J.G., Rich T.C.G. Native and alien invasive plants: more of the same? Ecography 1995; 18:390-402.

Tilman D. The importance of mechanisms of the interspecific interaction. Am Nat 1987; 129:769-774.

Tilman D. *Plant Strategies and the Dynamics and Structure of Plant Communities*, Princeton, New Jersey: Princeton University Press, 1988

Tilman D. Community invasibility, recruitment limitation, and grassland biodiversity. Ecology 1997; 78:81-92.

Tilman D. The ecological consequences of changes in biodiversity: a search for general principles. Ecology 1999; 80:1455-1474.

van der Putten W.H., Mortimer S.R., Hedlund K., van Dijk C., Brown V.K., Lepš J., Rodriguez-Barrueco C., Roy J., Diaz Len T.A., Gormsen D., Korthals G.W., Lavorel S., Santa Regina I., Smilauer P. Plant species diversity as a driver of early succession in abandoned fields: a multi-site approach. Oecologia 2000; 124:91-99.

Vitousek P.M., D'Antonio C.M., Loope, L.L., Westbrooks R. Biological invasions as global environmental change. Am Sci 1996; 84:468-478.

Wade M. "Predicting plant invasions: making a start." In *Plant Invasions: Studies from North America and Europe*, J.H. Brock, M. Wade, P. Pyšek, D. Green. Leiden: Backhuys Publ., 1997; pp. 1-18.

Weiher E., Keddy P.A. "Assembly rules as general constraints on community composition." In *Ecological Assembly Rules: Perspectives, Advances, Retreats*, E. Weiher, P.A. Keddy. Cambridge, UK: Cambridge University Press, 1999; pp. 251-271.

Weiher E., van der Werf A., Thompson K., Roderick M., Garnier E., Eriksson O. Challenging Theophrastus: a common core list of plant traits for functional ecology. J Veg Sci 1999; 10:609-620.

Williamson M. "Mathematical models of invasion." In *Biological Invasions: A Global Perspective*, J.A. Drake, H.A. Mooney, F. di Castri, R.H. Groves, F.J Kruger, M. Rejmánek, M. Williamson, eds. Chichester, UK: John Wiley & Sons, 1989; pp. 329-350.

CHAPTER 3

CROP-ASSOCIATED WEEDS
THE STRATEGY FOR ADAPTATION

T. TOMINAGA[1] & Y. YAMASUE[2]

[1]*Kyoto Prefectural University, Kyoto 619-0244 and* [2]*Kyoto University, Kyoto 606-8502, Japan*

Some weeds are always associated with a particular crop and inhabit only the cropland, rarely expanding into other habitats. They have adapted to the habitats which man disturbs for the crop cultivation. Such weeds are the product of evolution under the specific selection pressures imposed by the crop cultivation over many years. Over time, they developed strategies adapted to the cultural practices or environments encountered throughout their life cycles that enable them to survive and even dominate there for generations. These cultural practices include field operations such as hand-weeding, but also crop harvest, threshing, and winnowing.

Evolution of weeds in general has been discussed by many authors, including Bunting (1960), Barrett (1983), Harlan (1992), and Ladizinsky (1998). In this chapter, we describe how two weeds are adapted to their associated crops: *Lolium temulentum* L. in wheat and barley, and *Echinochloa oryzicola* Vasing. in lowland rice. In particular, we discuss the vegetative and grain mimicry of *L. temulentum* against crops, and the vegetative mimicry and physiological traits *E. oryzicola* exhibits to survive in lowland rice.

LOLIUM TEMULENTUM

The genus *Lolium* comprises eight species (Terrell, 1968), including *L. perenne* L. and *L. multiflorum* Lam., important forage grasses widely cultivated in temperate regions of the world. These two species are diploid (2n=14), although some tetraploid commercial cultivars have been bred from them. *L. rigidum* Gaud.(2n=14) is a polymorphic weed found in many winter crops. These three species are wind-pollinated allogamous plants.

The other five species, *L. remotum* Schrank, *L. subulatum* Vis., *L. persicum* Boiss. and Hohen ex Boiss., *L. canariense* Steud. and *L. temulentum* L., are all diploid autogamous weeds. *L. temulentum* (darnel) is frequently found in wheat and barley fields in which cereal crops are cultivated using indigenous cropping systems. Darnel is an important genetic resource of self-compatible genes (Thorogood and Hayward, 1992), especially since *L. perenne* and *L. multiflorum* are self-incompatible allogamous species.

Inderjit (ed.), Weed Biology and Management, 47-63.

Distribution

Darnel is native to the Mediterranean region and has been dispersed worldwide together with cereal crops such as wheat and barley (Holm et al., 1977). It has a long history as a weed accompanying cereal cultivation and is even mentioned in the Bible. Its grains - the term "grain" is used here to refer to the caryopsis of darnel and of cereal crops - * have been found among cereal grains from 3200 B.C. (McCreery, 1979) and along with emmer wheat chaff in Egyptian tombs from around 2000 B.C. (Täckholm and Täckholm, 1941).

In modern agricultural systems, however, the weed is becoming rare, due to the introduction of improved cereal crop cultivars and herbicides and to the recent development of improved seed cleaning technologies. In our field surveys from 1995 to 2001, the weed was rarely found in western Europe, but was abundant in the areas of South and West Asia and North Africa where indigenous wheat and barley cultivation is still practiced (Figure 1). It is likely that modern agricultural systems are creating new selection pressures that endanger this weed highly adapted to indigenous cereal cultivation systems.

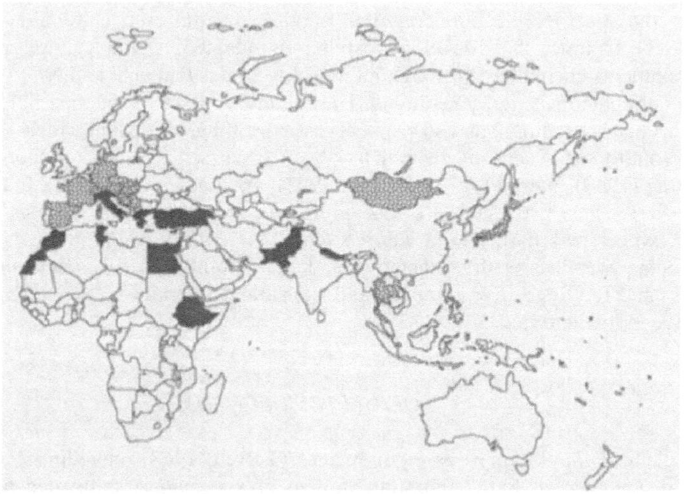

Figure 1. Distribution of darnel found in our field surveys from 1995 to 2001. Black shading indicates countries where darnel was found in cereal fields, and dotted shading indicates countries where darnel was not found.

---

* The term "grain" is used here to refer to caryopsis of darnel and of cerel crops.

Grain Mimicry

Darnel grain morphology is very similar to those of wheat and barley grain. Furthermore, darnel grains are exceptionally large and heavy compared with those of other weed species. The 100-grain weight of darnel, 1.64 g, is eight to nine times greater than those of other *Lolium* weeds, e.g. 0.18 to 0.22 g, for *L. perenne*, *L. multiflorum* and *L. rigidum* (Hayashi and Tominaga, unpublished), and even greater than those of millets such as foxtail millet (*Setaria italica* Beauv.), Japanese millet (*Echinochloa utilis* Ohwi et Yabuno) and proso millet (*Panicum miliaceum* L.). Selection for darnel grains of similar size and weight as those of cereal grains has occurred and darnel grains cannot be separated from cereal grains by hand or by winnowing. The large and heavy grains also enable darnel seedlings to grow rapidly after germination, allowing them to compete well with cereal crops (Harlan, 1992).

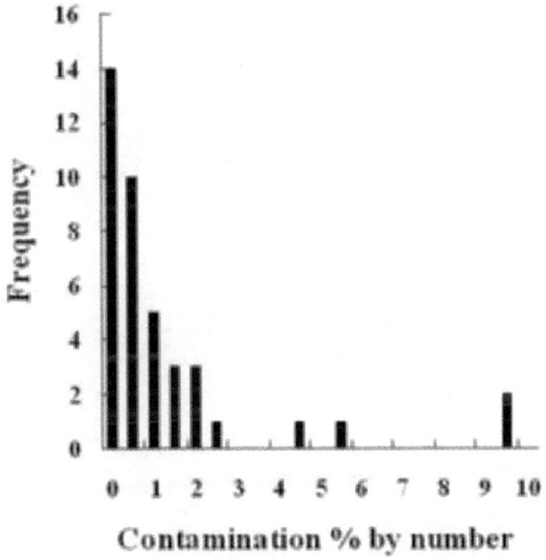

Figure 2. Frequency of darnel grain contamination in wheat and barley grain stocks sampled in a local market in Ethiopia (Tominaga and Fujimoto, 2000).

Darnel grains are frequently present in wheat and barley grain stored in farm granaries or found in local markets in South and West Asia and in North Africa. Darnel contamination of barley grain stocks in an Ethiopian village market, for example, was found to be as high as 9.4% (Figure 2). Darnel is harvested along with cereal crops because, like cereal crops, it does not have grain shattering habit. In areas of indigenous cereal grain production, where wheat and barley are cleaned by winnowing and by hand, darnel grains remain in wheat and barley grain stocks because of this grain mimicry.

Darnel grains are either awned or awnless, and the awnless is dominant over the awned (Jenkin, 1954; Senda and Tominaga, 2003). In Delvinaki, Greece, the awnless form was found in bread wheat and macaroni wheat, while the awned one was found to be associated with barley and oats (Kobayashi and Sakamoto, 1985). In this village, wheat was cleaned with sieves for human consumption, but barley and oats were not, because they were intended for livestock. Kobayashi and Sakamoto (1985) hypothesized that sieving eliminated the awned grains of darnel, because the length of the awned grain was far greater than the mesh size of the sieves, but allowed the awnless grains to remain in wheat grain stocks.

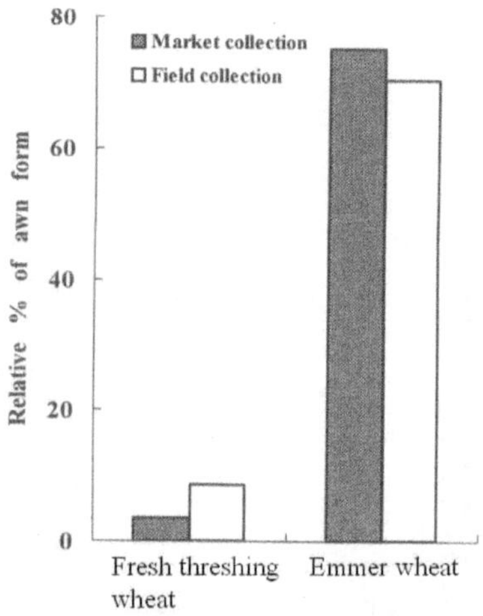

Figure 3. Association of the awn and awnless forms of darnel with free threshing wheat and emmer wheat.

In Malo, Ethiopia, awned darnel is found in emmer wheat, and awnless darnel is generally associated with bread wheat, macaroni wheat and rivet wheat (Figure 3). In this village, grain cleaning is done by winnowing and subsequent hand removal of contaminants. Emmer wheat has non-free threshing grains, and the other three crops have free threshing grains. Awned darnel's grain morphology is similar to that of emmer wheat grain, and the awnless darnel grain resembles the free threshing grains of bread, macaroni and rivet wheat. Separating awned darnel grains from emmer wheat grains is difficult, as is separating awnless grains from free threshing wheat grains; thus grain mimicry controls the close association of the two forms of darnel with the wheat species. In this village, rotation of cereals (wheat or barley) with pulses (peas or broad beans) and fallow is practiced. Because of darnel's grain mimicry and non-shattering habit, the darnel seed bank is found not in the soil but in farmers' seed stocks.

Wild gramineous plants usually have awns or other dispersal organs on their grains. Domestication has led to a loss of such dispersal organs in cereal crops because man disperses their grains (Ladizinsky, 1998). The presence or absence of awns in darnel may have evolved in association with grain cleaning methods as well as with the non-free threshing or free threshing characteristics of the model cereal crop. These varied selection pressures on the presence or absence of awns on darnel grain, including the methods of grain cleaning, the types of cereals cultivated in a given area, and the types of agricultural systems employed locally, lead to a multifaceted relationship between cereal crops and darnel.

In Ethiopia, where wheat and barley have a large genetic diversity (Harlan, 1992) and cereal spikes are sometimes colored black to dark purple, darnel spikes are also colored black to dark purple (Figure 4). In this case, darnel has developed mimicry not only of grain morphology and size but of color as well. Darnel's exceedingly specialized mimicry of wheat and barley in Ethiopia results from its intimate association with indigenous wheat and barley cultivation, accomplished exclusively by hand without the use of any machines or chemicals.

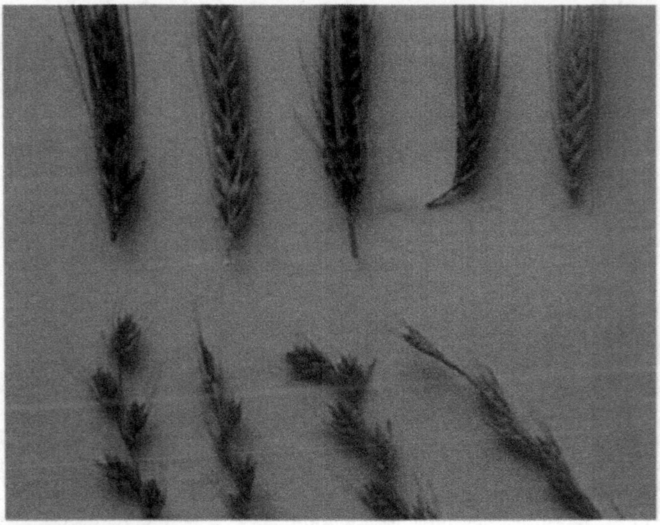

Figure 4. Colored barley spikes (top) and darnel spikes (bottom) in Ethiopia.

Life Cycle

Germination

Most weed seeds are dormant for several months after maturing in order to avoid germinating under conditions unfavorable for their growth (Baker, 1965). Darnel grains, however, do not show dormancy (Holm et al., 1977) and easily germinate about one month after harvest (Tominaga and Ishiyama, 2001). When stored under conditions simulating a farm granary at temperatures ranging from 23.2 °C to 26.2 °C, about 60% of the darnel grains germinated after one month of storage, and more than 91% of the grains had germinated by three months (Figure 5). In contrast, grains of

L. rigidum from wheat fields did not germinate at all after three month of storage, and only 2% of the grains germinated after four months. When the grains of the two species were buried at a depth of 5 cm in soil, about 80% of the darnel grains germinated after one month, and all grains germinated after two months. Forty percent of the *L. rigidum* grains from wheat fields germinated after being buried for one month, and 90% of the grains germinated after four months.

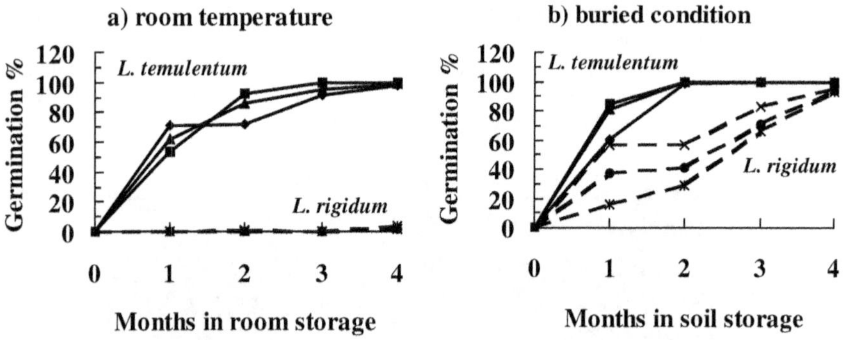

Figure 5. Germination of darnel (*L. temulentum*, solid lines) and *L. rigidum* (broken lines) grains under a) room temperature and b) buried conditions (redrawn from Tominaga and Ishiyama, 2001). Three accessions of each species were used in this experiment.

At sowing time, darnel grains are scattered on the fields along with the cereal grains, and darnel seedlings compete with cereal seedlings for light, water and minerals. Late emerging seedlings do not compete well with cereal seedlings and are removed from the population.

Growth and Heading

Darnel also resembles wheat and barley in seedling morphology. In indigenous wheat and barley cultivation systems, hand-weeding is usually done a month after seeding. In the seedling stage, crop mimicry enables darnel to escape hand-weeding. Darnel and wheat or barley crops in the same field show similar growth curves and head (show spikes) at almost the same time (Figure 6). Taller darnel plants may be removed by hand-weeding, and shorter ones do not compete well with the crops. This growth characteristic also seems to have resulted from repeated selection through hand-weeding and harvesting. Darnel mimics indigenous wheat or barley landraces in both its life cycle and seedling morphology.

Geographical Differentiation within Ethiopia

Fukui (1971) collected data on the local names for free threshing wheat in Ethiopia. He found that farmers did not distinguish the three free threshing wheat species, namely bread wheat, macaroni wheat and rivet wheat, from one other and considered them a single folk species. The local names for wheat were classified into three groups. Each of the three name groups had a distinct geographical distribution, and Fukui (1971) hypothesized that free threshing wheat was introduced to Ethiopia via three different routes.

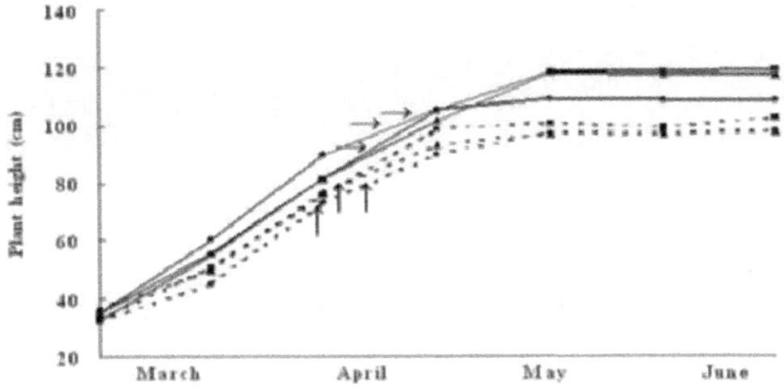

Figure 6. Changes in plant height of wheat (solid lines) and of darnel (broken lines) from the same field (Hayashi and Tominaga, unpublished). Both horizontal and vertical arrows show the heading time of each accession.

The local name for darnel also differs from area to area and fell into three groups, as in the case of free threshing wheat. Around Addis Ababa, free threshing wheat is called "sinde" and darnel is "inkurdad"; around Sodo, wheat is "qamade" and darnel is "bulassi"; and around Malo, wheat is "giste" and darnel is "leesho" (Figure 7). The geographical distribution of the three local name groups for darnel coincided with those for free threshing wheat. Our RAPD and UPGMA cluster analysis based on Nei's genetic distances (Nei, 1972) using data with accesssions from this area showed that darnel corresponding to each local name group formed different clusters and was genetically differentiated (Figure 8). Humans brought not only the plant seeds, both cereal crops and darnel, but also their original local names with them. These results demonstrate that darnel was introduced to Ethiopia with three separate free threshing wheat landraces (Tominaga, 2002).

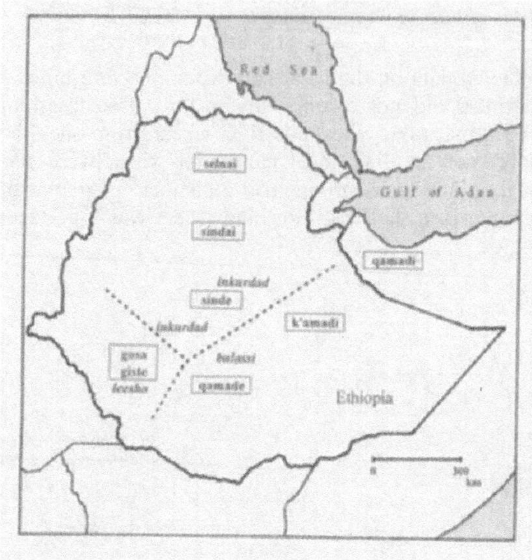

Figure 7. Distribution of local names for free threshing wheat (boxes) and darnel (italicized) in Ethiopia (adding our darnel data to Fukui, 1971).

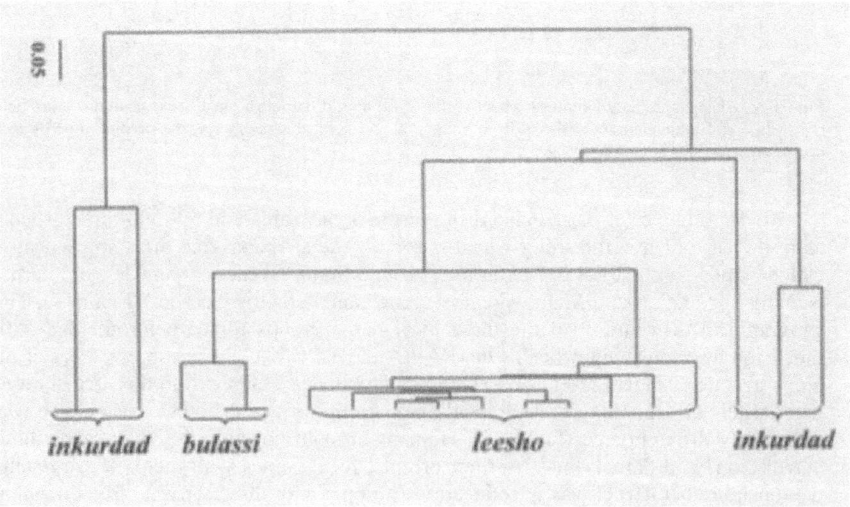

Figure 8. Genetic differentiation of darnel in Ethiopia (redrawn from Senda and Tominaga, 2003).

Agricultural Systems Allowing Darnel to Persist

Through harvesting, threshing, winnowing, storage and sowing with cereal crops such as wheat and barley, darnel, with its large and non-dormant grains, has developed grain mimicry. Mimicry of the cereal crops' life cycles has been selected for by hand-weeding under indigenous agricultural systems. Darnel is widely distributed throughout the world, but the prevalence of this species has diminished following the introduction of modern agricultural systems.

ECHINOCHLOA ORYZICOLA

E. oryzicola Vasing. (syn: *E. phyllopogon* Stapf ex Kessenko, *E. crus-galli* Beauv. var. *oryzicola* Ohwi, a barnyardgrass) is a summer annual and a progenitor of *E. crus-galli* Beauv., which is an allohexaploid and ranked among the World's worst weeds (Yabuno, 1966; Holm et al., 1977). *E. oryzicola* is found in lowland rice fields of East Asia, South East Asia, Europe and North America, and its occurrence other than in the rice fields has never been reported. Here, various adaptive characters of the weed to lowland rice are described, confining the species in Japan where rice has been cultivated for a long time with the traditional system and rarely rotated with other crops.

E. oryzicola was introduced to Japan along with rice in its pre-historic age (Maekawa, 1943). The native name of an *Echinochloa* weed appears in the oldest Japanese anthology compiled before 800 A.C., and a weed, which is likely *E. oryzicola*, is listed in an agronomic book in 1847. Rice farmers of the country are characterized as being very punctual about following traditional farming methods, and the cultural practices have not been largely changed until recent mechanization and pesticide usage, including herbicides. A farmer's diary that was written more than a hundred years ago indicated that he went to the rice fields more than 40 times for hand-weeding during a growing season, which is extended about four months. *E. oryzicola* is physiologically tolerant in seed germination and seedling growth under submergence, but also evolved many noteworthy characters to invade a lowland rice field of the country.

Seed Dormancy and Germination

As Baskin and Baskin (1998) indicated for many annual weeds, seeds (spikelets) of *E. oryzicola* also exhibit annual changes in their dormancy state in the soil-buried seed bank of the crop filed. Seeds shattered from mother plants in late summer are innately dormant and so never germinate in autumn. They gradually awake from dormancy during the period from winter to early spring (Figure 9).

Seeds still dormant in spring are forced to become nondormant by the anaerobic soil conditions created by the field flooding (Yamasue and Ueki, 1987). In the soil-buried seed bank, nondormant seeds (defined by Harper (1977) as being in a state of enforced dormancy), always appear at the season when rice farming starts. It starts in early spring with seedling preparation in the nursery bed and with flooding in the field where the seedlings are to be transplanted. Rice seedlings are transplanted into the flooded field from early May to late June (Figure 9). Most of the nondormant seeds buried at shallow soil depths of a few centimeters germinate during the period following rice transplanting. The remaining nondormant seeds, particularly those buried deep in the soil, do not germinate but are induced again into dormancy in the summer, and remain in the seed bank available for future germination. In this way, the weed seeds germinate coincidentally with the period from late spring to early summer when rice farmers begin field preparation.

For a plant with submergence tolerance, a flooded soil like that of a lowland rice field is rather favorable for survival since there is sufficient water and a small daily temperature fluctuation. Also, the number of other weed species that competes with it is restricted in the field. However, oxygen is deficient in this flooded soil. Oxygen concentration is lower than the theoretical value of its solubility in pure water at 20 °C, *ca* 9 ppm (v/v), because of oxygen consumption by chemical oxidation of soil organic matter and microbial respiration (Takai, 1978).

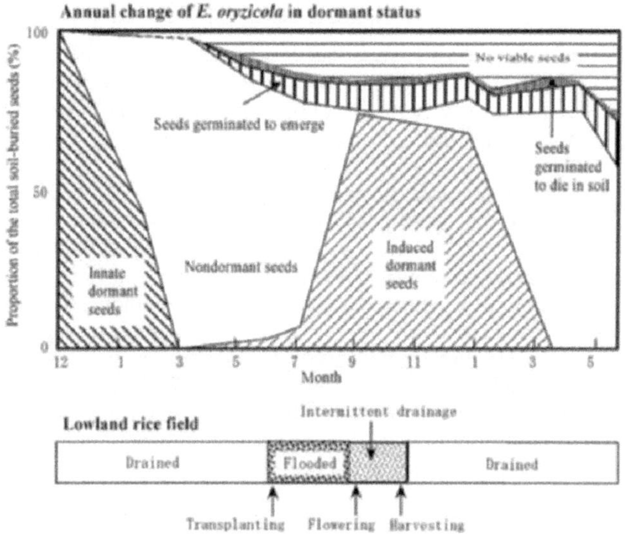

Figure 9. Annual change of dormancy status of soil-buried *E. oryzicola* seeds following to seed shed from mother plants (Modified from Miyahara, 1972 with permission). The change is compared to soil conditions of a lowland rice field of Japan.

Similarly to rice, seed germination of *E. oryzicola* is metabolically adapted to submerged conditions (Kataoka and Kim, 1978; Rumpho and Kennedy, 1981; Kennedy et al., 1983; Yamasue et al., 1987). The seeds are capable of germinating under flooded conditions and even in pure nitrogen. We compared the percent germination of nondormant seeds of *E. oryzicola*. and *E. crus-galli* var. *praticola* under conditions ranging from pure nitrogen to 21% oxygen. The latter species is a weed confined to upland habitats such as roadsides and open lands. Seeds of this species did not germinate under conditions with less than 3% oxygen, but seeds of *E. oryzicola* showed 100% germination regardless of oxygen concentration. Seeds germinating in a nitrogen atmosphere rapidly increased alcohol dehydrogenase activity and concomitantly produced a large amount of CO_2 and ethanol. The molar ratios of ethanol to CO_2 production were near unity in nondormant seeds (Table 1). That is, nondormant seeds of *E. oryzicola* appeared to respire mainly through anaerobic alcohol fermentation. Fermentation produces far smaller amounts of energy than the conventinal aerobic respiration, but it is compensated by the larger seed weight (3-5mg per seed) relative to *E. crus-galli* var. *praticola* (about 1mg per seed), seeds of which germinate by the aerobic respiration. When submerged, germinating seeds of *E. oryzicola* elongate the coleoptile and develop the primary leaf above the surface of the water, but those of *E. crus-galli* var. *praticola* cannot survive (Yabuno, 1966; Yamasue et al., 1989).

Table 1. Germination percentage and ethanol (EtOH) production of nondormant and dormant seeds of *E. oryzicola* during imbibition at 30°C in the light under the ambient atmosphere of air and nitrogen (Yamasue and Ueki, 1987).

	48 hours			72 hours		
Ambient conditions	Germination (%)	EtOH (nmol/ seeds)	EtOH/ CO_2 (ratio)	Germination (%)	EtOH (nmol/ seeds)	EtOH/ CO_2 (ratio)
			Nondormant seeds			
Air	0	0.226	0.8	100	0.568	0.8
N_2	0	0.290	1.2	100	0.604	0.9
			Dormant seeds			
Air	0	0.008	0.1	0	0.028	0.2
N_2	0	0.186	1.2	0	0.290	1.1

Vegetative Mimicry and Simultaneous Heading with the Crop

E. oryzicola mimics the rice plant throughout its development from seedling to heading (Figure 10). The weed so closely resembles rice that even a veteran farmer can mistake it for a rice plant; it thus evades hand-weeding. Yabuno (1966) first suggested that the vegetative mimicry of this weed was selected by intensive hand-weeding in lowland rice. Later, many weed biologists took it as a representative example to show vegetative mimicry of a weed, that evolved when its associates with a specific crop (Barrett, 1983; Radosevich and Holt, 1984; Harlan, 1992; Ladizinsky, 1998). Although the weed is botanically distinguishable from rice by the absence of a leaf ligule, there is no doubt that mimicry during vegetative growth helps the weed escape from hand-weeding, which has been persistently practiced for many years in the lowland rice fields of Japan.

Photoperiod-sensitive heading time in *E. oryzicola* is another character selected by hand-weeding. In lowland rice, both rice and the weed begin heading about the same time in late summer. The weed is a short-day plant, as is rice.

Figure 10. Vegetative mimicry of *E. oryzicola* (right) and rice (left) (Yamasue, 2001).

Japanese local rice cultivars differ from each other in photoperiod sensitivity and exhibit a geographical cline in time of heading or flowering (Oka, 1958); the photoperiod sensitivity increases as the latitude decreases. Heading time of the local forms of the weed also exhibit a geographical cline that parallels that of the rice cultivars (Morinaga and Nagamatsu, 1942; Yabuno, 1966; Nakatani et al., 1998) (Figure 11). The best time for the weed to start heading in order to evade hand-weeding, is coincident with the time of rice heading. Though plant mimicry helps the weed escape being removed during vegetative growth, one cannot mistake the weed for rice after heading. However, rice farmers are reluctant to walk in the rice paddy to do weeding during heading (Harlan, 1992) because it may disturb crop growth and pollination. Weeding after rice heading is not effective because the weed has already begun shattering seeds.

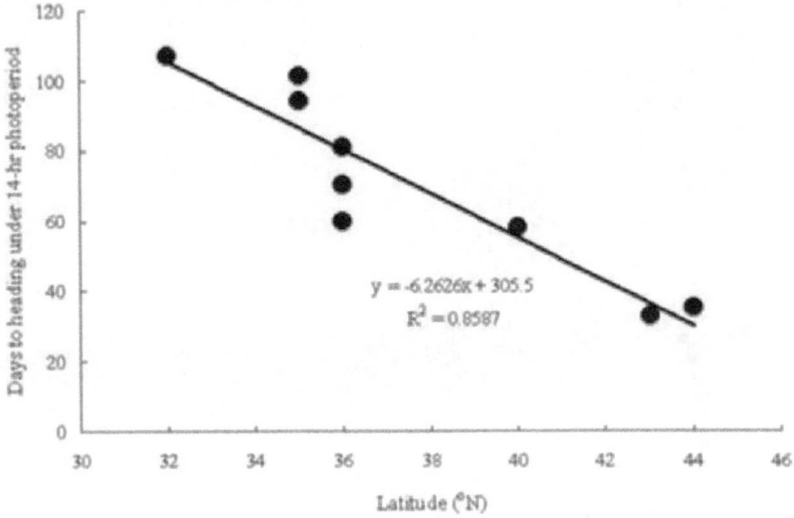

Figure 11. Geographical cline of *E. oryzicola* strains collected in Japanese lowland rice fields at different latitudes (Nakatani et al., 1998). The number of days from transplanting to heading was compared under photoperiod at 14 hours.

Competition with the Crop

E. oryzicola is highly competitive against rice and causes a large crop yield loss (Chisaka, 1977). This aggressiveness is partly conferred by its phenotypic plasticity in plant height (Yamasue, 2001). We studied its competitive ability by planting rice and *E. oryzicola* in mixed stands with a 1:1 substituted design of 20 x 20 cm spacing (Yamasue et al., 1997a, b). In the mixed stands, rice and weed increased in plant height at a similar constant rate until mid season, but thereafter, the weed increased in height more rapidly. The crop and weed attained heights of 110 and 140 cm at crop harvest time, respectively. Upper leaves of the weed occupied most of the stand canopy (Figure 12) and reduced the relative light intensity received by the rice leaves to 40 to 60 %.

In the farmer's fields we frequently observe that irrespective of the rice cultivar planted, a few upper leaves and panicles of *E. oryzicola* always become distinct above the crop canopy around the time of heading. We also observed in a field experiment that when planted together with gramineous plants of different heights, the final height of the weed ranged from 115 to 150 cm and was always a little taller than the neighboring plants (Table 2). This phenotypic plasticity of *E. oryzicola* in height is of significant contribution when the weed competes with the crop for light interception.

Modern rice cultivars tend to have no such phenotypic plasticity; there is very little height difference between weeded and unweeded fields.

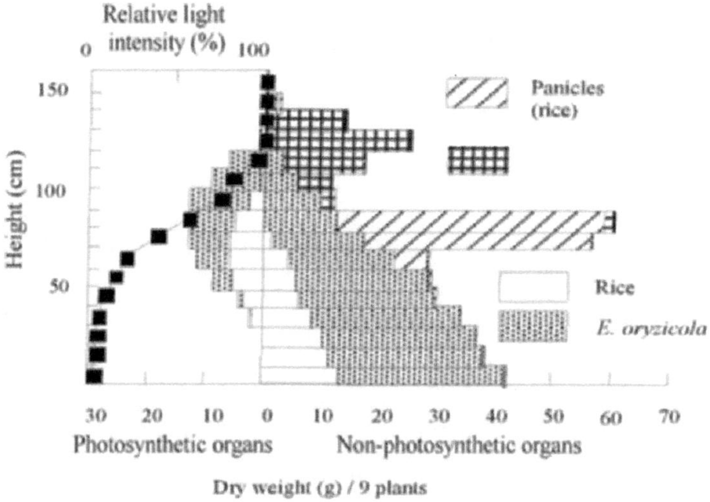

Figure 12. Productive structure of rice and E. oryzicola in a mixed stand (Yamasue et al., 1987). The crop and weed were planted with a 1:1 substituted design with 20 cm spacing between plants. The structure was obtained just before harvesting.

Seed Shattering Habit and Dormancy

Though darnel has neither seed (grain) shattering habit nor dormancy, *E. oryzicola* distinctly has both seed shattering habit and dormancy. When rice farmers conduct hand-weeding after crop heading, many seeds of *E. oryzicola* are shattered onto the soil surface. The shattering habit is important for the weed to disperse seeds and supply them to the seed bank in soil. The habit of *E. oryzicola* is obvious when compared with a cultivated form of this weed. The cultivated form was found in 1994 in Yunnan Province of China (Yabuno, 1996). The weedy form develops abscission layers in the spikelet rachis before shattering, but the cultivated form develops no particular structures at a comparable region of the spikelet rachis (Hirosue et al., 2000).

The seeds shattered onto the soil surface are innately dormant and buried in the soil by cultivation. Their longevity in a field is estimated to be about 10 years (Miyahara, 1972). During this span, seeds annually change dormancy state as described above (Figure 9). When seeds in the seed bank are dormant from late summer to early winter, the field is usually drained and the soil is exposed to aerobic conditions. Then, dormant seeds absorb larger amount of oxygen and have a higher activity of the Krebs cycle enzyme and cytochrome oxidase than nondormant seeds (Yamasue et al., 1987; Fukao et al., 2003), indicating that they slowly respire through the conventional aerobic pathway to maintain viability. When the seeds are released from dormancy and germinate under flooding, they respire through anaerobic alcohol fermentation. This respiratory transition is ecologically adaptive when we consider the oxygen tension in flooded paddy fields.

Table 2. Plant height of *E. oryzicola* as affected by four neighboring plants (From Yamasue et al., 1997a and unpublished data)[a]

Plants	Neighboring Plants Plant height (cm)	*E. oryzicola* Plant height (cm)
Rice (cv Nipponbare)[a]	72	116
Rice (cv Raja lele)[a]	117	121
E. crus-galli var. *formosensis*[a]	153	158
Rice (cv Yamada-nishiki)[b]	116	122
Rice (cv Nurasakinomai)[b]	134	160

[a] A trial was conducted in 1995 at The Kyoto Experimental Farm of our University by transplanting a strain of the weed and two rice cultivars with a substitutive design at a ratio of 1:1, 20-cm spacing between seedlings. *E. crus-galli* var. *formosensis* were also planted with *E. oryzicola* in rows with 20-cm spacing. Transplanting of all the seedlings was carried out at the 5 to 6 leaf stage on June 16. Data are means of five plants ± SD.
[b] Naturally occurring *E. oryzicola* in a lowland rice field.

Two additional findings on seed dormancy of *E. oryzicola* are noteworthy. Firstly, seed weight of *E. oryzicola* continuously varies among those produced from the same panicle; 1.5 to 3.5 mg per seed. The weight is linked to the degree of dormancy; smaller seeds have the greater degree of dormancy (Yoshioka et al., 1985). It partially causes the uneven germination and emergence of the weed in a lowland rice field, and contributes the weed to escape from a single weeding. Secondly, calcium cyanamide ($CaCN_2$), a nitrogen fertilizer, is effective in breaking dormancy (Ishida and Yamasue, 1997). When the fertilizer is applied after harvesting rice, many seedlings unseasonably emerge in the fall and are killed in the succeeding winter. Therefore, it is likely that seed dormancy is an important trait for the weed to avoid germination in the fall just before a hostile winter. There is a period in the fall after rice harvest when soil temperatures and moisture are sufficient for germination after enough precipitation. In the temperate region, this fall period with suitable conditions for the germination becomes shorter as the latitude increases. It follows then, that the degree of seed dormancy decreases as the latitude increases (Yamasue et al., 1981). But both the seed dormancy mechanism of *E. oryzicola* and the dormancy-breaking mechanism of cyanamide are still unknown.

CONCLUSION

Strategies of crop-associated weeds for adaptation to the cropping system have been described for two examples: Darnel in wheat and barley, and *E. oryzicola* in lowland rice. Both darnel and *E. oryzicola* are restricted to the cropland habitat. Each of the two weeds evolved various characters adapted to the indigenous or traditional cropping systems of a region or country. The selection pressure of hand-weeding evolves vegetative mimicry in both of them. However, differences in seed or grain shattering habit and dormancy between these two weed species are noteworthy. Darnel, which has neither grain shattering habit nor dormancy, developed various types of the grain mimicry in size, awns and color by the crop seed cleaning practices, which differ from region to region and between the two cereal crops. The grain mimicry is necessary for the weed grains to be harvested and sown together with the crop seeds. Farmer's stock of the cereal grains works as a bank to preserve the weed grains as an impurity. On the other hand, *E. oryzicola*, which has both seed shattering habitat and dormancy, developed only the vegetative mimicry to rice. The weed seeds are readily dispersed by the shattering habit and reserved for a prolonged period in a persistent seed bank in the soil by the dormancy.

In modern agriculture, manual weeding and seed cleaning are replaced with herbicides and machines. Those characteristics of modern agriculture are creating new selection pressures that endanger weed species adapted to the traditional cropping system. Darnel is now becoming a threatened species. The dominant *Echinochloa* weeds in lowland rice of Japan is changing from *E. oryzicola* to *E. crus-galli* var. *crus-galli*. Although *E. crus-galli* var. *crus-galli* has neither plant mimicry nor heading photoperiodic sensitivity parallel to that of rice, it is more competitive against rice and less sensitive to herbicides because of the large plant stature and aggressiveness in growth.

REFERENCES

Baker H.G. "Characteristics and modes of origin of weeds." In *The Genetics of Colonizing Species,* Herbert G. Baker, G. Stebbins, ed. Oxford; Academic Press, 1965; pp.147-168.

Baskin C.C., Baskin J.M. *Seeds.* San Diego, CA: Academic Press, 1998.

Bunting A.H. "Some reflections on the ecology of weeds" In *The Biology of Weeds,* John. L. Harper, ed. Oxford; Blackwell Sci. Publ., 1960; pp. 11-26.

Barrett C.H.S. Crop mimicry in weeds. Eco Bot 1983; 37:255 – 282.

Chisaka H. "Weed damage to crops: yield loss due to weed competition." In *Integrated Control of Weeds,* J. D. Fryer, S. Matsunaka, ed. Tokyo; Univ. Tokyo Press, 1977; pp.1-16.

Fukao T., Ida S., Rumpho M.E., Kennedy R.A., Yamasue Y. Differential gene expression of the α-chain of mitochondrial H^+-transporting ATP synthase between dormant and non-dormant seeds of paddy *Echinochloa* weeds. Weed Biol Manage 2003; 3:15-20.

Fukui K. The local names of cultivated plants in Ethiopia. Kikan-Jinruigaku 1971; 2:3-83

Harlan J.R. *Crops and Man.* Madison, WI: American Society for Agronomy and Crop Science Society of America, 1992.

Harper J.L. *Population Biology of Plants.* London: Academic Press, 1977.

Hirosue T., Yamasue Y., Yabuno T. Shattering habit and dormancy of spikelets in a cultivated form of *Echinochloa oryzicola* recently found in China. Weed Res 2000; 40: 449-456

Holm L.G., Plucknett D.L., Pancho J.V., Herberger J.P. *The World's Worst Weeds.* Honolulu, HI: University Press of Hawaii, 1977.

Ishida R., Yamasue Y. Effect of cyanamide on seed dormancy and respiratory enzyme activity of *Echinochloa oryzicola* Vasing. Proceedings of the 16th Asian-Pacific Weed Sci Soc Conf.; 1997 September 8-12; 89-91; Kuala Lumpur, Malaysia: The Asian-Pacific Weed Sci. Society, 1997.

Jenkin T.J. Interspecific and intergeneric hybrids in herbage grasses. VIII. *Lolium loliaceum, Lolium remotum* and *Lolium temulentum,* with references to '*Lolium canadense*'. J Genet 1954; 52: 318-331.

Kataoka T., Kim S.Y. Oxygen requirement for seed germination of several weeds. Weed Res (Japan) 1978; 23:9-12.

Kennedy R.A., Rumpho M., VanderZee V. Germination of *Echinochloa crus-galli* (barnyardgrass) seeds under anaerobic conditions: respiration and response to metabolic inhibitors. Plant Physiol 1983; 72:787-794

Kobayashi H., Sakamoto S. Association with cereals of awned and awnless darnel (*Lolium temulentum* L.). Weed Res (Japan) 1985; 30 (suppl): 89-90.

Ladizinsky G. *Plant evolution under Domestication.* Netherlands: Kluwer Academic Press, 1998.

Maekawa F. Prehistoric-naturalized plants to Japan. Acta Phytotaxonomica et Geobot 1943; 13:274-279.

McCreery D.W. Flotation of the Bab edh-Dhra and Numeria Plant Remains. Ann Am School Orient Res 1979; 46: 165-169.

Miyahara M. Physiological and ecological studies on dormancy of the seeds of *Echinochloa crus-galli* Beauv. var. *oryzicola* Ohwi, a paddy field weed. J Cent Agr Expt Sta 1972; 16:1-62.

Morinaga T., Nagamatsu T. Genecological studies on the wild Japanese barnyardgrass in the lowland rice field. I. The heading date of Japanese strains of Japanese barnyardgrass. Ikushu kennkyu 1942; 1: 116-122.

Nakatani K., Tanisaka T., Yamasue Y. Geographical variation in heading photoperiodic sensitivity of *Echinochloa oryzicola* Vasing., an obligate weed in rice. J Weed Sci Tech 1998; 43:108-113.

Nei M. Genetic distance between populations. Ame Natu 1972; 106: 283-292.

Oka H. Photoperiodic adaptation in rice. Phyton 1958; 11: 153-160.

Radosevich S.R., Holt J.S. *Weed Ecology.* New York: John Willey & Sons., 1984.

Rumpho M., Kennedy R.A. Anaerobic metabolism in germinated seeds of *Echinochloa crus-galli* (barnyard grass): metabolites and enzyme studies. Plant Physiol 1981; 68: 165-168.

Senda T. , Tominaga T. Inheritance mode of the awnlessness of darnel (*Lolium temulentum* L.). Weed Biol Manage 2003; 3: 41-43.

Täckholm, V., Tackholm, G. Flora of Egypt. Fouad I University 1941; 1:301-311.

Takai Y. The principal behaviors of a soil under waterlogged conditions. Agric Horti 1978; 18:12-24.

Terrell E.E. A taxonomic revision of the genus *Lolium.* USDA Techn Bull 1968; 392:1-65.

Thorogood D., Hayward M.D. Self compatibility in *Lolium temulentum* L.: its genetic control and transfer into *L. perenne* L. and *L. multiflorum* Lam. Heredity 1992; 68: 71-78.

Tominaga T. Association of darnel (*Lolium temulentum* L.) with wheat and barley cultivation. Breed Res 2002; 52 (suppl): 68-71.

Tominaga T., Fujimoto T. Evolution of the diversity in darnel (*Lolium temulentum* L.) in Malo, Ethiopia. J Weed Sci Tech 2000; 45 (suppl): 176-177.

Tominaga T., Ishiyama T. Differences in seed dormancy of darnel (*Lolium temulentum* L.) collected from different habitat. J Weed Sci Tech 2001; 46 (suppl): 148-149.

Yabuno T. Biosystematic study of the genus *Echinochloa.* Jap J Bot 1996; 19:277-323.

Yamasue Y., Koda S., Ueki K., Matsunaka S. Variation in growth, seed dormancy and herbicide susceptibility among strains of *Echinochloa oryzicola* Vasing. Weed Res (Japan) 1981; 26:6-13.

Yamasue Y., Ueki K., Chisaka H. Seed dormancy and germination of *Echinochloa oryzicola* Vasing.: an observation through respiration and several enzyme activities, Weed Res (Japan) 1987; 32: 188-197.

Yamasue Y., Ueki K. Seed dormancy and germination of *Echinochloa oryzicola* Vasing.: substantive evidences for alcohol fermentation at early germination. Weed Res (Japan) 1987; 32: 263-267.

Yamasue Y., Asai Y., Ueki K., Kusanagi T. Anaerobic seed germination for the habitat segregation in *Echinochloa* weeds. Japan J Breed 1989; 39: 159-168.

Yamasue Y., Murayama H., Inoue H., Matsui T., Kusanagi T. Growth analysis of rice and *Echinochloa oryzicola* Vasing. in mixed stands. J Weed Sci Tech 1997a; 42:366-373.

Yamasue Y., Murayama H., Inoue H., Matsui T., Kusanagi T. Productive structures of rice and *Echinochloa oryzicola* Vasing. in mixed stands. J Weed Sci Tech 1997b; 42:357-365.

Yamasue Y. Strategy of *Echinochloa oryzicola* Vasing. for survival in flooded rice. Weed Biol Manage 2001; 1: 28-36.

Yoshioka T., Yamasue Y., Ueki K. Seed ecological studies in relation to the asynchronous emergence of *Echinochloa oryzicola* Vasing. 1. variations in weight and ripening date of seeds among setting positions within a plant. Weed Res (Japan) 1985; 30: 58 – 64.

CHAPTER 4

EVOLUTIONARY BIOLOGY OF THE FOXTAIL (*SETARIA*) SECIES-GROUP

JACK DEKKER

Weed Biology Laboratory, Agronomy Department, Iowa State University, Ames, IA 50011, USA

The foxtails (*Setaria* spp.) are one of the worst weed groups interfering with U.S. and world agriculture and land management (Holm et al., 1977, 1979, 1997), therefore one of the world's most successful terrestrial plant organisms. The success of the foxtails is to a large extent due to their intimate evolutionary relationship with humans, disturbance, agriculture, and land management. The invasive threat of this weed group continues due to its increasing geographic range of distribution, the appearance of new species and biotypes, and the increase of herbicide resistant variants. The invasive and colonizing ability of the weedy foxtails is buffered against change by the accumulation of dormant seed in soil seed banks, a trait permitting enduring occupation of a locality.

Frequently more than one *Setaria* species co-exist together in a locality; possibly allowing a more complete exploitation of resources left available by human disturbance and management. Several weedy foxtail species often coexist in a single field (most commonly with *S. viridis*), each exploiting a slightly different "opportunity space" or niche. The genus *Setaria* also contains the important world crop foxtail millet whose geographic distribution and evolutionary history is intimately connected with the weedy foxtails. As such, a review of the weedy foxtails should appropriately begin at the level of the *Setaria* species-group (a group of closely related species, usually with partially overlapping ranges; Lincoln et al., 1998) (Table 1; Pohl, 1951, 1966; Rominger, 1962; Wang et al., 1995a).

Reviews of *Setaria* species have been published previously (Dekker, 2003), notably two in the Biology of Canadian Weeds series (*S. pumila* and *S. verticillata*, Steel et al., 1983; *S. viridis*, Douglas et al., 1985). The definitive taxonomic treatment is Rominger (1962). This review is presented from the perspective of the evolutionary ecology of the foxtails. Foxtails are weeds because of their interactions with humans in situations where disturbance, crop production and land management occur.
Weed management and the activities of humans are primary selection forces shaping and determining the conditions of existence of weedy plants. The traits and adaptations that weeds acquire in this struggle are the immediate problems managers confront. Therefore, weed management is the management of weed selection, adaptation and evolution. The goal of this review of the weedy *Setaria* species-group is to provide the phylogenetic, life and evolutionary histories with which foxtail weed management might be advised.

Inderjit (ed.), Weed Biology and Management, 65-113.

Table 1. The foxtail (*Setaria*) species-group; [1]common name, Weed Science Society of America nomenclature (WSSA, 1984).

Common Name[1]	Genus	Specific Epithet	Sub-spp.	Authority
Giant foxtail	*Setaria*	*faberii*	-	Herrm.
Yellow foxtail	*Setaria*	*pumila*	-	(Poir.) Roem. & Schult.
Knotroot foxtail	*Setaria*	*geniculata*	-	(Lamarck) Beauv.
Bristly foxtail	*Setaria*	*verticillata*	-	(L.) Beauv.
Green foxtail	*Setaria*	*viridis*	*viridis*	(L.) Beauv.
Foxtail millet	*Setaria*	*viridis*	*italica*	

PHYLOGENESIS

The evolutionary history and line of descent of the *Setaria* species-group, its phylogenesis, is the product of underlying genetic mechanisms and preadaptations that constrain its future evolutionary change. The genotypic and phenotypic characteristics of the weedy foxtails provide the basis for understanding what they are now, and what they might be in the future.

Genotypic Variation

The genetic constitution of individual plants in a community is the primary determinate of what foxtails do. The taxonomy, genetics, hybridization and speciation of the foxtail genotype are presented as both consequences and determinants of evolutionary change.

Taxonomy

Setaria spp. were first described by Linnaeus (1753). World-wide there are 125 foxtail species divided among several subgenera (Hubbard, 1915; Rominger, 1962). The genus *Setaria* (the foxtails) is widely distributed in tropical, sub-tropical and temperate parts of the world and is of significant agricultural importance as crops (cultivated grains, perennial forage grasses) and as noxious weeds (Prasada Rao et al., 1987; Rominger, 1962). The taxonomy of the genus is very complex, and an accurate classification has been confounded by the high degree of overlapping morphological characters both within and between species, and the diverse polyploidy levels within the genus. There is evidence that the genus *Setaria* is actually divided into two groups, or clades (Doust and Kellogg, 2002). This reassessment of the genus would place *S. pumila* and *S. geniculata* into a common (*S. pumila*) clade, and *S. viridis* (both subspecies), *S. faberi* and *S. verticillata* in another (*S. viridis*).

The genus *Setaria* belongs to the tribe Paniceae, subfamily Panicoideae and family Poaceae (Pohl, 1978). There are 43 species of *Setaria* in North America segregated into 3 subgenera: *Ptychophyllum* (6 spp.), *Paurochaetium* (10 spp.), and *Setaria* (27 spp) (Rominger, 1962). Of these, 25 are native to North America (including the weedy *S. geniculata*), 10 have centers of origin in South America, 8 species are adventives from the Old World, including the five weedy *Setaria* species of the temperate regions of the northern hemisphere, including North America and Eurasia (Table 1).

The origins of *S. geniculata* are of particular note. *Setaria geniculata*, a perennial, is the only weedy foxtail native to the New World, and very closely resembles *S. pumila*, an annual species of Eurasian origin (Rominger, 1962; Wang et al., 1995b). *S. geniculata* and *S. pumila* are frequently misidentified by weed managers. The primary morphological differences between *S. geniculata* and *S. pumila* are found in the spikelet length and lower floret, palea, and the presence of short rhizomes (Hitchcock, 1971). *S. geniculata* has a high degree of sterilization in the fascicle, unlike any other species native to North America, but similar to tropical *Setaria* species. It has been suggested that the cause of this enigma was a very ancient dispersal event westward from the Old to New World (Rominger, 1962).

Both *S. viridis* (foxtail millet) and *S. pumila* (korali) have been domesticated and cultivated as crops (de Wet et al., 1979). Linnaeus (1753) was the first to recognize foxtail millet and *S. viridis* as two independent species, *Panicum italicum* (foxtail millet) and *P. viride* (*S. viridis*). *Setaria viridis,* sub-species *italica* (L.) Beauv. (foxtail millet), is a minor world cereal crop grown across Eurasia and Africa. Foxtail millet (*S. viridis*, subsp. *italica*) and *S. viridis* (*S. viridis*, subsp. *viridis*) are subspecies of *S. viridis*, and not separate species, and have continuous and overlapping genetic variation, evidence of the weedy origins of the crop (Prasada Rao et al., 1987; Wang et al., 1995a). *S. viridis* subspp. *viridis* and *italica* are interfertile (Darmency et al., 1987a; Li, 1934; Li et al., 1935; Prasada Rao et al, 1987; Takahashi and Hoshino, 1934; Till-Bottraud et al., 1992; Willweber-Kishimoto, 1962). Prasada Rao et al. (1987) suggested that both taxa be considered as subspecies of *S. italica* (*S. viridis*, *S. italica*, subsp. *viridis*; foxtail millet, *S. italica*, subsp. *italica*). This incorrect nomenclature implies the primacy of the crop over that of its wild *Setaria* ancestors (see interelationship between wild-weed-crop *Setaria* below). Several racial taxonomies below the subspecies level have been proposed for foxtail millet based on morphology rather than genetic distances (see domesticated foxtail millet variants below).

Genetics and Hybridization

Molecular Genetics. DNA of *S. viridis* subsp. *italica* has several distinguishing features: low nuclear DNA content (haploid, C-value, 1C: 0.82 pg), low proportion of repetitive DNA (30%), and the presence of a high melting point component (4.7-6.7% of the genome) (Lakshmi and Ranjekar, 1984; Siva Raman et al., 1986). The total proportion of repetitive DNA was the lowest value in the Poaceae family at the time it was reported (Lakshmi and Ranjekar, 1984). Additionally, this low repetitive DNA value provides genomic evidence of the tropical origins of the genus *Setaria.* The authors also suggest that the high melting (G + C rich) component has a functional role in genomic adaptations to specific environments (i.e. dry, nutrient poor, high light). It has been suggested this high thermal stability in foxtail millet may indicate very little to no base sequence mismatch, a rare observation, or that this stability maintains internal homogeneity by 'cross over' fixation (Smith, 1976).

A major part of the foxtail millet genome (85%) consists only of long interspersed repeated DNA sequences (greater than 1.5 kilo base pairs; Siva Raman et al., 1986).

The genome organization patterns in plants are related to the dynamics of DNA (Thompson and Murray, 1981). The dynamics of DNA is defined in terms of the rate of turnover of the genome: low rates of turnover give rise to predominantly long interspersed DNA sequences, as in foxtail millet (Flavell, 1980). Genome size can be maintained by this slow turnover, in which repeated DNA sequences are not subjected to mutations, deletions or base substitutions at a very rapid rate (Siva Raman et al., 1986).

The significance of genome size in phylogeny is not understood, the so-called 'C-value paradox' (Orgel and Crick, 1980). Cavalier-Smith (1978) asserts there is a good correlation between low C-values and strong r-selection, selection for weedy qualities. Dawkins (1999) suggests that intragenomic selection pressure may lead to a decrease or elimination of "junk" DNA (untranslated introns) and therefore smaller genome size in colonizing, invasive species like foxtails.

Cytogenetics. The nuclear genome of *Setaria* spp. typically has a complement of 9 chromosomes in its monoploid form (gametophytic egg, pollen; n=9=x). Sporophytic tissue within the genus includes the polyploid series of 2x (*S. viridis* subsp. *italica* and *viridis*), 4x (*S. faberi, S. pumila, S. geniculata, S. verticillata*), 6x (*S. verticillata*), 8x (*S. pumila, S. geniculata*) possibilities. Aneuploidy (chromosome number of individual not exact multiple of typical haploid set for species) in the genus has been reported, but is considered insignificant. A wide range of chromosome complements have been reported within a particular foxtail species. Whether this is due to misidentification, or a product of the complex taxonomy of the genus, is not apparent. If these reports are true, they raise important questions about intra-specific fertility, and may indicate reproductive barriers within species. Within each foxtail species, reports of nuclear chromosome numbers are: *S. viridis* subspp. *viridis* and *italica*: diploid, 2n=18 (Avdulov, 1931; Fairbrothers, 1959; Gupta and Singh, 1977; Kishimoto, 1938; Li et al., 1935, 1945; Willweber-Kishimoto, 1962); *S. faberi*: tetraploid (2n=36) (Fairbrothers, 1959; Kishimoto, 1938; Li et al., 1942; Willweber-Kishimoto, 1962), allotetraploid (Till-Bottraud et al., 1992; Li et al., 1942); *S. geniculata*: a polyploid complex (Rominger, 1962), tetraploid (2n=36) (Gupta and Singh, 1977; Rominger, 1962), octaploid (2n=72) (Brown, 1948; Osada, 1989; Rominger, 1962); *S. pumila*: diploid (2n=18) (de Wet, 1954), a polyploid complex, tetraploid (4x, 2n=36) (Avdulov, 1931; Gupta and Singh, 1977; Khosla and Sharma, 1973; Kishimoto, 1938; Krishniswami and Rangaswami Ayyangar, 1935; Willweber-Kishimoto, 1962), octaploid (8x, 2n=72) (Brown, 1948; Khosla and Sharma, 1973; Rominger, 1962), aneuploidy, reports of hypertetraploidy (n=22; n=19+0-3B) in different Indian populations (Khosla and Sharma, 1973); *S. verticillata*: diploid, 2n=18 (de Wet, 1954; Pohl, 1962; Osada, 1989), autotetraploid (Till-Bottraud et al., 1992), 2n=36 (Avdulov, 1931; Gupta and Singh, 1977; Osada, 1989; Rominger, 1962; Till-Bottraud et al., 1992; Willweber-Kishimoto, 1962), hexaploid (n=27) (Gupta and Singh, 1977; Khosla and Sharma, 1973; Osada, 1989). Genetic maps of foxtail millet have been constructed based on RFLP markers (Wang et al., 1998; Devos et al., 1998).

Asexual reproduction is not a common mode of reproduction in weedy foxtails. The only weedy foxtail relying on vegetative asexual reproduction is *S. geniculata*, a perennial species with short, branched, knotty rhizomes (Rominger, 1962). Agamospermy (seed formation without fertilization) in *S. viridis* has been noted

(Mulligan and Findlay, 1970), and has also been suggested for *S. pumila* and *S. verticillata* (Steel et al., 1983). Apomixis has also been reported in other *Setaria* species within the *Setaria* subgenus (Chapman, 1992a; Emery, 1957). Emery (1957) reports both sexual (autogamy) and apomictic seed production in *S. leucopila*, *S. machrostachya* and *S. texana*, a highly unusual condition. Giant and *S. pumila* tillers readily root in soil when cut, separated from the plant, and buried in moist soil, an important trait allowing weedy foxtails to reestablish themselves after cultivation and mowing (Barrau, 1958; Santlemannn et al., 1963; Schreiber, 1965b).

Hybridization. Pollen and gene flow in the weedy foxtails can take the form of intra-specific (autogamy, self-fertilization; allogamy, outcrossing) or inter-specific hybridization (introgression between different foxtail species). The weedy foxtails are primarily a self-pollinated species (Mulligan and Findlay, 1970; Pohl, 1951). Wind pollination (anemophily) is the mode in those rare circumstances of outcrossing (Pohl, 1951), with pollen probably moving at most a few hundred meters when introgression does occur (Nguyen and Pernes, 1985). The combination of predominantly self pollination, with a very small amounts of out-crossing to provide new genetic combinations, is a weedy adaptation that has provided the foxtails much of their success (Allard, 1965; Baker, 1974; Simpson, 1990). This combination of reproductive modes has allowed the foxtails to colonize a new area with as little as one seed, to compete well in that specific niche, to maintain an enduring occupation of that niche, as well as provide new variants to invade new niches.

Intra-Specific Hybridization. Although weedy *Setaria* spp. normally are self-pollinated, natural outcrossing between plants does occur, and is an important source of new variants to the several species. In *S. viridis*, spontaneous outcrossing rates among plants in the field have been observed to be between 0 and 7.6% of the autogamous rates: 0.2-2.8 (Darmency et al., 1987a); less than 2% natural outcrossing in continuous *S. viridis* populations (Prasada Rao et al., 1987); for *S. viridis* subsp. *viridis* and subsp. *italica*, between 0-2.2%, with bagging panicles together it could reach 4% (Till-Bottraud et al., 1992); between 0.59-2.26% (Takahashi and Hoshino, 1934); for subsp. *italica*, 5.6-7.63% with plants 30cm apart (Li et al., 1934, 1935); and between 0 and 9.86% (average 0.7%) (Mac Vicar and Parnell, 1941). *S. viridis* subsp. *viridis* mated with subsp. *italica* had 0.3-0.6% outcrossing with some sterile seed when plants were 17 cm apart (Till-Bottraud et al., 1992). With this cross, there were an increased number of later-flowering hybrids amongst the crop cultivars. Crosses of these two subspecies yielded hybrids 50% sterile although meiotic chromosome behavior was regular (Li et al., 1945).

Interspecific Hybridization. Introgression of pollen between weedy species is a rare event, but when it does occur it can have very important consequences. When fertilization between species does occur, the progeny are almost always sterile. When fertile progeny are produced, the new genetic variant is usually unfit for competitive conditions it encounters in natural ecosystems.

Most of our information about interspecific hybrids of the foxtails comes from controlled experimentation aimed at producing these types of hybrids. Crosses between *S. viridis* subsp. *viridis/italica* (n=9) and *S. verticillata* (n=9) have lead to naturally occurring sterile hybrids in French foxtail millet fields (Poirier-Hamon and

Pernes, 1986). Till-Bottraud et al. (1992) obtained seed only with the wild parent as female; they observed a 0.5% hybridization rate. The hybrids were all triploid (2n=27=3x) and almost totally sterile. The F_2 progeny resembled the female *S. verticillata* parent and were tetraploid, 2n=36. Stace (1975) found a naturally occurring *S. viridis*, subsp. *viridis* and *S. verticillata* hybrid called *S.* X *ambigua* Guss. (2n=36; Clayton, 1980) (syn. *S.* X *ambigua* Guss emend. Koyama; *S.* X *decipiens* Schimp. (Osada, 1989)) throughout southern Europe. Although attempted, no natural or artificial crosses between *S. viridis* (n=9) and *S. pumila* (n=18) succeeded (Till-Bottraud et al., 1992; Willweber-Kishimoto, 1962). Willweber-Kishimoto (1962) successfully pollinated both subspecies (*viridis*/*italica*) of *S. viridis* (n=9) with *S. faberi* (n=18; tetraploid) and hybrid seed was formed. Only in the subsp. *viridis* cross was fertile seed found. Li et al. (1942) also successfully crossed these species, but most crosses were almost completely sterile. Willweber-Kishimoto (1962) successfully pollinated *S. pumila* (n=18) with *S. faberi* (n=18), no hybrid seed were formed.

Speciation and Reproductive Barriers between Setaria *Species.* New foxtail species can be formed by several processes (e.g. mutation, hybridization, polyploidization, etc.), but most of foxtail biology provides reproductive barriers ensuring such events occur very rarely. Many factors contribute to form partial reproductive barriers between the foxtail species, a genetic condition favoring introgression and gene flow at very low levels withing the species-group and wild-crop-weed complex (Darmency et al., 1987a, b; de Wet, 1966; Harlan, 1965; Harlan et al., 1973). Panicle height differences, and times of fertilization (pollen shed; stigma receptivity), both prevent hybridization events from occurring between many populations. The often greater height of foxtail millet (*S. viridis* subsp. *italica*) favors weed pollination by the crop in natural stands (Willweber-Kishimoto, 1962). Reproductive barriers between the foxtail species does not occur at the level of pollen germination and pollen tube growth in the stigma in any combination of pollen and stigma among *S. italica, S. viridis, S. faberi, S. pumila,* and *S. pallide-fusca* (Willweber-Kishimoto, 1962).

Speciation by polyploidization (either alloploidy or autoploidy) may be an important means by which new foxtail species were formed. Based on chromosomal data, Khosla and Sharma (1973) speculate polyploidization at various levels has played an active role in speciation and delimiting taxa in *Setaria*. Polyploidization of the more diverse and ancient *S. viridis* may have been the genesis of the specialized and less diverse *Setaria* spp. set (giant, yellow, bristly, knotroot) of specialized and less diverse *Setaria* spp. Allotetraploid forms of *S. faberi* and *S. pumila* have been explained as ancient crosses of *S. viridis* with and unknown diploid species (Li et al., 1942; Kholsa and Singh, 1971; Till-Bottraud et al., 1992). Li et al. (1942) concluded that *S. faberi* is a product of *S. viridis* and an unknown *Setaria* sp., followed by a chromosome doubling (polyploidization) event, forming an alloploid species. Relatively homogeneous population allozyme data suggest this polyploidization event in *S. faberi* was a relatively recent evolutionary event (Wang et al., 1995b). Genetic analysis has suggested that some variants of *S. verticillata* (n=18) may be the product of chromosome doubling of *S. viridis*, autotetraploidy (Till-Bottraud et al., 1992)

Geographical barriers to inter-species hybridization in current times are much less important that previously due to the profound increase in human exchanges around world, especially with agricultural grain trade and human travel.

Phenotypic Variation

The phenotype (total observable structural and functional properties) of an individual foxtail plant is determined by its genetic constitution, and varies when observed under different environmental and temporal conditions. Genetic composition is the "blueprint" for foxtail fitness, but selection acts on the individual phenotype. One of the most striking observations of foxtail behavior is the utilization of phenotypic heterogeneity within a single genotype or individual, rather than genetic diversity among the individuals of a population, as a means of exploiting a locality (Scheiner, 1993; Wang et al., 1995a, 1995b). This phenotypic heterogeneity takes the form of phenotypic plasticity and somatic polymorphism in many of its most important traits, especially during reproduction. Plastic responses to the immediate environment allow an individual plant to maximize its fitness to the resources available at a fixed locality, avoiding premature mortality of the individual and/or its parts.

The set of all possible foxtail phenotypes defines its niche, the reaction norm, the ecological role of the *Setaria* species-group in a community. The coordinates of the many parameters representing the condition of existence of the foxtails can be conceptualized as a multi-dimensional space, the foxtail niche. Most of the phenotype space of the foxtails has not been characterized. Only the most obvious behaviors and morphologies have been described (e.g. color morphology, herbicide resistance), and these are usually the characters that are most apparent in management situations. What is known of the *Setaria* species-group phenotype space can be described as variation in morphology, physiology, geography and domestication.

Morphological Variants

Often the most striking morphological differences among foxtails are similar in terms of allelic variation (e.g., compare *S. viridis* subsp. *viridis* var. *pachystachys* and *S. viridis* subsp. *italica* race *maxima),* while genotypic variation can be wide within nearly identical morphologies (e.g., compare Old World *S. pumila* and New World *S. geniculata*). Additionally, many of the morphological variants described as biotypes are in fact extreme examples of continuous characters (e.g. leaf coloration).

S. viridis varies greatly in morphology and growth characteristics. A number of morphological variants of *Setaria viridis* have been reported, including named varieties: *S. viridis* var. *viridis*; *S. viridis* var. *weinmanni*; *S. viridis* var. *ambigua*; *S. viridis* var. *major*, giant green; *S. viridis* var. *robusta-alba*, robust white; *S. viridis* var. *robusta-purpurea*, robust purple; *S. viridis* var. *pycnocoma*; and *S. viridis* var. *pachystachys* (Hubbard, 1915; Kawano and Miyake, 1983; Schreiber and Oliver, 1971). The taxonomic validity of formally recognized morphological varieties within *S. viridis* varieties has been questioned because they do not differ isozymatically from typical forms of *S. viridis* (Wang et al., 1995a). It was observed that these varieties how no genetic differentiation from the common type, and were completely embedded in common *S. viridis*, *S. viridis* var. *viridis* allozyme principal component analysis (PCA) plots. Confusion with identification is highlighted by the observation that presence or absence of hairs on upper leaf surface, normally thought of as diagnostic for *S. faberi*, are found in both *S. viridis* (var. *major*, giant green) and *S. faberi* (green giant; Wang et al., 1995a, b).

Abnormal forked, divided or segmented panicles in *S. viridis* have been observed, although these traits apparently are not heritable. Reproductive developmental mutants have been observed in which spikelets proliferate as green scales, named S. *viridis*, var. *vivipara* (Dore and McNeill, 1980).

Seed (spikelet) morphology and shape varies widely among individual foxtail species and genotypes. These variations may play an important role in interactions with the soil environment determining their behavior in soil seedbanks, especially at the soil particle-seed interface affecting the uptake of water and gases into the foxtail seed placental pore (see dormancy below). The heterogeneity in seed shape has neither been characterized nor has its role been hypothesized.

S. geniculata and *S. pumila*, natives to two different continents, were found to be so similar morphologically that the 24 accessions studied by Wang et al., 1995b were initially identified as *S. pumila.* Allozyme data proved to be diagnostic in the identification of these cryptically different species. Wide morphological and physiological variation among accessions of yellow and *S. faberi* were observed (Santelman and Meade, 1961).

Physiological Variants

The very rapid increase in frequency of herbicide resistant variants in populations is the most apparent behavioral variation among *Setaria* phenotypes observed in management situations in the last 25 years (e.g., Holt and LeBaron, 1990; Reschly et al., 1996; Ritter et al., 1989; Santelmann and Meade, 1961; Stoltenberg and Wiederholt, 1995; Tranel and Dekker, 1996; Thornhill and Dekker, 1993; Volenberg et al., 2001; Wiederholt and Stoltenberg, 1996). The mechanisms of resistance present in weedy foxtail populations includes exclusion and metabolism (e.g. Thornhill and Dekker, 1993; Wang and Dekker, 1995). In many instances these traits (not from *Setaria* sources) have been incorporated into crop cultivars (Dekker and Duke, 1995; Dekker and Comstock, 1993).

Physiological variation in dormancy and germinability induced in seed during embryogenesis is one of the most important phenotypic traits leading to colonizing ability and is discussed in detail below (see seed dormancy below; Dekker et al., 1996).

Variation in drought tolerance among the foxtails has been observed (Blackshaw et al., 1981a; Manthey and Nalewaja, 1982, 1987; Taylorson, 1986). Potentially salt-tolerant genotypes of *S. viridis* var. *pachystachys* have been observed along the seacoasts of Japan (Kawano, 1983; personal observation, 1992 and 2000, data not reported). Salt tolerance exists in other *Setaria* species (Chapman, 1992b). These studies indicate there may be an intimate physiological and morphological relationship between drought tolerance, salt tolerance and seed dormancy. This is most apparent in extreme habitats where foxtail millet remains an important crop and cultivated cereal (e.g. Central Asia including Afghanistan, India, sub-Saharan Africa). "Primitive" variants of foxtail millet (e.g race *moharia*) are grown in marginal areas of Asia (e.g. Afghanistan) and are well adapted to high elevation, semi-arid conditions on poor soil (de Wet et al., 1979). These variants yield a crop in areas with as little as 12.5 cm of rainfall per annum.

Geographic Variants
Locally adapted populations of *Setaria* occur throughout its range of distribution, yet almost nothing is known of this type of adaptation or how this local adaptation occurs. Wang et al. (1995a, b) observed north-south genotypic variation in North American populations of *S. viridis* and *S. geniculata*. Norris and Schoner (1980) found differences in *S. pumila* seed dormancy and germination requirements among five *S. pumila* accessions from California, Iowa, Pennsylvania, Connecticut, and Massachusetts. A more complete presentation of global geographic variation is presented in population genetic structure below.

Domesticated Foxtail Millet Variants
The principal *Setaria* domesticate is *S. viridis* subspecies *italica* (foxtail millet), but *S. pumila* (korali) has also been domesticated (Wang et al., 1995a; de Wet et al., 1979). The *Setaria* millets are characterized as wild-weed-crop complexes, as are all cereals (Harlan, 1965). Despite the fact that several racial taxonomies below the subspecies level have been proposed for the foxtail millets, continuous hybridization has resulted in a wide variety of types with overlapping morphology and genetic constitution.

Giant *S. viridis*, *S. viridis* var. *major* Pospichal, is a weed of Eurasia and the U.S. corn belt, and is believed to be a hybridization product of foxtail millet and *S. viridis* (Pohl, 1951, 1966; Rominger, 1962). This variant is robust (1.5-2.5 m height), with larger panicles (to 20 cm), compared to other *S. viridis* variants. It also has larger, lobed panicle branch fascicles (to 4 cm).

Hubbard (1915) found foxtail millet to be exceedingly variable, and describes in detail 3 subspecies, 5 varieties, 6 subvarieties, and 6 formae; as well as many different cultural names for the different variants. "Primitive" races of foxtail millet resemble giant *S. viridis* inflorescences, while more "highly evolved" (domesticated) races possess extremely large panicles (de Wet et al., 1979). A more recent classification of foxtail millet recognizes 2 or 3 different morphological races based on plant and panicle structure (Prasada Rao et al., 1987; de Wet et al., 1979).

Race moharia. These variants are 25-100 cm tall, with 5-52 (ave. 9) culms per plant. They have small (5-20 cm long), erect or slightly nodding panicles with short (1-3 cm long) fascicle branches on which the spikelets are arranged in clusters (Prasada Rao et al., 1987; de Wet et al., 1979). The setae are well developed. They differ from weedy variants by having spikelets that persist after maturity, aiding harvest. Korenicke and Werner (1885a, b) recognize 3 varieties based on setae and seed color: *metzgerii*, *mite*, and *atrum*. More advanced cultivars are believed to be derived from this race.

Race maxima. Variants in this race are 45-100 cm tall, 1-8 (ave. 2) unbranched culms, and large, pendulous panicles (to 35 cm long, 8 cm wide) with elongated secondary (fascicle) branches (2-6 cm long) forming lobed clusters (Prasada Rao et al., 1987; de Wet et al., 1979). Race *maxima* is highly variable (Dekaprelevich and Kasparian, 1928; Kernicke and Werner, 1885a, b). Gritzenko (1960) divides the race into three overlapping complexes: Manchurian, Mongolian and Korean.

Race indica. Race *indica* are intermediate to the other two races, but are unique due to their domestication in India (Prasada Rao et al., 1987). Plants have 1-25 (ave. 3) culms, with panicles 6-30 cm long whose fascicle branches are loosely arranged along the main axis.

An alternative organization is provided by Fukunaga et al. (1997), who reported five restriction fragment length polymorphisms (RFLP) phenotypes from among 117 Eurasian landraces of foxtail millet, types I-V. Most were from types I-III. Type 1, the most common, was found in temperate Eurasia, including Taiwan, Philippines, Nepal, Indonesia. Type II was found on Taiwan, Phillipines and Japan of that continuous Asian-Pacific island chain. Type III was mainly from South Asia (India, Nepal and Bangladesh). The rarer type IV came from Japan, Korea, Taiwan, Thailand and southern Pakistan. Another rarer type V came from southern Pakistan and Primorskaya province of Russia.

Korali, domesticated *S. pumila* is an endemic kharif crop grown in India (Bor, 1960; de Wet et al., 1979). It is also gathered as a cereal from wild plants due to its large seeds (2-3 mm).

LIFE HISTORY AND PHENOLOGY

The life history of the weedy foxtail species is the product of the interaction between the individual genotype and the environment. This complex developmental process is distinguished by both discrete and continuous events. The *Setaria* life history includes singular, threshold, often irreversible events (e.g., ferilization, abscission, germination), which are especially useful as experimental landmarks with which to make comparisons. The annual life cycle of most foxtail species emphasizes the short time frame within which intense selection pressure on this process occurs. The foxtail life history includes reproduction, seed life in the soil and vegetative growth.

Reproduction

The reproductive phase of the foxtail life history includes fertilization, embryogenesis and seed abscission and dispersal.

Fertilization

The life cycle of the foxtail seed begins with the initiation of flower primordia within panicle (inflorescence, seedhead) tissues in the culm of the vegetative parent plant. As the vegetative foxtail plant develops, the first events of the reproductive phase of plant development are initiated in the shoot meristem. Some meristems make the shift from leaf growth to producing cells that will eventually form the inflorescence and its constituent flowers. The panicle emerges from the culm and continues development, followed by the extrusion of the anthers, fertilization and the beginning events of embryogenesis. The fertilization process in the foxails involves floral and inflorescence morphology, as well as temporal and morphological patterns of flowering.

Panicle, Fascicle and Spikelet Morphology. The morphology of *Setaria* inflorescences consists of organs and tissues in organized in hierarchical nested structural sets: tiller culm, panicle, fascicle, spikelet and floret. The first externally visible event in the reproductive phase is the emergence of the immature panicle from the subtending leaf collar of its tiller. The most basal spikelets of the panicle are immediately adjacent to the leaf collar at that time (Dekker et al., 1996). In *S. faberi* the tiller culm supporting the panicle does not elongate further from the leaf collar until flowering has proceeded to the basal end of the inflorescence. Once flowering has proceeded basipedally to the bottom of the panicle the culm elongates and extends out of the leaf collar. This axis elongation is one of two means by which the light microenvironment in the panicle changes during flowering, fertilization and embryogenesis. The second mechanism, and its possible role in dormancy induction, involves fascicle spacing in the inflorescence (see below).

Panicle Architecture. The inflorescence, panicle, in the subgenus *Setaria* is usually narrow, terminal on the culm, very dense, cylindrical, and spicate, with very short branches (fascicles) only a few mm long (Rominger, 1962; Willweber-Kishimoto, 1962; Naryanaswami, 1956). The fascicles are spirally arranged around the main axis, each bearing a number of branchlets.

Fascicle Architecture. The fascicle is the panicle branch, consisting of 1 to many spikelets and a cluster of setae (bristles; Rominger, 1962). Each foxtail millet fascicle bears four to six or more spikelets amid a cluster of setae in a complicated system of branching (Naryanaswami, 1956). Spikelets are clustered on the rachis diverged from a main axis of a panicle, for most *Setaria* species these clusters are dense and the panicle looks spike-like ("foxtails") or lobed (foxtail millet and some weed-crop morphological intergrades). The number of fertile spikets per fascicle within each giant or *S. viridis* panicle is variable and more plastic in response to environmental conditions (Clark and Pohl, 1996). *S. pumila* panicle morphology differs from that of green or *S. faberi*: only a single spikelet is found in each *S. pumila* fascicle. In the mature inflorescence, the number of fascicles per unit length of panicle axis decreases from the distal to the basal end (Dekker et al., 1996). Flowering branch density along *S. faberi* panicle axes ranged from 16 fascicles/cm at the distal end to 5 fascicles/cm at the basal end. This arrangement of fascicles is a second mechanism allowing greater light penetration to those flowers at the basal end (see dormancy induction below).

Spikelet Architecture. The basic unit of the *Setaria* inflorescence is the dorsally compressed, two-flowered spikelet disarticulating below the glumes and subtended by 1 to several bristle-like setae (Hitchcock, 1971; Rominger, 1962). The spikelet consists of the rachilla, a sterile or staminate floret below, and a perfect floret above, and three empty glumes (Li et al., 1935; Willweber-Kishimoto, 1962). For *S. viridis*, *S. faberi* and *S. verticillata* the lower floret is entirely degenerated. In *S. pumila*, the lower floret has no pistil but three well developed anthers which open 3-7 days later than first floret (Willweber-Kishimoto, 1962).

The perfect or fertile floret above is often referred to as the fruit, grain or seed. This seed is indurate, usually transversely wrinkled, lemma and palea (hull) of similar marking and texture, which tightly encloses the caryopsis within at maturity (see dormancy regulation below for caryopsis morphology). The degree of rugosity of the lemma is a valuable taxonomic diagnostic character. Rugosity varies from smooth and shiny in foxtail millet, to finely ridged in *S. viridis*, to very coarse rugose seed in other foxtail species (Rominger, 1962). This rugosity may play a role in soil-seed contact relevant to seed germination and seedling emergence.

Foxtail spikelets are subtended by one-to-several setae that persist even when the spikelets are shed from the parent plant (Prasada Rao et al., 1987). The setae represent the stalks of abortive spikelets (Hofmeister, 1868; Chase, 1937). The number of these sterile branchlets varies by foxtail species (1-3, *S. viridis*, *S. faberi*, *S. verticillata*; 6 or more, *S. pumila*, *S. geniculata*; Hitchcock, 1971). The one-seeded spikelet in members of the subgenus *Setaria* falls off entire because it disarticulates from the pedicel (rachilla) and includes the unequal, membranaceous, glumes (weedy, seed-shattering, types; Darmency et al., 1987b; Rominger, 1962). These combined structures may provide protection to the caryopsis from predators before incorporation into the soil seed bank (alternative role in water wicking to placental pore in soil). The foxtail millet crop seed disarticulates above the glumes at shedding, a trait strongly associated with domestication.

Spikelets for *Setaria* spp. are 1.0-4.5 mm long, usually 2-3 mm, and vary both within and between species (Rominger, 1962). *S. viridis* seed are 1.8-2.4 mm long (var. giant *S. viridis*, 1.8-2.2 mm; var. robust purple and robust white, 2.0-2.4 mm), 1.0-1.3 mm wide, 1.8-2.2 mm in length, and weigh 0.71-0.87 mg (var. giant *S. viridis*, 1.08 mg; var. *pachystachys* 0.87 mg; var. robust purple 1.27 mg; var. robust white 1.39 mg). *S. faberi* seed are 2.5-3.0 mm in length, and weigh between 2.00-2.169 mg (Schreiber and Oliver, 1971; Kawano and Miyake, 1983). *S. pumila* seed are 2.5-3.4 mm long, 1.5-2.2 mm wide, 1.0-1.5 mm thick, and weigh between 2.7-4.5 mg (Kawano and Miyake, 1983; Povilaitis, 1956; Steel et al., 1983; Stevens, 1932; Schreiber and Oliver, 1971). *S. verticillata* spikelets are 1.8-2.3 mm long, 1.0-1.5 mm wide, 0.6-1.3 mm thick, and weigh between 1.1-1.3 mg (Steel et al., 1983).

Patterns of Flowering and Fertilization. The pattern of flowering and seed fertilization is a complex process occurring on several different spatial and temporal scales: within and among individual panicles on the same plant, during the annual growing season, and during a daily or diurnal period.

Pattern of Flowering in the Panicle. Panicles emerge usually after the summer solstice, and flowering commences and continues after that time into the autumn. There is considerable variation in the patterns of flowering, tiller panicle emergence, embryogenesis and abscission among foxtail inflorescences. Time of seedling emergence, as well as the photoperiod and temperature coinciding with an individual plant's seasonal growth period, are major determinants of these flowering phenomena. Under controlled environmental conditions (Haar, 1998; 16:8h light:dark diurnal; 25° C) found flowering within individual *S. faberi* panicles lasted 15-50 days (primary (1°), 15 days; secondary (2°), 30-35 days; tertiary (3°), 25-50 days). Foxtail flowering and seed production (anthesis of first seed until abscission of last seed on an individual panicle) lasted 45-75 days (1°, 45-50 days; 2°, 55-70 days; 3°, 40-70 days). North Dakota *S. viridis* flowering began in mid-July and the first seeds were mature 2 weeks later (Stevens, 1960). Seed formation has been observed in panicles before they emerged from the boot (culm leaf tissue surrounding the emerging seedhead), precocious autogamy (Steel et al., 1983).

In foxtail millet (*S. viridis* subsp. *italica*), flower blooming starts 5 days after the panicle emerges from the boot (Li et al., 1935). Depending on environmental conditions, the foxtail millet seedhead will bloom over 12-15 days. The flowers open gradually. As the glumes spread, the anthers and pistil push themselves out of the seedhead, sometimes simultaneously. In weedy foxtails, sometimes the flowers never open to the outside and fertilize themselves within the enclosing lemma and palea (cleistogamy; Dekker et al., 1996, data not reported). Pollen fertility in *S. viridis*, subsp. *italica* was ca. 95% (Willweber-Kishimoto, 1962). Anthers begin to dehisce after they are fully extruded. After dehiscence, the flower begins to close up, leaving only the shrivelled anthers and the tip of the stigma outside (Kishimoto, 1938). The order of blooming of the spikelets in the seedhead is very regular, from the top proceeding to the bottom of the head (in both the main culm and tiller panicles).

The flowering pattern in *S. faberi* is different than that reported for foxtail millet. Two patterns of flowering along the inflorescence were observed (Dekker et al., 1996). The first fertile spikelet in each fascicle consistently flowered first. The flowering of these first spikelets of the fascicle was first apparent in the distal 30-40% section of the panicle. From that first section, flowering continued in two directions at the same time, in one direction toward the proximal end and in another direction toward the distal end of the axis. Flowering of these spikelets reached completion at the distal end first, followed by those at the basal end. A second pattern of flowering was observed among the other fertile spikelets within individual fascicles along the axis. This second pattern was very complex and no consistent observations of its nature along the axis were possible. Schreiber (1965a) erroneously reported that a strict basipedal flowering order occurs in *S. faberi.*

Pattern of Seasonal Flowering. The seasonal environment influences all aspects of panicle development, including the eventual dormancy state of the seed when shed. Little information is available about the role of moisture and other environmental factors, but the effect of radiation (photoperiod and temperature) on foxtail development has been characterized to some extent.

Photoperiod. Foxtails that emerge in the spring appear to initiate flowering and fertilization in response to the shortening daylight (lengthening dark period) photoperiods after the summer solstice. Long daylengths appear to prolong these developmental responses, while continuous light markedly delayed panicle production and inhibited flowering, in all the weedy and crop *Setaria* spp. (Fabian, 1938; King, 1952; Peters and Yokum, 1961; Peters et al., 1963; Santelman et al., 1963; Schreiber, 1965a; Vanden Born, 1971). The time from planting to the first *S. faberi* and *S. viridis* panicle appearance increased as the photoperiod length increased (Schreiber and Oliver, 1971; Schreiber, 1965a, b). Panicles appeared and developed sooner when grown in shorter, compared to longer, photoperiods (Nieto-Hatem, 1963). The time from planting until flowering is highly variable and is dependent on the diurnal photoperiod (Schreiber and Oliver, 1971; Stevens, 1960). The optimal photoperiod for this response may vary with locally adapted populations at different latitudes, especially with the widely distributed *S. viridis* (Vanden Born, 1971). Triazine resistant biotypes of *S. viridis* flowered earlier than susceptible variants (Ricroch et al., 1987).

Temperature and humidity. Blooming is negatively correlated with temperature and positively with relative humidity (Li et al., 1935). The photoperiodic response of flowering in the weedy foxtails appears to be quantitatively modified by temperature. Within certain limits, higher temperatures may hasten the first appearance of *Setaria* flowers, especially for those grown in longer photoperiods (Ricroch et al., 1987; Schreiber and Oliver, 1971). Almost no *S. viridis* panicles developed in any photoperiod at 50°C; when they did there was no viable seed. Denser (seed number per unit axis length) than normal panicles developed in plants grown at 60°C, and seed maturation was relatively slower at that temperature, in all photoperiods.

Pattern of Diurnal Flowering. Foxtail flowering usually occurs during specific time periods of the day, but this varies with local environmental conditions. In *S. viridis,* subsp. *italica*, there appears to be two periods of maximum flower opening during the day: 0600-0800 h and 2200-0000 h (Li, 1935), 0400 to 0800 h and 1900 to 2300 h (Rangaswami Ayyangar et al., 1933), with no blooming between 1200 and 1800h, and 0600 to 0700 and 2200 to 2300 h (Willweber-Kishimoto, 1962). These studies suggest that *Setaria* blooming is an endogenously regulated rhythm and the two maxima peaks occur, and those only in the dark. The average time of flower opening and closing is 70 min. (Kishimoto, 1938).

Embryogenesis

Embryogenesis begins with anthesis and fertilization of the new embryo, and ends when the embryo enveloped with parental tissues (i.e. lemma, palea, caryopsis coat) becomes physiologically separated from the parent plant (abscission). During this time many events occur, including the maturation of the embryo, the formation and hardening of the endosperm, and induction of dormancy.

Formation of the Seed. Three genomes cooperate and interact to form the *Setaria* seed or spikelet: zygote (2n), endosperm (3n) and the parent (2-8n). The parent contributes sporophytic tissues to the seed that both nuture and protect the developing embryo, as well as provide significant contributions to dormancy and control of post-abscission germination timing (e.g. seed envelopes such as the caryopsis coat), and physical protection in the soil.

At pollination a male pollen grain adheres to the sticky female stigma surface and germinates by forming a tube that grows down through the style of the flower to reach the small micropyle opening of the female egg nucleus. One of the two surviving sperm cell nuclei fuses with the egg nucleus to form a diploid zygote; the other nucleus unites with the two haploid polar nuclei to form a triploid endosperm. The zygote develops to form the embryo, which draws upon the surrounding endosperm tissue for nourishment (Simpson, 1990).

Embryogenesis, under controlled environmental conditions (Haar, 1998; 16:8h light:dark diurnal; 25° C), lasted 8-15 days (1° panicle, 8-10 days; 2° panicle, 15 days; 3° panicle, 15 days), consistent with previous reports in different controlled environment studies in the Netherlands (Dekker et al., 1996).

Induction of Seed Dormancy. The production of seeds with different germination requirements is a function of genotype, plant architecture and the environmental conditions during their life history. Individual foxtail panicles on a single parent plant produce a diverse array of seeds, each with potentially different after-ripening (AR;

e.g. time at 4° C, moist, dark) requirements for germination (heteroblasty; Atchison, 2001; Dekker et al., 1996; Haar, 1998). Earlier fertilized seeds (both intra- and inter-panicle) are relatively more dormant than later developing seeds. Haar (1998) and Atchison (2001) has shown that seed produced early in the season (August) is more dormant than seed produced in September, which is again more dormant than October seed. Although mean temperatures and photoperiod length decrease during the seed rain period (August-November), these environmental cues to dormancy induction are dynamic. Others have observed that decreasing temperature and photoperiod length result in increased dormancy in seeds (Schrieber, 1965a; Simpson, 1990; Mohamed et al., 1985).

Atchison (2001) observed that foxtail seed maturing early in the growing season, and grown in relative shade, produce seed with considerably more dormancy than nearby plants in full sun. But, the effect of shade is lost in seed maturing late in the growing season. Axis elongation and fascicle spacing are two mechanisms differentially modifying the light microenvironment of individual spikelets (Dekker et al., 1996). This differential light reception may be a mechanism by which different levels of dormancy are induced in individual florets of a panicle: relatively greater light levels are associated with relatively reduced dormancy in later maturing seeds on a panicle. Greater light penetration is correlated with relatively less dormancy in the more widely spaced basal florets, while the greater height and light above the soil surface (and competitors) is correlated with less dormant florets maturing later on an individual panicle (Atchison, 2001; Dekker et al., 1996; Haar, 1998)

Other environmental conditions experienced by a developing foxtail panicle can influence the dormancy induced in its seeds. Some information about the influence of temperature, light quantity and injury is available, but little is known of the influences of nutrition, light quality and moisture. Foxtail plants subjected to mowing subsequently produced seed with relatively more dormancy the later the main culm was cut, regardless of plant age or emergence time (Schreiber, 1965b). Fawcett and Slife (1978) observed that foxtail plants treated with 2,4-D produced seed with less dormancy compared to an untreated control. The influence of environment on dormancy in other species is available (e.g. Baskin and Baskin, Ch. 8, 1998; Simpson, 1990).

Seed Abscission and Dispersal

Weedy foxtail seeds become physiologically independent when mature, but complete separation from the parent plant does not occur until the pedicel degrades and spikelet dispersal can occur. The quantity, quality and time of dispersal of seeds is variable.

Abscission. Seed abscission is the physiological event marking the beginning of the seed rain period, the time of seed dispersal. Abscission is a singular, threshold event in the life history of a new foxtail plant, but it is only a partial separation of the new and parental generations: parental, apoplastic tissues surround the seed symplast, affecting its ability to germinate. These parental tissues of the new seed provide a delaying, conservative, bet-hedging strategy for the more readily germinable embryo within the seed. Abscission is an experimentally important time, allowing direct comparison of seeds in terms of the effects of events and treatments before and after this time (e.g., differences in dormancy among seed shed or subsequent seedling emergence). Experimentally and physiologically abscission is discrete (Dekker et al., 1996).

Dispersal. Seeds of weedy foxtail seeds separate (shatter) from the parent plant soon after abscission, although the time from maturity of the abscission cell layer until separation, dispersal varies based on spikelet stalk deterioriation and wind. Seeds of the crop foxtail millet have been selected in the domestication process to not shatter readily, allowing near complete harvest by the grower. *Setaria* seed dispersal occurs by gravity, animal (e.g., birds), wind, water and human (e.g., farm machinery) vectors (Bor, 1960; Ridley, 1930; Steel et al., 1983; Wilson, 1980). The retrorsely barbed seta of S. verticillata easily stick to animals and humans facilitating their dispersal (Bor, 1960). *S. viridis* and *S. pumila* seeds are considered the most important weed species in the U.S. in terms of food value to wildlife predators (Martin et al., 1961).

Seed Rain Periodicity. The foxtail seed rain period in the north temperate regions of the world occur generally from July through December, depending on seasonal environmental conditions, notably the time of soil freezing and frost at the end of the season (Atchison, 2001; Haar, 1998; Peters et al., 1963; Stevens, 1960).

Seed Productivity. The numbers and yield of seed produced by a weedy foxtail plant is highly plastic, and strongly dependent on biomass accumulation, plant tillering and fascicle architecture (Haar, 1998; Kawano and Miyake, 1983). Foxtail plant biomass is strongly dependent on available resources, competition from neighbors, and the time of seedling emergence. In general, plant size is greater for plants emerging earlier in the growing season than later. For example, seed number per panicle decreased as emergence date increased: May, 920 seeds/panicle; July, 600 seeds per panicle; August, 20 seeds/panicle (Schreiber, 1965b). Reports of seed productivity therefore vary: *S. viridis*: 350-12,000 seeds/panicle; giant green, 4000-6000 seeds/panicle; *S. faberi*: 2000 seeds per panicle; *S. pumila*: 180 seeds per panicle, 3-47 panicles/plant, 540-8460 seeds/plant; *S. verticillata*: 120 seeds/panicle, 120-5000 seeds per plant (Peters et al., 1963; Rominger, 1962; Slife, 1954; Steel et al., 1983; Vanden Born, 1971).

Seed number per panicle, panicle length and seed number per unit panicle length (seed density) were found to vary among foxtail species (giant, green, yellow), panicle types (1°, 2°, 3°) and field locations (Haar,1998). *S. faberi* had longer panicles than *S. viridis*. *S. viridis* panicles had greater number of seed and higher seed density than *S. pumila*. Earlier developing panicle types were always greater than or similar to the later developing panicle type for all these seed productivity parameters. This study reported higher numbers of seed produced by foxtail plants than generally reported, emphasizing the importance of collection methods. Many reports of foxtail seed productivity are based on seed collection at one or few times in the season. The seed rain period of an individual *S. faberi* (abscission of first to last seed) ranged from 35-65 days (1° panicle, 40 days; 2° panicle, 40-55 days; 3° panicle, 35-65 days) under controlled environmental conditions (16:8h light:dark diurnal; 25° C; Haar,1998). Because the foxtail seed rain is continuous over a weeks-to-months period, artifacts are introduced by infrequent collection methods and inefficient capture techniques.

Attempts to provide experimental seed productivity estimates (e.g. relationship between panicle length and seed number per unit length measurements) have produced mixed results (Barbour and Forcella, 1993; Defelice et al., 1989; Fausey et al., 1997; Haar, 1998). The number of fertile spikelets per fascicle within a giant or *S. viridis* panicle is somewhat plastic, while only a single fertile spikelet is found in each *S. pumila* fascicle (Clark and Pohl, 1996). Floret abortion in response to the environment can lead to different numbers of fertile spikelets per fascicle in the more plastic green and *S. faberi*, relative to that in the more fixed fascicle structure in *S. pumila*. Estimates of seed productivity are more likely to be accurate for *S. pumila* compared to green or *S. faberi* for this reason.

Seed Color. The color of seeds shed by foxtail plants is highly variable, from very light brown (almost white) to black, with hues of brown, green and red possible (Schreiber, 1965b). The color of foxtail seeds is a function of genotypic, as well as environmental and maturity, factors. There exists conflicting evidence correlating seed color with seed dormancy (darker seed more dormant), although color may be associated with viability (immature, sterile or incomplete seed green and/or lighter colored than others of the same plant) (Moore and Fletchall, 1963; Stanway, 1971).

Seeds In the Soil

As annual species, seeds are the form of the foxtail plant in which they spend the longest time of their life cycle. Once buried in the soil, the risk of mortality to seeds may be lower than in their vegetative form, but seeds do not reproduce. The portion of their life history spent in the soil as a seed is a trade-off between decreased risk of mortality and potential fitness. One of the most important traits foxtails possess for invasiveness, colonizing ability and enduring occupation of a locality is the strategic timing of emergence from the soil for subsequent reproduction. The selection pressures exerted on the seed, and the resulting adaptations to the soil environment, are critical to the seed's ability to remain inactive, yet alive in the soil for long periods of time until environmental conditions are more favorable for germination, seedling establishment, growth and production of seed. Eventual success in the soil phase of the weedy foxtail life history is based on the heterogeneity of dormancy states in individual seeds produced by individual plants, and how this diversity results inevitably in the formation of soil seed banks. The diverse seeds interact with their individual microenvironments in the soil, resulting in the consequential patterns of seedling emergence, mortality or seed longevity.

Seed Dormancy and Germination

Setaria seed dormancy is a complex phenomena, and much is unknown about what causes its heterogeneity and how the environment influences that heterogeneity. Past experimentation reveals some information of the nature of dormancy state heterogeneity as well as how environment influences the diversity of seed entering the seed bank.

Heteroblasty. Individual foxtail panicles on a single parent plant produce a diverse array of seeds, each with potentially different dormancy states (heteroblasty). These dormancy states are experimentally revealed by the duration required for them to after-ripen and germinate (e.g. time at 4° C, moist, dark; stratification; Atchison, 2001; Dekker et al., 1996; Haar, 1998). The production of foxtail seeds with different germination requirements is a function of genotype, plant architecture and the environmental conditions during the reproductive phase of their life history.

Earlier fertilized seeds (both intra- and inter-panicle) on the same plant are relatively more dormant than later developing seeds. The first seeds on an individual panicle were shown to possess relatively greater dormancy (greater after-ripening requirement) than the last seeds maturing on the same panicle (Haar, 1998). Additionally, 1° panicles produce seeds with relatively greater dormancy than those produced on 2°, and again on 3°, panicles of the same parent plant.

Atchison (2001) has shown that seeds produced early in the season (August) are more dormant than seed produced in September, which are again more dormant than October seed. The temporal pattern of seed production (e.g., August to November) is roughly the pattern of 1° to 3° panicle seed productivity, on plants emerging in the spring and growing for a full season (Haar, 1998). Significant heterogeneity in dormancy states among seeds shed by a single plant allow these species to subsequently emerge at appropriate times within a cropping season and in different years (Dekker et al., 1996; Forcella et al., 1997).

Many reports have observed that seeds shed by foxtail plants are all dormant (*S. pumila*: Norris and Schoner, 1980; Peters and Yokum, 1961; Povilaitis, 1956; *S. verticillata*: Lee, 1979), but closer observation has revealed differences in dormancy among seeds collected (Kollman, 1970; Schreiber, 1965a; Taylorson, 1986; *S. pumila*: Nieto-Hatem, 1963; *S. viridis*: Martin, 1943; Vanden Born, 1971; *S. faberi*: Fletchall, 1963; Stanway, 1971). Morphological variants of *S. viridis* (robust white, robust purple and giant green) do not appear to possess seed dormancy (Schreiber, 1977). Triazine resistant *S. faberi* seed were reported to be inherently more dormant than susceptible biotypes (Tranel and Dekker, 1996, 2002), consistent with observations in other species (Mapplebeck et al., 1982).

Vivipary (premature germination) in *S. viridis* plants have been observed (Dore and McNeill, 1980; Hubbard, 1915). This pre-harvest sprouting is often associated with unusual environmental conditions of excessive atmospheric moisture during embryo maturation (King, 1983). It is also the least dormant seed in a continuous frequency distribution of dormancy states of individuals shed by a parent plant.

Dormancy, Germination and Environment. The inherent dormancy state of a foxtail seed interacts with the immediate soil environment it encounters after abscission and dispersal, which determines its subsequent behavior (germination, seedling emergence, mortality or extended life in the soil). The influence of environmental factors (light, temperature, gases and seed damage) on seeds in the soil changes with time.

Light and Germination. The influence of light on foxtail seed germinability is, at best, small. What is known of the effect of light is equivocal. Some studies indicate foxtail germination is not controlled by light (Nakamura, 1962; Taylorson, 1982, 1986). Others conclude light promotes germination (Povalitis, 1956; Holm and Miller, 1972). Taylorson reported that anesthetic chemicals can stimulate phytochrome-induced *S. faberi* germination (Hendricks and Taylorson, 1978; Taylorson, 1980/81, 1982). Emergence from depth in the soil indirectly supports a positive (Dawson and Bruns, 1962) and negative (Waldron, 1904) influence of light on foxtail seed germination.

Temperature and Germination. The influence of temperature on foxtail seed dormancy changes with time and development. The responses invoked in foxtail seeds by temperature is complex, and requires a knowledge of the extent and heterogeneity in dormancy among the individuals of the population at the time it is being tested. Often the same temperatures can have opposite effects on seeds depending on their dormancy state. The range of temperatures stimulating foxtail seed germination widens with after-ripening, as more individuals become highly germinable (Norris and Schoner; 1980).

Low temperature (e.g. 5° C) and adequate moisture (stratification) is one of the most effective and consistent ways to after-ripen dormant foxtail seed (Anderson, 1968; Atchison, 2001; Dekker et al., 1996; Nieto-Hatem, 1963; Povilaitis, 1956; Stanway, 1971; Vanden Born, 1971). Once fully after-ripened these same low temperatures can enforce dormancy in the same seed until temperatures capable of supporting the germination process are attained (i.e. 10-15° C or more). The duration (days to weeks) of the stratification treatment necessary to overcome dormancy is a function of the amount and variability of dormancy in the foxtail population (Atchison, 2001). Dormancy has also been overcome with after-ripening under dry conditions, but with considerably longer time periods than when moist, typically more than a year (Povilaitis, 1956; Stanway, 1971; Thornhill, 1997; Vanden Born, 1971). Very low temperature (e.g. -20 to -50° C) and aerobic conditions can slow after-ripening almost completely, yet retain seed viability (Dekker, personal observation, data not reported).

Temperatures of 20-25 °C can rapidly promote germination of non-dormant and near-complete after-ripened seeds, but dormant seeds at those temperatures after-ripen very slowly or not at all. Optimal temperatures for foxtail seed germination are a function of the amount and heterogeneity of dormancy states among seed evaluated. As such, there is no single optima for any naturally occurring group of weedy foxtail seed. Optimal foxtail seed germination temperatures have been reported (Banting et al., 1973; Blackshaw et al., 1981a; Vanden Born, 1971; Martin, 1943; Lauer, 1953; King, 1952; Norris and Schoner, 1980; Manthey and Nalewaja, 1987), although few have characterized the extent or variability in dormancy among individuals of the population tested (Atchison, 2001). Non-dormant *S. viridis* seeds readily germinated at temperatures from 15 to 35°C, although it took longer to reach maximum germination at the lower temperatures (Blackshaw, 1981a; Vanden Born, 1971). In other reports temperature optima for *S. viridis* seed germination is between 25-35°C (Martin, 1943) or 20-30° C (Banting et al., 1973; Blackshaw, 1981a; Vanden Born, 1971). *S. pumila* germination maxima varied between 20°C (MA, IA) and 25°C (CA, CT, PA), depending on the geographic biotype tested (Norris and Schoner, 1980). *S. verticillata* germination occurred at temperatures of less than 25°C to greater than 40°C (Lauer, 1953). Germination may occur at a faster rate in green compared to *S. pumila*, and be less affected by extremes in temperature (i.e. 10°C to 30°C or greater), suggesting higher temperatures favor green over *S. pumila* (Manthey and Nalewaja, 1987).

High temperatures (e.g. 30 °C or greater) can either stimulate non-dormant seed to germinate or induce secondary dormancy. In other situations these high temperatures can speed after-ripening and release from dormancy, possibly due to heat damage and cracking of gas-tight foxtail seed envelopes (Taylorson and Brown, 1977). "Accelerated after-ripening" occurred in green, yellow and *S. faberi* by exposing seeds to 50°C for up to 14 days. High temperatures and moisture can induce secondary dormancy in mature, highly germinable, after-ripened *S. faberi* seed (Taylorson, 1982, 1986), or have no affect on dormancy (Vanden Born, 1971).

Daily alternating temperatures increased germination of foxtail seeds when compared to germination in static temperature conditions (Anderson, 1968; James, 1968; Sells, 1965). Maximal foxtail seed germination occurred with daily alternating temperatures between about 20 and 35°C, compared to other alternating temperature or constant temperature regimes (King, 1952).

Moisture and Germination. Adequate moisture is a requirement for foxtail seed germination, and foxtail germination exhibits a high level of tolerance to variable soil moisture conditions (Manthey and Nalewaja, 1987; King, 1952; Taylorson, 1986). Moisture stresses of several kinds can both stimulate germination and induce dormancy. Alternate wetting and drying of soil stimulated *S. faberi* germination (King, 1952; Taylorson, 1986). Moderate, temporary water stress on *S. faberi* seed from the beginning of imbibition, as well as slow hydration in a humid atmosphere, caused some dormant *S. faberi* seeds to germinate (Taylorson, 1986). Taylorson (1986) also noted that a prolonged water stress could induce secondary dormancy. *S. viridis* germination is more tolerant of moisture stress than *S. pumila*, which suggests that *S. viridis* seed may germinate more rapidly than yellow in soils with limited moisture (Manthey and Nalewaja, 1987). This could allow *S. viridis* to dominant *S. pumila* in areas under moisture stress. Temperature and moisture may have an interactive effect on the germination of *S. viridis*. As water potential decreased, the effect of temperature became greater (Blackshaw et al., 1981a). *S. viridis* seed germination was inhibited completely at 15, 20 and 25C with a water potential of -7.8 bars. As water potential was decreased the effect of temperature became greater, indicating that soil moisture had a greater effect than temperature on *S. viridis* germination.

Gases and Germination. The gaseous components of agricultural soils influencing seed behavior include oxygen, nitrogen, carbon dioxide, water vapor and volatile metabolites such as carbon monoxide and ethylene (Siegel et al., 1962; Siegel and Siegel, 1987; Corbineau and Côme, 1995). The gaseous environment of the soil surface layers within which seeds germinate is influenced by the structural characteristics of the soil (texture, drainage, aggregation, crusting, and cracking), chemical qualities of the gases, and biological activity within the soil (Pareja and Staniforth, 1985; Parej et al., 1985).

Oxygen is required for germination by almost all plant seeds (Bewley and Black, 1994; Simpson, 1990), but exceptions have been observed (Corbineau and Côme, 1995). Both oxygen and carbon monoxide affect foxtail seed behavior in the soil. *S. faberi* seed germination was markedly affected by O_2 concentration (in N_2) above and below that of air (20% O_2): the largest increase in germination (from 37 to 60%)

occurred between 20-25% O_2; between 0-10% O_2 germination increased from 0-30%; and surprisingly germination at 10 and 20% O_2 was similar (Dekker and Hargrove, 2002). These observations reveal an asymmetrical response to incremental changes in O_2 above and below that typically found in agricultural soils. Carbon monoxide had opposite effects on *S. faberi* germination in air depending on concentration, stimulation and inhibition: germination increased from 37 to 56% with the addition of 1% CO, but decreased from 37 to 14% with 75% added CO. An explanation may be that there are two separate effects of CO, each occurring in different physiological systems of dormant seeds at the same time. At high concentrations (75%) in air CO inhibited seed germination, probably by inhibiting mitochondrial respiration. But low CO concentrations (0.1 or 1%) in air stimulated seed germination. It was not apparent which physiological system(s) CO and O_2 affected. It seems unlikely that CO-stimulated germination arises from effects on the respiratory apparatus, but may be a consequence of CO interactions with an as yet uncharacterized physiological factor in the seed. Nitrogen plays a neutral role in its affects on foxtail germination (Dekker and Hargrove, 2002), while the effects of carbon dioxide on foxtail seeds in the soil are unknown.

Seed Damage and Germination. Disrupting foxtail seed envelopes and bypassing natural gas-tight barriers resulted in increased germination (Dekker et. al., 1996; Heise, 1941; Nieto-Hatem, 1963; Kollman, 1970; Peters and Yokum, 1961; Rost, 1975; Stanway, 1971; Taylorson and Brown, 1977). Stimulation of foxtail germination may also result from cracks and damage to the seed envelopes subjected to high temperature (Taylorson and Brown, 1977).

Foxtail Soil Seed Banks

The foxtails infest vast areas due to their ability to form long-lived soil seed banks, the source of all future annual weed infestations (Buhler et al., 1997a; Dekker, 1999). These seed banks are reservoirs of highly heterogeneous seed (Dekker et al., 1996; Dekker, 1997), poised to inflict heavy crop losses. A key trait possessed by weedy foxtails in the soil is their ability to precisely time seed germination over large temporal scales (hours to decades; Atchison, 2001; Forcella et al., 1992, 1997). Significant heterogeneity in dormancy states among seeds shed by a single plant allow these species to emerge at appropriate times within a cropping season and in different years (Dekker et al., 1996). Soil seed banks consisting of a diverse collection of foxtail species and genotypes, each contributing a heterogeneous collection of dormancy phenotypes, reveals a hedge-betting strategy for adaptation to changing conditions within agroecosystems (e.g., Cohen, 1966; Philippi and Seger, 1989).

The foxtail seed banks formed by individuals with diverse dormancy states results in several observed behaviors (Figure 1). Seedling emergence occurs over the entire season, with the majority in the spring and early summer (Atchison, 2001; Forcella et al., 1992, 1997). Seed become after-ripened in the cool, moist conditions of the spring, but only a minority fraction germinate and emerge every year. Foxtail seeds can become ready for germination (a germination "candidate") in after-ripening conditions (cool, moist, dark), but fail to germinate without a change to somewhat different environmental conditions specifically favorable to germination (warm, moist). The higher soil temperatures of summer reverse the earlier after-ripening effects and secondary (2°) dormancy is induced (Atchison, 2001). Taken together these environmental conditions result in an annual cycling between dormant and highly germinable states within the heterogeneous seeds remaining in the soil seed bank (Baskin and Baskin, 1985; Baskin and Baskin, Ch. 4, 1998).

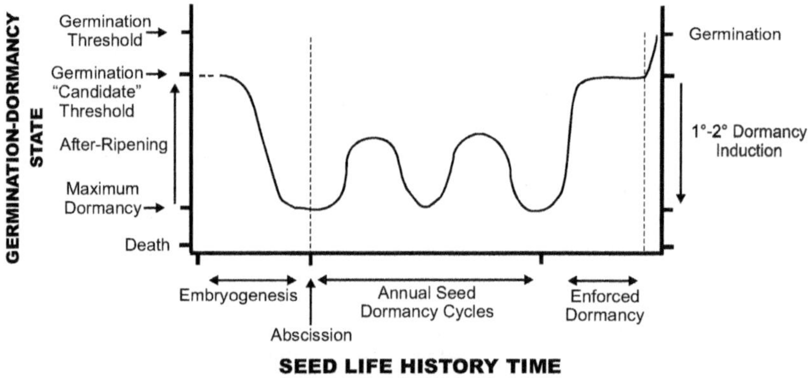

Figure 1. Schematic model of life history changes in individual foxtail seed germination-dormancy states with time from embryogenesis to germination or death; dashed (---) vertical lines represent irreversible life history events; axes not to scale (see Dekker and Hargrove, 2002).

Regulation of Foxtail Seed Behavior. The wide geographic range of adaptation, heterogeneity in dormancy phenotypes, and genotypic diversity raise the question of what mechanisms in foxtail seeds drive seed behaviors (e.g., seed dormancy, germination, or seedling emergence). There is evidence that soil water and oxygen play a unifying role in regulating both the biogeographic distribution of *Setaria*, as well as the responses of individual seeds in a soil microsite (Dekker, 2000; Dekker and Hargrove, 2002; Dekker et al., 2001).

The morphology of foxtail seeds also provides important clues about what environmental factors limit germination and maintain dormancy. The foxtail seed symplast (embryo, endosperm, and aleurone layer) is enveloped by several layers controlling its behavior, notably the hull (lemma, palea) and the caryopsis coat (Dekker et al., 1996; Rost, 1971, 1972, 1973, 1975; Rost and Lersten, 1973). The caryopsis coat is filmy, oily to touch, water- and gastight, and continuous except at the placental pore opening on the basal end of the seed. The mature foxtail seed is capable of freely imbibing water and dissolved gases, but entry is restricted and regulated by the placental pore tissues, notably membrane control by the transfer aleurone cell layer (TACL; Rost and Lersten, 1970). Gases entering the moist seed symplast must be dissolved in the imbibed water passing through the narrow placental pore and TACL. The importance of gases entering dry seed is unknown. The morphology of foxtail seeds strongly suggests that seed germination is restricted by water availability in the soil and by the amount of oxygen dissolved in water reaching the inside of the seed symplast to fuel metabolism. Oxygen solubility in the water entering the symplast is an inverse function of the diurnally, seasonally, and annually changing soil temperature. The control by this oxygen-limited, gastight morphology is supported by observations of increased germination when foxtail seed envelopes, including the caryopsis coat, are punctured (Dekker et al., 1996; Stanway, 1971), as well as by the stimulatory effects of oxygen and carbon monoxide (Dekker and Hargrove, 2002).

Based on these observations, biogeographic distribution, as well as individual foxtail seed behavior, appears to be regulated by the amount of oxygen dissolved in water taken into the seed over time. A robust seed regulatory system could result from the interaction of the TACL which restricts O_2-H_2O entry into the seed, and an oxygen scavenging system which delays the transduction of O_2 in the symplast (Dekker and Hargrove, 2002). When adequate amounts of water and oxygen reach the embryo, sufficient energetic equivalents are generated to support germination metabolism (Bewley and Black, 1994). When inadequate amounts of oxygen reach the embryo, dormancy is maintained or secondary dormancy is induced (Corbineau and Côme, 1995). This model of foxtail seed behavior is schematically represented in Figure 2 (see Dekker and Hargrove, 2002). The global regions that possess soils with suitable qualities of fertility, moisture, temperature, aeration, and disturbance are extensive (e.g., Eurasia, North America) and coincide with foxtail distribution.

The Environmental Signal Cueing Foxtail Seed Behavior. What is the signal to which foxtail seeds respond? Given that foxtail seed behavior is regulated by the amount of oxygen dissolved in water taken into the seed over time, the signal stimulating foxtail behavior in an individual seed is oxygen mass per water volume of symplast (caryopsis) per time (e.g. hour, day): mass O_2^{-1} volume H_20^{-1} time^{-1} seed^{-1} (oxy-hydro time; Dekker et al., 2001). Therefore, the internal oxygen mass is a function of the oxygen solubility at the ambient temperature of the seed. The O_2 mass is also a function of symplast H_20 volume of the seed, which varies by foxtail species and biotype as well as by soil water content (%). Changes over relatively short time periods (e.g. hours) will result in oxygen mass changes inside a seed depending on whether the temperature increases (increasing oxygen solubility; decreasing diffusion rates) or decreases (decreased oxygen solubility; increased diffusion rates). These daily soil temperature changes could produce two different types of O_2-H_2O signals: a "chronic" signal due to O_2 solubility as a function of the daily mean temperature, and an "acute" signal due to relatively rapid O_2 solubility changes as a function of daily soil temperature changes. Each day in the spring the sun warms the moist, cool, nightime seed in the soil resulting in a rapid decrease in O_2 solubility. This warming causes a pulse of de-gassed oxygen to be released in the seed symplast where it's ability to diffuse out of the narrow placental pore rapidly is impeded. The large temperature differences over the course of a day that occur in the spring and early summer result in frequent, daily pulses of de-gassed oxygen in the seed, a phenomena we call the "Big Spring Oscillator" (Dekker, 2000), the proposed acute H_20-O_2 signal.

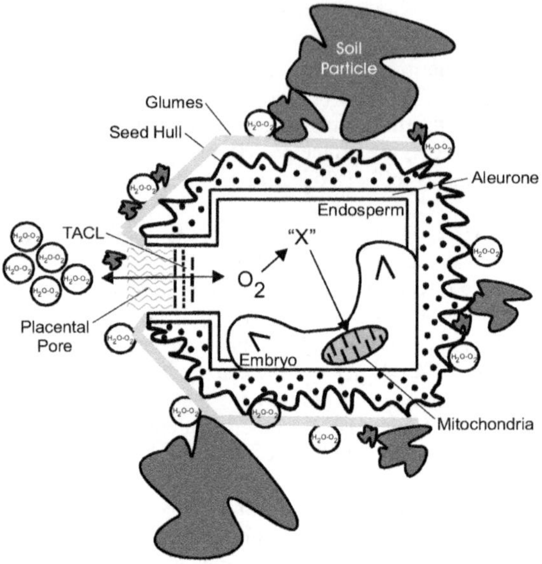

Figure 2. Schematic model of hydrated, weedy foxtail (*Setaria* spp.) seed parts involved in dormancy and germination and the signal transduction pathway of water-oxygen (H_20-O_2) movement from the outside environment; soil-seed contact via the glumes and seed hull; through the placental pore and the transfer aleurone cell layer (TACL; membrane) at the basal end of the seed, into the endosperm, hypothesized oxygen scavenging factor "X", aleurone, embryo, and mitochondria (see Dekker and Hargrove, 2002).

Seed Longevity in the Soil. The length of time that Setaria seeds are able to survive in the soil is quite variable, but mortality decreases with depth of burial. The maximum period of survival in the soil varies by species: *S. pumila*: typical, 13 years; maximum, 30 years (Darlington, 1951; Dawson and Bruns, 1975; Kivilaan and Bandurski, 1973; Toole and Brown, 1946); *S. viridis*: maximum 10-39 years (Burnside et al., 1981; Dawson and Bruns, 1975; Thomas et al., 1986; Toole and Brown, 1946); *S. verticillata*: maximum, 39 years (Toole and Brown, 1946). More typically, the majority of foxtail seed live in the soil for much shorter time periods. In one study in Iowa *S. faberi*, 86-95% of the buried portion of seed from one *S. faberi* population remained one year after dispersal (*S. viridis*, 87-92%; *S. pumila*, 81-91%), while 75-80% and 62-69% remained 2 and 3 years after dispersal, respectively (Atchison, 2001). Predation of foxtail seed prior to burial is undoubtably very high, therefore the percent of the total seed rain remaining in the soil after these time periods is probably much lower. In general, green and S. pumila seed on the surface lost viability sooner than buried seed (Banting et al., 1973; Thomas et al., 1986). The rate of survival of foxtail seed is longer under uncultivated conditions (Stoller and Wax, 1974; Waldron, 1904). Burial of foxtail seed increases both the level of dormancy, viability, and the longevity, of those seeds, possibly due to decreased oxygen at greater depths or within soil aggregates (Banting et al., 1973; Dekker and

Hargrove, 2002; James, 1968; Pareja and Staniforth, 1985; Pareja et al., 1985; Stoller and Wax, 1974). Dawson and Bruns (1975) showed precipitation had no effect on *S. viridis* seed longevity in the soil.

Vegetative Growth

The life history of foxtail plant vegetative growth and development, from seedling emergence to senesence have been reviewed previously (Douglas et al., 1985; Steele et al., 1983; see also phenotype section above). The growth and development of foxtail plants has been described and modelled (e.g., Fausey et al., 1997; Forcella and Banken, 1996; Orwick et al., 1978; Swinton et al., 1994a, 1994b), as well as the spatial variability of those plants in the field (Colbach et al., 2000a, 2000b).

With the threshold event of seedling emergence, the life history of an individual foxtail plant can take a wide variety of paths. The size and fecundity of any individual foxtail plant is a plastic growth response to the environmental conditions (light, water, nutrients, temperature) it experiences after it emerges from the soil. These plastic responses to environment can stimulate large differences in plant stature and fitness. For example, long day length photoperiods, little competition from neighbors, early seedling emergence and full season growth, ample nutrients and water, in full sunlight, can result in large *S. faberi* plants attaining 2 m height with multiple tillers and panicles, producing thousands of seeds. At the other extreme, a small *S. viridis* plant will result from a seedling emerging at the very end of the season, when the light portion of the photoperiod is very short, attaining the untillered height of 5 cm, with 2-3 leaves, producing a single panicle with one seed.

The size and branching that an individual foxtail plant attains after seedling emergence varies widely depending on environmental conditions to which the genotype is exposed. Tiller branch development begins with the emergence of the seedling main culm, the primary (1°) tiller. With adequate time and resources this is followed by branches forming at the lower nodes of the main culm, the secondary (2°) tiller. Tertiary (3°) tillers arise at nodes of the secondary tillers, as do subsequent branches (e.g., quanternary, 4°, tillers). Development of secondary, tertiary and subsequent tillers begins at the lower nodes of the culm from which they arose. Giant and *S. pumila* tillers readily root in soil when cut, separated from the plant, and buried in soil, an important trait allowing weedy foxtails to reestablish themselves after cultivation and mowing (Barrau, 1958; Santlemann et al., 1963; Schreiber, 1965b).

Seedling Emergence

Periodicity of Seedling Emergence. The periodicity of foxtail seedling emergence is complex and difficult to predict with any degree of specificity in a particular field. In general, *Setaria* seeds are capable of germinating and emerging anytime the soil is unfrozen. Most foxtail seedlings emerge in the early part of the growing season of a locality (e.g. April-June in the northern hemisphere; Atchison, 2001; Buhler et al., 1997b; Forcella et al., 1992, 1997): *S. viridis* (Banting et al., 1973; Chepil, 1946; Dawson and Bruns, 1962); *S. pumila* (Dawson and Bruns, 1962, 1975; Manthey and Nalewaja,1987; Martin, 1943; Sells, 1965; Steel et al., 1983; Stoller and Wax, 1974); *S. verticillata* (Martin, 1943). Seedlings that emerge in this early part of the season have both the greatest likelihood of mortality from agricultural practices (e.g. tillage, herbicides) as well as the greatest seed producing potential after a full season of growth.

As the soils warm during mid-season (e.g., June-July), seedling emergence decreases as secondary (2°; summer) dormancy is induced (Atchison, 2001). Seedlings that emerge just after layby, when agricultural practices cease until harvest, have a significantly reduced risk of density-independent mortality but face a shorter growth period to produce seeds as well as greater competition from established neighbors.

Low numbers of seedlings emerge for the remainder of the season until the soil freezes (e.g. December; Atchison, 2001). Seedlings that emerge late produce less seed due to less time for growth and the hastening of the onset of reproduction due to shortening photoperiod. They also often face less competition from neighbors due to late season crop senescence, and the fewer seeds produced can ensure a minimum hedge bet for the species for enduring occupation of that locality. Seeds from these late emerging plants are usually less dormant than earlier ones, and are more likely to be the first to emerge in the following season (Atchison, 2001; Haar, 1998).

The timing and numbers of seedling that germinate and emerge from the soil seed bank in an individual field is a function of the inherent dormancy of the seeds (species, dormancy state heterogeneity), seed age, and the environmental conditions of the particular season that influence the seeds. The age structure of seeds (year of dispersal) in the seed bank changes with time. The least dormant seeds are the first to after-ripen and germinate, leaving the more dormant for emergence in subsequent years. These older, more dormant seeds, are a long-term hedge bet for continuing occupation of a locality when no new seed are shed in a year. When these older foxtail seed (e.g. 2-3 years after dispersal) do emerge, they are more likely to emerge later in the growing season than younger, less dormant seed (e.g. 0-1 years after dispersal; Atchison, 2001). With knowledge of these factors (dormancy states, environmental H_20-O_2 signals) it may be possible to predict foxtail emergence in a locality with precision (Dekker et al., 2001).

Depth of Foxtail Seed Germination from the Soil. Foxtail spp. seedlings emerge from relatively shallow depths in the soil, usually between 1-5 cm depth. Emergence is greatest at 1.5-2.5 cm and declines with depth to a maximum of about 7.5-14 cm (Buhler, 1995; Buhler and Mester, 1991; Dawson and Bruns, 1962; Gregg, 1973; Waldron, 1904). Emergence of foxtail spp. does not usually occur on the soil surface except in the debris of no-till systems, an indication that light may not play an important role in seed germination. The soil depths from which the maximum S. faberi emergence occurs is a function of the tillage system, the greater the depth of tillage the deeper and more varied the seed placement and seedling emergence: 0-2 cm (no-till), 1-3 cm (disk), 1-4 cm (chisel plow), 1-5 cm (moldboard plow) (Mester and Buhler, 1986; Yenish et al., 1992).

Atchison (2001) observed that *S. faberi* depth of emergence is a function of age and dormancy state of the seed: in the first year after dispersal seed emerged from shallower depths (0-5 vs. 5-10 cm), but older, more dormant, seed emergence was similar from these two depths. Others have correlated depth of emergence and seed weight, surmising that greater seed weight indicates more seed reserves (Dawson and Bruns, 1962; King, 1952). Oxygen content decreases with soil depth, inhibiting after-ripening and germination at greater depths and enforcing dormancy (Dekker and Hargrove, 2002; James, 1968).

Growth and Environment

The weedy foxtails are very competitive in cropping fields (e.g. Blackshaw et al., 1981b; Dryden and Whitehead, 1963; Harrison et al., 1985; Morrison et al., 1981; Rahman and Ashford, 1972; Staniforth, 1965). Changes in the plant community and resource availability associated with crop rotations, and changes in crop rotation, can affect the competitive ability of weedy foxtails (e.g., Alex et al., 1972; Dunham et al., 1958; Laudien and Koch, 1972; Schreiber, 1965b). Resource competition in agroecosystems by foxtails includes that for light, soil nutrients and pH, water and temperature

Light. Foxtail plant size decreases with decreases in light and competition with neighbors (e.g., Bubar and Morrison, 1984; Santlemannn et al., 1963; VandenBorn, 1971). The changing photoperiod during the growing season also has important effects on growth and development, especially in combination with other growth requisites (Orwick and Schreiber, 1975; Schreiber and Orwick, 1978; VandenBorn, 1971). Foxtails elongate in response to light competition, but bristley foxtail is more sensitive to shading, a possible reason it is less prevalent in cropping situtations compared to green and *S. pumila* (Lee and Cavers, 1981). Foxtails have C_4 leaf-structural type of photosynthetic metabolism, with 'classical NADP-ME' type biochemistry (Black et al., 1969; Elmore and Paul, 1983; Hattersley and Watson, 1992; Ku et al., 1974). Despite this compensatory metabolism, photosynthesis in water-stressed *S. faberi* plants was enhanced with enriched carbon dioxide atmospheres (Sionit and Patterson, 1985). Although triazine resistant biotypes of other species have been shown to have reduced seed yield relative to susceptible variants (LeBaron and Gressel, 1982), it is not clear this photosynthetic variant results in decreased growth and yield in *S. viridis* (Ricroch et al., 1987). Pleiotropic effects resulting from the dynamic reorganization of a plant with the mutant psbA chloroplast gene may modify the plants growth and competitiveness in different environmental conditions (e.g. competitive advantage of resistance in stressful conditions; Dekker, 1993; Dekker and Sharkey, 1992).

Soil nutrients and pH. Nutrient availabilty, soil fertility and soil pH affect the size of foxtail plants and their growth and competitive ability with other plants of a community (intra- and inter-specific competition; e.g., Chambers and Holm, 1965; Hume, 1982; Kuzin, 1973; Moyer and Dryden, 1977; Moyer et al., 1979; Orwick and Schreiber, 1979; Schreiber and Orwick, 1978; Weaver and Hamill, 1985). Different foxtail species frequently co-habitat the same field, exploiting overlapping niches in the available nutrient resource space of a particular production system. The foxtail species that dominates can vary depending on the conditions of the locality. *S. faberi* dominated green and *S. pumila* in the same Indiana, U.S.A. field when grown together (Schreiber, 1977). Over time the other foxtail species were not displaced, but their relative proportion to *S. faberi* in the field changed. In less fertile conditions, *S. pumila* can dominate and displace *S. faberi*, possibly by greater root growth in *S. pumila* compared to giant when nitrogen fertility is low (Schreiber and Orwick, 1978).

Water and Temperature. Foxtail plant size decreases (tillering, leaf production, height, shoot and root biomass) with decreasing available water (e.g. Alex et al., 1972; Nadeau, 1983; Nadeau and Morrison, 1983; Orwick and Schreiber, 1975). Water stress reduces photosynthetic rates in *S. faberi* (Sionit and Patterson, 1985). Foxtail growth and development is dependent on temperature (e.g., Schreiber, 1965a).

Mortality. Foxtail plant mortality from density-dependent (e.g., effects of neighbors) and density-independent (e.g., predation, parasitism, catastrophe) causes are important sources of weed losses (Douglas et al., 1985; Harper, 1977; Steele et al., 1983). Foxtail mortality is poorly understood, and characterization and quantification of foxtail mortality losses are one of most experimentally difficult life history events to document. A more complete understanding of the forces in agroecosystems that regulate mortality could provide a sustainable approach to weed management.

RECENT EVOLUTIONARY HISTORY

The foxtails are a very ancient group of plants, and their evolutionary history is characterized by invasion and colonization followed by selection and local adaptation for enduring occupation of a site.

History of Global Invasion

There appears to be five major phases, or waves, of foxtail invasiveness in earth's history. The tropical genus *Setaria* dispersed to Eurasia from its probable first home in Africa. The first invading species was probably *S. viridis* or *viridis*-like diploid annual (Rominger, 1962; Stapf and Hubbard, 1930).

In the second phase, *S. viridis* weedy progenitor species spread over Eurasia as a wild colonizing species. With domestication it then spread both as a weed (Li et al., 1942, 1945; Wang et al., 1995a, b) and a crop (foxtail millet, S. *viridis*, subsp. *italica*; Wang et al., 1995a, b; de Wet, 1995; Fukunaga et al., 1997, 2002; korali, *S. pumila*). Foxtail millet was a crop in China 6000 years B.P. (Before Present) (Cheng, 1973; Gao and Cheng, 1988; Ho, 1975; Kawase and Sakamoto, 1984; Li and Wu, 1996), while its use as a crop in Europe dates to 3500 B.P. (Dembinska, 1976; de Wet and Harlan, 1975; Kuuster, 1984; Neuweiler, 1946). *S. pumila* seed were gathered from wild plants and later cultivated in India (de Wet, 1992). Other weedy foxtail species arose from *S. viridis*-like ancestors (Li et al., 1942, 1945; Rominger, 1962; Willweber-Kishimoto, 1962; Williams and Schreiber, 1976; Prasada Rao et al., 1987). These polyploid specialists probably included *S. pumila* and *S. verticillata*.

In the third wave, the weedy foxtails spread to the New World at two different times: pre-Columbian (before ca. 1500 A.D.) and post-Columbian invasions (after ca. 1500 A.D.). The pre-Columbian origins of *S. geniculata*, the only weedy foxtail native to the New World, suggest an ancient dispersal event westward from the Old World (Africa) to New World (Rominger, 1962). A species of *Setaria* was the oldest cultivated cereal in the Americas, its origins dating to 8000 B.P. (Callen, 1965, 1967; Smith, 1968). Prehistoric migration of *Setaria* could date from when the ancestors of contemporary native Americans first crossed the Bering Strait land connecting Asia and North America (Beringia; approximately 10,000 B.P.). The *S. pumila* invasion of the Americas came primarily came from Asia (Wang et al., 1995b). The post-Columbian dispersal of weedy foxtails by Eurasian human emigration over the last 500 years is most likely the major source of the weedy foxtail floral introductions we see in North American agroecosystems today (Wang et al., 1995a, b). For example, *S. viridis* had invaded Canada by 1821 (Rousseau and Cinq-Mars, 1969). The foxtails have expanded their range since their introduction to North America in the last several hundred years as a consequence of global trade and human-mediated dispersal (Hafliger and Scholz, 1980).

The fourth wave may have been determined recently by an extremely rare allo(tetra)polyploidization event giving rise to a fertile hybridization product in southern China, *S. faberi* (Pohl, 1951, 1962; Wang et al., 1995b). *S. faberi* was first introduced to North America in the 1920's near New York City (Fairbrothers, 1959). Soon thereafter it was found in Philadelphia (1931) and Missouri (1932). From this early colonization *S. faberi* spread rapidly in the eastern and midwestern states (Fernald, 1950; Rominger, 1962; Reed, 1970). More recently, *S. faberi* has spread northward into Ontario in the 1970's (Dore and McNeill, 1980; Warwick, 1990). The largest increase and spread of *S. faberi* occurred across the North American corn and cereal growing regions in post-World War II era. The introduction and adoption of the herbicide 2,4-D to control dicotyledonous weeds created an opportunity space for, or population shift to, grassy weeds like the foxtails (Pohl, 1951; Slife, 1954). Introduction of other new selective herbicides, as well as expansion of maize and soybean production in North America, accelerated these trends (Warwick, 1990). In little over a half century, *S. faberi* has colonized a large part of the most fertile, disturbed areas of North America (Hafliger and Scholz, 1980).

Currently we are experiencing a fifth invasive phase, from the late 1970's through the 2000's, with the appearance of many different types of herbicide resistant biotypes of foxtail (e.g. Holt and LeBaron, 1990; LeBaron and Gressel, 1982; Ritter et al., 1989; Stoltenberg and Wiederholt, 1995; Thornhill and Dekker, 1993). Concurrent with this is the incremental spread of *S. faberi* and *S. pumila* in the northern prairies of North America.

Geographic invasion can simultaneously be viewed from the perspective of genetic invasion: *Setaria* polyploidization. Polyploid foxtails more specifically adapted to niche opportunities in agroecosystems include those four weedy foxtails with more narrow geographic distributions than diploid *S. viridis*: knotroot, yellow, *S. verticillata*, and more recently *S. faberi*. The 4th geographic invasive wave was also the 4th genetic-polyploidization, weed speciation event.

Colonization and Biogeography

Geographical Distribution

The genus *Setaria* is cosmopolitan, and can be found on every land mass in the world except the polar regions (Prasada Rao et al., 1987; Rominger, 1962). *Setaria* is particularly well represented in Africa, Asia and South America, with the greatest concentration in numbers of species in tropical Africa (Stapf and Hubbard, 1930). In the western hemisphere, *Setaria* has its center of distribution in Brazil, and radiates poleward in both directions (Rominger, 1962). The weedy foxtails are primarily temperate species but are widely distributed between 45° S and 55° N latitudes (Holm et al., 1977, 1997; Wang et al., 1995a, b). They are found in every state in the continental U.S. and every province in Canada (Lorenzi and Jeffery, 1987). The Eurasian foxtails are rare in the southwest U.S., Mexico and more southerly, tropical regions.

The center of origin of the original wild Eurasian *Setaria* sp. is disputed (compare e.g., Prasada Rao et al., 1987; Rominger, 1962; Simpson, 1990). The large number of *Setaria* spp. in Africa (74 of 125), as well as genomic evidence of tropical origins, are strong arguments that Africa is the center of origin of this genus (Lakshmi and Ranjekar, 1984; Stapf and Hubbard, 1930). From these probable tropical origins, *Setaria* spp. adapted to a wider range of environmental conditions and habitats. Speciation and adaptation followed this spread in geographic distribution to subtropical, and then temperate regions. Grassy members of other species of same tribe, Paniceae, probably originated in, and adapted to, tropical and subtropical ecological zones of the world (alternating wet and dry environments) (Simpson, 1990). Li (1942) emphasized the primacy of *S. viridis* subsp. *viridis* as the progenitor species of these weedy, temperate adapted types. He concluded from detailed cytological evidence that two genomic components comprise the chromosome complements in *Setaria* spp., his "A" and "B genoms". *S. viridis* subsp. *viridis* was the source of the "A genom" in the many *Setaria* spp. The source of his "B genom" is unknown. For example, *S. faberi* may be the result of hybridization of the A and B "genoms", associated with a chromosome polyploidization (alloploidy) event.

S. viridis. *S. viridis* is primarily a temperate species, but widely distributed between 45 °S and 55 °N latitudes (Holm et al., 1977, 1979; Hafliger and Scholz, 1980). It is one of the most widely distributed weedy foxtail species both globally and in the U.S. Globally it ranges from North America, through Central America to parts of South America; from Europe to northern Africa, and from east Asia to south Asia and Australia (Hafliger and Scholz, 1980). It is found in every state in the continental U.S. and every province of Canada, but is rare in southeast U.S. (e.g., Georgia, Florida) (Lorenzi and Jeffery, 1987).

Foxtail millet origins as a crop are ancient in Asia, Europe, India, and possibly Africa and Central and South America, but in North America in this century its culture has been limited (Wheeler, 1950). Today it is grown as a crop in southeastern Europe, South and Central Asia, and China. It is also grown in North and South America, Australia and parts of Africa as a minor cereal, especially as caged bird feed (Gao and Chen, 1988; Kawase and Sakamoto, 1984; Prasada Rao et al., 1987; de Wet et al., 1979). It remains an important crop only in parts of India, Afghanistan, Central Asia, Manchuria, Korea and the Georgian Republic. Foxtail millet, race *moharia*, was once grown widely in southern Europe, but now is an important crop only in Hungary, the Czech and Slovakian Republics, the mountains of southwestern Russia, Pakistan and in Afghanistan. All three complexes of race *maxima* are grown in Transcaucasian Russia. The Manchurian complex is grown in northeastern China, Manchuria and eastern Russia; the Mongolian complex in northeastern China, Mongolia and adjacent Russia; and the Korean complex in Korea and adjacent Central China. Race *indica* is grown in India, including central India and Sri Lanka.

S. faberi. *S. faberi* occurs in central Europe, Russia, the Middle East, eastern Asia, and North America (Holm et al., 1977, 1979; Hafliger and Scholz, 1980). The range of *S. faberi* is more restricted in North America than *S. viridis*, and is primarily found in the northern, central and eastern corn belts (Ontario, Iowa and Missouri east to the Atlantic coast). *S. faberi* probably originated in China (Pohl, 1951).

S. pumila. *S. pumila* is widely distributed in the United States and Canada, Argentina and Uruguay, Africa, the Middle East, Europe, Asia, Australia, and many Pacific Islands (Holm et al., 1977, 1979; Hafliger and Scholz, 1980). Korali (millet) is grown as crop in India, including near Poona and Southern India, and seeds from wild plants are collected across southern Asia (Bor, 1960; Prasada Rao et al., 1987; de Wet et al., 1979). *S. pumila* with very large spikelets and panicles has been observed by the author in 2000 being used as a forage grass in upland fields on Tsushima Island, between Korea and Japan.

S. verticillata. *S. verticillata* is reported to be a native to the tropical and temperate Old World (Bor, 1960; Hsu, 1975). It is distributed throughout Africa, North America, Europe, the Near East, west Asia, south Asia, east Asia, Australia (Holm et al., 1977, 1979; Hafliger and Scholz, 1980).

S. geniculata. The only significant weedy foxtail adventive from the new world is *S. geniculata*. *S. geniculata* occurs throughout the Americas, southern Aftrica, most of Europe, the Middle East, Asia, Australia, and several Pacific Islands (Holm et al., 1977, 1979; Hafliger and Scholz, 1980). In the United States, *S. geniculata* accurs along the Atlantic coast from Massachusetts to Florida, the Gulf coast from Florida to Texas, and the Carolinas westward to Kansas and California (Rominger, 1962). *S. geniculata* is also more localized in distribution than *S. viridis*, being found primarily in the southeast quarter of continental U.S. The center of origin of the weed *S. geniculata* is reported to be Central America or northern South America. It is closely related to several *Setaria* species in tropical Africa, the suspected center of origin of the genus (Stapf and Hubbard, 1930; Rominger, 1962). The close resemblance of *S. geniculata* (New World origin) to *S. pumila* (Old World origin) represents a phylogenetic as well as geographic enigma (Rominger, 1962).

Population Genetic Structure

One of the important aspects of population biology is population genetic structure, which forms the basis of plant spatial and temporal organization. Knowledge of population genetic structure has practical implications (Barrett and Husband, 1990). It can be used to reconstruct the historical process of invasion, migration and colonization, and provide insights into the ecological persistence and evolutionary potential of populations in new habitats, leading to a better understanding of weedy adaptation. It may also provide a foundation for understanding the spatial structure of individuals within an individual field.

Genetic Diversity in the Weedy Foxtails. The pattern of genetic diversity within an individual foxtail species is characterized by unusually low intra-population genetic diversity, and unusually high genetic diversity between populations, compared to an "average" plant (Hamrick and Godt, 1990; Wang et al., 1995a, b). These two patterns of population genetic structure appear to typify introduced, self-pollinated weeds that are able to rapidly adapt to local conditions after invasion and colonization (e.g., Brown and Marshall, 1981; Rice and Jain, 1985; Barrett and Richardson, 1986; Barrett, 1988; Barrett and Shore, 1989).

Hamrick and Godt (1990) developed a framework for interpretation of genetic diversity in plants in which diversity estimates for 449 species were summarized and an attempt was made to find correlations between the amount and distribution of allozyme diversity and various life history features. In comparison to an "average" plant species, all four weedy foxtail (green, yellow, giant and knotroot) species contain low to exceedingly low amounts of total genetic variation. Among the four species studied, *S. geniculata* exhibits the highest diversity levels, followed by *S. viridis*, *S. pumila*, and finally *S. faberi*, which is nearly monomorphic as a species. These low diversity estimates are especially noteworthy given the geographic breadth of sampling of each of the four species studied, which in all cases included accessions from at least three different continents (Wang et al., 1995a, b).

Although relative genetic diversity within each of the several foxtail species is very low, differences between homogeneous populations is high, indicating a strong tendency for local adaptation by a single genotype. The limited diversity contained within weedy foxtails is partitioned among populations as alternative homozygous genotypes. Heterozygosity was rarely detected, suggesting strong inbreeding in nearly all populations of the four species, either in the generalized sense or as a specific consequence of self-pollination. Apomixis in the foxtails cannot be ruled out (Emery, 1957). Nearly all populations were found to consist of a single multilocus genotype. Foxtail populations are homogeneous but are differentiated from one another by fixation of different alleles at one or more loci. Best expressed as the coefficient of population differentiation (G_{ST}), which is much higher in the weedy foxtails (yellow, 0.73; knotroot, 0.91; green, 0.65) than in the "average" plant species (0.22; Hamrick and Godt, 1990) (Wang et al, 1995b). In contrast to this indication of population divergence (specialized adaptation), a common multilocus genotype of both green and *S. pumila* occurred in many accessions from widely separated geographic areas (general adaptation).

Genetic Pattern of Colonization and Regional Differentiation. In addition to the potential selective forces responsible for low genetic diversity and high population differentiation, stochastic forces have played a major role in shaping foxtail population genetic structure. Perhaps the most important phenomenon, in this respect, has been a history of genetic bottlenecks associated with founder events, genetic drift, and natural selection. The effects of these stochastic and selective forces are evident in the allozyme data (Wang et al., 1995a, 1995b).

The founder effect was observed in *S. viridis* to a certain degree. *S. viridis* accessions from North America had reduced allelic richness compared to those of Eurasia: fewer unique alleles, lower percentage of polymorphic loci, and fewer alleles per locus. Genetic drift probably has occurred in *S. viridis*, as indicated by the many fixed alleles in North American accessions. On the other hand, heterozygosity was equal between Eurasian and North American accessions, and Nei's genetic identity between them was 1.0. Multiple introductions of *S. viridis*, in the absence of local adaptation, should have produced a random, mosaic pattern of geographic distribution among North American accessions. Instead, a strong intra-continental differentiation was observed in *S. viridis* populations, both in Eurasia and North America (Jusuf and Pernes, 1985; Wang et al., 1995a). *S. viridis* populations in North America were genetically differentiated into northern and southern groups separated on either side of a line at 43.5 °N latitude. The northern type was less variable than southern type. This regional divergence suggests that natural selection has partitioned *S. viridis* along a north-south gradient. Populations with genotypes more suitable to northern conditions probably flourished there, while those more suited to southern conditions had a fitness advantage in the south.

These results were interpreted to imply that the present patterning among *S. viridis* populations in North America is the consequence of multiple introductions into the New World followed by local adaptation and regional differentiation.

The geographic distance (from global to local) separating foxtail populations does not indicate the genetic distance separating them. The size of the geographic range from which populations were sampled was not an accurate indicator of the extent of genetic diversity found among populations from that region. The degree of genetic differentiation of *S. viridis* populations at the state, county and farm level showed little hierarchical patterning (Wang et al., 1995a). The genetic diversity of *S. viridis* populations from a particular local geographic area was probably largely determined by both the number of independent introductions to that area, and the intensity and duration of natural selection pressures at those sites.

S. pumila populations are genetically clustered into overlapping Asian, European, and North American groups (Wang et al., 1995b). *S. pumila* populations from the native range (Eurasia) contained greater genetic diversity and a higher number of unique alleles than those from the introduced range (North America). Within Eurasia, Asian populations have greater genetic diversity than those from Europe; more unique alleles found in Asian populations. These results were interpreted as indicating the there have been numerous introductions of *S. pumila* from Eurasia to North America. The Asian cluster was most diverse, although it represented the smallest numbers of samples, indicating *S. pumila* originated in Asia, not Europe. The bulk of introductions of *S. pumila* to North America are from Asia.

The pattern of *S. pumila* genetic variability in North American was unexpected (Wang et al., 1995b). Within the distinct cluster from North America, nearly the entire diversity of *S. pumila* appears to be encompassed by accessions from Iowa, whereas populations collected from the other North American locations were nearly monomorphic for the same multilocus genotype. In this respect, it is significant or coincidental that this pattern was repeated in the diversity data for *S. viridis*, also a native of Eurasia (Wang et al., 1995a). Iowa possesses a surprising genetic diversity in the foxtails. All five weedy foxtail species are present: bristly, giant, green, knotroot and yellow. Typically two or more foxtail species occur in same field at the same time. It was also noted that Iowa is the center of the north-south agro-ecological gradient in North America, perhaps leading to greater environmental heterogeneity.

Despite originating on different continents, the genetic diversity patterns for *S. geniculata* parallel those for *S. pumila* and *S. viridis*: greater genetic diversity was observed in accessions from the New World than in those from the introduced range (Eurasia) (Wang et al., 1995b). These data most likely reflect genetic bottlenecks associated with sampling a limited number of founding propagules and the history of multiple introductions from the Americas to Eurasia. The population genetic structure of *S. geniculata* consists of three nearly distinct clusters, groups from Eurasia, northern United States, southern United States. Accessions from Eurasia and North America are approximately equally diverse genetically. Within North America, *S. geniculata* accessions were strongly differentiated into southern and northern groups at about the Kansas-Oklahoma border (37 °N latitude); indicating greater genetic differentiation within North American populations than between North American and Eurasian populations.

S. faberi contains virtually no allozyme variation. Fifty of the 51 accessions surveyed were fixed for the same multilocus genotype (Wang et al., 1995b).

Genetic Variation and Evolutionary Success of Colonizers. The population genetic structure of many widely distributed, introduced, self-pollinating weed species clearly indicates that a high level of genetic variation is not a prerequisite for successful colonization and evolutionary success (Allard, 1965; Barrett and Shore, 1989). Two contrasting, adaptive, strategies are hypothesized to explain weedy adaptation and the success of colonizers: genetic polymorphism with the development of locally adapted genotypes ("specialists"), and phenotypic plasticity for the development of "general purpose" genotypes (“generalists”) adaptable to a wide range of environmental conditions (Baker, 1965, 1974; Bradshaw, 1965; Barrett and Richardson, 1986). Foxtail population genetic structure itself allows some insight into the dichotomy of generalism versus specialization. Foxtails possess both generalists and specific strategy types. A key observation is that although a single, multilocus genotype predominates or is fixed in all populations, not all multilocus genotypes are equally prevalent within individual weedy foxtail species (Wang et al., 1995a).

The most striking example of this is in *S. pumila*, where the most common multilocus genotype was found in 53 (of 94 evaluated) accessions from widely separated geographic areas in Europe (Belgium, Czech Republic), Asia (Manchuria, China, Turkey), and North America (Ontario, Canada; eastern U.S.: MD, PA; western US: WA, WY; south U.S.: AK; north and midwest U.S.: IA, IN, KS, MN, ND, WI) (Wang et al., 1995b). Overlaying this pattern of homogeneity were other, less abundant *S. pumila* genotypes, each with a more narrow geographical and ecological distribution.

The population genetic structure of *S. viridis* suggests that this species also possesses both generally adapted and specially adapted genotypes. Many *S. viridis* populations are genetically strongly differentiated (e. g., northern and southern North America), while other populations remain identical. The most widely distributed *S. viridis* genotype (fixed multilocus genotype) occurred in 25 (of 168 evaluated) accessions from six countries, from both the Old World and the New World (Wang et al., 1995a). This common *S. viridis* genotype was found in many geographic locations, in very different ecological conditions: the midwestern U.S. corn Belt areas of Indiana, Minnesota, and North Dakota; the eastern U.S. corn Belt of Maryland; the western U.S. agricultural valley of Washington; the highlands of Ontario, Canada; the lowlands of Belgium in northwest Europe; the plains of Bohemia, Czechoslovakia in central Europe; the waste areas interspersed between limestone outcroppings in the Akiyoshi Dai National Park, Honshu, Japan; and between cracks in a cement pier over the bay in Yokohama Harbor, Honshu, Japan. Interestingly, this common genotype has not yet been found in Iowa, despite the diversity of *S. viridis* populations found in this state.

These observations reveal a complex hedge-betting strategy by individual foxtail species, a strategy that balances general adaptation with the additional niche opportunities available with specialization: the ratio of general and special genotypes available in a species for invasion. The ratio of generalist to specialist genotypes was quite different within yellow, knotroot and *S. viridis* (Wang et al., 1995a, b).

Selection and Local Adaptation

The primary selection pressure the foxtails have adapted to is large, periodic disturbances in their habitat. In agroecosystems this typically includes tillage and other disturbances of the soil, grazing by herbivores, and fires. In more recent times it also includes selection by herbicides. The foxtails are plants adapted as colonizers of recently disturbed ecosystems. Most foxtail species have an annual life cycle, and are primarily self-pollinating. This reproductive trait allows them to establish themselves in new, freshly disturbed habitats when only a few individuals (as little as one) are present or there is an absence of pollinators (Mulligan and Findlay, 1970). These successful original colonizers then are able to rapidly increase their numbers with genetically identical individuals as well adapted to that area as the founder plant(s) (Baker, 1974). The accumulation of dormant seed into soil seed banks ensures their enduring occupation of that locality.

These processes of invasion, colonization and enduring occupation are only possible because unused or under-utilized resources left available as a consequence of disturbance. Resources are left available in agroecosystems when crop production practices such as tillage and weed control eliminate non-crop species, creating an "opportunity space", or niche, ripe for exploitation. The colonizing wild species within the genus *Setaria* before the advent of agriculture were most probably adapted to the non-human disturbance opportunities then available. Pre-agriculture wild *Setaria* was poised to take advantage of the vast new opportunity spaces created by agriculture. Selection and adaptation among pre-existing traits resulted in shifts within and among populations, favoring more fit genotype individuals. The range and quality of agricultural habitats expanded rapidly with time and provided opportunity for weedy selection and adaptation for several *Setaria* species within both discrete overlapping niches. This evolutionary process continues today, the most probable phylogenetic basis for the foxtail species-group.

The vast geographic range, as well as the genotypic, and phenotypic diversity, of the foxtails provides individuals in this group the ability to readily change and exploit any changes that occur in the opportunity space. Changes in the environment (short and long-term), crops and crop rotations, crop production practices (e.g. planting, tillage) and weed management tactics (e.g. herbicides) are all strong selection pressures causing shifts in weed populations that are better adapted to exploit resources left available.

Weed Population Shifts

Most changes is foxtail population shifts are difficult to experimentally document, with the exception of shifts in populations caused by herbicides. The most obvious and dramatic weed population shift in the last few decades has been the shift to herbicide resistant biotypes (e.g., Holt and LeBaron, 1990; Reschly et al., 1996; Ritter et al., 1989; Santelmann and Meade, 1961; Stoltenberg and Wiederholt, 1995; Tranel and Dekker, 1996; Thornhill and Dekker, 1993; Volenberg et al., 2001; Wiederholt and Stoltenberg, 1996). Herbicides cause shifts in weed populations by both direct (e.g., selection and enrichment of resistant biotypes in local populations) and indirect mechanisms (e.g., decrease in dicotyledonous weed populations with 2,4-D use and consequent increase in foxtail populations). Foxtail population shifts in seedling emergence timing have been more difficult to demonstrate experimentally (Atchison, 2001). Anecdotally, many weed managers have observed increased weed problems later in the season since the introduction and increased use of herbicides since the late 1940's. Changes in tillage practices in the late 20th century provided selection pressure for changes in the location and soil depth from which foxtail seeds germinate and emerge (Buhler, 1995; Mester and Buhler, 1986; Yennish et al., 1992).

Species Niche Differentiation

The geographic distribution of the several weedy foxtail species provides an initial insight into the niches they occupy. More than one *Setaria* species often co-exist together in a field, exploiting resources left available in somewhat different ways and times. *S. viridis* is the most ancient and widespread weedy foxtail species, and is often the most commonly found species when more than one is found in a single field. The polyploid weedy foxtails are less widely distributed than the diploid *S. viridis*, and are often more specialized. Weedy foxtail biogeographical distribution is associated soils that possess a wide range of moisture, gas, and nutrient regimes with distinctive seasonal and diurnal temperature fluctuations. These seasonal temperature-water cycles are correlated with their cyclical germination behavior (e.g., Dekker et al., 2001; Forcella et al., 1992, 1997).

The geographic distribution of the foxtails has been expanding in North America in terms of species present, new habitats, and population density in occupied areas (Alex, 1972; Blackshaw et al., 1981a; Dexter et al., 1981; Manthey and Nalewaja, 1987; Oliver and Schreiber, 1971; Pohl, 1951; Schreiber, 1977; Vanden Born, 1971; Warwick, 1990). For example, in the Canadian prairies differential responses to soil moisture and temperature may be the cause of green and *S. pumila* spread and geographic distribution. *S. viridis* germination is more tolerant of moisture stress than *S. pumila*, which could allow *S. viridis* to dominant *S. pumila* in areas under moisture stress (Manthey and Nalewaja, 1987). Germination may occur at a faster rate in green compared to *S. pumila*, and be less affected by extremes in temperature (i.e. 10°C to 30°C or greater), suggesting higher temperatures favor green over *S. pumila* (Blackshaw et al., 1981a; Manthey and Nalewaja, 1987). *S. faberi* dominated *S. pumila* when grown together in the same Indiana, U.S.A. field (Schreiber, 1977). Under less fertile conditions *S. pumila* can dominate and displace *S. faberi* by greater root growth when nitrogen availability is low (Schreiber and Orwick, 1978). *S. verticillata* vegetative growth is sensitive to shading, may make it less competitive in cropping situations (Lee and Cavers, 1981). Changes in cropping practices and herbicide usage have been responsible for *S. faberi* invasion and spread northward from midwestern U.S. since 1960 (Warwick, 1990).

Domestication

Annual species in the *Setaria* genus are one of the most ancient wild-weed-crop species-groups in human experience. As such, they have had a very long time to spread, diversify, evolve, and adapt to a wide range of conditions. Crucial to the ability of foxtails to invade and adapt to local conditions has been their domestication. Several foxtail species have been domesticated by humans. Foxtail domestication surprisingly appears to have occurred at a similar time in both Eurasia (6000 B.P.) and the Americas (8000 B.P.). Wild colonizing foxtail species adapted to pre-agricultural disturbances were first domesticated by gathering seeds from natural stands. Later, these seeds were sown purposefully and harvested. Domestication and utilization by gathering and sowing continues today in the 21st century.

Interelationship between Wild-Weed-Crop Setaria. Selection criteria in crop breeding and domestication are an important cause of weedy foxtail adaptation: uniform and rapid germination (non-dormancy), uniform population maturity, increased yield (larger endosperm; larger panicles), improved harvestability (non-shattering: spikelets remain on panicle at maturity; spikelet disarticulation above the glumes), loss of natural seed dispersal ability, and altered nutritional quality (Harlan et al., 1973; Prasada Rao et al., 1987). The intentional selection criteria used in foxtail breeding and domestication has often resulted in unintentional consequences. For example, selection for uniform and rapid germination following planting has resulted in smoother and thinner S. viridis seed hulls (lemma, palea).

Intentionally and unintentionally, weedy and domestication traits have been exchanged by wild, weedy and crop variants of the same species in production fields since agriculture began, forming wild-weed-crop complexes (Darmency et al., 1987; Harlan, 1965; Harlan et al., 1973). Interactions in the history of this *Setaria* complex include wild-weed, wild-crop, weed-crop, and wild-weed-crop. Hybridization and gene flow between wild, weedy and crop *Setaria* in the same fields has occurred since the first wild relatives were affected by human agricultural activities. The subsequent co-evolutionary consequences of domestication, selection and adaptation has lead to the genotypes that dominate fields today. The history of the weedy foxtails is the history of the foxtail millets.

Genotypes of domesticated, wild and weedy *Setaria* relatives are not completely isolated by the disruptive selection in domestication. Sterility barriers between crop and wild *Setaria* variants have evolved (e.g. China), providing an additional isolating mechanism (with self-pollination) in areas where they occupy the same habitat. Li et al. (1945) found 50% of *S. viridis* subspecies (foxtail millet, *S. viridis*) hybrids were sterile, evidence of the development of partial reproductive isolating barriers. The partial reproductive isolating barriers between crop and weed variants becomes a positive advantage for weedy success when a new variant acquires weedy traits and invades the human-managed habitat (e.g., *S. viridis* var. *major*; Darmency et al., 1987b). Human selection accelerates these phenotypic and adaptive changes in the crop, while disruptive selection maintains variant type unity in the wild-weed-crop complex (Prasada Rao et al., 1987). These observations also have implications for the release of transgenic *Setaria* crops (Till-Bottraud et al., 1992).

Weedy *S. viridis* subsp. *viridis* is reported to always occur where *S. viridis* subsp. *italica* is cultivated (Willweber-Kishimoto, 1962). The relatively large amount of genetic heterogeneity between and amongst *S. viridis* subsp. *viridis* and subsp. *italica*, and their mutual interfertility, indicates that over time numerous introgressions between them have occurred (Nguyen Van and Pernes, 1985). Populations of the weedy and crop *S. viridis* derived from the same region have stronger genetic similarities than do those from different regions. Allozyme analysis of North American and Eurasian *S. viridis* populations revealed that weedy (subsp. *viridis*) and crop (subsp. *italica*) taxons from the same region were more alike (closer genetic distances) than were weedy types from the two different regions, or the crop types from the two regions (Li et al., 1945; de Wet et al., 1979; Jusuf and Pernes, 1985; Nguyen Van and Pernes, 1985; Darmncy et al., 1987a; Till-Bottraud et al., 1992).

Center of Origin. The center of origin of the foxtail millet crop *S. viridis*, subsp. *italica* in Eurasia is not clear. Some believe it to be northern India and the regions adjacent to the northern Indian frontier (Koernicke and Werner, 1885a, b; Prasada Rao, 1987; Sakamoto, 1987; Werth, 1937; Willweber-Kishimoto, 1962). Other sources argue that it was first domesticated in southeastern Europe, Egypt and Africa (Becker-Dillingen, 1927). The domestication of wild Setaria is very ancient and most likely occurred independently in many different Eurasian locations (de Wet et al., 1979; Fukunaga et al., 1997; Prasada Rao et al., 1987). Variants from Afghanistan and India were found to be completely different restriction fragment length polymorphism phenotypes, indicating foxtail millet from these two areas originated independently (Fukunaga et al., 1997).

A more recent analysis provides compelling evidence for 5 geographic landraces of foxtail millet, each characterized by specific combinations of RFLP, rDNA and intraspecific hybrid sterility types (Fukunaga et al., 2002). The 5 geographic landraces are from Europe, Afghanistan and vicinity and East Asia, and the "tropical group" from both India and surrounding regions, as well as Taiwan and the Philippine Islands. The most diversity was found within Chinese populations as is likely one of the original sites of this crops domestication, although multiple origins here and elsewhere cannot be ruled out (Harlan, 1975).

The wild, progenitor *Setaria* spp. that adapted after multiple introductions to Eurasia after leaving its African home was most likely similar to *S. viridis* subsp. *viridis* (Kihara and Kishimoto, 1942; Koernicke and Werner, 1885a, b; Li, 1934; Li et al., 1942, 1945; Prasada Rao et al., 1987; Rominger, 1962; Werth, 1937; Willweber-Kishimoto, 1962). These introductions were closely associated with multiple crop domestication events on several continents (de Wet, 1979; Jusuf and Pernes, 1985; Nguyen Van and Pernes, 1985). The processes of migration, domestication, hybridization, differentiation and adaptation have been very active over time within and between these two *S. viridis* subspecies and have obscured their origins. Exchange of germplasm for last 2000 years between Europe and Asia via southern Russia has masked the evolutionary history of foxtail millet and its centers of variability (de Wet et al., 1979).

More problematical are the origins of cultivated *Setaria* in the New World. The oldest known cultivated cereal in the Americas was a species of *Setaria* cultivated almost 8000 B.P. (Callen, 1965, 1967; de Wet, 1992). If this ancient date is correct, it suggests earlier origins than that of Eurasia from the available evidence, antecedents dating to the early colonization of the Americas from Asia over the Bering Strait land bridge at that time. It also suggests an explanation of the genetic similarity enigma of yellow and *S. geniculata* (see taxonomy above; Rominger, 1962).

Antiquity of Cultivation. The oldest known cultivated cereal in the Americas was a species of Setaria, cultivated in the tropical highlands of Mexico among other locations (Callen, 1965, 1967; de Wet, 1992; Smith, 1968). It was found in the oldest (almost 9000 B.P.) occupied layers of the Valley of Mexico. The use of this foxtail millet was replaced about 7000 B.P., when it disappeared with the adoption of maize culture. Subsequently, it had a resurgence in use when it was harvested from weed plants invading maize fields. Its use in northwestern Mexico dates to 6000 B.P. by peoples occupying the Ocampo caves. Grain size increased significantly as a consequence of domestication selection during the 2000 years it was cultivated in Tamaulipas, but it never lost its natural seed dispersal ability (Callen, 1967).

Foxtail millet, *S. viridis* subsp. *italica*, was domesticated as long ago as 6000 years B.P. in Eurasia (Cheng, 1973; Naeiri and Belliard, 1987). Foxtail millet was used as a cereal in China 6000 years B.P., and has been grown in China for about 5,000 years, making it one of the oldest cultivated cereals of China (Cheng, 1973; Gao and Cheng, 1988; Ho, 1975; Kawase and Sakamoto, 1984; Wheeler, 1950). Archaeological evidence of the abundance foxtail millet in China includes storage jars filled with husks of foxtail millet at Pan-p'o areas from about 4000 B.C., its association with farming implements of the Yang-shao culture in the loess highlands of China, and other Yang-shao archaeological sites in Shensi and Kansu provinces of China (Chang, 1973; Ho, 1969; Nai, 1963). Foxtail millet was grown with *Panicum miliaceum* (proso millet) by proto-Chinese groups in southern Shansi province of China around 7000 B.P. The name of the legendary ancestor of the Chou tribe of the loess hills of China (Hou Chi) means "Lord of the Millets" (Ho, 1975)

In Japan, *Setaria* millet was the primary grain during the Jomon (cord mark) period (ending 300 B.C.). During the Yayoi period (300 B.C.-300 A.D.) it was slowly replaced by rice culture, a phenomena that was occurring throughout Asia at that time. Foxtail millet culture had a resurgence during the Kamakura (1185-1333 A.D.) and Muromachi (1576-1600 A.D.) periods, coinciding with the imposition of a tax on rice culture (but not millet). A tax on the millets after that time saw the decline in its culture.

In Africa, a panicle of *S. verticillata* was found in a garland excavated at El Lahum, and dates to the 12th Egyptian dynasty (1991-1784 B.C.) (Dixon, 1969).

Foxtail millet grains were found in European neolithic sites, and its use as a cereal in Europe dates to 3600 B.P. (Dembinska, 1976; Helbaek, 1960; Neuweiler, 1946; de Wet and Harlan, 1975). Foxtail millet was found in the debris of Stone Age Swiss lake dwellings and Austrian agricultural sites dating from 3000 B.P. (Heer, 1886; Rominger, 1962; Werth, 1937). Archaeological evidence indicates foxtail millet was gathered as a wild cereal in northern Europe, but cultivated in southern Europe (van Zeist, 1970). However culture of foxtail millet as a crop during the Bronze Age only occurred in Alpine southeastern Europe.

Cultivation and Utilization. Foxtail millet is a minor world crop. It is grown as a crop and seeds from wild and weed plants are also gathered. The reasons why it is not a major human food crop are historical and agronomic. Foxtail millet is not traded on world markets, but is traded locally. *Setaria* millets, like other millet species of world (e.g. *Panicum miliaceum*, proso millet; *Digitaria* sp., Africa, "hungry rice"; *Echinochloa esculentum*; *Fagopyron* sp.; *Eleucine coracana*, finger millet (Japan, "Shikoku millet"); *Pennisetum americanum*; Coix sp.), have not been the focus of modern plant breeding, crop improvement, biotechnology or international trade.

Several species of *Setaria* are extensively collected as wild cereals: African savannas in times of scarcity (*S. phacelata*; Dalziel, 1937; Upper Volta (*S. pallidifusca*; Chevalier, 1913; *S. verticillata* (Chevalier, 1932); by Australian aborigines (Bulmer, 1964); India in times of scarcity (Bor, 1960); and by forest tribes in the Philippines (*S. palmifolia*; Monsalud et al., 1966)

Foxtail millet (*S. viridis*) use as human food includes steamed grain as a rice substitute, cooked grains mixed with pulse crops, ground flour for bread, pasta (noodles), boiled for porridge or mash, baked as unleaven bread, and vegetatively propagated and the edible shoots used as a green vegetable (Barrau, 1958; Kadkol and Swaminathan, 1955; Monsalud et al., 1966; de Wet et al., 1979). In Russian Georgia the sprouted grain is used to make beer, and its mash is fermented and distilled (fxka) (Rozhevitz and Shishkin, 1934). It is used as an animal feed for caged birds and domesticated fowl, and the stalks and straw are fed to cattle. Compared to other cereals, foxtail millet is lower in protein content, and higher in fat, mineral content and indigestible fiber (Anderson and Martin, 1949).

Korali (*S. pumila*) is harvested as a cereal in India and other locations (de Wet et al., 1979). *S. pumila* with very large panicles and seeds has been observed by the author being used as a forage grass in fertile upland fields on Tsushima Island, between Korea and Japan. Hill tribes of the eastern and western Ghats of India are one of the few groups extant sowing and cultivating Korali as a cereal (de Wet, 1992).

SUMMARY

The *Setaria* (foxtail) wild-crop-weed species-group is one of the most successful terrestrial plant complexes on earth. The genotypic and phenotypic biodiversity of this species-group has conferred on it the ability to invade, colonize, adapt and endure in a wide range of disturbed habitats in temperate, tropical, and sub-tropical regions. Over 100 *Setaria* species were distributed throughout the world before the advent of agriculture.

About 6000 to 8000 years ago annual *Setaria* species were first domesticated, creating the conditions in which wild, pre-agricultural, colonizing foxtails could adapt and thrive in the newly created agroecosystems. These domestication events occurred first in the Americas, specifically in Mexico. A *Setaria* species was the first cultivated cereal in the western hemisphere. At about the same time *Setaria viridis* was first domesticated in Eurasia, probably in China.

Domestication of *Setaria* created the new selective conditions in which *Setaria* weeds could evolve from preadapted wild colonizing species. The partial reproductive isolating barriers thus created between self-pollinated wild, crop and weed foxtail species provided an ideal genetic condition in which introgression of highly adapted weedy traits could thrive, improve and gain protection. Polyploidization of *S. viridis*-like, diploids that invaded Eurasia from Africa instantly created new foxtail species. These polyploidization events occurred regularly in the foxtail's evolutionary history, providing specialized species capable of exploiting niches adjacent and overlapping that of the progenitor diploid species.

Typical of self-pollinated weeds able to rapidly adapt to local conditions after invasion and colonization, the population genetic structure of the weedy foxtails is characterized by unusually low intra-population, and unusually high inter-population, genetic diversity compared to an average plant species. Foxtails possess both generalists and specific strategy types: *S. viridis* and *S. pumila* have a common multi-locus genotype in widely separated geographic locations around the world, as well as less abundant genotypes with more narrow geographic and ecological distribution.

For example, many green and *S. geniculata* populations within North America are genetically differentiated into north-south latitudinal groups.

S. viridis has one of the smallest plant nuclear genomes. Intragenomic selection pressure may eliminate "junk" DNA (untranslated introns), leading to smaller genome size in colonizing, invasive species like the foxtails. *S. viridis* has one of the lowest total proportion of repetitive DNA in the Poaceae family, giving rise to slow rates of turnover of the genome in which repeated DNA sequences are not subjected to mutations, deletions or base substitutions at a very rapid rate. The presence of a high melting point DNA component of the *S. viridis* genome may indicate very little to no base sequence mismatch, a rare observation, and suggesting a functional role in genomic adaptations to specific environments.

Evolution of a complex seed dormancy system has given the foxtails the ability to precisely time seedling emergence on time scales from hours to decades, providing them with the ability for enduring occupation of a locality. Seed dormancy state heterogeneity among individual seeds induced during embryogenesis has inevitably led to the formation of soil seed banks. Foxtail soil seed banks are the source of all future annual weed infestations in a locality, spreading the risk of continued development over time, a bet-hedging strategy balancing mortality against future fitness.

Phenotypic plasticity in growth and development has allowed the foxtails to optimize the match between the immediate resources available and vegetative size and reproductive output, avoiding premature mortality of the individual or its parts. The foxtails are endowed with the ability to tolerate high concentrations of salt in the soil, low soil water, tillage, extremes in temperature, and the ability to resist herbicides.

The success of the foxtails is due to the intentional and unintentional consequences of weed management for over 8000 years. Therefore, the foxtail management systems of the future will continue to be founded on management of human-imposed selection forces in agroecosystems.

REFERENCES

Alex J.F., Banting J.D., Gebhardt J.P. Distribution of *Setaria viridis* in western Canada. Can J Plant Sci 1972; 52:129-138.

Allard R.W. "Genetic systems associated with colonizing ability in predominantly self-pollinated species". In *The Genetics of Colonizing Species*, H. G. Baker, G. L. Stebbins, eds. New York, NY: Academic Press, 1965.

Anderson R.N. *Germination and Establishment of Weeds for Experimental Purposes*. WSSA Handbook. Geneva, NY: W.F. Humphrey Press, 1968.

Anderson J.H., Martin J.H. World production and consumption of millet and sorghum. Econ Bot 1949; 3:265-288.

Avdulov N.P. Karyo-systematische untersuchungen der familie Gramineen. Bull Appl Bot Suppl 1931; 43. Leningrad.

Atchison B. Relationships between foxtail (*Setaria* spp.) primary dormancy at abscission and subsequent seedling emergence. M.S. Thesis 2001, Iowa State University, Ames.

Baker H. G. "Characteristics and modes of origin of weeds". In *The Genetics of Colonizing Species*, H. G. Baker, G. L. Stebbins, eds. New York, Academic Press, 1965, pp. 147-172.

Baker H.G. The evolution of weeds. Ann Rev Ecol Syst 1974; 5:1-24.

Banting J. D., Molberg E.S., Gebhardt J.P. Seasonal emergence and persistence of *S. viridis*. Can J Plant Sci 1973; 53:369-376.

Barbour J.C., Forcella F. Predicting seed production by foxtails (*Setaria* spp.). Proc North Cent Weed Sci Soc 1993; 48:100.

Barrau J. Subsistence agriculture in Melanesia. Bernice P. Bishop Museum, Hawaii Bull 1958; 219.

Barrett S. C. H. "Genetics and evolution of agricultural weeds". In *Weed Management in Agroecosystems: Ecological Approach*, M. Altieri, M. Liebman, eds. Boca Raton, FL: CRC Press, 1988; pp. 58-75.

Barrett S. C. H., Husband B.C. "The genetics of plant migration and colonization". In *Plant Population Genetics, Breeding and Genetic Resources*, A. H. D. Brown, M. T. Clegg, A. L. Kahler, B. C. Weir, eds. Sunderland, MA: Sinauer, 1990, pp. 255-277.

Barrett S. C. H., Richardson B.J. "Genetic attributes of invading species". In *Ecology of Biological Invasions*, R. H. Groves, J. J. Burdon, eds. Canberra: Australian Academy of Science, 1986, pp. 21-33.

Barrett S. C. H., Shore J.S. "Isozyme variation in colonizing plants". In *Isozymes in Plant Biology*, D. E. Soltis, P. S. Soltis, eds. Portland, OR: Dioscorides, 1989, pp. 106-126.

Baskin J.M., Baskin C.C. The annual dormancy cycle in buried weed seeds: a continuum. Bioscience 1985; 35:492-498.

Baskin C. C., Baskin J. M. *Seeds-Ecology, Biogeography, and Evolution of Dormancy and Germination.* New York: Academic Press, 1998.

Becker-Dillingen, J. Berlin: Handbuch des Getreidebaues 1927.

Bewley J. D. Black, M. *Seeds Physiology and Germination.* New York, NY: Plenum Press, 1994.

Black C.C., Chen T.M., Brown R.H. Biochemical basis for plant competition. Weed Sci 1969; 17:338-344.

Blackshaw R. E., Stobbe E. H., Shayewich C. F., Woodbury W. 1981a. Influence of soil temperature and soil moisture on *S. viridis* (*Setaria viridis*) establishment in wheat (*Triticum aestivum*). Weed Sci 29:179-184.

Blackshaw R. E., Stobbe E.H., Sturko A.R.W. Effect of seeding dates and densities of *S. viridis* (*Setaria viridis*) on the growth and productivity of spring wheat (*Triticum aestivum*). Weed Sci 1981b; 29:212-217.

Bor N.L. *Grasses of Burma, Ceylon, India and Pakistan.* Int. Monogr Pure and Appl Biol Vol. 1. London: Pergamon Press, 1960.

Bradshaw A. D. Evolutionary significance of phenotypic plasticity in plants. Adv Gen 1965; 13: 115-155.

Brown W.V. A cytological study in the Gramineae. Am J Bot 1948; 35:382-385.

Brown A. H. D., Marshall D.R. "Evolutionary changes accompanying colonization in plants". In *Evolution Today*, G. C. E. Scudder, J.L. Reveal, eds. Pittsburgh, PA: Hunt Institute for Botanical Documentation, Carnegie-Mellon University, 1981, pp. 351-363.

Bubar C.J., Morrison I.N. Growth responses of green and yellow foxtail (*Setaria viridis* and *Setaria lutescens*) to shade. Weed Sci 1984; 32:774-780.

Buhler D.D. Influence of tillage systems on weed population dynamics and management in corn and soybeans in the central USA. Crop Sci 1995; 35:1247-1258.

Buhler D.D., Hartzler R.G., Forcella F. Implications of weed seed and seed bank dynamics to weed management. Weed Sci 1997a; 45:329-336.

Buhler D.D., Hartzler R.G., Forcella F., Gunsolus J.L. Relative emergence sequence for weeds of corn and soybeans. Iowa St. Univ. Extension Publ. SA-11 1997b.

Buhler D.D., Mester T.C. Effect of tillage systems on the emergence depth of giant foxtail (*Setaria faberi*) and *S. viridis* (*Setaria viridis*). Weed Sci 1991; 39:200-203.

Buhlmer R. Edible seeds and prehistoric stone mortars in the highlands of New Guinea. Man 1964; 64:147-150.

Burnside O.C., Fenster C.R., Evetts L.L., Mumm R.F. Germination of exhumed weed seed in Nebraska. Weed Sci 1981; 29:577-586.

Callen E.O. Food habits of some Pre-Columbian Mexican Indians. Econ Bot 1965; 19:335-343.

Callen E.O. The first New World cereals. Amer Antiquity 1967; 32:535-538.

Cavalier-Smith T. Nuclear volume control by nucleoskeletal DNA, selection for cell volume and cell growth rate, and the solution to the DNA C-value paradox. J Cell Sci 1978; 34:247-278.

Chambers E.E., Holm L.G. Phosphorus uptake as influenced by associated plants. Weeds 1965; 13:312-314.

Chapman G.P. "Apomixis and evolution". In *Grass Evolution and Domestication*, G.P. Chapman, ed. Cambridge, UK: Cambridge Univ. Press, 1992a, pp. 138-155.

Chapman G.P. "Domestication and its changing agenda". In *Grass Evolution and Domestication*, G.P. Chapman, ed.. Cambridge, UK: Cambridge Univ. Press, 1992b, pp. 316-337.

Chase A. *First Book of Grasses.* San Antonio, TX: 2nd ed., 1937.

Cheng K. Radio carbon dates from China: some initial interpretations. Curr Anthro 1973: 14: 525-528.

Chepil W.S. Germination of weed seeds. I. Longevity, periodicity of germination and vitality of seed in cultivated soil. Sci Agric 1946; 26:307-347.

Chevalier A. Etudes sur la flore de l'Afrique centrale Francaise. Paris: 1913.

Chevalier A. Resources vegetales du Sahara et de ses confins nord et sud. Paris: Mus. Hist. Nat. Lab. d'Agron. Coloniale Bull., 1932.

Clark L.G., Pohl R.W. *Agnes Chase's first book of grasses.* Washington and London: 4th ed. Smithsonian Institution Press, 1996.

Clayton W. D. "*Setaria*". In *Flora Europa*, T.G. Tutin, V.H. Heywood, N.A. Burges, D.M. Moore, D.H. Valentine, S.M. Walters, D.A. Webb, eds. Cambridge, UK: Cambridge University Press, 1980, pp. 263-264.

Cohen D. Optimizing reproduction in a randomly varying environment. J Theore Biol 1966; 12: 119-129.

Colbach N., Dessaint F., Forcella F. Evaluating field-scale sampling methods for the estimation of mean plant densities of weeds. Weed Res 2000a; 40:411-430.

Colbach N., Forcella F., Johnson G.A. Spatial and temporal stability of weed populations over five years. Weed Sci 2000b; 48:366-377.

Corbineau F., Côme D. "Control of seed germination and dormancy by the gaseous environment". In *Seed Development and Germination*, J. Kigel, G. Galili, eds. New York, New York: Marcel Dekker, 1995.

Dalziel J.M. *The Useful Plants of West Africa*. London: Crown Agents, 1937.

Darlington H. T. The seventy year period of Dr. Beal's seed vitality experiment. Am J Bot 1951; 38:379-381.

Darmency H., Ouin C., Pernes J. Breeding foxtail millet (*Setaria italica*) for quantitative traits after interspecific hybridization and polyploidization. Genome 1987a; 29:453-456.

Darmency H., Zangre G.R., Pernes J. The wild-weed-crop complex in *Setaria*: a hybridization study. Genetica 1987b; 75:103-107.

Dawkins R. *The extended phenotype. The long reach of the gene*. New York: Oxford Univ. Press, 1999.

Dawson J. H., Bruns V.F. Emergence of barnyardgrass, green and yellow foxtail seedlings form various soil depths. Weeds 1962; 10:136-139.

Dawson J. H., Bruns V.F. Longevity of barnyardgrass, green and yellow foxtail seed in soil. Weed Sci 1975; 23:437-440.

Defelice M.S., Brown W.B., Aldrich R.J, Sims B.D, Judy D.T, Guethle D.R. Weed control in soybeans (*Glycine max*) with reduced rates of postemergence herbicides. Weed Sci 1989; 37:365-374.

Dekaprelevich L.L., Kasparian A.S. A contribution to the study of foxtail millet (*Setaria italica* P.B. *maxima* Alf.) cultivated in Georgia (western Transcaucasia). Bull Appl Bot Plt Breed 1928; 19:533-572.

Dekker J. Pleiotropy in triazine resistant *Brassica napus*: leaf and environmental influences on photosynthetic regulation. Zeitschrift Naturforschung 1993; 48:283-287.

Dekker J. Weed diversity and weed management. Weed Sci 1997; 45:357-363.

Dekker J. Soil weed seed, seed banks and weed management. J Crop Prod 1999; 2(1):139-166. Simultaneously published in book format as *Expanding the Context of Weed Management*, D. Buhler, ed., New York: Haworth Press Inc., 1999.

Dekker J. "Emergent weedy foxtail (*Setaria* spp.) seed germinability behavior". In *Seed Biology: Advances and Applications*, M. Black, K.J. Bradford, J. Vasquez-Ramos, eds. Wallingford, UK: CAB International, 2000; pp. 411-423.

Dekker, J. The foxtail (*Setaria*) species-group. Weed Sci 2003; 51: In Press.

Dekker J., Comstock, G. Ethical, environmental and ecological considerations in the release of herbicide resistant crops. Agric Human Values 1993; 9:31-44.

Dekker J., Dekker B.I., Hilhorst H., Karssen C. Weedy adaptation in *Setaria* spp.: IV. Changes in the germinative capacity of *S. faberi* embryos with development from anthesis to after abscission. Am J Bot 1996; 83:979-991.

Dekker J., Duke S.O. Herbicide resistance in field crops. Adv Agron 1995; 54:69-116.

Dekker J., Hargrove M. Weedy adaptation in *Setaria* spp.: V. Effects of gaseous atmosphere on giant foxtail (*Setaria faberii*) (Gramineae) seed germination. Am J Bot 2002; 89:410-416.

Dekker J., Lathrop J., Atchison B., Todey D. The weedy *Setaria* spp. phenotype: how environment and seeds interact from embryogenesis through germination. Proceedings of the British Crop Protection Conference-Weeds 2001; 65-74.

Dekker J., Sharkey T.D. Regulation of photosynthesis in triazine resistant and susceptible *Brassica napus*. Plant Physiol 1992; 98:1069-1073.

Dembinska M. Wild corn plants gathered in the 9th to 13th centuries in light of paleobotanical materials. Folia Quaternaria 1976; 47:97-103.

Devos K.M., Wang Z.M., Beales J., Sasaki T., Gale M.D. Comparative genetic maps of foxtail millet (*Setaria italica*) and rice (*Oryza sativa*). Theor Appl Genet 1998; 96:63-68.

de Wet J.M.J. Chromosome numbers of a few South African grasses. Cytologia 1954; 19:97-103.

de Wet J.M.J. The origin of weediness in plants. Proc of the Okla Acad Sci 1966.

de Wet J.M.J. "The three phases of cereal domestication". In *Grass Evolution and Domestication*, G.P. Chapman, ed.. Cambridge, UK: Cambridge Univ. Press, 1992, pp. 176-198.

de Wet J.M.J. "Foxtail millet *Setaria italica*". In *Evolution of Crop Plants*, J. Smartt, N.W. Simmonds, eds. Essex, UK: Longman Scientific and Technical, 2nd ed., 1995, pp. 170-172.

de Wet J., Harlan J. Weeds and domesticates: evolution in the man-made habitat. Econ Bot 1975; 29:99-107.

de Wet J.M.J., Oestry-Stidd L.L., Cubero J.I. Origins and evolution of foxtail millets. J Agric Trop Bot Appl 1979; 26:54-64.
Dexter A.G., Nalewaja J.D., Rasmussen D.D., Buchli J. 1981. Survey of wild oats and other weeds in North Dakota. 1978 and 1979. North Dakota Agric Exp Stn Res Rep No 79, 1981.
Dixon D.M. "A note on cereals in ancient Egypt". In *The Domestication and Exploitation of Plants and Animals*, J.P.Ucko, F.W. Dimbleby, eds. Chicago, IL: Aldine, 1969, pp. 131-142.
Dore W.G. and McNeill J. *Grasses of Ontario*. Hull, Quebec, Canada: Res. Branch, Agric. Canada, Monograph 26, 1980.
Douglas B.J., Thomas A.G., Morrison I.N., Maw M.G. The biology of Canadian weeds. 70. *Setaria viridis* (L.) Beauv. Can J Pl Sci 1985; 65:669-690.
Dryden R.D., Whitehead C.W. The effect of TCA on *S. viridis* in competition with cereals. Can J Pl Sci 1963; 43:451-456.
Dunham R.S., Robinson R.G., Anderson R.N. Crop rotation and associated tillage practices for controlling annual weeds in flax and reducing weed seed populations of the soil. Tech. Bull. 230. Minn Agric Exp Sta, 1958.
Elmore C.D, Paul R.N. Composite list of C_4 weeds. Weed Sci 1983; 31:686-692.
Emery W.H.P. A study of reproduction in *Setaria macrostachya* and its relatives in the southwestern United States and Northern Mexico. Bull Torrey Bot Club 1957; 84:106-121.
Fabian I. Beitrage zum Lang- und Kurztagsproblem. Zeitschr Bot 1938; 33:305-357.
Fairbrothers D.E. Morphological variation of *Setaria faberii* and *S. viridis*. Brittonia 1959; 11:44-48.
Fausey J.C., Kells J.J., Swinton S.W., Renner K.A. Giant foxtail (*Setaria faberi*) interference in non-irrigated corn (*Zea mays*). Weed Sci 1997; 45:256-260.
Fawcett R.S., Slife F.W. Effects of 2,4-D and dalapon on weed seed production and dormancy. Weed Sci 1978; 26: 543-547.
Fernald M.L. *Gray's Manual of Botany*. 8th ed. New York: American Book Co., 1950.
Flavell R. The molecular characterization and organization of plant chromosomal DNA sequences. Annu Rev Plant Physiol 1980; 31:569-596.
Forcella F., Banken K.R. Relationships among green foxtail (*Setaria viridis*) seedling development, growing degree days, and time of nicosulfuron application. Weed Tech 1966; 10:60-67.
Forcella F., Wilson R.G., Dekker J., Kremer R.J., Cardina J., Anderson R.L., Alm D., Renner K.A., Harvey R.G., Clay S., Buhler D.D. Weed seed bank emergence across the corn belt. Weed Sci 1997; 45:67-76.
Forcella F., Wilson R.G., Renner K.A., Dekker J., Harvey R.G., Alm D.A., Buhler D.D., Cardina J.A. Weed seedbanks of the U.S. corn belt: magnitude, variation, emergence, and application. Weed Sci 1992; 40:636-644.
Fukunaga K., Domon E., Kawase K. Ribosomal DNA variation in foxtail millet, *Setaria italica* (L.) P. Beauv., and a survey of variation from Europe and Asia. Theor Appl Genet 1997; 95:751-756.
Fukunaga K., Wang Z., Kato K., Kawase M. Geographical variation of nuclear genome RFLPs and genetic differentiation of foxtail millet, *Setaria italica* (L.) P. Beauv. Genet Res Crop Evol 2002; 49:95-101.
Gao M. J., Chen J.J. Isozymic studies on the origin of cultivated foxtail millet. Acta Agro Sin 1988; 14: 131-136.
Gregg W. Ecology of the annual grass *Setaria lutescens* on old fields of the Pennsylvania Piedmont. Proc Nat Acad Natural Sci Philadelphia 1973; 124:135-196.
Gritzenko R.J. Chumiza (Italian millet) taxonomy (*Setaria italica* (L.) P.B. subsp. *maxima* Alef.). Bull Appl Bot Genet Plt Breed 1960; 32:145-182.
Gupta P.K., Singh R.V. Variations in chromosomes and flavonoids in *Setaria* Beauv. The Nucleus 1977; 20:167-171.
Haar J. M. Characterization of foxtail (*Setaria* spp.) seed production and giant foxtail (*Setaria faberii*) seed dormancy at abscission. Ph.D. dissertation, Iowa State University, Ames, Iowa, USA, 1998.
Hafliger E., Scholz H. Grass weeds 1-weeds of the subfamily Panicoideae. Basle, Switzerland: Documenta Ciba-Geigy, 1980.
Hamrick J.L., Godt M.J.W. "Allozyme diversity in plant species". In *Plant Population Genetics, Breeding and Genetic Resources*, A. H. D. Brown , M. T. Clegg, A. L. Kahler, B. S. Weir, eds. Sunderland, MA: Sinauer, 1990, pp. 43-63.
Harlan J.R. The possible role of weed races in the evolution of cultivated plants. Euphytica 1965; 14:173-176.
Harlan J.R. *Crops and Man*. Madison, WI: Am Soc Agro Crop Sci Soc, 1975.
Harlan J.R., de Wet J.M.J., Price E.G. Comparative evolution of cereals. Evolution 1973; 27:311-325.
Harper J.L. *Population Biology of Plants*. New York: Academic Press, 1977.
Harrison S.K., Williams C.S, Wax L.M. Interference and control of giant foxtail (*Setaria faberi*) in soybeans (*Glycine max*). Weed Sci 1985; 33:203-208.
Hattersley P.W., Watson L. "Diversification of photosynthesis". In *Grass Evolution and Domestication*, G.P. Chapman, ed. Cambridge, UK: Cambridge Univ. Press, 1992, pp. 38-116.

Heer O. Pflanzen die Pfahlbaut. Zurich: Neujahrblat der Naturforschenden Gesellsch. 1886.
Heise A. C. Germination of *S. viridis* seeds. Proc Assoc Off Seed Anal 1941; 33:43-44.
Helback H. "The paleobotany of the Near East and Europe". In *Prehistoric Investigations in Iraq Kurdistan*, Braidwood, R.J., B. Howe, eds. Chicago: Chicago University Press, 1960.
Hendricks S.B., Taylorson R.B. Dependence of phytochrome action in seeds on membrane organization. Plant Physiol 1978; 61:17-19.
Hitchcock A.S. *Manual of the grasses of the United States*. 2nd Ed., Vol. 2. New York, NY: Dover Public., 1971.
Ho P.-T. The loess and the origin of Chinese agriculture. Amer Hist Rev 1969; 75:1-36.
Ho P.-T. *The Cradle of the East*. Univ. Chicago, IL: Chicago Press, 1975.
Hofmeister W. Allgemeine morphologie der Gewachse. Leipzig: Handbuch der Physiologischen Botanik, Vol. 1, Pt. 2, 1868.
Holm L. G., Plucknett D.L., Pancho J.V., Herberger J.P. *The World's Worst Weeds - Distribution and Biology*. Honolulu, HI: The East-West Food Institute, 1977.
Holm L., Doll J., Holm E., Pancho J., Herberger J. *World weeds: Natural Histories and Distribution*. New York: John Wiley & Sons, Inc., 1997.
Holm L., Pancho J.V., Herberger J.P., Plucknett D.L. *A Geographic Atlas of World Weeds*. New York: John Wiley & Sons, 1979.
Holm R. E., Miller M.R. Weed seed germination responses to chemical and physical treatment. Weed Sci 1972; 20:150-153.
Holt J.S., LeBaron H.M. Significance and distribution of herbicide resistance. Weed Tech 1990; 4:141-149.
Hsu C. *Taiwan grasses*. Taipei: Taiwan Prov. Educ. Assoc., 1975.
Hubbard F.T. A taxonomic study of *Setaria* and its immediate allies. Am J Bot 1915; 2:169-198.
Hume L. The long-term effects of fertilizer application and three rotations on weed communities in wheat (after 21-22 years at Indian Head, Saskatchewan). Can J Plant Sci 1982; 62:741-750.
James A. L. Some influences of soil atmosphere on germination of annual weeds. Ph.D. dissertation. Iowa State University, Ames, Iowa, USA, 1968.
Jusuf M., Pernes J. Genetic variability of foxtail millet (*Setaria italica* Beauv.) electrophoretic studies of five isozyme systems. Theor Appl Genet 1985; 71:385-391.
Kadkol S.B., Swaminathan M. The nutritive value of Italian millet (*Setaria italica*). Sci Cult 1955; 20:340-341.
Kawase K., Sakamoto S. Variation, geographical distribution and genetical analysis of esterase isozymes in foxtail millet, *Setaria italica* (L.) P. Beauv. Theor Appl Gen 1984; 67: 529-533.
Kawano S., Miyake S. The productive and reproductive biology of flowering plants. X. Reproductive energy allocation and propagule output of five congeners of the genus *Setaria* (Gramineae). Oecologia 1983; 57:6-13.
Khosla P.K., Sharma M.L. Cytological observations on some species of *Setaria*. Nucleus 1973; 26:38-41.
Khosla P.K., Singh S. Cytotaxonomical investigations on *S. pumila* Beauv. complex. Abst., 1st All India Cong Cytol Genet 1971; 15.
Kihara H., Kishimoto E. Bastarde Zwishen *Setaria italica* und *S. viridis*. Bot Mag Tokyo 1942; 56:62-67.
King L. J. Germination and control of the *S. faberi* grass. Contr Boyce Thompson Inst 1952; 16:469-487.
King R.W. "The physiology of pre-harvesting sprouting-a review". In *3rd International Symposium. Pre-harvest sprouting in cereals*, J.E. Kruger, D.E. Laberge, eds. Boulder, CO: Westview Press, 1983, pp. 11-21.
Kishimoto E. Chromosomenzahlen in den Gattungen *Panicum* und *Setaria*, part I. Chromosomenzahlen einier *Setaria*-Arten. Cytologia 1938; 9:23-27
Kivilaan A., Bandurski R.S. The ninety year period for Dr. Beal's seed viability experiment. Am J Bot 1973; 60:140-145.
Koernicke F., Werner H. Handbuch des Getreidesbaues; Vol. I. Berlin: Verlag von Paul Parey, 1885a.
Koernicke F., Werner H. Handbuch des Getreidesbaues; Vol. 2, Die Sorten und der Anbau des Getreides. Berlin: Verlag von Paul Parey, 1885b .
Kollman G. E. Germination-dormancy and promoter-inhibitor relationships in *Setaria lutescens* seeds. Ph.D. dissertation. Iowa State Univ. Ames, IA, 1970.
Krishniswami N., Rangaswami A.G.N. Chromosome numbers in some *Setaria* species. Cur Sci 1935; 3(11):559-560.
Ku S.B., Gutierrez M., Edwards G.E. Localisation of the C_3 and C_4 pathways of photosynthesis in the leaves of *Pennisetum purpureum* and other C_4 species. Insignificance of phenol oxidase. Planta 1974; 119:267-278.
Kuuster H. "Neolithic plant remains from Eberdingenhochdorf, southern Germany". In *Plants and Ancient Man (Studies in Paleoethnobotany)*, W.V. Van Zeist, W.A. Caparoe, eds. Rotterdam: A.A. Bakame, Rotterdam; 1984.

Kuzin I.N. The effect of *S. viridis* in spring wheat crops with mouldboardless cultivation on the availability of nitrogen, phosphorus and potassium. Agrokimiya 1973; 10:130-131.

Lakshmi S., Ranjekar P.K. Novel molecular features of millet genomes. Indian J Biochem Biophys 1984; 21:299-303.

Laudien H., Koch W. Some comments on the ecology and distribution of *Echinochloa crus-galli* (L.), *Digitaria sanguinalis* (Scop.) and *Setaria* spp. in the Federal Republic of Germany, Notiziario sulle Malatie della Piante (1972) No. 86. Univ. Hohenheim, German Federal Republic, 1972, pp. 149-162.

Lauer E. Uber die Keimtemperatur von Acherrunkrautern und deren Einfluss auf die Zusammensetzung von Unkrautgesellschaften. Flora Allg. Bot Stg 1953; 140:551-595.

LeBaron H.M., Gressel J. *Herbicide Resistance in Plants.* New York: John Wiley & Sons, Inc., 1982.

Lee S.M. The distribution and abundance of three species of *Setaria* Beauv. around London Canada, with particular reference to the effects of shade. M.Sc. Thesis, Univ. of Western Ontario, London, Ont., 1979.

Lee S.M., Cavers P.B. The effects of shade on growth, development and resource allocation patterns of three species of foxtail (*Setaria*). Can J Bot 1981; 59:1776-1786.

Li C.H., Pao W.K., Li H.W. Interspecific crosses in *Setaria.* J Hered 1942; 33:351-355.

Li H.W. Studies in millet breeding methods. Bul Col Agr Honan Univ 1934; 2:1-22 (in Chinese).

Li H.W., Li C.H., Pao W.K. Cytological and genetical studies of the interspecific cross of the cultivated foxtail millet, *Setaria italica* (L.) Beauv. and the green foxtail millet *S. viridis* L. J Amer Soc Agron 1945; 37:32-54

Li H.W., Meng C.J., Liu T.N. Problems in the breeding of millet (*Setaria italica* (L.) Beauv.). J Amer Soc Agron 1935; 27:963-970.

Li Y., Wu S.Z. Traditional maintenance and multiplication of foxtail millet (*Setaria italica* (L.) P. Beauv.) landraces in China. Euphytica 1996; 87:33-38.

Lincoln R., Boxshall G., Clark P. *A Dictionary of Ecology, Evolution and Systematics.* 2nd Ed., UK: Cambridge Univ. Press, UK, 1998.

Linnaeus, C. *Species Plantarum.* Stockholm, 1753.

Lorenzi H. J., Jeffery L.S. *Weeds of the United States and their Control.* New York: Van Nostrand Reinhold Co., 1987.

MacVicar R.M., Parnell, H.R. The inheritance of plant colour and the extent of natural crossing in foxtail millet. Scient Agric 1941; 22:80-84.

Manthey D. R., Nalewaja J.D. Moisture stress effects on foxtail seed germination. Proc North Central Weed Control Conf 1982; 37:52-53.

Manthey D.R., Nalewaja J.D. Germination of two foxtail (*Setaria*) species. Weed Technol 1987; 1:302-304.

Mapplebeck L.R., Sousa Machado V., Grodzinski B. Seed germination and seedling growth characteristics of atrazine-susceptible and resistant biotypes of *Brassica campestris.* Can J Plant Sci 1982; 62:733-739.

Martin J. N. Germination studies of the seeds of some common weeds. Proc Iowa Acad Sci 1943; 50:221-228.

Martin A.C., Zim H.S., Nelson A.L. *American Wildlife and Plants. A Guide to Wildlife Food Habits.* New York: Dover Publications Inc., 1961.

Mester T. C., Buhler D.D. Effects of tillage on the depth of giant foxtail germination and population densities. Proc North Cent Weed Control Conf 1986; 41:4-5.

Mohamed H.A., Clark J.A., Ong C.K. The influence of temperature during seed development on the germination characteristics of millet seed. Pl Cell Environ 1985; 8: 361-362.

Monsalud M.R., Tongaran A.L., Lopez F.R., Lagrimas M.Q. Edible wild plants in Philippine forests. Philippine J Sci 1966; 95:431-561.

Moore D.J., Fletchall O.H. Germination-regulating mechanisms of giant foxtail (*Setaria faberii*). Res Bull. Missouri Ag Exp Sta No 829, 1963.

Morrison I.N., Maurice D., Bubar C.J. The relative competitive ability of green and *S. pumila* in wheat and their response to shade. Abstracts, XIII Int Bot Cong, Sydney, Australia, 1981.

Moyer J.R., Dryden R.D. Effects of combined applications of triallate or trifluralin with solution nitrogen on wheat, wild oats and *S. viridis.* Can J Plant Sci 1977; 57:479-484.

Moyer J.R., Dryden R.D., Chow P.N.P. Effect of barban and flamprop methyl with solution nitrogen on wheat, wild oats and *S. viridis.* Can J Plant Sci 1979; 59:351-356.

Mulligan G.A., Findlay J.N. Reproductive systems and colonization in Canadian weeds. Can J Bot 1970; 48:859-860.

Nadeau L.B. The effect of water stress on the root growth of green foxtail and yellow foxtail. M.Sc. thesis, The Univ. of Manitoba, Winnepeg, Manitoba, 1983.

Nadeau L.B., Morrison I.N. Root development of two *Setaria* species under different soil moisture regimes. Asp Appl Biol 1983; 4:125-134.

Naeiri Y., Belliard J. Le millet *Setaria italica* une plante a redecouvrir. J Agric Trop Bot Appl 1987; 34:65-87.

Nai H. Archaeology in New China. Antiquity 1963; 37:176-185.

Nakamura S. Germination of grasses. Proceedings Intl Seed Testing Assoc 1962; 27:710-729.

Narayanaswami S. Structure and development of the caryopsis in some Indian Millets. VI. *Setaria italica.* Bot Gaze 1956; 118:112-122.

Neuweiler E. Nachtrage urgeschichtlicher Pflanzen. Vierteljahrsschr. Naturf Ges Zurich 1946; 91:122-236.

Nguyen Van F., Pernes J. "Genetic diversity of foxtail millet (*Setaria italica*)". In *Genetic Differentiation and Dispersal in Plants*, P. Jacquard, ed. NATO ASI Series, Vol. G5. Berlin: Springer-Verlag, 1985, pp. 113-128.

Nieto-Hatem J. Seed dormancy in *Setaria lutescens* Ph.D. dissertation. Iowa State Univ. Ames, IA, 1963.

Norris R. F., Schoner C.A. Jr. Yellow foxtail (*Setaria lutescens*) biotype studies: dormancy and germination. Weed Sci 1980; 28:159-163.

Oliver L.R., Schreiber M.M. Differential selectivity of herbicides on six *Setaria* taxa. Weed Sci 1971; 19:428-431.

Orgel L.E., Crick F.H.C. Selfish DNA: the ultimate parasite. Nature 1980; 284:604-607.

Orwick P.L., Schreiber M.M. Differential root growth of four *Setaria* taxa. Weed Sci 1975; 23:364-368.

Orwick P.L., Schreiber M.M. Analysis of nonstructural carbohydrates in redroot pigweed (*Amaranthus retroflexus*) and robust foxtail (*Setaria viridis* var. *robusta*) throughout the growing season. Weed Sci 1979; 27:374-379.

Orwick P.L., Schreiber M.M., Holt D.A. Simulation of foxtail (*Setaria viridis* var. *robusta-alba, Setaria viridis* var. *robusta-purpurea*) growth: the development of SETSIM. Weed Sci 1978; 26:691-699.

Osada, T. *Illustrated grasses of Japan*. Tokyo: Heibonsha Ltd. Publishers, 1989.

Pareja M. R., Staniforth D.W. Seed-soil microsite characteristics in relation to weed seed germination. Weed Sci 1985; 33: 190-195

Pareja M. R., Staniforth D.W., Pareja G.P. Distribution of weed seed among soil structural units. Weed Sci 1985; 33: 182-189.

Peters R.A., Meade J.A., Santelmann P.W. Life history studies as related to weed control in the northeast. 2. Yellow foxtail and giant foxtail. Agric Exp Sta, pp. 18, Univ. Rhode Island, Kingston, 1963.

Peters R.A., Yokum H.C. Progress report on a study of the germination and growth of yellow toxtail (*Setaria glauca* (L.) Beauv.) Proc Northeast Weed Control Conference 15:350-355, 1961.

Philippi T., Seger J. Hedging one's bets, revisited. Trends Ecol Evol 1989; 4: 41-44.

Pohl R.W. The genus *Setaria* in Iowa. IA St J Sci 1951; 25:501-508.

Pohl R.W. Notes on green foxtail and giant foxtail (Gramineae). Brittonia 1962; 14:210-213.

Pohl R.W. The grasses of Iowa. Iowa St Jour Sci 1966; 40:341-373.

Pohl R.W. *How to know the grasses*, 3rd Edit., Dubuque, IA: Wm. C. Brown Co. Publishers, 1978.

Poirier-Hamon S., Pernes J. Instabilite chromosomique dans les tissus somatiques des descendants d'un hybride interspecifique *Setaria verticillata* (P. Beauv.), *Setaria italica* (P. Beauv.). R Acad Sci. Paris 1986; 302:319-324.

Povilaitis B. Dormancy studies with seeds of various weed species. Proceedings Intl Seed Testing Assoc 1956; 21:87-111.

Prasada Rao K.E., de Wet J.M.J., Brink D.E., Mengesha, M.H. Intraspecific variation and systematics of cultivated *Setaria italica*, foxtail millet (*Poaceae*). Econ Bot 1987; 41:108-116.

Rahman A., Ashford R. Control of *S. viridis* in wheat with trifluralin. Weed Sci 1972; 20:23-27.

Rangaswami Ayyangar G.N., Narayanan T.R., Seshadri Sarma P. Studies in *Setaria italica* (Beauv.), the Italian millet. Part 1. Anthesis and pollination. Indian J Agr Sci 1933; 3:561-571.

Reed C.F. *Selected Weds of the United States*. U.S. Dept. Agric. Agric. Handbook 366, 1970.

Reschly B., Dekker J.H., Stoltenberg D.E. Comparison of seed germinability between acetyl coenzyme A carboxylase inhibitor resistant and susceptible giant foxtail. Agro Abst 1996; 6.

Ricroch A., Mousseau M., Darmency H., Pernes J. Comparison of triazine-resistant and -susceptible cultivated *Setaria italica*: growth and photosynthetic capacity. Plant Physiol Biochem 1987; 25:29-34.

Rice K., Jain S.K. "Plant population genetics and evolution in disturbed environments". In *The Ecology of Natural Disturbance and Patch Dynamics*, S. T. A. Pickett, P. A. White, eds. New York: Academic Press, 1985, pp. 287-303.

Ridley H.N. *The Dispersal of Plants throughout the World.* Ashford, Kent, U.K.: L. Reeve and Co., 1930.

Ritter R.L., Kaufman L.M., Monaco T.J., Novitsky W.P., Moreland D.E. Characterization of triazine-resistant giant foxtail (*Setaria faberi*) and its control in no-tillage corn (*Zea mays*). Weed Sci 1989; 37:591-595.

Rominger J.M. *Taxonomy of Setaria (Gramineae) in North America*. Illinois Biol Monogr 29, 1962.

Rost T. L. Structural and Histochemical investigations of dormant and non-dormant caryopsis of *Setaria lutescens* (Gramineae). Ph. D. Thesis, Iowa State University, Ames, 1971.

Rost T.L. Ultrastructure and physiology of protein bodies and lipids from hydrated dormant and non-dormant embryos of *Setaria lutescens*. Am J Bot 1972; 59:607-616.

Rost T. L. The anatomy of the caryopsis coat in mature caryopses of the yellow foxtail grass (*Setaria lutescens*). Bot Gaz 1973; 134:32-39.

Rost T. L. The morphology of germination in *Setaria lutescens* (Gramineae): The effects of covering structures and chemical inhibitors on dormant and non-dormant florets. Ann Bot 1975; 39:21-30.

Rost T. L., Lersten N.R. A synopsis and selected bibliography of grass caryopsis anatomy and fine structure. Iowa State J Res 1973; 48:47-87.

Rost T. L., Lersten N.R. Transfer aleurone cells in *Setaria lutescens* (*Gramineae*). Protoplasma 1970; 71: 403-408.

Rousseau C., Cinq-Mars L. Les plantes introduites du Quebec. Jeune Sci 1969; 7:163-222.

Rozhevitz R.U., Shishkin B.K. "Gramineae". In *Flora Unionas Rerumpublicarum Sovieticarum Socialisticarum, Vol. 2*, K. Komarov, ed. Leningrad, 1934.

Sakamoto S. Origin and dispersal of common millet and foxtail millet. JARQ 1987; 21:84-89.

Santlemann P. W., Meade J.A. Variation in morphological characteristics and dalapon susceptibility within the species *Setaria lutescens* and *S. faberi*. Weeds 1961; 9:406-410.

Santlemann P. W., Meade J.A., Peters R.A. Growth and development of yellow foxtail and giant foxtail. Weeds 1963; 11:139-142.

Schreiber M. M. Development of giant foxtail under several temperatures and photoperiods. Weeds 1965a; 13:40-43.

Schreiber M.M. Effect of date of planting and stage of cutting on seed production of giant foxtail. Weeds 1965b; 13:60-62.

Schreiber M. M. Longevity for foxtail taxa in undisturbed sites. Weed Sci 1977; 25:66-72.

Schreiber M.M., Oliver L.R. Two new varieties of *Setaria viridis*. Weed Sci 1971; 19:424-427.

Schreiber M.M., Orwick P.L. Influence of nitrogen fertility on growth of foxtail (*Setaria*) taxa. Weed Sci 1978; 26:547-550.

Scheiner S.M. Genetics and evolution of phenotypic plasticity. Annu Rev Ecol Syst 1993; 24:35-68.

Sells G.D. CO_2-O_2 ratios in relation to weed seed germination. Ph.D. Thesis, Iowa State University, Ames, 1965.

Siegel S. M., Renwick G., Rosen L.A. Formation of carbon monoxide during seed germination and seedling growth. Science 1962; 137: 683-684.

Siegel S. M., Siegel B.Z. Biogenesis of carbon monoxide: Production by fungi and seed plants in the dark. Phytochemistry 1987; 26: 3117-3119.

Simpson G.M. *Seed Dormancy in Grasses*. Cambridge, UK: Cambridge Univ. Press, 1990.

Sionit N., Patterson D.T. Responses of C_4 grasses to atmospheric CO_2 enrichment. II. Effect of water stress. Crop Sci 1985; 25:533-537.

Siva Raman L., Gupta V.S., Ranjekar P.K. DNA sequence organization in the genomes of three related millet plant species. Plant Mol Biol 1986; 6:375-388.

Slife F.W. A new *Setaria* species in Illinois. Proc North Cent Weed Cont Conf 1954; 11:6-7.

Smith C.E. Jr. The new world centers of origin of cultivated plants and the archaeological evidence. Econ Bot 1968; 22:253-266.

Smith G.P. Evolution of repeated DNA sequences by unequal crossover. Science 1976; 191:528-535.

Stace C.A. *Hybridization and the Flora of the British Isles*. London: Academic Press, 1975.

Staniforth D.W. Competitive effects of three foxtail species on soybeans. Weeds 1965; 13:191-193.

Stanway V. Laboratory germination of giant foxtail, *Setaria faberii* Herrm. at different stages of germination. Proc Assoc Off Seed Anal 1971; 61:85-90.

Stapf O., Hubbard C.K. "*Setaria*". In *Flora of tropical Africa 9*, Prain, ed., London.

Steel M.G., Cavers P.B., Lee, S.M. The biology of Canadian weeds. 59. *Setaria glauca* (L.) Beauv. and *S. verticillata* (L.) Beauv. Can J Plant Sci 1983; 63:711-725.

Stevens O.A. The number and weight of seeds produced by weeds. Am J Bot 1932; 19:784-794.

Stevens O.A. Weed development notes. Res Rep 1, pp.22, North Dakota Agric Exp Sta, 1960.

Stoller E. W., Wax L.M. Dormancy changes fate of some annual weed seeds in the soil. Weed Sci 1974; 22:154155.

Stoltenberg D.E., Wiederholt R.J. Giant foxtail (*Setaria faberi*) resistance to arlyoxyphenoxypropionate and cyclohexanedione herbicides. Weed Sci 1995; 43:527-535.

Swinton S., Sterns J., Renner K., Kells J. Estimating weed-crop interference parameters for weed management models. Research Report 538. Michigan Agricultural Experiment Station, Michigan State University, East Lansing, MI, 1994.

Swinton S.M., Buhler D.D., Forcella F., Gunsolus J.L., King R.P. Estimation of crop yield loss due to interference by multiple weed species. Weed Sci 1994; 42:103-109.

Takahashi N., Hoshino T. Natural crossing in *Setaria italica* (Beauv.). Proc Crop Sci Soc, Japan 1934; 6:3-19.

Taylorson R. B. Anesthetic release of seed dormancy- An Overview. Israel J Bot 1980/1981; 29:273-280.
Taylorson R. B. Anesthetic effects on secondary dormancy and phytochrome response in *Setaria faberi* Hermm. Plant Physiol 1982; 70:882-886.
Taylorson R. B. Water stress induced germination of giant foxtail (*Setaria faberi*) seeds. Weed Sci 1986; 34:871-875.
Taylorson R. B., Brown M.M. Accelerated after-ripening for overconing seed dormancy in grass weeds. Weed Sci 1977; 25:473-476.
Thomas A. G., Banting J.D., Bowes G. Longevity of *S. viridis* seeds in a Canadian prairie soil. Can J Plant Sci 1986; 66:189-192.
Thompson W.F., Murray M.G. "The nuclear genome: structure and function". In *The Biochemistry of Plants. A Comprehensive Treatise Vol. 6. Proteins and Nucleic Acids*, A. Marcus, ed. New York: Academic Press, New York, 1981.
Thornhill R. The effect of temperature and moisture on germination of giant foxtail (*Setaria faberi*) seeds, caryopses and embryos. M.S. Thesis, Iowa State University, Ames, 1997.
Thornhill R., Dekker J. Mutant weeds of Iowa: V. S-triazine resistant giant foxtail (*Setaria faberii* Hermm.). J Iowa Acad Sci 1993; 100:13-14.
Till-Bottraud I., Reboud X., Brabant P., Lefranc M., Rherissi B., Vedel F., Darmency H. Outcrossing and hybridization in wild and cultivated foxtail millets: consequences for the release of transgenic crops. Theor Appl Genet 1992; 83:940-946.
Toole E.H., Brown E. Final results of the Duvel buried seed experiment. J Agric Res 1946; 72:201-210.
Tranel D.M., Dekker J.H. Differential seed germinability in triazine resistant and susceptible giant foxtail (*Setaria faberi*). NCWSS Proceedings 1996; 51:165.
Tranel D.M., Dekker J. Differential seed germinability in triazine-resistant and -susceptible giant foxtail (*Setaria faberii*). Asian J Pl Sci 2002; 1:334-336.
Vanden Born W. H. Green foxtail: seed dormancy, germination and growth. Can J Plant Sci 1971; 51:53-59.
Van Zeist W. Prehistoric and early historic food plants in the Netherlands. Palaehistoria 1970; 14:41-173.
Volenberg D.S., Stoltenberg D.E., Boerboom C.M. Biochemical mechanism and inheritance of cross-resistance to acetolactate synthase inhibitors in giant foxtail. Weed Sci 2001; 49:635-641.
Waldron L.R. Weed studies. Vitality and growth of buried weed seed. N Dakota Agric Exp Sta Bull 1904; 62:439-446.
Wang R.L., Dekker J. Weedy adaptation in *Setaria* spp.: III. Variation in herbicide resistance in *Setaria* spp. Pestic Biochem Physiol 1995; 51:99-116.
Wang R.L, Wendel J., Dekker J. Weedy adaptation in *Setaria* spp.: I. Isozyme analysis of the genetic diversity and population genetic structure in *S. viridis*. Am J Bot 1995a; 82(3):308-317.
Wang R.L, Wendel J., Dekker J. Weedy adaptation in *Setaria* spp.: II. Genetic diversity and population genetic structure in *S. glauca*, *S. geniculata* and *S. faberi*. Am J Bot 1995b; 82(8):1031-1039.
Wang Z.M., Devos K.M., Liu C.J., Wang R.Q., Gale M.D. Construction of RFLP-based maps of foxtail millet, *Setaria italica* (L.) P. Beauv. Theor Appl Genet 1998; 96:31-36.
Warwick S. I., Thompson B.K., Black L.D. Life history and allozyme variation in populations of the weed species *Setaria faberi*. Can J Bot 1987; 65:1396-1402.
Warwick S.I. Allozyme and life history variation in five northwardly colonizing North American weed species. Pl Syst Evol 1990; 169:41-54.
Weaver S., Hamill A. Effects of soil pH on competitive ability and leaf nutrient content in corn (*Zea mays* L.) and three weed species. Weed Sci 1985; 33:447-451.
Weed Science Society of America. Composite list of weeds. Weed Sci 1984; 32, supplement 2.
Werth E. Zur Geographie und Geschichte der Hirsen Berlin. Angew Bot 1937; 19:42-88.
Wheeler W.A. *Forage and Pasture Crops*. New York: D. Van Nostrand Co. Inc., 1950.
Wiederholt R.J., Stoltenberg D.E. Abscence of differential fitness between giant foxtail (*Setaria faberi*) accessions resistant and susceptible to acetyl-coenzyme A carboxylase inhibitors. Weed Sci 1996; 44:18-24.
Williams R.D., Schreiber M.M. Numerical and chemotaxonomy of the green foxtail complex. Weed Sci 1976; 24:331-335.
Willweber-Kishimoto E. Interspecific relationships in the genus *Setaria*. Contrib Bio Kyoto Univ 1962.
Wilson R.G. Jr. Dissemination of weed seeds by surface irrigation water in western Nebraska. Weed Sci 1980; 28:87-92.
Yenish J.P., Doll J.D., Buhler D.D. Effects of tillage on vertical distribution and viability of weed seed in soil. Weed Sci 1992; 40:429-433.

CHAPTER 5

AQUATIC WEEDS

S. SIDORKEWICJ, M.R. SABBATINI[1], O.A. FERNÁNDEZ[1] & J.H. IRIGOYEN[1]

Departmento de Biología, Bioquímica y Farmacia, [1]Departamento de Agronomía and Centro de Recursos Naturales Renovables de la Zona Semiárida; Universidad Nacional del Sur, 8000 Bahía Blanca, Argentina

The welfare of an ever-increasing human population throughout the world, depends in part upon man's competence to manage continental freshwater ecosystems. Nowadays, these freshwater resources include natural water bodies (e.g., rivers, streams, lakes, ponds), and the construction around the earth of a wide range of standing and flowing waters, such as thousands of reservoirs of all sizes and millions of kilometres of irrigation and drainage channel networks. These natural and man-made aquatic ecosystems provide an excellent environment for aquatic plant growth.

Aquatic plants include numerous emergent, submerged, and floating-leaved macrophytes and periphytic and phytoplanktonic algae. They are ecologically important in most freshwater ecosystems. As primary producers they constitute at least part of the base of most aquatic food chains. In the process of photosynthesis they release oxygen needed to meet the respiration needs of aquatic biota. They regulate sediment quality and nutrient cycling, and provide bioarchitectural variety, and habitats for a variety of other aquatic and terrestrial organisms. However, when aquatic plant growth reaches high abundance, it can be a major problem affecting the suitability of the water for urban use, agriculture, industry and recreation.

By definition, plants which are harmful to human activities are classified as weeds; this terminology can also be applied to plants growing in aquatic habitats, i.e. aquatic weeds are nuisance plants that may impair the diversity and scale of services we have come to anticipate from our surface freshwater resources (e.g. Pieterse, 1990). It is interesting to note that a high percentage of studies on the subject of aquatic vegetation are connected with weed problems. The reason for this is that the utilization of water, one of the earth's most fundamental resources needed by man, is closely associated to the damage which can be caused by the presence of some types of aquatic plants. The economic losses incurred by public services, farmers, industry, fisheries and recreation, are severe enough to require active management of aquatic plants in many freshwater systems.

Despite the existance of *ca* 700 species of aquatic macrophytes, only a few species are the cause of most of the prevailing aquatic weed problems today. Nevertheless, these few species may take over huge areas of continental water bodies worldwide. A list of the 20 most important aquatic macrophyte weeds on a global scale is included in Table 1.

Inderjit (ed.), Weed Biology and Management, 115-135.

Table 1. Aquatic macrophyte weed species most troublesome around the world.

Submerged	Emergent	Free-floating
Potamogeton pectinatus	*Alternanthera philoxeroides*	*Eichhornia crassipes*
Hydrilla verticillata	*Echinochloa crus-galli*	*Pistia stratiotes*
Elodea spp.	*Phragmites australis*	*Salvinia molesta*
Myriophyllum spicatum	*Typha* spp.	*Azolla* spp.
Myriophyllum aquaticum	*Scirpus* spp.	*Lemna* spp.
Ceratophyllum demersum	*Panicum* spp.	
Vallisneria americana		
Najas spp.		
Zannichellia palustris		
Filamentous algae		
(*Cladophora* spp.,		
Enteromorpha spp.,		
Vaucheria spp.)		

A reasonable understanding of why a given sort of aquatic plant community dominates a specific freshwater system is important and generally may help indicate the system's ecological integrity, as well as the sustainability of the services we must anticipate from it.

PROBLEMS CAUSED BY AQUATIC WEEDS

In most instances an aquatic weed problem is directly associated with an excessive growth of one or a few aquatic species that interferes with the human utilization of a given freshwater resource. The obstruction caused by aquatic weeds may block pump intakes in hydroelectric complexes. Aquatic weeds also represent one of the biggest concerns in the proper management of water in irrigation and drainage channels - the abundance of aquatic vegetation increases the frictional resistance to water flow, restraining water availability and increasing water level in the system. Often this results in overbank spill and flooding, as well as the increase of deposition of suspended sediments. In the irrigation district of the lower valley of the Río Colorado, southern Argentina, for example, *Chara contraria* and *Potamogeton pectinatus* cause the malfunctioning of drainage channels which causes impaired land drainage, and problems of soil salinity in cultivated lands (Sabbatini et al., 1986; Fernández et al., 1987).

In the tropical and subtropical regions, the extremely high densities of free-floating aquatic vegetation may cause serious blockage problems in navigation waterways; and they are a recurrent nuisance in stagnant or slow flowing water. This is the case of the free floating mats of *Eichhornia crassipes,* a South American species which through its prodigious rate of increase has colonized tropical water systems throughout the tropics (and some cooler areas, as far north for example as southern Portugal in Europe), constituting the "world's worst aquatic weed" (Gopal and Sharma, 1981; Cook, 1990; Moreira et al., 1998a; Pieterse, 1998; Hill, 1999). A similar problem may be caused by free-floating ferns of the genera *Salvinia,* and others.

The problem of aquatic weeds in rice paddy fields is one of the main causes of reduction in crop yields of rice in tropical and subtropical countries; the fall in production due to the presence of aquatic weeds in Argentina was estimated from 35 to 70 per cent (Fernández et al., 1990), a situation that has not changed much nowadays.

Aquatic weeds may also massively colonize fish ponds, shallow lakes and reservoirs, affecting fisheries, industries and recreation. Dense growths of aquatic plants may provide habitats favourable to vectors of water-borne diseases, such as the snails that transmit the species of *Schistosoma*, which cause a chronic parasitic disease in tropical countries.

In numerous situations, excessive aquatic plant growth is one of the main factors degrading water quality. This is a crucial issue in circumstances in which the purity of water source is of paramount importance, as it is in the case for human domestic consumption or certain industries. For instance, 61% of the drinking water supplies coming from about 12,000 lakes in the United States and Canada were impaired in 1983 by the extreme abundance of algae, cyanobacteria and macrophytes, affecting several million people (American Water Works Association, 1987; Welsh and Lindell, 1992; Lewis and Wang, 1997). There are several species of cyanobacteria that under bloom growth conditions are known to produce enough amount of toxins able to cause illness to man and even death to domestic and wild animals (Gangstad and Cardarelli, 1990; Carmichael, 1992). As a recent example, the 350.000 people living in and around the city of Bahía Blanca, Argentina (38° 41' S, 62° 15' W), were deprived in 1999 of drinking water for several months by the presence of cyanobacterial blooms of the genera *Anabaena* and *Microcystis.* These species heavily colonized the 3000 ha Paso de las Piedras reservoir, which constitutes the source of domestic water to the local population (Fernández and Vázquez, 2001). Large local petrochemical and fertilizer industries using the same water supply also had difficulties. In most of these situations, purification of water to restore its properties to an acceptable level of utilization may demand a series of costly additional manipulations (filtration, detoxification, other chemical and physical treatments).

ECOLOGICAL FACTORS GOVERNING AQUATIC PLANT GROWTH

The biology of aquatic vascular plants has been treated in detail by Sculthorpe (1967), including two chapters dealing with aquatic weeds. The biology and ecology of some of the world's most troublesome aquatic weeds have been reviewed by Cook (1990), Spencer and Bowes (1990) and Wade (1990).

Aquatic weeds have evolved a great deal of adaptations to grow in continental aquatic systems. Nearly all of them are characterized by a rapid vegetative growth, and most by this potential to reproduce by sexual and asexual means. A typical example of a range of survivorship strategies is given by several species of the genus *Potamogeton*. In effect, they ensure the presence of the most successful genotypes in a given aquatic environment by simultaneous clonal reproduction by rhizomes, tubers and turions, while the abundance of seeds of sexual origin gives the opportunity for the creation of new genotypes open to explore changing environments. For example, the reproductive potential of *Potamogeton pectinatus*, in alliance with its large phenotypic plasticity (Yeo, 1965; Kantrud, 1990), allowed this species to achieve extraordinary success in invading a wide range of natural and artificial freshwater ecosystems, making it one of the most menacing aquatic plant species world-wide.

The influence of some of the environmental factors affecting physiological processes crucial to aquatic plant growth (photosynthesis and respiration for example), varies greatly for emergent, floating or submerged plants. In this chapter we emphasise the significance of components associated with the aquatic submerged freshwater environment. While soil type, lack of water, temperature stress and grazing are important factors in terrestrial environments, aquatic ecosystems are more influenced by water depth and velocity, inorganic carbon supply, light availability and the physico-chemical properties of water. One of the greater distinctive attributes between aqueous and aerial environments is the slowness of solute diffusion through water, in particular oxygen and the dissolved carbon molecules (CO_2, HCO_3^-) primarily used for photosynthesis by submerged plants.

The availability of dissolved inorganic carbon (DIC) for photosynthesis is a potential limiting factor for submerged leaves. They are faced with a carbon dioxide diffusion rate that is 10^4 times slower in water than in air. To cope with the problem of DIC accessibility, some submerged plants have evolved a range of physiological, morphological and anatomical adaptation, such as the use of bicarbonate ions, photosynthetic pathways involving C_4 acid metabolism, the presence of thin finely divided leaves to increase diffusion rate, root uptake of CO_2 from sediments and carbon fixation at night (Murphy, 1995).

The stress caused by a low level of light availability may influence the growth and distribution of submerged plants. Light attenuation increases with water depth, preventing the occurrence of submerged macrophytes in deep waters; a mat of floating species may further limit light penetration, turbidity is frequently one of the main factors limiting submerged plant growth through reduced light availability. The case of the irrigation district of the lower valley of the Río Colorado, Argentina, illustrates this point (Fernández et al., 1998). Within the irrigation complex substantial changes in water clarity and macrophyte abundance have occurred during recent years in the irrigation canals. Prior to the construction of a major dam 700 km upstream the irrigation area, the high load of suspended sediments resulted in a very turbid and reddish colour of the water (giving the name to the river). Irrigation water derived from the river was also very turbid, preventing submerged vegetation growth at all in primary or secondary irrigation canals. Since the closure of the dam (1989), river turbidity has declined markedly due to the settlement of the clay sediment load in the reservoir. This, in turn resulted in an increase of light penetration, rapidly followed by an increase of submerged macrophytes in most of the 5000 km irrigation network, dominated by *P. pectinatus.* As a consequence, weed problems became a real management issue in irrigation canals in the early 1990s. The drainage network of the system already had a long history of aquatic weed problems. Around the mid-1980s, fish of the species *Cyprinus carpio* (common carp) were accidentally introduced into these channels. The sediment-browsing activities of the fish increased the turbidity of the water with a concomitant reduction in submerged plant biomass (Fernández et al., 1998; Sidorkewicj et al., 1998).

Submerged macrophytes species have evolved a series of physiological and morphological adaptation to cope with the lower light availability of aquatic habitats. Morphologically these adaptations comprise thin and finally dissected leaves and chloroplasts in the outer cell layers, frequently associated with low light compensation points. The development for some species of a dense upper canopy close to the water surface maximizes the interception of solar energy, and deprives deeper-growing competitor species of light.

One source of mineral ions for aquatic plant's development is the nutrient medium in which the plants are bathed, since they can use nutrients from water by direct foliar absorption. Nevertheless, sediment appears to be the most important source of nutrients for rooted macrophytes (Barko and Smart, 1981; Barko et al., 1991). Several species would be able to exploit both sources. The availability of nutrient supplies is a topic of maximum concern in the management of aquatic weeds, since the enrichment of terrestrial aquatic habitats by human actions in the catchment is one of the main causes favouring macrophytes and algal blooms conditions. This issue is treated in more detail in the next section.

As a general conclusion concerning the environmental factors controlling aquatic plant growth, for an effective management of aquatic weeds, it is meaningful to consider the full range of physico-chemical water attributes that may influence the type of plant community colonizing a given freshwater ecosystem. This includes: turbidity, salinity, pH, hardness, nutrients (mainly phosphorus and nitrogen), dissolved oxygen and inorganic carbon, biochemical oxygen demand, water velocity and depth. Modification of any of these environmental parameters can bring about changes in the type of vegetation dominating a specific water body. New biological interactions may also be the reason of significant modification on aquatic flora, via competition with other plant species, diseases and herbivory. Examples of aquatic weed management by the interaction with other organisms are discussed in some detail later in this chapter.

AQUATIC PLANT COMMUNITIES AND INTERACTION WITH THE LAND SURFACE

Type of aquatic vegetation growing in rivers, streams, lakes or man-made waterways and reservoirs depends ultimately on the water properties. This in turn is the outcome of the kind of terrestrial eco-region (soil type, relief, vegetation, other organisms) from which water drains into the system, known as the river basin, watershed or catchment. A catchment can be hydrologically defined as a limited three-dimensional surface of land in which rainfall gives rise during a short period of time (hours) to surface runoff following the normal discharge pathway of the relief, and to a slow subsurface flow (months) of infiltrated water. These water movements, flowing inward from the natural boundaries of the catchment, carry and dissolve inorganic and organic components in their path. Whatever their extent, catchments are critical components in determining the quality and quantity of water caught by surface water systems.

Nowadays, a range of anthropogenic activities within catchments can strongly influence the physico-chemical properties of natural and man-made freshwater systems. For example, pollution by waste water discharge from human settlements may increase metals, detergents and acidification of water; deforestation, overgrazing and agriculture can cause soil erosion with the eroded soil eventually ending up in water bodies; the prevalent heavy contribution of agrochemicals for crops is responsible in many parts of the world of nutrients increase in the water. The above problem for the domestic water supplies to the city of Bahía Blanca, due to the abundance of algae, appeared to be related, as in many other places throughout the world, to increasing inputs of water containing high-nutrient concentrations (especially phosphorus and nitrogen), as a result of agricultural activities within the catchment (Marchena et al., 2000; Sabbatini et al., 2002).

In many (if not all) cases, the issue of aquatic weed management can only be properly faced within the context of anthropogenic impacts within a catchment. This is because the origin of aquatic weed problems may be essentially a consequence of what is going on the neighbouring land, rather than in the water system itself. Therefore, the solution may not be restricted only to direct weed control actions applied to target nuisance species, but ought to be coupled with an appropriate land use and management within the catchment. The development of an integrated "catchment-water system" ecological assessment program, can provide information about the rate of export of allochthonous materials towards a river, stream, lake or reservoir. Furthermore, data on the complex physical and chemical interaction between land and water can increasingly be predicted from aquatic plant community information (or other freshwater biota), by means of biomonitoring systems to indicate the state of ecological health of the system (e.g., Caffrey, 1986; Ali et al., 1999; Janauer, 2001; Sabbatini et al., 2002).

MANAGEMENT OF AQUATIC WEEDS

If the sustainability of freshwater resources has to be protected, then in many systems active control of aquatic weeds cannot be avoided. This situation reflects the present social and economic importance of the problem. However, this does not mean eradication. It is hard to conceive of natural aquatic continental habitats without plants; and in many instances the ecological integrity of the water system itself may depend on their presence. Consequently, the problem cannot be as simple as killing aquatic weeds, and the question is to determine the acceptable density of weeds which can be tolerated in a given system (at which stage, by definition, they can no longer be considered as "weeds" *per se*) to ensure the stability of the water resource. Often this becomes a question of how we can manipulate the ecosystem in order to favour a most desirable aquatic plant community. Therefore, in most cases aquatic weed management programmes do not aim at the total removal of nuisance species, but rather the decrease of plant biomass to levels low enough to allow a proper functioning of the aquatic system. The essence of the problem may differ radically according to the type of services we demand from each specific water resource, the core concern will be quite different if we are dealing with drainage and irrigation channels, natural or man-made lakes, or rivers and streams.

There are three potential ways of controlling the excessive growth of undesirable aquatic vegetation: physical, chemical and biological. Other strategies, however, are also used such as the manipulation of the environment or the combination of different alternatives in an integrated approach.

Physical Control and Environmental Manipulation

Physical control by manual and mechanical methods is the oldest strategy of dealing with aquatic weeds. The subject was extensively examined by Wade (1990). Direct manual removal of vegetation, or the employment of hand-held tools such as cutting, forks, hoes, scythes, floating ropes and nets, although still widely practised in many places where labour is plentiful, are generally restricted to isolated group of plants, small ponds and water courses, and the marginal populations in lakes and channels. Their efficiency is low, demands a prolonged period for weed removal, and the operator may be exposed to important disease risks, as for example filariasis, malaria, and schistosomiasis.

The extensiveness of the aquatic weed problems in heavily infested large water bodies of the world has resulted in the fabrication of a large variety of machines devised to reduce aquatic plant biomass. They fall into two main categories: those aimed at only cutting and removing stands of vegetation, and those that also dredgeout part of the sediment (Wade, 1990). Cutting and harvesting machines mounted on boats are regularly used for the removal of submerged and floating vegetation in lakes, rivers and channel maintenance. Some of them are able to partially or totally execute the operations of cutting, shredding, harvesting and dumping the plants out of the water.

A large range of bank-based cutting mechanisms, hydraulic dragline excavators, rakes, and grab-buckets, are widely used against submerged and emergent vegetation worldwide. Some of these machines can also excavate part of the bottom sediment. In southern Argentina a common method for the control of massive infestations of *Potamogeton illinoensis* in main irrigation channels is the use of heavy chains with sharp blades that are dragged upstream by two tractors, one on each bank, tearing plant foliage along the way (Dall′Armellina et al., 1996; Bezic, 2001). Due to rapid regrowth, chaining may be repeated up to seven times in the main channels during a growing season to ensure proper water supply to irrigated crops (Murphy, 1995). During winter, the channels remain dry and plant survivorship is dependent on possession of an underground rhizomatous system, which also provides for regrowth after removal of foliage by the chaining operations (Dall′Armellina et al., 1996). Similar methodologies are used in other parts of the world.

There are a number of advantages in using machinery for aquatic weed control, such as the potential to solve rapidly at any time a weed problem independently of species phenology, and the control of weed populations at a chosen level in mixed plant stands or multi-purpose water use. However, there are some inherent drawbacks. These include: incomplete coverage of the infested area, capital costs of machinery can be high, and there is a need for proper bank facilities to permit the use of heavy equipment. In addition the survivorship of plant propagules buried in the sediment bank may remain unaffected by clearance operations, and submerged and underground plant fragmentation can facilitate reinfestation and weed dispersion to other parts of the system. As a result, regrowth following a mechanical clearance operation may be rapid, and may demand frequent repetition of the treatments with concomitant effects on labour and operational costs (Wade, 1990).

Methodologies that involve environmental alterations that interfere with weed growth have been used to manage aquatic vegetation in many places. Water level manipulations are often employed in small ponds and irrigation systems. Control is the result of dehydration of plant tissue due to its exposure to the air. For example, short-term drawdown (4-5 days) during the summer in the warm and dry irrigation district of Río Dulce in central Argentina (27°25′ S, 63°50′ W), causes quick desiccation of the aboveground biomass of the submerged vegetation due to high air temperature and low humidity, bringing an improvement in water flow once the levels are restored to normal. This procedure is inevitably followed by rapid plant regrowth, however, due to its low cost, the method can be repeated as many times as necessary during the growing season (Eng. M. T. Sobrero, pers. comm.).

A methodology that has been tested for submerged weed management is the use of black plastic, or geotextile, lining that cover the bottom of the irrigation channels avoiding the growth of rooting species. In one trial carried out in winter 1998 in irrigation districts of southern Argentina, a 7-m-wide black plastic liner was positioned along 20 km of the bed of an irrigation channel. Although this did give rapid initial relief from the weed problem, the long-term effect was poorer because sediment accumulated over the plastic, and vegetation colonized the new system and expanded rapidly.

The control of blue-green algal (cyanobacterial) blooms affecting drinking water is best done by reducing the quantities of nutrients that reach the water system. This may include the study of the catchment and the water system as single ecological unit, with the aims of preventing large quantities of soil and agrochemicals from entering the water body, and providing better treatment of urban and industrial sewage before its discharge into the system. Since these algal blooms occur in standing waters, reduction of algal growth is often possible by artificial mixing or aeration. The resulting turbulence produced prevents the phytoplankton from obtaining the amount of light needed for rapid growth (Moss, 1998). The strategy also helps in reducing fast growth of free-floating species given that they absorb nutrients directly from the water medium.

Chemical Control

Herbicides can offer an effective and rapid method of aquatic weed control. They are important aids especially in the maintenance of (i) waterbodies in areas with warm temperatures and long growing seasons invaded by floating and submerged weeds, as for example several water bodies severely affected by *Hydrilla verticillata* and *Eichhornia crassipes* in south and south east-Asia (Gopal, 1990), and the state of Florida in the United States (Anderson and Steward, 1990; Haller et al., 1990), (ii) open channel networks in semiarid irrigation districts usually affected by submerged and emergent aquatic weeds, as for example in Australia and New Zealand (Bowmer, 1979), the western United States (Nordone et al., 1996) or southern Argentina (Fernández et al., 1987), (iii) drainage channels in lowland areas usually invaded by submerged weeds and algae, as for example in England (Miles, 1976; Sabbatini and Murphy, 1996), and (iv) enhanced recreational uses of lakes and ponds for sport fishing, swimming, sailing, etc.

Herbicides are powerful tools, which require knowledge and understanding to be used safely (Petta, 2002). In terms of toxicity to non-target aquatic organisms, herbicides may have both direct and indirect effects (Murphy and Barrett, 1990). Direct effects could include the damage produced by certain herbicide (at concentrations likely to result in the water from use at their designed application rate) on fish, non-target vegetation and other non-target organisms. For example, aquatic herbicides such as acrolein, copper sulphate and dichlobenil, may have severe effects on fish populations at certain concentrations (Tooby et al., 1974; Bartley and Hattrup, 1975; Bohl et al., 1982). The indirect effects are related to the *in situ* decay and death of the aquatic plants as a result of herbicide treatment. A typical example of the indirect effect could be the fast and massive fish killing that typically happens as a consequence of the decomposition of dense stands of submerged aquatic weeds. This is due to the rapid oxygen depletion. This can happen even though the chemical was applied at a concentration non-toxic to the fish. An example is a major roach (*Rutilus rutilus*) mortality which occurred approximately 45 days after dichlobenil application to a British lake, when daytime dissolved oxygen levels fell to less than 5% saturation (Tooby and Spencer-Jones, 1978).

Despite the clear negative impact that herbicide may produce in water bodies, it is important to state that when they are used with caution, they are an effective way to improve water quality in many types of waterbodies, especially when there are no other practical alternatives for weed control. A good example was reported by Olaleye et al. (1993), who showed that diquat and terbutryn applied in a Nigerian waterway, Kofawei Creek, not only succeeded in killing an *Eichhornia crassipes* infestation without causing any fish mortality, but also improved fish species composition qualitatively and quantitatively in the treated sites.

The application of herbicides to water bodies for weed control is a much more restricted event than other forms of general weed control. For example, in the United States no product can be labelled for aquatic use if it poses more than a one in a million chance of causing significant damage to human health, the environment or wildlife resources (Madsen, 2000). Moreover, the high costs of investment involved and relatively small market, have caused herbicide companies to cease pursuing the registration of aquatic herbicides in recent years, it can take $ 20-40 million and 8-12 years to complete laboratory and field testing to finally register the product, with no guarantee of a return on investment (Getsinger, 1991). The result of this situation is that only a limited spectrum of existing chemicals is available for aquatic weed control, with little or no new research underway on developing new products.

Aquatic herbicides are chemicals specifically formulated to be used for the control of several types of aquatic plants (free-floating, emergent, rooted with floating leaves, submerged and algae). They can be used in static and flowing water bodies as ponds, lakes, rivers, canal systems, etc. Conventional spraying techniques are effective against emergent or floating foliage but ineffective against submerged weeds or algae. Because of differences in the biology of aquatic and terrestrial plants, and the environments where they live, herbicide efficacy is lower in aquatic situations. In particular this is due to the diluting and dispersing effects of the water surrounding, and sometimes covering the weeds. For effective control of submerged weeds the lethal concentration of the chemical has to be introduced into the water and remain in contact with the above-ground plant tissues long enough to be absorbed by the plants. In static water systems, this can be solved by applying the product in doses calculated in relation to the water volume for a determined area. In flowing waters, the herbicide may be added at a constant rate to maintain a given concentration for the required period, so that a pulse of herbicide moves downstream, being absorbed by the target weeds as it passes (Bowmer and Smith, 1984). Special formulations, such as slow-release granules or viscous gel formulations (for example, diquat- alginate) which stick to weeds, aim to enhance the particular uptake properties of individual chemicals (Barrett and Murphy, 1982). The aim of these formulations is to transport the herbicide to the weed and release it in a way that enhances uptake and performance (Killgore, 1984). The aquatic herbicides most commonly used around the world (Murphy and Barrett, 1990; Madsen, 2000; Lewis, 2002) are listed below.

1. 2,4-D: often called the "hormonal" herbicide because its effect on plants mimics those of natural auxinic growth-regulating plant hormones. 2,4-D is commonly used against free-floating weeds such as *Eichhornia crassipes*.
2. Dichlobenil: a non-selective and translocated herbicide. It is used for a wide range of submerged and floating-leaved species.
3. Bipyridinium herbicides (diquat and paraquat): water-soluble and non-selective. They act by contact to disrupt the integrity of plant cell membrane. They are rapidly absorbed by submerged aquatic plants and widely used in static waters, such as ponds and lakes.
4. Endothall: a contact herbicide usually used for submerged weeds, which acts by inactivating plant protein synthesis.
5. Fluridone: a systemic herbicide, mainly used against emergent and floating species. It inhibits carotenoid biosynthesis preventing chlorophyll synthesis in the plant. Fluridone is used mainly for submerged weeds like *Hydrilla* and *Myriophyllum*.
6. Glyphosate: a non-selective and systemic herbicide that disrupts synthesis of phenylalanine. It is widely used against species with floating and emergent tissues.
7. Triclopyr: a selective plant growth regulator used in lakes and slow-flowing waters.
8. Acrolein: a very toxic and explosive compound that destroys enzyme systems and cell membranes. Its use in aquatic systems is strongly restricted due to its high toxicity to both plants and animals, but it is still used in some countries for the control of submerged weeds in channel networks of artificial irrigated areas.
9. Copper: a plant cell toxicant mainly used as an algicide. Due to its toxicity to fish and invertebrates, many countries have restricted its use to waters without fish or where they have secondary importance in the functioning of the water system.
10. Triazines (terbutryne and simazyne): photosynthesis inhibitors that are used for control of submerged and free-floating weeds and algae in water courses, reservoirs, and fish ponds. They are quite persistent in water and hydrosoils.

Madsen (2000) has reported that the future of aquatic herbicides may include (a) the application of plant growth regulators, which reduce plant elongation rather than cause plant death, and (b) integrated control, where herbicides are used in conjunction with other management techniques to improve their effectiveness, discussed further in this chapter.

Biological Control

In general, the same basic principles of biological control can be applied for aquatic macrophytes as for terrestrial weeds. There are three alternatives for this purpose (Pieterse, 1990):

1. use of selective organisms (i.e., organisms which attack one or only a few species),

2. use of non-selective organisms (i.e., organisms which attack all, or nearly all, weed species present), and
3. use of competitive plant species (i.e., plants which compete with a weed species for one or more critical growth factors).

When using selective organisms for aquatic weeds, the strategy most extensively applied is classic biocontrol, namely, the use of one or more exotic agents that attack the weed in its native range but do not occur in the region where the plant has been introduced.

Compared with other control methods, biological control takes longer to yield results because of its slowness. For this reason, its use is generally recommended for maintenance of low plant densities than for application in emerging situations when the degree of plant colonisation is high. Sometimes it is useful to apply another control technique to rapidly decrease weed densities, such as a mechanical or chemical clearance, and subsequently introduce a natural enemy to maintain low aquatic plant biomass. The costs of biological control can be higher in the short term than chemical or mechanical approaches are, but tend to be lower in the longer term (Doeleman, 1989; Quimby and Birdsall, 1995)

Several studies have been performed on biocontrol of aquatic weeds, in a variety of lotic and lentic environments. The organisms more extensively used for this purpose are arthropods and herbivorous fish along with naturally occurring pathogens, and there are also references for the use of other animals and competitive plants.

Arthropods as Biocontrol Agents of Aquatic Weeds

The best results in aquatic weed control have been obtained using arthropods by means of the classic approach. The first big success occurred in 1964, when a chrysomelid beetle, *Agasicles hygrophila*, was released in USA to control *Alternanthera philoxeroides* (DeLoach et al., 1989; Harley and Forno, 1990; Julien et al., 1995). Following the success in controlling the weed, this chrysomelid was released in other countries such as Australia, Thailand, New Zealand and China (Harley and Forno, 1990; Napompeth, 1990; Wan and Wang, 1991). Other natural enemies have been studied and released in these countries against *A. philoxeroides*, but none of them has matched the results obtained with *A. hygrophila*.

Another successful case was the biocontrol of *Eichhornia crassipes*. Outstanding results in the control of this aggressive species have been obtained with two South American curculionids of the same genus, *Neochetina eichhorniae* and *N. bruchi*. These weevils were successfully introduced into at least 20 countries of America, Asia, Africa and Oceania (Jayanth, 1988; Harley and Forno, 1990; Cilliers, 1991a). *Neochetina bruchi* was also used in western Argentina, from where it is native, to control *E. crassipes* in the Dique Los Sauces, constituting the most successful event of biocontrol in this country (Alvarado, 1996; Cordo and Zapater, 1996). Other insects that seem to have a biocontrol potential against *E. crassipes* are the pyralid moth *Niphograpta albiguttalis* (= *Sameodes albiguttalis*), and the mite *Orthogalumna terebrantis*. There are several reports on their use alone or in combination with *Neochetina* spp. (Axelsen and Julian, 1988; Jayanth and Visalakshy, 1989; Harley, 1990).

The third most successful biocontrol event was the control of *Salvinia molesta*, a floating aquatic fern that has colonised most tropical and subtropical regions outside its native range (South America). This weed is efficiently controlled by the curculionid *Cyrtobagous salviniae*, which was first released at Lake Moondarra, Australia, in 1980 (Harley and Forno, 1990). Following the impressive success achieved, the weevil was introduced and released in several places, mainly from Australia, Asia and Africa, with excellent results (Axelsen and Julian, 1988; Cilliers, 1991b; Jayanth and Singh, 1993; Pieterse et al., 2002).

For other genera and species of aquatic plants, several natural enemies have been used or are yet under test, and some of them are promising as biological control agents. The most promising agent to be used against *Pistia stratiotes* is up to now *Neohydronomous affinis* (Coleoptera: Curculionidae), which has been demonstrated to significantly reduce *Pistia* infestations in Africa (Cilliers, 1991c; Cilliers et al., 1996) and Australia (Harley et al., 1990).

Research on the use of arthropods for biological control of *Hydrilla verticillata* and *Myriophyllum* spp. is a more recent matter. For *Hydrilla* control, the most promising agent is the curculionid weevil *Bagous affinis* (Bennett and Buckingham, 1991; Godfrey and Anderson, 1994), the pyralid moth *Parapoynx diminutalis* (Buckingham and Bennett, 1989; Napompeth, 1991), and the ephydrid fly *Hydrellia pakistanae* (Krishnaswamy and Chacko, 1990). In the case of *Myriophyllum spicatum*, good results have been reported in some places with the chironomid *Cricotopus myriophylli* (Macrae et al., 1990) and the curculionid *Euhrychiopsis lecontei* (Creed and Sheldon, 1993, 1995). For *M. aquaticum*, a plant native to South America that has gained international importance as a weed over the last several years, the first biocontrol agent used was *Lysathia* sp. (Coleoptera: Chrysomelidae). This agent, released in South Africa in 1994, successfully established and reached population densities that severely affected the growth of *M. aquaticum*; however the weed infestations recovered when beetle numbers decline, and new releases would be required to ensure continuous weed control (Cilliers et al., 1999).

Biological Control of Aquatic Weeds by Means of Pathogens

Many microorganisms have been evaluated for biocontrol of aquatic weeds since the 1970s (Joye, 1990). However, the agents most commonly considered for this purpose have been fungi, which have a high suitability in terms of pathogenic efficacy and feasibility of production, formulation and delivery (Charudattan, 1990).

The ability of plant pathogens of weeds to infect nontarget plants and even other organisms (fish, crustaceans, molluscs, etc.) is a matter of concern, and efforts must be concentrated to find organisms that provide the greatest benefit against the target weeds with the least risk to nontarget species (Quimby and Birdsall, 1995; TeBeest, 1996). This is particularly important with fungi that have broad host ranges. Techniques such as development of auxotrophs and strain selection are being exploited to decrease environmental risks (Quimby and Birdsall, 1995).

There are several fungi that have been reported as having biocontrol potential against aquatic weeds, but the most studied case is the use of *Cercospora piaropi* (= *C. rodmanii*) to control *Eichhornia crassipes*. *C. piaropi* was discovered in 1973 in Florida (Charudattan, 1990), and up to now it appears as the most promising pathogen to control this weed. The reasons are: its high specificity, absence of toxic effects to fish, high efficacy to infect *E. crassipes* mainly under conditions that favoured a reduced growth rate of the plant, and the excellent results obtained when combined with other pathogens or insects (Charudattan, 1990).

Notwithstanding the scientific knowledge that has been gained for the potential use of fungi against aquatic weeds, to date there has been no successful development of bioherbicides for a large-scale use. According to Forno and Cofrancesco (1993), bioherbicide development has been limited by technology development and, further, the industrial or commercial sector is hesitant to market a product that is restricted in its effect (i.e., attacks only one aquatic weed). This may not be a cost-effective approach due to the costly registration requirements of mycoherbicides and due to the small size of the potential market.

The effectiveness of pathogens may be affected synergistically by other pathogens, arthropods or even chemical compounds, enabling better results than when using only one technique. Novel approaches for bioherbicidal control of weeds lie in the use of a synergistic multiple-pathogen strategy, in which various host-specific fungal plant pathogens are combined and applied inundatively for the control of several weeds (Sonoda et al., 2000; Vincent and Charudattan, 2000).

Biological Control of Aquatic Weeds with Fish

There are several freshwater fish species that feed mainly on plants, mostly from families Cichlidae and Cyprinidae, and some of these have been tested as biocontrol agents of aquatic weeds. However, the only species that is extensively used for this purpose is the Chinese grass carp or white amur, *Ctenopharyngodon idella*. Several biological features of this fish give it a high utility in biocontrol programmes. It is a voracious feeder on aquatic and even terrestrial plants with a relatively low selectivity; it starts to consume vegetation at a length of 2 cm; is relatively long-lived; is tolerant to a wide range of water temperature (about 2-35 °C), pH (5-9) and oxygen content (< 1 mg l^{-1} to supersaturation); and generally does not breed naturally outside its native habitat (Opuszynski, 1967; Van der Zweerde, 1990; Winfield and Townsend, 1991; Billard, 1995). Due to these characteristics, grass carp are used worldwide as a non-native (and not-target-specific) biocontrol agent that must be released again each time the weed population erupts (Randall and Tu, 2001), though once introduced the fish can persist several years in the system even in the absence of reproduction.

White amur has been introduced in more than 25 countries of Africa, America, Asia, Europe and Oceania, mostly during the 1960s and 1970s (Wohlfarth, 1995). The risk of reproduction of this fish species outside its native range, the river systems of north-eastern China, is said to be low because suitable environmental conditions for successful spawning and larval development are not readily available (Van der Zweerde, 1990). Nevertheless, spontaneous reproduction has been documented in some riverine systems of Hungary/Yugoslavia, Japan, Mexico, Philippines, Taiwan, the USA and Russia (Wohlfarth, 1995). For this reason, many countries have restricted the introduction and release of this species and have generated considerable research into methods for biotechnological production of sterile fish (hybrids and triploids), mainly in the USA (Cassani and Caton, 1985; Billard, 1995; Hulata, 1995). The best results have been obtained with the sterile triploids, which indicates a promising future in aquatic weed control programmes (Spencer, 1994; Blackwell and Murphy, 1996). However, care must be taken when implementing a biocontrol programme of weeds with grass carp, because several factors must be considered. The vegetation type and biomass, fish size and age, stocking rate, possible fish losses by predation, adult dispersion and time of the year the fish are introduced, are among the factors that need to be examined to obtain good results (Fowler, 1978; Bain, 1989; Pine et al., 1990).

A successful degree of aquatic weed control with *C. idella* was achieved in several regions in the world. In Europe, grass carp is currently used in approximately 1% of the watercourses of the Netherlands, with satisfactory results in the great majority of cases (Murphy et al., 1990). Attempts have been made for the use of this fish as a biocontrol tool in other countries with limited success (Robinson, 1986; Robson, 1986; Moreira et al., 1998b).

C. idella has generally had a limited success in controlling the growth of aquatic weeds in most Asian countries, owing to its nonpreference on some of the more troublesome taxa present: *Eichhornia crassipes*, *Salvinia* spp, and emergent weeds (Gopal, 1990). Good results, however, have been reported locally in some countries (Wijaya et al., 1991; Katano and Gomi, 2000).

In Africa, grass carp was introduced for weed control in Egyptian irrigation systems, where more than 82% of canals and drains are infested by all types of aquatic weeds (Khattab and El-Gharably, 1986). Excellent results were obtained, not only in weed control (Bakry, 1994), but also indirectly in controlling the intermediate hosts of *Schistosoma* owing to the diminution of vegetation coverage (Khattab and El-Gharably, 1986).

The introduction of grass carp into Australia has encountered strong restrictions. In New Zealand, several studies have been carried out to evaluate its potential for weed control since 1971. Although the investigations showed successful results, final permission to allow releases in the field was withheld, as a consequence of an accidental escape suffered in 1984 (Brumley, 1991).

In North America, *C. idella* was first introduced to the U.S. waters for aquatic weed control in 1963, generating public controversy about the possible ecological consequences (Leslie et al., 1983). The later development of a sterile triploid strain of grass carp allowed its introduction and use for experimentation in places where dense infestations of aquatic weeds were present. Triploids became legal for use since 1992 (Blackwell and Murphy, 1996), and good results have been obtained in several states, mainly to control *Hydrilla* (Bain, 1989; Anderson, 1990). In the USA the fish has a promising future in weed control, as a single tool or in integrated management programmes (Spencer, 1994).

In Central and South America, reports on the use of *C. idella* for biocontrol purposes are scarce. In Argentina, grass carp has been proposed to control submerged weeds in the hydroelectric scheme of El Nihuil, Mendoza. The fish has been reported by DeLoach et al. (1989) as locally used to control aquatic weed infestations in some irrigation systems of northern and central parts of the country, and at the Parana Delta, but detailed information on this topic is not available. A programme for the use of grass carp to control aquatic weeds in the irrigation system of the lower valley of the Río Negro, in southern Argentina, started in 1993. The results of the studies with *C. idella* under natural conditions showed a high degree of biocontrol after two months of fish introduction suggesting a stocking rate lower than 100 kg ha^{-1} was enough for local conditions and for a prolonged fish presence in the area (Dall' Armellina et al., 1998, 1999).

The common carp, *Cyprinus carpio*, is another cyprinid fish that has received considerable attention, mainly as a result of the negative impacts observed on macrophytes following its introduction to new habitats outside its native range in Eastern Asia (McCrimmon, 1965; Fletcher et al., 1985). There are several reports of the adverse effects of this fish on submerged vegetation, primary as a result of its benthic feeding habit. Due to physical disturbance of the bottom sediments, this species causes a turbidity increase in the water column, affecting submerged plant survivorship, mainly in shallow water bodies (Roberts et al., 1995; Sidorkewicj et al., 1996). Another indirect effect is the uprooting of the vegetation as a consequence of plant bed destruction (Robel, 1961; Crivelli, 1983). Under certain conditions, as for example high fish density, fish stress and scarcity of preferred food items, plant consumption has also been observed (Sibbing, 1988; Sidorkewicj, 1998).

In Argentina, the possibility of using common carp for aquatic weed control in drainage channels of the irrigation complex of the Río Colorado has been evaluated. Common carp was introduced into this country at the beginning of the 20th century, and presumably ingressed accidentally into this irrigation system around the mid-1980s, now being a naturalised constituent of the fish fauna of the channels (Sidorkewicj et al., 1998; López Cazorla and Sidorkewicj, 2002). Experimental studies performed by Sidorkewicj et al. (1998) in drainage channels of the area, using 500-2000 juvenile fish ha^{-1}, have demonstrated that common carp had a significant effect in reducing the aboveground biomass of submerged weed species. A 40-100% biomass reduction with respect to the untreated control was registered after 3-4 months of fish stocking. In the same experiments, turbidity was also increased up to 60-160 NTU, depending on the fish density. These results indicated that, rather than being considered a nuisance, manipulation of common carp populations in the mentioned system could be valuable as the basis of a biological control strategy for managing channel weed problems, at a lower cost than conventional control regimes. It is remarkable that, although *C. carpio* is not actually manipulated within a biocontrol program, its presence in the drainage channels maintains a low density of submerged weeds (Fernández et al., 1998).

Other Methodologies used in Biological Control of Aquatic Weeds

Several studies have been carried out to investigate the possibilities of biocontrol of aquatic weeds with other types of non-selective organisms, as for example snails and herbivorous mammals.

Snails of the species *Marisa cornuarietis* and *Pomacea canaliculata* (Fam. Ampullariidae), which are voracious feeders on aquatic vegetation, have been considered for use in biocontrol programmes (Blackburn et al., 1971; Okuma et al., 1994).

The West Indian Manatee, *Trichechus manatus*, is a sirenid mammal that occurs naturally in coastal rivers as well as the sea, from southeastern USA to northeastern South America, and eats great amounts of floating and rooted vegetation. Early studies showed its great potential in keeping irrigation and drainage channels clear (Bertram and Bertram, 1962). More recently, it has been proposed for biocontrol of aquatic weeds in Guyana (Haigh, 1991). However, experiments carried out in Florida concluded that the use of manatees is a not practical alternative because of the great number of animals that would be necessary to control extensive weed infestations (Etheridge et al., 1985).

Other semi-aquatic mammals that have been proposed as biocontrol agents are the South American rodents *Hydrochoerus hydrochaeris* (capybara) and *Myocastor coypus* (coypu). However, their introduction into areas outside their native range is not recommended because they burrow into the banks of water bodies and may feed on crops (Pieterse, 1990).

The use of plant competition has also been indicated as a possible biological control alternative against aquatic weeds. Amongst the strategies proposed are planting shade trees along the stream banks or the restoration of native riparian vegetation to reduce light (Nichols, 1991; Bunn et al., 1998), and the addition of fertilizers to water to stimulate phytoplankton, which compete with macrophytes for one or more critical growth factors (Pieterse, 1990).

Integrated Weed Management

The results of many investigations and experience in different parts of the world have demonstrated that an integrated weed management program is the best alternative to control aquatic weeds in the long term (Fernández, 1982). Successful vegetation management is based on defined management objectives followed by careful assessment of the needs of the site (Garnett, 2002). The achievement of these programs is recommended when different strategies could improve the result obtained by a single procedure, and producing less damage to the environment.

Using a multivariate approach, Sabbatini et al. (1998) identified in irrigation systems of southern Argentina four functional aquatic plant groups, occurring in different sets of disturbance and stress conditions. They found that the most troublesome submerged weeds belonged to groups well adapted for disturbance-tolerance. Thus, we should not expect a reduction in its abundance in the long term as a consequence of weed control procedures based especially in the destruction of its above-ground biomass.

The developing of innovative channel management regimes for two irrigation systems of semi-arid Argentina (CEC, 1996), showed the advantage of integrating biological control measures based on the use of fish with other management options. Several recommendations arose, however, from the study, one of them being that the manipulation of water turbidity for weed management purposes does have quite some potential when combined with (initial) other measures, like mowing. However, some decisions have to be taken by the channel network management authorities for implementation. For example, since biological control of weeds by fish is incompatible with chemical control using acrolein (a strong aquatic biocide), an official prohibition of the chemical is necessary for the use of this herbicide in channels were fish populations are manipulated.

ACKNOWLEDGEMENTS

Authors thank two anonymous referees for helpful criticism of the manuscript.

REFERENCES

Ali M.M., Murphy K.J., Abernethy V.J. Macrophyte functional variables v. species assemblages as predictors of trophic status in flowing waters. Hydrobiologia 1999; 415:131-138.

Alvarado L. Situación actual del control biológico en Argentina y posibilidades de desarrollo de actividades conjuntas en el Cono Sur. Anais do Simposio de Controle Biológico (V SICONBIOL), Foz do Iguaçu, Brasil, 1996; pp. 384-393.

American Water Water Works Association. Current methodology for the control of algae in surface reservoirs. American Water Works Association Research Foundation, Denver, Co. USA, 1987.

Anderson L.W.J. "Aquatic weed problems and management in North America: (a) Aquatic weed problems and management in the western United States and Canada." In *Aquatic Weeds. The Ecology and Management of Nuisance Aquatic Vegetation*, A. H. Pieterse, K. J. Murphy, eds. Oxford Science Publications, 1990; pp. 371-391.

Anderson L.W.J., Steward K.K. "Aquatic weed problems and management in North America." In *Aquatic Weeds. The Ecology and Management of Nuisance Aquatic Vegetation*, A. H. Pieterse, K. J. Murphy, eds. Oxford Science Publications, Oxford, 1990; pp. 371-405.

Axelsen S., Julian C. Weed control in small dams. Part II: Control of salvinia, azolla and water hyacinth. Queensland Agric J 1988; 114: 291-298.

Bain M.B. Sterile grass carp may control hydrilla in Guntersville Reservoir. Highlights of Agricultural Research, Alabama Agricultural Experiment Station 1989; 36: 3-6.

Bakry M.F. The improvement of hydraulic efficiency of irrigation systems with particular reference to biological control method. Irri Drain Syst 1994; 8:123-133.

Barko J.W., Smart R.M. Sediment-based nutrition of submerged macrophytes. Aqua Bot 1981; 10:339-352.

Barko J.W., Gunnison S.R., Carpenter S.R. Sediment intteraction with submerged macrophyte growth and community dynamics. Aquat Bot 41:41-65.

Barrett P.R.F., Murphy K.J. The use of diquat-alginate for weed control in flowing waters. Proceedings EWRS 6th Symposium Symposium on Aquatic Weeds 1982:200-208.

Bartley T.R., Hattrup A.R. Acrolein residues in irrigation water and effects on rainbow trout. Report REC-GRC-75-8, Bureau of reclamation, Denver, Engineering and Res Center, 1975, p. 19.

Bennett C. A., Buckingham G.R. Laboratory biologies of *Bagous affinis* and *B. laevigatus* (Coleoptera: Curculionidae) attacking tubers of *Hydrilla verticillata* (Hydrocharitaceae). Ann Entomol Soc Am 1991; 84:420-428.

Bertram G. C. L., Bertram C. K. R. Manatees of Guiana. Nature 1962; (London) 196: 1329.

Bezic C. R Estrategias de supervivencia de *Potamogeton illinoensis* Morong. MSc Thesis, Universidad Nacional del Sur, Bahía Blanca, Argentina, 2001, 217 p.

Billard R. "The major carps and other cyprinids." In *Production of Aquatic Animals. Fishes,* C. E. Nash, A. J. Novotny, eds. World Animal Sciences C8, The Netherland: Elsevier, 1995, pp.21-55.

Blackburn R.D., Sutton D.L., Taylor T. M. Biological control of aquatic weeds. J Irri Drain Division, Proceedings of the American Society of Civil Engineers 97 (IR3), 1971, 421-432.

Blackwell B.G., Murphy B.R. Low-density triploid grass carp stockings for submersed vegetation control in small impoundments. J Freshwater Ecol 1996; 11:475-484.

Bohl M., Wagner H., Hoffman R. Long-term investigations with periodic applications of cooper sulphate as a herbicide in trout ponds. Proceedings of the EWRS 6th Symposium on Aquatic Weeds, 1982, 244-254.

Bowmer K.H. Management of aquatic weeds in Australian irrigation systems. Proceedings 7th Asian-Pacific Weed Science Society Conference 1979, 219-222.

Bowmer K.H., Smith G.H. Herbicides for injection into flowing water: acrolein and endothal-amine. Weed Res 1984; 25:449-459.

Brumley A. R. Cyprinids of Australasia. cyprinid fishes: systematics, biology and exploitation. fish and fisheries Series 3, In I. J. Winfield, J. S. Nelson, eds. London: Chapamn and Hall, 1991, pp. 264-283.

Buckingham G. R., Bennett C.A. Laboratory host range of *Parapoynx diminutalis* (Lepidoptera: Pyralidae), an Asian aquatic moth adventive in Florida and Panama on *Hydrilla verticillata* (Hydrocharitaceae). Environ Entomol 1989; 18:526-530.

Bunn S.E., Davies P.M., Kellaway D.M., Prosser I.P. Influence of invasive macrophytes on channel morphology and hydrology in an open tropical lowland stream, and potential control by riparian shading. Freshwater Biol 1998; 39:171-178.

Caffrey J.M. "Macrophytes as biological indicators of organic pollution in Irish rivers." In *Biological Indicators of Pollution*. D. H. S. Richardson, ed. Royal Irish Academy, Dublin, 1986, 77-87.

Carmichael W.W. A status report on planktonic cyanobacteria (Blue -green algae) and their toxins. USEPA 600/R-92/079, 1992, 141 pp.

Cassani J. R., Caton W. E. Induced triploidy in grass carp, *Ctenopharyngodon idella* Val. Aquaculture 1985; 46: 37-44.

CEC. Biological management of irrigation channel weed problems in irrigated semi-arid agriculture. Final Report to Commission of the European Communities. IHE, Delft, The Netherlands, 1996, 91 pp.

Charudattan R. "Biological control of aquatic weeds: (c) Biological control of aquatic weeds by means of fungi." In *Aquatic Weeds. The Ecology and Management of Nuisance Aquatic Vegetation.* A.H. Pieterse, K. J. Murphy, eds. Oxford Science Publications, 1990, 186-201.

Cilliers C. J. Biological control of water hyacinth, *Eichhornia crassipes* (Pontederiaceae), in South Africa. Agric Ecosyst Environ 1991*a*; 37:207-217.

Cilliers C. J. Biological control of water fern, *Salvinia molesta* (Salviniaceae), in South Africa. Agric Ecosyst Environ 1991*b*; 37:219-224.

Cilliers C.J. Biological control of water lettuce, *Pistia stratiotes* (Araceae), in South Africa. Agric Ecosyst Environ 1991*c*; 37:225-229.

Cilliers C. J., Zeller D., Strydom G. Short- and long-term control of watre lettuce (*Pistia stratiotes*) on seasonal water bodies and on a river system in the Kruger National Park, South Africa. Hydrobiologia 1996; 340:173-179.

Cilliers C. J., Olckers T., Hill M.P. Biological control of parrot's feather, *Myriophyllum aquaticum* (Vell.) Verdc. (Haloragaceae) in South Africa. Biological Control of Weeds in South Africa, African Entomology Memoir 1999; N° 1: 113-118.

Cook C.D.K. "Origin, autecology, and spread of some of the world's most troublesome aquatic weeds." In *Aquatic Weeds: The Ecology and Management of Nuisance Aquatic Vegetation,* A. H. Pieterse, K. J. Murphy, eds. Oxford Science Publications, 1990, pp.31-38.

Cordo H.A., Zapater M. C. Control biológico de malezas en la Argentina: progresos y situación actual, a dos décadas de su comienzo. Proceedings of the III Mesa Redonda de Control Biológico en el Neotrópico, Río de Janeiro, Brazil, 1996, 9-15.

Creed R.P. Jr., Sheldon S.P. The effect of feeding by a North American weevil, *Euhrychiopsis lecontei*, on eurasian watermilfoil (*Myriophyllum spicatum*). Aquatic Botany 1993; 45 (2-3): 245-256.

Creed R.P. Jr., Sheldon S. P. Weevils and watermilfoil: did a North American herbivore cause the decline of an exotic plant? Ecol Appl 1995; 5:1113-1121.

Crivelli A.J. The destruction of aquatic vegetation by carp. Hydrobiologia 1983; 106: 37-41.

Dall'Armellina A., Bezic C. R., Gajardo O. A. Propagation and mechanical control of *Potamogeton illinoensis* Morong in irrigation canals in Argentina. J Aquat Plant Manage 1996; 34:12-16.

Dall'Armellina A., Bezic C., Gajardo O. Aquatic weed control with herbivorous fish in southern Argentina. Fish density adjust. Proceedings of the 10th EWRS Symposium on Aquatic Weeds, Lisbon, Portugal, 1998, 303-306.

Dall'Armellina A., Bezic C. R., Gajardo O. A. Submerged macrophyte control with herbivorous fish in irrigation channels of semiarid Argentina. Hydrobiologia 1999; 415: 265-269.

DeLoach C., Cordo H. A., Crouzel I. S. Control Biológico de Malezas. El Ateneo, 1989, 266 pp.

Doeleman J. A. Biological control of *Salvinia molesta* in Sri Lanka: an assessment of costs and benefits. ACIAR- Technical Reports Series N° 12, 1989, 14 pp.

Etheridge K., Rathburn G. B., Powell J. A., Kochman H. I. Consumption of aquatic plants by the West Indian Manatee. J Aquat Plant Manage 1985; 23: 21-25.

Fernández O. A. Manejo integrado de malezas. Planta Dahina (Brazil) 1982; 5 (2): 69-79.

Fernández O. A., Vazquez R.J. El caso de las floraciones algales del dique Paso Piedras. La Ingeniería para la Comunidad, Colegio de Ingenieros de la Provincia de Buenos Aires, Argentina, 2001; 9: 4-7.

Fernández O. A., Irigoyen J. H., Sabbatini M. R., Brevedan R. E. Aquatic plant management in drainage canals of southern Argentina. J Aquat Plant Manage 1987; 25: 65-67.

Fernández O. A., Sutton D. L., Lallana V.H., Sabbatini M. R., Irigoyen J. H. "Aquatic weed problems and management in South and Central América." In *Aquatic Weeds. The Ecology and Management of Nuisance Aquatic Vegetation.* A.H. Pieterse, K.J. Murphy, eds.) Oxford University Press, 1990, 406-425.

Fernández O. A., Murphy K. J., López Cazorla A., Sabbatini M. R., Lazzari A., Domaniewski J. C., Irigoyen J. H. Interrelationships of fish and channel environmental conditions with aquatic macrophytes in an Argentine irrigation system. Hydrobiologia 1998, 380 (1/3): 18-25.

Fletcher A.R., Morison A.K., Hume D.J. Effects of carp, *Cyprinus carpio* L., on communities of aquatic vegetation and turbidity of waterbodies in the Lower Goulburn River Basin. Aust J Marine Freshwater Res 1985; 36:311-327.

Forno I. W., Cofrancesco A. F. New frontiers in biocontrol. J Aquat Plant Manage 1993; 31:222-224.

Fowler M. C. The biological control of water weeds. Seventh Report, Agricultural Research Council Weed Research Organization, 1978, 85-89.

Gangstad E.O., Cardarelli N.F. "The relation between aquatic weeds and public health." In *Aquatic Weeds. The Ecology and Management of Nuisance Aquatic Vegetation*, A. H. Pieterse, K. J. Murphy, eds. Oxford University Press, 1990, pp. 85-90.

Garnett R.P. Integrated aquatic vegetation management with glyphosate herbicide. Proceedings of the 11th EWRS International Symposium on Aquatic Weeds, Moliets et Maâ, France, 2002, 355-358.

Getsinger K.D. Chemical Control Technology: History and overview. Miscellaneous Paper A-91-3, US Army Engineer Waterways Experiment Station, Vicksburg, Mississippi, 1991, pp. 197-200.

Godfrey K. E., Anderson L. W. J. Feeding by *Bagous affinis* (Coleoptera: Curculionidae) inhibits germination of hydrilla tubers. Florida Entomol 1994; 77:480-488.

Gopal B. "Aquatic weed problems and management in Asia." In *Aquatic Weeds. The Ecology and Management of Nuisance Aquatic Vegetation.* A.H. Pieterse, K.J. Murphy, eds. Oxford Science Publications, 1990, 318-340.

Gopal B., Sharma K.P. Water hyacinth (*Eichhornia crassipes*): the most troublesome weed of the world. Hindosia Publishers, Delhi, India, 1981, 128.

Haigh M.D. The use of manatees for the control of aquatic weeds in Guyana. Irri Drain Syst 1991; 5:339-349.

Haller W. T., Fox A. M., Shilling D. G. Hydrilla control program in the upper St Johns River, Florida, USA. Proceedings of the 8th EWRS Symposium on Aquatic Weeds, Sweden, 1990, 111-116.

Harley K.L.S. The role of biological control in the management of water hyacinth, *Eichhornia crassipes*. Biocont News Informat 1990; 11:11-22.

Harley K.L.S., Forno I.W. "Biological control of aquatic weeds: (b) Biological control of aquatic weeds by means of arthropods." In *Aquatic Weeds. The Ecology and Management of Nuisance Aquatic Vegetation*. A. H. Pieterse, K. J. Murphy, eds. Oxford Science Publications, 1990, 177-186.

Harley K.L.S., Kassulke R.C., Sands D.PA., Day M.D. Biological control of water lettuce, *Pistia stratiotes* (Araceae) by *Neohydronomus affinis* (Coleoptera: Curculionidae). Entomophaga 1990; 35:363-374.

Hill M. P. The world's worst aquatic weed. *Pesticide Outlook* 1999; 10 (2): 58-61.

Hulata G. A review of genetic improvement of the common carp (*Cyprinus carpio* L.) and other cyprinids by crossbreeding, hybridization and selection. Aquaculture 1995; 129: 143-155.

Janauer G.A. Macrophytes and the classification of the ecological status in rivers and lakes. In S. Bäck and K. Kartunnen (eds.), Classification of ecological status of lakes and rivers. Report TemaNord 2001, Copenhagen. 584:20 - 22.

Jayanth K.P. Biological control of water hyacinth in India by release of the exotic weevil *Neochetina bruchi*. Curr Sci 1988; 57:968-970.

Jayanth K. P., Visalakshy P. N. G. Establishment of the exotic mite *Ortholagumna terebrantis* Wallwork on water hyacinth in Bangalore, Ind J Biol Cont 1989; 3:75-76.

Jayanth K.P., Singh S.P. Biological control of aquatic weeds. Indian Farm 1993; 42:25-30.

Joye G. F. Biocontrol of *Hydrilla verticillata* with the endemic fungus *Macrophomina phaseolina*. Plant Dis 1990; 74:1035-1036.

Julien M.H., Skarratt B., Maywald G.F. Potential geographical distribution of alligator weed and its biological control by *Agasicles hygrophila*. J Aquat Plant Manage 1995; 33:55-60.

Kantrud H.A. Sago pondweed (*Potamogeton pectinatus* L.): A literature review. United States Department of the Interior, Fish and Wildlife Service, Resource Publication 176. Washington D. C., 1990, 90 pp.

Katano M., Gomi K. Effect of carp and aigamo duck setting on growth, yield, and root system of paddy rice, one case study at Hakusui village of Aso district in Kumamoto prefecture. Proceedings of School of Agriculture, Kyushu-Tokai University, 2000; 19: 79-89.

Khattab A. F., El-Gharably Z. A. Management of aquatic weeds in irrigation systems with specisl reference to the problem in Egypt. Proceedings of the EWRS/AAB 7th Symposium on Aquatic Weeds, Loughborough, England, 1986, 199-206.

Killgore K. J. Field methods to measure aquatic plant treatment method efficacy. Miscellaneous Paper A-84-3, US Army Engineer Waterways Experiment Station, Vicksburg, Mississippi, 1984.10 pp.

Krishnaswamy S., Chacko M. J. *Hydrellia* spp. (Diptera: Ephydridae) attacking *Hydrilla verticillata* in south India. Entomophaga 1990; 35:211-216.

Leslie A.J., Nall L.E., Van Dyke J.M. Effects of vegetation control by grass carp on selected water quality variables in four Florida lakes. Transact Am Fish Soc 1983; 112:777-787.

Lewis G. "Aquatic Environments." In GA Pest Control Handbook, Special Bulletin 28. P. Guillebeau, ed. Cooperative Extension Service, The University of Georgia, Athens, USA, 2002, 18 pp

Lewis M.A., Wang W. "Water quality and aquatic plants." In *Plants for Environmental Studies*. W. Wang, J.W. Gorsuch, J.S. Hughes, eds. Lewis Publication, Boca Raton, New York, 1997, pp. 141-175.

López Cazorla A.C., Sidorkewicj N. S. Fish fauna from the irrigation system of the Lower Valley of the Colorado River, Argentina. Ecology, Environment and Conservation 2002; 8:1-8.

Macrae I.V., Winchester N.N., Ring R.A. Feeding activity and host preference of the milfoil midge, *Cricotopus myriophylli* Oliver (Diptera: Chironomidae). J Aquat Plant Manage 1990; 28:89-92.

Madsen J.D. Advantages and disadvantages of aquatic plant management techniques. LakeLine 2000; 20:22-34.

Marchena J.A., Sidorkewicj N.S., Sabbatini M.R., Freije R.H., Asteasuain R.O., Irigoyen J.H., Fernández O.A. Evaluación de la calidad del agua en la cuenca alta del Río Sauce Grande, Argentina. Proceedings V Encuentro de Química Analítica y Ambiental, Talca, Chile, 2000, 15.

McCrimmon H.R. Carp in Canada. Bull Fish Res Board Canada. Ottawa, 1965; 165:1-93.

Miles W.G. Land drainage and weed control. Proceedings Conference on Aquatic Herbicides, British Crop Protection Concil, Monograph 1976; 16: 7-13.

Moreira I., Catarino L., Monteiro A., Ferreira T. Gestão das infestantes aquáticas. Estudos e Documentos de Trabalho N° 4, 1998a, 60 pp.

Moreira I., Ferreira T., Monteiro A., Catarino L., Vasconcelos T. Aquatic Weeds and its management in Portugal: insights for the 10th Symposium. Proceedings of the 10th EWRS Symposium on Aquatic Weeds, Lisbon, Portugal, 1998b, 3-9.

Moss B. *Ecology of Fresh Waters. Man and Medium.* Blackwell Scientific Publications, 2nd edition, Oxford, 1998, 417 pp.

Murphy K.J. Aquatic Weeds. Encycl Environ Biol 1995; 1:71-80.

Murphy K.J., Barrett P.R.F. Chemical control of aquatic weeds. In *Aquatic Weeds. The Ecology and Management of Nuisance Aquatic Vegetation.* A.H. Pieterse, K.J. Murphy, eds. Oxford Science Publications, 1990, 136-173.

Murphy K.J., Robson T.O., Arsenovic M., van der Zweerde W. Aquatic weed problems and management in Europe. In *Aquatic Weeds. The Ecology and Management of Nuisance Aquatic Vegetation.* A. H. Pieterse, K.J. Murphy, eds. Oxford Science Publications, 1990, pp. 295-317.

Napompeth B. Country report: Thailand. Biological control of weeds in Thailand. BIOTROP- Special Publication 1990; 38: 23-36.

Napompeth B. Aquatic weed management by biological methods in Thailand. BIOTROP- Special Publication 1991; 40: 51-61.

Nichols S.A. The interaction between biology and the management of aquatic macrophytes. Aquat Bot 1991; 41:225-252.

Nordone A.J., Matherly R., Bonnivier B., Doane R., Caravello H., Paakonen S., Parent R. The mobility and degradation of acrolein in agricultural canals treated with Magnacide H herbicide. Chemosphere 1996; 32:807-814.

Okuma M., Tanaka K., Sudo S. Weed control method using apple snail (*Pomacea canaliculata*) in paddy fields. Weed Res (Tokyo) 1994; 39:114-119.

Olaleye V. F., Akintunde E. A., Akinyemiju O. A. Effect of a herbicidal control of water hyacinth (*Eichhornia crassipes* Mart.) on fish composition and abundance in the Kofawei Creek, Ondo State, Nigeria J Environ Manage 1993; 38:85-97.

Opuszynski K. Comparison of temperature and oxygen tolerance in grass carp (*Ctenopharyngodon idella* Val.), silver carp (*Hypophthalmichthys molitrix* Val.), and mirror carp (*Cyprinus carpio* L.). Ekologia Polska A 1967; 17:385-400.

Petta J. F. Chemical toxicity and environmental fate: an evaluation of aquatic pesticide use and risk assessment. Proceedings of the 11th EWRS International Symposium on Aquatic Weeds, Moliets et Maâ, France, 2002, 379-380.

Pieterse A. H. "Biological control of aquatic weeds: (a) Introduction to biological control of aquatic weeds." In *Aquatic Weeds. The Ecology and Management of Nuisance Aquatic Vegetation.* A.H. Pieterse, K.J. Murphy, eds. Oxford Science Publications, 1990, pp. 174-177.

Pieterse A. H. The dramatic advance of water hyacint (*Eichhornia crassipes*) in Africa in recent decades and options for its management: is further spread inevitable? Proceedings of the 10th EWRS International Symosium on Aquatic Weeds, Lisbon, Portugal, 1998, 279-282.

Pieterse A. H., Kloff S., Dieme C., Diouf S., Ndao I., Sarr K., Diawara Y., Kettunen M., Tarvainen A., Janauer G., Exler N., Hellsten S. Effective biological control of *Salvinia molesta* in the Senegal River by means of the weevil *Cyrtobagous salviniae*. Proceedings of the 11th EWRS International Symposium on Aquatic Weeds, Moliets et Maâ, France, 2002, 381-384.

Pine R.T., Anderson L. W.J., Hung S.S.O. Control of aquatic plants in static and flowing water by yearling triploid grass carp. J Aquat Plant Manage 1990; 28:36-40.

Quimby P. C. Jr., Birdsall J. L. 1995. "Fungal agents for biological control of weeds: classical and augmentative approaches." In *Novel Approaches to Integrated Pest Management.* R. Reuveni, ed. Lewis Publishers, Boca Raton, Florida, 1995, 293-308.

Randall J. M., Tu M. "Biological control." In *Weed Control Methods Handbook: Tools and Techniques for Use in Natural Areas*, M. Tu, C. Hurd, J. M. Randall, eds. Wildland Invasive Species Program, Nature Conserv 2001; pp. 4.1-4.24.

Robel R.J. The effect of carp populations on the production of waterfowl food plants on a western waterfowl marsh. Transactions of the Twenty-sixth North American Wildlife Conference 1961;26:147-159.

Roberts J., Chick A., Oswald L., Thompson P. Effect of carp, *Cyprinus carpio* L., an exotic benthivorous fish, on aquatic plants and water quality in experimental ponds. Marine Freshwater Res 1995; 46:1171-1180.

Robinson G.W. Weed control in land drainage channels in England and Wales with reference to conservation. Proceedings of the EWRS/AAB 7th Symposium on Aquatic Weeds, Loughborough, England, 1986, 257-262.

Robson T.O. Aquatic plant management problems in Europe. Proceedings of the EWRS/AAB 7th Symposium on Aquatic Weeds, Loughborough, England, 1986, 263-269.

Sabbatini M.R., Murphy K.J. Submerged plant survival strategies in relation to management and environmental pressures in drainage channel habitats. Hydrobiologia 1996; 340: 191-195.

Sabbatini M.R., Irigoyen J.H., Fernández O.A. Phenology and biomass dynamics of *Chara contraria* A. Braun ex Kütz. in drainage channels of a temperate irrigation area in Argentina. Proceedings of the EWRS/AAB 7th Symposium International on Aquatic Weeds, Loughborough, England, 1986, 285-289.

Sabbatini M.R., Murphy K.J., Irigoyen J.H. Vegetation-environmental relationships in irrigation channel systems of southern Argentina. Aquat Bot 1998; 62:119-133.

Sabbatini M.R., Sidorkewicj N.S., Murphy K.J., Marchena J.A., Irigoyen J.H., Fernández O.A. Macrophytes, bioseston and periphyton as water quality indicators in the Sauce Grande river catchment, Argentina. Proceedings of the 11th EWRS International Symposium on Aquatic Weeds, Moliets et Maâ, France, 2002, pp. 251-254.

Sculthorpe C.D. *The Biology of Aquatic Vascular Plants*. Edward Arnold, London, 1967, 610 pp.

Sibbing F.A. Specializations and limitations in the utilization of food resources by the carp, *Cyprinus carpio*: a study of oral processing. Environ Biol Fishes 1988; 22:161-178.

Sidorkewicj N. S. Efecto de *Cyprinus carpio* L. (Pisces: Cyprinidae) sobre la vegetación sumergida y la calidad del ambiente acuático. PhD Thesis, Universidad Nacional del Sur, Bahía Blanca, Argentina, 1998, 268 pp.

Sidorkewicj N. S., López Cazorla A. C., Fernández O. A. The interaction between *Cyprinus carpio* L. and *Potamogeton pectinatus* L. under aquarium conditions. Hydrobiologia 1996; 340: 271-275.

Sidorkewicj N. S., López Cazorla A. C., Murphy K. J., Sabbatini M. R., Fernández O. A., Domaniewski J. C. J. Interaction of common carp with aquatic weeds in Argentine drainage channels. J Aquatic Plant Manage 1998; 36: 5-15.

Sonoda R. M., Singh M., Charudattan R., Chandramohan S. Multiple-pathogen strategy: a novel approach for bioherbicidal control of several weeds. Proceedings of the Third International Weed Science Congress, Foz do Iguaçu, Brazil, 2000, 182-183.

Spencer D. F. Estimating the impact of triploid grass carp on sago pondweed in the Byrnes Canal: implications for biological control in northern California irrigation systems. Ecol Model 1994; 72:187-204.

Spencer W., Bowes G. "Ecophysiology of the world's most troublesome aquatic weeds." In *Aquatic Weeds. The Ecology and Management of Nuisance Aquatic Vegetation*, A. H. Pieterse, K. J. Murphy, eds. Oxford Science Publications, 1990, 39-73.

TeBeest D.O. Biological control of weeds with plant pathogens and microbial pesticides. Adv Agron 1996; 56:115-137.

Tooby T.E., Spencer-Jones D.H. The fate of the aquatic herbicide dichlobenil in hydrosoil, water and roach (*Rutilus rutilus* L.) following treatment of areas of a lake. Proceedings of the EWRS 5th Symposium on Aquatic Weeds, Amsterdam, The Netherlands,1978, 323-331.

Tooby T.E., Durbin F.J., Rycroft R.J. Accumulation and elimination of residues of the aquatic herbicide dichlobenil in two species of British freshwater fish. Proceedings of the EWRS 4th Symposium on Aquatic Weeds, 1974, 323-331.

Van der Zweerde W. "Biological control of aquatic weeds: (d) Biological control of aquatic weeds by means of phytophagous fish." In *Aquatic Weeds. The Ecology and Management of Nuisance Aquatic Vegetation*, A. H. Pieterse, K. J. Murphy, eds. Oxford Science Publications, 1990, 201-221.

Vincent A. C., Charudattan R. Evaluation of a combination of two pathogens as a potential bioherbicide for *Eichhornia crassipes* [Mart.] Solms-Laub. under field conditions. Proceedings of the Third International Weed Science Congress, Foz do Iguaçu, Brazil, 2000, pp. 217-218.

Wade P. M. 1990. "General biology and ecology of aquatic weeds." In *Aquatic Weeds. The Ecology and Management of Nuisance Aquatic Vegetation*, A. H. Pieterse, K. J. Murphy, eds. Oxford, Science Publications, 1990, 17-30.

Wan F.H., Wang R. Achievements of biological weed control in the world and its prospects in China. Chinese J Biol Control 1991; 7:81-87.

Welsh E.B., Lindell T. *Ecological effects of wastewater*. Ellis Horwood, New York, 1992, 425 pp.

Wijaya F., Suwignyo P., Taufan M. Practical implementation of aquatic weed management in Situ Leutik, Darmaga campus, Bogor, Indonesia. BIOTROP- Special Publication 1991; N° 40: 117-132.

Winfield I.J., Townsend C.R. "The role of cyprinids in ecosystems." In *Cyprinid Fishes: Systematics, Biology and Exploitation. Fish and Fisheries,* I.J. Winfield, J. S. Nelson, eds. Series 3, Chapman and Hall, London, 1991, pp. 552-571.

Wohlfarth, G.W. The common carp and Chinese carps. In *Conservation of Fish and Shellfish Resources: Managing Diversity*. J. E. Thorpe, G. A. E. Gall, J. E. Lanan, C. E. Nash, eds. Academic Press Ltd., 1995, 138-176.

Yeo R.R. Life history of sago pondweed. Weeds 1965; 13:314-321.

CHAPTER 6

AGROECOLOGICAL BENEFITS FROM WEEDS

NICHOLAS JORDAN & CHRISTINE VATOVEC

Agronomy and Plant Genetics Department and Minnesota Center for Community Genetics, University of Minnesota-Twin Cities, MN 55108, USA

We begin by stating a basic premise: that weed science must be fundamentally concerned, like all of agricultural science, with the imperative of *sustainability*. This concern underlies the mounting questions about the profitability, long-term efficacy, and environmental safety of the discipline's historical focus: herbicidal management of weeds (i.e., "plants out of place"). Herbicide-based approaches are currently relied upon in managed ecosystems of many sorts: e.g., farms, forests, and waterways (Cousens and Mortimer, 1995; Sheley et al., 1997). Our own assessment is that economically, ecologically, and agronomically sustainable weed management systems cannot be attained by current concepts and means of weed management. In addition to concerns for agricultural sustainability, there is rapidly-growing recognition that biological invasions by plants are a troublesome component of *global change* (i.e., large-scale detrimental changes in ecological systems that provide "life-support" functions to humanity; Daily, 1997; Mack et al., 2000). Weed scientists are actively considering how they can contribute to the scientific and social response to this problem. The challenge is to expand the scope of weed science from current emphases on herbicides and field-crop agroecosystems to the larger work of coping with invasive plants in ecosystems of all kinds, including areas that are managed to preserve their "natural" qualities. Below, we argue that progress towards more expansive and sustainable notions of weed management will require new theories of weeds and weed management that are based on a broader view of the agroecological roles of weeds.

A critical step in such a development of weed management is a conceptual shift to explicit *acceptance* of the presence of weed communities in managed ecosystems. When practical, eradication of invasives is a wise strategy during initial phases of invasion episodes (Simberloff, 2001). However, eradication of many or most invasive plants is practically unattainable in most managed ecosystems. Also, ecosystems in which multiple invasive organisms occur commonly experience problematic secondary effects as a result of eradication efforts focused on individual invasive species, such as compensatory increases in abundance of other invasives (Zavaleta et al., 2001). These realities make plain the limitations of current weed-management strategies, which aim implicitly at eradication of particular problem species (Liebman et al., 2001).

Inderjit (ed.), Weed Biology and Management, 137-158.

Moreover, we believe that at least some weeds should be regarded with appreciation, rather than seen only as an inevitable problem of managed ecosystems. Broadly, there is rapidly growing awareness (Swift and Anderson, 1993; CAST, 1999) that biodiversity has vitally important effects on productivity, resource efficiency, resilience and other properties of agroecosystems and other managed ecosystems. A growing body of evidence indicates that weeds can provide a variety of beneficial ecological effects. Therefore, we suggest that at least some weeds should be seen as part of the *beneficial* or *protective* biota of managed ecosystems. The beneficial biota is that fraction of the total biodiversity of an ecosystem that provides some ecological service (Swift and Anderson, 1993). The protective biota (Pankow et al., 1991) is that fraction of biodiversity that helps protect the integrity of valued ecosystem properties, e.g., nutrient cycling (Woodmansee, 1984).

We propose that weed management should aim to steer the dynamics of the weed flora so as to *reduce harm* and *increase benefits* resulting from this flora. We define such management as part of the *ecological restoration* of the plant community of a managed ecosystem - i.e., efforts to increase the level of ecological services and amenities that are sustainably produced by an ecosystem. This *restoration* concept of weed management builds on recently-proposed integrative concepts of weed management. For example, Swanton et al. (1993) proposed that weed management should be approached as the management of ecological succession. Liebman and Dyck (1993) and Liebman et al. (2001) have proposed that higher weed community diversity is an important management objective. We propose a simple and general goal of weed management: *amelioration* of weed communities, by management that aims to promote the abundance of weed species that provide ecological services. For weeds, this goal can be pursued by two strategies. First, management actions can be used to direct the dynamics of the weed flora so as to increase the frequency of more beneficial genotypes within species, and more beneficial species within the weed flora as a whole. Second, management can be used to manipulate the factors - e.g., applications of resources, organisms, disturbances, and different stresses - that structure the agroecosystem, so as to increase benefits and reduce harmful effects of a given weed flora. The most obvious realization of this latter strategy is management action that aims to establish weed populations in spatial and temporal "patches" in which the weeds provide a significant net benefit, while aiming to minimize weed abundance beyond such patches.

There is additional justification for assessing the agroecological roles of weeds: in a broad range of cropping systems, farmers view weeds as an integral part of a functional agroecosystem, not just as a problem. For example, a survey of farmers in Minnesota (Vatovec, unpublished) revealed that these farmers were aware of a variety of beneficial uses of weeds including their use as self-sowing cover and green manure “crops”, trap “crops” for insects, soil quality indicators, competitors of other more harmful weeds, forage for livestock, and for direct human consumption. Similarly, a case study in southeastern Mexico (Chacon and Gliessman, 1982) reported farmers leaving certain weeds in crop fields based on a classification system that regarded non-crop plants as either “mal monte” (bad weeds) or “buen monte” (good weeds) depending on their direct human use potential as food, medicine, ceremonial materials and teas, and their effects on crops and soil. Interviews with these farmers revealed that 81% believed “buen monte” beneficially affected soil temperatures; 40% thought that these weeds contributed nitrogen to crops and improved soil tilth. We believe that the time has come to carefully test farmer perceptions of ecological benefits of weeds.

In what follows, we examine relevant evidence and propose a research agenda to test the proposition that weeds can be a useful component of the beneficial biota of field-crop agroecosystems. To begin, we survey current evidence and claims regarding ecological benefits provided by weeds; we conclude that there is a broad range of suggestive evidence, but in most instances definitive evidence is lacking. Next, we present an integrative hypothesis concerning the potential agroecological roles of weeds, namely, that weeds are necessary elements of *ensembles of plant diversity* that work together with organisms at other trophic levels to produce a range of agroecological services. According to this hypothesis, weeds serve, via various ecological interactions, to maintain populations or communities of beneficial organisms at levels that suffice to provide ongoing ecological services. Therefore, we argue that weed management should be conceptualized as part of a broader program of spatial and temporal deployment of plant diversity in order to effectively and efficiently realize a broad range of ecological services from the total biodiversity of agroecosystems. Finally, we propose a set of provisional guidelines for fostering the *ecological restoration* of weed communities in particular. Throughout, we aim to integrate a wide range of ecological evidence and theory to suggest what agroecological roles of weeds might be *possible*. We believe such "forecasting" is necessary to help agriculturalists envision what might be possible in the more biologically-rich agroecosystems that must be developed in the near future (Jordan, 2002; Uphoff, 2002).

ECOLOGICAL BENEFITS PROVIDED BY WEEDS

There are a variety of mechanisms by which weeds might provide ecological benefits in cropping systems. Only a few critical studies have been done on these mechanisms; we review evidence from this work and highlight important knowledge gaps. We organize our discussion in four sections: weed effects on soil quality and fertility, weed effects on herbivorous arthropods and their natural enemies, weeds functioning as "cover crops", and a section of miscellaneous claimed benefits.

Weed Effects on Soil Quality and Fertility

In some circumstances, weeds can reduce soil erosion, slow nutrient loss, increase soil organic matter, improve nitrogen levels and conserve soil water. Weeds that have proportionally deeper rooting systems relative to co-occurring crops may be capable of significant net transfer of nutrients from deep soil layers not available to such crops (Cocannouer, 1950; quoted by Patriquin, 1986). Such weeds may also have a valuable ability to reduce soil erosion, as seen in species such as *Digitaria* spp. (crabgrasses) that have very extensive root systems (Holzner, 1982). There are no critical studies that quantify the effects of such nutrient transfer mechanisms, but increases in soil-surface calcium and magnesium levels were noted in association with increased abundance of certain deep-rooted weed species (Patriquin et al., 1986). In a related study of the same farm, Patriquin (1986) proposed that weeds made an important contribution to internal recycling of nitrogen. In the nitrogen budget estimated for the farm, weeds (mainly growing during fallow periods) were found to protect between 10 and 20 kg N/ha per year from loss via leaching. This farm practiced small grain and poultry production, and had minimal external N inputs. Therefore, the farm depended on efficient

N cycling because of minimal external inputs, and it was estimated that the prevention of N leaching provided by weeds provided N sufficient for ca. 1 ton/ha of grain output. Grain yields on the farm averaged 25% lower than local conventional farms, but profits were higher because of decreased input costs. Therefore, weeds played an important role in increasing resource efficiency via improving nitrogen recycling.

Of course, most studies of weed-crop interactions have shown negative effects on crops (Radosevitch et al., 1997). However, substantial crop benefits from weed association have been shown in a few cases. For example, Kapoor and Ramakrishnan (1975) reported that a leguminous weed enhanced growth of associated wheat. Dry weight yield per wheat plant increased from 8g when grown in a clean field, to 25g when grown along with a density of 800 weed (*Trigonella polycerata*), trigonella plants/m^2 (wheat plants grown at 20/ m^2). Likewise in jhum (shifting cultivation) cropping systems in northeast India, weeds are regarded as nutrient sources. In this situation, a variety of weeds are early colonizers after site preparation, and likely act as nutrient sinks, rapidly assimilating nutrients made available after the slash and burn process, and preventing nutrient losses from the cropping system. Swamy and Ramakrishnan (1988) reported that in this system, traditional weed-tolerant practices have little negative on crop yield, and can conserve up to 20% of soil resources. For example, in plots following a 20 year jhum cycle, a loss of nitrate N (NO_3-N) in run-off water from traditionally weeded fields was 4.3 kg/ha/yr, compared to 5.3 kg/ha/yr in weed-free plots, and 2.8 kg/ha/yr in non-weeded plots. Similar trends were also observed in P, K, Ca and Mg. Furthermore, water losses in the traditional weeding system were significantly lower than under weed-free regimes. The jhum farmer leaves ca. 20% of weed biomass standing in fields to act as nutrient stores; other weeds are uprooted and applied as a rapidly decomposing mulch (Ramakrishnan, 1992). Finally, some non-leguminous weeds may have potential to improve nitrogen fertility via nitrogen fixation by rhizosphere bacteria (Patriquin, 1986); possibly, such effects could be substantially enhanced in soils managed to encourage beneficial interactions between weeds and soil biota.

Weeds can also play an important ecological role via their associations with beneficial soil organisms. Colonization of bare soil by plants is usually accompanied by an increase in soil microbial biomass (Wardle, 1992). Hence, weed density may be an important factor affecting abundance of soil biota. In addition, weeds increase plant community diversity, and may remedy the low plant species diversity typical of high-input cropping systems that is thought to generally increase incidence of harmful soil pests and diseases (De Ruiter et al., 1995; Brussaard, 1997). Wardle's (1995) assessment of studies involving the effects of herbicides on soil organisms supports this view of weeds as sponsors of soil biological activity. For example, herbicides negatively impacted total soil microflora in 45% of studies of where weeds were present, as compared to negative effects in only 20% of studies were weeds were absent, thus revealing more direct effects of herbicdes on microflora. In particular, 35% of studies showed negative herbicidal effects on fungal biomass when weeds were present, whereas no negative herbicide effects were observed in absence of weeds.

Abundance of soil arthropods also increased strongly in response to weed abundance (Wardle, 1995). For example, in a study comparing insects in living mulch, weedy, and crop-only plots, significantly more ground predators were found in the weed and mulch plots than in the clean cultivated plots, and specialized herbivore densities were significantly reduced in weed and mulch plots (Altieri et al., 1985). Ants are thought to have beneficial effects on soil quality, via water and soil movement; and ant density has been shown to increase with weed density (Lobry de Bruyn, 1999). In this study, approximately 70% of ants caught in pitfall traps over a ten-week period were in weedy plots, 12% were in clover living-mulch plots, and 18% in crop-only plots. In a three-year study of the effects of over-wintering habitats for predators of cereal aphids, the number of spiders increased from $32.3/m^2$ to $170.6/m^2$ in habitats of non-controlled weeds (Thomas et al., 1992). Cardwell et al. (1994) found that the total number of carabid beetles in uncropped and conservation headlands was significantly higher than in fully sprayed crop fields and headlands. Similarly, carabid abundance was significantly higher in crop fields adjacent to headlands than in those adjacent to conservation and fully sprayed headlands combined. These studies suggest that weedy areas can act as refuges for some predatory groups, providing a source for colonization of adjacent cropped areas (Altieri and Letourneau, 1984; Wardle, 1995).

There is an important special case of the relationship between weeds and beneficial soil organisms: certain weed species have been found to form a symbiosis with arbuscular mycorrhizal fungi (AMF) and thus are hosts to these beneficial organisms (Jordan et al., 2000a). These fungi have many beneficial effects in agroecosystems (Bethlenfalvay, 1992; Smith and Read, 1997; Douds and Millner, 1999; Dodd, 2000). These include effects on crops such as improved water relations, pest and disease resistance, tolerance to root pathogens, tolerance to drought and adverse environmental conditions, and enhanced nutrient uptake. AMF act antagonistically against non-host weed species of a variety of families (Jordan et al., 2000a) and thus may be capable of broad-spectrum weed biocontrol. More broadly, mycorrhizal networks may be important to soil food webs that support soil biota including macroinvertebrates (Rabatin and Stinner, 1989). While critical studies of effects of weed diversity and abundance *per se* on AMF are few in number, it is clear that reduced plant diversity reduces the density and diversity of AMF (Rabatin and Stinner, 1989). For example, twenty AMF species were found in prairie as compared to only four to six AMF species in winter-wheat fields (Daniels et al., 1983). Therefore, reduced diversity and abundance of weed floras, particularly of host species, as is typical of low-diversity cropping systems (Liebman and Dyck, 1993) may reduce diversity and abundance of AMF and concomitant loss of ecological benefits (Rabatin and Stinner, 1989).

There is some suggestive evidence that certain weed-AMF relationships may be important in within-season interspecific nutrient redistribution processes. These processes are potentially of major importance in cropping systems in which nutrient-use efficiency is a critical attribute, for economic and or environmental reasons. AMF can redistribute nutrients from dying or senescing plant to others that are at an earlier stage in growth (Smith, 1997). In the particular case of weeds, several studies have shown that small quantities of N can be mobilized from senescing weeds to crop plants via AMF connections, raising the possibility that these weeds could function as short-term nutrient sinks (Bethlenfalvay et al., 1992 a, b) that are accessible to crops later in growth. Weed senescence might be induced by management (e.g., by sublethal herbicide doses; Bethlenfalvay et al., 1996 a, b; Rejon et al., 1997), or naturally, as in the case of winter-annual weeds.

To date, only modest levels of nutrient transfer have been observed. However, in theory, resource transfer from a senescing host plant to another via AMF will occur only when transfer produces benefits to the AMF (Wilkinson, 1998; Fitter et al., 1998). Conversely, when nutrient retention rather than "investment" in symbiotic plants will increase the benefit obtained from that nutrient by the fungus, then minimal nutrient transfer is expected. Such retention might be favored in situations where the plant-

AMF symbiosis might be disrupted, e.g, in situations of intensive tillage, or when modern crop cultivars that are poor hosts to AMF are present. There is evidence, from studies of N transfer from a legume to a non-leguminous crop, that certain AMF taxa are substantially more effective in resource redistribution than others (Martensson et al., 1998). Appropriate management practices might serve to increase the frequency and resource-transfer efficiency of such AMF taxa (Perry, 1998), increasing within-season nutrient transfer from crops to weeds and thus improving nutrient efficiency in crop production.

Resource transfer is likely to be particularly significant when there is sufficient spatial heterogeneity of soil resources to cause growth limitations due to local deficits of resources. In clonal plants, these limiting effects can be alleviated by within-clone resource transfer. In these cases, when different portions of a physiologically-integrated clonal plant are subject to limiting deficits of different resources, reciprocal transfer of "surplus" resources between these portions can enable abundant production across the entire clone. Very large (e.g., 500%) increases in total growth can result. Obviously, clonal plants are strongly selected for such resource transfers, perhaps much more so than in the case of plant-AMF systems, where conflicts of interest between AMF and plants can limit nutrient transfer, as discussed above (Wilkinson, 1998). However, these results suggest that large productivity gains could result from resource transfer. This potential mechanism is of particular importance given that reduced-tillage, reduced-input agroecosystems may develop considerable patchiness in the distribution of soil resources (Patriquin, 1986; Bever et al., 1997). Conceivably, large gains in resource-efficiency in addition to productivity could be gained by these sorts of AMF-mediated resource transfers. Weeds may figure in these transfers, if certain weed species function as self-sowing intercrops that provide some ecological benefit as companion crops to a main crop.

Taken together, the above considerations provide substantial support for Altieri's (1991) assertion that "systems with higher levels of weed biomass are conducive to supporting a rich and diverse soil biota..." However, more quantification of the ability of weeds to provide soil quality and fertility benefits is needed. Specifically, evaluations are needed of the quantitative capacity of weeds to perform nutrient capture and scavenging, to provide physical protection of soil from erosion and other processes that degrade soil quality, and to sponsor ecological processes in soil food webs that improve efficiency of nutrient cycling and maintain soil tilth. These are urgently needed to extend and test the current narrow base of evidence regarding the importance of weeds to soil quality and fertility.

Weed Effects on Herbivorous Arthropod Pests and their Natural Enemies

Weeds may have a number of direct and indirect effects on herbivorous arthropod pests. The preponderance of evidence suggests that low-diversity cropping systems promote instability among populations of herbivorous arthropods and their natural enemies (Altieri, 1999). In these cropping systems, lack of alternative food sources, shelter, hibernation sites, and breeding and nesting areas negatively affect native predators (Altieri and Whitcomb, 1979). Diversification of cropping systems often results in reduced populations of herbivorous arthropods (Altieri, 1999), as the result of "ecological control": a complex multi-trophic interaction that is an emergent property of the biodiversity of an agroecosystem as a whole. Ecological control arises when the structure and function of the agroecosystem causes sufficient densities of pest predators and parasitoids to coincide in time and space with their prey (Barbosa and Benrey, 1998). In particular, vegetation that creates habitat for a diverse and abundant complex of invertebrate predators can be highly effective at regulating populations of herbivorous arthropods (Barbosa and Wratten, 1998). Weeds can contribute to ecological control by effects on the spatial and temporal distribution of plant diversity and abundance in the agroecosystem. Weeds can act by providing a 'sink' that attracts pests away from crops, or by serving as a 'source' that increases abundance of natural enemies, or by acting to visually or olfactorally confuse pests by increasing diversity of habitat within fields, or by modifying microclimate variables so that these are less favorable to herbivorous pests (Bugg, 1992). From this point of view, weeds and herbivorous pests are seen as necessary and interactive elements of ecological control mechanisms, problematic only when sufficient population densities coincide with vulnerable phases of crop development (Bugg, 1992).

Of course, such beneficial effects must be balanced against harm caused by weeds by other mechanisms. A thorough review of interactions between weeds, arthropod pests and pest natural enemies (Norris and Kogan, 2000) recognized that weeds can contribute to ecological control, but pointed out that weeds can compromise ecological control in some circumstances by providing alternative food sources or refuges for pests (Jervis et al., 1993). However, most studies of weed effects on arthropod pests have *not* addressed agroecosystems where the spatial and temporal distribution of weeds was managed to provide beneficial effects on ecological control while minimizing interference with crop growth and management (Bugg, 1992). Such agroecosystems are the most relevant systems in which to assess the role of weeds in ecological control; using weeds for this purpose does not imply a simplistic notion of generally lax control of weeds. Rather, weed management for ecological benefit should have a spatial dimension: weeds should be managed within a landscape to provide weed species of choice in the place where they are desired for specific beneficial attributes. As well, a temporal dimension is necessary: weeds should be managed to produce desirable attributes at a specific time.

For example, management of common knotweed (*Polygonum aviculare* L.) in California, USA illustrates a case where farmers aim to manage the weed's spatial and temporal distribution to good advantage. In this situation, the weed served as a *de facto* cover crop in permanent-bed vegetable production (Bugg et al.,1987). Common knotweed flowers provided nectar to generalist predators including bigeye bugs (*Geocoris* spp.), and also attracted an abundant supply of supplementary foods such as thrips, spider mites, a psyllid, and an aphid. In the fall as the weed senesced, adjacent crops such as radishes apparently benefited from the dispersal of bigeye bugs. Bigeye bug densities in radish flats adjacent to common knotweed (mean density of 0.33 bugs per plant in the first period, 1.11 during the second period) were significantly higher than flats adjacent to other weed species (0.06 bugs per plant during both periods; Bugg et al., 1987). This case also reveals the complex roles of weeds in such pest management schemes: common knotweed was suspected of sometimes providing an overabundance of food, thereby attracting beneficial predators away from crops during the growing season.

At this point, the critical issue regarding weed interactions with herbivorous arthropods and their natural enemies is whether practically significant beneficial effects can be obtained from weed patches that sufficiently localized in space and time to minimize harmful effects of weeds, caused by via direct and indirect interference with crop production, and otherwise. Critical studies of beneficial effects in such situations are very few; these are needed in a variety of cropping systems in situations where weed populations are being deliberately managed for their ecological benefits.

Weeds as Self-Sowing "Cover Crops"

Certain weed species might be appropriately seen as 'self-sowing' cover crops. Beneficial ecological effects of cover crops are well-documented (Lal et al., 1991; Bugg, 1992) and include conservation of soil, water and organic matter (Ditsch and Alley, 1991) and pest-management benefits. Many weeds could potentially play similar functional roles. For example, in a conservation tillage cropping system in the USA (Thompson, 1991), certain early-emerging weeds were thought to provide a strongly suppressive effect on later-emerging problematic weed species; in the farmer's view, these "self-sowing cover crops" substantially augmented the weed-control effect of a deliberately sown cover crop. In this situation, early-emerging species, in combination with the sown cover crop, were thought to substantially reduce the abundance of later-emerging species that were difficult to control in that cropping system. Thus, these early-emerging weed species were appreciated as valuable components of overall integrated weed management. Moreover, the self-perpetuating qualities of weeds might alleviate some of the difficulties with timely establishment that often complicate use of cover crops. Weeds *are* weeds because they are able to grow and proliferate in soils with varying fertility, and they are able to survive in very harsh environmental conditions. Furthermore, weeds can decompose very rapidly as a result of low C:N ratios, thereby providing an important form of available organic matter (Wardle, 1995) and N. Therefore, certain weeds might be valuable components of cover-cropping, especially if the aim is a multi-species cover-crop "polyculture" that can provide multiple ecological benefits.

Miscellaneous Benefits from Weeds

A considerable miscellany of other benefits have been reported by various observers, although many of these have not received any critical evaluations (Patriquin, 1986). These claims include beneficial effects on storage and edible-use quality of both root and fruit crops, reduction of allelopathic effects, provision of a "biological tillage" effect and other forms of "ecological engineering" of soils such as reducing drought-susceptibility of shallow-rooted crops, serving as effective "companion crops" (i.e., providing a positive neighbor effect of some sort) to certain crop species, and providing a beneficial phase (a "weedy fallow" year) in diversified crop rotation schemes.

Many weed species that are close relatives of crops have been used in crop breeding programs, e.g., as sources of resistance to diseases and environmental stress factors (Holzner, 1982). Weeds can also serve as useful bioindicator tools, medicinal herbs, and forage or food sources (Holzner, 1982), particularly in traditional cropping systems of developing countries. For example, weeds serve as a food source for indigenous peoples of the Mexican Sierras and other cultures during the periods before crop plants mature, and during years when crops are destroyed by hail storms (Bye, 1981).

Finally, when wild or feral relatives of crops are present as weeds, e.g., in traditional cropping systems of developing countries (Altieri, 1988) hybridization between weeds and crops often occurs (Ellstrand et al., 1999). Possibly, this weed-to-crop gene flow can benefit the crop via introducing weed traits that improve crop adaptation to local conditions, although no careful examination of this possibility has yet been made. For example, in Mexico farmers reportedly encourage teosinte-corn hybrids as part of in-situ seed production processes (Altieri, 1995).

SUSTAINING COMPLEX SYMBIOSES: A CONCEPTUAL FRAMEWORK FOR ANALYSIS OF ECOLOGICAL BENEFITS FROM WEEDS

Under certain circumstances weeds can provide ecological benefits by a variety of pathways. Here we wish to focus attention on a pathway that appears to deserve greater emphasis and analysis. Essentially, we propose that weeds can complement the planned diversity and spatial-temporal extent of plant communities of agroecosystem plant communities (e.g.., "cash" and "cover" crops), such that substantial increases occur in ecological services provided by these communities. Our broad hypothesis is that certain weeds exert strong qualitative effects on interactions between agroecosystem plant communities and beneficial biota at other tropic levels. Beneficial organisms whose diversity, abundance and provision of beneficial effects are plausibly affected by interactions with weeds (Altieri, 1995) include (at a minimum) soil biota (Boucher, 1990) arthropods (Bugg, 1992), and beneficial microorganisms on leaf surfaces (Kinkel, 1997). To illustrate and develop this hypothesis, we will focus on an important particular case: the role of weeds in interactions between plant communities and beneficial soil biota.

Close and mutually-beneficial interactions occur between plant communities and a range of soil biota, including mycorrhizal and other endosymbiotic fungi, and a wide variety of beneficial bacteria including rhizobial, P-solubilizing, metal-chelating, pathogen-antagonistic and other types (Boucher, 1990). Increasingly, such ecological relationships are recognized as important in plant-community dynamics and ecosystem functioning in a wide range of ecosystems (Perry et al., 1989; Perry, 1995; Bever et al., 1997; Watkinson, 1998). Specifically, it is clear that interactions with soil biota can strongly influence both the species composition and diversity of plant communities, and therefore these interactions may serve to maintain an array of plant species that provides ecological services in an agroecosystem. These interactions gain additional importance because of their potential role in agroecological restoration of ecosystems from low levels of plant and soil-biota diversity. This importance results from the potential for these relationships to be self-organizing and self-sustaining, via positive feedback between plant communities and soil biota, (i.e., the reciprocal promotion of population growth; Bever et al., 1997; Bever, 2002). Cost-effective restoration of beneficial soil biota might be possible via a strategy of establishing conditions that favor these self-organizing plant-soil biota relationships. The self-organizing dynamics of these relationships could then be harnessed to establish plant and soil biota that provide increased ecological services to the agroecosystem.

The significance of weeds in this scenario of restoration and maintenance of ecological services lies in the *structure of relationships* between plant communities and soil biota. Commonly, multiple plant species within a community share a common set of beneficial soil biota, and in effect cooperate in the maintenance of that biota (Perry et al., 1989; Perry, 1995; Bever et al., 1996). The resulting set of relationships can be seen as a *complex plant-soil biota symbiosis*. These symbioses provide an important example of positive interactions between species within ecosystems (Callaway, 1995; Leigh and Rowell, 1995; Wilson, 1997).

If a complex symbiosis is important to the development and maintenance of a plant community, then in principle that symbiosis will be important to any provision of ecological services by that community. An important ecological service that is probably linked to complex symbioses is maintenance of soil quality. For example, this linkage would arise if complex symbiosis was key to the ongoing activity of soil biota that produce soil tilth, such as mycorrhizal fungi (Miller and Jastrow, 1992). Symbioses may also figure in regulation of nutrient cycling, via plant-community effects that enforce a close coupling of microbially-mediated nutrient mineralization and plant activity and uptake patterns (Woodmansee, 1984; Patriquin, 1986; Swift, 1997). Growing evidence supports the notion that this coupling is achieved via strong plant-community effects on microbial communities, e.g., by effects on physical factors limiting microbial activity, as well as other mechanisms. Finally, resistance to plant invasion is now conceptualized as being strongly affected by the spatial and temporal distribution of limiting resources (Davis et al., 2000). In turn, such resource distributions are strongly determined by corresponding distributions of plant activity (Crawley, 1989), which are related to community diversity (Tilman, 1997; Naeem et al., 2000) and successional stage (Rejmanek, 1989). Both diversity (Bever et al., 1997; Van der Heijden, 1998a; Watkinson, 1998; Klironomos et al., 2000) and successional dynamics are likely to be related to the functioning of complex symbioses. Mycorrhizal fungi, for example, have been shown to be capable of significant effects on plant community composition and dynamics (Grime et al., 1987; Allen et al., 1989; Francis and Read, 1994; Zobel et al., 1997; Van der Heijden, 1998b; Watkinson, 1998; Hartnett and Wilson, 1999). In particular, these fungi have recently been shown to exert strong antagonistic effects on certain problematic invasive plants that are non-hosts for the fungi (Francis and Read, 1995). Thus, complex symbioses are plausibly important to many of the ecological services associated with plant diversity (Swift and Anderson, 1993; Swift et al., 1996) in agroecosystems.

Below, we focus on the role of weeds in these symbioses, to illustrate how weeds can function as integral and necessary elements of ensembles of plant diversity that interact with beneficial organisms. After stating some hypotheses about these ecological roles of weeds, we review supporting evidence and suggest research priorities needed to further explore these hypotheses. We will focus on a relatively well-studied class of soil biota, the AMF, that appear to play a central role in plant-soil biota interactions, via their role in structuring the rhizosphere, the biologically-active zone surrounding plant roots (Smith and Read, 1997).

Our core hypothesis is that certain elements of weed communities are essential for *maintenance of mutualism* (i.e. a mutually-beneficial ecological relationship) in complex plant-soil biota symbioses, and therefore to the resultant provision of ecological services. We speak of maintenance of mutualism because mutual benefit is not the only outcome of plant-soil biota interactions or of other symbioses involving plants: various forms of parasitic relationships are also well-known (Johnson et al., 1992; Johnson, 1993; Bronstein, 1994; Saikkonen et al., 1998; Herre et al., 1999). In fact, most soil biota cannot be unambiguously classifed as beneficial or parasitic to their plant hosts (Boucher, 1990); relations of plants with AMF provide a good example. A single plant species can host as many as 13 species of AMF (Helgason et al., 1999). When examined singly, plant growth responses to various co-infecting AMF can range from strongly positive to strongly negative, suggesting a spectrum of AMF roles from highly beneficial to strongly parasitic. Similarly, AMF growth responses to different plant hosts can vary widely (Bever et al., 1996; Smith and Read, 1997). Moreover, often there is non-reciprocity in these relationships: in these cases, the most beneficial AMF taxon for a particular plant species is *not* most strongly benefited by interaction with that species (Bever et al., 1996; Bever, 2002).

These patterns suggest that plant-soil symbioses are complex on at least two levels: by virtue of the densely-woven interactive web linking multiple plant and AMF species, and also by the complexity of the pattern of beneficial and parasitic relationships across the interactive web. We hypothesize that certain weeds serve, by participation in these complex symbioses, to maintain important mutualisms that are embedded with these symbioses. By maintaining these mutualisms, these weeds would serve to maintain the *agronomic integrity* of a complex symbiosis: i.e., its ability to provide ecological services on an ongoing basis. For example, weeds may be necessary to maintain the mutualistic quality of AMF-crop relationships. Below, we review the modest but noteworthy body of relevant evidence, which is consistent with the hypothesis that reductions in the diversity and abundance of weeds that host AMF can cause those AMF to become much more parasitic on crops that they infect. Specifically, this hypothesis is supported by demonstrations that reductions in agroecosystem plant diversity, or in weed diversity and abundance *per se*, can compromise the integrity of these complex symbioses.

First, AMF interactions with host weeds may be important to maintaining the *diversity and abundance* of AMF communities. Interactions between host weeds and certain AMF taxa can be important to maintaining these taxa in soil communities. When AMF-host weeds were removed by handweeding, AMF communities declined in diversity by 50% (as estimated by spore-based methods of assessing AMF diversity; Feldmann and Boyle, 1999). Given the strong patterns of plant species by AMF species interaction that have been observed or hypothesized in several different plant communities (Sieverding, 1991; Bever et al., 1996; Van der Heijden et al., 1998 a, b; Klironomos et al., 2000) (and plant-genotype by AMF interaction; Schultz et al., 2001), it is certainly plausible that AMF host-weeds maintain AMF diversity and abundance by providing high-quality hosts to certain AMF, or by increasing the temporal and spatial range of high-quality hosts.

AMF communities also can respond *functionally*, in terms of their effects on plant performance, to a decline in weed diversity and/or abundance. Taxonomically-depleted AMF communities that resulted from experimental removal of host weed species were considerably less beneficial to crop growth. Maize biomass production was reduced by 28% as a result of host-weed removal (Feldmann and Boyle, 1999). This experiment is the only direct assessment of the effect of weeds on the agronomic outcome of the crop-AMF symbiosis. However, there is supporting evidence from several other agroecosystems. In a tropical soybean cropping system, certain weed species permitted much higher spore production of the most effective AMF taxa for promoting plant growth than did soybean (Pellet and Sieverding, 1986), suggesting a similar beneficial role of these weeds to that observed by Feldmann and Boyle (1999). Higher overall AMF spore densities were associated with higher weed densities in a temperate maize-soybean cropping system (Kurle and Pfleger, 1994) and with overwintering dandelion (*Taraxacum officinale*) densities in a temperate wheat cropping system (Kabir and Koide, 2000). Although none of these latter studies assessed whether any agronomic benefits were associated with the observed effects on AMF spore densities, there are several studies that show that increased crop diversity (i.e., crop rotation versus monoculture) increases agronomic benefits provided by AMF (Johnson et al., 1992; Scullion et al., 1998; Eason et al., 1999).

It is not clear why crop growth in these studies was sensitive to reduction in AMF diversity and/or abundance. However, as noted above, individual plant hosts can form mycorrhizae with many AMF taxa (e.g., > 10; Helgason et al., 1999). Such symbiotic complexity has also been observed in other symbioses, e.g., corals (Herre et al., 1999). The functional importance of this complexity is unknown (Herre et al., 1999), but a multiplicity of co-occurring symbionts may broaden the range of beneficial effects provided by AMF (Wilkinson, 1998). Beyond well-known effects on mineral nutrition, AMF are also known to provide many other direct benefits to hosts (e.g., stress resistance, protection against some pathogens and herbivores). These manifold benefits may be sensitive to AMF community diversity.

Reductions in plant diversity may also degrade the quality of mutualisms by causing qualitative changes in the functioning of individual AMF taxa. Removal of AMF-host weeds caused rapid changes within the AMF taxa that survived after weeding treatments that removed all host weeds, so that these AMF could form mycorrhizae only with maize (Feldmann and Boyle, 1999). In pot-culture experiments, maize biomass production at 42 days after planting was reduced by 42 % by these surviving AMF taxa. Any of the observations of plant-diversity effects on functional qualities of AMF (Johnson et al., 1992, 1997; Scullion et al., 1998; Eason et al., 1999) may result in part or in whole from such changes in the functioning of individual AMF taxa.

In the above discussion, we have considered how weeds might provide certain ecological services via interaction with soil biota within the context of complex plant-soil symbioses. The broader implication is that weeds can provide other ecological services via participation in other complex ecological interactions, in which mutualistic and antagonistic interactions are also intertwined. A likely case of such a complex interactive system is the nexus of weeds, crops, herbivorous arthropods, and their natural enemies. In these interactive networks – as for plant-soil systems – weeds may be critical to maintaining the integrity of these networks by adding diversity and broadening the spatial and temporal dimensions of the plant-community component of the network. For example, weeds play such a role in the case of common knotweed noted above, in which the weed provided nectar to generalist predators and attracted arthropod prey species, thus maintaining predator populations that later dispersed into adjacent crops (Bugg et al., 1987). If playing roles in complex interactive networks prove to be central to ecological benefits of weeds, then the establishment and maintenance of such networks will be an important goal of agroecological restoration.

ECOLOGICAL RESTORATION OF WEED FLORAS

Our objective in this section is to suggest *guiding principles* for an approach to weed management that aims at a form of ecological restoration, namely to increase ecological benefits provided by weeds, while minimizing problematic effects of this flora. We propose that a fundamental goal of weed science research is to define principles for the design and management of agroecosystems that will be *self-organizing* to a considerable extent (Perry, 1995). Such an agroecosystem will respond to appropriate management with a trajectory that is driven by positive interactions among populations of weeds and other biota. Such dynamics are desirable because they will be, by nature, resilient to disturbances from weather events and variable impacts of agronomic operations, and also will not require a large amount of information to guide its facilitation via management. For example, maintaining a sufficient level of host-plant diversity in agroecosystems may in turn maintain mycorrhizal fungi communities that will have substantial negative effects on problematic non-host weeds and substantial positive effects on ecologically-beneficial weed species. These fungal communities would increase an agroecosystem's ecological resilience and resistance to invasion by undesirable weed species. Thus, it is plausible that certain patterns of spatial and temporal deployment of cash crop, cover crop and weed diversity will have strong self-organizing properties, greatly aiding agroecosystem management.

As mentioned in the begining, we envision a restored weed flora as a succession of weed "patches" that appear at opportune locations in space and time. Moreover, we envision that the *practice and process* of restoration entails determining how to tailor farming operations - including but certainly not limited to specific weed-management operations - so as to have the effect of "confining" the weed community largely to situations where the weeds produce net ecological benefits.

The value of manipulating spatial and temporal placement of weed patches for ecological benefit is well illustrated by the use of weed patches to guide the population processes (Andow and Hidaka, 1989) of arthropod pests and their natural enemies (Bugg, 1992). By considering factors such as relative effects of weed patches on size, dispersal, and resilience to disturbance of populations of both herbivorous arthropods and their natural enemies, it is possible to reduce pest abundance in crops during vulnerable periods of crop growth and development (Bugg, 1992; Barbosa and Benrey, 1998; Barbosa and Wratten, 1998). Manipulations of temporal distributions involve modifying weed phenology (by stimulating weed emergence or by destroying weed patches) to attract pests and beneficials only at opportune times. For example, weed patches that have attracted pests might be destroyed at the most vulnerable pest life-cycle stages, or permitted to flower at times that will pests away from crops (Bugg, 1992). Spatial manipulations include planting weed strips as barriers to pest dispersal, or as refuge for natural enemies (Bugg et al., 1987).

Such tailoring of the distribution of weed patches is necessarily highly site specific and highly dependent on the *knowledge and resources of farmers*. However, as noted in the introduction, there will also be a need for scientists to better define the potential benefits of weeds in cropping systems. Efforts should address current systems, but it is more important to look forward to beneficial roles of weeds in future cropping systems that are more highly dependent on ecological services from biodiversity. Scientists also need to determine how weeds can enable agroecological restoration, and how specific management practices affect the dynamics of restoration. Other professionals (e.g., extension workers and farmer advisors; Jordan et al., 2003) also have very important roles to play in developing the necessary knowledge base. Based on present knowledge, we propose the following as important principles in designing and implementing a weed restoration program:

- *The Spatial and Temporal Location of Weeds is Determined by the Set of Ecological Benefits that is Desired from Weeds in a Particular Cropping System*

First, the particulars of a farming operation will determine where, when and how weeds might feasibly provide ecological benefits. For example, certain weeds may be desired to improve the nutritional quality of corn harvested for silage production, or to suppress other less desirable weeds, or to support specific beneficial organisms, or to as components of permanent vegetated walkways in horticultural crop production.

- *The Farm (or Other Managed Ecosystem) - as a Whole - is the Instrument of Restoration*

This principle highlights the importance of farmers as designer-operators of agroecosystems that will restore – rather than degrade – their weed communities. A farm-specific restoration process should begin by identifying the magnitude and spatial/temporal location of stressors that may cause degradation of the weed community: examples of relevant factors include low levels of diversity in the cropping system (Liebman and Dyck, 1993), or high levels of factors that damage mycorrhizal fungi communities, including synthetic fertilizer inputs and soil disturbance.

Then, stress factors, desired benefits, and management constraints must be considered together to decide how a cropping system can be "fine-tuned" for restorational purposes. Fine-tuning refers to modification of farming practices so as to achieve *agroecological selectivity* (Fernholz, 1992; Bender, 1993; Liebman et al., 2001): promoting growth and reproduction of desirable weeds (and crops) and disadvantaging undesirable species. For example, early-emerging weeds and ridge-tillage cultivation operations were used to select against large-seed dicotyledonous weeds that were problematic in a low-input conservation-tillage cropping system (Thompson, 1991). In general, the sequencing of cash and cover crops and the nature and timing of fertilizer, irrigation and tillage application all have strong effects on weed community dynamics (Liebman et al., 2001). Therefore, each of these management factors may afford some opportunity to fine-tune the agroecosystem for restorational purposes. For instance, manipulations of nutrient cycling by nutrient application and tillage – e.g., use of organic N sources or high C residues to reduce early-season nitrogen availability – may have strong effects in reducing the growth and reproduction of certain weed species that are problematic because they are highly-competitive with crops at relatively low densities (Dyck and Liebman, 1995). It is likely that the spatial and temporal placement of these nutrient-cycling interventions will be critical to their effect on weed restoration (Nordell, 1993).

Practical methods that working farmers can use to design their farms (e.g., "farmscaping"; Bugg, 1992; Pickett and Bugg, 1998) and to monitor results of weed restoration efforts are critically needed. One valuable approach is careful observation and mapping of spatial and temporal patterns of relevant attributes of the farm. *Pattern analysis* (Conway and Barbier, 1990) is a systematic approach to such mapping. One important objective in mapping is to consider when and how weeds provide a function that appears necessary to other significant biota in the agroecosystem, such as provision of habitat (Zavaleta et al., 2001). Observation, mapping, and fine-tuning of management practices must encompass both the within-farm scale (e.g., addressing weeds in field margins or in permanent walkways) and the scale of the landscape that embeds that farm (e.g., addressing hedgerows, or other non-field landscape elements; Fry, 1994).

- *Situating Weeds along Ecological Boundaries may be an Important Design Consideration in Determining the Advantageous Location of Weed Patches*

Ecological boundaries (so-called ecotones at various scales) are locations where fluxes of organisms, energy and materials are concentrated and controlled (Risser, 1995). Many agroecosystem properties are affected by the functional properties of these interface locations (Perry, 1995), including pest dynamics and nutrient cycling (Letourneau, 1998; Laurance, 2000). For example, situating weed patches at field margins may increase landscape connectivity, allowing a variety of beneficial organisms to disperse more readily and thus helping to sustain and increase populations of these species (Fry, 1994).

- *Refuges from various Disturbances may Help Maintain Desirable Weed Species*

Studies of a variety of plant communities show that refuges from disturbance are key to preservation of ecosystem functioning in response to disturbance (Perry, 1995). Of course, agroecosystems are subject to many forms of disturbance on an ongoing basis: e.g., tillage, harvesting, nutrient inputs, etc. Refuges preserve "biological legacies": complements of biological diversity that provide for ecosystem resilience, resistance or recovery after disturbance (Perry, 1995). Legacies persist through disturbance events only if some form of protective refuge exists in the ecosystem, within which necessary biodiversity is sufficiently sheltered from disturbance to be able to exert desired effects on ecosystem functioning can continue after disturbance. Accordingly, an important consideration in ecological restoration of farms is the provision of refugia from disturbance. In agroecosystems, study of refugia and biological legacies is just beginning, but examples exist in the concept of 'beetle banks' (Pretty, 1995, p. 110), in use of field margin vegetation (Bugg, 1992), and in ridge tillage, zone tillage and other conservation tillage methods that provide semipermanent protection from soil disturbances. Nentwig et al. (1998) argue for the importance of refuges – which they term "ecological compensation areas" – that maintain natural areas within and between agroecosystems. They advocate use of deliberately-sown weed strips in refuges, and envision that older strips could serve biological legacies that aid establishment of new strips.

- *Ongoing Monitoring, and 'Data-Poor' Adaptive Management are Necessary to Guide Restoration*

In principle, successful restoration of the weed flora in a particular agroecosystem is measured by a temporal trend of increasing ecological benefits from weeds and decreasing harmful effects. Unfortunately, accurate direct measurement of agronomic costs and benefits of the weed flora of a particular agroecosystem is prohibitively costly. Therefore, we believe that the only viable approach is to use practical on-farm monitoring techniques along with *data-poor* and *adaptive management* (Johannes, 1998). These management modes aim to develop "precautionary" styles of management that seek to gain adequate (not necessarily short-term maximum) production while also protecting valuable biodiversity from degradative stressors, so that ecological services can be provided by this biodiversity on an ongoing basis. By definition, data-poor guidelines require only moderate amounts of quantitative information about the agroecosystem to guide their application. Such guidelines typically take the form of 'rules of thumb' for the spatial and temporal application of various management techniques (Johannes, 1998). In the ideal, these rules of thumb will support the operation of self-regulating ecological dynamics that maintain the desired balance between production and protection of valuable biodiversity. For example, current evidence on the role of weeds in complex symbioses with AM fungi suggests a 'rule of thumb' that weed control practices should aim to leave modest residual populations of AMF-host weeds will maintain AMF communities that are strongly beneficial to crop growth, whereas lowering the density of these species to very low levels may cause substantial loss of beneficial AMF effects on crop production. robably, farmer experience and ecological research can be used to develop some archetypal rules of thumb (e.g., restoration guidelines proposed above). However, these guidelines will need to be tailored for the particularities of particular farming operations. Therefore, practical monitoring techniques are needed, which could be applied by farmers, perhaps with support from management consultants. Such techniques have been developed for soil quality monitoring (Roming et al., 1996). Weed monitoring could proceed by observing and mapping patterns of weeds and noting their abundance on an ongoing basis. There must also be some means of assessing benefits associated with weed patches, probably through various easily-observed indirect measures – e.g., noting trends in soil quality attributes, or abundance of natural enemies of herbivorous pests. It is also necessary to have practical ways of assessing harmful effects of weeds. These monitoring techniques should be used to assess the outcome of applying rules of thumb to design a weed restoration program for a farming operation, and to modify management and agroecosystem design accordingly. Of course, this interplay between management actions guided by rules of thumb and evaluation of management outcomes by monitoring is the essence of adaptive management.

Can We Find Examples of Successful Ecological Restoration of Weed Communities?

There is little direct scientific evidence for or against the notion that successful ecological restoration of weeds is possible. However, it is a commonplace observation among farmers using organic and biological production methods that these methods lead to some amelioration of weed communities – i.e., reduced problems and increased benefits (Patriquin, 1986; Andow and Hidaka, 1989). Certainly, we would predict that some aspects of organic/biological production would exert a restorative effect on weeds. In particular, the nutrient management strategies used on these farms, particularly the use of organic N sources (Dyck and Liebman, 1995), use of mulches, cover crops and careful application of tillage in order to increase the diversity and abundance of beneficial soil biota, and diversification of cropping system and weed management tactics all have the potential to be restorative to weed communities. Therefore, we predict that careful study of the functional and compositional dynamics of weed communities on these farms would indicate a restorational trend. However, we are aware of no formal studies of these dynamics on organic/biological farms.

In a comparison of weed flora in organic and conventional winter wheat fields, Moreby et al. (1994) determined that there were indeed differences between the weeds of these farming regimes. There were three times as many weed species per 0.25 m^2 in organic fields than in conventional fields. Broad-leaved species made up the bulk of this diversity difference, and their mean percent coverage also differed, with organic fields having nearly 6% cover from broadleaf weeds during the first year of the study, as compared to 0.44% in conventional fields. The second year of the study showed an even more striking difference with broadleaf weeds making up 12% cover in organic fields, and only $< 1\%$ in conventional fields. This evidence supports the hypothesis that organic management regimes can restore weed community diversity. Studies in a variety of ecosystems show that once ecological degradation has occurred (e.g., loss of mechanisms for control of nutrient and hydrological cycling; Carpenter and Cottingham, 1997) , then restoration of ecological integrity may be very slow; i.e, degradation may not be a readily reversable process (Swift et al., 1996).

CONCLUSIONS

Taken together, direct evidence from agroecosystems, relevant theory, and applicable evidence from other ecosystems strongly encourages the notion that weeds can provide substantial ecological benefits. Certainly, there is need for a deeper and broader understanding of what ecological benefits could be received from weeds in return for deliberate management with this goal in mind, and whether and when these benefits will outweigh negative effects of such weeds. Clearly, deliberate management of weeds for ecological benefits will be complicated and site-specific, no less so than management of other forms of agroecosystem biodiversity, and it will require adaptive and experimental management. However, development of resource-efficient, environmentally benign agroecosystems is critical to the human future, and no stone can be left unturned in the search for the keys to such agroecosystems. We strongly recommend that scientists and farmers work in partnership on the important challenge of deliberate management of weed communities for ecological benefits.

REFERENCES

Allen M.F., Allen E.B., Friese C.F. Response of the non-mycotrophic plant *Salsola kali* to invasion by vesicular-arbuscular mycorrhizal fungi. New Phytol 1989; 111:45-49.

Altieri M.A. The ecological role of biodiversity in agroecosystems. Agricult Ecosys Environ 1999; 74:19-31.

Altieri M.A. *Agroecology: The Science of Sustainable Agriculture*. 2nd ed. Boulder, CO: Academic Press, 1995.

Altieri M.A. *Biodiversity and Pest Management in Agroecosystems*. Binghampton, NY: Food Products Press, 1994.

Altieri M.A. "The influence of adjacent habitats on insect populations in crop fields." In *Biodiversity and Pest Management in Agroecosystems*, M. Altieri, ed. New York: Haworth Press, 1994; pp. 109-130.

Altieri M.A. How can we best use biodiversity in agroecosystems? Outlook Agr 1991; 20:15-23.

Altieri M.A. "The impact, uses, and ecological role of weeds in agroecosystems." In *Weed Management in Agroecosystems: Ecological Approaches*, M.A. Altieri, M. Liebman, eds. Boca Raton, FL: CRC Press, Inc., 1988; pp. 1-8.

Altieri M.A., Letourneau D.K. Vegetation diversity and insect pest outbreaks. Crit Rev Plant Sci 1984; 2:131-69.

Altieri M.A., Whitcomb W.H. Manipulation of insect populations through seasonal disturbance of weed communities. Prot Ecol 1979; 1:185-202.

Altieri M.A., Wilson R.C., Schmidt L.L. The effects of living mulches and weed cover on the dynamics of foliage- and soil-arthropod communities in three crop systems. Crop Prot 1985; 4:201-213.

Andow D.A., Hidaka K. Experimental natural history of sustainable agriculture: syndromes of production. Agricult Ecosys Environ 1989; 27:447-462.

Barbosa P., Benrey B. "The influence of plants on insect parasitoids: implications for conservation biological control." In *Conservation Biological Control*, P. Barbosa, ed. New York, NY: Academic Press, 1998; pp. 55-82.

Barbosa P., Wratten S.D. "The influence of plants on invertebrate predators: implications to conservation biological control." In *Conservation Biological Control*, P. Barbosa, ed. New York, NY: Academic Press, 1998; pp. 83-100.

Bender J. *Future Harvest*. Lincoln, NE: University of Nebraska Press, 1993.

Bethlenfalvay G.J. "Mycorrhizae and crop productivity." In *Mycorrhizae in Sustainable Agriculture*, G.J. Bethlenfalvay, R Linderman, eds. ASA Special Publication No. 54, Madison, WI: American Society of Agronomy, 1992a; pp. 1-27.

Bethlenfalvay G.J. "Preface." In *Mycorrhizzae in Sustainable Agriculture*, G.J. Bethlenfalvay, R. Linderman, eds. ASA Special Publication No. 54, Madison, WI: American Society of Agronomy, 1992b; pp. viii-xiii.

Bethlenfalvay G.J., Mihara K.L., Schreiner R.P., McDaniel H. Mycorrhizae, biocides and biocontrol: 1. Herbicide-mycorrhiza interactions in soybean and cocklebur treated with bentazon. Appl Soil Ecol 1996a; 3:197-204.

Bethlenfalvay G.J., Schreiner R.P., Mihara K.L., McDaniel H. Mycorrhizae, biocides and biocontrol: 2. Mycorrhizal fungi enhance weed control and crop growth in a soybean-cocklebur association treated with the herbicide bentazon. Appl Soil Ecol 1996b; 3:205-214.

Bever J.D. Host-specificity of AM fungal population growth rates can generate feedback on plant growth. Plant Soil 2002; 244:281-290.

Bever J.D., Morton J.B., Antonovics J., Schultz P.A. Host-dependent sporulation and species diversity of arbuscular mycorrhizal fungi in a mown grassland. J Ecol 1996; 84:71-82.

Bever J.D., Westover K.M., Antonovics J. Incorporating the soil community into plant-population dynamics-the utility of the feedback approach. J Ecol 1997; 85:561-573.

Boucher D.H. "Beneficials in agricultural soils." In *Agroecology*, C.R. Carroll, J.H. Vandermeer, P.M. Rosset, eds. NY: McGraw-Hill, 1990; pp. 329-340.

Boucher D.H. (ed.) *The Biology of Mutualism: Ecology and Evolution*. Oxford, UK: Oxford University Press, 1985.

Bronstein J.L. Conditional outcomes in mutualistic interactions. Trends Ecol Evolut 1994; 9:214-217.

Brussaard L. Biodiversity and ecosystem functioning in soil. Ambio 1997; 26:563-570.

Brussaard L. Soil fauna, guilds, functional groups and ecosystem processes. Appl Soil Ecol 1998; 9:123-135.

Bugg R.L. Using cover crops to manage arthropods on truck farms. HortSci 1992; 27:741-745.

Bugg R.L., Ehler L.E., Wilson L.T. Effect of common knotweed (*Polygonum aviculare*) on abundance and efficiency of insect predators of crop pests. Hilgardia 1987; 55:1-53.

Bye R.A. Qualites-ethnoecology of edible greens-past, present and future. J Ethnobiol 1981; 1:109-114.

Callaway R.M. Positive interactions among plants. Bot Rev 1995; 6:306-349.

Cardwell C., Hassall M., White P. Effects of headland management on carabid beetle communities in Breckland cereal fields. Pedobiologia 1994; 38:50-62.

Carpenter S.R., Cottingham K.L. Resilience and restoration of lakes. Conserv Ecol 1997; 1:2. Available on-line with updates at http://www.consecol.org/vol1/iss1/art2.

Chacon J.C., Gliessman S.R. Use of the "non-weed" concept in traditional tropical agroecosystems of south-eastern Mexico. Agro-ecosystems 1982; 8:1-11.

Cocannouer J.A. *Weeds: Guardians of the Soil*. Old Greenwich, CT: Devin-Adair, 1950.

Conway G.R, Barbier E.B. *After the Green Revolution: Sustainable Agriculture for Development*. London: Earthscan Publications, 1990.

Council for Agricultural Science and Technology. *Benefits of Biodiversity*. Ames, IA: CAST, 1999.

Cousens R, Mortimer M. *Dynamics of Weed Populations*. Cambridge, UK: Cambridge University Press, 1995.

Crawley M.J. "Chance and timing in biological invasions." In *Biological Invasions: A Global Perspective*, J.A. Darke, H.A. Mooney, F. di Castri, R.H. Groves, F.J. Kruger, M. Rejmanek, M. Williamson, eds. Chichester, UK: John Wiley and Sons, 1989; pp. 407-424.

Crossley D.A., House G.J., Snider R.M., Snider R.J., Stinner B.R. "The positive interactions in agroecosystems." In *Agricultural Ecosystems—Unifying Concepts*, R. Lowrance, B.R. Stinner, G.J. House, eds. New York: John Wiley, 1984; pp. 73-82.

Daily G.C. *Nature's Services: Societal Dependence on Natural Ecosystems*. Washington, DC: Island Press, 1997.

Daniels H.B.A., Bloom J. Vesicular-arbuscular mycorrhizal fungi associated with native tall grass prairie and cultivated winter wheat. Can J Botany 1983; 61:2140-2146.

Davis M.A., Grime J.P., Thompson K. Fluctuating resources in plant communities: a general theory of invasibility. J Ecol 2000; 88:528-534.

De Ruiter P.C., Neutel A.M, Moore J.C. Energetics, patterns of interaction strengths, and stability in real ecosystems. Science 1995; 269:1256-1260.

Ditsch D.C., Alley M.M. Nonleguminous cover crop management for residual N recovery and subsequent crop yields. J Fert Issues 1991; 8:6-13.

Dodd J.C. The role of arbuscular mycorrhizal fungi in agro- and natural ecosystems. Outlook Agr 2000; 29:55-62.

Doersch R.E., Buhler D.D. *Controlling Weeds in Conservation Tillage Corn Production*. Ext Pub 3425, University of Wisconsin-Madison, 1989.

Douds D.D., Millner P.D. Biodiversity of arbuscular mycorrhizal fungi in agroecosystems. Agricult Ecosys Environ 1999; 74:77-93.

Dyck E., Liebman M. Crop-weed interference as influenced by a leguminous or synthetic fertilizer nitrogen source. II. Rotation experiments with crimson clover, field corn and lambsquarters. Agricult Ecosys Environ 1995; 56:109-120.

Eason W.R., Scullion J.R., Scott E.P. Soil parameters and plant responses associated with arbuscular mycorrhizas from contrasting grassland management regimes. Agricult Ecosys Environ 1999; 73:245-255.

Ellstrand N.C., Prentice H.C., Hancock J.F. Gene flow and introgression from domestic plants into their wild relatives. Annu Rev Ecol Syst 1999; 30:539-564.

Feldmann F., Boyle C. Weed-mediated stability of arbuscular mycorrhizal effectiveness in maize monocultures. J Appl Botany 1999; 73:1-5.

Fernholz C. *Sustainable Management Practices for the Nineties*. Madison, MN: A Frame Press, 1992.

Fitter A.J., Graves J.D., Watkins N.K., Robinson D., Scrimgeour C. Carbon transfer between plants and its control in networks of arbuscular mycorrhizas. Funct Ecol 1998; 12:406-412.

Francis R., Read D.J. The contributions of mycorrhizal fungi to the determination of plant community structure. Plant Soil 1994; 159:11-25.

Francis R., Read D.J. Mutualism and antagonism in the mycorrhizal symbiosis, with special reference to impacts on plant community structure. Can J Botany 1995; 73(Suppl. 1):1301-1309.

Fry G.L.A. "The role of field margins in the landscape." In *Field Margins: Integrating Agriculture and Conservation*, N. Boatman, ed. BCPC Monograph No. 58, United Kingdom: BCPC, 1994; pp. 31-46.

Grime J.P., Mackey J.M.L., Hillier S.H., Read D.J. Floristic diversity in a model system using experimental microcosms. Nature 1987; 328:420-422.

Hartnett D.C., Wilson G.W.T. Mycorrhizae influence plant community structure and diversity in tallgrass prairie. Ecology 1999; 80:1187-1195.

Helgason T., Daniell T.J., Husband R., Fitter A.H., Young J.P.Y. Ploughing up the wood-wide web? Nature 1999; 394:431.

Herre E.A., Knowlton N., Mueller U.G., Rehner S.A. The evolution of mutualisms: exploring the paths between conflict and cooperation. Trends Ecol Evolut 1999; 14:49-53.

Holzner W. "Concepts, categories and characteristics of weeds." In *Biology and Ecology of Weeds*, W. Holzner, N. Numata, eds. The Hague: Dr. W. Junk Publishers, 1982; pp. 3-20.

Hutchings M.J., Wijesinghe D.K. Patchy habitats, division of labour and growth dividends in clonal plants. Trends Ecol Evolut 1999; 12:390-394.

Jackson W., Piper J. The necessary marriage between ecology and agriculture. Ecology 1989; 70:1591-1593.

Jervis M.A., Kidd N.A., Fitton M.G., Huddleston T., Dawah H.A. Flower visiting by hymenopteran parisitoids. J Nat Hist 1993; 27:67-105.

Johannes R.J. The case for data-less marine resource management: examples from tropical nearshore finfisheries. Trends Ecol Evol 1998; 13:243-246.

Johnson N.C. Can fertilization of soil select less mutualistic mycorrhizae? Ecol Appl 1993; 3:749-757.

Johnson N.C., Copeland P.J., Crookston R.K., Pfleger F.L. Mycorrhizae: possible explanation for yield decline with continuous corn and soybean. Agron J 1992; 84:387-390.

Johnson N.C., Grahan J.H., Smith F.A. Functioning of mycorrhizal associations along the mutualism-parasitism continuum. New Phytol 1997; 135:575-585.

Jordan N. "Agroecological restoration: sustaining production with biodiversity." In *The Farm as a Natural Habitat*, D. Jackson, L. Jackson, eds. Covelo, CA: Island Press, 2002; pp. 155-186.

Jordan N., Becker R., Gunsolus J., White S., Damme S. Knowledge networks: an avenue to ecological management of invasive weeds. Weed Sci 2002; 51:271-277.

Jordan N., White S., Gunsolus J., Becker R., Damme S. "Learning groups developing collaborative learning methods for diversified, site-specific weed management: a case study from Minnesota, USA." In *Cow Up a Tree: Knowing and Learning for Change in Agriculture*, M. Cerf, D. Gibbon, B. Hubert, R. Ison, J. Jiggins, M. Paine, J. Proost, N. Röling, eds. Paris: INRA, 2000; pp. 85-95.

Jordan N.R., Zhang J., Huerd S. Arbuscular-mycorrhizal fungi: potential roles in weed management. Weed Res 2000; 40:397-410.

Jordan N., Becker R., Gunsolus J., White S., Damme S. Knowledge networks: an avenue to ecological management of weeds. Weed Sci 51:271-277.

Kabir Z., Koide R.T. The effect of dandelion or a cover crop on mycorrhiza inoculum potential, soil aggregation and yield of maize. Agricult Ecosys Environ 2000; 78:167-174.

Kapoor P., Ramakrishnan P.S. Studies on crop-legume behaviour in pure and mixed stands. Agro-Ecosys 1975; 2:61-74.

Kinkel L.L. Microbial population dynamics on plant surfaces. Annu Rev Phytopathol 1997; 35:327-347.

Klironomos J.N., McCune J., Hart M., Neville J. The influence of arbuscular mycorrhizae on the relationship between plant diversity and productivity. Ecol Lett 2000; 3:137-141.

Krebs J.R., Wilson J.D., Bradbury R.B., Siriwardena G.M. The second silent spring? Nature 1999; 400:611-612.

Kurle J.E., Pfleger F.L. Arbuscular mycorrhizal fungus spore populations respond to conversions between low-input and conventional management practices in corn-soybean rotation. Agron J 1994; 86:467-475.

Lal R.E., Regnier E., Eckert D.J., Edward S.W.M., Hammond R. "Expectations of cover crops for sustainable agriculture." In *Cover Crops for Clean Water*, W.L. Hargrove, ed. Ankeny, IA: Soil and Water Conservation Society, 1991; pp. 1-11.

Laurance W.F. Do edge effects occur over large spatial scales? Trends Ecol Evolut 2000; 15:134-135.

Leigh E.G., Rowell T.E. The evolution of mutualism and other forms of harmony at various levels of biological organization. Ecologie 1995; 26:131-158.

Letourneau D.K. "Conservation biology: lessons for conserving natural enemies." In *Conservation Biological Control*, P. Barbosa, ed. San Diego,CA: Academic Press, 1998; pp. 9-38.

Liebman M., Dyck E. Crop rotation and intercropping strategies for weed Management. Ecol Appl 1993; 3:92-122.

Liebman M., Mohler C.L., Staver C.S. *Ecological Management of Agricultural Weeds*. Cambridge, UK: Cambridge University Press, 2001.

Lobry de Bruyn L.A.L. Ants as bioindicators of soil function in rural environments. Agricult Ecosys Environ 1999; 74:425-441.

Mack R.N., Simberloff D., Lonsdale W.M., Evans H., Clout M., Bazzaz F.A. Biotic invasions: causes, epidemiology, global consequences, and control. Ecol Appl 2000; 10:689-710.

Martensson A.M., Rydbergi Vestberg G.M. Potential to improve transfer of N in intercropped systems by optimising host-endophyte combinations. Plant Soil 1998; 205:57-66.

Matson P.A., Parton W.J., Power A.G., Swift M.J. Agricultural intensification and ecosystem properties. Science 1997; 277:504-509.

Miller R.M., Jastrow J.D. "The role of mycorrhizal fungi in soil conservation." In *Mycorrhizae in Sustainable Agriculture*, G.J. Bethlenfalvay, R Linderman, eds. ASA Special Publication No. 54, Madison, WI: American Society of Agronomy, 1992; pp. 29-44.

Moreby S.J., Aebischer N.J., Southway S.E., Sotherton N.W. A comparison of the flora and arthropod fauna of organically and conventionally grown winter wheat in southern England. Ann Appl Biol 1994; 125:13-27

Naeem S.J., Knops M.H., Tilman D., Howe K.M., Kennedy T., Gale S. Plant diversity increases resistance to invasion in the absence of covaring extrinsic factors. Oikos 2000; 91:97-108.

Neher D.A., Barbercheck M.E. "Diversity and function of soil mesofauna." In *Biodiversity in Agroecosystems*, W.W. Collins, C.O. Qualset, eds. Boca Raton, FL: CRC Press, 1998; pp. 27-47.

Nentwig W., Frank T., Lethmayer C. "Sown weed strips: artificial ecological compensation areas as an important tool in conservation biological control." In *Conservation Biological Control*, P. Barbosa, ed. New York, NY: Academic Press, 1998; pp. 133-154.

Newsham K.K., Fitter A.H., Watkinson A.R. Multi-functionality and biodiversity in arbuscular mycorrhizas. Trends Ecol Evolut 1995b; 10:407-411.

Nordell E. A few long furrows on horsedrawn tillage. Small Farmers' J; 1993:17:17-23.

Norris R.F., Kogan M. Interactions between weeds, arthropod pests, and their natural enemies in managed ecosystems. Weed Sci 2000; 48:94-158.

Pankow W., Boller T., Wiemken A. The significance of mycorrhizas for protective ecosystems. Experientia 1991; 47:391-394.

Patriquin D.G. "Biological husbandry and the "nitrogen problem." In *The Role of Microorganisms in a Sustainable Agriculture*. Selected papers from The Second International Conference on Biol. Agric., University of London, Wye College, Wye, Kent, UK. J.M. Lopez-Real, R.D. Hodges, eds. Berkamsted, : A.B. Academic, 1986; pp. 81-103.

Patriquin D.G., Hill N., Baines D., Bishop M., Allan G. Observations on a mixed farm during the transition to biological husbandry. Biol Agric Hortic 1986; 4:69.

Pellet D., Sieverding. E. "Host preferential multiplication of fungal species of the endogonaceae in the field, demonstrated with weeds." In *Physiological and Genetical Aspects of Mycorrhizae*. V. Gianinazzi-Pearson, S. Gianinazzi, eds. Paris: INRA, 1986; pp. 555-557.

Perry D.A. Self-organizing systems across scales. Trends Ecol Evolut 1995; 10:241-244.

Perry D.A. A moveable feast: the evolution of resource sharing in plant-fungus communities. Trends Ecol Evolut 1998; 13:432-434.

Perry D.A., Amaranthus M.A., Borchers J.G., Borchers S.L., Brainerd R.E. Bootstrapping in ecosystems. Bioscience 1989; 39:230-237.

Perry D.A., Borchers J.G., Borchers S.L., Amaranthus M.P. Species migrations and ecosystem stability during climate change: the belowground connection. Conserv Biol 1990; 4:266-274.

Pickett C.H., Bugg R.L. (eds.) *Enhancing Biological Control: Habitat Management to Promote Natural Enemies of Agricultural Pests*. Berkeley, CA: University of California Press, 1998.

Pimentel D., McLaughlin L., Zepp A., Lakitan B., Kraus T., Kleinman P., Vancini F., Roach W.J., Graap E., Keeton W.S. Environmental and economic effects of reducing pesticide use: a substantial reduction in pesticides might increase food costs only slightly. Bioscience 1991; 41:402-409.

Pokarzhevskii A.D., Krivolutskii D.A. Problems of estimating and maintaining biodiversity of soil biota in natural and agroecosystems:a case study of chernozem soil. Agricult Ecosys Environ 1997; 62:127-133.

Pretty J.N. *Regenerating Agriculture: Policies and Practice for Sustainability and Self-reliance*. London, UK: Earthscan, 1995.

Rabatin S.C., Stinner B.R. The significance of vesicular-arbuscular mycorrhizal fungal-soil macroinvertebrate interactions in agroecosystems. Agricult Ecosys Environ 1989; 27:195-204.

Radosevich S., Holt J., Ghersa C. *Weed Ecology—Implications for Management.* New York: John Wiley and Sons, 1997.

Ramakrishnan P.S. *Shifting Agriculture and Sustainable Development: An Interdisciplinary Study From Northeastern India.* UNESCO-MAB Series, Paris, Parthenon, Carnforth, 1992.

Rejmanek M. "What attributes make some species more invasive?" In *Biological Invasions: A Global Perspective*, J.A. Darke, H.A. Mooney, F. di Castri, R.H. Groves, F.J. Kruger, M. Rejmanek, M. Williamson, eds. Chichester, UK: John Wiley and Sons, 1989; pp. 369-388.

Rejon A, Garcia-Romera I, Ocampo J.A., Bethlenfalvay G.J. Mycorrhizal fungi influence competition in a wheat-ryegrass association treated with the herbicide diclofop. Appl Soil Ecol 1997; 7:51-57.

Risser P. The status of the science examining ecotones. Bioscience 1995; 45:318-325.

Roming D.E., Garlynd M.J., Harris R.F. *Farmer-Based Assessment of Soil Quality: A Soil Health Scorecard.* SSSA Special Publication 49. Madison, WI: Soil Science Society of America, 1996.

Rosemeyer M.E., Gliessman S.R. "Modifying traditional and high-input agroecosystems for optimization of microbial symbioses: a case study of dry beans in Costa Rica." In *Biotic Diversity in Agroecosystems*, M.G. Paoletti, D. Pimentel, eds. Amsterdam: Elsevier, 1992; pp. 61-70.

Saikkonen K., Faeth S.H., Helander M., Sullivan T.J. Fungal endophytes: a continuum of interactions with host plants. Annu Rev Ecol Syst 1998; 29:319-343.

Schlult P.A., Miller R.M., Jastrow J.D., Rivetta C.V., Bever J.D. Evidence of a mycorrhizal mechanism for the adaptation of *Andropogon gerardii* (Poaceae) to high- and low-nutrient prairies. Am J Bot 2001; 88:1950-1656.

Scullion J., Eason W.R., Scott E.P. The effectivity of AMF from high-input conventional and organic grassland and grass-arable rotations. Plant Soil 1998; 204:243-254.

Shaver C.P. "Knowledge, science and practice in ecological weed management: farmer-extensionist-scientist interactions." In *Ecological Management of Agricultural Weeds*, M. Liebman, C.L. Mohler, C.S. Staver, eds. Cambridge, UK: Cambridge University Press, 2001; pp. 99-138.

Shea K., NCEAS Working Group on Population Management. Management of populations in conservation, harvesting and control. Trends Ecol Evolut 1998; 13:371-375.

Sheley R.L., Svejcar T.J., Maxwell B.D. A theoretical framework for developing successional weed management strategies on rangeland. Weed Technol 1996; 10:766-773.

Shipley P.R., Meisinger J.J., Decker A.M. Conserving residual corn fertilizer nitrogen with winter covercrops. Agron J 1992; 84:869-876.

Sieverding E. *Vesicular-arbuscular Mycorrhizal Management in Tropical Agroecosystems.* Deutsche Gesellschaft fur Technische Zusammenarbeit, Eschborn, Federal Republic of Germany, 1991.

Simberloff D. Eradication—preventing invasions from getting off the ground. WSSA Abstracts 2001; 41:137.

Smith W.E., Read D.J. *Mycorrhizal Symbiosis*. 2nd ed. San Diego, CA: Academic Press, 1997.

Swamy P.S., Ramakrishnan P.S. Ecological implications of traditional weeding regimes under slash and burn agriculture (jhum) in northeastern India. Weed Res 1988; 28:127-136.

Swanton C.J., Clements D.R., Derksen D.A. Weed succession under conservation tillage: a hierarchical framework for research and management. Weed Technol 1993; 7:286-297.

Swift M.J. "Biological management of soil fertility as a component of sustainable agriculture: perspectives and prospects with particular reference to tropical regions." In *Soil Ecology in Sustainable Agricultural Systems*, L. Brussard, R. Ferrera-Cerrato, eds. Boca Raton, FL: CRC Press, 1997; pp. 91-112.

Swift M.J., Anderson J.M. "Biodiversity and ecosystem function in agricultural systems." In *Biodiversity and Ecosystem Function*, E.D. Schultz, H.A. Mooney, eds. Berlin, Germany: Springer-Verlag, 1993; pp. 15-41.

Swift M.J., Vandermeer J., Ramakrishnan P.S., Anderson J.M., Ong C.K., Hawkins B.A. "Biodiversity and agroecosystem function." In *Functional Roles of Biodiversity-A Global Perspective*, H.A. Mooney, J.H. Cushman, E. Medina, O.E. Sala, E.-D. Schultz, eds. New York: John Wiley and Sons, Inc., 1996; pp. 261-294.

Thomas M.B., Wratten S.D., Sotherton N.W. Creation of "island" habitats in farmland to manipulate populations of beneficial arthropods: predator densities and species composition. J Appl Ecol 1992; 29:524-531.

Thompson D. *Thompson On-Farm Research*. Greenbelt, MD: Wallace Institute, 1991.

Tilman D. Community invasibility, recruitment limitation, and grassland biodiversity. Ecology 1997; 78:81-92.

Van der Heijden M.G.A., Boller T., Weimken A., Sanders I.R. Different arbuscular mycorrhizal fungal species are potential determinants of plant community structure. Ecology 1998a; 79:2082-2091.

Van der Heijden M.G.A., Klironomos J.N., Ursic M., Moutoglis P., Streitwolf-Engel R., Boller T., Wiemken A., Sanders I.R. Mycorrhizal fungal diversity determines plant biodiversity, ecosystem variability and productivity. Nature 1998; 396:69-72.

Vandermeer J. The ecological basis of alternative agriculture. Annu Rev Ecol Syst 1995; 26:201-224.

Vatovec C.M. A Weed Survey of the Farmers in the Sustainable Farming Association of Minnesota, Unpublished, 2001.

Wardle D.A. A comparative assessment of factors which influence microbial biomass carbon and nitrogen levels in soil. Biol Rev 1992; 67:321-358.

Wardle D.A. Impacts of disturbance on detritus food webs in agro-ecosystems of contrasting tillage and weed management practices. Adv Ecol Res 1995; 26:105-184.

Watkinson A.R. The role of the soil community in plant population dynamics. Trends Ecol Evolut 1998; 13:171-177.

Whitson T.D. (ed.) *Weeds of the West*. 5th ed. Jackson, WY: Pioneer of Jackson Hole, 1996.

Wilkinson D.M. The evolutionary ecology of mycorrhizal networks. Oikos 1998; 82:407-410.

Wilson D.S. Biological communities as functionally organized units. Ecology 1997; 78:2018-2024.

Woodmansee R. "Comparative nutrient cycles of natural and agricultural ecosystems: a step toward principles." In *Agricultural Ecosystems Unifying Concepts*, R. Lowrance, B.R. Stinner, G.J. House, eds. New York, NY: John Wiley, 1984; pp. 145-156.

Zavaleta E.S., Hobbs R.J., Mooney H.A. Viewing invasive species removal in a whole-ecosystem context. Trends Ecol Evolut 2001; 16:454-459.

Zobel M., Moora M., Haukioja E. Plant coexistence in the interactive environment: arbuscular mycorrhiza should not be out of mind. Oikos 1997; 78:202-208.

CHAPTER 7

RISING CARBON DIOXIDE AND WEED ECOLOGY

LEWIS H. ZISKA
Alternate Crop and Systems Laboratory, USDA-ARS,
Beltsville, MD 20705 USA

Documented and projected changes in the concentration of atmospheric carbon dioxide (CO_2) and other gases suggest potential changes in climate stability that could negatively impact human systems. This would include negative impacts on agricultural crops and associated weeds. Worldwide crop production losses from weeds are estimated at 12%, with a 25% loss in undeveloped agronomic systems (Parker and Fryer, 1975). Climatic outcomes regarding crops and weed/crop interactions have been addressed in several seminal reviews (Patterson and Flint, 1980; Patterson, 1993, 1995 a, b; Bunce and Ziska, 2000). The current review acknowledges the importance of such climatic variables, but recognizes the uncertainty of future climatic change. At present there is greater consensus on the direct effect of CO_2 on plant systems (Cure and Acock, 1986; Kimball, 1983, 1993); consequently, the principal focus of the current review is to integrate what is known regarding the direct effects of rising CO_2 on weed biology.

OVERVIEW OF RISING CO_2

Based on modern records of CO_2 data obtained from the Mauna Loa observatory, atmospheric CO_2 concentration has already risen from 315 µatm in the mid 1950s to approximately 370 µatm today (Keeling and Whorf, 2001). The Mauna Loa data represent the longest continuous direct record of measured atmospheric CO_2 for the 20th century. As the global demand for energy and agricultural land increases, deforestation and fossil fuel burning will continue to be anthropogenic sources of atmospheric CO_2. The present rate of increase, approximately 1.5 µatm $year^{-1}$ is expected to continue with concentrations approaching 600 µatm by the end of the 21st century (Intergovernmental Panel on Climate Change, (IPCC) scenarios IS 92e and IS 92 a, respectively in Schimel et al., 1996).

Measurements at Mauna Loa are part of the carbon dioxide information analysis center (cdiac.esd.ornl.gov/home.html) which monitors background atmospheric CO_2. This monitoring network samples air at high elevations away from anthropogenic sources. However, for most areas where people live, or crops grow, there should be macro-environmental differences in CO_2 related to the amount of industrialization, wind speed, dark respiration that influence both the average daily CO_2 as well as the diurnal CO_2 pattern. For example, data obtained for the city of Phoenix show the presence of what the researchers describe as an "urban CO_2 dome" (Idso et al., 1998, 2001). CO_2 data on four separate transects from the outskirts of the city to the city center in January of 1998 and again in July of 1998 show a consistent increase in CO_2

Inderjit (ed.), Weed Biology and Management, 159-176.

concentration from 370 to 555 μatm. Similarly, recent global data comparing ground-level CO_2 daily averages show much higher values for metropolitan centers near Washington, D.C. and Sydney, Australia (~ 460 μatm) than other rural locations (Ziska et al., 2001). This suggests that atmospheric CO_2 increases anticipated for the globe as a whole may already be occurring in some areas as a result of urbanization.

On a global basis, the increase in atmospheric CO_2 has been accompanied by increases in other radiation trapping gases including methane (CH_4) at a rate of 0.9% per year; nitrous oxide (N_2O) at a rate of 0.25% per year; and, chlorofluorocarbons (CFCs) at a rate of 4% per year (Ashmore, 1990; MacCracken et al., 1990). Recent IPCC reports based, in part, on a National Academy of Sciences assessment indicate that the rise of these "greenhouse" gases could lead to a 3-12 °C increase in global surface temperatures with subsequent effects on climate, although the degree of temperature/climatic conditions remains uncertain (Cushman, 1999).

DIRECT EFFECTS OF CO_2 ON PLANT FUNCTION

While the extent of future temperature increases remains undetermined, there is acknowledged consensus regarding the direct impact of increasing CO_2 on plant physiology. Elevated CO_2 concentrations stimulate net photosynthesis in C_3 plants by increasing the CO_2 concentration gradient from air to leaf and by reducing the loss of CO_2 through photorespiration. Because oxygen competes with CO_2 for active sites of the enzyme ribulose-bisphosphate carboxylase/oxygenase (Rubisco), increasing CO_2 (relative to O_2) stimulates photosynthesis and reduces carbon lost via photorespiration. Because the competition between O_2 and CO_2 for active sites is temperature sensitive (i.e., favoring O_2 with increasing temperature), the stimulation of net photosynthesis by elevated CO_2 should increase as the temperature increases. For plants that rely solely on the C_3 photosynthetic pathway, (~95% of all plant species), increasing CO_2 and temperatures associated with climate change should be favorable for increased growth. Alternatively, plants with the C_4 photosynthetic pathway (~4% of all known plant species) have an internal mechanism for concentrating CO_2 around Rubisco; therefore increases in external CO_2 concentration should have little effect on net photosynthesis in C_4 plants (Bowes, 1996; Ghannoum et al., 2000). However, one of the most consistent responses of both C_3 and C_4 species to elevated CO_2 is a decrease in stomatal conductance (Morison, 1998; Eamus, 1991; Bunce, 1998).

The decrease in stomatal conductance can result in significant increases in leaf transpiration efficiency (CO_2 assimilated / H_2O transpired) or water use efficiency (dry matter obtained / H_2O transpired). Hence, under water-limiting conditions, elevated CO_2 should result in significant increases in photosynthesis and biomass for both C_3 and C_4 species.

WHAT ABOUT ACCLIMATION?

In the short-term (hours/days), photosynthetic rates of C_3 plants grown at elevated CO_2 are generally greater than the same plants grown at ambient CO_2. However, over time, plants grown at elevated CO_2 may demonstrate a reduction in photosynthetic rate per unit leaf area relative to plants grown at ambient CO_2 when measured at a common CO_2 concentration. This phenomenon, termed photosynthetic acclimation, or down-regulation, was initially thought to occur in response to restricted root volumes associated with plants in small pots (cf Arp, 1991; Thomas and Strain, 1991). However, the appearance of acclimation has been confirmed for a variety of plant species even under field conditions (cf Bunce, 1995). The mechanistic basis for the occurrence of acclimation is unclear, but may be related to an imbalance between carbon sinks and sources which result in sugar-induced feedback changes in the transcription of Rubisco m-RNA (see Sheen, 1990; Sage, 1994; Van Oosten et al., 1994; Moore et al., 1999). In turn, such source-sink imbalances may be related to environmental limitations (e.g. temperature, water, nutrients) that are sub-optimal with a subsequent reduction in new sink development and decreased photosynthetic rates at elevated CO_2 (Patterson and Flint, 1982; Pettersson and McDonald, 1994; Poorter, 1998). However, acclimation to increased CO_2 is not a universal response.

CO_2 AND WEED BIOLOGY

Weeds, strictly speaking, do not exist in nature. The term, "weed" is an anthropogenic concept used to describe a type of plant that is generally recognized as objectionable or undesirable to human activities. However, among such undesirable plants there are some general biological similarities, including colonization of disturbed habitats, vigorous growth, prodigious seed production, and seed longevity (Baker, 1974).

One of the most recognized undesirable characteristics of weedy species is interference in crop production. Human selection of desirable crop species has led to inadvertent selection of other species that simulate or mimic a particular crop (e.g., commercial and wild oat). Therefore, while crops are likely to benefit from the ongoing increase in atmospheric CO_2, growth of weedy competitors is also likely to be stimulated. To date, the majority of studies to determine the impact of rising CO_2 have been conducted for crop and tree species (Poorter, 1993; Kimball, 1983, 1993; Curtis and Wang, 1998); however, sufficient studies on agronomic weeds exist so that relative comparisons can be made.

Overall, C_3 and C_4 crops show a range of response from 1.10 to 2.43x and 0.98 to 1.24x, respectively while C_3 and C_4 weeds show a range of response from 0.95-2.72x and 0.6 to 1.61x, respectively, with a projected doubling of ambient atmospheric CO_2 (See reviews by Patterson and Flint, 1990; Patterson, 1993; Patterson et al., 1999). The range of stimulation for some C_4 weeds, even under optimal water, is somewhat surprising given the expected response of the C_4 pathway to increasing CO_2. Differences in subtype or degree of CO_2 "leakiness" within the bundle sheath do not provide a general explanation to the relative stimulation among C_4 species to elevated CO_2 and further mechanistic information is needed (Ziska et al., 1999a; Ziska and Bunce, 1999; Ghannoum et al., 1998). Overall, the greater range of responses observed for weeds at elevated CO_2 is consistent with the suggestion of Treharne (1989) that weeds have a greater physiological plasticity and genetic diversity relative to crop species.

Another undesirable characteristic of some weedy species is related to alteration of natural systems. In this instance, plant invaders may become dominant to such an extent that natural species are displaced or eliminated. Millions of acres of once healthy, productive rangelands, forests and riparian areas, have been overrun by noxious or invasive weeds with subsequent environmental and economic costs. Initial work by Sasek and Strain (1988, 1989, 1990) on two invasive species, Kudzu

(*Pueraria lobata*) and Japanese honeysuckle (*Lonicera japonica*) showed a significant response to increasing atmospheric CO_2 (e.g., 2.5x for kudzu from 350 to 1000 μatm CO_2, Sasek and Strain, 1990). Recent work on the response of six invasive species, Canada thistle (*Cirsium arvense*), field bindweed (*Convolvulus arvensis*), leafy spurge (*Euphorbia esula*), perennial sowthistle (*Sonchus arvensis*), spotted knapweed (*Centaurea maculosa*) and yellow starthistle (*Centaurea solstitialis*) to future, elevated CO_2 demonstrated a range of responses (overall average of 1.46x), with the largest relative response observed for Canada thistle (1.72x relative to 380 μatm) (Figure 1A). Interestingly, the growth response of these same species to the increase in atmospheric CO_2 that occurred during the 20th century (i.e. 285 to 380 μatm) was much higher, averaging 2.1x (Figure 1B) with Canada thistle again having the strongest response (2.8x). The relative stimulation of these invasives is approximately 3x greater than any other species examined over the same range of CO_2 concentrations (Sage, 1995) and suggests the possibility that a strong response to recent increases in atmospheric CO_2 may be a common characteristic among invasive species.

Plants may also be deemed weeds based on their negative impact on public health. Such impacts may occur through allergenic reactions, skin irritations, mechanical injury, or internal poisoning (Ziska, 2001a). To date, little is known regarding the impact of rising atmospheric CO_2 on such species. One exception is common ragweed (*Ambrosia artemisiifolia*). Ragweed is a ubiquitous weed of agricultural crops and vacant lots. Like most wind pollinated (anemophilous) plants, ragweed sheds enormous quantities of pollen in order to achieve reproductive success. Allergies or "hay fever" are a reaction to inhalation of these airborne pollens. A large skin-test survey demonstrated that approximately 29 million Americans are ragweed sensitive (Gergen et al., 1987).

Increasing CO_2 during the 20th century (280 to 370 μatm.) resulted in a significant stimulation of leaf photosynthesis (2.7x), biomass (1.4x) and pollen production (2.3x). The response of ragweed to future CO_2 (600-700 μatm) increases also indicated a significant response to pollen production (1.6-1.9x) above ambient CO_2 levels (Ziska and Caulfield, 2000; Wayne et al., 2002). Although additional data, particularly under field conditions are needed, these preliminary studies indicate a potential direct effect of CO_2 on human health.

HOW DO ENVIRONMENTAL LIMITATIONS MODIFY THE CO_2 ENHANCEMENT EFFECT IN WEEDY SPECIES?

As temperatures increase, larger stimulation of photosynthetic rates by elevated CO_2 should occur at higher temperatures based on photorespiratory carbon loss (Long, 1991); however, this is not always observed. For velvetleaf (*Abutilon theophrasti*), increasing day/night temperature with a doubling of CO_2 can either decrease leaf area and biomass (Ackerly et al., 1992; Coleman and Bazzaz, 1992) or have no effect (Tremmel and Patterson, 1993). Similarly, CO_2 enrichment and temperature did not interact for two C_4 weed species barnyardgrass and goosegrass (*Echinochloa crus-galli* and *Eleusine indica*, respectively) (Potvin and Strain, 1985). Alternatively, dry matter in spurred anoda (*Anoda cristata*) increased at 700 μatm when day/night temperatures increased from 26/17 to 32/23 °C (Patterson et al., 1988). At present, there is little unequivocal evidence for significant differences in response to CO_2 with increasing temperatures. With respect to sub-optimal temperatures (i.e. from 0-15°C), data on the modification of CO_2 enhancement in weedy species is almost non-existent. One exception is work by Bunce (2000) with dandelion (*Taraxacum officinale*). Bunce states, "It is premature to conclude that low temperatures will necessarily reduce the relative stimulation of photosynthesis caused by rising atmospheric carbon dioxide." Overall, theoretical limitations based on biochemical models have generally assumed no growth temperature effects on carboxylation kinetics (V_{Cmax}) and/or no limitation on potential rate of electron transport (J_{max}); however, recent studies with velvetleaf indicate that long-term adaptation to growth temperature may adjust both parameters, lowering the temperature sensitivity of CO_2-induced photosynthetic stimulation (Bunce, 2000; Ziska, 2001b).

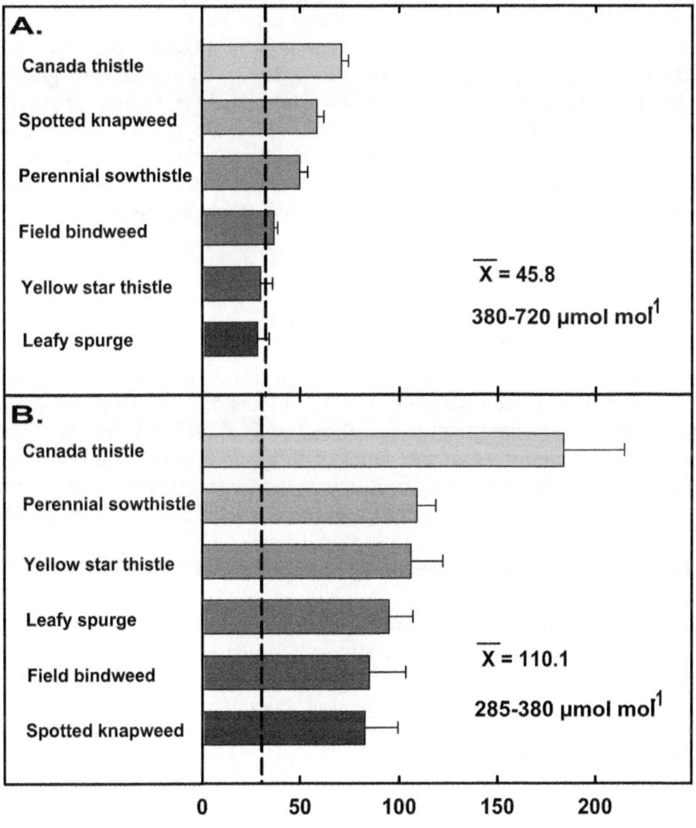

Figure 1. A. The average relative increase (%) determined from the ratio of dry weight at elevated (719 µatm), relative to ambient (380 µatm) CO_2 averaged over three experimental runs for each of six invasive weeds at a common harvest date of 54 days after sowing. Dotted line is the average response of plants as reported by Kimball (1983). B. Same as A., but for CO_2 of 284 and 380 µatm CO_2. Dotted line is the published response of plants form 270-380 µatm (Sage, 1995). Bars are ± SE.

Potential increases in global temperature may be accompanied by changes in both amount and frequency of precipitation. However, because of the indirect effect of CO_2 on stomatal aperture, even under water limiting conditions, elevated CO_2 can still stimulate plant photosynthesis and growth (Patterson, 1986; Choudhuri et al., 1990). For some C_4 weeds, increased photosynthesis and growth at elevated CO_2 may only occur under dry conditions, presumably due to increased water use efficiency (WUE, the ratio of carbon gained to water lost) and reduced water stress. Although water shortages should not limit the response to elevated CO_2, no assessment on CO_2 response under flooded conditions is available for weedy species.

Sub-optimal amounts of nutrients are frequently cited as inducing photosynthetic acclimation to elevated CO_2 (see Poorter,1998; Pettersson and McDonald, 994). For lambsquarters, low N availability did reduce photosynthetic response to CO_2, but only at low temperatures (Sage et al., 1990). Similarly, growth response to CO_2 was unaffected by N level in wild radish (*Raphanus raphanistrum*) (Jablonski,1997), and sicklepod (*Cassia obtusifolia*) (Patterson and Flint, 1982). However, the growth response to elevated CO_2 was reduced, but not eliminated, in showy crotalaria (*Crotalaria spectabilis*) (Patterson and Flint, 1982). Overall, it is reasonable to suppose a reduction in the relative photosynthetic and growth response of weeds to elevated CO_2 is likely under extreme nutrient deficiencies. However, such deficiencies may be less relevant to agronomic situations where nutrients are usually optimal for crop growth.

WEED/CROP COMPETITION

It is sometimes assumed that because different plant species do not compete for CO_2 directly, CO_2 is less important than competition for nutrients or water (Radosevich et al., 1997). However, any resource that affects the growth of an individual alters its ability to compete with individuals of the same or different species (Patterson and Flint, 1990). Hence, competition not only occurs in response to limited resources, but also occurs when species respond differently to resource enhancement. For example, in weed/crop interactions, additions of supra-optimal level of N can increase weedy competition and reduce crop yields because weeds utilize a given resource (N) more efficiently than crops (Vengris et al., 1955; Appleby et al., 1976; Carlson and Hill, 1985). Analogous to the N response, differential inter and intra-specific responses to CO_2 have been observed for the increase in atmospheric carbon dioxide which has already occurred during the 20th century (Sage, 1995) and that projected for the end of the 21st century (Poorter, 1993). Although, in general, the relative effect of increasing CO_2 is greater for C_3 than C_4 species, species-specific responses demonstrate a wide range of relative enhancement within C_3 and C_4 species (e.g., Patterson and Flint, 1990).

It is the difference in relative stimulation between C_3 and C_4 species that is most applicable to weed/crop interactions. It is clear that a disproportionate number of weeds have C_4 metabolism. For example, among the "worlds worst weeds" 14 out of 18 are C_4 (Holm et al., 1977) whereas of the 86 plant species that make up 95% of the world food supply, only five are C_4 (Patterson, 1995a). Because the C_4 photosynthetic pathway is overly represented in troublesome weedy species, many experiments and most reviews concerned with weed competition and rising CO_2 have reported on C_3

crop-C_4 weed interactions (Alberto et al., 1996; Patterson et al., 1984; Patterson 1986, 1993; Patterson and Flint, 1990) ("Troublesome" refers to weeds that result in a significant reduction in crop yield or quality; Bridges, 1992). Increasing CO_2 increased the crop-weed (i.e., the C_3/C_4) biomass ratio in all of these studies (Bunce and Ziska, 2000). Based on the relative proportion of C_3 and C_4 photosynthesis among crops and weeds, many global models have assumed less yield loss due to weedy competition as atmospheric CO_2 increases (Rosenzweig and Hillel, 1998).

However, the idea that crops are fundamentally C_3 and weeds C_4, and that weed competition will consequently decrease with rising atmospheric CO_2 is overly simplistic. Clearly there are C_4 crops of economic and nutritional importance (e.g., corn (*Zea mays*), grain sorghum (*Sorghum bicolor*), pearl millet (*Pennisetum americanum*), sugarcane (*Saccharum officinarum*)), and many important C_3 weeds (e.g. lambsquarters, wild oat (*Avena fatua*), field bindweed, Canada thistle). Crop-weed interactions vary significantly by region; consequently depending on temperature, precipitation, soil, etc. C_3 and C_4 crops may interact with C_3 and C_4 weeds (Bridges, 1992). For example in the U.S. 4 out of 5 of the most troublesome weeds for soybean (*Glycine max*) are C_3 (Bridges, 1992) whereas on a global basis the most troublesome weeds for soybean are predominantly C_4 (Holm et al., 1977). Furthermore, many of the studies which have included both weed and crop species have been analyzed in the context of weed/crop competition even though species were grown individually. Some examples include reductions in weed/crop biomass ratios at elevated (CO_2) in itchgrass (*Rottboellia exaltata*) and soybean (C_4/C_3 comparison) at elevated (CO_2) (Patterson and Flint, 1980), and barnyardgrass, goosegrass and southern crabgrass (*Digitaria ciliaris*) (all C_4 species) relative to soybean (C_3) (Patterson, 1986). While such studies are necessary to examine species responses, competitive outcomes are difficult to predict on the basis of the response of isolated species (see Bazzaz and Carlson, 1984; Zangerl and Bazzaz, 1984; Bazzaz and Garbutt, 1988 *inter alia*).

To date only a few studies have examined the response of crop and weed species to elevated CO_2 in competition. These investigations fall into two broad categories, those where the photosynthetic pathway differs, and those where the pathway is the same. With one exception, those studies with differing pathways have contrasted C_3 crops with C_4 weeds (Table 1). Increasing CO_2 increased the crop/weed ratio in all of these studies, which is consistent with the relative stimulation of the C_3 and C_4 photosynthetic pathways to elevated CO_2. The exception was grain sorghum (C_4 crop) with common cocklebur (*Xanthium strumarium*, C_3 weed) (Ziska, 2001c). This comparison is significant because cocklebur is listed as a troublesome weed in nine states in the southern or central United States (Bridges, 1992). In this study plant relative yield (PRY) increased significantly for common cocklebur and decreased significantly for sorghum in competitive mixtures at elevated CO_2 (Ziska, 2001c). With respect to competitive comparisons for the same photosynthetic pathway, no competitive comparisons are available for C_4 crops and weeds, even though studies of individual C_4 crops and weeds have indicated a greater relative stimulation to CO_2 for weedy species (Ziska and Bunce, 1997). For comparisons of C_3 crops and weeds, weed/crop biomass was stimulated in all cases where weed and crop emerged simultaneously (Table 1). In a study comparing lambsquarters and sugarbeet (*Beta vulgaris*), the competitive advantage of sugarbeet at elevated CO_2 was attributed to late emergence of the weed species within the experiment (Houghton and Thomas, 1996).

While these studies have compared the relative stimulation of biomass between crops and weeds, only a single investigation has quantified the relative change in yield loss from weedy competition at elevated CO_2 in the field (Ziska, 2000). In this study, soybean was grown at ambient and elevated CO_2 with and without the presence of two weeds, lambsquarters (C_3) and redroot pigweed (*Amaranthus retroflexus*, C_4) at a density of two weeds per meter of row. In a weed-free environment, elevated CO_2 resulted in a significant increase in vegetative dry weight and seed yield at maturity for soybean (33 and 24%, respectively) which is consistent with other studies (Baker et al., 1989). However, for lambsquarters the reduction in soybean seed yield relative to the weed-free condition increased from 28 to 39% with elevated CO_2, with a 65% increase in the average vegetative dry weight of lambsquarters. In contrast, for pigweed, soybean seed yield losses diminished with elevated CO_2 from 45 to 30%, with no change in vegetative dry weight of pigweed. Interestingly, the presence of either weed negated the ability of soybean to respond either vegetatively or reproductively to enhanced CO_2 (Ziska, 2000). This is of interest since almost all CO_2 enhancement studies involving crops do not consider weed/crop interactions (Kimball, 1983, 1993).

HOW DO ENVIRONMENTAL LIMITATIONS (TEMPERATURE, WATER AND NUTRIENTS) ALTER THE CO_2 EFFECT ON WEED/CROP COMPETITION?

As with field evaluations of seed loss, only a single experiment has examined the interaction between temperature, CO_2 and crop/weed competition (Alberto et al., 1996). In this phytotron study, an assessment of competitive changes based on PRY and replacement series diagrams comparing a C_3 crop (rice, *Oryza sativa*) and a C_4 weed (cockspur, barnyardgrass, *Echinochloa glabrescens*) indicated that rice was favored over the weed with elevated CO_2 (594 μatm) at day/night temperatures of 27/21 °C; but that at day/night temperatures of 37/29 °C, a greater reduction in PRY in rice was observed relative to the C_4 weed. This suggested that concurrent changes in CO_2 and temperature could still favor a C_4 weed over a C_3 crop. Although no other studies have examined the role of temperature and CO_2 in weed/crop competition, there are a number of potential interactive effects. Greater increases in night-time relative to day time temperatures for example, could decrease seed production to a greater extent in crop relative to weed species (e.g., cowpea; Ahmed et al., 1993) with a differential effect on sink strength, photosynthetic enhancement and competition. In addition, differential response of seed emergence to temperature could influence weed/crop competition at elevated CO_2 (e.g., Houghton and Thomas, 1996). Weeds of tropical origin show a large stimulation to small changes in air temperature (Flint et al., 1984; Flint and Patterson, 1983; Patterson, 1993), but it is unclear if a greater synergy with rising CO_2 would be anticipated for such weeds relative to warm season crops. Another effect of temperature would be a potential alteration in the composition of weed/crop mixtures since increasing temperature would change ranges of both agronomic (Patterson et al., 1979) and noxious weeds (Patterson, 1995a). Such range changes could indirectly alter weed/crop competition to elevated CO_2 by increasing the ratio of C_4 weeds (via northward expansion) to C_3 crops (Patterson, 1993, 1995 a, b, Patterson et al., 1999).

Table 1. Competition studies of crop and weed species as a function of CO_2.

Weed	Crop	Favored?	Environment	Reference
1. Different Photosynthetic Pathway				
Sorghum halepense (C_4)	Meadow fescue	Crop	Glasshouse	Carter and Peterson (1983)
Sorghum halepense (C_4)	Soybean	Crop	Chamber	Patterson et al. (1984)
Echinochloa glabrescens (C_4)	Rice	Crop	Glasshouse	Alberto et al. (1996)
Paspalum dilatatum (C_4)	Grasses	Crop	Chamber	Newton et al. (1996)
Grasses (C_4)	Lucerne	Crop	Field	Bunce (1993)
Amaranthus retroflexus (C_4)	Soybean	Crop	Field	Ziska (2000)
Xanthium strumarium (C_3)	Sorghum	Weed	Glasshouse	Ziska (2001c)
2. Same Photosynthetic Pathway				
Chenopodium album (C_3)	Soybean	Weed	Field	Ziska (2000)
Taraxacum officinale (C_3)	Lucerne	Weed	Field	Bunce (1995)
Plantago lanceolata (C_3)	Grasses	Weed	Chamber	Newton et al. (1996)
Taraxacum and *Plantago* (C_3)	Grasses	Weed	Field	Potvin and Vasseur (1997)

Whether drought alters the competitive relationships between weeds and crops at elevated CO_2 is not known. Studies that have examined the interactive effects of increasing CO_2 and water stress on annual weed and crops in competition are not available. For a given photosynthetic pathway, crops and weeds have relatively similar responses to drought (Patterson, 1995b), although the overall impact of weeds may be reduced because of decreased growth of both crops and weeds in response to water availability (Patterson, 1995b). Although competition was not determined directly, the proportion of weed biomass increased with CO_2 to a similar extent in wet and dry treatments in a pasture mixture (Newton et al., 1996).

As with the interactive effects of CO_2 and water availability, there is no study addressing the interaction between increasing CO_2 and nutrient availability for crop/weed competition. Presumably as with water stress, reductions in growth induced by nutrient availability may lessen the impact of weeds on crop yield (Patterson, 1995b). Although nutrient availability can modify the enhancement response to elevated CO_2 in individual weeds and crops, the relative effect on weed/crop competition will be species specific and will depend, in part, on soil/plant interactions including degree of rhizobial symbiosis, mycorrhizal interactions, etc. Schenk et al. (1997) suggested that elevated CO_2 could affect competitive outcomes between ryegrass (*Lolium* spp.) and white clover (*Trifolium repens*) depending on the nitrogen supply.

IMPACT OF ELEVATED CO_2 ON CHEMICAL WEED MANAGEMENT

Clearly, climatic change associated with increasing atmospheric CO_2 can potentially alter abiotic variables such as temperature, wind speed, soil moisture, and atmospheric humidity. Alteration of such variables can influence the effectiveness of herbicide applications (Muzik, 1976). In addition, these same environmental variables can affect crop injury due to herbicide application. A recent economic evaluation based on anticipated climate change suggested that increasing temperature increased pesticide cost variance for corn, potatoes and wheat, while decreasing it for soybean (Chen and McCar, 2001).

Although the effect of such variables on herbicide efficacy has been well studied, can rising CO_2 directly alter chemical weed control? Theoretically, efficacy could be reduced in one of two ways: either the amount of herbicide entering the plant is reduced, or the physiological action of the herbicide is impaired in some way. Carbon dioxide induced changes that could, potentially, reduce foliar absorption include stomatal closure, reductions in stomatal number and changes in leaf or cuticular thickness. Decreased stomatal conductance with increasing CO_2 could also reduce transpiration and uptake of soil-applied herbicides. The duration of time weeds are susceptible to herbicide treatment could also be affected because at elevated CO_2, time spent in the seedling stage (usually the time of greatest effectiveness in post-emergence, foliar applied herbicides) could be reduced if CO_2 stimulates growth. In addition, for perennial weeds, increasing CO_2 could stimulate greater below ground growth (e.g., rhizomes, tubers, roots), making herbicidal control of such weeds more difficult and costly.

Alternatively, there are a number of CO_2-induced physiological changes, specific for each herbicide. For example for glyphosate, CO_2-induced reductions in protein content per gram of tissue could result in less demand for aromatic amino acids.

Since glyphosate inhibits aromatic amino acid production through the shikimic acid pathway, this could influence the toxicity of the herbicide. Another example is glufosinate, a herbicide that inhibits glutamine synthetase with a subsequent decrease in glutamine and serine. Lack of these amino acids inhibits the transanimation of glyoxylate into glycine during photorespiration. In previous studies when photorespiration is suppressed (either by increasing CO_2 or decreasing O_2), the efficacy of glufosinate is reduced (LaCuesta et al., 1992; for a C_3 vs C_4 comparison see Wendler et al., 1993).

Although much remains to be known regarding the impact of rising CO_2 on herbicide efficacy, some initial studies have been conducted. In a greenhouse experiment, herbicide efficacy was determined by examining whether a doubling of CO_2 altered glyphosate tolerance in two weedy species, redroot pigweed and lambsquarters (Ziska et al., 1999b). Tolerance was determined by measuring growth (biomass) following spraying at 0 (control, 0.112 kg ai ha^{-1} (10% of recommended rate) and 1.12 kg ai ha^{-1} (100% or recommended rate) over a 14 day period in four separate trials which differed in time of application. Independently of CO_2, growth of the C_4 weed, redroot pigweed was significantly reduced and eliminated altogether at glyphosate application rates of 0.112 and 1.12 kg ai ha^{-1}, respectively. In contrast to the ambient CO_2 treatment, an application rate of 0.112 kg ai ha^{-1} had no effect on growth, and a 1.12 kg ai ha^{-1} reduced, but did not eliminate growth in elevated CO_2 grown lambsquarter. Glyphosate tolerance did increase with plant size at the time of application; however, differences in glyphosate tolerance between CO_2 treatments could not be explained by size alone. Similarly for a C_3 perennial weed, quackgrass (*Elytrigia repens*), glyphosate tolerance increased for two different cohorts for application rates of 2.24 kg ai ha^{-1} (100% of recommended rate) in climate- controlled glasshouse (Ziska and Teasdale, 2000).

The basis for the increased tolerance to glyphosate at elevated CO_2 in these experiments is unclear. Short-term switching of the quackgrass to the elevated CO_2 condition prior to spraying did not increase tolerance, suggesting that stomatal closure did not play a factor (Ziska and Teasdale, 2000). Specific evaluations of uptake, allocation and physiological interaction are required before a mechanistic basis for increased glyphosate tolerance at elevated CO_2 can be established for these species.

Recent unpublished data for Canada thistle, a noxious agronomic weed grown at elevated CO_2 under field conditions for a two year period indicated significant increases in root biomass (+70-150%) relative to the ambient condition (Figure 2). However, in contrast to the ambient CO_2 treatment, glyphosate and glufosinate applications had no effect on shoot and root regrowth, respectively, when Canada thistle was grown at the elevated CO_2 concentration (i.e. shoot regrowth for glyphosate in 2000, and root regrowth for glufosinate in 2001 did not differ from unity, Figure 3).

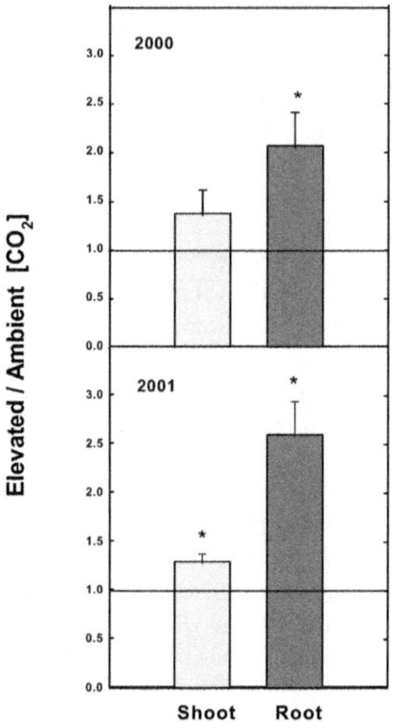

Figure 2. Shoot and root biomass at elevated (771 μatm) relative to ambient (421 μatm) CO_2 for field grown Canada thistle in 2000 and 2001 at the time of herbicide application. Shoot biomass refers to all above ground herbaceous material. Root biomass was obtained from soil subsamples (2.36 L in volume, to a depth of 30 cm). * indicates a significant difference relative to unity (i.e. no CO_2 treatment effect). Bars are ±SE.

For this study, visual shoot symptoms did not differ as a function of CO_2 concentration. Rather, CO_2-induced changes in herbicide tolerance were consistent with the idea that greater CO_2 could stimulate greater below-ground growth in perennial weeds, making herbicidal control more costly.

It can be argued that CO_2-induced changes in herbicide tolerance are irrelevant given the rate of atmospheric CO_2 increase (i.e. other herbicides will be developed in the future). However, herbicide use can persist over decades (e.g. 2,4-D) coinciding with significant increases in atmospheric CO_2 (e.g. 310-372 ϒatm from 1950 to 1999). Given the investment of large companies in genetically modified crops and their associated herbicides, it seems more likely that use of current herbicides will persist for decades. Furthermore, as previously stated, background CO_2 is not uniform and can differ from urban to rural areas. For example, ragweed control in an urban setting (i.e. higher CO_2 and temperature) may be less effective than that for a farm setting at similar application rates (Ziska and Caulfield, 2000). Obviously chemical control can still be obtained if additional sprayings occur, or if concentration increases, but this could, potentially, alter the economic and environmental costs.

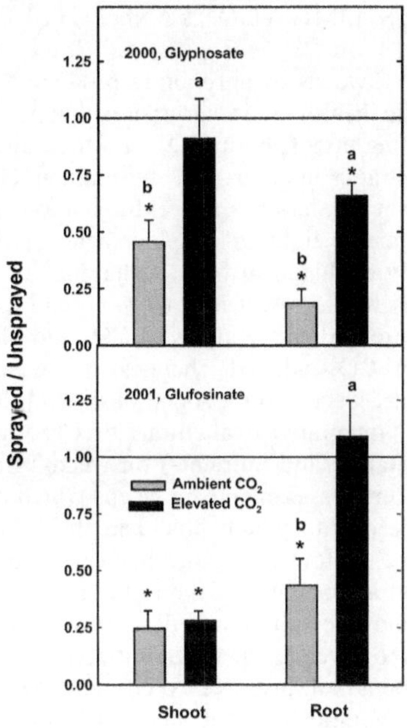

Figure 3. Ratio of sprayed to unsprayed biomass for shoots and roots grown at the CO_2 given in Figure 2 six weeks after treatment with recommended rates of either glyphosate (2000) or glufosinate (2001). * indicates a significant difference relative to unity (i.e. no CO_2 treatment effect). Different letters indicate a CO_2 treatment effect for either shoot or roots. Bars are $\pm$SE.

ELEVATED CO_2 AND BIOLOGICAL CONTROL

In theory, elevated CO_2 could alter the efficacy of any biocontrol agent by changing the growth, development and reproductive output of any selected weed. Effects of CO_2 on increasing the C:N ratio could also affect biocontrol by changing the behavior, growth and reproductive cycle of weedy herbivores (Gutierrez, 2000).

ELEVATED CO_2 AND MECHANICAL CONTROL

Tillage is widely regarded as a means of mechanical weed control in agronomic systems. However, if increasing atmospheric CO_2 results in stimulation of root and rhizome growth, (e.g. Rogers et al., 1994), then mechanical tillage may lead to additional plant propagation and subsequent difficulty in control of such weeds (e.g., Canada thistle).

UNCERTAINTIES, LIMITATIONS AND FUTURE DIRECTION

Given the importance of weeds to agronomic production, species diversity (i.e. invasive weeds) and public health, it is surprising that so few data are available assessing the impact of rising atmospheric CO_2. Furthermore, most of the data that are available are based on studies in controlled environment chambers or glasshouses. This represents a significant limitation since extrapolation of such studies to field environments may differ due to light or edaphic factors (Ghannoum et al., 1997). Although weeds are usually excluded in field evaluations of CO_2 sensitivity in crop species, inclusion of weeds would result in a more certain assessment of crop/weed interactions. Overall, a more realistic analysis of CO_2 and plant biology should also include: quantification of CO_2-induced changes in secondary compounds and potential allelopathic effects, weed crop associations which include perennial crops, C_4 weed/crop interactions, integrated evaluations of CO_2 and other environmental changes (e.g. water, temperature and nutrients) on weedy growth, reproduction and potential crop losses, long-term assessments on weeds which could alter public health (e.g. poison ivy) and experiments which elucidate the basis and degree of CO_2-induced changes in chemical, biological and mechanical weed control.

The ongoing increase in atmospheric carbon dioxide concentration, and potential changes in temperature and precipitation will almost certainly alter the growth, reproduction and location of weedy species. Initial results regarding changes in weed/crop competition, spread of invasive weeds and weed-mediated changes in public health suggest unfavorable consequences in human systems. Overall, our current understanding regarding the impact of rising CO_2 with respect to weed establishment, growth, reproduction and competition is limited. Yet, the environmental and economic costs of not understanding these impacts, and the appropriate control measures, may be substantial. To that end, it is hoped that this paper will serve as a guide for interested researchers in assessing the importance of past and current increases in atmospheric CO_2 to weed biology; and will elucidate key areas where additional information is needed.

REFERENCES

Ackerly D.D., Coleman J.S., Morse S.R., Bazzaz F.A. CO_2 and temperature effects on leaf area production in two annual plant species. Ecology 1992; 73:1260-1269.

Ahmed F.E., Hall A.E., Madore M.A. Interactive effect of high temperature and elevated carbon dioxide concentration on cowpea (*Vigna unguiculata* (L.) Walp.). Plant Cell Environ 1993;16:835-842.

Alberto A.M., Ziska L.H., Cervancia C.R., Manalo P.A. The influence of increasing carbon dioxide and temperature on competitive interactions between a C_3 crop, rice (*Oryza sativa*) and a C_4 weed (*Echinochloa glabrescens*). Aust J Plant Physiol 1996; 23:795-802.

Appleby A.P., Olson P.D., Colbert D.R. Winter wheat yield reduction from interference by Italian ryegrass. Agron J 1976; 68:463-466.

Arp W.J. Effects of source-sink relations on photosynthetic acclimation to elevated CO_2, Plant Cell Environ 1991; 14:869-875.

Ashmore M. The greenhouse gases. Trends Ecol Evol 1990; 5:296-297.

Baker H.G. The evolution of weeds. Ann Rev Ecol Sys 1974; 5:1-24.

Baker J.T., Allen L.H. Jr., Boote K.J., Jones P., Jones J.W. Response of soybean to air temperature and carbon dioxide concentration. Crop Sci 1989; 29:98-105.

Bazzaz F.A., Carlson R.W. The response of plants to elevated CO_2. I. Competition among an assemblage of annuals at two levels of soil moisture. Oecologia 1984; 62:196-198.

Bazzaz F.A., Garbutt K. The response of annuals in competitive neighborhoods: effects of elevated CO_2. Ecology 1988; 69:937-946.

Bowes G. "Photosynthetic responses to changing atmospheric carbon dioxide concentration." In *Photosynthesis and the Environment.* Neil R. Baker, ed. Dordrecht, Netherlands: Kluwer Publishing, 1996; pp 387-407.

Bridges D.C. *Crop Losses Due to Weeds in the United States.* Weed Science Society of America, Champaign, IL, USA, 1992.

Bunce J.A., Ziska L.H. "Crop ecosystem responses to climatic change: crop/weed interactions." In *Climate Change and Global Crop Productivity,* KR Reddy, HF Hodges, eds. New York, NY, CABI Publishing 2000; pp 333-348.

Bunce J.A. Growth, survival, competition, and canopy carbon dioxide and water vapor exchange of first year alfalfa at an elevated CO_2 concentration. Photosynthatica 1993; 29:557-565.

Bunce J.A. Long-term growth of alfalfa and orchard grass plots at elevated carbon dioxide. J Biogeo 1995; 22:341-348.

Bunce J.A. Effects of humidity on short-term responses of stomatal conductance to an increase in carbon dioxide concentration. Plant Cell Environ 1998; 21:115-120.

Bunce J.A. Acclimation to temperature of the response of photosynthesis to increased carbon dioxide concentration in *Taraxacum officinale.* Photosyn Res 2000; 64:89-94.

Carlson H.L., Hill J.E. Wild oat (*Avena fatua*) competition with spring wheat: effects of nitrogen fertilization. Weed Sci 1985; 34:29-33.

Carter D.R., Peterson K.M. Effects of a CO_2 enriched atmosphere on the growth and competitive interaction of a C_3 and C_4 grass. Oecologia 1983; 58:188-193.

Chen C.C., McCar, B.A. An investigation of the relationship between pesticide usage and climate change. Clim Change 2001; 50:475-487.

Choudhuri U.N., Kirkham M.B., Kanemasu E.T. Carbon dioxide and water level effects on yield and water use of winter wheat. Agron J 1990; 82:637-641.

Coleman J.S., Bazzaz F.A. Effects of CO_2 and temperature on growth and resource use of co-occurring C_3 and C_4 annuals. Ecology 1992; 73:1244-1259.

Cure J.D., Acock B. Crop responses to carbon dioxide doubling: a literature survey. Agri Forest Meteorol 1986; 38:127-145.

Curtis P.C., Wang X. A meta-analysis of elevated CO_2 effects on woody plant mass, form and physiology. Oecologia 1998; 113:299-313.

Cushman R.M. "Global climatic-change modeling and monitoring." In *Encyclopedia of Environmental Science.* D. E. Alexander, R. W. Fairbridge, eds. Dordrecht: Kluwer Academic Publishers, 1999; pp 101-132.

Eamus D. The interaction of rising CO_2 and temperature with water-use-efficiency. Plant Cell Environ 1991; 14:843-852.

Flint E.P., Patterson D.T., Mortensen G.H., Beyers J.L. Temperature effects on growth and leaf production in three weed species. Weed Sci 1984; 32:655-663.

Flint E.P., Patterson D.T. Interference and temperature effects on growth in soybean (*Glycine max*) and associated C_3 and C_4 weeds. Weed Sci 1983; 31:193-199.

Gergen P.J., Turkeltaub P.C., Kovar M.C. The prevalence of allergic skin test reactivity to eight common aeroallergens in the U.S. population: results from the second National Health and Nutrition Examination survey. J Allergy Clinical Immun 1987; 80, 669-679.

Ghannoum O., von Caemmerer S., Barlow E.W.R., Conroy J.P. The effect of CO_2 enrichment and gas exchange of a C_3 (*Panicum laxum*) and a C_4 (*Panicum antidotale*) grass. Aust J Plant Physiol 1997; 24:227-237.

Ghannoum O., Siebke K., von Caemmerer S., Conroy J.P. The photosynthesis of young C_4 Panicum leaves is not C_3-like. Plant Cell Environ 1998; 21:1123-1131.

Ghannoum O., von Caemmerer S., Ziska L.H., Conroy J.P. The growth response of C_4 plants to rising atmospheric CO_2 partial pressure: a reassessment. Plant Cell Environ 2000; 23:931-942.

Gutierrez, A.P. "Crop ecosystem responses to climatic change: pests and population dynamics". In *Climate Change and Global Crop Productivity,* KR Reddy, HF Hodges, eds. New York: CABI Publishing, 2000; 353-370.

Holm L.G., Plucknett D.L., Pancho J.V., Herberger J.P. *The Worlds Worst Weeds. Distribution and Biology.* Honolulu, HI, University of Hawaii Press, 1977.

Houghton S.K., Thomas T.H. Effects of elevated carbon-dioxide concentration and temperature on the growth and competition between sugar beet (*Beta vulgaris*) and fat-hen (*Chenopodium album*). Asp App Biol 1996; 45:197-204.

Idso C.D., Idso S.B., Balling Jr. R.C. The urban CO_2 dome of Phoenix, Arizona. Phys Geo 1998; 19:95-108.

Idso C.D., Idso S.B., Balling Jr. R.C. An intensive two week study of an urban CO_2 dome in Phoenix Arizona, USA. Atmos Environ 2001; 35:995-1000.

Jablonski L.M. Responses of vegetative and reproductive traits to elevated CO_2 and nitrogen in *Raphanus* varieties. Can J Bot 1997; 75:533-545.
Keeling C.D., Whorf T.P. "Atmospheric CO_2 records from sites in the SIO air sampling network." In *Trends: A Compendium of Data on Global Change.* Carbon Dioxide Information Analysis Center, Oak Ridge National Laboratory, U.S. Department of Energy, Oak Ridge, Tenn., U.S.A 2001.
Kimball B.A. Carbon dioxide and agricultural yield: an assemblage and analysis of 430 prior observations. Agron J 1983; 75:779-788.
Kimball B.A. Effects of increasing atmospheric CO_2 on vegetation. Vegetatio 1993; 104/105:65-75.
LaCuesta M., Munoz-Rueda A., Gonzales-Murua C., Sivak M.N. Effect of phosphinothricin (glufosinate) on photosynthesis and chlorophyll fluorescence emission by barley leaves illuminated under photorespiratory and non-photorespiratory condition. J Exp Bot 1992; 43:159-165.
MacCracken M.C., Budyko M.I., Hecht A.D., Izrael Y.A. *Prospects for Future Climate.* Chelsea, MI, USA, Lewis Publishers Inc. 1990.
Moore B.D., Cheng S.H., Sims D., Seemann J.R. The biochemical and molecular basis for photosynthetic acclimation to elevated atmospheric CO_2. Plant Cell Environ 1999; 22:567-582.
Morison J.I.L. Stomatal response to increased atmospheric CO_2. J Exp Bot 1998; 49:443-452.
Muzik T.J. "Influence of environmental factors on toxicity to plants" In *Herbicides: Physiology, Biochemistry, Ecology*, L.J. Audus, ed. New York: Academic Press, 1976; pp 102-131.
Newton P.C.D., Clark H., Bell C.C., Glasglow E.M. Interaction of soil moisture and elevated CO_2 on the above-ground growth rate, root length density and gas exchange of turves from temperate pasture. J Exp Bot 1996; 47:771-779.
Parker C., Fryer J.D. *Weed control problems causing major reductions in world food supplies.* FAO Plant Protection Bull 1975; 23:83-95.
Patterson D.T. Responses of soybean (*Glycine max*) and three C_4 grass weeds to CO_2 enrichment during drought. Weed Sci 1986; 34:203-210.
Patterson DT Implications of global climate change for impact of weeds, insects and plant diseases. Internat Crop Sci 1993; 1:273-280.
Patterson D.T. Weeds in a changing climate. Weed Sci 1995a; 43:685-701.
Patterson D.T. Effects of environmental stress on weed/crop interactions. Weed Sci 1995b; 28:71-75.
Patterson D.T., Flint E.P. Potential effects of global atmospheric CO_2 enrichment on the growth and competitiveness of C_3 and C_4 weed and crop plants. Weed Sci 1980; 28:71-75.
Patterson D.T., Flint E.P. Interacting effects of CO_2 and nutrient concentration. Weed Sci 1982; 30:389-394.
Patterson D.T., Flint E.P. "Implications of increasing carbon dioxide and climate change for plant communities and competition in natural and managed ecosystems." In *Impact of Carbon Dioxide, Trace Gases and Climate Change on Global Agriculture.* Kimball, B.A., Rosenburg N.J., Allen L.H., Jr., eds. ASA Special Publication No. 53, Madison, WI: American Society of Agronomy, 1990; pp 83-110.
Patterson D.T., Meyer C.R., Flint E.P., Quimby Jr P.C. Temperature responses and potential distribution of itchgrass (*Rottboellia exaltata*) in the United States. Weed Sci 1979; 27:77-82.
Patterson D.T., Flint E.P., Beyers J.L. Effects of CO_2 enrichment on competition between a C_4 weed and a C_3 crop. Weed Sci 1984; 32:101-105.
Patterson D.T., Highsmith M.T., Flint E.P. Effects of temperature and CO_2 concentration on the growth of cotton (*Gossypium hirsutum*), spurred anoda (*Anoda cristata*), and velvetleaf (*Abutilon theophrasti*). Weed Sci 1988; 36:751-757.
Patterson D.T., Westbrook J.K., Joyce R.J.V., Lingren P.D., Rogasik J. Weeds, insects and diseases. Clim Change 1999: 43:711-727.
Pettersson R.P., McDonald J.S. Effects of nitrogen supply on the acclimation of photosynthesis to elevated CO_2. Photosyn Res 1994; 39:389-400.
Poorter H. Interspecific variation in the growth response of plants to an elevated ambient CO_2 concentration. Vegetatio 1993; 104/105: 77-97.
Poorter H. Do slow-growing species and nutrient-stressed plants respond relatively strongly to elevated CO_2? Glob Change Biol 1998; 4:693-697.
Potvin C., Strain B.R. Effects of CO_2 enrichment and temperature on growth in two C_4 weeds, *Echinochloa crus-galli*, and *Eleusine indica.* Can J Bot 1985; 63:1495-1499.
Potvin C., Vasseur L. Long-term CO_2 enrichment of a pasture community. Ecology 1997; 78:666-677.
Radosevich S., Holt J., Ghersa C. "Weed responses to CO_2." In *Weed Ecology: Implications for Management*, Radosevich, S., Holt, J., Ghersa, C., eds., New York: John Wiley and Sons, 1997; pp 241-245.
Rogers H.H., Runion G.B., Krupa S.V. Plant responses to atmospheric CO_2 enrichment with emphasis on roots and the rhizosphere. Environ Poll 1994; 83:155-189.

Rosenzweig C., Hillel D. "Effects on weeds, insects and diseases." In *Climate Change and the Global Harvest*, Rosenzweig, C, Hillel, D., eds. Oxford University Press, UK 1998; pp. 101-122.
Sage R.F., Sharkey T.D., Pearcy R.W. The effect of leaf nitrogen and temperature on the CO_2 response of photosynthesis in the C_3 dicot *Chenopodium album*. Austral J Plant Physiol 1990; 17:135-148.
Sage R.F. Acclimation of photosynthesis to increasing atmospheric CO_2: The gas exchange perspective. Photosyn Res 1994; 39:351-368.
Sage R.F. Was low atmospheric CO_2 during the Pleistocene a limiting factor for the origin of agriculture? Glob Change Biol 1995; 1: 93-106.
Sasek T.W., Strain, B.R. Effects of carbon dioxide enrichment on the growth and morphology of Kudzu (*Pueraria lobata*), Weed Sci 1988; 36:28-36.
Sasek T.W., Strain B.R. Effects of carbon dioxide enrichment on the expansion and size of Kudzu (*Pueraria lobata*) leaves. Weed Sci 1989; 37:23-28.
Sasek T.W., Strain B.R. Implications of atmospheric CO_2 enrichment and climatic change for the geographical distribution of two introduced vines in the USA. Climat Change 1990; 16:31-51.
Schenk U., Jager H.J., Weigel H.J. The response of perennial ryegrass/white clover swards to elevated atmospheric CO_2 concentrations. I. Effects on competition and species composition and interaction with N supply. New Phytol 1997; 135:67-79.
Schimel D., Alves D., Enting I., Heimann M., Joos F., Raynaud D., Wigley T., Prather M., Der Went R., Ehhalt D., Fraser P., Sanhueze E., Zhou X., Jonas P., Charlson R., Rodhe H., Sadasivan S., Shine K.P., Fouquart Y., Ramasawamy V., Solomon S., Srinivasan J., Albritton D., DerWent R., Isaksen I., Lal M., Wuebbles D. "Radiative forcing of climate change." In *Climate Change 1995: The Science of Climate Change*. Houghton J.T., Meira-Filho L.G., Callander B.A., Harris N., Kattenberg A., Maskell K. eds. Cambridge, UK, Cambridge University Press, 1996; pp 23-45.
Sheen J. Metabolic repression of transcription in higher plants. The Plant Cell 1990; 2:1027-1038.
Thomas R.C., Strain B.R. Root restriction as a factor in photosynthetic acclimation of cotton seedlings grown in elevated carbon dioxide. Plant Physiol 1991; 96:627-634.
Treharne K. "The implications of the 'greenhouse effect' for fertilizers and agrochemicals." In *The Greenhouse Effect and UK Agriculture*, Bennett R.C., ed., CAS Paper 19, Ministry of Agriculture, Fisheries and Food, UK, 1989; pp 67-78.
Tremmel D.C., Patterson D.T. Responses of soybean and five weeds to CO_2 enrichment under two temperature regimes. Can J Plant Sci 1993; 73:1249-1260.
Van Oosten J.J., Wilkins D., Besford R.T. Regulation of the expression of photosynthetic nuclear genes by high CO_2 is mimicked by carbohydrates: A mechanism for the acclimation of photosynthesis to high CO_2? Plant Cell Environ 1994; 17:913-923.
Vengris J., Colby W.G., Drake, M. Plant nutrient competition between weeds and corn. Agron J 1955; 47:213-216.
Wayne P., Foster S., Connolly J., Bazzaz F., Epstein P. Production of allergenic pollen by ragweed (*Ambrosia artemisiifolia L.*) is increased in CO_2-enriched atmospheres. Ann Allergy Asthma Immunol 2002; 88:279-282.
Wendler D.J., Haber E., Keddy C. Effect of phosphinothricin (glufosinate) on photosynthesis and photorespiration of C_3 and C_4 plants. Photosyn Res 1993; 24:55-62.
Zangerl A.R., Bazzaz F.A. The response of plants to elevated CO_2. II. Competitive interactions among annual plants under varying light and nutrients. Oecologia 1984; 62:412-417.
Ziska L.H., Bunce J.A. Influence of increasing carbon dioxide concentration on the photosynthetic and growth stimulation of selected C_4 crops and weeds. Photosyn Res 1997; 54:199-208.
Ziska L.H., Bunce J.A. Effect of elevated carbon dioxide concentration at night on the growth and gas exchange of selected C_4 species. Aust J Plant Physiol 1999; 26:71-77.
Ziska L.H., Caulfield F.A. Rising CO_2 and pollen production of common ragweed (*Ambrosia artemiisifolia L.*), a known allergy-inducing species: implications for public health. Austral J Plant Physiol 2000; 27:893-898.
Ziska L.H., Teasdale J.R. Sustained growth and increased tolerance to glyphosate observed in a C_3 perennial weed, quackgrass (*Elytrigia repens*), grown at elevated carbon dioxide. Aust J Plant Physiol 2000; 27:159-166.
Ziska L.H., Sicher R.C., Bunce J.A. The impact of elevated carbon dioxide on the growth and gas exchange of three C_4 species differing in CO_2 leak rates. Physiol Plant 1999a; 105:74-80.
Ziska L.H., Teasdale J.R., Bunce J.A. Future atmospheric carbon dioxide may increase tolerance to glyphosate. Weed Sci 1999b; 47:608-615.
Ziska L.H. The impact of elevated CO_2 on yield loss from a C_3 and C_4 weed in field-grown soybean. Glob Change Biol 2000; 6:899-905.
Ziska L.H. My view: weeds and public health. Weed Sci 2001a; 49:437-438.

Ziska L.H. Growth temperature can alter the temperature dependent stimulation of photosynthesis by elevated carbon dioxide in *Abutilon theophrasti*. Physiol Plant 2001b; 111:322-328.
Ziska L.H. Changes in competitive ability between a C_4 crop and a C_3 weed with elevated carbon dioxide. Weed Sci 2001c; 49:622-627.

CHAPTER 8

MOLECULAR ASPECTS OF HOST-PARASITE INTERACTIONS

OPPORTUNITIES FOR ENGINEERING RESISTANCE TO PARASITIC WEEDS

JAMES H. WESTWOOD

Virginia Tech, Department of Plant Pathology, Physiology, and Weed Science, 410 Price Hall, Blacksburg, VA 24061-0331, U.S.A

Within the discipline of weed science there are few areas in which a fundamental knowledge of weed biology offers more promise for increasing control than that of parasitic weeds. One reason is that the close physiological relationship between parasite and host offers many potential points of intervention in which knowledge of the biology of the interaction could be used to control the parasite. In contrast to autotrophic weeds, parasitic plants interact with crops through direct physical and physiological contact as tissues of the parasite invade the host plant. Although autotrophic weeds can profoundly impact neighboring crops through competition for resources (light, water, mineral nutrients) and allelopathic interactions, the ability of the crop to respond and protect itself is mediated by the environment and thus complicated by many factors.

Parasitic weeds are in direct interaction with the crop, and in many cases are totally dependent, so they are vulnerable to resistance mechanisms produced by the crop. This dependence offers the opportunity for intervention by plant scientists, including the potential that altering genes or gene expression in the host crop could lead to significant improvements in the host's ability to withstand parasitic attack. This chapter will discuss the life cycles of parasitic weeds and recent advances in understanding the biology of parasitism that offer opportunities for using genetic engineering strategies to enhance control of these weeds. Discussion will focus on root parasites of the *Orobanche* and *Striga* genera, for which the body of physiological and molecular data are relatively large.

PARASITIC WEEDS

Orobanche and *Striga* spp. are parasitic plants that live on the roots of a host plant. Hosts for *Orobanche* spp. include many important dicotyledonous crops from such families as the Solanaceae, Fabaceae, Apiaceae, Brassicaceae and Asteraceae. Hosts for *Striga* spp. are monocotyledonous species and a few Fabaceae (Parker and Riches, 1993). *Orobanche* spp. are holoparasites, i.e., lack photosynthetic capacity, and thus

Inderjit (ed.), Weed Biology and Management, 177-198.

rely on their hosts for all reduced carbon substrates, water, and mineral nutrients. *Striga* spp. are hemiparasites and posses approximately normal photosynthetic capacity, although they are also able to derive a large portion of their reduced carbon needs from the host (Press, 1995). These parasitic species are notorious for their habit of attacking crops and are capable of inflicting high yield losses. Together, they present a major limitation to agriculture in regions where they are found, which are predominantly Africa, southern Europe, and the Middle East (Oerke et al., 1994; Parker and Riches, 1993).

Control of parasitic weeds is perhaps one of the most difficult challenges in weed science. The intimate association of these weeds with crop roots and the fact that they are hidden below ground during most of their life cycle makes mechanical control at an early stage impossible. Removing shoots of emerged parasites can help reduce reproduction and replenishing of the parasite seed bank, but does not avoid crop yield losses because most damage is inflicted before parasite emergence from the soil.

A wide variety of chemical methods have been tried for *Orobanche* control, but most have disadvantages (Foy et al., 1989). Stimulating suicidal seed germination with ethylene is effective for *Striga* control (Babiker et al., 1993), but not for *Orobanche*, and requires considerable expense. Soil sterilization with methyl bromide or metham sodium is effective, but also is expensive and potentially harmful to the environment. Translocatable herbicides, such as glyphosate, are effective at low rates, but the window of selectivity is small in crops that do not have a high tolerance to the herbicides (discussed in detail below). Refinements of herbicide application techniques, such as seed dressing with imazethapyr (Garcia-Torres et al., 1996) or addition of sulfonylurea herbicides to sprinkler irrigation water (Kleifeld et al., 1996), offer improvements in *Orobanche* control in specific crops and cropping systems.

The optimal long-term strategy for controlling parasitic weeds is through development of resistant crop genotypes, enabling reduced labor, less expense, increased cropping choices, and reduced input of chemicals that may be harmful to the environment. However, despite many years of work by plant breeders, parasite-resistant varieties of most affected crops are not available. The major barrier to successful breeding for resistance to parasitic weeds has been identification of germplasm with resistance traits that can be transferred to agronomically acceptable crop varieties. Of all the crops affected by parasitism, the best examples of resistance to *Orobanche* have been obtained with sunflower (*Helianthus annuus*) and broadbean (*Vicia faba*) (Cubero et al., 1994), and for resistance to *Striga* have been with sorghum (*Sorghum bicolor*) and cowpea (*Vigna unguiculata*) (Ejeta et al., 1991).

The parasite life cycle is an example of adaptive specialization designed to synchronize growth of the parasite with availability of a suitable host (Figure 1). At each step of the life cycle a host-parasite interaction occurs which, if disturbed, could result in failure of the parasitic association. Indeed, it is a marvel of biology that the parasites are so successful given the complexity of this process. It is possible to envision multiple mechanisms of resistance by the host, including the following: reduced germination stimulant production, mechanical barriers to penetration, inhibition of parasite digestive enzymes, phytoalexin production, active post-attachment resistance mechanisms (e.g., hypersensitive response), altered hormone supply to the parasite, insensitivity to parasite-derived toxins, or avoidance of parasites through root growth habit (Haussmann et al., 2000). Not all of these mechanisms have been documented with parasitic weeds, but they illustrate the rich variety of physiological processes that are potential targets for modification.

The mechanisms of host resistance to parasites with the most experimental support, and those that will be discussed in this chapter, are indicated in the boxes in Figure 1. Our understanding of parasitic plant-host interactions remains limited, with little molecular-level characterization of either the host responses to parasitism or the mechanisms of parasite penetration, establishment, and growth. However, recent efforts in this area show promise that it may soon be possible to identify specific genes involved in key steps in the interaction. The application of genetic engineering strategies to develop crops engineered for *Orobanche* resistance will have a large impact on control of parasitic weeds.

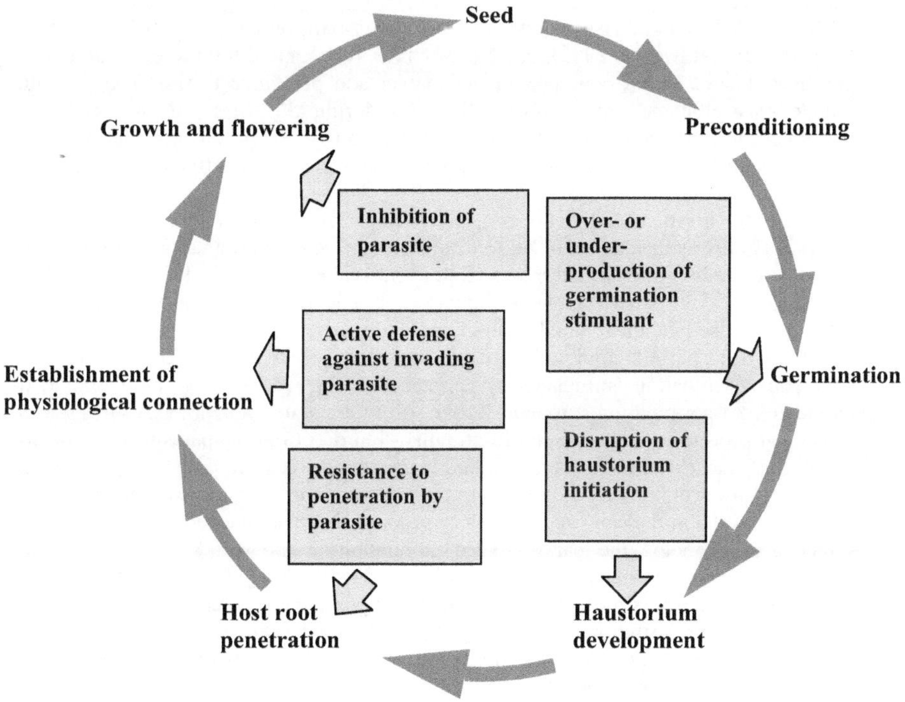

Figure 1. Life cycle of a root parasite on a plant, including boxes indicating potential points of intervention with genetically engineered crops.

PARASITE GERMINATION SIGNALING

The first step in host-parasite interaction occurs at the time of seed germination. Root parasites such as *Orobanche* and *Striga* have complex mechanisms for detecting the presence of a suitable host root (Joel et al., 1995a; Press et al., 1990; Stewart and Press, 1990; Yoder, 1999). Parasite seed germination is controlled by what is essentially a double-check system by which the parasite determines both that conditions are favorable for growth and that an appropriate host plant is near. The first requirement is preconditioning, which is a one to three week period when the seed is exposed to moisture at an optimal temperature (Kebreab and Murdoch, 1999). During this time, changes occur in the seed that allow it to detect and respond to a chemical germination signal emitted from host roots. This chemical signal comprises the second condition for germination, and the ability to specifically detect a host by the chemicals released from the host roots has long been one of the most fascinating aspects of parasitic plant biology, and has lead to much research on germination stimulants, their mechanisms of action, and their potential utility in controlling parasitic weeds.

Seeds of *Striga* and *Orobanche* are extremely small, ranging from 0.2 to 0.3 mm long (Parker and Riches, 1993). Such seeds have very limited food reserves and once germinated have only about a week to contact and parasitize a host. The parasite radicle may elongate only a few millimeters during this time, so any seed that germinates too far from a host root has no chance to live. Such "suicidal" germination is the objective for some control strategies and has been used effectively in certain cases.

The plant hormone, ethylene, triggers germination in *Striga* species (Babiker et al., 1993) and was injected into soil to trigger mass suicidal germination to rid the soil of parasite seeds in the United States (Eplee and Langston, 1991). This approach has not been applied to other regions of the world because the quarantine and eradication program as used in the United States was extremely expensive. However, much effort has been put into finding affordable equivalent procedures based on natural or synthetic germination stimulants. Approaches along these lines have included application of germination stimulants to the soil in inundative quantities, or using trap crops that produce a large amount of stimulant but do not allow parasite attachment. Neither of these methods have worked particularly well in practice. Synthetic stimulants are prohibitively expensive to manufacture and need to be relatively stable to remain active in the soil long enough to produce the desired effect. Trap crops are effective in reducing the parasite seed population in the soil but are inefficient because several years of growing trap crops are needed to significantly reduce a soil seed bank that harbors millions of parasite seeds per square meter. High amounts of stimulant present during preconditioning inhibited parasite seed germination (Hsiao et al., 1979), suggesting that very high stimulant-producers may be protected by repressing germination of parasite seeds near their roots.

A second strategy involves crops that do not produce germination stimulants. This has been the basis for some of the best characterized examples of parasite resistance. The sorghum line SRN-39 is a low stimulant producer and shows increased resistance to *Striga* (Vogler et al., 1996). This ability to avoid detection by the parasite would not directly cause a decrease in the numbers of parasite seeds in the soil, but if maintained over time could have this effect through mortality of dormant seeds in the soil.

Both high- and low-stimulant approaches are ideal candidates for employing genetic engineering strategies against parasitic weeds. In theory the modification of germination stimulant production is the simplest stage in the life cycle to address because a single host metabolic step could be either blocked to prevent stimulant production or overexpressed to increase stimulant production. However, this requires detailed knowledge of plant metabolism leading to synthesis of germination stimulants, and this information has been difficult to obtain. Indeed, even the identification of germination stimulants has been - and continues to be - a very challenging task.

The first naturally occurring stimulant to be isolated was strigol (Figure 2) from cotton, a non-host that happens to produce high quantities of this stimulant (Cook et al. 1966, 1972). Subsequently, compounds with stimulant activity were isolated from host plants: sorgoleone (Chang et al., 1986) and sorgolactone (Hauck et al., 1992) from sorghum, alectrol from cowpea (Müller et al., 1992), and orobanchol from red clover (*Trifolium pratense*) (Yokota et al., 1998). All of these compounds, except sorgoleone, share a sesquiterpene lactone structure with the lactone bridge between the C and D rings being the major structural component associated with stimulant activity (Figure 2) (Mangnus and Zwanenburg, 1992).

Synthetic analogs of these compounds have been produced, notably GR-24 (Figure 2), which stimulate germination of *Orobanche* and *Striga* seeds and are useful in laboratory studies (Mangnus et al., 1992). In addition, many synthetic precursors of strigol that contain the lactone feature have stimulant activity on both *Striga* and *Orobanche* species (Vail et al., 1990). Unfortunately, information on other germination stimulants or stimulants from the many other host plants have been very difficult to obtain. Part of this difficulty is due to the extremely low quantities of stimulant produced by the hosts. Strigol, for example, stimulates germination at concentrations on the order of 10^{-12} M. Furthermore, sorgolactone, alectrol, orobanchol, and strigol all contain a lactone bridge that is key to the activity of the molecule, but which is also highly unstable, leading to further problems in purification and identification of these compounds.

Other compounds capable of stimulating germination are the plant growth regulators zeatin, abscisic acid, and kinetin, which all showed some activity on preconditioned *S. asiatica* seeds (Hsiao et al., 1988). However, overexpression of these compounds for the purpose of stimulating parasite seed germination would be problematic in that any gain in parasite control would likely be offset by adverse physiological effects on the host plant.

The best approach for using stimulant production to control parasites is by reducing the quantity of natural stimulants produced by host plants. Promising leads in this direction come from genetics and biochemistry, which are leading to identification of important genes in the synthesis and exudation of stimulant compounds. Hope for a simple resolution to the problem comes from low-stimulant sorghum lines such as SRN-39 (Vogler et al., 1996), Framida, 555, and SRN 6496 (Ramaiah et al., 1990) in which the low stimulant production is controlled by a single, recessive gene.

The prospect of greatly reducing stimulant production through modification of a single gene is exciting, but this trait also appears to be controlled somewhat by additive gene action, so the genetics may not be simple. In fact, at least three germination stimulants are exuded from sorghum roots: sorgolactone, strigol, and sorgoleone (Chang et al., 1986; Hauck et al., 1992; Siame et al., 1993). The greatest contributor of stimulant capacity to this mixture is believed to be sorgolactone, and while the gene controlling the major stimulant effect in SRN-39 is likely related to the production of this compound, the exact identity of the responsible gene remains unknown. Interactions of minor genes involved in stimulant production and variations in stimulant sensitivity by parasite strains from different regions further complicate the situation (Haussmann et al., 2001).

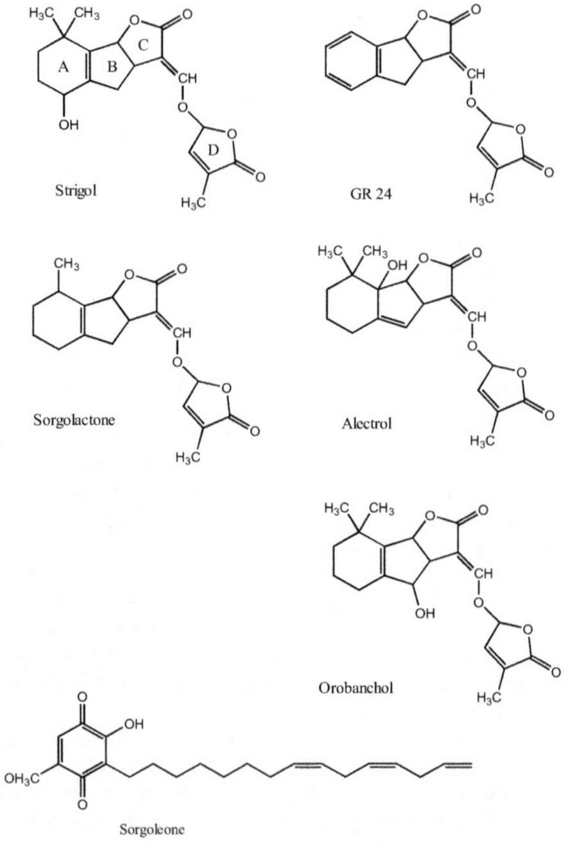

Figure 2. Chemical structures of some germination stimulants.

In contrast to *Striga*, little research has been done to characterize low-stimulant varieties of *Orobanche* hosts. Rather, it has been shown that levels of stimulant production are often not predictive of crop resistance to *Orobanche* parasitism. Goldwasser et al. (1997) found that *Orobanche*-resistant purple vetch stimulated greater germination of *O. aegyptiaca* than common vetch genotypes that were more sensitive to the parasite.

The difficulty in identifying the gene(s) responsible for low stimulant production may soon be alleviated. Expanding knowledge of grass genomics, notably for rice but also sorghum and maize (Draye et al., 2001), offers hope for accelerated identification of such genes in sorghum and other grass crops within the next several years. A similar approach is the use of *Arabidopsis thaliana* to simplify the identification of mutant plants with alterations in germination stimulant synthesis. *Arabidopsis* is a suitable host for certain *Orobanche* species, and it is capable of acting as a host through all important parasite growth stages from stimulating germination through flowering of the parasite (Goldwasser et al., 2000a; Westwood, 2000).

Research with *Arabidopsis* has already demonstrated that flavonoid compounds are not required for host-parasite communication with *Orobanche* species (Westwood, 2000). More importantly, *Arabidopsis* mutants are currently being screened to identify lines lacking stimulant production. Promising initial results from screens have been reported (Denev et al., 2001; Goldwasser and Yoder, 2001). The high number of single copy genes in *Arabidopsis* makes it likely that researchers will identify low expressing mutants and the gene responsible for this trait. Of course, the germination stimulant(s) from *Arabidopsis* may not be the same as those from affected crops. However, the structural similarity among germination stimulants from sorghum, cowpea, and clover suggests that *Arabidopsis* produces a similar compound, which will be of considerable value in finding homologous genes in major crops.

Biochemical approaches are also contributing to identification of the possible biosynthetic pathways involved in stimulant production. Because germination stimulants closely resemble the structure of sesquiterpene lactones, Denev et al. (2001) studied whether roots of host plants (tobacco, *Arabidopsis*, sunflower) producing different amounts of germination stimulants have corresponding differences in enzymatic activities converting the radio-labeled universal sesquiterpene precursor farnesyl diphosphate to a sesquiterpene product. Only sunflower showed trace enzymatic activities capable of converting farnesyl diphosphate to the products, trans-nerolidol and germacrene A, but the activities were not correlated with stimulant production. These researchers also report that inhibition of mevalonic acid formation in *Arabidopsis* and tobacco did not effect stimulant production, whereas inhibition of the alternative methylerythritol phosphate pathway in plastids decreased germination stimulant levels by 2.5 times. These observations suggest that the mevalonate pathway is not the primary source of precursors for germination stimulants. Such biochemical approaches narrow down the possible stimulant pathways and help focus research efforts.

As a final complication to the problem of parasite seed germination, evidence suggests that resistant sunflower lines may exude inhibitors of *Orobanche* germination or radicle growth. This was demonstrated by discoloration of *O. cernua* radicles in the presence of roots of a resistant variety, but not by a susceptible variety (Serghini et al., 2001). These resistant plants produced more of two coumarins, scopolin and ayapin, in response to parasitism as compared to the susceptible line. Furthermore, these compounds were excreted from the roots where they could act in the rhizosphere. This research raises the possibility that high-inhibitor crop varieties could be developed.

DEVELOPMENT OF THE HAUSTORIUM

After germination, the next critical step in the life of the parasite is development of a haustorium. The haustorium is a unique anatomical structure that forms the physical and physiological connection between parasite and host, and that differentiates parasitic from non-parasitic plants (Kuijt, 1969). The haustorium develops from the tip of the radicle in *Orobanche* and *Striga* species, but may originate from mature tissues in other parasites. In root parasites, haustorial initiation is triggered by a chemical signal distinct from the germination stimulant. This transition of the parasite radicle to a haustorium has been well characterized in *Striga* and the facultative root parasite *Triphysaria versicolor*, where the change is marked by a pronounced alteration in root morphology consisting of swelling and generation of hair-like protuberances (Albrecht et al., 1999; Riopel and Timko, 1995; Smith et al., 1990; Timko et al., 1989). In *Orobanche*, this transition is less prominent but appears to be an analogous event because the radicle apex enlarges and develops papillate cells at the point of host root contact (Joel and Losner-Goshen, 1994).

Haustorium induction factors (HIFs) are derived from host roots, but they are not produced in the absence of the parasite. Rather, wounded roots or roots in the presence of parasites emit the HIFs, leading to the hypothesis that the HIFs are components of host tissues, e.g., cell wall fragments, that are released in response to lytic activity produced by the parasite. This amounts to a form of chemical radar in which the parasite emits an enzyme that cleaves components of the host cell wall, which then diffuse away from the host root in a concentration gradient that provides the parasite with a signal that a host root is nearby. However, where this has been studied in *Striga* spp., there is a notable absence of detectable enzymatic activity that would account for this mechanism (Kim et al., 1998). Rather a model involving H_2O_2 exudation from the parasite has been proposed in which the H_2O_2 is responsible for liberating the signal molecules from the host root (Keyes et al., 2000, 2001).

HIFs include 2,6-dimethoxy-*p*-benzoquinone (DMBQ), which is the best characterized of the phenolic compounds (including quinones, flavonoids, and *p*-hydroxy acids) thought to be responsible for this signaling (Estabrook and Yoder, 1998; Lynn and Chang, 1990). The mechanism by which these compounds trigger the chain of events leading to haustorial initiation has been suggested to depend on the redox potential of the compound, as only those compounds within a certain redox range were effective stimulants. The elucidation of this mechanism led to the discovery of a haustorial initiation inhibitor, cyclopropyl-*p*-benzoquinone (CPBQ), an analog of DMBQ (Smith et al., 1996). CPBQ specifically and irreversibly blocked the induction process, suggesting that specific receptors for HIFs exist in the parasite.

Implications of these findings for control of parasitic weeds through manipulation of haustoria formation include three potential mechanisms. First, as proof of the role of H_2O_2 in cleaving of DMBQ and similar compounds from the host roots, Keyes et al. (2001) added catalase to a medium containing cell wall preparations and demonstrated that removal of H_2O_2 was sufficient to block haustorial induction. Engineering of plant roots to protect themselves from the activity of H_2O_2 by exudation of catalase may enable them to avoid stimulating haustoria formation in parasites.

Likewise, clearer understanding of the synthesis and mechanism of action of the haustorial inhibitor, CPBQ, may enable engineering of plants enhanced in the synthesis and exudation of this inhibitor. Further investigation into this mechanism may suggest other natural or synthetic inhibitors of this receptor. Host-produced haustorial inhibitors would comprise a particularly elegant defense mechanism wherein the parasite would germinate but be unable to initiate attachment.

A final potential mechanism would be to alter the host root composition to avoid production of compounds that liberate the HIFs. Evidence for the possibility of such a mechanism comes from studies of species recognition in haustorial initiation. Formation of haustoria by the facultative root parasite, *Triphysaria*, was greatest when roots of an unrelated host were present, less when another species of *Triphysaria* was present, and least in the presence of its own species (Yoder, 1997). This suggests fundamental differences between species that reflects recognition of self, but also suggests that roots of the parasite are protected in some way from producing HIFs. Although manipulating the cell wall composition of host plants is a daunting task, further investigation of the mechanisms involved will lead to a clearer idea of the differences that are important in this process.

Among the questions that will have to be answered to engineer parasite resistance at the level of haustorial induction is the precise source(s) of the HIFs. For example, flavonoids have been suggested as sources of HIFs, but *Arabidopsis* plants completely lacking in flavonoids are just as susceptible to parasitism as wild-type plants (Westwood, 2000). Variation among parasite species will also be a factor, and heritable differences in sensitivity to DMBQ have been shown both among and within species of *Triphysaria* (Jamison and Yoder, 2001). Further genetic studies of this trait will improve the chances of understanding the phenomenon of haustorial induction.

Haustoria formation is generally a less specific recognition event than germination stimulation. For example many parasite species germinate only in response to a relatively narrow range of hosts, but once germinated they can form haustoria and attach to a wider range of species. Nevertheless, this is an interesting area of investigation and recent progress provides encouragement that this could lead to practical solutions for parasite control.

HOST RESPONSE TO PENETRATION AND ESTABLISHMENT

The haustorium must attach to and penetrate the host root. This involves physical and chemical interactions between the two plants, which further complicates efforts to distinguish the processes involved. Much of our knowledge about this stage comes from anatomical and physiological studies of susceptible and resistant responses, which offer clues as to potential mechanisms that could be used to develop resistant plants.

The primary haustorium of *Orobanche* and *Striga* develops from the radicle tip, and in *Striga* is accompanied by elongation of papillae. This process in both species involves a secreted, mucilaginous substance that adheres the parasite to the host root surface (Joel and Losner-Goshen, 1994; Olivier, 1996). In the Scrophulariaceae, pectins are implicated in the adhesion of papillae to the host root surface (Neumann et al., 1999). The haustorium then grows intrusively to penetrate the host root by pushing between cells. Digestive enzymes secreted by the parasite aid in this process, and studies of germinating *Orobanche* seedlings in the presence of various substrates indicate that the parasite secretes pectin methylesterase (PME) (Ben-Hod et al., 1993), polygalacturonase, and endocellulase, but not exocellulase, β-glucosidase or xylanase (Ben-Hod et al., 1993; Losner-Goshen et al., 1998; Shomer-Ilan, 1993).

These observations have been supported *in situ* by immunocytochemical studies that demonstrate the presence of PME in *Orobanche* haustorial cells and in the apoplast of adjacent host tissue (Losner-Goshen et al., 1998). This study also used histochemical staining and immunolabeling of pectins to demonstrate a decrease in methylated pectins in the host middle lamella near the parasite intrusive cells. This is in contrast to parasitism of cowpea by *S. gesnerioides*, in which little degradation of host root pectin occurred (Reiss and Bailey, 1998). Cytochemical evidence of host cell wall degradation was determined by labeling cellulosic β-1,4-glucans in sorghum parasitized by *S. hermonthica.* Loosening and fragmenting of host cell wall fibrils was observed slightly distant from the immediate parasite intrusive cells, suggesting diffusion of digestive enzymes, such as cellulases or hemicellulases, from the parasite (Olivier et al., 1991).

Haustoria of different parasite species vary greatly in anatomy, depending on the needs of the parasite (Kuijt, 1977). For example, haustoria of hemiparasitic Scrophulariaceae species, including *Striga* and *Triphysaria* species discussed here, form only xylem connections, while holoparasites such as *Orobanche* species form sieve elements that appear to be active in the transfer of resources from host phloem (Dörr, 1996; Heide-Jørgensen and Kuijt, 1995). However, it is likely that many of the same processes are involved in establishing these connections, and reports of host-parasite interactions indicate a high level of similarity, especially concerning host resistance responses. The host resistance responses suggested include production of phytoalexins and mechanical barriers to inhibit haustorial penetration, a hypersensitive-like response in the host root that results in localized death of host cells, and post-attachment mechanisms.

Phytoalexins and Mechanical Barriers

The host root endodermis presents the greatest structural barrier to haustorial penetration, although haustorial growth may also be arrested in the cortex. Even in a susceptible interaction, *S. asiatica* attacking sorghum readily traverses the cortex in 1-2 days, but then pauses for 3-4 days before breaching the endodermis to form vascular connections (Hood et al., 1998). Parasitism of wild grass species by *S. asiatica* failed after root penetration, which was attributed to inability to penetrate the endodermis (Kuiper et al., 1998). A similar response characterized vetch (*Vicia* spp.) resistance to *O. aegyptiaca* (Goldwasser et al., 2000b). In some resistant host varieties, the endodermis may be traversed and vascular connections established, but subsequent haustorium development may be limited. *S. hermonthica*, parasitizing the resistant sorghum variety Framida, had a restricted haustorium that did not fully develop, which led to decreased uptake of photosynthates from the host and thus significantly less growth of the parasite (Arnaud et al., 1999). This may cause death of the parasite or restrict its growth such that the shoot fails to emerge, thereby limiting damage to the host.

Several studies report the accumulation of electron-dense material in resistant roots where parasite growth was arrested in the root cortex or at the barrier of the endodermis (Dörr et al., 1994; Neumann et al., 1999; Olivier et al., 1991). As the haustorium expands within the host root it crushes some of the host cortical cells, creating a wound site. In many cases this is accompanied by accumulation of glycoproteins and phenolic compounds. Parasitism of susceptible hosts by Scrophulariaceae species lead host roots to accumulate hydroxyproline-rich glycoproteins (Neumann et al., 1999), which are associated with wall strengthening. Sorghum parasitized by *S. hermonthica* accumulated electron dense material, presumably rich in phenolic compounds, in association with cortical cells and xylem (Olivier et al., 1991). In this study, the timing and tissue in which this material appeared, e.g., early during parasitism as opposed to later in the xylem vessels, correlated to resistant or susceptible varieties. In the nonhost marigold (*Tagetes erecta*) cortical cells in advance of the haustorium had denser cytoplasm than observed in susceptible species, suggesting greater metabolic activity as part of a defense response. For vetch resistant to *O. aegyptiaca*, accumulation of a reddish substance in the apoplast between the parasite and cells of the resistant species was correlated with increased levels of phenolics, lignin, and peroxidase activity compared to susceptible species (Goldwasser et al., 1999; Goldwasser et al., 2000b). In addition, higher baseline activity of phenylalanine ammonia lyase occurred in these resistant plants. In all of these studies it is not clear that the induction of lignin and phenolic biosynthetic pathways is directly related to the resistance mechanism, or accompanies other, as yet unidentified, responses that have a more direct role in effective defense. Accumulation of colored material in association with resistance in a *S. hermonthica*-sorghum interaction was attributed to a final phase in resistance, after parasite development had been effectively stopped (Arnaud et al., 1999).

To increase understanding of these defense responses, it is important to investigate the underlying molecular events in the host-parasite interaction. This has been possible through use of transgenic plants expressing promoter:reporter fusions that allow visualization of gene expression patterns (Weissenborn et al., 1995; Westwood et al., 1998). *O. aegyptiaca* stimulated expression of the locally-induced defense-specific gene, *HMG2*, in a manner similar to that reported for this promoter in wounded or elicited tissue. Likewise, Joel and Portnoy (1998) found that this same parasite species induced expression of *PRB-1b*, a basic pathogenesis-related (PR) protein expressed in response to such diverse stimuli as ethylene, salicylic acid (SA), tobacco mosaic virus (TMV), and darkness (Sessa et al., 1995). This gene is jasmonic acid (JA) dependent and wound inducible (Eyal et al., 1992), rather than being controlled by SA.

Together the induction of *HMG2* and *PRB-1b* indicate that host plants recognize *O. aegyptiaca* as an assault, and that specific defense signaling pathways may be involved in the host response to parasitism. For example, the induced expression of HMGR in response to wounding (Weissenborn et al., 1995) suggests that a significant aspect of the stimulus is the wound caused by the parasite haustorium as it establishes itself inside the host root. To expand this analysis, expression of other genes with known roles in wounding and/or pathogen defense situations have been examined. These studies indicate that metabolic intermediates of the mevalonic acid pathway are diverted from branch pathways leading to products used in normal growth and cell maintenance functions toward pathways leading to the production of phytoalexins (Griffitts et al., 2001). Interestingly, genes encoding PR proteins that are mediated through SA and the systemic acquired resistance (SAR) pathway do not appear to be expressed in host roots parasitized by *O. aegyptiaca* (Westwood, unpublished results). This suggests a model in which *O. aegyptiaca* may trigger defenses associated with JA and wounding, but not SA-mediated signaling pathways. It will be important to determine the host defense pathways that do or do not respond to parasitism in order to identify gene targets for engineering an effective defense response.

Hypersensitive Response

Some cases of host resistance to parasitism are accompanied by death of host root tissue immediately surrounding the haustorium, suggesting the hypersensitive response (HR) as another mechanism of resistance. HR has been best characterized in gene-for-gene interactions, in which a pathogen carrying an avirulence (*Avr*) gene invades a host carrying a corresponding resistance (*R*) gene (Bent, 1996). The result is a strong defense response involving production of phytoalexins, PR proteins, and ultimately the death of cells around the point of pathogen ingress. HR is also associated with initiation of a translocated signal involved in SAR, which leads to a condition of heightened defense readiness in plant parts distant from the site of initial pathogen perception (Hunt and Ryals, 1996). The effect of such a response against parasitic plants should be highly effective because the parasite relies on a physiological continuum with living host tissue.

HR was proposed as a mechanism of defense in cowpea variety 58-57 resistance to *S. gesnerioides* (Lane et al., 1993). This variety exhibited necrosis of the root tissue around the haustorium, with *S. gesnerioides* seedling death 3-4 days after host root penetration. This necrosis was observed frequently in the resistant variety but was only rarely seen in susceptible interactions. Interestingly, *S. gesnerioides* seeds from different regions differed in ability to elicit this response. *S. gesnerioides* from Burkina Faso could not attach due to the resistance response, whereas *S. gesnerioides* from Mali parasitized this variety. Although HR may occur in plant-pathogen interactions that do not involve gene-for-gene mechanisms, this is evidence for host recognition of a factor specific to the strain of parasite.

Better evidence for *R* gene involvement in parasite resistance comes from *S. asiatica* interaction with the non-host, marigold. Resistance in this interaction is characterized by cessation of haustorium penetration in the host root cortex, which is associated with necrosis of the cortical cells bordering the haustorium. Analysis of gene expression in these host roots identified a gene, named *NRSA-1*, with homology to *R* genes that was induced during the unsuccessful parasitic attack (Gowda et al., 1999). The protein encoded by this gene shares homology with *R* gene products N, NL25, RPP5, L6, and M, and contains a domain similar to that of the toll/interleukin-1 receptor (TIR). This indicates that the *NRSA-1* protein is localized in the cytoplasm. However, the lack of structural domains such as LRR (leucine-rich repeats) or NBS (nucleotide binding sites) distinguishes it from other *R* gene products and suggests that it may not be involved in pathogen recognition, but rather play a part in signal transduction following detection of the pathogen.

Expression of *NRSA-1* was induced by treatment with methyl jasmonate (Me-JA) but not by SA or mechanical wounding (Gowda et al., 1999). The mediation of parasite defense response signaling by Me-JA, but not SA, is consistent with observations of JA-associated gene expression as described above. A surprising aspect of *NRSA-1* expression is that it is induced systemically in response to *Striga* parasitism, as shown by an accumulation of *NRSA-1* mRNA in untreated roots and shoots of parasitized plants. This is interesting because little evidence currently exists for systemic defense against root parasites and such a mechanism offers the possibility that a systemic signal in response to an initial parasitic event could predispose the entire host root system to rapidly respond to subsequent parasites. While the role of *NRSA-1* in defense against parasitic plants has yet to be determined, this research demonstrates the potential value of identifying resistance genes that could be manipulated to confer resistance on currently susceptible host species.

In addition to the cases discussed above, host tissue necrosis around parasite attachment sites has been reported frequently (Goldwasser et al., 1999; Serghini et al., 2001). However, as noted previously, it is difficult to distinguish cause and effect relationships. The necrosis and discoloration observed may be the result of another defense response that results in death of the parasite, with subsequent discoloration and necrosis associated with accumulation of secondary compounds involved in wound repair. More research will be needed to understand these cases of resistance at the molecular level.

Post-Attachment Mechanisms

A parasite that has made successful connections to the host vascular tissue may still fail to survive on resistant hosts. During this critical time the parasite must acquire nutrients from the host and expand its contact with host vascular tissue. This is a complex interaction in which the two plants wage a war of nutrient acquisition within the host tissues. Important steps in this process for hemiparasites include formation of connections to the host xylem (Dörr, 1997), while in holoparasites cytoplasmic connections between host and parasite sieve elements have been documented (Dörr, 1996).

Parasites capture water and other resources by establishing a strong osmotic gradient relative to the host, a process that is aided by the parasite's ability to accumulate inorganic ions and synthesize polyhydric alcohols such as mannitol that contribute to a lower water potential than that of the host (Graves, 1995; Harloff and Wegmann, 1993; Press, 1995; Robert et al., 1999). The ability of *Orobanche* to act as a strong sink has been demonstrated by its ability to accumulate compounds present in the host phloem, ranging from sugars (Aber et al., 1983) to translocated herbicides (Müller and Distler, 1991; Nandula et al., 1999).

Blocking Parasite Acquisition of Nutrients

Some host crop varieties are susceptible to parasitism but yield as well as non-parasitized plants. This phenomenon has been attributed to the host's ability to keep a greater proportion of photosynthates for its own reproduction rather than being lost to the parasite. Although this ability to tolerate parasitism may not be considered true resistance, it can reduce parasite growth.

An example of post-attachment resistance was suggested by Lane et al. (1993) as a second mechanism of cowpea resistance to *S. gesnerioides* (in addition to the HR discussed above). This resistance was characterized by variety B301, which allowed initial tubercle formation, but tubercle growth was arrested at a young stage. The mechanism for this inhibition was not determined.

A similar resistance was observed in the sorghum variety Framida, which may limit the growth of *S. hermonthica* by restricting the parasite ability to gain host resources (Arnaud et al., 1999). The specific basis of this resistance was not identified as research suggested the involvement of multiple mechanisms. However, ^{14}C-labeling studies demonstrated that even established parasites growing on Framida were only able to withdraw a fraction of the photosynthates compared to parasites on a susceptible variety. Similar studies of radiolabeled host photosynthates showed that an *O. cumana*-resistant sunflower hybrid, LR1, allowed less ^{14}C into parasite tubercles (Labrousse et al., 2001). Again, the mechanism for limitation of carbon movement was unclear, but the authors proposed that this could be attributed to occluding materials observed in host vessels adjacent to the haustorium, suggesting that the host may seal off the flow of water and/or nutrients to the parasite.

The elaborate interplay of parasite and host physiology at this stage confuses attempts to identify resistance mechanisms. It is probable that several additional resistance responses operate in addition to mechanical blocking of parasite nutrition. The relative osmotic balance between host and parasite could play a role, with hosts having a greater capacity for generating lower water potentials than an attached parasite, thus limiting translocation into the parasite. Alterations in host hormone balance have also been suggested as an impact induced by the parasite, so resistance to this perturbation might enable the host to withstand parasitism. Production of phytoalexins could also be important at this stage, assuming they are able to enter the parasite. In this capacity, phytoalexins would act as mobile, selective toxins to the parasite, an idea that has been proven to be effective using herbicides.

Engineered Herbicide Resistance

One of the best, although unintended, applications of genetically engineered herbicide-resistant crops is their potential use in controlling parasitic weeds. Very few chemical options exist for parasitic weed control, primarily due to the lack of selectivity between crop and parasite (Foy et al., 1989). Thus, most herbicide applications risk either injury to the crop or failure to control the weed. However, the advent of herbicide-resistant crops has changed this situation, at least in principle. Crops that can withstand a normal use rate of broad-spectrum, translocatable herbicides become powerful tools in controlling parasitic weeds because the herbicide can be applied to the host and allowed to translocate to the parasite where it exerts its toxic effect. For example, glyphosate applied to glyphosate-resistant oilseed rape (*Brassica napus*) effectively controlled *O. aegyptiaca* (Joel et al., 1995b). Normally glyphosate would be toxic to both plant species, but when the host is resistant the herbicide translocates through the host and accumulates in the parasite tubercle, which is a very strong sink.

This principle has also been shown for other herbicides, including chlorsulfuron- and asulam-resistant tobacco (Joel et al., 1995b). An important point is that the mechanism of herbicide resistance cannot be due to enhanced herbicide metabolism in the host, because this will lead to detoxification of the herbicide before it reaches the parasite.

One current barrier to adoption of herbicide-resistant crop technology is the limited availability of crops with the engineered resistance genes. Apart from maize, which is an important host for *Striga* spp., these traits are not being incorporated into the subsistence and cash crops that are targets of parasitic plants in Africa, the Middle East, and other parts of the world. However, this situation will likely improve as the technology becomes more widespread and government and nonprofit organizations work to protect key crops. One example is the effort being put into creating asulam-resistant potato lines that can be grown in *Orobanche*-infested areas (Surov et al. 1998).

Herbicide-resistant crop technology should become common for improving crops in regions affected by parasites, but is not without disadvantages. Compared to parasite resistant crop varieties, this technology will require higher input costs and management skills associated with herbicide application. However, a promising innovation is the application of herbicides directly to crop seeds, such that the herbicide is absorbed in to the germinating seedling (Díaz-Sánchez et al., 2002; Kanampiu et al., 2002). Any parasites attacking such protected seedlings receive a lethal dose of the herbicide. Innovative application methods such as this will encourage use of herbicide resistant crop technology, but as herbicide use increases, the probability of herbicide resistance developing in the parasite populations will also increase (Gressel et al., 1996). Careful management will be required to sustain the benefits of this technology.

PARASITE VARIABILITY AND HOST RESISTANCE

An important aspect of host resistance to parasitism is recognition of the heterogeneous nature of the parasite species, including the presence of different races within species. The crop that has received the most attention in this respect is sunflower, in which resistance to *O. cumana* Wallr. (syn. *O. cernua* Loefl.) was first obtained in 1916 by Russian researchers (reviewed in Alonso, 1998). However this resistance was not durable over time or geography, as resistant sunflower varieties were susceptible when challenged with new 'races' of the parasite. This led to a cycle of success and failure in which subsequent new resistant varieties were eventually found to be susceptible to *O. cumana* from different regions and each new parasite race had increased virulence.

Currently at least six *O. cumana* races are recognized, designated A, B, C, D, E, and F. Genetic studies indicate that resistance to some of these races is controlled by single dominant genes (Ish-Shalom-Gordon et al., 1993; Sukno et al., 1999), and a gene-for-gene resistance has been suggested whereby sunflower genes *Or1*, *Or2*, *Or3*, etc. confer resistance to the corresponding pathotype of *O. cumana* (Alonso, 1998). However the genetics is not so simple, as other studies have indicated involvement of multiple genes and maternal effects in sunflower resistance, all of which are confounded by environmental influences and parasites with different virulence levels (Alonso et al., 1996; Dominguez et al., 1996). Several mechanisms for the resistance in sunflower have been suggested, including increased lignification, hypersensitivity, and increased phytoalexin production (Alonso, 1998), but few studies have been able to draw a clear causal correlation between host plant responses and parasite death.

Genetic markers can be used to study resistance, which offers hope that genes responsible for these traits will be cloned and identified. The sunflower *Or5* resistance gene has been linked to a sunflower RFLP map (Lu et al., 1999), while two genes controlling resistance to *S. gesnerioides* in cowpea have been linked to AFLP markers (Ouédraogo et al., 2001). This latter work can begin to sort out the complexities present in the five races of *S. gesnerioides* currently recognized in West Africa.

Although identification of parasite races provides an interesting point for asking questions about the specificity of host-parasite interactions, it also raises daunting challenges for creating parasite-resistant crops. For example, the high level of genetic diversity present in *Striga* and *Orobanche* species (Pieterse and Verkleij, 1991) could lead to rapid breakdown in engineered resistance mechanisms, just as it presents problems for resistance obtained through conventional means. Gressel et al. (1996)

have argued that use of herbicides acting at the acetolactate synthase target site in parasitic weeds could select for herbicide-resistant parasitic weeds in 2 to 6 years. This is based on assumptions of a high frequency of resistance in the population and a high selection pressure with the herbicide.

An analogous situation may be envisioned for host plants engineered for parasite resistance based on any single mechanism. The extremely high number of *Striga* or *Orobanche* seeds in the soil presents an abundant source of genetic variation, and high reproduction rates will rapidly amplify the numbers of any survivors that overcome the resistance mechanism. For example, for resistance to result in depletion of the soil seed bank, Kebreab and Murdoch (2001) calculated that the resistance mechanism would have to be 99.6% effective. This model assumes that no other control measures were used, and was calculated for *Orobanche* species assuming an average of 2,500 seeds produced per seed capsule. The model should also apply to *Striga*, which shares a remarkable fecundity.

As researchers look for resistance genes, whether naturally occurring or novel strategies, consideration should be given to methods for delaying the inevitable breakdown of the resistance mechanism by the parasite. Of course, one way is to implement additional control strategies consistent with integrated weed control ideals that deter parasite growth such as hand weeding or cultural practices. In addition, it should be possible to incorporate multiple resistance mechanisms into a host plant. For instance, a trait that minimizes stimulation of parasite seed germination could be combined with an active defense response, which should provide a more durable resistance because the low induction of parasite germination would decrease the number of parasite seeds that germinate, thus reducing the number that confront, and perhaps overcome, the resistance mechanism.

CONCLUSIONS

Scientific research is advancing to the point that it will soon be possible to manipulate gene expression and gene products with a high degree of precision. Understanding metabolic pathways and signal transduction mechanisms may further allow researchers to significantly enhance plant responses to pathogens. It is relatively easy to envision ways in which crop interactions with parasitic weeds could be modified, perhaps even by single gene traits. As we learn more about the fundamental processes involved in this interaction, from germination and haustorial initiation to host defense pathways, we will have several potential targets for enhancing resistance. In the larger context of weed biology, this can serve as a model from which to explore other crop modifications that would enhance plant interactions with their environment, including autotrophic weeds. It is possible to imagine studies of host-parasite signaling leading to progress in enhancing the allelopathic ability of crops, or studies of resource allocation contributing to increased crop competitiveness.

ACKNOWLEDGEMENTS

Research on parasitic weeds leading to preparation of this chapter has been supported by USDA NRICGP award 2001-35320-10900, Research Grant No. IS-3048-98 from BARD, The United States - Israel Binational Agricultural Research and Development, and USDA Hatch Project no. VA-135657.

REFERENCES

Aber M., Fer A., Sallé G. Etude du transfert des substances organiques de l'hôte (*Vicia faba*) vers le parasite (*Orobanche crenata* Forsk.). Z Pflanzenphysiol Bd 1983; 112:297-308.

Albrecht H., Yoder J.I., Phillips D.A. Flavonoids promote haustoria formation in the root parasite *Triphysaria versicolor*. Plant Physiol 1999; 119:585-591.

Alonso L.C. Resistance to *Orobanche* and resistance breeding: a review. Proceedings of the Fourth International Workshop on *Orobanche*; 1998 September 23 - 26; Albena, Bulgaria. Near General Toshevo: Institute for Wheat and Sunflower "Dobroudja", 1998.

Alonso L.C., Fernandez-Escobar J., Lopez G., Rodriguez-Ojeda M.I., Sallago F. New highly virulent sunflower broomrape (*Orobanche cernua* Loefl.) pathotypes in Spain. Advances in Parasitic Plant Research; 1996 April 16 - 18; Cordoba, Spain. Cordoba: Junta de Andalucia, 1996.

Arnaud M.-C., Véronési C., Thalouarn P. Physiology and histology of resistance to *Striga hermonthica* in *Sorghum bicolor* var. *Framida*. Aust J Plant Physiol 1999; 26:63-70.

Babiker A.G.T., Ejeta G., Butler L.G., Woodson W.R. Ethylene biosynthesis and strigol-induced germination of *Striga asiatica*. Physiol Plant 1993; 88:359-365.

Ben-Hod G., Losner D., Joel D.M., Mayer M. Pectin methylesterase in calli and germinating seeds of *Orobanche aegyptiaca*. Phytochemistry 1993; 32:1399-1402.

Bent A.F. Plant disease resistance genes: function meets structure. Plant Cell 1996; 8:1757-1771.

Chang M., Netzly D.H., Butler L.G., Lynn D.G. Chemical regulation of distance: characterization of the first natural host germination stimulant for *Striga asiatica*. J Am Chem Soc 1986; 108:7858-7860.

Cook C.E., Whichard L.P., Turner B., Wall M.E., Egley G.H. Germination of witchweed (*Striga lutea* Lour.): isolation and properties of a potent stimulant. Science 1966; 154:1189-1190.

Cook C.E., Whichard L.P., Wall M.E., Egley G.H., Coggon P., Luhan P.A., McPhail A.T. Germination stimulants. II. The structure of strigol - a potent seed germination stimulant for witchweed (*Striga lutea* Lour.). J Am Chem Soc 1972; 94:6198-6199.

Cubero J.I., Pieterse A.H., Khalil S.A., Sauerborn J. Screening techniques and sources of resistance to parasitic angiosperms. Euphytica 1994; 73:51-58.

Denev I., Pereira A., Verstappen F., Bouwmeester H. Biosynthesis of *Orobanche* germination stimulants. 7th International Parasitic Weed Symposium; 2001 June 5 - 8; Nantes, France. Nantes: Faculté des Sciences, 2001.

Díaz-Sánchez J., López-Martínez N., López-Granados F., De Prado R., García-Torres L. Absorption, translocation, and fate of herbicides in *Orobanche cumana*-sunflower system. Pest Biochem Physiol 2002; 74:9-15.

Dominguez J., Melero-Vara J.M., Refoyo A. Virulence groups of *Orobanche cernua* in Andalusia, (Southern Spain). Advances in Parasitic Plant Research; 1996 April 16 - 18; Cordoba, Spain. Cordoba: Junta de Andalucia, 1996.

Dörr I. New results on interspecific bridges between parasites and their hosts. Advances in Parasitic Plant Research; 1996 April 16 - 18; Cordoba, Spain. Cordoba: Junta de Andalucia, 1996.

Dörr I. How *Striga* parasitizes its host: a TEM and SEM study. Ann Bot 1997; 79:463-472.

Dörr I., Staack A., Kollmann R. Resistance of *Helianthus* to *Orobanche* - histological and cytological studies. Biology and management of *Orobanche*, Proceedings of the Third International Workshop on *Orobanche* and related *Striga* research; 1993 November 8 - 12; Amsterdam, The Netherlands. Amsterdam: Royal Tropical Institute, 1994.

Draye X., Lin Y.-R., Qian X., Bowers J.E., Burow G.B., Morrell P.L., Peterson D.G., Presting G.G., Ren S., Wing R.A., Paterson A.H. Toward integration of comparative genetic, physical, diversity, and cytomolecular maps for grasses and grains, using the sorghum genome as a foundation. Plant Physiol 2001; 125:1325-1341.

Ejeta G., Butler L.G., Hess D.E., Vogler R.K. Genetic and breeding strategies for *Striga* resistance in sorghum. Proceedings of the 5th International Symposium of Parasitic Weeds; 1991 June; Nairobi, Kenya. Nairobi: CIMMYT, 1991.

Eplee R., Langston M. Integrated control methodology for *Striga* in the USA. Proceedings of the 5th International Symposium of Parasitic Weeds; 1991 June; Nairobi, Kenya. Nairobi: CIMMYT, 1991.

Estabrook E.M., Yoder J.I. Plant-plant communications: rhizosphere signaling between parasitic angiosperms and their hosts. Plant Physiol 1998; 116:1-7.

Eyal Y., Sagee O., Fluhr R. Dark-induced accumulation of a basic pathogenesis-related (PR-1) transcript and a light requirement for its induction by ethylene. Plant Mol Biol 1992; 19:589-599.

Foy C.L., Jain R., Jacobsohn R. Recent approaches for chemical control of broomrape (*Orobanche* spp.). Rev Weed Sci 1989; 4:123-152.

Garcia-Torres L., Jurado-Exposito M., Castejon-Munoz M., Lopez-Granados F. Herbicide-treated crop seeds for control of *Orobanche* spp. Advances in Parasitic Plant Research; 1996 April 16 - 18; Cordoba, Spain. Cordoba: Junta de Andalucia, 1996.

Goldwasser Y., Hershenhorn J., Plakhine D., Kleifeld Y., Rubin B. Biochemical factors involved in vetch resistance to *Orobanche aegyptiaca*. Physiol Mol Plant Pathol 1999; 54:87-96.

Goldwasser Y., Kleifeld Y., Rubin B. Variation in vetch (*Vicia* spp.) response to *Orobanche aegyptiaca.* Weed Sci 1997; 45:756-762.

Goldwasser Y., Plakhine D., Yoder J.I. *Arabidopsis thaliana* susceptibility to *Orobanche* spp. Weed Sci 2000a; 48:342-346.

Goldwasser Y., Plakhine D., Kleifeld Y., Zamski E., Rubin B. The differential susceptibility of vetch (*Vicia* spp.) to *Orobanche aegyptiaca*: anatomical studies. Ann Bot 2000b; 85:257-262.

Goldwasser Y., Yoder J.I. *Arabidopsis thaliana* induction of *Orobanche* seed germination. 7th International Parasitic Weed Symposium; 2001 June 5 - 8; Nantes, France. Nantes: Faculté des Sciences, 2001.

Gowda G.S., Riopel J., Timko M.P. *NRSA-1*: a resistance gene homolog expressed in roots of non-host plants following parasitism by *Striga asiatica* (witchweed). Plant J 1999; 20:217-230.

Graves, Jonathan D. "Host-plant responses to parasitism." In *Parasitic Plants*, M. C. Press, J. D. Graves, eds. London: Chapman & Hall, 1995.

Gressel J., Segel L., Ransom J.K. Managing the delay of evolution of herbicide resistance in parasitic weeds. Internat J Pest Manage 1996; 42:113-129.

Griffitts A.A., Cramer C.L., Westwood J.H. Characterization of host plant responses to parasitization by *Orobanche aegyptiaca.* 7th International Parasitic Weed Symposium; 2001 June 5 - 8; Nantes, France. Nantes: Faculté des Sciences, 2001.

Harloff H.J., Wegmann D. Evidence for a manitol cycle in *Orobanche ramosa* and *Orobanche crenata*. J Plant Physiol 1993; 141:513-520.

Hauck C., Müller S., Schildknecht H. A germination stimulant for parasitic flowering plants from *Sorghum bicolor*, a genuine host plant. J Plant Physiol 1992; 139:474-478.

Haussmann B.I.G., Hess D.E., Omanya G.O., Reddy B.V.S., Welz J.G., Geiger H.H. Major and minor genes for stimulation of *Striga hermonthica* seed germination in sorghum, and interaction with different *Striga* populations. Crop Sci 2001; 41:1507-1512.

Haussmann B.I.G., Hess D.E., Welz H.-G., Geiger H.H. Improved methodologies for breeding *Striga*-resistant sorghums. Field Crops Res 2000; 66:195-211.

Heide-Jørgensen H.S., Kuijt J. The haustorium of the root parasite *Triphysaria* (Scrophulariaceae), with special reference to xylem bridge ultrastructure. Am J Bot 1995; 82:782-797.

Hood M.E., Condon J.M., Timko M.P., Riopel J.L. Primary haustorial development of *Striga asiatica* on host and nonhost species. Phytopathology 1998; 88:70-75.

Hsiao A.I., Worsham A.D., Moreland D.E. Factors affecting conditioning and germination of witchweed (*Striga asiatica* [L.] Kuntze) seed under laboratory conditions. Second International Symposium on Parasitic Weeds; 1979 July 16 - 19; Raleigh. Raleigh: North Carolina State University, 1979.

Hsiao A.I., Worsham A.D., Moreland D.E. Effects of chemicals often regarded as germination stimulants on seed conditioning and germination of witchweed (*Striga asiatica*). Ann Bot 1988; 62:17-24.

Hunt M.D., Ryals J.A. Systemic acquired resistance signal transduction. Crit Rev Plant Sci 1996; 15:583-606.

Ish-Shalom-Gordon N., Jacobsohn R., Cohen Y. Inheritance of resistance to *Orobanche cumana* in sunflower. Phytopathology 1993; 83:1250-1252.

Jamison D.S., Yoder J.I. Heritable variation in quinone-induced haustorium development in the parasitic plant *Triphysaria*. Plant Physiol 2001; 125:1870-1880.
Joel D.M., Kleifeld Y., Losner-Goshen D., Gressel J. Transgenic crops against parasites. Nature 1995b; 374:220-221.
Joel D.M., Losner-Goshen D. The attachment organ of the parasitic angiosperms *Orobanche cumana* and *O. aegyptiaca* and its development. Can J Bot 1994; 72:564-574.
Joel D.M., Portnoy V.H. The angiospermous root parasite *Orobanche* L. (Orobanchaceae) induces expression of a pathogenesis related (PR) gene in susceptible tobacco roots. Ann Bot 1998; 81:779-781.
Joel, D.M., Steffens, Matthews J.C., David E. "Germination of weedy root parasites." In *Seed Development and Germination*, J. Kigel, G. Galili, eds. New York: Marcel Dekker, Inc., 1995a.
Kanampiu F.K., Ransom J.K., Friesen D., Gressel J. Imazapyr and pyrithiobac movement in soil and from maize seed coats to control *Striga* in legume intercropping. Crop Prot 2002; 21:611-619.
Kebreab E., Murdoch A.J. A model of the effects of a wide range of constant and alternating temperatures on seed germination of four *Orobanche* species. Ann Bot 1999; 84:549-557.
Kebreab E., Murdoch A.J. Simulation of integrated control strategies for *Orobanche* spp. based on a life cycle model. Expl Agric 2001; 37:37-51.
Keyes W.J., O'Malley R.C., Kim D., Lynn D.G. Signaling organogenesis in parasitic angiosperms: xenognosin generation, perception, and response. J Plant Growth Regul 2000; 19:217-231.
Keyes W.J., Taylor J.V., Apkarian R.P., Lynn D.G. Dancing together. Social controls in parasitic plant development. Plant Physiol 2001; 127.
Kim D., Kocz R., Boone L., Keyes W.J., Lynn D.G. On becoming a parasite: evaluating the role of wall oxidases in parasitic plant development. Chem Biol 1998; 5:103-117.
Kleifeld Y., Goldwasser Y., Herzlinger G., Plakhine D., Golan S., Chilf T. Selective control of *Orobanche aegyptiaca* in tomato with sulfonylurea herbicides. Advances in Parasitic Plant Research; 1996 April 16 - 18; Cordoba, Spain. Cordoba: Junta de Andalucia, 1996.
Kuijt, J. *The Biology of Parasitic Flowering Plants.* Berkeley: University of California Press, 1969.
Kuijt J. Haustoria of phanerogamic parasites. Ann Rev Phytopathol 1977; 17:91-118.
Kuiper E., Groot A., Noordover E.C.M., Pieterse A.H., Verkleij J.A.C. Tropical grasses vary in their resistance to *Striga aspera*, *Striga hermonthica*, and their hybrids. Can J Bot 1998; 76:2131-2144.
Labrousse P., Arnaud M.C., Serieys H., Bervillé A., Thalouarn P. Several mechanisms are involved in resistance of *Helianthus* to *Orobanche cumana* Wallr. Ann Bot 2001; 88:859-868.
Lane J.A., Bailey J.A., Butler R.C., Terry P.J. Resistance of cowpea [*Vigna unguiculata* (L.) Walp.] to *Striga gesnerioides* (Willd.) Vatke, a parasitic angiosperm. New Phytol 1993; 125:405-412.
Losner-Goshen D., Portnoy V.H., Mayer A.M., Joel D.M. Pectolytic activity by the haustorium of the parasitic plant *Orobanche* L. (Orobanchaceae) in host roots. Ann Bot 1998; 81:319-326.
Lu Y.H., Gagne G., Grezes-Besset B., Blanchard P. Integration of a molecular linkage group containing the broomrape resistance gene *Or5* into an RFLP map in sunflower. Genome 1999; 42:453-456.
Lynn D.G., Chang M. Phenolic signals in cohabitation: implications for plant development. Annu Rev Plant Physiol Plant Mol Biol 1990; 41:497-526.
Mangnus E.M., Stommen P.L.A., Zwanenburg B. A standardized bioassay for evaluation of potential germination stimulants for seeds of parasitic weeds. J Plant Growth Regul 1992; 11:91-98.
Mangnus E.M., Zwanenburg B. Tentative molecular mechanism for germination stimulation of *Striga* and *Orobanche* seeds by strigol and its synthetic analogues. J Agric Food Chem 1992; 40:1066-1070.
Müller F., Distler B. Translocation of glyphosate in the host/parasite system *Vicia faba* and *Orobanche crenata*. Progress in *Orobanche* research: Proceedings of the International Workshop on *Orobanche* Research; 1989 August 19-22; Obermarchtal, FRG. Tübingen: Eberhard-Karls-Universität, 1991.
Müller S., Hauck C., Schildknecht H. Germination stimulants produced by *Vigna unguiculata* (L.) Walp. cv Saunders Upright. J Plant Growth Regul 1992; 11:77-84.
Nandula V.K., Foy C.L., Orcutt D.M. Glyphosate for *Orobanche aegyptiaca* control in *Vicia sativa* and *Brassica napus*. Weed Sci 1999; 47:486-491.
Neumann U., Vian B., Weber H.C., Sallé G. Interface between haustoria of parasitic members of the Scrophulariaceae and their hosts: a histochemical and immunocytochemical approach. Protoplasma 1999; 207:84-97.
Oerke E.-C., Dehne H.-W., Schonbeck F., Weber A., *Crop Production and Crop Protection. Estimated Losses in Major Food and Cash Crops.* Amsterdam: Elsevier, 1994.
Olivier A. La relation entre le *Striga hermonthica* et ses hôtes: une synthese. Can J Bot 1996; 74:1119-1137.

Olivier A., Benhamou N., Leroux G.D. Cell surface interactions between sorghum roots and the parasitic weed *Striga hermonthica*: cytochemical aspects of cellulose distribution in resistant and susceptible host tissues. Can J Bot 1991; 69:1679-1690.

Ouédraogo J.T., Maheshwari V., Berner D.K., St-Pierre C.-A., Belzile F., Timko M.P. Identification of AFLP markers linked to resistance of cowpea (*Vigna unguiculata* L.) to parasitism by *Striga gesnerioides*. Theor Appl Genet 2001; 102:1029-1036.

Parker C., Riches C.R. *Parasitic Weeds of the World: Biology and Control.* Wallingford, Oxon: CAB International, 1993.

Pieterse A.H., Verkleij, J.A.C. "Genetic variability in *Orobanche* (broomrape) and *Striga* (witchweed) and its implications for host crop resistance breeding." In *Ecological Responses to Environmental Stresses,* J. Rozema, J. A. C. Verkleij, eds. The Netherlands: Kluwer Academic Publishers, 1991.

Press, Malcolm C. "Carbon and nitrogen relations." In *Parasitic Plants*, M. C. Press and J. D. Graves, eds. London: Chapman & Hall, 1995.

Press M.C., Graves J.D., Stewart G.R. Physiology of the interaction of angiosperm parasites and their higher plant hosts. Plant, Cell and Environ 1990; 13:91-104.

Ramaiah K.V., Chidley V.L., House R.L. Inheritance of *Striga* seed-germination stimulant in sorghum. Euphytica 1990; 45:33-38.

Reiss G.C., Bailey J.A. *Striga gesnerioides* parasitising cowpea: development of infection structures and mechanisms of penetration. Ann Bot 1998; 81:431-440.

Riopel J.L., Timko M. P. "Haustorial initiation and differentiation." In *Parasitic Plants,* M. C. Press, J. D. Graves, eds. London: Chapman & Hall, 1995.

Robert S., Simier P., Fer A. Purification and characterization of mannose 6-phosphate reductase, a potential target for the control of *Striga hermonthica* and *Orobanche ramosa*. Aust J Plant Physiol 1999; 26:233-237.

Serghini K., Pérez de Luque A., Castejón-Muñoz M., García-Torres L., Jorrín J.V. Sunflower (*Helianthus annuus* L.) response to broomrape (*Orobanche cernua* Loefl.) parasitism: induced synthesis and excretion of 7-hydroxylated simple coumarins. J Exp Bot 2001; 52:2227-2234.

Sessa G., Yang X.-Q., Raz V., Eyal Y., Fluhr R. Dark induction and subcellular localization of the pathogenesis-related PRB-1b protein. Plant Mol Biol 1995; 28:537-547.

Shomer-Ilan A. Germinating seeds of the root parasite *Orobanche aegyptiaca* Pers. excrete enzymes with carbohydrase activity. Symbiosis 1993; 15:61-70.

Siame B.A., Weerasuriya Y., Wood K., Ejeta G., Butler L.G. Isolation of strigol, a germination stimulant for *Striga asiatica*, from host plants. J Agric Food Chem 1993; 41:1486-1491.

Smith C.E., Dudley M.W., Lynn D.G. Vegetative/parasitic transition: control and plasticity in *Striga* development. Plant Physiol 1990; 93:208-215.

Smith C.E., Ruttledge T., Zeng Z., O'Malley R.C., Lynn D.G. A mechanism for inducing plant development: the genesis of a specific inhibitor. Proc Natl Acad Sci USA 1996; 93:6986-6991.

Stewart G.R., Press M.C. The physiology and biochemistry of parasitic angiosperms. Annu Rev Plant Physiol Plant Mol Biol 1990; 41:127-151.

Sukno S., Melero-Vara J.M., Fernández-Martínez J.M. Inheritance of resistance to *Orobanche cernua* Loefl. in six sunflower lines. Crop Sci 1999; 39:674-678.

Surov T., Aviv D., Aly R., Joel D.M., Goldman-Guez T., Gressel J. Generation of transgenic asulam-resistant potatoes to facilitate eradication of parasitic broomrapes (*Orobanche* spp.), with the *sul* gene as the selectable marker. Theor Appl Genet 1998; 96:132-137.

Timko M.P., Florea C.S., Riopel J.L. "Control of germination and early development in parasitic angiosperms." In *Recent Advances in the Development and Germination of Seeds*, R. B. Taylorson, eds. New York: Plenum Press, 1989.

Vail S.L., Dailey O.D., Blanchard E.J., Pepperman A.B., Riopel J.L. Terpenoid precursors of strigol as seed germination stimulants of broomrape (*Orobanche ramosa*) and witchweed (*Striga asiatica*). J Plant Growth Regul 1990; 9:77-83.

Vogler R.K., Ejeta G., Butler L.G. Inheritance of low production of *Striga* germination stimulant in sorghum. Crop Sci 1996; 36:1185-1191.

Weissenborn D.L., Denbow C.J., Laine M., Lang S., Yang Z., Yu X., Cramer C.L. HMG-CoA reductase and terpenoid phytoalexins: molecular specialization within a complex pathway. Physiol Plant 1995; 93:393-400.

Westwood J.H. Characterization of the *Orobanche-Arabidopsis* system for studying parasite-host interactions. Weed Sci 2000; 48:742-748.

Westwood J.H., Yu X., Foy C.L., Cramer C.L. Expression of a defense-related 3-hydroxy-3-methylglutaryl CoA reductase gene in response to parasitization by *Orobanche* spp. Mol Plant-Microbe Interact 1998; 11:530-536.

Yoder J.I. A species-specific recognition system directs haustorium development in the parasitic plant *Triphysaria* (Scrophulariaceae). Planta 1997; 202:407-413.

Yoder J.I. Parasitic plant responses to host plant signals: a model for subterranean plant-plant interactions. Curr Opinion Plant Biol 1999; 2:65-70.

Yokota T., Sakai H., Okuno K., Yoneyama K., Takeuchi Y. Alectrol and orobanchol, germination stimulants for *Orobanche minor*, from its host red clover. Phytochemistry 1998; 49:1967-1973.

CHAPTER 9

ALTERED HERBICIDE TARGET SITES

IMPLICATIONS FOR HERBICIDE-RESISTANT WEED MANAGEMENT

MICHAEL J. CHRISTOFFERS, VIJAY K. NANDULA, LEMMA W. MENGISTU & CALVIN G. MESSERSMITH

Department of Plant Sciences, North Dakota State University, Fargo, ND 58105, U.S.A.

Herbicides have been important tools for weed management ever since the introduction of auxin-analog herbicides in the mid-1940s. While herbicides continue to be viable weed control options, herbicide-resistant weeds are an ever-increasing concern for world agriculture (Heap, 1997; Morrison and Devine, 1994; Shaner, 1995). Herbicide resistance is defined as the inherited ability of a plant to survive and reproduce following exposure to a dose of herbicide normally lethal to wild-type plants (Anonymous, 1998). Selection for rare resistant biotypes in weed populations is believed to be responsible for the evolution of most herbicide-resistant weed populations. Recurrent selection pressure is provided by season-after-season use of herbicides that do not control the resistant biotype(s). These herbicides are most likely those with the same or similar mode of action. Herbicide resistance has been documented in 98 dicot and 67 monocot weed species worldwide, totaling 275 species-by-mode-of-action combinations to date (Heap, 2003).

Several biological mechanisms, alone or in combination, may be responsible for herbicide resistance in weeds. These mechanisms include expression of resistant or overexpression of wild-type herbicide-target-site proteins, modified metabolism including enhanced deactivation or reduced activation of herbicide molecules, and altered herbicide absorption, translocation, or sequestration. The most common mechanism appears to be alteration of target sites, although this conclusion may partly be due to research favoring the study of altered target sites over other, more complicated mechanisms.

Target-site resistance has been extensively reviewed (Devine and Eberlein, 1997; Devine and Shukla, 2000; Powles and Holtum, 1994; Preston and Mallory-Smith, 2001), and we refer the reader to these articles for details not covered in this chapter. Our discussion will focus on naturally occurring herbicide-target-site alterations in weeds, emphasizing potential implications for weed management.

Inderjit (ed.), Weed Biology and Management, 199-210.

TARGET-SITE RESISTANCE CLASSIFICATION

Confirmation of target-site resistance in weeds is commonly a two-step process. Herbicide resistance is first confirmed by comparing the response of putatively resistant plants to known susceptible biotypes (Beckie et al., 2000). This is typically accomplished by treating soil-grown whole plants with a range of herbicides to demonstrate lack-of-control consistent with resistance. Alternative methods, such as petri-dish assays of seedling growth-reduction in the presence of herbicide, also may be used to confirm resistance (Frie, 1995; Moss, 2000; Murray et al., 1996).

The second step in classifying target-site herbicide resistance is to confirm that resistance is indeed due to target-site alteration. Whole-plant resistance screening assays may give some indication of the resistance mechanism (Moss, 2000), but preferably the activity of target enzymes from putative resistant and susceptible plants in the presence of active herbicide is evaluated. Enzyme assays are typically conducted under *in vitro* conditions and often use crude enzyme preparations with minimal purification. The activity of enzymes under these conditions can depend on experimental conditions such as extraction buffer pH or assay temperature, and optimization of conditions is essential to prevent inaccurate results. For example, Shukla et al. (1997), based on the results of carefully planned experiments, recommended that buffer conditions used for extraction of acetyl-CoA carboxylase (ACCase) from plant tissue be carefully considered to accurately detect target-site resistance. Similar precautions might be warranted in other target-site assay systems. An additional limitation of enzyme assays for confirmation of target-site herbicide resistance is that the target proteins for many herbicides are not known.

Information gained from herbicide resistance research, including whole-plant and enzyme studies, have clearly established that expression of target-site resistance often varies from biotype to biotype. Cross-resistance is the ability of a single resistance mechanism, which may be coded by one or more genes, to confer resistance to herbicides of more than one chemical class (Hall et al., 1994). For example, a specific alteration of acetolactate synthase (ALS) may confer resistance to sulfonylurea and imidazolinone classes of ALS-inhibiting herbicides (Tranel and Wright, 2002). This is cross-resistance because resistance to the two herbicide classes are the result of a single mechanism. Target-site alterations often confer cross-resistance, but not always. Weeds with target-site resistance may also vary in response to individual herbicides within a chemical class, whether or not they are cross-resistant. Negative cross-resistance, or supersensitivity, occurs when a resistant biotype is more susceptible to other classes of herbicide than wild-type biotypes (Gressel, 1991). Negative cross-resistance appears to be relatively rare, but may often go undetected due to high wild-type susceptibility in most herbicide treatment experiments.

Weeds that resist herbicides from different chemical classes due to more than one mechanism are considered to have multiple resistance (Hall et al., 1994). Multiple resistance includes the alteration of more than one target site, such as unrelated alterations in ALS and in the D1 protein of photosystem II (PS II), providing separate resistance to inhibitors of ALS and PS II, respectively (Foes et al., 1999). Multiple resistance likely evolves through separate selection processes (Gunsolus, 1993).

TARGET-SITE ALTERATIONS

Several distinct molecular alterations conferring herbicide resistance have been identified among genes encoding target-site proteins. By definition, herbicide-resistant target proteins are not the wild-type and are generally considered to be mutant forms. The most likely source of altered target sites is the spontaneous mutation of wild-type genes into resistant forms (Jasieniuk et al., 1996), although direct evidence to support the assumption that target-site resistance genes are descendents of susceptible forms is not always available. In this review, we will consider mutation to be the source of resistance when differences in herbicide response among biotypes is the result of target-site DNA sequence polymorphism.

Identification of target-site alterations has been facilitated by work not only with naturally-occurring resistant weeds but also with resistant crops and model organisms including laboratory-derived resistant strains. The aim of this article is to review target-site resistance alterations that have been discovered among natural weed populations. Additional artificially-derived resistance mutations or those known among non-weedy species will be excluded from this discussion, and the reader is referred to other reviews for this information (Devine and Eberlein, 1997; Powles and Holtum, 1994). Unfortunately, many resistant weed biotypes are studied in detail without identification of the responsible alteration, even when target-site resistance is suspected and possible gene mutations are known. The number of target-site alterations known to confer resistance in weeds will no doubt continue to grow with further research.

There are currently six target sites for which resistance alterations at the DNA level have been found among weeds (Table 1). These target sites are the D1 protein of PS II, ALS, ACCase, tubulin, 5-enolpyruvylshikimate-3-phosphate synthase (EPSPS), and phytoene desaturase (PDS). The following is a general overview of resistance typical of specific target-site mutations with the intent of illustrating mutation-dependent resistance variation.

Table 1. Naturally-occurring target-site herbicide resistance substitutions among weedy plant species. S = Susceptible; R = Resistant.

Gene[a]	S[b]	R	Weed	Reference
ACCase	Ile_{1769}	Leu	*Alopecurus myosuroides*	Brown et al. (2002); Délye et al. (2002a)
	Ile_{1769}	Leu	*Avena fatua*	Christoffers et al. (2002)
	Ile_{1769}	Leu	*Avena sterilis*	A. Tal, pers. commun.
	Ile_{1769}	Leu	*Lolium rigidum*	Tal and Rubin (2002); Zagnitko et al. (2001)
	Ile_{1769}	Leu	*Setaria viridis*	Délye et al. (2002b); Zhang and Devine (2000)

Continued.

Table 1. Continued.

Gene[a]	S[b]	R	Weed	Reference
α-tubulin	Thr_{239}	Ile	*Eleusine indica*	Anthony et al. (1998); Yamamoto et al. (1998)
	Met_{268}	Thr	*Eleusine indica*	Yamamoto et al. (1998)
ALS	Ala_{122}	Thr	*Amaranthus hybridus*	cited in Tranel and Wright (2002)
	Ala_{122}	Thr	*Solanum ptycanthum*	Milliman et al. (2000); cited in Tranel and Wright (2002)
	Ala_{122}	Thr	*Xanthium strumarium*	Bernasconi et al. (1995)
	Pro_{197}	Ala	*Brassica tournefortii*	Boutsalis et al. (1999)
	Pro_{197}	Ala	*Kochia scoparia*	Guttieri et al. (1995)
	Pro_{197}	Ala	*Lindernia dubia*	Uchino and Watanabe (2002)
	Pro_{197}	Ala	*Raphanus raphanistrum*	Tan and Medd (2002)
	Pro_{197}	Arg	*Kochia scoparia*	Guttieri et al. (1995)
	Pro_{197}	Gln	*Kochia scoparia*	Guttieri et al. (1995)
	Pro_{197}	Gln	*Lindernia micrantha*	Uchino and Watanabe (2002)
	Pro_{197}	Gln	*Lindernia procumbens*	Uchino and Watanabe (2002)
	Pro_{197}	His	*Lactuca serriola*	Guttieri et al. (1992)
	Pro_{197}	Ile	*Sisymbrium orientale*	Boutsalis et al. (1999)
	Pro_{197}	Leu	*Amaranthus retroflexus*	Michel et al. (2001); Sibony et al. (2001)
	Pro_{197}	Leu	*Galium spurium*	Horsman and Devine (2000)
	Pro_{197}	Leu	*Kochia scoparia*	Guttieri et al. (1995)
	Pro_{197}	Leu	*Salsola iberica*	cited in Devine and Eberlein (1997)
	Pro_{197}	Ser	*Amaranthus blitoides*	Michel et al. (2001)
	Pro_{197}	Ser	*Kochia scoparia*	Guttieri et al. (1995)
	Pro_{197}	Ser	*Lindernia dubia* var. *major*	Uchino and Watanabe (2002)
	Pro_{197}	Ser	*Lindernia micrantha*	Uchino and Watanabe (2002)
	Pro_{197}	Ser	*Lindernia procumbens*	Uchino and Watanabe (2002)
	Pro_{197}	Thr	*Kochia scoparia*	Guttieri et al. (1992, 1995)
	Pro_{197}	Thr	*Raphanus raphanistrum*	Tan and Medd (2002)
	Ala_{205}	Val	*Xanthium strumarium*	Woodworth et al. (1996)
	Trp_{574}	Leu	Numerous	Tranel and Wright (2002)
	Ser_{653}	Asn	*Amaranthus rudis*	Patzoldt and Tranel (2001)
	Ser_{653}	Thr	*Amaranthus retroflexus*	McNaughton et al. (2001); cited in Tranel and Wright (2002)

Table 1. Continued.

Gene[a]	S[b]	R	Weed	Reference
	Ser_{653}	Thr	*Amaranthus powellii*	McNaughton et al. (2001); cited in Tranel and Wright (2002)
	Ser_{653}	Thr	*Amaranthus rudis*	Patzoldt and Tranel (2001)
EPSPS	Pro_{106}	Ser	*Eleusine indica*	Baerson et al. (2002b)
PDS	Arg_{304}	Cys	*Hydrilla verticillata*	F. E. Dayan, pers. (commun.); Michel et al. (2002)
	Arg_{304}	His	*Hydrilla verticillata*	F. E. Dayan (pers. commun.); Michel et al. (2002)
	Arg_{304}	Ser	*Hydrilla verticillata*	F. E. Dayan (pers. commun.); Michel et al. (2002)
*psb*A	Val_{219}	Ile	*Amaranthus powellii*	Dumont and Tardif (2002)
	Val_{219}	Ile	*Kochia scoparia*	L. W. Mengistu, unpublished data
	Val_{219}	Ile	*Poa annua*	Mengistu et al. (2000)
	Ser_{264}	Gly	Numerous	Trebst (1996)
	Ser_{264}	Thr	*Portulaca oleracea*	Masabni and Zandstra (1999)

[a]Abbreviations: ACCase, multifunctional plastidic acetyl-CoA carboxylase; ALS, acetolactate synthase; EPSPS, 5-enolpyruvylshikimate-3-phosphate synthase; PDS, phytoene desaturase.
[b]To account for gene sequence inter- and intra-specific variation, amino acid residue positions are standardized to the following: ACCase, *Triticum aestivum* (GenBank accession no. AF029895); α-tubulin, *Eleusine indica* (AJ005598); ALS, *Arabidopsis thaliana* (AY042819); EPSPS, predicted mature enzyme of *Eleusine indica* (AJ417034); PDS, the initial cloned variant of *Hydrilla verticillata* (F. E. Dayan, pers. commun.); *psb*A, *Arabidopsis thaliana* (X79898).

D1 Protein of Photosystem II

Inhibitors of PS II block photosynthetic electron transport and prevent the reduction of $NADP^+$ required for CO_2 fixation. The herbicidal activity of PS II-inhibitors is due to oxidative stress generated from the electron transport block, leading to destruction of lipid and chlorophyll molecules, and the PS II reaction center (Gronwald, 1994). Herbicides inhibiting PS II include those of the triazine, triazinone, urea, benzothiadiazole, and several other chemical classes.

Resistance to triazine herbicides in higher plants was first observed in 1968 (Ryan, 1970), and is now documented in 64 weed species worldwide (Heap, 2003). Weed biotypes with target-site resistance to PS II-inhibitors typically have a chloroplast genome mutation on the *psb*A gene, leading to an altered D1 protein (Hirshchberg and McIntosh, 1983). A Ser_{264} to Gly substitution is responsible for conferring resistance in almost all of these cases (see Gronwald, 1994) (Table 1). The phenotype associated with the Ser_{264} to Gly substitution generally has high resistance to triazine and moderate resistance to triazinone herbicides. Negative cross-resistance to a few herbicides, including bentazon, a benzothiadiazole, has been reported (Fuerst et al., 1986; Oettmeier, 1999).

Other *psbA* substitutions include Ser_{264} to Thr in a resistant biotype of common purslane (*Portulaca oleracea*) (Masabni and Zandstra, 1999) (Table 1). Ser_{264} to Thr seems to confer high resistance to triazines, similar to the Ser_{264} to Gly substitution, and also results in high resistance to ureas, whereas Ser_{264} to Gly plants have little to no urea resistance. Moderate resistance to metribuzin, a triazinone herbicide, and diuron, a urea herbicide, is associated with a Val_{219} to Ile substitution in annual bluegrass (*Poa annua*) (Mengistu et al., 2000) (Table 1). This same Val_{219} to Ile substitution has also been reported in resistant Powell amaranth (*Amaranthus powellii*) (Dumont and Tardif, 2002), and resistant kochia (*Kochia scoparia*) (L. W. Mengistu, unpublished data).

Resistance to PS II-inhibiting herbicides is complicated by the observation that resistance genes are not always expressed as expected. For example, common groundsel (*Senecio vulgaris*) plants carrying the Ser_{264} to Gly substitution are not always resistant, most likely due to chloroplast DNA polymorphism (Frey et al., 1999).

Acetolactate Synthase

ALS, also known as acetohydroxyacid synthase (AHAS), is the first common enzyme in the synthesis of branched-chain amino acids (Shaner and Singh, 1997). ALS is inhibited by herbicides including those of the sulfonylurea, imidazolinone, and triazolopyrimidine classes. Substitutions involving various residues within ALS are known to provide resistance to one or more classes of ALS-inhibitors in higher plants (Tranel and Wright, 2002). Substitutions for at least five of these residues have been identified among resistant weeds (Table 1).

ALS substitutions of Val for Ala_{205} and Leu for Trp_{574} provide resistance to sulfonylurea, imidazolinone, and triazolopyrimidine herbicides (Devine and Eberlein, 1997; Devine and Shukla, 2000). These residue positions, and the ones that follow, refer to the mouse-ear cress (*Arabidopsis thaliana*) ALS enzyme (GenBank accession no. AY042819). A lower level of resistance is typical of Ala_{205} to Val compared to the high resistance conferred by Trp_{574} to Leu.

Mutation of Pro_{197} confers high level resistance to sulfonylurea herbicides, with cross-resistance varying from susceptibility to moderate resistance for imidazolinone herbicides and low to high resistance to triazolopyrimidine herbicides. Cross-resistance involving mutations of Pro_{197} appear dependent on specific substitutions. Resistance to imidazolinones is characteristic of Ala_{122} to Thr, while susceptibility to sulfonylurea and triazolopyrimidine herbicides is maintained. Substitution of Thr for Ser_{653} also confers resistance to imidazolinones but not sulfonylureas (Tranel and Wright, 2002).

Acetyl-CoA Carboxylase

ACCase is responsible for production of malonate from acetyl-CoA, the first committed step of fatty acid synthesis (Devine, 1997). Aryloxyphenoxypropionate and cyclohexanedione herbicides inhibit plastidic ACCase of most grasses. Although inhibition of fatty acid synthesis is the presumed cause of death, not all evidence supports this conclusion and the exact mode-of-action for ACCase-inhibitors may not be fully understood (Devine and Shimabukuro, 1994). Regardless, it is quite apparent that mutations within the plastidic ACCase target site is a mechanism by which weeds may resist ACCase-inhibiting herbicides. Identification of these mutations, however, has been relatively slow.

Only recently has their been enough supporting evidence to associate an Ile to Leu substitution with resistance (Zagnitko et al., 2001) (Table 1). This substitution is at position 1769 based on the plastidic ACCase sequence of wheat (*Triticum aestivum*) (GenBank accession no. AF029895). The Ile_{1769} to Leu substitution is associated with high resistance to sethoxydim and low to moderate resistance to other cyclohexanedione and aryloxyphenoxypropionate herbicides. Other candidate substitutions for aryloxyphenoxypropionate resistance have been identified (Christoffers et al., 2000), but their significance toward resistance is still unclear.

Tubulin

Polymerization of heterodimers consisting of α and β-tubulin is required for production of microtubules necessary for spindle-fiber formation and cell division. Herbicides including dinitroanilines inhibit this polymerization by binding to tubulin heterodimers (Anthony and Hussey, 1999). The two substitutions determined to be associated with resistance to these herbicides in weeds were found within α-tubulin of goosegrass (*Eleucine indica*) (Anthony et al., 1998; Yamamoto et al., 1998) (Table 1). Despite early evidence for herbicide resistance associated with β-tubulin (Vaughn and Vaughan, 1990), resistance alterations have not been found among the β-tubulin genes of resistant goosegrass. Substitutions detected within α-tubulin of goosegrass differ in the resistance they confer to dinitroanilines; Thr_{239} to Ile confers high level resistance while Met_{268} to Thr confers moderate level resistance.

5-Enolpyruvylshikimate-3-Phosphate Synthase

Glyphosate is a herbicide of major importance for world agriculture. The target-site of glyphosate is EPSPS, an important enzyme for production of the aromatic amino acids phenylalanine, tryptophan and tyrosine. Only one alteration of the EPSPS protein has been found as likely responsible for target-site resistance to glyphosate in weeds (Table 1). A Pro to Ser substitution was found in the EPSPS of glyphosate-resistant goosegrass, corresponding to position 106 of the predicted mature goosegrass EPSPS enzyme (Baerson et al., 2002b).

Currently, altered EPSPS has not been identified as a major resistance mechanism among glyphosate-resistant biotypes of other weed species, including rigid ryegrass (*Lolium rigidum*) (Baerson et al., 2002a) and horseweed (*Conyza canadensis*) (Van Gessel, 2001).

Phytoene Desaturase

PDS is an enzyme of the carotenoid pathway, catalyzing the desaturation of phytoene into phytofluene and ζ-carotene. Additional carotenoids are produced from ζ-carotene, and herbicidal inhibition of PDS leads to photobleaching due to lack of carotenoid protection. Three substitutions involving Arg_{304} of PDS have been found among herbicide-resistant biotypes of the aquatic weed hydrilla (*Hydrilla verticillata*) (F. E. Dayan, pers. commun.; Michel et al., 2002) (Table 1). Substitution of Cys, His, or Ser for Arg_{304} provides resistance to the PDS-inhibitor fluridone. Variation in the level of fluridone resistance and cross-resistance to other PDS-inhibitors appears to be dependent on the specific substitution.

SUMMARY OF TARGET-SITE ALTERATIONS

Three observations are revealed through a survey of the known target-site alterations for herbicide resistance in weeds. First, the typical resistance alteration is a single point mutation resulting in substitution of an amino acid residue, i.e., a missense mutation. Only in one case of ALS-inhibitor resistant eastern rocket (*Sisymbrium orientale*) has the missense mutation been due to substitution of two nucleotides within the same codon (Boutsalis et al., 1999). Second, mutations within the target-site genes are not distributed at random, but rather substitution of a limited number of residues results in resistance. Third, the variety of substitutions leading to resistance varies from site to site. For example, all possible substitutions of Pro_{197} that may result from mutation of a single nucleotide have been found among ALS-inhibitor resistant kochia (Guttieri et al., 1995). Conversely, only the substitution of Leu for Trp_{574} has been reported at this residue among resistant weeds, with other possibilities resulting in inactive ALS (Bernasconi et al., 1995).

These three observations are consistent with target-site modifications that provide herbicide insensitivity while maintaining biological function sufficient for life. Alterations that disrupt herbicide/target site interaction but severely disrupt the biological function of the target protein, although theoretically more common, would not be selected in the field due to overwhelming lack of fitness. A finite number of alterations would therefore be expected to result in problematic target-site resistance in the field, thus limiting the number of target-site alterations to consider when characterizing individual resistance cases.

HERBICIDE-RESISTANT WEED MANAGEMENT

What is the best way to manage resistant populations and/or slow down the rate at which they occur? Several recommendations have been made (Anonymous, 2002), but here we consider those most related to target-site resistance. Two primary recommendations are to annually rotate herbicides based on target site and/or apply mixtures of herbicides that control similar weed species through interaction with different target sites. These recommendations are designed to limit selection pressure for resistance and slow its occurrence.

Classification of herbicides based on target site is fundamental to these resistance management strategies. Both the Herbicide Resistance Action Committee and the

Weed Science Society of America have established such classification systems (Retzinger and Mallory-Smith, 1997). Resistance management based on target-site is gaining strength as evidenced by mandatory labeling of herbicides in Australia with their target-site classification (Shaner and Leonard, 2001), and voluntary labeling guidelines developed in Canada (Anonymous, 1999) and the United States (Anonymous, 2001). Recent voluntary pesticide labeling guidelines announced by the U.S. Environmental Protection Agency state that pesticide mode-of-action/target site is the "cornerstone of most resistance management programs" (Anonymous, 2001).

Should resistance management based on target site be as widely accepted as it currently is? Cross-resistance and even resistance to individual herbicides varies within target-site categories as illustrated above. This fact is well established but perhaps under-appreciated. Non-target-site resistance is even less amenable to management based on herbicide target sites. For instance, metabolism-based mechanisms may provide cross-resistance to herbicides with different target sites (Preston and Mallory-Smith, 2001).

Annually rotating herbicides and/or applying herbicide mixtures may underutilize viable herbicides while overutilizing alternative herbicides. Re-evaluations concerning the general applicability of these strategies have raised questions about their overall benefit (Friesen et al., 2000; Pannell and Zilberman, 2001). The cost of implementing a resistance management program prior to appearance of resistance may be unnecessarily burdensome in some cases. Implementation of a rotation and/or mixture program may also require at least a reduction in use of a grower's preferred method of weed control and may also require crop rotation adjustments that deviate from a grower's preference.

An additional and accepted resistance management recommendation is to monitor trends or changes in weed populations (Anonymous, 2002). To implement weed population monitoring to its fullest potential requires a case-by-case approach, including the means to identify resistance problems early, to quickly confirm resistance after a lack-of-control, and to predict the likely success or failure of alternative weed control measures. Central to these goals is the ability to quickly diagnose and classify potential or realized resistance problems based on resistant weed biology.

Target-site herbicide resistance alterations are significant to case-by-case weed management for two reasons. First, target-site mutations are the direct cause of many cases of resistance and may form a basis for herbicide response prediction. Second, continuing advances in molecular genetic techniques will greatly facilitate practical genetic testing for resistant weed diagnosis. A re-prioritization of resistance management strategies may be required to take full advantage of these new technologies.

ACKNOWLEDGMENT

This material is based upon work supported by the Cooperative State Research, Education, and Extension Service, U.S. Department of Agriculture, under Agreement No. 98-34361-6831.

REFERENCES

Anonymous. "Herbicide resistance" and "herbicide tolerance" defined. Weed Technol 1998; 12:789.

Anonymous. Voluntary Pesticide Resistance-Management Labelling based on Target Site/Mode of Action. Regulatory Directive DIR99-06. Ottawa: Pest Management Regulatory Agency, Health Canada, 1999.

Anonymous. Notice to Manufacturers, Formulators, Producers and Registrants of Pesticide Products. Pesticide Registration Notice 2001-5. Washington, DC: Office of Pesticide Programs, U.S. Environmental Protection Agency, 2001.

Anonymous. HRAC Guidelines - How to Minimize Resistance Risks and How to Respond to Cases of Suspected and Confirmed Resistance. Herbicide Resistance Action Committee. Online. Internet. January 29, 2002. Available at www.plantprotection.org/HRAC/.

Anthony R.G., Hussey P.J. Dinitroaniline herbicide resistance and the microtubule cytoskeleton. Trends Plant Sci 1999; 4:112-116.

Anthony R.G., Waldin T.R., Ray J.A., Bright S.W.J., Hussey P.J. Herbicide resistance caused by spontaneous mutation of the cytoskeletal protein tubulin. Nature 1998; 393:260-263.

Baerson S.R., Rodriguez D.J., Biest N.A., Tran M., You J., Kreuger R.W., Dill G.M., Pratley J.E., Gruys K.J. Investigating the mechanism of glyphosate resistance in rigid ryegrass (*Lolium rigidum*). Weed Sci 2002a; 50:721-730.

Baerson S.R., Rodriguez D.J., Tran M., Feng Y., Biest N.A., Dill G.M. Glyphosate-resistant goosegrass. Identification of a mutation in the target enzyme 5-enolpyruvylshikimate-3-phosphate synthase. Plant Physiol 2002b; 129:1265-1275.

Beckie H.J., Heap I.M., Smeda R.J., Hall L.M. Screening for herbicide resistance in weeds. Weed Technol 2000; 14:428-445.

Bernasconi P., Woodworth A.R., Rosen B.A., Subramanian M.V., Siehl D.L. A naturally occurring point mutation confers broad range tolerance to herbicides that target acetolactate synthase. J Biol Chem 1995; 270:17381-17385.

Boutsalis P., Karotam J., Powles S.B. Molecular basis of resistance to acetolactate synthase-inhibiting herbicides in *Sisymbrium orientale* and *Brassica tournefortii*. Pestic Sci 1999; 55:507-516.

Brown A.C., Moss S.R., Wilson Z.A., Field L.M. An isoleucine to leucine substitution in the ACCase of *Alopecurus myosuroides* (black-grass) is associated with resistance to the herbicide sethoxydim. Pestic Biochem Physiol 2002; 72:160-168.

Christoffers M.J., Berg M.L., Messersmith C.G. Analysis of acetyl-CoA carboxylase gene sequences from fenoxaprop-P-resistant wild oat biotypes. Proc North Cent Weed Sci Soc 2000; 55:67.

Christoffers M.J., Berg M.L., Messersmith C.G. An isoleucine to leucine mutation in acetyl-CoA carboxylase confers herbicide resistance in wild oat. Genome 2002; 45:1049-1056.

Délye C., Calmès É., Matéjicek A. SNP markers for black-grass (*Alopecurus myosuroides* Huds.) genotypes resistant to acetyl CoA-carboxylase inhibiting herbicides. Theor Appl Genet 2002a; 104:1114-1120.

Délye C., Wang T., Darmency H. An isoleucine-leucine substitution in chloroplastic acetyl-CoA carboxylase from green foxtail (*Setaria viridis* L. Beauv.) is responsible for resistance to the cyclohexanedione herbicide sethoxydim. Planta 2002b; 214:421–427.

Devine M.D. Mechanisms of resistance to acetyl-coenzyme A carboxylase inhibitors: a review. Pestic Sci 1997; 51:259-264.

Devine M.D, Eberlein C.V. "Physiological, biochemical and molecular aspects of herbicide resistance based on altered target sites." In *Herbicide Activity: Toxicology, Biochemistry and Molecular Biology*. R.M. Roe, J.D. Burton, R.J. Kuhr, eds. Amsterdam: IOS Press, 1997, pp. 159-185.

Devine, M.D., Shimabukuro R.H. "Resistance to acetyl coenzyme A carboxylase inhibiting herbicides." In *Herbicide Resistance in Plants: Biology and Biochemistry*, S.B. Powles, J.A.M. Holtum, eds. Boca Raton, FL: Lewis Publishers, 1994, pp. 141-169.

Devine M.D., Shukla A. Altered target sites as a mechanism of herbicide resistance. Crop Protect 2000; 19:881-889.

Dumont M., Tardif F.J. Resistance to linuron in a Powell amaranth population. Abstr Weed Sci Soc Am 2002; 42:49-50.

Foes M.J., Liu L., Vigue G., Stoller E.W., Wax L.M., Tranel P.J. A kochia (*Kochia scoparia*) biotype resistant to triazine and ALS-inhibiting herbicides. Weed Sci 1999; 47:20-27.

Frey J.E., Müller-Schärer H., Frey B., Frei D. Complex relation between triazine-susceptible phenotype and genotype in the weed *Senecio vulgaris* may be caused by chloroplast DNA polymorphism. Theor Appl Genet 1999; 99:578-586.

Frie J.B. Wild Oat (*Avena fatua*) Resistance to ACCase-Inhibiting Herbicides in North Dakota and Western Minnesota. M.S. Thesis. Fargo, ND: North Dakota State University, 1995.

Friesen L.J.S., Ferguson G.M., Hall J.C. Management strategies for attenuating herbicide resistance: untoward consequences of their promotion. Crop Protect 2000; 19:891-895.

Fuerst E.P., Arntzen C.J., Pfister K., Penner D. Herbicide cross-resistance in triazine-resistant biotypes of four species. Weed Sci 1986; 34:344-353.

Gressel J. "Why get resistance? It can be prevented or delayed." In *Herbicide Resistance in Weeds and Crops,* J.C. Caseley, G.W. Cussans, R.K. Atkin, eds. Oxford, U.K.: Butterworth-Heinemann Ltd., 1991, pp. 1-25.

Gronwald J.W. "Resistance to photosystem II inhibiting herbicides." In *Herbicide Resistance in Plants: Biology and Biochemistry,* S.B. Powles, J.A.M. Holtum, eds. Boca Raton, FL: Lewis Publishers, 1994, pp. 27-60.

Gunsolus J.L. Herbicide Resistant Weeds. North Cent Reg Ext Pub 468. St. Paul, MN: Minnesota Extension Service, University of Minnesota, 1993.

Guttieri M.J., Eberlein C.V., Mallory-Smith C.A., Thill D.C., Hoffman D.L. DNA sequence variation in domain A of the acetolactate synthase genes of herbicide-resistant and -susceptible weed biotypes. Weed Sci 1992; 40:670-676.

Guttieri M.J., Eberlein C.V., Thill D.C. Diverse mutations in the acetolactate synthase gene confer chlorsulfuron resistance in kochia (*Kochia scoparia*) biotypes. Weed Sci 1995; 43:175-178.

Hall L.M., Holtum J.A.M., Powles S.B. "Mechanisms responsible for cross resistance and multiple Resistance." In *Herbicide Resistance in Plants: Biology and Biochemistry*, S.B. Powles, J.A.M. Holtum, eds. Boca Raton, FL: Lewis Publishers, 1994; pp. 243-261.

Heap I.M. The occurrence of herbicide-resistant weeds worldwide. Pestic Sci 1997; 51:235-243.

Heap I.M. The International Survey of Herbicide Resistant Weeds. Online. Internet. May 6, 2003. Available at www.weedscience.com.

Hirschberg J., McIntosh L. Molecular basis of herbicide resistance in *Amaranthus hybridus*. Science 1983; 222:1346-1349.

Horsman G.P., Devine M.D. Molecular basis of ALS resistance in a biotype of false cleavers (*Galium spurium* L.). Abstr Weed Sci Soc Am 2000; 40:5.

Jasieniuk M., Brûlé-Babel A.L., Morrison I.N. The evolution and genetics of herbicide resistance in weeds. Weed Sci 1996; 44:176-193.

Masabni J.G., Zandstra B.H. A serine-to-threonine mutation in linuron-resistant *Portulaca oleracea*. Weed Sci 1999; 47:393-400.

McNaughton K.E., Lee E.A., Tardif F.J. Mutations in the ALS gene conferring resistance to group II herbicides in redroot pigweed (*Amaranthus retroflexus*) and green pigweed (*A. powellii*). Abstr Weed Sci Soc Am 2001; 41:97.

Mengistu L.W., Mueller-Warrant G.W., Liston A., Barker R.E. *psb*A mutation (valine$_{219}$ to isoleucine) in *Poa annua* resistant to metribuzin and diuron. Pest Manag Sci 2000; 56:209-217.

Michel A., Dayan F.E., Netherland M.D., Duke S.O., Scheffler B.E. Discovery of Natural Resistance to PDS-Inhibitors in Higher Plants. 10th IUPAC International Congress on the Chemistry of Crop Protection; August 4-9; Basil, 2002.

Michel A., Wagner J., Haas H., Hurle K. Characterization of resistance to ALS-inhibitors in three *Amaranthus* species. Abstr Weed Sci Soc Am 2001; 41:125.

Milliman L.D., Riechers D.E., Simmons F.W., Wax L.M. Two biotypes of eastern black nightshade that are resistant to ALS-inhibiting herbicides. Proc North Cent Weed Sci Soc 2000; 55:86.

Morrison I.N., Devine M.D. Herbicide resistance in the Canadian prairie provinces: five years after the fact. Phytoprotection 1994; 75 (Suppl.):5-16.

Moss S.R. The Rothamsted rapid resistance test for detecting herbicide-resistance in annual grass weeds. Abstr Weed Sci Soc Am 2000; 40:42-43.

Murray B.G., Friesen L.F., Beaulieu K.J., Morrison I.N. A seed bioassay to identify acetyl-CoA carboxylase inhibitor resistant wild oat (*Avena fatua*) populations. Weed Technol 1996; 10:85-89.

Oettmeier W. Herbicide resistance and supersensitivity in photosystem II. Cell Mol Life Sci 1999; 55:1255-1277.

Pannell D.J., Zilberman D. "Economic and sociological factors affecting growers' decision making on herbicide resistance." In *Herbicide Resistance and World Grains*, S.B. Powles, D.L. Shaner, eds. Boca Raton, FL: CRC Press, 2001; pp. 251-277.

Patzoldt W.L., Tranel P.J. ALS mutations conferring herbicide resistance in waterhemp. Proc North Cent Weed Sci Soc 2001; 56:67.

Powles S.B., Holtum J.A.M. *Herbicide Resistance in Plants: Biology and Biochemistry*. Boca Raton, FL: Lewis Publishers, 1994.

Preston C., Mallory-Smith C.A. "Biochemical mechanisms, inheritance, and molecular genetics of herbicide resistance in weeds." In *Herbicide Resistance and World Grains*, S.B. Powles, D.L. Shaner, eds. Boca Raton, FL: CRC Press, 2001, pp. 23-60.

Retzinger E.J. Jr., Mallory-Smith C. Classification of herbicides by site of action for weed resistance management strategies. Weed Technol 1997; 11:384-393.

Ryan G.F. Resistance of common groundsel to simazine and atrazine. Weed Sci 1970; 18:614-616.

Shaner D.L. Herbicide resistance: Where are we? How did we get here? Where are we going? Weed Technol 1995; 9:850-856.

Shaner D.L., Leonard P. "Regulatory aspects of resistance management for herbicides and other crop protection products." In *Herbicide Resistance and World Grains*, S.B. Powles, D.L. Shaner, eds. Boca Raton, FL: CRC Press, 2001, pp. 279-294.

Shaner D.L., Singh B.K. "Acetohydroxyacid synthase inhibitors." In *Herbicide Activity: Toxicology, Biochemistry and Molecular Biology*, R.M. Roe, J.D. Burton, R.J. Kuhr, eds. Amsterdam: IOS Press, 1997, pp. 69-110.

Shukla A., Dupont S., Devine M.D. Resistance to ACCase-inhibitor herbicides in wild oat: evidence for target site-based resistance in two accessions from Canada. Pestic Biochem Physiol 1997; 57:147-155.

Sibony M., Michel A., Haas H.U., Rubin B., Hurle K. Sulfometuron-resistant *Amaranthus retroflexus*: cross-resistance and molecular basis for resistance to acetolactate synthase-inhibiting herbicides. Weed Res 2001; 41:509-522.

Tal A., Rubin B. A point mutation in the ACCase gene is responsible for the resistance to ACCase inhibiting herbicides in *Lolium rigidum*. 12th European Weed Research Society Symposium; 2002 June 24-27; Wageningen, The Netherlands.

Tan M.K., Medd R.W. Characterization of the acetolactate synthase (ALS) gene of *Raphanus raphanistrum* L. and the molecular assay of mutations associated with herbicide resistance. Plant Sci 2002; 163:195-205.

Tranel P.J., Wright T.R. Resistance of weeds to ALS-inhibiting herbicides: what have we learned? Weed Sci 2002; 50:700-12.

Uchino A., Watanabe H. Mutations in the acetolactate synthase genes of sulfonylurea-resistant biotypes of *Lindernia* spp. Weed Biol Manage 2002; 2:104-109.

VanGessel M.J. Glyphosate-resistant horseweed from Delaware. Weed Sci 2001; 49:703-705.

Vaughn K.C., Vaughan M.A. "Structural and biochemical characterization of dinitroaniline-resistant *Eleusine*." In *Managing Resistance to Agrochemicals*. ACS Symposium Series 421, M.B. Green, H.M. LeBaron, W.K. Moberg, eds. Washington, DC: American Chemical Society, 1990; pp. 364-375.

Woodworth A., Bernasconi P., Subramanian M., Rosen B. A second naturally occurring point mutation confers broad-based tolerance to acetolactate synthase inhibitors. Plant Physiol 1996; 111:S105.

Yamamoto E., Zeng L., Baird W.V. α-tubulin missense mutations correlate with antimicrotuble drug resistance in *Eleusine indica*. Plant Cell 1998; 10:297-308.

Zagnitko O., Jelenska J., Tevzadze G., Haselkorn R., Gornicki P. An isoleucine/leucine residue in the carboxyltransferase domain of acetyl-CoA carboxylase is critical for interaction with aryloxyphenoxypropionate and cyclohexanedione inhibitors. Proc Natl Acad Sci USA 2001; 98:6617-6622.

Zhang X.Q., Devine M.D. A possible point mutation of plastidic ACCase gene conferring resistance to sethoxydim in green foxtail (*Setaria viridis*). Abstr Weed Sci Soc Am 2000; 40:33.

CHAPTER 10

HERBICIDE RESISTANCE
MOVING BEYOND SIMPLE EXPLANATIONS

L.J. SHANE FRIESEN[1] & J. CHRISTOPHER HALL[2]

Department of Environmental Biology, University of Guelph, Guelph, Ontario, Canada, N1G 2W1; [1]Present address: Western Australian Herbicide Resistance Initiative, School of Plant Biology, University of Western Australia, Crawley WA, Australia, 6009. [2]Corresponding author

Herbicide resistance is the inherited ability of a plant to survive and reproduce following its exposure to a dose of herbicide normally lethal to the susceptible biotype (Heap, 2003). Herbicide-resistant populations develop from small numbers of resistant plants within susceptible populations following selection with a herbicide(s) (Maxwell and Mortimer, 1994; Friesen et al., 2000). After years of herbicide selection, a population of weeds may shift from predominantly susceptible to predominantly resistant, and may become an agronomic problem.

The phenomenon of herbicide resistance is a global one (Darmency and Gasquez, 1990) affecting an estimated 210,000 fields worldwide (Heap, 2003) and has the potential to substantially (1) increase the costs of crop production (Beckie et al., 1999), (2) limit the types of crops that may be grown, (3) lower the expected crop yields, (4) limit the types of herbicides that may be used, and/or (5) lower the resale value of crop land. Furthermore, herbicide resistance has occurred with most of the chemistries commonly used with an associated total of 272 weed biotypes from 163 species (Heap, 2003). Frequencies of resistance development have increased exponentially (Heap, 2003) despite the promotion of management strategies for more than three decades (Hall and Friesen, 1999), suggesting that current management strategies and herbicide resistance prediction models (Gressel and Segel, 1978, 1990; Putwain and Mortimer, 1989; Maxwell et al., 1990; Mortimer et al., 1993; Gardner et al., 1998; Richter and Zwerger, 1999; Cavan et al., 2000; Cavan and Moss, 2001), upon which many of the current strategies are based, may not be effective. It has been proposed that this may be partially due to poor estimates of the relevant underlying processes and characteristics that influence the development of herbicide-resistant populations (Friesen et al., 2000). In order to improve our understanding of herbicide resistance and to develop new models that can more accurately predict the evolution and severity of new herbicide-resistant populations, these underlying factors must be re-examined. More effective resistance management strategies could be derived from these new models.

Inderjit (ed.), Weed Biology and Management, 211-225.

The plant factors that influence the appearance and intensity of herbicide resistance are those that contribute to the initial frequencies of resistance, i.e., before selection occurs, amplification of resistance through selection, and perpetuation of resistance. These factors may differ among weed species and weed populations and thus may explain why the appearance of resistance is different among weed species and weed populations.

Because it is believed that herbicide-resistant plants develop through random mutation events and thus are present in a susceptible population before a herbicide is ever used, the number of resistant plants present in a population when a herbicide is first applied will ultimately affect the rate at which herbicide-resistant populations develop (Diggle and Neve, 2001). This number is known as the initial frequency of resistance (Jasieniuk et al., 1996; Diggle and Neve, 2001) and is affected by a number of factors including the frequency of mutations leading to herbicide resistance, the levels at which resistance alleles are expressed and their probability for survival in the gene pool, i.e., allele fitness, and the speed and breadth of resistance allele dissemination, i.e., allele distribution.

Herbicide selection has the effect of amplifying the initial frequency of resistance alleles in a population. Factors that are associated with the selection pressure for resistance and that are innate characteristics of plant species include generation time, seed dormancy period, and level of susceptibility to a herbicide. They also include allele fitness and allele distribution, which are not necessarily innate charecteristics of a plant species.

The relationships of generation time, dormancy period, and level of susceptibility to the development of herbicide-resistant populations have been discussed (Gressel and Segel, 1978, 1990; Maxwell et al., 1990; Gasquez, 1997; Shaner, 1997). In general, it has been concluded that the rate and intensity of herbicide resistance development is directly related to the selection pressures. According to this tenet, herbicide-resistant populations are more likely to develop in plant species exhibiting a relatively short generation time and low dormancy period, and in species that are highly susceptible to the herbicide. Mutation frequency, allele fitness, and allele distribution have been previously identified to influence the evolution of herbicide resistance by several authors but are yet to be comprehensively discussed (e.g., Gressel and Segel, 1978, 1990; Putwain and Mortimer, 1989; Maxwell et al., 1990; Mortimer et al., 1993; Ghersa et al., 1994a; Gould, 1995; Jasieniuk et al., 1996; Martinez-Ghersa et al., 1997; Gardner et al., 1998; Richter and Zwerger, 1999; Cavan et al., 2000; Cavan and Moss, 2001; Diggle and Neve, 2001).

In this chapter, we discuss (1) mutation frequency for resistance, (2) resistance allele fitness, and (3) resistance allele distribution, three important factors that may substantially influence the appearance and intensity of herbicide resistance but which have heretofore been addressed perfunctorily in resistance evolution models. Furthermore, we examine how and why herbicide-resistant populations are developing with broad-spectrum resistance from the overuse of a single herbicide group.

MUTATION FREQUENCY FOR HERBICIDE RESISTANCE

Gene mutation is an important process that occurs naturally and is the only natural way in which alternative alleles, e.g., resistance alleles, at a gene locus are produced

(Shorrocks, 1978). Mutations may occur in several different ways, many of which may not result in a change in the amino acid sequence. Those that have been demonstrated include (1) missense mutations, where there is a change at one or more nucleotide base pairs, (2) nonsense mutations, where codon changes result in mRNA molecules that encode shorter or longer polypeptides, (3) frameshift mutations, where nucleotides are deleted or added resulting in a reading frame that encodes a different amino acid sequence downstream of the insertion or deletion, (4) mutation of gene regulators, which control gene activation and repression, and (5) silent mutations, where a mutation does not result in a change in the encoded amino acid (Spiess, 1989).

Scientists have attempted to measure mutation frequencies in plants since the early 1900s (Bartlett, 1915a, b; DeVries, 1915, 1918). At that time, a dichotomy of ideologies about origins of species had developed. For example, DeVries (1915) argued that species may evolve from "mass mutation" events and therefore by sudden leaps, in opposition to the Darwinian notion of speciation through slow and imperceptible changes (DeVries, 1915). DeVries' argument was based on the fact that some species, especially *Oenothera* species (evening primrose), have relatively high mutation frequencies of leaf, stem, and floral traits, ranging from 0.05% to 2%.

Two values that have been cited as estimates of mutation frequency for herbicide resistance in the literature (Saari et al., 1994; Jasieniuk et al., 1996) are those for resistance to chlorsulfuron in Arabidopsis (*Arabidopsis thaliana*) (Haughn and Somerville, 1987) and to primisulfuron in tobacco (*Nicotiana tabacum*) (Harms and DiMaio, 1991). Haughn and Somerville (1987) treated Arabidopsis seeds with the mutagen ethyl methane sulfonate, allowed the seeds to germinate, grow to maturity, and set seed. Progeny plants were subsequently screened for resistance to chlorsulfuron, and the level of measured resistance was adjusted for the estimated affect of the mutagen as a way to back calculate what the mutation frequency might be in a nonmutagenized population (Haughn and Somerville, 1987). From this experiment, they estimated that the spontaneous frequency of mutation to chlorsulfuron resistance in Arabidopsis was approximately 1×10^{-9}. This value may not accurately reflect the actual mutation frequency for resistance because the mutagen they used only produces GC to AT transitions, and thus all possible mutations leading to chlorsulfuron resistance were not represented in their mutagenized populations. Harms and DiMaio (1991) screened tobacco cell lines with primisulfuron and calculated a nonmutagenized resistance mutation rate of 2.7×10^{-8} per cell generation. This value is likely lower than the rates of mutation that would be measured in the gametes of whole tobacco plants because many of the processes that are responsible for natural mutation events in plants, like crossover events (Griffiths et al., 1993), occur primarily during meiosis (Gould, 1995) and therefore would not have been measured in their experiments.

Genetic diversity among plant populations and species has been reported by many authors (e.g., Darmency and Gasquez, 1983; Price et al., 1983, 1985; Darmency and Aujas, 1987; Chauvel and Gasquez, 1994; Wang and Dekker, 1995; Allphin et al., 1998; Andrews et al., 1998; Cavan et al., 1998; Chung and Chung, 2000; Gitzendanner and Soltis, 2000; Pappert et al., 2000; Rutledge et al., 2000; Crawford et al., 2001; Stankiewicz et al., 2001; Volis et al., 2001; techniques reviewed in Jasieniuk and Maxwell, 2001). The above authors report the variation in specific gene loci or protein sequences among chosen populations and some have attempted to associate levels of variation with the herbicide resistance phenotype (Darmency and

Gasquez, 1983; Price et al., 1983, 1985; Andrews et al., 1998; Cavan et al., 1998; Rutledge et al., 2000; Stankiewicz et al., 2001). Their techniques may prove useful in determining the origin or source of a mutation, or in estimating initial frequencies of resistance. However, because initial frequencies of resistance are dependent on mutation frequency, allele fitness, and allele distribution, they do not provide a measure of mutation frequency. Some authors assume that initial frequencies of resistance are equal to mutation frequencies (e.g., Putwain and Mortimer, 1989; Cavan and Moss, 2001), and therefore neglect the effects of allele fitness and allele distribution on the initial frequencies of resistance. However, these characteristics may also significantly affect the initial appearance and intensity of herbicide-resistant weed populations.

The idea that mutations conferring herbicide resistance occur in weeds when or because herbicides are applied, i.e., responsive mutations, is controversial and not generally accepted (Diggle and Neve, 2001). Although some work has been done with the weed *Chenopodium album* that suggests this possibility (Gasquez et al., 1985; Bettini et al., 1987; Tremolieres et al., 1988; Darmency and Gasquez, 1990; reviewed in Darmency, 1994), there has been no clear demonstration that this process occurs in this species. Some herbicides have been associated with genetic mutation, however. For example, the *s*-triazine herbicides are known to induce mutations in some plant species (reviewed in Jasieniuk et al., 1996). Furthermore, the herbicide alachlor and the plant growth regulator maleic hydrazide induce a significant increase in the frequency of sister chromatid exchange in cultured human (*Homo sapiens*) lymphocytes (Ribas et al., 1996). Maleic hydrazide also induces a high frequency of somatic mutations in tobacco leaves and chromosome aberrations in *Vicia faba* roots (Gichner et al., 2000).

The effects of stress on genetic variation in other organisms has been well documented and reveals a clear positive relationship between the two, i.e., that genetic variation increases with stress (reviewed in Hoffmann and Parsons, 1991). This has also been observed with plants, but not for mutations that lead to herbicide resistance (Marcotrigiano, 2000; Garces et al., 2001; Turpeinen et al., 2001; Queitsch et al., 2002). Thus, although it is clear that stress, including treatment with some herbicides, may induce mutations, there is yet little evidence to support the idea that responsive mutations occur because of herbicide application and thereby accelerate the evolution of herbicide resistance in weeds.

One factor regarding gene mutation frequency that has not been considered in herbicide resistance prediction models is weed species ploidy level (e.g., Gressel and Segel, 1978; Putwain and Mortimer, 1989; Gressel and Segel, 1990; Maxwell et al., 1990; Mortimer et al., 1993; Gardner et al., 1998; Richter and Zwerger, 1999; Cavan et al., 2000; Cavan and Moss, 2001). Because mutations occur equally at any one of the gene copies present within a genome, the rate of acquiring resistance mutations could, therefore, be directly related to the number of gene copies, which is in turn determined by ploidy level. It could be argued that a hexaploid weed species (6N) containing six copies of a gene might have a frequency of acquiring herbicide resistance mutations three times higher than that of a diploid weed species (2N) containing only two copies of the gene. Thus, the ploidy level of a weed species may also be an important factor to consider when attempting to predict herbicide resistance evolution.

Current prediction models either assume that the mutation rate for herbicide resistance is 1×10^{-6} gametes per locus per generation (Maxwell et al., 1990; Gardner et al., 1998; Richter and Zwerger, 1999; Cavan et al., 2000; Cavan and Moss, 2001), or do not consider mutation frequency (Gressel and Segel, 1978; Putwain and Mortimer, 1989; Gressel and Segel, 1990; Mortimer et al., 1993). The source of 1 x 10^{-6} was not referenced in most of these papers. Maxwell et al. (1990) referenced a chapter by Georghiou and Taylor (1986); in that chapter, the authors do not give a value for mutation frequency for resistance. Rather, they state that the frequency of homozygous insecticide resistance alleles for a specific gene locus typically ranges from 10^{-4} to 10^{-8}. The source for this frequency is not given. It is likely that the value of 1 x 10^{-6} used in prediction models was derived from an average of mutation frequencies measured for numerous unrelated gene loci in other, predominantly non-plant, organisms.

Merrell (1981) compiled known mutation frequencies that produce dominant alleles at specific gene loci in Drosophila (*Drosophila melanogaster*), house mouse (*Mus musculus*), corn (*Zea mays*), human, and *Escherichia coli*. In Drosophila, house mouse, corn, and human the mutation frequencies for the tested loci range in orders of magnitude from 10^{-4} to 10^{-7} gametes per locus per generation. Mutation frequencies for loci tested in *E. coli* range from 10^{-6} to 10^{-9} genes per cell division (Merrell, 1981). There was a considerable amount of variation among gene loci in the same organism, for example, ranging from 10^{-4} to 10^{-6} gametes per locus per generation for the loci tested in corn. Contrary to the assumptions made in prediction models that mutation frequencies for resistance are homogenous among species and among gene loci of the same species, these data suggest that they may be substantially heterogeneous among both plant species and gene loci. Possible reasons for differences in frequency of mutation among gene loci have been described by Merrell (1981), and include the ideas that (1) large genes have a higher probability of mutation than small genes, (2) the rates of DNA repair may be different among gene loci, and (3) the mutation frequency of specific gene loci may be affected by other genes, e.g., mutator or antimutator genes. These ideas could be interpreted to suggest that some genes encoding herbicide target sites may be more mutable than others, and may explain why herbicide resistance develops more quickly to certain target-site chemistries than to others.

Mutation for herbicide resistance is a complex process that has not been considered fully in prediction models. True values of mutation frequency for resistance have not been measured to our knowledge for any weed species or weed population. Moreover, it is highly improbable that the frequencies of mutation for herbicide resistance are uniform among weed species or gene loci within the same species, as is assumed in prediction models. In order to better understand how herbicide resistance occurs and to predict when it will develop, accurate values for this parameter are required.

HERBICIDE RESISTANCE ALLELE FITNESS

The term "allele fitness" is used to describe those factors that affect the level at which an allele is expressed and the probability of its survival in a population. These include factors such as the genome in which the mutation occurs, i.e., nuclear or cytoplasmic, negative cross-resistance or decreased catalytic efficiencies associated with the resistance allele, and level of outcrossing. Jasieniuk et al. (1996) discussed nuclear versus cytoplasmic inheritance in the context of the appearance and spread of herbicide resistance and concluded that nuclear inheritance of a resistance gene is more conducive to the development of herbicide-resistant populations. This makes sense considering that if resistance alleles are inherited maternally their potential distribution is severely limited, having essentially no dissemination by pollen. Indeed, most mechanisms of herbicide resistance that have been determined are conferred by nuclear-inherited genes (reviewed in Jasieniuk et al., 1996).

The survivability of resistance alleles is also associated with any negative cross-resistance or catalytic inefficiencies that they may confer, and has been reviewed by several authors (Holt and Thill, 1994; Warwick and Black, 1994; Jasieniuk et al., 1996; Friesen et al., 2000; Diggle and Neve, 2001). In general, large phenotypic fitness differences are not characteristic of most types of herbicide resistance described to date, with the exception of some triazine-resistant biotypes (Jasieniuk et al., 1996; Diggle and Neve, 2001). It is important to note that if there is a significant fitness cost associated with a resistance allele, its frequency will decrease over time in the absence of herbicide selection, and, conversely, if there is not a significant fitness cost associated with a resistance allele, i.e., it is a neutral or nearly neutral mutation, its relative frequency in the gene pool is expected to remain relatively constant (Kimura and Crow, 1964; Kimura, 1968; Kimura and Ohta, 1973; Lewontin, 1974).

There is an inverse relationship between the level of outcrossing and the appearance of herbicide resistance conferred by recessive alleles, and a positive relationship between the level of outcrossing and the appearance of herbicide resistance conferred by dominant alleles (Gressel and Segel, 1990; Maxwell et al., 1990; Martinez-Ghersa et al., 1997; Cavan and Moss, 2001; reviewed in Gould, 1995; Jasieniuk et al., 1996). Resistance conferred by recessive alleles will only be expressed in the homozygous state, and because an individual in a plant population is more likely to become homozygous recessive for a specific gene locus in highly selfing species, the probability of recessive alleles causing herbicide resistance is related to the proportion of offspring that are derived through selfing, thus the probability is higher at lower outcrossing rates (Jasieniuk et al., 1996). Dominant resistance alleles will be phenotypically expressed in the heterozygous state, so the appearance of resistance caused by them will depend on the breadth of their distribution, and the speed with which they are disseminated. The breadth of resistance allele distribution will be more extensive and their speed of dissemination will be faster in plant species that exhibit high levels of outcrossing.

Most of the published model simulations assume a constant level of outcrossing and/or do not test the influence of different outcrossing rates (Gressel and Segel, 1978; Putwain and Mortimer, 1989; Gressel and Segel, 1990; Maxwell et al., 1990; Mortimer et al., 1993; Gardner et al., 1998; Richter and Zwerger, 1999; Cavan et al., 2000). Simplifying by making initial assumptions such as these is a necessary starting point when building a model. It should be clearly understood, however, that predictions may not be valid for other weed species or weed populations of the same species that exhibit substantially different rates of outcrossing than the value used in a particular model. Because levels of outcrossing are known or have been estimated for many problem weeds, it should be relatively straightforward to incorporate these values into future prediction models.

A relatively new model by Cavan and Moss (2001) has been used to simulate the influence of different levels of outcrossing on the development of resistance conferred by dominant alleles in wild oat (*Avena fatua*) by using hypothetical outcrossing rates ranging from 0.1% to 100%. Although their model determined that in this species the rate of resistance development was relatively insensitive to the level of outcrossing, this conclusion appears counterintuitive for most weed species given that higher levels of outcrossing should promote the dissemination of resistance alleles in the absence of herbicide selection.

Ultimately, a perfect model would consider both cytoplasmic and nuclear inheritance, all levels of outcrossing, and all levels of resistance allele survivability. This would enable the modeller to predict when resistance is likely to develop on a particular field given the type of weeds present and the cropping practices employed.

HERBICIDE RESISTANCE ALLELE DISTRIBUTION

A herbicide resistance allele may be naturally disseminated via pollen, seeds, or through vegetative propagation. These processes are collectively described as "gene flow" and have been thoroughly discussed by many authors (Maxwell et al., 1990; Ghersa et al., 1994a; Ghersa et al., 1994b; Matthews, 1994; Maxwell and Mortimer, 1994; Mulugeta et al., 1994; Saari et al., 1994; Gould, 1995; Stallings et al., 1995; Jasieniuk et al., 1996; Martinez-Ghersa et al., 1997; Jenczewski et al., 1999; Diggle and Neve, 2001; Franssen et al., 2001; Messeguer et al., 2001). Prediction models are usually based on the assumption that gene flow is completely random (Gressel and Segel, 1978; Gressel and Segel, 1990; Maxwell et al., 1990; Mortimer et al., 1993; Jasieniuk et al., 1996; Martinez-Ghersa et al., 1997; Richter and Zwerger, 1999; Diggle and Neve, 2001). Furthermore, it is presumed that resistance alleles follow the Castle-Hardy-Weinberg law and that the initial frequencies of resistance alleles among weed populations are relatively homogeneous. Initial frequencies of resistance conferred by single dominant alleles are usually given in prediction models as approximately 1×10^{-6} (Gressel and Segel, 1982; Maxwell et al., 1990; Mortimer et al., 1993; Jasieniuk et al., 1996; Martinez-Ghersa et al., 1997; Gardner et al., 1998; Richter and Zwerger, 1999; Cavan et al., 2000; Cavan and Moss, 2001).

The assumption of completely random mating may not be applicable to weed populations (Gould, 1995; Diggle and Neve, 2001). For example, many of the world's major weeds are highly self fertilizing and do not mate randomly (reviewed in Jasieniuk et al., 1996). But, even with a highly outcrossing weed species, the range of pollination and of seed and vegetative dispersal is usually relatively small (Darmency and Gasquez, 1990; Gould, 1995) and, furthermore, may not follow Castle-Hardy-Weinberg equilibrium with respect to pollination due to inter-sibling fertilization (Gould, 1995). This suggests that the initial frequencies of herbicide resistance alleles may not follow the Castle-Hardy-Weinberg law, that they may be highly heterogeneous among weed populations, and therefore, that they should not be assumed to be consistent among populations or individual fields for the purposes of resistance modelling. Moreover, self and inter-sibling fertilization will have the effect of concentrating neutral or nearly neutral resistance alleles in weed populations (reviewed in Gould, 1995). These points may explain why the initial frequencies of resistance in some populations are significantly higher than expected by mutation frequency alone (Price et al., 1983; Matthews and Powles, 1992).

Actual herbicide resistance allele distribution patterns are not known, but will likely vary from being relatively homogeneous to highly heterogeneous among weed populations. This variation will affect the frequency of resistance alleles present within a population both before selection occurs, and in the years when the herbicide is not used after selection has begun. Therefore, to improve the power of prediction models, it is imperative that the patterns of allele distribution are determined.

NON-SELECTED HERBICIDE RESISTANCE

The evolution of multiple herbicide resistance selected with the overuse of a single herbicide group has been observed for over a decade. This type of resistance has been identified in populations of *Lolium rigidum* (Christopher et al., 1991), *Avena fatua* (Morrison et al., 1995), and *Galium spurium* (Hall et al., 1998). These populations in particular are all resistant to at least two different herbicide target-site chemistries, one or more of which had never been previously used. In addition to these confirmed cases, there is a preponderance of *L. rigidum* populations (Tardif et al., 1997) and a substantial proportion of *A. fatua* populations (Beckie et al., 1999) that are resistant to multiple herbicide chemistries, but whose field histories have not been characterized. It is likely that some of these populations developed in the same way as those described by Christopher et al. (1991) and Morrison et al. (1995).

Because multiple resistance in the *L. rigidum*, *A. fatua*, and *G. spurium* populations described above is associated with selection by one herbicide group (Christopher et al., 1991; Morrison et al., 1995; Hall et al., 1998, respectively), it is highly unlikely that each of the respective target-site enzymes has mutated to become herbicide insensitive. Rather, it is more likely that resistance developed from an alteration to a pathway that is common for all of the herbicide chemistries to which resistance developed. In theory, such a mutation may enable a population of weeds to resist multiple herbicide chemistries, even if the population has only been exposed to some of them.

Aside from multiple, independent target-site mutations, there are four well described types of mutations in plants that may confer resistance to multiple herbicide chemistries: (1) decreased absorption, (2) decreased translocation, (3) increased detoxification, and (4) increased sequestration of the herbicides. Decreased absorption and translocation are not considered major mechanisms by which herbicide resistance evolves in weeds (Preston and Mallory-Smith, 2001), and therefore the most likely mechanisms are increased herbicide detoxification and sequestration. Resistance in the *L. rigidum* population described by Christopher et al. (1991) has been determined to be due to enhanced detoxification (Holtum et al., 1991; Christopher et al., 1992; reviewed in Powles and Preston, 2002). Christopher et al. (1991, 1992) liken the metabolic resistance of this weed population to the general metabolic tolerance of wheat (*Triticum aestivum*) to these herbicides. The *A. fatua* and *G. spurium* populations identified by Morrison et al. (1995) and Hall et al. (1998), respectively, have been partially characterized (Friesen, 2001; Van Eerd et al., 2001, respectively), although no significant differences in herbicide absorption, translocation, or detoxification were identified. Thus, the mechanism of multiple-chemistry herbicide resistance in these populations is still unknown.

Herbicide sequestration has been hypothesized to be the mechanism of resistance to the bypyridyl herbicides in several weed species (reviewed in Preston and Mallory-Smith, 2001), but few mechanisms of resistance in weeds for other herbicide chemistries have been associated with sequestration. There is evidence that suggests this mechanism may be likely for many herbicides other than the bipyridylliums. Therefore, sequestration may provide an explanation as to how herbicide resistance has developed to numerous unrelated target-site chemistries without enhancement of the detoxification pathways in *A. fatua* and *G. spurium*.

One way by which herbicides may potentially be sequestered in weeds is via across-membrane transporters. Across-membrane transporters have been extensively characterized in mammalian and microbial systems (Theodoulou, 2000) and have been shown to move drugs and other toxic compounds (Holyoak et al., 1999; Andrade et al., 2000; Del Sorbo et al., 2000; Hamamoto et al., 2001; Nishino and Yamaguchi, 2001; Sakamoto et al., 2001; Schoonbeek et al., 2001; Vermeulen et al., 2001). They have also been associated with the development of resistance to a multitude of previously effective, unrelated drug chemistries (Bain and LeBlanc, 1996; Kolaczkowski et al., 1996; Loe et al., 1996a; Loe et al., 1996b; Loe et al., 2000; Hayashi et al., 2001; Leslie et al., 2001). Five families of across-membrane drug transporters have been identified by sequence homology: (1) the major facilitator family, (2) the small multidrug resistance family, (3) the resistance nodulation cell division family, (4) the multidrug and toxic compound extrusion family, and (5) the ATP-binding cassette (ABC) family (Nishino and Yamaguchi, 2001).

Because of their known importance in other organisms, researchers have recently begun investigating ABC transporters in plants (Theodoulou, 2000). The ABC protein family is the largest known (Henikoff et al., 1997), and most of its members function as ATP-dependent pumps, i.e., transporters (Theodoulou, 2000; Sanchez-Fernandez et al., 2001). Some have also been associated with ion channels and are involved with channel regulation (Theodoulou, 2000). There are a reported 129 open reading frames capable of encoding ABC transporters in Arabidopsis, which on a gene for gene basis is relatively high, e.g., more than double that identified to date in humans (Sanchez-Fernandez et al., 2001; Anonymous, 2002). The quantity in other vascular plants has not been established (Sanchez-Fernandez et al., 2001), but should become available soon for species whose genomes have recently been sequenced, e.g., rice (*Oryza sativa*).

Plant ABC transporters have been associated with the movement of herbicide conjugates (Gaillard et al., 1994; Klein et al., 1996; Lu et al., 1997; Klein et al., 1998; Lu et al., 1998; Rea et al., 1998; Rea, 1999; Klein et al., 2000; Martinoia et al., 2000; Klein et al., 2001; Liu et al., 2001), which for many herbicides are often the first metabolites formed in a series of detoxification steps (Edwards, 1997; Hall et al., 2001), and which may retain phytotoxicity. They have also been associated with natural auxin transport in Arabidopsis (Noh et al., 2001).

Therefore, there appears to be the potential for herbicide sequestration pathways to exist in plants, which could have broad substrate/herbicide specificities, and thus could be associated with resistance to unrelated herbicide chemistries. Such pathways have not yet been demonstrated to confer resistance to multiple herbicide chemistries in weeds, but may explain why resistance has mysteriously developed to unrelated herbicides without any identified changes in the rates of herbicide detoxification, as in several *A. fatua* and *G. spurium* populations.

CONCLUSIONS

The evolution of herbicide-resistant weed populations is a complex process that involves a dynamic interplay between many weed and agroecological factors. Included in these are the mutation frequency for resistance, resistance allele fitness, and resistance allele distribution. Although it is clear that mutation frequency, allele fitness, and allele distribution may have a substantial influence on the appearance and intensity of herbicide resistance, this potential influence has not been fully characterized and should be studied.

The identification of resistance in weed populations to herbicide chemistries that were not previously applied to those populations appears to contradict the theory of resistance evolution. However, this type of resistance still seems to be associated with selection by a herbicide. The most likely mechanisms of resistance in these populations are enhanced detoxification and/or sequestration. Either of these pathways could theoretically explain how herbicide resistance develops to multiple unrelated chemistries via the use of one of them. Therefore, this type of resistance may actually not be in conflict with the theory of herbicide resistance evolution.

There are several commonly recommended management strategies for attenuating herbicide resistance, many of which are based on conclusions derived from prediction models (reviewed in Friesen et al., 2000). In this chapter we have shown that, with respect to mutation frequencies for resistance, resistance allele fitness, and resistance allele distribution, the current prediction models often rely on assumptions and/or estimates that have not been validated. Future models need to address these characteristics more thoroughly and, ultimately, should use real, measured values.

ACKNOWLEDGEMENTS

Sincere thanks to Dr. Michael D. McLean and Mr. Kurt C. Almquist for their helpful discussions, and also to Dr. McLean for his thorough editing of the early versions of this manuscript.

REFERENCES

Allphin L., Windham M.D., Harper K.T. Genetic diversity and gene flow in the endangered dwarf bear poppy, *Arctomecon humilis* (Papaveraceae). Am J Bot 1998; 85:1251-1261.

Andrade A.C., Van Nistelrooy J.G., Peery R.B., Skatrud P.L., De Waard M.A. The role of ABC transporters from *Aspergillus nidulans* in protection against cytotoxic agents and in antibiotic production. Mol Gen Genet 2000; 263:966-977.

Andrews T.S., Morrison I.N., Penner G.A. Monitoring the spread of ACCase inhibitor resistance among wild oat (*Avena fatua*) patches using AFLP analysis. Weed Sci 1998; 46:196-199.

Anonymous. Human ATP-binding cassette transporters. Nutrition, metabolism and genomics group. Wageningen University, The Netherlands. Online. Internet. January 24 2002; Available www.humanabc.org/.

Bain L.J., LeBlanc G.A. Interaction of structurally diverse pesticides with the human *MDR1* gene product P-glycoprotein. Toxicol Appl Pharmacol 1996; 141:288-298.

Bartlett H.H. Additional evidence of mutation in *Oenothera*. Bot Gaz 1915a ; 59:81-123.

Bartlett H.H. Mass mutation in *Oenothera pratincola*. Bot Gaz 1915b; 60:425-456.

Beckie H.J., Thomas A. G., Legere A., Kelner D. J., Van Acker R. C., Meers S. Nature, occurrence, and cost of herbicide-resistant wild oat (*Avena fatua*) in small grain production areas. Weed Technol 1999; 13:612-625.

Bettini P., McNally S., Sevignac M., Darmency H., Gasquez J., Dron M. Atrazine resistance in *Chenopodium album*. Plant Physiol 1987; 84:1442-1446.

Cavan G., Moss S. Managing the risks of herbicide resistance in wild oat. Weed Sci 2001; 49:236-240.

Cavan G., Biss P., Moss S. R. Localized origins of herbicide resistance in *Alopecurus myosuroides*. Weed Res 1998; 38:239-245.

Cavan G., Cussans J., Moss S. R. Modelling different cultivation and herbicide strategies for their effect on herbicide resistance in *Alopecurus myosuroides*. Weed Res 2000; 40:561-568.

Chauvel B., Gasquez J. Relationships between genetic polymorphism and herbicide resistance within *Alopecurus myosuroides* Huds Heredity 1994; 72:336-344.

Christopher J.T., Powles S.B., Holtum J.A.M. Resistance to acetolactate synthase-inhibiting herbicides in annual ryegrass (*Lolium rigidum*) involves at least two mechanisms. Plant Physiol 1992; 100:1909-1913.

Christopher J.T., Powles S. B., Holtum J.A.M., Liljegren D.R. Cross-resistance to herbicides in annual ryegrass (*Lolium rigidum*). II. Chlorsulfuron resistance involves a wheat-like detoxification system. Plant Physiol 1991; 95: 1036-1043.

Chung M.G., Chung M.Y. Levels and partitioning of genetic diversity in populations of *Eurya japonica* and *Eurya emarginata* (Theaceae) in Korea and Japan. Int J Plant Sci 2000; 161:699-704.

Crawford D.J., Ruiz E., Stuessy T.F., Tepe E., Aqeveque P., Gonzalez F., Jensen R.J., Anderson G.J., Bernardello G., Baeza C.M., Swenson U., O M. S. Allozyme diversity in endemic flowering plant species of the Juan Fernandez Archipelago, Chile: Ecological and historical factors with implications for conservation. Am J Bot 2001; 88:2195-2203.

Darmency H. "Genetics of herbicide resistance in weeds and crops." In *Herbicide Resistance in Plants: Biology and Biochemistry*, S. B. Powles, J.A.M. Holtum, eds. Boca Raton, FL: CRC Press. 1994, pp. 263-298.

Darmency H., Aujas C. Character inheritance and polymorphism in a wild oat (*Avena fatua*) population. Can J Bot 1987; 65: 2352-2356.

Darmency H., Gasquez J. Appearance and spread of triazine resistance in common lambsquarters (*Chenopodium album*). Weed Technol 1990; 4:173-177.

Darmency H., Gasquez J. Interpreting the evolution of a triazine-resistant population of *Poa annua* L. New Phytol 1983; 95:299-304.

Del Sorbo G., Schoonbeek H., De Waard M.A. Fungal transporters involved in efflux of natural toxic compounds and fungicides. Fungal Genet Biol 2000; 30:1-15.

DeVries H. Mass mutations and twin hybrids of *Oenothera grandiflora* Ait. Bot Gaz 1918; 65:377-422.

DeVries H. The coefficient of mutation in *Oenothera biennis* L. Bot Gaz 1915; 59:169-196.

Diggle A. J., Neve P. "The population dynamics and genetics of herbicide resistance – a modeling approach." In *Herbicide Resistance and World Grains*, S. B. Powles, D.L. Shaner, eds. Boca Raton, FL: CRC Press. 2001, pp. 61-100.

Edwards R. "Glutathione transferases and herbicide metabolism and selectivity." In *Weed and Crop Resistance to Herbicides*, R. De Prado, J. Jorrin, L. Garcia-Torres, eds. Dordrecht, The Netherlands: Kluwer Academic Publishers. 1997, pp. 109-116.

Franssen A.S., Skinner D.Z., Al-Khatib K., Horak M.J., Kulakow P.A. Interspecific hybridization and gene flow of ALS resistance in *Amaranthus* species. Weed Sci 2001; 49:598-606.

Friesen L.J.S. Examination of multiple-group herbicide resistance in two wild oat populations. M.Sc. Thesis. University of Guelph, Guelph, Ontario, Canada. 2001.

Friesen L.J.S., Ferguson G.M., Hall J.C. Management strategies for attenuating herbicide resistance: Untoward consequences of their promotion. Crop Prot 2000; 19: 891-895.

Gaillard C., Dufaud A., Tommasini R., Kreuz K., Amrhein N., Martinoia E. A herbicide antidote (safener) induces the activity of both the herbicide detoxifying enzyme and of a vacuolar transporter for the detoxified herbicide. FEBS Letters 1994; 352:219-221.

Garces H., Durzan D., Pedroso M.C. Mechanical stress elicits nitric oxide formation and DNA fragmentation in *Arabidopsis thaliana*. Ann Bot 2001; 87:567-574.

Gardner S.N., Gressel J., Mangel M. A revolving dose strategy to delay the evolution of both quantitative vs major monogene resistances to pesticides and drugs. Int J Pest Manage 1998; 44:161-180.

Gasquez J. "Genetics of herbicide resistance within weeds. Factors of evolution, inheritance and fitness." In *Weed and crop resistance to herbicides*, R. De Prado, J. Jorrin, L. Garcia-Torres, Eds. Kluwer Academic Publishers, Dordrecht, The Netherlands. 1997, pp. 181-190.

Gasquez J., Mouemar A. A., Darmency H. Triazine herbicide resistance in *Chenopodium album* L.: occurrence and characteristics of an intermediate biotype. Pestic Sci 1985; 16:392-396.

Georghiou G. P., Taylor C. E. Factors influencing the evolution of resistance. In *Pesticide resistance: Strategies and tactics for management*. National Academy Press, Washington, D.C. 1986.

Ghersa C.M., Martinez-Ghersa M.A., Brewer T.G., Roush M.L. Selection pressures for diclofop-methyl resistance and germination time of Italian ryegrass. Agron J 1994a; 86:823-828.

Ghersa C.M., Martinez-Ghersa M.A., Brewer T.G., Roush M.L. Use of gene flow to control diclofop-methyl resistance in Italian ryegrass (*Lolium multiflorum*). Weed Technol 1994b; 8:139-147.

Gichner T., Menke M.,. Stavreva D. A, Schubert I. Maleic hydrazide induces genotoxic effects but no DNA damage detectable by the Comet assay in tobacco and field beans. Mutagen 2000; 15:385-389.

Gitzendanner M. A., Soltis P. S. Patterns of genetic variation in rare and widespread plant congeners. Am J Bot 2000; 87:783-792.

Gould F. Comparisons between resistance management strategies for insects and weeds. Weed Technol 1995; 9:830-839.

Gressel J., Segel L.A. Modelling the effectiveness of herbicide rotations and mixtures as strategies to delay or preclude resistance. Weed Technol 1990; 4:186-198.

Gressel J., Segel L.A. Interrelating factors controlling the rate of appearance of resistance: the outlook for the future. In *Herbicide Resistance in Plants*, H. M. LeBaron, J. Gressel, eds. Tornoto: John Wiley & Sons. 1982, pp. 325-348.

Gressel J., Segel L.A. The paucity of plants evolving genetic resistance to herbicides: possible reasons and implications. J Theor Biol 1978; 75:349-371.

Griffiths A.J.F., Miller J.H., Suzuki D.T., Lewontin R.C., Gelbart W.M. *An Introduction to Genetic Analysis*. 5th edition. W. H. Freeman and Company. New York, NY. 1993, pp. 50-86.

Hall J.C., Friesen L.J.S. Generally accepted weed management strategies – after three decades, how effective have they been? Symposium lecture at the XIVth International Plant Protection Congress (IPPC): *Plant Protection Towards the Third Millennium – Where Chemistry Meets Ecology*; 1999 July 25-30; Jerusalem, Israel.

Hall J.C., Wickenden J.S., Yau K.Y.F. "Biochemical conjugation of pesticides in plants and microorganisms: an overview of similarities and divergences." In *Pesticide Biotransformation in Plants and Microorganisms: Similarities and Divergences*, J. C. Hall, R. E. Hoagland, R. M. Zablotowicz, Eds. ACS Symposium Series 777. American Chemical Society, Washington, DC. 2001, pp. 89-118.

Hall L. M., Stromme K. M., Horsman G. P., Devine M. D. Resistance to acetolactate synthase inhibitors and quinclorac in a biotype of false cleavers (*Galium spurium*). Weed Sci 1998; 46:390-396.

Hamamoto H., Nawata O., Haswgawa K., Nakaune R., Lee Y. J., Makizumi Y., Akutsu K., Hibi T. The role of the ABC transporter gene *PMR1* in demethylation inhibitor resistance in *Penicillium digitatum*. Pestic Biochem Physiol 2001; 70:19-26.

Harms C.T., DiMaio J.J. Primisulfuron herbicide-resistant tobacco cell lines. Application of fluctuation test design to *in vitro* mutant selection with plant cells. J Plant Physiol 1991; 137:513-519.

Haughn G., Somerville C.R. "Selection for herbicide resistance at the whole-plant level." In *Biotechnology in Agricultural Chemistry*, H. M. LeBaron, R. O. Mumma, R. C. Honeycutt, J. H. Duesing, eds. Wahington, DC: American Chemical Society. 1987, pp. 98-107.

Hayashi K., Schoonbeek H-J., Sugiura H., De Waard M.A. Multidrug resistance in *Botrytis cinerea* associated with decreased accumulation of the azole fungicide oxpoconazole and increased transcription of the ABC transporter gene *BcatrD*. Pest Biochem Physiol 2001; 70:168-179.

Heap I. International survey of herbicide-resistant weeds. Online. Internet. February 23, 2003. Available www.weedscience.org.

Henikoff S., Greene E.A., Pietrokovski S., Bork P., Attwood T.K., Hood L. Gene families: the taxonomy of protein paralogs and chimeras. Science 1997; 278:609-614.

Hoffman A.A., Parsons P.A. "Effects of stress on genetic variation." In *Evolutionary Genetics and Environmental Stress*. New York: Oxford University Press Inc. 1991.

Holt J.S., Thill D.C. "Growth and productivity of resistant plants." In *Herbicide Resistance in Plants: Biology and Biochemistry*, S. B. Powles, J.A.M. Holtum, eds. Boca Raton, FL: CRC Press.1994, pp. 299-316.

Holtum J.A.M., Hausler R.E., Devine M.D., Powles S.B. Cross-resistance to herbicides in annual ryegrass (*Lolium rigidum*). III. On the mechanism of resistance to diclofop-methyl. Plant Physiol 1991; 97:1026-1034.

Holyoak C. D., Bracey D., Piper P. W., Kuchler K., Coote P. J. J Bacteriol 1999; 181:4644-4652.

Jasieniuk M., Maxwell B.D. Plant diversity: new insights from molecular biology and genomics technologies. Weed Sci 2001; 49:257-265.

Jasieniuk M., Brule-Babel A.L., Morrison I.N. The evolution and genetics of herbicide resistance in weeds. Weed Sci 1996; 44:176-193.

Jenczewski E., Prosperi J-M., Ronfort J. Evidence for gene flow between wild and cultivated *Medicago sativa* (Leguminosae) based on allozyme markers and quantitative traits. Am J Bot 1999; 86:677-687.

Kimura M. Genetic variability maintained in a finite population due to mutational production of neutral and nearly neutral isoalleles. Genet Res 1968;11:247-269.

Kimura M., Crow J. F. The number of alleles that can be maintained in a finite population. Genetics 1964; 49:725-738.

Kimura M., Ohta T. The age of a neutral mutant persisting in a finite population. Genetics 1973; 75: 199-212.

Klein M., Martinoia E., Weissenbock G. Directly energized uptake of beta-estradiol 17-(beta-D-Glucuronide) in plant vacuoles is strongly stimulated by glutathione conjugates. J Biol Chem 1998; 273:262-270.

Klein M., Martinoia E., Hoffmann-Thoma G., Weissenbock G. A membrane-potential dependent ABC-like transporter mediates the vacuolar uptake of rye flavone glucuronides: Regulation of glucuronide uptake by glutathione and its conjugates. Plant J 2000; 21:289-304.

Klein M., Martinoia E., Hoffmann-Thoma G., Weissenbock G. The ABC-like vacuolar transporter for rye mesophyll flavone glucuronides is not species-specific. Phytochemistry 2001; 56:153-159.

Klein M., Weissenbock G., Dufaud A., Gaillard C., Kreuz K., Martinoia E. Different energization mechanisms drive the vacuolar uptake of a flavonoid glucoside and a herbicide glucoside. J Biol Chem 1996; 271:29666-29671.

Kolaczkowski M., van der Rest M., Cybularz-Kolaczkowska A., Soumillion J-P., Konings W.N., Goffeau. A. Anticancer drugs, ionophoric peptides, and steroids as substrates of the yeast multidrug transporter Pdr5p. J Biol Chem 1996; 271:31543-31548.

Leslie E. M., Deeley R. G., Cole S. P. C. Toxicological relevance of the multidrug resistance protein 1, MRP1 (ABCC1) and related transporters. Toxicology 2001;167: 3-23.

Lewontin R. C. "The paradox of variation." In *The Genetic Basis of Evolutionary Change*. New York: Columbia University Press. 1974.

Liu G., Sanchez-Fernandez R., Li Z-S., Rea P. A. Enhance multispecificity of *Arabidopsis* vacuolar multidrug resistance-associated protein-type ATP-binding cassette transporter, AtMRP2. J Biol Chem 2001; 276:8648-8656.

Loe D.W., Deeley R.G., Cole S.P.C. Verapamil stimulates glutathione transport by the 190-kDa multidrug resistance protein 1 (MRP1). J Pharmacol Exp Ther 2000; 293:530-538.

Loe D.W., Almquist K.C., Cole S.P.C., Deeley R.G. ATP-dependent beta-estradiol 17-(beta-D-Glucuronide) transport by multidrug resistance protein (MRP). J Biol Chem 1996a; 271:9683-9689.

Loe D.W., Almquist K.C., Deeley R.G., Cole S.P.C. Multidrug resistance protein (MRP)-mediated transport of leukotriene C_4 and chemotherapeutic agents in membrane vesicles. J Biol Chem 1996b; 271:9675-9682.

Lu Y-P., Li Z-S., Rea P. A. *AtMRP1* gene of *Arabidopsis* encodes a glutathione *S*-conjugate pump: isolation and functional definition of a plant ATP-binding cassette transporter gene. Proc Natl Acad Sci USA. 1997.

Lu Y-P., Li Z-S., Drozdowicz Y.M., Hortensteiner S., Martinoia E., Rea P.A. AtMRP2, and *Arabidopsis* ATP binding cassette transporter able to transport glutathione *S*-conjugates and chlorophyll catabolites: Functional comparisons with AtMRP1. Plant Cell 1998; 10:267-282.

Marcotrigiano M. Herbivory could unlock mutations sequestered in stratified shoot apices of genetic mosaics. Am J Bot 2000; 87:355-361.

Martinez-Ghersa M.A., Ghersa C.M., Bila-Aiub M.M., Satorre E.H., Radosevich S.R. Evolution of resistance to diclofop-methyl in ryegrass (*Lolium multiflorum*): investigation of the role of introgression with related species. Pestic Sci 1997; 51:305-308.

Martinoia E., Massonneau A., Frangne N. Transport processes of solutes across the vacuolar membrane of higher plants. Plant Cell Physiol 2000; 4(:1175-1186.

Matthews J.M. "Management of herbicide-resistant weed populations." In *Herbicide Resistance in Plants*, S. B. Powles, J. A. M. Holtum, eds. Boca Raton, FL: CRC Press. 1994, pp. 317-335.

Matthews J.M., Powles S.B. Aspects of the population dynamics of selection for herbicide resistance in *Lolium rigidum* (Gaud.). Proc. First Int. Weed Control Congr, 1992.

Maxwell B.D., Mortimer A.M. "Selection for herbicide resistance." In *Herbicide Resistance in Plants: Biology and Biochemistry*, S. B. Powles, J. A. M. Holtum, eds. Boca Raton, FL: CRC Press. 1994.

Maxwell B.D., Roush M.L., Radosevich S.R. Predicting the evolution and dynamics of herbicide resistance in weed populations. Weed Technol 1990; 4:2-13.

Merrell D.J. "Mutation." In *Ecological Genetics*. University of Minnesota Press, Minneapolis, MN. 1981.

Messeguer J., Fogher C., Guiderdoni E., Marfa V., Catala M.M., Baldi G., Mele E. Field assessments of gene flow from transgenic to cultivated rice (*Oryza sativa* L.) using a herbicide resistance gene as tracer marker. Theor Appl Genet 2001; 103:1151-1159.

Morrison I.N., Bourgeois L., Friesen L., Kelner D. Betting against the odds: the problem of herbicide resistance. Presented at the 2nd Biennial Western Canada Agronomy Workshop; 1995, July 5-7. Red Deer, AB.

Mortimer A.M., Ulf-Hansen P.F., Putwain P.D. "Modelling herbicide resistance - a study of ecological fitness." In *Achievements and Developments in Combating Pesticide Resistance*, I. Denholm, A. L. Devonshire, D. W. Hollomans, Eds. Elsevier, Amsterdam. 1993, pp. 148-164.

Mulugeta D., Maxwell B.D., Fay P.K., Dyer W.E. Kochia (*Kochia scoparia*) pollen dispersion, viability and germination. Weed Sci 1994; 42:548-552.

Nishino K., Yamaguchi A. Analysis of a complete library of putative drug transporter genes in *Eschericia coli*. J Bacteriol 2001; 183:5803-5812.

Noh B., Murphy A.S., Spalding E.P. Multidrug resistance-like genes of *Arabidopsis* required for auxin transport and auxin-mediated development. Plant Cell 2001; 13:2441-2454.

Pappert R.A., Hamrick J.L., Donovan L.A. Genetic variation in *Pueraria lobata* (Fabaceae), an introduced, clonal, invasive plant of the southeastern United States. Am J Bot 2000; 87:1240-1245.

Powles S.B., Preston C. Herbicide cross resistance and multiple resistance in plants. Online. Internet. March 26, 2002. Available www.plantprotection.org/HRAC/.

Preston C., Mallory-Smith C.A. "Biochemical mechanisms, inheritance, and molecular genetics of herbicide resistance in weeds." In *Herbicide Resistance and World Grains*, S. B. Powles, D. L. Shaner, eds. Boca Raton, FL: CRC Press. 2001, pp. 23-60.

Price S.C., Hill J. E., Allard R.W. Genetic variability for herbicide reaction in plant populations. Weed Sci 1983; 31:652-657.

Price S.C., Allard R.W., Hill J. E., Naylor J. Associations between discrete genetic loci and genetic variability for herbicide reaction in plant populations. Weed Sci 1985; 33:650-653.

Putwain P. D., Mortimer A. M. The resistance of weeds to herbicides: Rational approaches for the containment of a growing problem. Proc. Brighton Crop Prot. Conf. – Weeds. British Crop Protection Council, Brighton, Great Britain. 1989, pp. 285-289.

Queitsch C., Sangster T. A., Lindquist S. Hsp90 as a capacitor of phenotypic variation. Nature. 2002; 417:618-624.

Rea P.A. MRP subfamily ABC transporters from plants and yeast. J Exp Bot 1999; 50:895-913.

Rea P.A., Li Z-S., Lu Y-P., Drozdowicz Y.M. From vacuolar GS-X pumps to multispecific ABC transporters. Ann Rev Plant Physiol Plant Mol Biol 1998; 49:727-760.

Ribas G., Surrales J., Carbonell E., Xamena N., Creus A., Marcos R. Genotoxicity of the herbicides alachlor and maleic hydrazide in cultured human lymphocytes. Mutagen 1996; 11:221-227.

Richter O., Zwerger P. "Temporal and spatial dynamics of herbicide resistance: a new approach combining population dynamics, population genetics and cellular automata." In *Human and Environmental Exposure to Xenobiotics*, A. A. M. Del Re, C. Brown, E. Capri, G. Errera, S. P. Evans, and M. Trivisan, eds. Proceedings of the XI symposium Pesticide Chemistry, Cremona, Italy. La Goliardica Pavese s.r.l. 1999, pp. 599-608.

Rutledge J., Talbert R.E., Sneller C.H. RAPD analysis of genetic variation among propanil-resistant and – susceptible *Echinochloa crus-galli* populations in Arkansas. Weed Sci 2000; 48:669-674.

Saari L. L., Cotterman J. C., Thill D. C. "Resistance to acetolactate synthase inhibiting herbicides." In *Herbicide Resistance in Plants: Biology and Biochemistry*, S. B. Powles, and J. A. M. Holtum, eds. Boca Raton, FL: CRC Press. 1994, pp. 83-140.

Sanchez-Fernandez R., Davies T. G. E., Coleman J. O. D., Rea P. A. The *Arabidopsis thaliana* ABC protein superfamily, a complete inventory. J Biol Chem 2001; 276:30231-30244.

Sakamoto K., Margolles A., van Veen H.W., Konings W.N. Hop resistance in the beer spoilage bacterium *Lactobacillus brevis* is mediated by the ATP-binding cassette multidrug transporter HorA. J Bacteriol 2001; 183:5371-5375.

Schoonbeek H., Del Sorbo G., De Waard M.A. The ABC transporter BcatrB affects the sensitivity of *Botrytis cinerea* to the phytoalexin resveratrol and the fungicide fenpiclonil. Mol Plant Microbe Interact 2001; 14:562-571.

Shaner D.L. "Herbicide resistance in North America: History, circumstances of development and current situation." In *Weed and Crop Resistance to Herbicides*, R. De-Prado, J. Jorrin, and L. Garcia-Torres, Eds. Kluwer Academic Publishers. Dordrecht, The Netherlands. 1997; pp. 29-38.

Shorrocks B. "Genetic diversity." In *The Genesis of Diversity*. Hodder and Stoughton Ltd. London, Great Britain, 1978.

Spiess E.B. "Mutation: Source of new variation." In *Genes in Populations*. John Wiley & Sons, Inc., New York, NY. 1989, pp. 361-390.

Stallings G.P., Thill D.C., Mallory-Smith C.A., Shafii B. Pollen-mediated gene flow of sulfonylurea-resistant kochia (*Kochia scoparia*). Weed Sci 1995; 43:95-102.

Stankiewicz M., Gadamski G., Gawronski S.W. Genetic variation and phylogenetic relationships of triazine-resistant and triazine-susceptible biotypes of *Solanum nigrum* – analysis using RAPD markers. Weed Res 2001; 41:287-300.

Tardif F. J., Preston C., Powles S.B. "Mechanisms of herbicide multiple resistance in *Lolium rigidum*." In *Weed and Crop Resistance to Herbicides*, R. De Prado, J. Jorrin, L. Garcia-Torres, eds. Dordrecht, The Netherlands: Kluwer Academic Publishers. 1997.

Theodoulou F. L. Plant ABC transporters. Biochim Biophys Acta 2000; 1465: 79-103.

Tremolieres A., Darmency H., Gasquez J., Dron M., Connan A. Variation of transhexadecenoic acid content in two triazine resistant mutants of *Chenopodium album* and their susceptible progenitor. Plant Physiol 1988; 86:967-970.

Turpeinen T., Tenhola T., Manninen O., Nevo E., Nissila E. Micosatellite diversity associated with ecological factors in *Hordeum spontaneum* populations in Israel. Mol Ecol 2001; 10:1577-1591.

Van Eerd L.L., Stephenson G.R., Hall J.C. Mechanism of quinclorac resistance in false cleavers (*Galium spurium*). Weed science society of America annual meeting – Abstract #238. Greensboro, North Carolina. 2001.

Vermeulen T., Schoonbeek H., De Waard M.A. The ABC transporter BcatrB from *Botrytis cinerea* is a determinant of the activity of the phenylpyrrole fungicide fludioxinil. Pest Manag Sci. 2001; 57:393-402.

Volis S., Mendlinger S., Turuspekov Y., Esnazarov U., Abugalieva S., Orlovsky N. Allozyme variation in Turkmenian populations of wild barley, *Hordeum spontaneum* Koch. Ann Bot 2001; 87:435-446.

Wang R-L., Dekker J. Weedy adaptation in *Setaria* spp. III. Variation in herbicide resistance in *Setaria* spp. Pestic Biochem Physiol 1995; 51:99-116.

Warwick S. I., Black L. D. Relative fitness of herbicide-resistant and susceptible biotypes of weeds. Phytoprotection 1994; 75:37-49.

CHAPTER 11

FATE AND PROPERTIES OF HERBICIDES IN TROPICAL SOILS

RUBEM S. OLIVEIRA JR., ANTONIO C. S. COSTA, & CÁSSIO A. TORMENA

Departamento de Agronomia, Universidade Estadual de Maringá – Av. Colombo 5790. 87020-900 – Maringá, PR – Brazil

DEFINING TROPICAL SOILS

Tropical Oxisols, Ultisols and Entisols are considered to present quantitatively a very simple mineralogy in the clay fraction, where three to four minerals show more than 90% dominance. Since in the tropics soils of other orders are also used for crop production, the result is a more diverse and complex mineralogy, which affects the chemical and physical properties of soil in a manner similar to that observed in any other region of the world.

The definition of tropical soils is not straightforward and differences among scientists still persist. Years ago most of the tropical soils were considered to be covered by Laterite soils (Mohr and van Baren, 1954; McNeil, 1964; Mohr et al., 1973). Early definitions took into account climatic delimitation. Temperature limits such as the 68° F (20 °C) (Dudal, 1979), or 70° F (21°C) (Sivakumar et al., 1992) have been proposed. Other climatic classification systems have used precipitation (Koeppen, 1936), precipitation and evaporation (Thornthwaite, 1948), and water availability (Hargreaves, 1971). A more suitable system proposed by Eswaran et al. (1992) defines tropical soils as those located between the 23°27' North and South parallels of the Tropic of Cancer and the Tropic of Capricorn, respectively, with an "Iso" soil regime temperature that represents 40% of the land on the surface of the earth. Otherwise the term Laterite has been applied to small regions where these soil morphological features are common, rather than to the majority of these soils. As observed from other latitudes, soils of many orders can be found in the tropical region (Table 1) in the soil taxonomy system (Sanchez and Logan, 1992). The Oxisols and Ultisols represent 43% of the tropical region, and if we add the Entisols (16%), Alfisols (15%) and Inceptsols (14%) orders, 88% are represented.

Inderjit (ed.), Weed Biology and Management, 227-249.

Table 1. Distribution of associations of soils between latitudes 23 1/2° N and S (Sanchez and Logan, 1992).

Predominant Soil Order	10^6 ha	%
Oxisols	833	23
Ultisols	749	20
Entisols	574	16
Alfisols	559	15
Inceptsols	532	14
Vertisols	163	5
Aridisols	87	2
Mollisols	74	2
Andisols	43	1
Histosols	36	1
Spodosols	20	1

MAJOR PROPERTIES OF TROPICAL AREAS AS RELATED TO HERBICIDE BEHAVIOR

The diversity of rocks, climate, landscapes, drainage conditions and organisms observed within the Tropic of Cancer and the Tropic of Capricorn, known as the torrid or tropical zone, are hints to recognize the influence of such diversity of soil forming factors and consequently the soils formed (Table 1). In Brazil, a continental complex surface area in Southern America with an area of more than eight million km^2, soils of all orders can be found within the Soil Taxonomy system with the exception of the Andisols. Therefore the mineralogy in the clay size fraction will vary from soils where the 2:1 clay minerals will dominate to soils where Fe and Al oxyhydroxides will be the major component (Figure 1). Differences in specific soil-forming factors will explain these variations, since in the northeastern part of the country semiarid climatic conditions dominate and in the Pantanal and deep southern areas poor drainage conditions are responsible for the low degree of weathering and slow removal of ions in solution that are suitable conditions for smectite and vermiculite diagenesis with different degrees of Al-interlayering.

The sustainable use of a soil should be based on the knowledge of its mineralogical, morphological, physical, chemical and biological characteristics. Soils of the tropical region have only recently been the subjects of systematic research (Sanchez, 1976; Uehara and Gillman, 1981; Theng, 1980; van Wambeke, 1992) and there are still gaps that remain in explaining the properties of such important group of soils.

Organic Matter

The total content of organic matter (OM) within soils in the tropics is no longer considered to be smaller than any other soil of the same order in the world (Greenland et al., 1992). Nevertheless tropical forest soils, when exposed to agricultural practices such as plowing, experience a rapid decrease in OM content that reaches values as low as 0.2% carbon content on the surface soils within less than a decade. The reduction in OM content can be due to increased soil erosion, faster OM mineralization and oxidation of soil organic carbon or smaller quantities of organic inputs in managed systems as compared to natural forests.

Figure 1. Distribution of the primary minerals in the clay fraction of soils from Brazil. (Modified from Melfi et al., 1979).

For most Oxisols and some Ultisols and Alfisols where there is a small increase in the clay content with depth and clay mineralogy is homogeneous throughout the profile there is, with depth, a good correlation between the effective cation exchange capacity (CEC) and the carbon content of these soils. This can be observed in Figure 2 using data (Embrapa, 1984) from several Oxisols from the State of Paraná-Brazil (Costa et al., 1999). The average value for the cation exchange capacity of the OM is around 450 $cmol_c$/kg organic carbon, while the clay size fraction, free of organic matter, has a CEC value slightly higher than 5 $cmol_c$/kg. Therefore for most of these soils the CEC is almost solely due to the carbon content (Table 2). This can be better understood if one observes the charge distribution of the minerals as determined using pure minerals. Kaolinite, the dominant clay mineral in the tropical soil zone, has the pH value in the point size zero charge (pH_{PZC}) (pH at which the mineral superficial charge is electrically neutral) close to 5 (Table 3); therefore it contributes a very small amount of negative charge mostly concentrated in the hydroxyls of the gibbsitic layer. Lim and Jackson (1980) determined values of CEC for several kaolinites varying from almost zero to close to one. These authors observed that CEC values for kaolinites greater than one are due to small amounts of 2:1 clay minerals rather than the kaolinites. The Fe and Al oxyhydroxides have very high pH_{PZC}, most of them above pH 6.5; therefore, they can be responsible for the Anion Exchange Capacity (AEC) of the soils.

Table 2. Total cation exchange capacity (CEC) and CEC associated to clay and organic matter fractions of soils from the State of Paraná-Brazil.

Soil	CEC ($cmol_c.kg^{-1}$)		
	Total	Clay	Organic Matter
Dusky Red Latosol	19.25	5.00	14.25
Dark Red Latosol	3.70	0.45	3.25
Cambisol	21.90	1.17	20.73
Organic Soil	69.00	8.00	60.00

Table 3. The pH values in the point of zero charge (pH_{PZC}) of some silicates and iron and aluminum oxides.

Minerals	Chemical Formula	pH_{PZC} [a]
	Silicates	
Montmorillonite	$(Na, Ca)_{0,33}$ $(Al, Mg)_2$ Si_4O_{10} $(OH)_2$	2.5
Kaolinite	Al_2 Si_2 O_5 $(OH)_4$	4.7-4.8
Quartz	SiO_2	2-2.9
Chlorite	(Mg, Fe^{2+}) 5 Al $(Si_3Al)O_{10}(OH)_8$	3
Illite	K $(Al, Mg, Fe)_2$ $(Si, Al)_4$ O_{10} $(OH)_2$ $H_2O)$	2.5
	Iron, Aluminum and Titanium Oxyhydroxides	
Goethite	α-FeOOH	8.2-8.4
Hematite	α-Fe_2O_3	8.4-8.5
Gibbsite	α-$Al(OH)_3$	5.0
Magnetite	Fe_3O_4	6.5
Maghemite	γ-Fe_2O_3	6.7
Rutile	TiO_2	6.5
Fe-amorphous	$Fe(OH)_3$	8.5

[a] Kaolinite – Sposito (1984); Montmorillonite, Quartz – Hunter (1981); Chlorite and Illite – Hussain et al. (1996); Goethite – Ikram et al. (1996); Magnetite – Sun et al. (1998); Maghemite – Garcell et al. (1998); Rutile – Fendorf and Fendorf (1996)

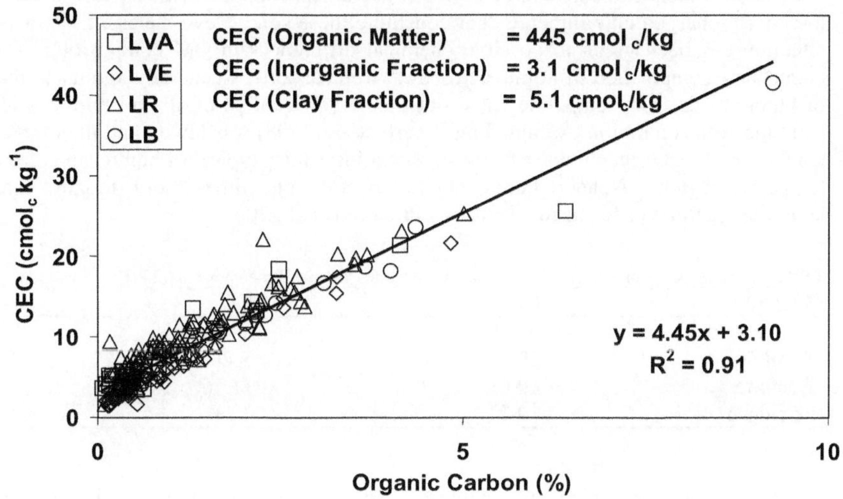

Figure 2. Correlation between carbon content and cation exchange capacity (CEC) of several Oxisols profiles from Campo Mourão-Brazil. LVA =Yellow red latosol; LVE = Dark Red Latosol, LR = Dusky Red Latosol; LB = Brownish Latosol. Above, average CEC of isolated fractions of these soils (Costa et al., 1999).

Role of Organic Matter on Herbicide Sorption in Tropical Soils

The content of soil organic matter (SOM) as humic material is the primary factor influencing the sorption of many herbicides, particularly for weak base and non-polar, non-ionic molecules. For weak acid herbicides, the pH is usually considered to be the main soil property affecting sorption, and therefore the fate of these herbicides. Even in this case, OM usually plays a relevant secondary role.

In non-tropical soils, at least 80% of the sorption of herbicides such as metribuzin, diuron and 2,4-D is directly connected to soil OC content (Mallawatantri and Mulla, 1992). Similar results have been shown for other herbicides and soil types in tropical conditions (Singh et al., 1989; Baskaran et al., 1996; Gerritse et al., 1996; Ahmad et al., 2001), and the dependence of sorption on OM is often high. For instance, a correlation analysis showed that atrazine sorption is significantly correlated with total OM and even more with pyrophosphate-extractable OM. In fact, the latter accounts for nearly 98% of the variation of the atrazine sorption in Tanzanian soils (Borggaard and Streibig, 1988a). For tropical soils, OM has been shown to be more important than pH in relation to sorption for weak acid herbicides such as imazaquin (Regitano et al., 2000); imazethapyr (Ahmad et al., 2001); metsulfuron-methyl (Bolan and Baskaran, 1996b; Ismail and Kalithasan, 1997) and 2,4-D (Bolan and Baskaran, 1996b; Vieira et al., 1999). In soils from Brazil, there are significant correlations between both OC content and soil CEC with sorption coefficients of sulfonylureas such as metsulfuron-methyl and sulfometuron-methyl but no significant correlation between pH and sorption of these herbicides was observed (Oliveira Jr. et al., 2001). Since the correlation of CEC to sorption depends on charges from OM, this is the most important soil characteristic affecting sorption of these herbicides.

The dependence of sorption on soil OM in tropical soils is obviously correlated to the poorly charged clay minerals that constitute these soils. Nevertheless, little or no attention has been given to possible chemical differences in OM composition. The elementary composition of humic acids from tropical soils is generally similar to that of humic acids from temperate soils, but the carbon content of all soil humic acids from the Amazon region examined by Rivero et al. (1998) was lower, and the oxygen content was higher than the average corresponding values typical of humic acids from temperate regions. Isolated humic acids can sorb nine times more atrazine than kaolinite, goethite or ferridrite (Table 4) (Procópio et al., 2001).

Table 4. Atrazine sorption coefficients in isolated substrates. Source: Procópio et al. (2001).

Substrate	Kd	Kf
Humic acid	91.89	87.31
Kaolinite	10.33	5.11
Goethite	10.08	11.16
Ferridrite	10.71	7.23

Despite its importance and increasing scientific understanding of the natural processes involved in SOM dynamics, there are several factors that limit the adoption of management strategies to maintain or increase soil OM in tropical ecosystems. A major constraint is often an inadequate supply of organic inputs due to competing uses for organic materials such as fodder, fuel and building materials (Prasad and Power, 1991). Difficulties imposed in handling organic inputs to soil have led to the assumption that management options for such organic inputs in tropical land-use systems are often limited by social and economic factors (Fernandes et al., 1997; Izac, 1997).

Mineralogy

Regarding the mineralogy of tropical soils, Schwertmann and Herbillon (1992) stated "*highly weathered soils of the tropics, such as Oxisols and Ultisols, often have a monotonous mineralogy... Among these minerals are 1:1 layer silicates of the kaolin group, oxides of Fe, Al and Ti and some resistant minerals.... such as moscovite and Al-hydroxy-interlayered vermiculite*." Considering the diversity of minerals within the clay size fraction, tropical soils are as complex as any other soil in the world, but in terms of proportions, minerals of the kaolin group, oxides of Fe and Ti and some Al-hydroxy interlayered 2:1 clay minerals will represent over 90% in most cases. Careful study, however, has shown that these minerals when exposed to very harsh environments display very unusual characteristics, which are important to the better management of these soils. Water retention, cations and anions and the reactivity of these minerals are very different, and literally hundreds of papers have been published using these isolated minerals. Otherwise, the properties of these minerals in soils either in separated size fractions or in the whole soil are poorly understood, and research is needed to clarify such topics.

Considering the soil in the landscape, changes within the clay mineralogy are always observed. Most of the profiles formed in the shoulder, backslope and footslope positions of the landscape present a more active clay mineralogy. They also develop a blocky to prismatic structure associated with the increase percentage in 2:1 clay minerals, still interlayered with Al and the accumulation of goethite with very high amounts of Al-substitution due to a higher water content. Changes on the concentration and type of clay minerals and iron oxides are very common, even in short distances within the same landscape (Curi and Franzmeer, 1984; Costa, 1996). Most of the time these changes are overlooked by agronomists and even by soil scientists in the field (Eswaran et al., 1992).

Effect of Al and Fe Oxyhydroxides on Herbicide Retention

The clay fraction of the soil is composed of both crystalline and amorphous minerals. Most of the charged and polar sorption sites are on the secondary minerals, the layer silicates. They provide most of the cation exchange capacity of many soils in the form of 1:1 and 2:1 layer silicates. In 1:1 layer silicates, exchange sites are located only on the exterior surfaces of the silicate sheets, while 2:1 layer silicate exchange sites can be located on both the exterior surfaces and interlamellar sites between silicate interlayers. Generally 2:1 layer silicates have the greater CEC of the two minerals. Amorphous minerals can also provide some hydrophobic sorption sites. Crystalline minerals such as quartz and feldspar typically contribute little to the sorption capacity of a soil (Harper, 1994).

As previously shown, Fe and Al-hydroxy clay minerals are important components of tropical soil mineralogy. Glyphosate sorption increases exponentially with the contents of Fe and Al oxides in Podzol soils from Australia, which led Gerritse et al. (1996) to suggest that glyphosate is particularly strongly sorbed to kaolinite. For sandy loam soils from Australia, which are low in organic matter and contain kaolinite and amorphous oxides, Borggaard and Streibig (1988b) found the amorphous Fe to have a high capacity for sorption of sulfonylureas at low pH, although the forces are relatively weak. Regitano et al. (2000) reported similar results for Brazilian soils, emphasizing the increased relative importance of Fe and Al oxide contents in imazaquin sorption under low OC contents. Comparison of imazaquin sorption in variable and permanent charged soils provided evidence that the principal sorption mechanism was hydrophobic partitioning. However, in positively charged soil sites, as on the surface of Fe and Al oxides, there is expressive imazaquin sorption due to ionic exchange and or ligand exchange, particularly in soils with low OC contents (Regitano et al., 1997).

Positive correlations were also found between Freundlich sorption coefficient (Kf) and either oxalate extractable Fe or pyrophosphate extractable Fe (Gerritse et al., 1996), indicating that adsorption takes place on the non crystalline and short range ordered iron oxides, and on the Fe-organic complexes. Poor correlation was found between Kf and extractable Fe consisting of both the crystalline and amorphous forms of iron ($R^2 < 0.3$) excluding the involvement of crystalline iron. Other authors also pointed out the high affinity of acidic organic compounds for the weakly crystallized Fe oxides. Pusino et al. (1997) reported a sorption coefficient (Kd) of imazapyr on an amorphous Fe oxide of the same order of magnitude as that on humic acid, confirming the role of oxalate extractable Fe. They also observed that adsorption on the iron oxide is strongly improved by a decrease in pH. This is a general finding of the studies on adsorption of acidic organic compounds with iron oxides which can be explained by an increase in net positive charge at the surface of iron oxides at decreasing pH, responsible for the adsorption of the anionic forms of the acids.

Maximum adsorption is usually obtained near the pKa of the acids where the increase in surface positive charge is counterbalanced by a decrease in negative charges of the acids. For most soils, the pKa values are usually lower than soil pH allowing relevant amounts of anionic forms available for adsorption on iron oxides when they are positively charged i.e. at pH values lower than their pH_{PZC}. The pH_{PZC} of pure iron oxides is close to 8 (Table 3) but in natural soils it depends on the characteristics of the surrounding medium such as electrolyte concentration of the soil solution, hydration state of the surface, and chemisorb anions. Iron oxides from tropical soils are likely to exhibit different PZC values, and this could be the reason the correlation between Kf and oxalate extractable iron is apparently not affected by pH.

Significant positive correlation between sorption and exchangeable Al in non-tropical soils has been explored less, although there are reports for soils from temperate areas such as Italy (Pusino et al., 1997), Canada (Huang et al., 1984), and USA (Johnson and Sims, 1993; Werkheiser and Anderson, 1996). In this case, the partial contribution to sorption from oxides is usually overwhelmed by the charge deriving from other clay minerals and OM.

Thus, evidence points to the importance of Fe and Al oxides in herbicide sorption, particularly in highly weathered, low OC content soils. This is probably even more important in subsurface soil layers, but no study has been reported that compares the effect of these oxides in superficial and subsuperficial tropical soil samples.

Chemistry

Quantitatively the tropical soils present a charge system dominated by pH-dependant colloids. Hence, their ion exchange capacity changes with pH. Kaolinite, Fe and Al oxyhydroxides and the organic matter present charges, positive or negative, which depend on the pH of the soil solution. The points of zero charge (pH_{PZC}) of these minerals are usually higher than 5 (Table 3), which determine a positive, or close to zero charge balance of the soils free of organic matter. Therefore these minerals comprise a system where most of the negative charge is associated with the organic matter and a very small amount with the surface hydroxyls on the kaolinite. The positive charges are associated with the presence of Fe and Al oxyhydroxides. The presence of permanent negative charges within the structure of the minor amounts of the 2:1 clay minerals is very small due to the increasing interlayering of Al^{3+} which, as weathering intensity increases within time and climatic conditions, replaces the basic cations (Ca^{2+}, Mg^{2+}, K^{+} and Na^{+}), decreasing their ability to swell cations and affect the physical behavior of these soils.

Acidity and Effects of Liming

The effect of soil pH on herbicide behavior is known and widely explored in literature. Briefly, pH affects more intensively the ionizable than the non-ionizable herbicides. Among the ionizable herbicides, two important groups are the herbicides that behave as weak acids and the herbicides that behave as weak bases. Weakly basic herbicides exist as either the protonated cation or the neutral species, depending on the pH of the system. Weakly acidic chemicals exist as either the undissociated molecule or the anion. Figure 3 shows two examples of such herbicides. The increased sorption of acidic herbicides such as 2,4-D with a decrease in pH result from an increase in the percentage of the undissociated (non-ionic) form of the herbicide. As the pH decreases, the proportion of atrazine cations in solution increases and hence the sorption increases.

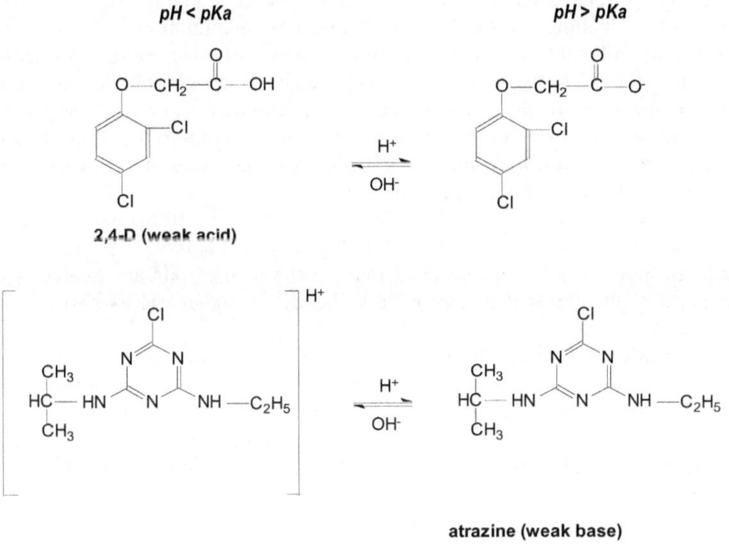

Figure 3. Ionization of a weak acid (2,4-D, pKa=2.8) and a weak base herbicide (atrazine, pKa=1.7).

It is not expected that pH can affect herbicide sorption in tropical soils differently than it does in any other soils. The sorption of non-polar, non-ionic herbicides such as linuron (Singh et al., 1989) and alachlor (Oliveira Jr. et al., 2001) appears to be independent of the soil pH, whereas the sorption of ionic pesticides increases when the pH decreases (Baskaran et al., 1996).

The combined effect of high weathering, acidic soils, low sorption capacity in clay minerals and low organic carbon contents usually found in tropical soils has led to the adoption of two agronomically significant techniques. The first is liming, usually performed to correct soil acidity and improve availability of mineral resources such as phosphorus and micronutrients. The second is the increasing trend to adopt minimum or no-till. Both have important effects on the fate and behavior of herbicides in these soils.

AGRONOMIC MANAGEMENT PRACTICES IN TROPICAL SOILS AFFECTING HERBICIDE BEHAVIOR

Effects of Liming

Pesticide disappearance from soil is recognized to include four main processes: volatilization, degradation, leaching and runoff. Liming can strongly affect all these important processes that control herbicide fate in soil, mainly because its potential effect of regulating concentrations of herbicides sorbed to soil or in solution.

Ahmad et al. (2001) compared 25 soils from Pakistan and Australia in relation to ametryn and imazethapyr sorption and pointed out that sorption of these herbicides in the alkaline soils from Pakistan was low and consequently there would be considerable risk of leaching and therefore groundwater contamination.

Evaluating the soil behavior of chlorsulfuron and triasulfuron applied to acidic soils of Australia, Blacklow and Pheloung (1992) detected more herbicide moving to lower layers of the soil profile in the higher pH environment. For these soils, the $CaCO_3$ treatment, which raised the pH of the surface soil from 4.9 to 6.5, caused more rapid leaching of triasulfuron from the 0 to 50 mm layer and more of both herbicides moved down the profile due to this treatment.

Several authors (Mersie and Foy, 1985; Brown, 1990; Hay, 1990) have detected increased mobility of sulfonylureas in soils with higher pH. It is likely, therefore, that the sulfonylurea herbicides will be leached more if the acidic soils are treated with lime, or in parts of the fields that may have soils with a higher pH as part of soil variability.

Since liming affects the availability of herbicides in soil solution, it is expected that it also influences degradation of these chemicals. Lab studies evaluating sorption and degradation of triazine herbicides in soils from western Australia (Walker and Blacklow, 1994) described increasing half-life values with increases in soil pH. Under similar conditions, degradation of simazine and atrazine herbicides was slower in limed soils (pH = 6.4, as compared to pH = 4.8 for non-treated soil) due to the higher pH obtained (Walker and Blacklow, 1995).

Liming is a normal agronomic practice used to overcome soil acidity in tropical soils. Changing soil pH greatly affects the behavior of herbicides in soil, especially ionic compounds. The increase in soil pH after liming usually decreases sorption. Herbicide that is leached from the soil surface lessens the duration of residual weed control and may contaminate groundwater.

Effects of Minimum Tillage on Herbicide Behavior

Agricultural systems and agronomic practices have undergone major changes, particularly in the last 10-15 years, leading to even greater dependence on chemical methods of pest control. Global problems such as soil erosion and eutrophication are posing serious threats to the sustainability of the modern agricultural systems. All around the tropical region of the world, land degradation in its various forms has been recognized as one of the most serious problems facing agriculture (Kim, 2000). Consequently, there has been a shift toward the use of minimum tillage practices in tropical agriculture in order to reduce loss in soil structure and erosion of soil particles, and to minimize nutrient runoff and subsequent eutrophication of surface waters. No-till systems are used on about 17.36 million hectares in Brazil in soybean, wheat, corn and sugarcane cropping and reforestation, and the area increases consistently every year (FEBRAPD, 2003).

The use of herbicides replaces cultivation as the primary means of weed control and facilitates the use of minimum- or no-till, which, together with crop rotations, contributes directly to environmental protection by preventing runoff losses of nutrients and other contaminants. On the other hand, the system depends heavily on chemicals. Because the application rates of pesticides for minimum tillage are higher than for conventional tillage (CT) systems, the potential loss from conservation tillage is greater. This loss would be significantly more important for those rainfall-runoff events occurring on, or shortly after, the day of pesticide application.

Locke and Bryson (1997) reviewed many aspects related to the interactions between herbicide and soil under minimum tillage. They consider minimum tillage as a general term describing several types of management practices, all of which exclude at least one major cultivation practice or minimize the intensity of a tillage operation. No-till (or zero tillage) is defined as a management practice that omits all operations of tillage, disking, or harrowing, and is one example of minimum tillage. In general, minimum or no-till is considered to be a technique that decreases soil erosion, soil crusting, the rate of surface evaporation and soil surface temperature by maintaining organic inputs on the soil surface (Wade and Sanchez, 1983; Prasad and Power, 1991; Ball-Coelho et al., 1993; Lal, 1997).

It is difficult to generalize the effects of tillage on herbicide persistence, dissipation, and movement because tillage influences so many soil characteristics. For instance, Koskinen and Clay (1997) reviewed literature on factors affecting atrazine fate in North Central U.S. soils, and concluded that differences in tillage practices have been reported to decrease, increase, or have no effect on herbicide dissipation. The effect of tillage on the amount of atrazine detected in surface runoff and leachate was also considered difficult to generalize. In most cases, environmental conditions during the study may play a more important role in the losses than tillage.

The increase in surface OM and plant residues on the soil surface affect the dynamics of weed emergence, the soil chemical, physical and biological properties and the behavior of herbicides applied to soil. Since the first topic is beyond the scope of this chapter, the following discussion will focus on the two latter aspects.

Minimum Tillage Effects on Soil Physics

Minimum tillage practices normally leave a layer of residue on the soil surface after the crop is harvested which impedes overland water flow, thereby enhancing the amount of water that filters through the soil profile. The rate of water infiltration is believed to be higher in no-till (NT) compared with the conventional-till (CT) systems, primarily because of the presence of macropores, small pores created by earthworms, root channels and cracks (Edwards et al., 1988; Wilson and Luxmoore, 1988; Rosenberg and McCoy, 1990). Consequently, concern has been raised that the use of more pesticides on soils that exhibit higher rates of water infiltration may accelerate leaching of pesticides into the subsurface waters underneath the agricultural croplands (Kenimer et al., 1987). Although the above argument seems logical, the overall assessment of pesticide fate and movement into groundwater under different soils, tillage methods, and climatic conditions is controversial. Recent studies have suggested that macropore flow in the NT systems is largely responsible for pesticide transport into shallow groundwater (Isensee et al., 1990; Gish et al., 1991). Other studies have shown that under certain conditions, rapid passage of water through macropores may bypass large proportions of the soil matrix. This should allow less time for interactions with the sorbed pesticide on the soil particles so that less pesticide is carried through the unsaturated zone and into the subsurface waters (Shipitalo et al., 1990; Fermanich and Daniel, 1991). Very little has been written on this subject for tropical soils. In some reports, effects of mulch on soil physical properties were confined mostly to the 0-5 cm depth, with a decreased bulk density and reduced penetration resistance (Lal, 2000).

Influence of Earthworms

Another effect described after adopting minimum tillage is the increase in the earthworm population and activity in soil. The influence of earthworms on the physical and chemical properties of soils has been well documented. Earthworms play an important role in nutrient cycling and in the development of the micromorphological structure for soil in which they are active (Mackay et al., 1983). Also, the importance of earthworms in the incorporation of plant residues and surface-applied agricultural materials such as pesticides has been under examination.

In no-till cropping soils, a large amount of soil material is ingested by earthworms and deposited as casts. It has been observed that earthworm casts are enriched with clay fractions, organic matter, and available nutrients (McKenzie and Dexter, 1987) compared to the surrounding soil. Under field conditions, the enrichment may be due to preferential feeding by earthworms in horizons of higher nutrient status, clay content, or organic matter (Tiwari, 1989). Enrichment of earthworm casts with organic matter and clay fractions is likely to affect the sorption and movement of surface-applied organic pesticides. Stehouwer et al. (1993) demonstrated that earthworm burrow linings contain higher levels of organic C than the bulk soil, resulting in increased atrazine sorption. Bolan and Baskaran (1996a) also found that earthworm casts had a higher pH than the source soil, resulting in a higher number of surface negative charges. The earthworm casts sorbed higher amounts of herbicides than the source soil (Table 5), mainly due to the increases in the amount of organic C and fine size fractions. The increased atrazine, 2,4-D and metsulfuron-methyl sorption by the casts resulted in decreased leaching. Sorption of atrazine was greater on the earthworm burrow linings, due to the enrichment in organic carbon and soluble organic carbon relative to bulk soil (Stehouwer et al., 1993, 1994). No intensive study has been conducted to date under tropical conditions to measure the effect of earthworms in soil-applied pesticide fate.

Table 5. Distribution of herbicides in the columns and leachates, and sorption coefficient expressed per unit of organic C, Koc. Adapted from Bolan and Baskaran (1996a).

Herbicides	Koc (adimensional) and distribution of herbicides (%)					
	Source soil			Earthworm casts		
	Koc	Column	Leachate	Koc	Column	Leachate
Atrazine	54.7	81.9	18.3	69.3	88.6	12.6
2,4-D	139.2	53.1	47.5	157.0	63.9	36.5
Metsulfuron-methyl	19.2	25.2	74.7	20.4	40.1	59.7

Interception by Straw, Degradation and Carryover

Crop residues and mulches may present a barrier, preventing herbicides from reaching the soil at the time of application. The retention of chemicals by different types of straw is obviously variable. Different types of straw may differ in relation to source, biomass production, decomposition kinetics and in relation to how they cover the soil surface.

There are few reports under field conditions. Rodrigues et al. (1998) applied different herbicides pre-emergence, soon after sowing, in areas with different amounts of oat residues on the soil surface. Using both bioassays and chromatographic analysis to detect herbicide residues, no leaching of trifluralin was detected after 20-mm irrigation in a soil covered with 6000 or 12000 kg/ha of oat residues. For atrazine, before irrigation, oat residues intercepted close to 85% of any rate applied. Irrigation with 20 mm after spraying was enough to leach nearly all atrazine to the soil surface (Fornarolli et al., 1998). In such cases, the main factors influencing herbicide behavior were water solubility and volatilization potential.

In a study that measured benzene, ethyl-benzene and 1, 2, 3-trichlorobenzene sorption (sorption characteristics similar to non-ionic herbicides) in a no-till corn soil, linear sorption coefficients for corn residues were 35 to 60 times greater than those for surface soil (Boyd et al., 1990). Sorption coefficients of the plant cuticles were estimated to be 5 to 15 times greater than the corn residues, indicating that this is the primary sorbing component. In the future, it may be possible and necessary to determine retention indices by different types of straw, similar to those currently adopted to characterize herbicide sorption to soil (Kd, Koc, Kf).

Tillage can also affect the kinetics of herbicide degradation in soils, but the interactions between tillage, herbicide persistence and phytotoxicity remain unclear.

Even though a no-till system can reduce soil erosion and often increases the sustainability of crop yield, in Australia the risk of damage to a subsequent susceptible crop makes many farmers reluctant to adopt no-till (Walker et al., 2000). Cultivation after sorghum harvest reduced atrazine residue in the surface soil in Northern South Wales (Australia) by 20-40%, depending on the initial rate of application. There was no associated increase in the 5-15 cm zone. Therefore, there was no distribution down the profile. Despite the reduction in atrazine residue, cultivation increased the severity of atrazine injury to wheat sown in one of the two soils used by Ferris et al. (1989).

Both degradation and leaching in soils covered by a crop canopy are likely to be less than in bare soils, as reported by Walker and Blacklow (1995), because of lower soil temperatures and less leaching. However, there is another approach to this question, based on the increased microbial activity in NT soils. NT systems, compared to CT, contributed to a faster dissipation of diclosulam in soil, decreasing chemical half-life from 87 to 67 days, which was attributed to the more expressive microbial activity in NT (Lavorenti et al., 2003).

WEATHER

Laboratory studies in which soils have been held under standardized conditions show that herbicide degradation rate and pathway are comparable between tropical and temperate soils. Under field conditions, pesticide degradation is likely to be faster than that reported for temperate climate regions (Lalah et al., 1994).

Langenbach et al. (2001) compared the fate of terbuthylazine in different soils and simulated weather conditions. Experiments with standardized climate conditions were compared under natural Brazilian climate conditions. They concluded that the influence of higher irrigation rates was evident in increasing leaching rates and higher vertical transport, both in sandy soil and organic soil. The expected effects of weather conditions in relation to mineralization and volatilization were more intense due to higher water supply rather than to higher temperatures (Table 6).

Nakagawa et al. (1995) described atrazine half-life of 56 days in a Humic Gley and of 54 days in a Deep Red Latosol from Brazil, similar to the values described for soils from temperate areas. Degradation studies were carried out under constant 25°C and 66% soil field capacity. Sorption was low and similar in both soils (Kf 0.66 – 0.74), which is also comparable to published values elsewhere (Francioso et al., 1992; Jenks et al., 1998; Koskinen and Clay, 1997; Sonon and Schwab, 1995).

In another study, the simulation of the effect of typical tropical climatic conditions on the persistence of atrazine led to the conclusion that atrazine degradation in hydroxyatrazine is related to increases in temperature and incubation time and the process involved is mainly chemical degradation; the half-life of atrazine in Brazilian soils decreased from 248.7 days at 10°C to 4.5 days at 40°C. Both incubation period and temperature increases led to more bound residues (Andréa et al., 1997).

Table 6. Fate of ^{14}C-terbuthylazine in microcosms with Brazilian soils under various conditions (in % of ^{14}C initially applied; mean of four microcosms). Source: Langenbach et al. (2001).

Process	Standard climatic conditions[a]			Brazilian summer conditions		
	Sandy soil	Clay soil	Organic soil	Sandy soil[b]	Organic soil, flooded[c]	Organic soil, dry[d]
Mineralization	0.09	0.02	0.053	0.008	0.012	0.07
Volatilization	1.45	0.10	0.58	0.25	0.01	0.01
Leaching	0.08	0.01	0.00	2.99	0.01	0.23
Transport in soil (% left at 0-1 cm depth)	40	47	82	24	45	70

[a] Air temperature 20 ± 1°C, relative air humidity 38 ± 2%, precipitation 60 mm/month.
[b] Air temperature 26.8 °C mean daily maximum, 20.9°C mean daily minimum, relative air humidity 60-80%, precipitation 300 mm/month.
[c] Air temperature 30.5 °C mean daily maximum, 24.6°C mean daily minimum, relative air humidity 60-80%, precipitation about 200 mm/month.
[d] Natural ambient Brazilian air temperature and natural Brazilian ambient air humidity in summer, precipitation 200 mm/month.

Langenbach et al. (2000) also concluded that the fate of atrazine in Brazilian Oxisol is not very different from that observed in other soils from temperate regions under similar rain and temperature conditions. Nevertheless, they proposed further research to evaluate the influence of heavy summer rain, summer sun irradiation and the movement of atrazine in sub-soil with high iron and aluminum content, conditions not found in temperate regions.

Dissipation in the topsoil, mobility within the soil profile and pesticide leaching were studied for a period of 28 days after application in a field experiment set up to evaluate the leaching potential of different pesticides in a Brazilian oxisol (Laabs et al., 2000). In contrast to temperate regions, the dissipation half-lives of pesticides were very short (<14 days), mainly due to enhanced degradation and volatilization losses. Nevertheless, significant vertical movement of metolachlor, atrazine, and simazine beyond 15 cm soil depth was found within the first four weeks after application, as well as substantial concentrations in lysimeter percolate at 35 cm soil depth. In spite of their fast dissipation under tropical climate conditions, the presence of herbicides in lysimeter percolate demonstrated that these substances possess a leaching potential.

Under lab conditions, the behavior of triazines in incubated (20°C and 33 or 66% of field capacity) samples of Western Australian soils was similar to that observed in soils elsewhere (Walker and Blacklow, 1994)

Few lab investigations have compared herbicide fate in tropical and temperate soils under similar experimental conditions. In Martinique soils, the sorption coefficients (Kf) of metolachlor in two different soils were 3.62 and 2.42 for a vertisol and a ferrisol, respectively. The influence of clay content and CEC seem to be the main factors, at times more important than organic carbon content, since differences in Kf values were not efficiently corrected by normalization of Kf to OC. Regarding leaching, it was found that, under field conditions, rainfall greatly influenced metolachlor migration. Comparison of the dissipation of metolachlor under tropical and Mediterranean conditions did not show substantial differences. Half-life was slightly longer in Mediterranean soils (Cooper and Zheng, 1994).

The interrelationships between herbicide sorption, availability in solution, and soil microbial biomass and activity form the basis of research to understand herbicide degradation in soil. The other variables which influence the rate of herbicide degradation and, hence, persistence in soil are soil temperature and soil moisture content. In the field, these are controlled by weather factors such as air temperature, rainfall and evaporation.

HOW TO OVERCOME LIMITED AVAILABLE DATA?

One of the most important constraints limiting the knowledge on the behavior of herbicides in tropical areas is the high cost associated with methodologies needed to detect and monitor these chemicals. To overcome this restriction, alternative techniques such as bioassays and computer-based modeling have been used.

Bioassays are usually considered to be a simple and versatile, although non-specific, method to detect and measure herbicide in soils. They can also provide an interesting tool to detect not only the active biological fraction of the herbicide, but also toxic metabolites. Table 7 summarizes a number of examples of bioassays developed to estimate parameters related to herbicide fate and behavior under tropical conditions.

Another simple approach to facilitate data generation is the development of empirical equations to express pesticide parameters based on soil properties. Many times, based on the strong correlation between sorption and OC contents for tropical soils, several empirical equations have been proposed as a rough estimation of soil sorption coefficients (Table 8), and may be useful to preview herbicide fate in these environments. Equations derived from empirical relationships are useful to develop simulation models. Selection of appropriate equations should be made on the basis of available data for a particular herbicide, and on the range of the input parameter values covered by each regression equation and on the chemical classes covered by each equation (Lyman, 1990). The most important advantage of simulating is the possibility of varying greatly the conditions and data for different scenarios and pesticides, assuming, for instance, different dynamics for herbicide degradation in different soil layers (Di et al., 2001).

Frequently, modeling studies seek to build models that predict the losses of a pesticide to the surrounding environment. Such predictions are required for a variety of purposes: to inform the most appropriate crop management practices, for regulatory use, and to aid understanding the factors affecting herbicide behavior in soil.

For simulation purposes, primary data on soil chemical and hydraulic properties, climate data, and the chemical and physical properties of the pesticides, are needed (Wagenet and Rao, 1990). A soil database for pesticide fate modeling should include soil properties at a number of depths in the soil profile and some characterization of spatial variability of soil properties, particularly the hydraulic properties. At present, there is very little work being done to develop appropriate soil databases in tropical countries, although initiatives are underway in countries such as Brazil and Australia (Kookana et al., 1998; Spadotto, 1999; Spadotto and Hornsby, 2000).

Most of the available chemical fate models have been developed in the United States or Europe, which often have a less variable climate and a smaller variety of agricultural soils than tropical areas. Also, models often incorporate empirical relationships based on data from a limited range of soil types. Empirical relationships must be validated locally and modified or adjusted if necessary. For instance, a comparison between calculated and measured soil temperatures revealed that the average between minimum and maximum air temperature, which gave good results as a basis for the simulation in HERBASYS model in Europe, resulted in soil temperatures usually 2 to 5 °C below the measured values in New Zealand, possibly due to the fact that the amount of energy given by radiation is higher in New Zealand and thus the soil is warmed up more quickly (Günther et al., 1992). Similar results could be expected for warmer tropical regions.

A worst-case approach is often adopted to test models because it requires very few circumstances to be considered, but the worst-case situation is generally not known *a priori*. A more realistic procedure is to develop a collection of representative sites for using the chemical and conduct a sensitivity analysis to determine the conditions under which there is an appreciable hazard to the environment. The greatest difficulty with this approach is the enormous variability in soil properties from one site to another, and even within sites (Gallant and Moore, 1993).

Validation of pesticide fate models in tropical developing countries conditions is limited at present because of the almost complete lack of detailed field studies concerning pesticide properties in these soils.

Table 7. Bioassays developed to study fate and behavior of herbicides in tropical soils.

Herbicide	Bioindicator	Purposes	Location and reference
Ametryn, diuron and oxyfluorfen	cucumber biomass	Persistence	Puerto Rico (Liu and Santiago-Córdova, 1991)
Atrazine	cucumber biomass	Retention by mulch	Brazil (Fornarolli et al., 1998)
Atrazine	oats dry biomass	Residual activity	South Africa (Reinhardt and Nel, 1993)
Atrazine	oats and turnip	Residual activity	Australia (Bowmer, 1991)
Bromacil, diuron and atrazine	cucumber, cowpea (% seedling survival)	Residual activity	India (Leela, 1984)
Chlorsulfuron and triasulfuron	lentil root growth	Residual activity	Australia (Blacklow and Pheloung, 1991)
Chlorsulfuron and triasulfuron	wheat seedlings shoot biomass	Movement, persistence and bioactivity	Australia (Blacklow and Pheloung, 1992)
Glufosinate-ammonium and imazapyr	rape	Persistence	Malaysia (Ismail and Ahmad, 1994)
Imazaquin	corn seedling root growth	Mobility and persistence	Nigeria (Evbuomwan et al., 1993)
Imazethapyr	rape leaf area and biomass	Leaching	South Africa (Van Wyk and Reinhardt, 2001)
Metsulfuron-methyl	long bean shoot biomass	Movement and residual activity	Australia (Ismail and Kalithasan, 1997)
Several	Several	Persistence and residual activity	Nigeria (Oyeniyi and Akinyemiju, 1990)
Sulfonylurea (various)	fresh weigh turnip; sunflower	Residual activity, water and leachate residues	New Zealand (Rahman et al., 1993)
Trifluralin	Italian ryegrass growth of seedlings	Bioactivity of different formulations	India (Coffman et al., 1984)

Table 8. Empirical equations to predict different sorption coefficients from soil properties in tropical areas. For all equations, the correlation coefficients (r^2) are ≥ 0.80.

Herbicide	Equation[a]	Origin of soils	Reference
2,4-D	Kd = 12.80-(2.21pH)+(0.66OC)+(0.73Al)	New Zealand	Baskaran et al. (1996)
25 non-ionic chemicals	Log Kom=0.52 log Kow+0.69	Australia	Briggs (1981)
Alachlor	Kd = 0.42+0.69OC	Brazil	Oliveira Jr. et al. (2001)
Atrazine	Kd=0.46+1.31OC	Brazil	Oliveira Jr. et al. (2001)
Atrazine	Kd = 0.967OC+0.419	Australia	Kookana et al. (1998)b
Atrazine	Kd=0.63+(0.45OC)-(0.18Al)	New Zealand	Baskaran et al. (1996)
Dicamba	Kd=0.05+0.06OC	Brazil	Oliveira Jr. et al. (2001)
Hexazinone	Kd=0.23+0.19OC	Brazil	Oliveira Jr. et al. (2001)
Metsulfuron-methyl	Kd=0.10+0.02OC	Brazil	Oliveira Jr. et al. (2001)
Metsulfuron-methyl	Kd = 1.12+(0.07OC)-(0.16pH)	New Zealand	Baskaran et al. (1996)
Simazine	Kd=-0.14+0.84OC	Brazil	Oliveira Jr. et al. (2001)
Sulfometuron-methyl	Kd=0.29+0.12OC	Brazil	Oliveira Jr. et al. (2001)

[a] Kd = soil sorption coefficient; Kow = octanol-water partition coefficient.
[b] Compilation from different authors.

CONCLUDING REMARKS

Herbicides are a broad class of bioactive compounds important for food and crop production. The safe development and use of these pesticides now encompasses a very wide range of disciplines including biochemistry of fate in plants, animals and soil, environmental chemistry, and risk assessment of formulations and residues. Although scientific knowledge is increasing, there is a lack of information on the behavior of a wide range of herbicides in tropical soils. Answers to questions on important processes such as herbicide desorption and hysteresis (Parkpian et al., 1998; Kookana et al., 1990), accumulation and availability of aged residues (Musumeci and Ostiz, 1994; Andréa and Wiendl, 1995a, 1995b; Regitano et al., 2001) and the effect of tropical climate on degradation (Racke et al., 1997), among others, are still lacking in the literature for tropical soils. Also, aspects of minimum tillage and liming in these soils have been shown to affect both the mobility and persistence of these chemicals. These unanswered questions represent a wide variety of opportunities for field and lab research work. Results may provide background to prioritize herbicides or chemical groups that should be evaluated under field conditions with regard to their leaching potential to ground water in tropical soils and needed information for their safe and effective use in tropical areas.

REFERENCES

Ahmad R., Kookana R.S., Alston A.M. Sorption of ametryn and imazethapyr in twenty-five soils from Pakistan and Australia. J Environ Sci Health B 2001; 36:143-160.

Andréa M.M., Wiendl F.M. Formation and release of bound residues of pesticides in two Brazilian soils. I. [^{14}C]-Lindane. Pesq Agropec Bras 1995a; 30:687-694.

Andréa M.M., Wiendl F.M. Formation and release of bound residues of pesticides in two Brazilian soils. II. [^{14}C]-Parathion. Pesq Agropec Bras 1995b; 30:695-700.

Andréa M.M., Matallo M.B., Tomita R.Y., Luchini L.C. Effect of temperature on dissipation of [^{14}C]-atrazine in a Brazilian soil. Pesq Agropec Bras 1997; 32:95-100.

Ball-Coelho B., Tiessen H., Stewart J.W.B., Salcedo I.H., Sampaio E.V.S.B. Residue management effects on sugarcane yield and soil properties in northeastern Brazil. Agron J 1993; 85:1004-1008.

Baskaran S., Bolan N.S., Rahman A., Tillman R.W. Pesticide sorption by allophanic and non-allophanic soils of New Zealand. New Zea J Agric Res 1996; 39:297-310.

Blacklow W.M., Pheloung P.C. Sulfonylurea herbicides applied to acidic sandy soils: a bioassay for residues and factors affecting recoveries. Aust J Agric Res 1991; 42:1205-1216.

Blacklow W.M., Pheloung P.C. Sulfonylurea herbicides applied to acidic sandy soils: movement, persistence and activity within the growing season. Aust J Agric Res 1992; 43:1157-1167.

Bolan N.S., Baskaran S. Characteristics of earthworms casts affecting herbicide sorption and movement. Biol Fertil Soils 1996a; 22:367-372.

Bolan N.S., Baskaran S. Biodegradation of 2,4-D herbicide as affected by its adsorption-desorption behaviour and microbial activity in soils. Aust J Soil Res 1996b; 34:1041-1053.

Borggaard O.K., Streibig J.C. Atrazine adsorption by some soil samples in relation to their constituents. Acta Agric Scand 1988a; 38:293-301.

Borggaard O.K., Streibig J.C. Chlorsulfuron adsorption by humic acid, iron oxides, and montmorillonite. Weed Sci 1988b; 36:530-534.

Bowmer K.H. Atrazine persistence and toxicity in two irrigated soils of Australia. Aust J Soil Res 1991; 29:339-350.

Boyd S.A., Xiangcan J., Lee J.F. Sorption of non-ionic organic compounds by corn residues from a no-tillage field. J Environ Qual 1990; 19:734-738.

Briggs G.G. Adsorption of pesticides by some Australian soils. Aust J Soil Res 1981; 19:61-68.

Brown H.M. Mode of action, crop selectivity, and soil relations of the sulfonylurea herbicides. Pestic Sci 1990; 29:263-281.

Coffman C.B., Gentner W.A., Shasha B.S. Herbicidal activity of controlled release formulations of trifluralin. Indian J Agric Sci 1984; 54:117-122.

Cooper J.F., Zheng S.Q. Behaviour of metolachlor in tropical and Mediterranean plain field conditions. Sci Total Environ 1994; 153:133-139.

Costa A.C.S., Torino C.A., Rak J.G. Capacidade de troca catiônioca dos colóides orgânicos e inorgânicos de latossolos do Estado do Paraná. Acta Sci 1999; 3:491-496.

Costa, Antonio Carlos S, Iron oxide mineralogy of soils derived from volcanic rocks in the Paraná river basin, Brazil. PhD Thesis. Columbus: Ohio State University, 1996.

Curi N., Franzmeier D.F. Toposequence of Oxisols from the central plateau of Brazil. Soil Sci Soc Am J 1984; 48:341-346.

Di H.J., Sparling G.P., Lee R., Magesan G.N. The effect of mineralisation rates of atrazine in surface and subsurface soils on its groundwater contamination potential. Aust J Soil Res 2001; 39:175-183.

Dudal, Rudi. "Soil related constraints to agricultural development in the tropics". In *Priorities for Alleviating Soil Related Constraints to Food Production in the Tropics*, Int. Rice Res. Inst., ed. Los Banos, Philippines: International Rice Research Institute, 1979, pp. 23-37.

Edwards W.M., Norton L.D., Redmond C.E. Characterizing macropores that affect infiltration into nontilled soils. Soil Sci Am J 1988; 52:483-487.

Embrapa – Empresa Brasileira de Pesquisa Agropecuária. Serviço Nacional de Levantamento e Conservação do Solo. Instituto Agronômico do Paranoia, *Levantamento de Reconhecimento dos Solos do Estado do Paraná*. Londrina, PR: EMBRAPA/IAPAR, 1984.

Eswaran, Hari; Beinroth, Fred H; Kimble, John "Soil diversity in the tropics: implications for agricultural development". In *Myths and science of soils of the tropics*, Rattan Lal, Pedro A Sanchez, eds. Madison, WI: Soil Sci. Soc. Am. (Special Publ., n. 29), 1992, pp. 1-16.

Evbuomwan F.O., Akinyemiju O.A., Akobundu, I.O. Persistence and mobility of imazaquin in a humid tropical soil. Plant Protect Quart 1993; 8:118-122.

FEBRAPD. Federação brasileira de plantio direto na palha. Brasil - Expansão da área cultivada em plantio direto. Available in http://www.febrapdp.org.br/pd_area_br.htm 20/02/2003.

Fendorf S., Fendorf M. Sorption mechanisms of lanthanum on oxide minerals. Clays Clay Miner 1996; 44:220-227.

Fermanich K.J., Daniel T.C. Pesticide mobility and persistence in microlysimeter soil columns from a tilled and no-tilled plot. J Environ Qual 1991; 20:195-202.

Fernandes E.C.M., Motavalli P.P., Castilla C., Mukurumbira L. Management control of soil organic matter dynamics in tropical land-use systems. Geoderma 1997; 79:49-67.

Ferris I.G., Felton W.L., Holland J.F., Bull M.S. Effects of tillage practice on the persistence of atrazine in two contrasting soils in northern New South Wales. Aust J Exp Agric 1989; 29:849-853.

Fornarolli D.A., Rodrigues B.N., Lima J., Valério M.A. Influência da cobertura morta no comportamento do herbicida atrazine. Planta Daninha 1998; 16:97-107.

Francioso O., Bak E., Rossi N., Sequi P. Sorption of atrazine and trifluralin in relation to the physico-chemical characteristics of selected soils. Sci Total Environ 1992; 123/124:503-512.

Gallant J.C., Moore I.D. Modelling the fate of agricultural pesticides in Australia. Agric Syst 1993; 43:185-197.

Garcell L., Morales M.P., Andres-Verges M., Tartaj P., Serna C.J. Interfacial and rheological characteristics of maghemite aqueous suspensions. J Coll Interf Sci 1998; 205:470-475.

Gerritse R.G., Beltran J., Hernandez F. Adsorption of atrazine, simazine, and glyphosate in soils of the Gnangara Mound, Western Australia. Aust J Soil Res 1996; 34:599-607.

Gish T.J., Isensee A.R., Nash R.G., Helling C.S. Impact of preferential transport of pesticides on water quality. Trans ASAE 1991; 34:1745-1753.

Greenland, Dennis J; Wild, Alan "Organic matter dynamics in soils of the tropics – from myth to complex reality." In *Myths and Science of Soils of the Tropics*, Rattan Lal, Pedro A Sanchez, eds. Madison, WI: Soil Sci. Soc. Am. (Special Publ., n. 29), 1992, pp. 17-33.

Günther P, James TK, Rahman A, Pestemer W. Adaptation of a german herbicide advisory (HERBASYS) to New Zealand conditions. Proceedings of the 45th New Zealand Plant Protection Conference; 1992; Lincoln: New Zealand Plant Protection Society, 1992.

Hargreaves, George H, Precipitation dependability and potential for agricultural production in Northeast Brazil. Logan: Utah State Univ/EMBRAPA., 1971 (Publ. 74-D159).

Harper S.S. Sorption-desorption and herbicide behavior in soil. Rev Weed Sci 1994; 6:207-225.

Hay J.V. Chemistry of sulfonylurea herbicides. Pestic Sci 1990; 29:247-261.

Huang P.M., Grover R., McKercher R.B. Components and particle size fractions involved in atrazine adsorption by soils. Soil Sci 1984; 138:20-24.

Hunter, Robert J, *Zeta Potential in Colloid Science*. London: Academic Press, 1981.

Hussain S.A, Demirci S., Ozbayoglu G. Zeta potential on three clays from Turkey and effects of clays on coal flotation. J Colloid Int Sci 1996; 184: 535-541.
Ikram U.H., Hhabib U.R., Shakirullah H. Interfacial properties of alpha-FeOOH, prepared by two different methods. J Chem Soc Pak 1996; 18:251-258.
Isensee A.R., Nash R.G., Helling C.S. Effect of conventional vs. no-tillage on pesticide leaching to shallow groundwater. J Environ Qual 1990; 19:434-440.
Ismail B.S., Ahmad A.R. Attenuation of the herbicidal activities of glufosinate-ammonium and imazapyr in two soils. Agric Ecosyst Environ 1994; 47:279-285.
Ismail B.S., Kalithasan K. Mobility of metsulfuron-methyl in tropical soils. Aust J Soil Res 1997; 35:1291-1300.
Izac A.M.N. Developing policies for soil carbon management in tropical regions. Geoderma 1997; 79:261-276.
Jenks B.M., Roeth F.W., Martin A.R., McCallister D.L. Influence of surface and subsurface soil properties on atrazine sorption and degradation. Weed Sci 1998; 46:132-138.
Johnson R.M., Sims J.T. Influence of surface and subsoil properties on herbicide sorption by Atlantic coastal plain soils. Soil Sci 1993; 155:339-348.
Kenimer A.L., Mostaghimi S., Young R.W., Dillaha T.A., Shanholtz V.O. Effects of residue cover on pesticide losses from conventional and no-tillage systems. Trans ASAE 1987; 30:953-959.
Kim KU. Weed management implication and trends of direct seeding in Asia. Proceedings of the Third International Weed Science Congress; 2000 June 6-11; Foz do Iguassu, Brazil. Oxford, MS: International Weed Science Society, 2000.
Koeppen, Wolfgang "Das geographische system der klimate." In *Handbuch der Klimatologie v.1, Part C*, Wolfgang Koeppen, ed. Berlin: Gebruder Borntrager, 1936.
Kookana R.S., Baskaran S., Naidu R. Pesticide fate and behavior in Australian soils in relation to contamination and management of soil and water: a review. Aust J Soil Res 1998; 36:715-764.
Kookana R.S., Gerritse R.G., Aylmore L.A.G. Effect of organic cosolvent on adsorption and desorption of linuron and simazine in soil. Aust J Soil Res 1990; 28:717-725.
Koskinen W.C., Clay S.A. Factors affecting atrazine fate in North Central U.S. soils. Rev Environ Contam Toxicol 1997; 151:117-165.
Laabs V., Amelung W., Pinto A., Altstaedt A., Zech W. Leaching and degradation of corn and soybean pesticides in an Oxisol of the Brazilian Cerrados. Chemosphere 2000; 41:1441-1449.
Lal R. Mulching effects on soil physical quality of an alfisol in western Nigeria. Land Degrad Develop 2000; 11:383-392.
Lal R. Soil degradative effects of slope length and tillage methods on alfisols in western Nigeria. 1. Runoff, erosion and crop response. Land Degrad Develop 1997; 8:201-219.
Lalah J.O., Acholla F.V., Wandiga S.O. Fate of C-14 p, p'-DDT in Kenyan tropical soils. J Environ Sci Health B 1994; 29:57-64.
Langenbach T., Schroll R., Paim S. Fate and distribution of ^{14}C-atrazine in a tropical oxisols. Chemosphere 2000; 40:449-455
Langenbach T., Schroll R., Scheunert I. Fate of the herbicide ^{14}C-terbuthylazine in Brazilian soils under various climatic conditions. Chemosphere 2001; 45:387-398.
Lavorenti A., Rocha A.A., Prata F., Regitano J.B., Tornisielo V.L., Pinto O.B. Comportamento do diclosulam em amostras de um Latossolo Vermelho Distroférrico sob plantio direto e convencional. Rev Bras Ci Solo 2003; 27:183-190.
Leela D. Studies on the persistence of herbicides in sandy loam soil. Indian J Hort 1984; 41:123-126.
Lim C.H., Jackson M.L. Kaolins- sources of differences in cation exchange capacities and cesium retention. Clays Clay Miner 1980; 2:223-229.
Liu L.C., Santiago-Córdova M. Persistence of three herbicides in a drip-irrigated banana field. J Agric Univ PR 1991; 75:19-23.
Locke M.A, Bryson C.T. Herbicide-soil interactions in reduced tillage and plant residue management systems. Weed Sci 1997; 45:307-320.
Lyman, Warren J. "Adsorption coefficient for soils and sediments". In *Handbook of Chemical Property Estimation Methods: Environmental Behavior of Organic Compounds*, Warren J Lyman, W. F. Reehl, D. H. Rosenblatt, eds. Washington, DC: American Chemical Society, 1990, pp. 4.1-4.33.
Ma C., Eggleton R.A. Cation exchange capacity of kaolinite. Clays Clay Miner 1999; 47:174-180.
Mackay A.D., Springett J.A., Syers J.K., Gregg P.H. Origin of the effect of earthworms on the availability of phosphorous in a phosphate rock. Soil Biol Biochem 1983; 15:63-73.
Mallawatantri A.P., Mulla D.J. Herbicide adsorption and organic carbon contents on adjacent low-input versus conventional farms. J Environ Qual 1992; 21:546-551.
McKenzie B.M., Dexter A.R. Physical properties of casts of the earthworm *Aporrecttodea rosea*. Biol Fert Soils 1987; 5:152-157.
McNeil M. Lateritic soils. Sci Amer 1964; 211:96-102.

Melfi A.J., Pedro G., Volkof B. Nature and distribution of ferrous compounds in Brazilian soils. R Bras Ci Solo 1979; 3:47-54.
Mersie W., Foy C.L. Phytotoxicity and adsorption of chlorsulfuron as affected by soil properties. Weed Sci 1985; 33:564-568.
Mohr, Edward C J; van Baren, Ferdinand A, *Tropical soils. A critical study of soil genesis as related to climate and vegetation*. London: The Royal Tropical Institute/Interscience, 1954..
Mohr E.C.J; van Baren, F.A., van Schuylenborgh J. *Tropical soils* 3rd ed. Netherlands: The Hague, 1973.
Musumeci M.R., Ostiz S.B. Binding of cypermethrin residue in Brazilian soils and its release by microbial activity. Rev Microbiol 1994; 25:216-219.
Nakagawa L.E., Luchini L.C., Musumeci M.R., Andréa M.M. Comportamento da atrazina em solos brasileiros em condições de laboratório. Pesq Agropec Bras 1995; 30:471-476.
Oliveira Jr R.S., Koskinen W.C., Ferreira F.A. Sorption and leaching potential of herbicides on Brazilian soils. Weed Res 2001; 41:97-110.
Oyeniyi A., Akinyemiju O.A. Use of bioassays for herbicide persistence studies in the humid tropics. Turrialba 1990; 40:265-271.
Parkpian P., Anurakpongsatorn P., Pakokong P., Patrick W.H. Adsorption, desorption and degradation of alpha-endosulfan in tropical soils of Thailand. J Environ Sci Health B 1998; 33:211-233
Prasad R., Power J.F. Crop residue management. Adv Soil Sci 1991; 15:205-251.
Procópio S.O., Pires F.R., Werlang R.C., Silva A.A., Queiroz M.E.L.R., Neves A.A., Mendonça E.S., Santos J.B., Egreja Filho F.B. Sorção do herbicida atrazine em complexos organominerais. Planta Daninha 2001; 19:391-400.
Pusino A., Petretto S., Gessa C. Adsorption and desorption of imazapyr by soil. J Agric Food Chem 1997; 45:1012-1016.
Racke K.D., Skidmore M.W., Hamilton D.J., Unsworth J.B., Miyamoto J., Cohen S.Z. Pesticide fate in tropical soils (Technical report). Pure Appl Chem 1997; 69:1349-1371.
Rahman A, James TK, Gunther P. Bioassays of soil applied herbicides. Proceedings of the International Symposium of Indian Society of Weed Science; 1993 November 18-20; Hisar. Hisar: Indian Society of Weed Science, 1993.
Regitano J.B., Alleoni L.R.F., Vidal-Torrado P., Casagrande J.C., Tornisielo V.L. Imazaquin sorption in highly weathered tropical soils. J Environ Qual 2000; 29:894-900.
Regitano J.B., Bischoff M., Lee L.S., Reichet J.M., Turco R.F. Retention of imazaquin in soil. Environ Toxicol Chem 1997; 16:397-404.
Regitano J.B., Tornisielo V.L., Lavorenti A., Pacovsky R.S. Transformation pathways of C^{14}-chlorotalonil in tropical soils. Arch Environ Contam Toxicol 2001; 40:295-302.
Reinhardt C.F., Nel P.C. Quantitative bioassays for monitoring the dissipation of atrazine in soil. S Afr J Plant Soil 1993; 10:58-62.
Rivero C., Senesi N., Paolini J., D´Orazio V. Characteristics of humic acids of some Venezuelan soils. Geoderma 1998; 81:227-229.
Rodrigues B.N., Lima J., Yada I.F.U., Fornarolli D.A. Influência da cobertura morta no comportamento deo herbicida trifluralin. Planta Daninha 1998; 16:163-175.
Rosenberg R.J., McCoy E.L Measurement of soil macropore air permeability. Soil Sci Am J 1990; 54:969-974.
Sanchez, Pedro A; Logan, Terry. "Myths and science about chemistry and fertility of soils in the tropics". In *Myths and Science of Soils of the Tropics*, Rattan Lal, Pedro A Sanchez, eds. Madison, WI: Soil Sci. Soc. Am. (Special Publ., n. 29), 1992, pp. 35-46.
Sanchez, Pedro A, *Properties and management of soils in the tropics*. New York: John Wiley and Sons, 1976.
Schwertmann, Udo; Herbillon, Adrien J "Some aspects of fertility associated with the mineralogy of highly weathered tropical soils." In *Myths and Science of Soils of the Tropics*, Rattan Lal, Pedro A Sanchez, eds. Madison, WI: Soil Sci. Soc. Am. (Special Publ., n. 29), 1992, pp. 47-60.
Shipitalo M.J., Edwards W.M., Dick W.A., Owens L.B. Initial storm effects on macropore transport of surface-applied chemicals in no-till soil. Soil Sci Am J 1990; 54:1530-1536.
Singh R., Gerritse G.R., Aylmore L.A.G. Adsorption-desorption behaviour of selected pesticides in some western australian soils. Aust J Soil Res 1989; 28:227-243.
Sivakumar, M.V.K., Manu, Andrew; Virmani, Surender M; Kanemasu, Edward T. "Relation between climate and soil productivity in the tropics." In *Myths and Science of Soils of the Tropics*, Rattan Lal, Pedro A Sanchez, eds. Madison, WI: Soil Sci. Soc. Am. (Special Publ., n. 29), 1992, pp. 91-120.
Sonon L.S., Schwab P. Adsorption characteristics of atrazine and alachlor in Kansas soils. Weed Sci 1995; 43:461-466.
Spadotto, Cláudio Antonio, *Sorption and transport of acidic pesticides in variable-charge soils*. Gainesville (FL): University of Florida, 1999. (PhD Thesis Soil And Water Science).

Spadotto CA, Hornsby AG. Sorption of acidic herbicides in variable-charge soils. Abstracts of the Third International Weed Science Congress; 2000 June 6-11; Foz do Iguassu, Brazil. Foz do Iguassu: International Weed Science Society. 2000.

Sposito, Garrison, *The Surface Chemistry of Soils*. New York: Oxford University Press, 1984.

Stehouwer R.C., Dick W.A., Traina S.T. Characteristics of earthworm burrow lining affecting atrazine sorption. J Environ Qual 1993; 22:181-185.

Stehouwer R.C., Dick W.A., Traina S.T. Sorption and retention of herbicides in vertically oriented earthworm and artificial burrows. J Environ Qual 1994; 23:286-292.

Sun Z.X., Su F.W., Forsling W., Samskog P.O. Surface characteristics of magnetite in aqueous suspensions. J Coll Interf Sci 1998; 197:151-159.

Theng, Benny K G, *Soils with Variable Charge*. Palmerston North: New Zealand Soc. Soil Sci., Offset Publ., 1980.

Thornthwaite C.W. An approach towards a rational classification of climate. Geogr. Ver. 1948; 38:55-94.

Tiwari S.C., Tiwari B.K., Mishra R.R. Microbial population, enzyme activities and nitrogen-phosphorous-potassium enrichment in earthworm casts and in the surrounding soils of a pineapple plantation. Biol Fert Soils 1989; 8:178-182.

Uehara, Goro; Gilmann, Gavin, *The Mineralogy, Chemistry and Physics of Tropical Soils with Variable Charge*. Westview Tropical Agric. Ser., No. 4. Boulder, CO: Westview Press, 1981.

van Wambeke, Arnand, *Soils of the tropics: properties and appraisal*. New York: McGraw Hill, Inc., 1992.

van Wyk L.J, Reinhardt C.F. A bioassay technique detects imazethapyr leaching and liming-dependent activity. Weed Technol 2001; 15:1-6.

Vieira E.M., Prado A.G.S., Landgraf M.D., Rezende M.O.O.R. Estudo da adsorção/dessorção do ácido 2,4 diclorofenoxiacético (2,4-D) em solo na ausência e presença de matéria orgânica. Química Nova 1999; 22:305-308.

Wade M.K., Sanchez P.A. Mulching and green manure applications for continuous crop production in the Amazon Basin. Agron J 1983; 75:39-45.

Wagenet, R.J., Rao P., Suresh C. "Modeling pesticide fate in soils." In *Pesticides in the Soil Environment: Processes, Impacts, and Modeling*, Hwei H. Cheng, ed. Madison, WI: Soil Sci. Soc. Am., 1990, pp. 351-399.

Walker S, Hollaway K, Osten V. Sulfonylurea persistence in various climates of Australia. Abstracts of the Third International Weed Science Congress; 2000 June 6-11; Foz do Iguassu, Brazil. Foz do Iguassu: International Weed Science Society. 2000.

Walker S.R., Blacklow W.M. Adsorption and degradation of triazine herbicides in soils used for lupin production in western Australia: Laboratory studies and a simulation model. Aust J Soil Res 1994; 32:1189-1205.

Walker S.R., Blacklow W.M. Movement and persistence of triazine herbicides in soils used for lupin production in western Australia: field measurements and interpretations with the aid of a simulation model. Aust J Soil Res 1995; 33:773-786.

Werkheiser W., Anderson S.J. Effect of soil properties and surfactant on primisulfuron sorption. J Environ Qual 1996; 25:809-814.

Wilson G.V., Luxmoore R.J. Infiltration, macroporosity, and mesoporosity distributions on two forested watersheds. Soil Sci Soc Am J 1988; 52:329-335.

CHAPTER 12

DIVERSIFIED WEED MANAGEMENT SYSTEMS

K. NEIL HARKER & GEORGE W. CLAYTON

Agriculture and Agri-Food Canada, Lacombe Research Centre, 6000 C & E Trail, Lacombe, Alberta, Canada T4L 1W1

Weed management has been the object of considerable attention and study ever since crops have been cultivated. Despite this prolonged effort and the more recent but significant activity of incredible herbicide technology, weeds still thrive in most global cropping systems. Like microorganisms that quickly adapt to and resist antibiotic attack, and insect populations that rapidly gain the ability to detoxify insecticides, weeds also adapt to herbicides. There are currently 156 weed species resistant to at least one herbicide mode of action group (Heap, 2002), and there is good reason to believe that number will continue to steadily increase. But it is probably our inability to unravel and exploit the intricacies of weed population ecology, rather than a lack of sufficient technology, that limits our ability to effectively manage weeds.

THE PERPETUAL PRESENCE OF WEEDS

Buhler (1999) lamented, "We seldom examine the causes of the perpetual presence of weeds." One might well ask why a weed like wild oat (*Avena fatua*), which has been the subject of almost constant attention and herbicide attack for the last few decades, continues to thrive. Ironically, it may be that we have focused too much on wild oat destruction as opposed to removing, or perhaps more realistically, varying the niche that wild oat thrives in. That niche may be as simple as crops seeded and harvested at relatively constant dates (Derksen et al., 1999). Beck (2002) suggests that rotations containing plants with the same growth patterns will lead to problems with weeds with similar growth patterns. Blackshaw (1994) noted that winter annual downy brome (*Bromus tectorum*) thrives in a repeated winter wheat rotation but is easily managed in cropping systems with growth habits other than winter wheat and downy brome. It is likely that the dominance of wild oat in many areas relates to the dominance of summer annual crops in those areas. Given year after year of the same practices, dominant weeds will increase sufficiently to thrive in simple cropping systems for many years (Buhler, 1999; Dekker, 1999). "A good rotation has diversity in plant types, planting dates, and harvest periods" (Beck, 2002). The ability of weed populations to adapt and thrive can be restricted by cropping system and weed management diversity.

Inderjit (ed.), Weed Biology and Management, 251-265.

THE DOMINANCE OF HERBICIDE TECHNOLOGY

Herbicides dominate weed management tools in most crop production systems (Buhler et al., 2000; Kropff and Walter, 2000). Continued improvements in the effectiveness and safety of herbicide technology will probably assure their dominance well into the current century. However, many consider that there has not been sufficient effort expended on herbicide alternatives, or on the integration of herbicide technology with alternative weed management techniques (Wyse, 1992). In 1991, Thill et al. concluded that weed scientists generally spend little time conducting or publishing integrated weed management research. Why is this so? Significant funding from the herbicide industry and constantly eroding public funding has influenced the lack of support for alternative weed management tools. Perhaps researchers and industry have simply become too enamored with the wonderful activity and specificity of herbicides. The simplicity of use and short-term cost-effectiveness of herbicides makes them difficult to ignore. Our focus in this chapter is not on alternatives to herbicides, but on diversifying the use of herbicides with other weed management tools. We contend that diversified use of current tools is as important as the search for new weed management tools.

WEED MANAGEMENT TOOL LONGEVITY

It is human nature to repeat a successful practice. This repetition is not problematic in the world of assembly lines, automata and machines where all "genotypes" are defined. But repeated practices are rapidly adapted to by living systems with inherently high levels of variation and diversity. Given this diversity, even in very small areas, weeds rapidly adapt to repeated practices and reduce once successful practices to failure. The lifespan of a given weed management practice is determined by how carefully it is employed. We commonly invoke the phenomena of the over-use of herbicide tools and subsequent weed resistance outbreaks to illustrate the above principle, but are often hesitant to apply this principle to less technological tools. Nevertheless, weed management tools other than herbicides can also be rendered ineffective if they are repeated in sequence and/or over-used (Gould, 1991; Holt, 1994).

WEED RESISTANCE TO HERBICIDES

The effective action of herbicides was unchallenged for many years after their discovery, but herbicide resistance has become a formidable challenge in some areas. At first, weed resistance was a rather minor problem in isolated areas (Hilton, 1957; Ryan, 1970; Switzer, 1957). But after a few years of rather infrequent detections, the familiar exponential pattern of confirmed resistant populations has been revealed for some of the major herbicide mode of action groups (Heap, 2002). It is clear that some mode of action groups are more susceptible to resistance buildup than others. Two of these groups (acetolactate synthase and acetyl-CoA carboxylase inhibitors) seem to have a higher frequency of natural resistance in weed populations. Weed populations resistant to these groups can be problematic after only a few years of continuous herbicide use (Heap, 2002). In Western Australia, rigid ryegrass (*Lolium rigidum*) resistance is thought to be one of the world's most significant resistance problems (Llewellyn and Powles, 2001). In western Canada, herbicide-resistant wild oat presents similar challenges across the Prairie Provinces (Beckie et al., 1999). In both cases, there are numerous populations of rigid ryegrass and wild oat exhibiting multiple resistances. But it also clear that the challenges associated with weed resistance have led researchers and crop producers to more carefully consider weed management strategies that are not so dependent on herbicides.

IMPACT OF HERBICIDE RESISTANCE ON WEED SCIENCE

Not surprisingly, in areas where weed resistance has presented the greatest challenges, there has been substantial new research emphasis on integrated weed management and diversified cropping systems. In Australia, where herbicide resistance problems are probably most severe, there are several institutions involved in alternative weed management system research. For example, the Western Australia Herbicide Resistance Initiative (http://www.wahri.agric.uwa.edu.au) involves researchers with emphases ranging from molecular biology and biochemistry to seed biology and population genetics, to resistance management and technology transfer. In response to the weed resistance problem in western Canada, and especially Manitoba, researchers at the University of Manitoba became early leaders in weed resistance detection, mapping and management (Bourgeois and Morrison, 1997; Heap et al., 1993; Shirtliffe et al., 2000a). In addition, weed resistance to triazine herbicides opened several productive avenues of important, basic research (Dekker and Burmester, 1992; Diesenhofer and Michel, 1989; Hirschberg et al., 1984; Machado et al., 1980). It is likely that the current effort in weed resistance research will continue to provide resistance management solutions as well as new discoveries with applications for agriculture and beyond.

Whether by design or out of necessity, there is no doubt that research on weed ecology has substantially increased in recent years (Table 1). From 1990 to 2000, of the total weed ecology and weed control articles, the proportion of weed ecology articles in "*Weed Research*" and "*Weed Science*" increased by more than 20%. Over the same period there were also a higher proportion of weed ecology articles written in "*Weed Technology*", which is mostly dedicated to the "hows" as opposed to the "whys" of weed science. Weed resistance and greater environmental awareness/concerns are at least partially responsible for renewed interest in weed ecology and diversified cropping systems. Given improved communication, networking, and perhaps a belated realization of the futility of overusing herbicides, Weed Scientists and Extension Specialists have probably also been responsible for renewed interest in weed ecology and more diversified weed management practices.

THE NECESSITY OF WEED DIVERSITY

Weeds are inherently different from pests such as insects. Whereas, insects can locate crop host plants in response to chemical or visual stimuli, weeds are mostly immobile or are dependent on less predictable and less controllable forces such as tillage, wind, water, or animals to scatter their propagules to desirable locations. Therefore, to a greater extent than insects, weeds must possess sufficient variability in a given location to respond to available niches. Weeds depend upon the genetic "memory" of their seed banks (Cavers, 1995) for a broad diversity of phenotypes that are available to exploit numerous environmental niches. Seed from successful plants in previous cropping systems are added to the seed bank and are then available to germinate and re-exploit the niche their parent plant once thrived in. Therefore, given variable environmental conditions and limited or unpredictable mobility, weed success over the long-term is dependent upon broad genotypic diversity.

Table 1. Number of articles in major weed journals with a primary focus on either weed control with herbicides or weed ecology in 1990 and 2000.

Year/Journal	Control[a]	Ecology[a]	% Ecology[b]
1990			
Weed Research	14	22	61
Weed Science	30	33	52
Weed Technology	91	24	21
Total	135	79	37
2000			
Weed Research	6	35	85
Weed Science	11	42	79
Weed Technology	64	20	24
Total	81	97	54

[a]These numbers are subject to the author's interpretation; some articles did not fit well into either category and were not included in the Table.
[b]Percent Ecology is calculated from the total number of weed control and weed ecology articles, it is not a percent of all articles in the journal.

Weed diversity is sufficient to counter a variety of simple, repeated crop production practices (Blackshaw, 1994; Buhler, 1999; Dekker, 1997; Harker et al., 2000). Indeed, the recent short-term, economic-survival thinking involved in low diversity, short crop rotation sequences have likely made it much easier for weeds to quickly adapt and thrive. Very effective herbicide tools allow growers to cope in such systems in the short-term. However, low diversity cropping systems are accompanied by pesticide dependency, and by definition disregard time-tested high diversity cropping systems that were necessary before herbicides were discovered. Although pre-herbicide-era weed management practices do not provide the rapid and rather spectacular effects of herbicides, their employment in various combinations can significantly augment judicious herbicide use, and in some cases, provide adequate weed management solutions in the absence of herbicides.

"BANDWAGONS" REDUCE DISCIPLINE DIVERSITY AND SUCCESS

Within a discipline, research emphasis on a particular subject ebbs and flows. The current emphasis on "invasive species" is important, but should not exclude continued emphasis in other weed science disciplines. "Copycat" research on favourite or in-vogue subjects must be avoided (Mortensen et al., 2000). Indeed, if we concentrate solely on invasive species and forget disciplines such as application technology, herbicide mode of action, and biological control, the lack of diversity in our studies will also be detrimental to our contest with weeds. Paradoxically, some would encourage diversity in weed management systems but discourage a diversity of weed science disciplines.

We should strive for weed management diversity, agribusiness diversity, expertise diversity, research partnership diversity, and general diversity of thought. If weed ecology and integrated weed management studies would not have been neglected in the 1950s through the 1980s, we would probably have much less weed resistance and a more balanced approach to current research. Some weed scientists discourage almost any research that involves herbicide efficacy. In 2000, a single article was published in Weed Science involving herbicide application technology (Wolf et al., 2000); in 2001 no application technology articles were published in Weed Science. Application technology research can lead to dramatic improvements in herbicide efficiency and reduced environmental impact, but these studies generally lack funding and ideological support, particularly in public institutions.

If we are to prosper in an uncertain future, we can only do so if we broaden our thinking. Important and progressive weed management insights are to be found in research involving weed ecology and biology, herbicide efficacy (Kropff and Walter, 2000), agronomy, transgenics, organic cropping systems, application technology, biological control, herbicide mode of action, environmental fate of herbicides, and a host of other related disciplines. If we can mimic the diversity found in natural systems, and let variation flourish in our weed science group, then perhaps we can flourish as a group in the future. We must not neglect any of our tools, technologies, systems, or research areas.

SINGLE VERSUS MULTIPLE TOOLS

Stressing diversity in weed management practices is more challenging than stressing single tools. The weed management concerns of crop producers are often so urgent that "quick-fix" solutions seem imperative. In response, research and extension personnel have often stressed single tools as opposed to packages of tools in entire cropping systems. Single tools are easily recommended, but, given weed diversity, are readily adapted to over the long-term (Buhler, 1999; Holt, 1994). Increased canola seeding rate can aid in weed suppression (O'Donovan and Newman, 1996), but could be even more effective against weeds when combined with an additional optimal factor such as competitive cultivar. For example, % dockage (weed seed and foreign material in the harvested seed) can be significantly reduced when a competitive canola cultivar such as InVigor 2153 is combined with higher seeding rates (Figure 1). Technologies and information require "stacking" or "pyramiding" into packages that combine several strategies for superior crop health and competition with weeds.

In another canola study, combining agronomic factors at optimal levels led to both yield and weed management benefits (Harker et al., 2003a). Combining the best cultivar, the highest seeding rate, and the earliest weed removal treatments led to a 41% yield increase compared to the combination of the weaker cultivar, the lowest seeding rate, and the latest time of weed removal (Figure 2). Perhaps just as important, combining the earliest weed removal treatment with the highest seeding rate not only led to lower weed biomass levels, but dramatically decreased weed biomass variability. From an integrated weed management point of view, it is important to restrict weed population variability (Liebman and Davis, 2000).

Beck (2002) reminds us that "successful crop production, regardless of the methods used, is a careful piecing together of numerous components into a system. Simply replacing one component with another is seldom successful". Focusing on crop competitiveness and health will lead producers to rely on packages of tools which include such things as sanitation (prevention of weed seed spread), low disturbance seeding systems, higher seeding rates, narrow crop rows, optimum fertilizer placement, and diverse crop rotations. These and other practices are common in integrated weed management systems (Boerboom, 1999; Swanton and Weise, 1991; Thill et al., 1991). A healthy, competitive crop is the key to weed management in any cropping system.

The complications of applying biological and knowledge-intensive management inputs (rotation, seeding dates, higher seeding rates, legumes, soil organisms, insects, etc.) versus the simple application of herbicides or inorganic fertilizers can be intellectually intimidating (Mortensen et al., 2000), but the former are more sustainable over the long term (Figure 3). A higher degree of knowledge is required for the successful integration of biological inputs since they are "living", more diverse, or less specific than single compounds and formulations of compounds. But the most economic and environmentally sustainable systems in the long-term are likely to include a combination of biological and technological inputs.

WEED MANAGEMENT ENCOMPASSES MORE THAN JUST WEEDS

Weed management cannot be successfully implemented if weeds are an exclusive focus. If ignored, diseases, insects or other pests can reduce crop health to the degree that the tools employed for weed management are negated. For example, a canola crop with root systems ravaged by root maggots (*Delia radicum* L.) or brown girdling root rot (*Rhizoctonia solani* Kuhn) will not be healthy enough to compete with weeds or even remain productive in the presence of relatively low weed populations. Similarly, a barley crop that was very competitive early in the growing season, but later compromised by leaf disease will allow more weeds to complete their life cycle and add their seeds back into the soil seed bank to flourish in the next rotational crop. Furthermore, if seed is carelessly planted too deep or too much fertilizer is placed with the seed, rapid crop emergence will be compromised and weed management will likely also be compromised.

Crop health and competitiveness demand that all aspects of a healthy crop are considered and implemented. The importance of agronomic factors such as crop rotation, adequate fertility, and high seeding rates to final crop yields may be less than their importance to the augmentation of weed management (Beck, 2002). Similarly, good disease and insect management, in some situations, may be more important for weed management, because of their impact on crop health, than for their direct effects on crop yield. Also, it is not enough to consider only the health of the current crop. Ironically, because weed management must be integrated with all other crops and crop production practices that influence the ecosystem (Swanton and Weise, 1991; Walker and Buchanan, 1982), the best thing a grower could do for weed management in a given crop may be in a previous rotational crop.

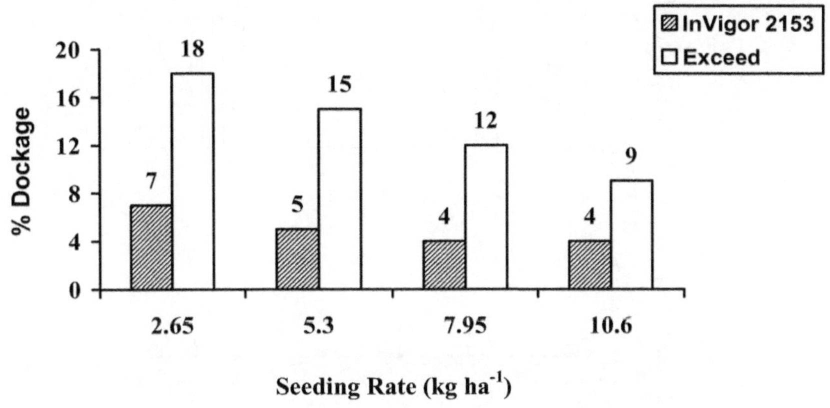

Figure 1. The relationship between canola (*Brassica napus*) cultivar and seeding rate on a weed management outcome (% dockage). Dockage is defined as the percent of harvested material that is not crop seed (weed seeds, crop stems and leaves, chaff, insects, etc.). 'InVigor 2153' is a vigorous, hybrid cultivar and 'Exceed' is a less competitive open-pollinated cultivar; both are tolerant to glufosinate. LSD (0.05) = 2. Data are from the AAFC Lacombe Research Centre, Lacombe, Alberta, Canada (Experiment 99013).

OPERATIONAL DIVERSITY – EXAMPLES

Operational diversity involves the modification of "normal" procedures for a given cropping practice. In the first example below, operational diversity pertains to planting earlier than normal. In the last example, operational diversity involves earlier than normal harvest. As will be seen, adding diversity to the operational side of cropping practices has obvious and less obvious impacts and advantages.

Fall-seeded or dormant-seeded canola has provided a new seeding dates option for canola growers in western Canada over the past few years. The primary risks of fall seeding are fall germination and/or frost damage to early-emerged canola the following spring. Benefits of fall-seeded canola include earlier flowering (avoidance of hot, dry periods during flowering), a longer flowering period, earlier maturity, reduced plant height, higher oil content and up to 38% higher yields (Kirkland and Johnson, 2000). Additional benefits of fall-seeded canola are less obvious than those mentioned above, but may be considerable in terms of diversified weed management systems.

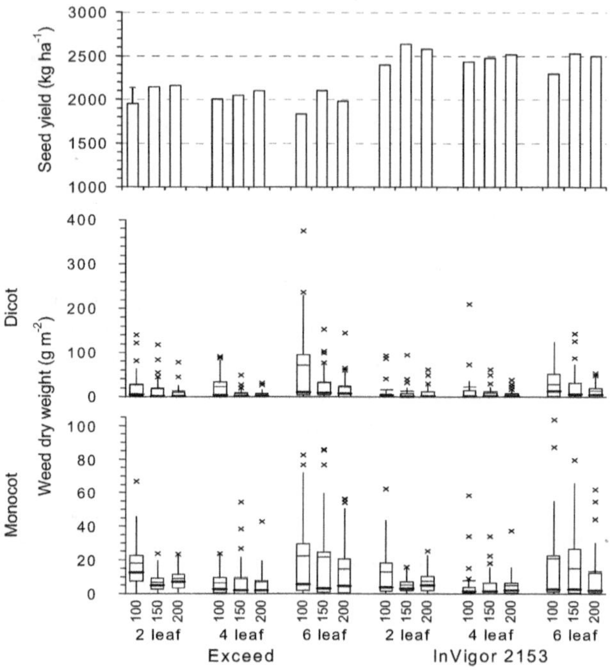

Figure 2. Seeding rate (seeds m^{-2}) and time of weed removal (canola leaf stage) effects on 'Exceed' and 'InVigor 2153' canola mean yields (TOP) at Lacombe (1998–2000) and Lethbridge (1999-2000), Alberta, Canada. LSD (0.05) on the first bar is for the three-way interaction. Dicot and monocot weed box-plots (MIDDLE & BOTTOM) are bounded by the first and third quartiles. 'Whiskers' extend 1.5 box lengths from the end of the box. Outliers (x) lie beyond the whiskers. Central tendency is indicated by the median (thick solid line) and the mean (thin solid line). Data are from AAFC, Lacombe, Alberta, Canada (F20014).

Weeds that have adapted to proliferate in crops seeded on conventional dates may be disadvantaged when required to compete with crops that emerge earlier in the season and are harvested earlier in the fall. Figure 4 shows canola phenology on June 26, 1999 when seeded in early November (1988), late April (1999), and in mid May (1999). On land subject to repeated seeding on standard dates, the bulk of the weed seed bank probably contains seeds that have adapted over many years to emerge just before or with crops seeded on those dates. It is likely that many weeds from these seed banks would emerge too late to effect serious interference with a fall-seeded canola crop. At harvest, weeds previously selected to mature and shatter seed just before mid May-seeded canola is harvested may not have sufficient time to complete their life cycle and replenish the seed bank in fall-seeded canola.

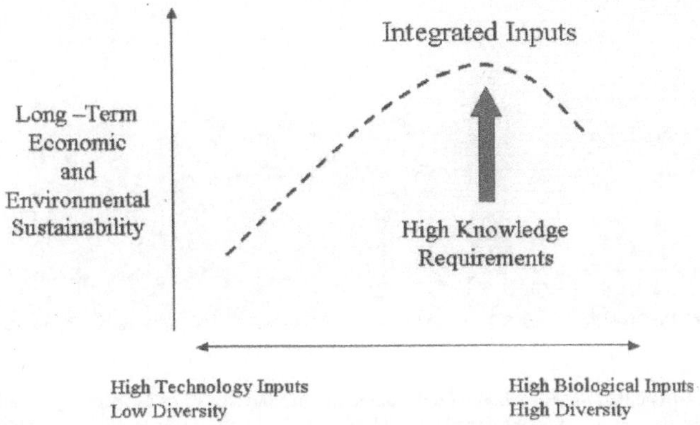

Figure 3. The relationship between high technology input, low diversity systems, and high biological input, high diversity systems on long-term environmental and economic sustainability in plant management systems. Effectively integrating inputs from both management philosophies requires the greatest knowledge.

The advantage of fall seeding is not because there is something inherently superior about fall seeding versus mid-May seeding, but because weeds adapted to May seeding will be subjected to operational diversity that leaves them less prepared for competition in crops seeded on different dates. Of course, if all seeding were shifted to an earlier date, weeds would also adapt to that practice. The key is not changing to one favored practice, but combining many different practices into a diverse cropping system to minimize weed niches (Dekker, 1999). Fall seeding allows crop managers to prevent continued weed adaptation by shifting the canola life cycle forward without a change in crop species.

Over the long-term, weeds might also adapt to repeated use of multiple-seeding dates (somewhat similar to multiple herbicide resistance). As is the case for any individual weed management practice, combining seeding date diversity with other weed management techniques would extend its term of effective use.

Operational diversity with potential for positive impacts on weed populations is also possible by harvest timing manipulation (Shirtliffe et al., 2000b). Figure 5 depicts spring wild oat populations after plots were subjected to various harvesting intervals (early-cut silage, normal-cut silage, and barley grain cut at maturity). Grain production in the absence of herbicides led to high wild oat populations over time. Hastening harvest timing reduced additions of viable wild oat to the soil seed bank and significantly reduced wild oat populations after two consecutive years of the treatment on the same plots (Harker et al., 2003b). Early-cut barley silage without herbicides led to wild oat populations as low or lower than normal-barley silage with half rates of herbicide, or lower than barley grain with full rates of herbicide. Because of the inherent genetic diversity of weed populations, it is certain that wild oat populations would adapt to earlier barley harvest should the practice be continually repeated. Weeds have even adapted to practices as seemingly foolproof as hand-weeding (Gould, 1991).

Figure 4. Stages of canola (*Brassica napus*) development and maturity when seeded in the fall of 1998 (November 5) and the spring of 1999 (April 29 and May 12). Photo taken on June 26, 1999, AAFC Lacombe Research Centre, Lacombe, Alberta, Canada (Experiment 99005).

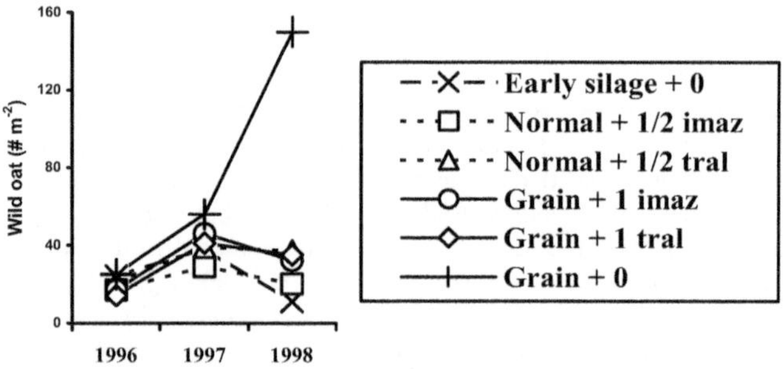

Figure 5. Effect of barley harvest dates on spring wild oat populations from 1996 to 1998 at Lacombe, Alberta, Canada. Barley grain was harvested at maturity. "Early" and "Normal" silage were harvested 1 week after heading (milk stage) and 1.5 to 2 weeks later at the soft dough stage, respectively. Half (1/2) or full (1) rates of two selective wild oat herbicides were included in some barley treatments (0 = no herbicide, imaz = imazamethabenz, tral = tralkoxydim). Treatments were applied to the same plots sequentially over the period of study.

ECOSYSTEM – LEAVING A FEW WEEDS

If repeated applications of herbicides such as glyphosate in glyphosate-resistant crops allow the removal of almost every weed in a field, and the crop represents 99.9% of the plant species in a given field, will other organisms in the ecosystem be affected? Should we be concerned about the diversity of weed flora (Cole, 1994)? What about interdependent avian, insect, soil macro fauna, and soil microbe food chains (Buhler et al., 2000; Taylor and Maxwell, 2001)? Are there weeds that some of these organisms require as a food substrate? What spectrum of organisms will remain when the only plant food substrate in large fields is essentially a single crop species (See Figure 6)? Are there microbial populations involved in nutrient cycling that will be adversely affected by lower ecosystem diversity? We know that legumes in rotations can have positive effects on soil microflora (Lupwayi et al., 1998), but should there also be species diversity in a given field? If ecosystem diversity is important in space and time (Cox and Wood 1999) should equal emphasis be placed in both areas? These questions are interesting, important, and, as yet, unanswered.

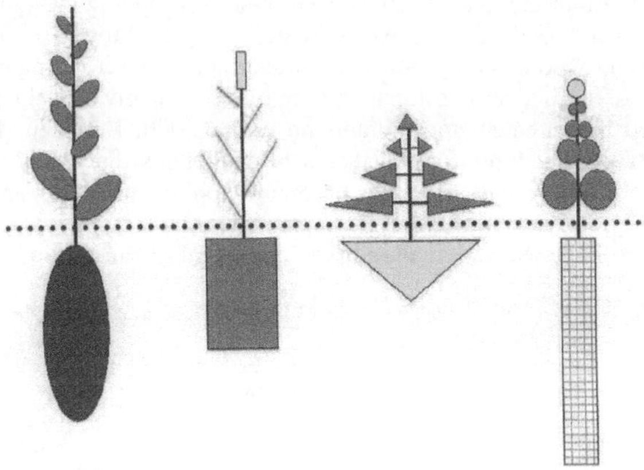

Figure 6. Simplified illustration of plants with diverse shoot and root systems. Diverse plant communities have shoots and roots (depicted by colour and shape) with unique attributes and composition. In a given area, shoots and roots of several species support a diverse community of plant and soil macro- and micro-fauna. The latter makes it more difficult for any single species of pest (fungi, bacteria, insect, nematode, weed, etc.) to dominate in the agroecosystem.

In support of "leaving a few weeds", we cite a recent canola study. Root maggots can infest and reduce yields in central Alberta canola fields (Dosdall et al., 1996). In several canola experiments we have observed less root maggot damage when weeds were removed at the 6-leaf as opposed to the 2-leaf stage of canola (Dosdall et al., 2003). Apparently, mated, gravid females will not lay eggs on host plants until they have made an average of four consecutive spiral flights into the crop canopy without landing on a non-host plant (Kostal and Finch, 1994). "Inappropriate landings" on weedy non-hosts probably reduced root maggot egg laying (Finch and Collier, 2000). It will be important to determine how many weeds are required and how long they need to remain in the canopy to significantly thwart egg laying. It will also be important to determine the extent of the compromise from leaving weeds in the crop in terms of crop losses due to early weed competition (Clayton et al., 2002; Harker et al., 2003a; Martin et al., 2001). These observations on root maggots clearly indicate that it is imprudent to implement weed management solutions without considering disciplines beyond weed science.

HIGHER SUCCESSION – TOWARDS INTERCROPS

Tolerating some weeds in a crop canopy has parallels to the practice of intercropping. Savory (1988) suggested that high weed population fluctuations (weed outbreaks) are characteristic of low successional communities (see Figure 7). Similarly, Dekker (1999) suggested that crop and weed management systems that leave few resources unexploited are more likely to have smaller, more easily managed seed banks. Generally, Weed Scientists have done little to exploit the effects of higher successional communities on weeds. Indeed, few have looked beyond simple monoculture cropping systems for weed management solutions. Admittedly, it is probably impossible to employ high (as opposed to higher) successional communities in our current large-scale cropping systems. Agriculture itself pushes plant community succession to early stages (Altieri, 1988). But we may be able to introduce some compromises. For example, intercropping mimics some of the attributes of higher successional communities in a given field and can be an effective weed management tool (Baumann et al., 2000; Liebman, 1988; Liebman and Davis, 2000). But intercropping leads to challenges that many do not want to face when the alternative is a simple herbicide application. Researchers and crop producers will look for new, and perhaps more challenging, weed management solutions as herbicide use and availability becomes more constrained.

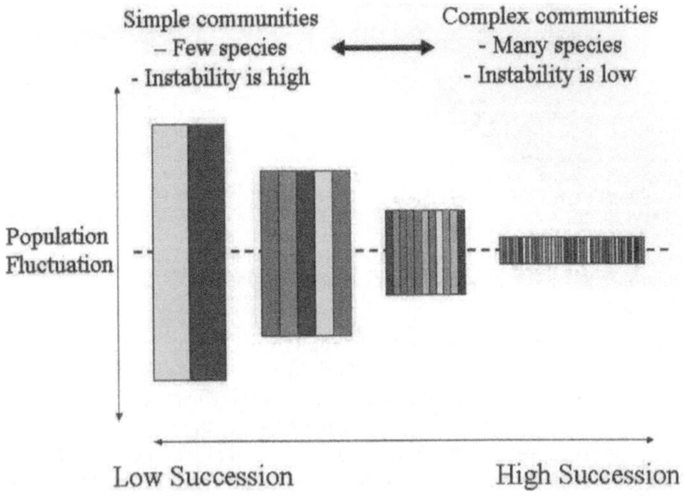

Figure 7. Illustration of the relationship between species diversity, succession and relative population stability in ecosystem communities. Species are represented by different coloured boxes and communities are represented by groups of boxes. Given a high degree of species diversity, there is little opportunity for a particular weed or other pest species to dominate the community. The illustration is highly simplified; a more realistic reflection of a very simple natural community would probably be much more diverse than the most complex community suggested in the diagram. Adapted from Savory (1988).

Despite the above, caution must be exercised in uncritically assuming that increased biodiversity will always increase stability in agroecosystems. Lenne and Wood (1999) point out that "simple communities may not be unstable". Furthermore, they suggest that ecological assumptions on diversity-stability, including those on pest management strategies, have not been borne out by experiments, observation, or models.

SUMMARY

Herbicide resistance, urban pressure to reduce herbicide use, and high input costs may push growers to adopt more diverse weed management systems to a greater degree than they, or many weed researchers, are comfortable with or prepared for. Research on higher diversity weed management systems is imperative. But systems studies require much longer time periods than more simple research projects (Mortensen et al., 2000).

There is a need for a greater knowledge of weed biology and ecology (Buhler et al., 2000; Mortensen et al., 2000). With this knowledge, we will be positioned to exploit the benefits of biological resources in cropping systems for our gain (Liebman and Davis, 2000). A focus on the tactical use of several tools (Boerboom, 1999; Liebman and Gallandt, 1997) versus individual tools, and on long-term population management versus immediate eradication, would increase the integration level of weed management (Cardina et al., 1999). Walker and Buchanon (1982) suggested that not only must weed management have a broader focus than weed control alone, but that it must be integrated with all other crops and crop production practices that influence the ecosystem. Diverse weed management systems will also ensure greater longevity for our most effective weed management tools - herbicides (Buhler, 1999; Llewellyn and Powles, 2001). We cannot depend on the continued release of new herbicide technology. Social, economic, and technological constraints have made new herbicide development difficult (Buhler et al., 2000).

Despite where weed management systems are at present, there are indications that the adoption of higher diversity systems will be greater in the future (Lenne and Wood, 1999). Issues such as herbicide resistance have forced growers to look beyond herbicide technology for weed management solutions. In Canada, a shift in public attitude from "environmental awareness to environmental action" (Swanton and Weise, 1991), may also force greater adoption of alternative weed management strategies in the future. The latter is also a common trend in many other nations.

Refocusing on the cultural systems that suppressed weeds before herbicides were available will help us move towards higher diversity systems which favour biological as well as technological inputs. It is no surprise that many of the cultural options important for integrated or alternative weed management systems (Boerboom, 1999; Nalewaja, 1999; Thill et al., 1994) mirror practices that are advocated for delaying or managing weed resistance to herbicides. We must be careful to consider all disciplines and even all former weed management practices as we face new challenges with weeds. Weeds fortunate enough to grow in simple, repeated cropping systems will continue to have little difficulty adapting and thriving in those systems.

REFERENCES

Altieri M.A. "The impact, uses, and ecological role of weeds in agroecosystems." In *Weed Management in Agroecosystems: Ecological Approaches*, M.A Altieri., M. Liebman, eds. Boca Raton, FL: CRC Press, 1988; pp.1-6.

Baumann D.T., Kropff M.J., Bastiaans L. Intercropping leeks to suppress weeds. Weed Res 2000; 40:359-374.

Beck D. No-till guidelines for the arid and semi-arid prairies. 2002; [Online] Available: http://www.dakotalakes.com/publications.htm [25 Mar 2002].

Beckie H.J., Thomas A.G., Legere A., Kelner D.J., Van Acker R.C., Meers S. Nature, occurrence, and cost of herbicide-resistant wild oat (*Avena fatua*) in small-grain production areas. Weed Technol 1999; 13:612-625.

Blackshaw R.E. Rotation affects downy brome (*Bromus tectorum*) in winter wheat (*Triticum aestivum*). Weed Technol 1994; 8:728-732.

Boerboom C.M. Nonchemical options for delaying weed resistance to herbicides in Midwest cropping systems. Weed Technol 1999; 13:636-642.

Bourgeois L., Morrison I.N. Mapping risk areas for resistance to ACCase inhibitor herbicides in Manitoba. Can J Plant Sci 1997; 77:173-179.

Buhler D.D. "Expanding the context of weed management." In *Expanding the Context of Weed Management*, D.D. Buhler, ed. New York: The Haworth Press, 1999; pp. 1-7.

Buhler D.D., Liebman M., Obrycki J.J. Theoretical and practical challenges to an IPM approach to weed management. Weed Sci 2000; 48:274-280.

Cardina J., Webster T.M., Herms C.P., Regnier E.E. "Development of Weed IPM: Levels of Integration for Weed Management." In *Expanding the Context of Weed Management*, D.D. Buhler, ed. New York: The Haworth Press, 1999; pp. 239-267.

Cavers P.B. Seed banks: memory in soil. Can J Soil Sci 1995; 75:11-13.

Clayton G.W., Harker K.N., O'Donovan J.T., Baig M.N., Kidnie M.J. Glyphosate timing and tillage system effects on glyphosate-tolerant canola (*Brassica napus*). Weed Technol 2002; 16:124-130.

Cole D.J. 1994. Introduction of herbicide-resistant crops. Pestic Outlook, June 1994; pp. 32-36.

Cox T.S., Wood D. "The nature and role of crop biodiversity". In *Agrobiodiversity: Characterization, Utilization and Management*, D. Wood, J.M. Lenne eds. New York, NY: CABI Publishing, 1999; pp. 35-57.

Dekker J.H., Burmester R.G. Pleiotropy in triazine-resistant *Brassica napus*. Ontogenetic and diurnal influences on photosynthesis. Plant Physiol 1992; 100:2052-2058.

Dekker J. Weed diversity and weed management. Weed Sci 1997; 45:357-363.

Dekker J. "Soil weed seed banks and weed seed management." In *Expanding the Context of Weed Management,* D.D. Buhler, ed. New York: The Haworth Press, 1999; pp. 139-166.

Derksen D.A., Harker K.N., Blackshaw R.E. Herbicide tolerant crops and weed population dynamics in western Canada. Proceedings of the Brighton Conference – Weeds; pp. 417-424, 1999.

Diesenhofer J., Michel H. The photosynthetic reaction center for the purple bacterium *Rhodopseudomonas viridis*. Science 1989; 245:1463-1473.

Dosdall L.M., Herbut M.J., Cowle N.T., Micklich T.M. The effect of seeding date and plant density on infestations of root maggots, *Delia* spp. (Diptera: Anthomyiidae), in canola. Can J Plant Sci 1996; 76:169-177.

Dosdall L.M., Clayton G.W., Harker K.N., O'Donovan J.T., Stevenson F.C. Weed control and root maggots: making canola pest management strategies compatible. Weed Sci 2003; 51:576-585.

Gould F. The evolutionary potential of crop pests. Am Sci 1991; 79:496-507.

Finch S., Collier R.H. Host-plant selection by insects – a theory based on "appropriate/inappropriate landings" by pest insects of cruciferous plants. Entomol Exp Appl 2000; 96:91-102.

Harker K.N., Baron V.S., Chanasyk D.S., Naeth M.A., Stevenson F.C. Grazing intensity effects on weed populations in annual and perennial pasture systems. Weed Sci 2000; 48:231-238.

Harker K.N., Clayton G.W., Blackshaw R.E., O'Donovan J.T., Stevenson F.C. Seeding rate, herbicide timing and competitive hybrid contribute to integrated weed management in canola (*Brassica napus).* Can J Plant Sci 2003a; 83:433-440.

Harker K.N., Kirkland K.J., Baron V.S., Clayton G.W. Early-harvest barley (*Hordeum vulgare*) silage reduces wild oat (*Avena fatua*) densities under zero tillage. Weed Technol 2003b; 17:102-110.

Heap I.M., Murray B.G., Loeppky H.A., Morrison I.N. Resistance to aryloxyphenoxypropionate and cyclohexanedione herbicide in wild oat (*Avena fatua*). Weed Sci 1993; 41:232-238.

Heap I. International survey of herbicide resistant weeds. 2002; [Online] Available: http://www.weedresearch.com/in.asp [25 March 2002].

Hilton H.W. Herbicide tolerant strains of weeds. Hawaiian Sugar Plant. Assoc. Annual Report 69., 1957.

Hirschberg J.A., Bleeker D.J., Kyle L. McIntosh L., Arntzen C.J. The molecular basis of triazine resistance in higher plant chloroplasts. Z Naturforsch 1984; 39c:412-419.

Holt J.S. Impact of weed control on weeds: new problems and research needs. Weed Technol 1994; 8:400-402.

Kirkland K.J., Johnson E.N. Alternative seeding dates (fall and April) affect *Brassica napus* canola yield and quality. Can J Plant Sci 2000; 80:713-719.

Kostal V., Finch S. Influence of background on host-plant selection and subsequent oviposition by the cabbage root fly (*Delia radicum*). Entomol Exp Appl 1994; 70:153-163.

Kropff M.J., Walter H. EWRS and the challenges for weed research at the start of a new millennium. Weed Res 2000; 40:7-10.

Lenne J.M., Wood D. "Optimizing biodiversity for productive agriculture." In *Agrobiodiversity: Characterization, Utilization and Management*, D. Wood, J.M. Lenne, eds. New York, NY: CABI Publishing, 1999; pp. 447-470.

Liebman, M. "Ecological suppression of weeds in intercropping systems: a review." In *Weed Management in Agroecosystems: Ecological Approaches,* M.A. Altieri, M. Liebman, eds. Boca Raton, FL: CRC Press, 1988; pp. 197-212.

Liebman M., Davis A.S. Integration of soil, crop and weed management in low-external-input farming systems. Weed Res 2000; 40:27-47.

Liebman M., Gallandt E.R. "Many little hammers: ecological management of crop-weed interactions." In *Ecology in Agriculture*, L. E. Jackson, ed. San Diego, CA: Academic Press, 1997; 291-343.

Llewellyn R.S., Powles S.B. High levels of herbicide resistance in rigid ryegrass (*Lolium rigidum*) in the wheat belt of western Australia. Weed Technol 2001; 15:242-248.

Lupwayi N.Z., Rice W.A., Clayton G.W. Soil microbial diversity and community structure under wheat as influenced by tillage and crop rotation. Soil Biol and Biochem 1998; 30:1733-1741.

Machado V.S., Beversdorf W., Switzer C. Potential for breeding triazine resistance into *Brassica* SPPS herbicide tolerance. Proceedings of the 1980 British Crop Protection Conf. – Weeds. pp. 855-860, Nov. 17-20, Brighton, UK, 1980.

Martin S.G., Van Acker R.C., Friesen L.F. Critical period of weed control in spring canola. Weed Sci 2001; 49:326-333.

Mortensen D.A., Bastiaans L., Sattin M. The role of ecology in the development of weed management systems: an outlook. Weed Res 2000; 40:49-62.

Nalewaja J.D. Cultural practices for weed resistance management. Weed Technol 1999; 13:643-646.

O'Donovan J.T., Newman J.C. Manipulation of canola (*Brassica rapa*) plant density and herbicide rate for economical and sustainable weed management. Proceedings of the 2nd International Weed Control Congress, Copenhagen, vol. 3:969-974, 1996.

Ryan G.F. Resistance of common groundsel to simazine and atrazine. Weed Sci 1970; 18:614-616.

Savory A. "Succession." In *Holistic Resource Management*, A. Savory, ed. Washington, DC: Island Press, 1988; pp. 73-89.

Shirtliffe S.J., Entz M.H., Van Acker R.C. Chaff collection reduces mechanical dispersal of wild oat. Weed Sci Soc Am Abst 38:39, 2000a.

Shirtliffe S.J., Entz M.H., Van Acker R.C. *Avena fatua* development and seed shatter as related to thermal time. Weed Sci 2000b; 48:555-560.

Swanton C.J., Weise S.F. Integrated weed management: the rationale and approach. Weed Technol 1991; 5:657-663.

Switzer C.M. The existence of 2,4-D resistant strains of wild carrot. Proc North Eastern Weed Control Conf 11:315-318, 1957.

Taylor R.L., Maxwell B.D. Indirect effects of herbicides on avian food resources and beneficial arthropods. Proceedings of the West. Soc. Weed Sci. Vol. 54, Abstract #71, page 49, 2001.

Thill D.C., Lish J.M., Callihan R.H., Bechinski E.J. Integrated weed management – a component of integrated pest management: a critical review. Weed Technol 1991; 5:648-656.

Thill D.C., O'Donovan J.T., Mallory-Smith C.A. Integrated weed management strategies for delaying herbicide resistance in wild oats. Phytoprotection (Suppl.) 1994; 75:61-70.

Walker R.H., Buchanan G.A. Crop manipulation in integrated weed management systems. Weed Sci (Suppl.) 1982; 30:17-24.

Wolf T.M., Harrison S.K., Hall F.R., Cooper J. Optimizing postemergence herbicide deposition and efficacy through application variables in no-till systems. Weed Sci 2000; 48:761-768.

Wyse D.L. Future of weed science research. Weed Technol 1992; 6:162-165.

CHAPTER 13

SOIL-IMPROVING PRACTICES FOR ECOLOGICAL WEED MANAGEMENT

ERIC R. GALLANDT

Department of Plant, Soil and Environmental SciencesUniversity of Maine, Orono, ME 04469-5722, USA

Weed management in industrial agroecosystems continues to rely heavily on the use of herbicides to minimize weed densities within a given crop growth period. Despite widespread contamination of surface and ground waters, human health concerns, and increasing incidence of herbicide resistance that have emerged over the past twenty years, and increasing research on the ecology of agricultural weeds, herbicide use remains pervasive. Initially, efforts to reduce reliance on herbicides focused on improving efficiency by reducing rates of active ingredient applied and substitution of herbicide applications with cultivation. Efforts to further improve the efficiency of herbicide- and cultivation-based weed management systems has prompted considerable research efforts describing weed density-crop yield loss relationships, providing threshold information which could guide decisions related to the implementation of these practices. Despite their contribution to reducing the herbicide load in agricultural systems, these strategies perpetuate a myopic perspective of weed management - the use of purchased inputs and equipment to reduce weed seedling densities.

Ecologically-based weed management recognizes that weeds are affected by, and respond to, a multitude of stresses within a cropping system (Liebman and Gallandt, 1997; Liebman et al., 2001). While conventional weed management efforts focus largely on weed seedling mortality, there are several other points within a weed's life cycle during which stresses may be imposed. Individually, it is unreasonable to expect that manipulation of a single important ecological process would be able to provide a level of weed control acceptable to a grower. However, the integration of ecological stresses at multiple points within a weed's life history, through cumulative and possible synergistic effects, offers opportunity to minimize reliance on herbicide applications and cultivation - moving these tactics from the first to the last line of defense.

Considerable progress has been made in manipulating the canopy environment of weeds and crops to the advantage of the latter. Competitive cultivars may be chosen. The crop density and spatial arrangement may be optimized to preempt weed access to light resources. Likewise, fertilizers may be selectively placed near the crop, a manipulation of the soil environment designed to disadvantage weeds. In the sections that follow, this approach is expanded to consider practices that may be used to improve or maintain soil quality. Aiming to develop agronomic practices congruent with both long-term farm sustainability and short-term needs for managing weeds and

Inderjit (ed.), Weed Biology and Management, 267-284.

maintaining crop yields, I start with the assumption that sustainable farming systems must maintain, and better yet improve, the quality of their soil resources.

Defined by Karlen et al. (1997), soil quality refers to the capacity of a soil to sustain plant and animal productivity, maintain or enhance water and air quality, and support human health and habitation. The principle management practices employed to maintain or improve soil quality include foremost a reduction in tillage-related soil disturbance. Because tillage encourages oxidation of soil organic carbon and destruction of soil aggregates, reduced tillage systems retain and thus accumulate soil organic matter. Furthermore, reducing tillage maintains macropores and surface residues which aide, respectively, in water infiltration and reducing soil erosion. Secondly, efforts are made to increase organic inputs, usually in the form of crop residues, composts, animal manures, municipal or industrial wastes. These inputs build and maintain soil carbon pools in various states of decomposition that contribute to the soil microbial activity and hence nutrient cycling, formation of aggregates, and ultimately the general improvement in "tilth" associated with high quality soils.

Particular tillage and amendment practices used may directly affect weed dynamics (Forcella et al., 1994). A change from moldboard plowing to no-till, for example, may change the structure of the weed community, encouraging proliferation of perennial and small-seeded annual grass species (Forcella et al., 1994). Densely planted cover crops protect the soil from erosion, are often very competitive with weeds, reducing weed biomass and seed production relative to areas where cover crops are not being grown. Raw animal manures often contain a considerable quantity of viable weed seed (Mt. Pleasant and Schlather, 1994). Furthermore, limited evidence suggests that resultant changes in soil quality may also affect weed dynamics, e.g., improving the ability of the crop to compete with weeds (Gallandt et al., 1998a). Based on examples like these, we previously identified three life-history processes, critical to the dynamics of annual weeds, that may be manipulated through practices used to improve soil quality and/or the resultant changes in the soil surface environment: (1) seed bank survival, (2) establishment, and (3) interspecific interference (Gallandt et al., 1999). These life-history processes, and the soil-quality-improving management practices that offer opportunity for manipulating them, are the focus of the discussion that follows.

CONSERVATION TILLAGE

Conservation tillage includes a wide range of agronomic practices that share a single effect - they result in 30% residue covering the soil after planting. In 1996 nearly 36% of U.S. planted acreage was under conservation tillage, a level similar to the previous couple of years, but a considerable increase over 26% of the 1989 acreage (Uri, 1999). Encouraged in response to greater understanding of the scope and impact of soil erosion, both on- and off-farm, planting strategies using less soil disturbance provide further benefits including increased soil organic matter, improved soil structure, and sequestration of greater levels of carbon (Uri, 1999). The use of particular practices to achieve this goal varies by region but includes, minimum- or reduced-tillage, ridge-till, strip- or zone-till, and zero- or no-till or direct seed cropping practices (Baker et al., 1996).

Conservation tillage systems generally have increased reliance on herbicides (Forcella et al., 1994), the magnitude of which is likely to be inversely related to the degree of soil disturbance. It is possible, however, to reduce herbicide dosage when either weed densities are low or weeds are small (e.g., DeFelice et al., 1989). Furthermore, there are circumstances where selective herbicides are either no longer effective (e.g., herbicide resistant wild oat, *Avena fatua*) or simply not available (e.g., jointed goatgrass, *Aegilops cylindrica*), which have inspired a search for management opportunities in the fundamental changes resulting from reduced disturbance. In particular, the concentration of weed seeds at the soil surface (Yenish et al., 1992), and the changes in the soil conditions within this strata, have motivated several research projects related to the ecological management of weeds in conservation, and in particular no-till systems. Seed predation, seed decay, and the probability of weed seedling establishment in conservation tillage systems, particularly no-till systems, are selected examples described below.

Seed Predation

Tillage operations, in general, have greater impact on larger than on smaller organisms (Kladivko, 2001). The macrofauna in this category includes both earthworms and carabid beetles, two important groups of invertebrate seed predators. Tillage also affects post-dispersal weed seed predation which may be responsible for substantial losses to the seed bank. Seed predators consumed 2.5 times more seeds in no-till soybean compared to a clean-tilled control over a five-week period (Brust and House, 1988). In Ontario, Cromar et al., (1999) found similar levels of seed predation in no-till and moldboard plowed treatments (32%), both being greater than chisel-plowed treatments (24%). Factors other than ground cover and the degree of soil disturbance apparently affect the foraging behavior of the ground-dwelling invertebrates that were responsible for greater than 75% of the predation measured. Based on published estimates of barnyardgrass (*Echinochloa crus-galli*) seed rain and potential predation rates measured in these experiments (2.9% per day), and conservative estimates of favorable conditions for predator activity, 82% of the weed seeds produced would be consumed by seed predators (Cromar et al., 1999). Germination generally accounts for the greatest decline in the weed seed bank. However, rates of loss by this mechanism are often dependent on satisfying complex dormancy requirements. Other causes of seed bank decline include seed mortality caused by extremes in environmental conditions (Lonsdale, 1993), age-related embryo death (Copeland and McDonald, 1995), and decay by microorganisms (Kremer, 1993). Combined mortality rates for these processes, however, are likely to be considerably less than the predicted effects of seed predators supporting the conclusion of Cromar et al., (1999) that seed predation may be the preeminent broad spectrum form of biological weed control, deserving consideration in the design of multi-tactic weed management programs.

Seed Mortality

Surface-residing weed seeds in no-till systems could be advantageous in controlling weeds, from both chemical and biological control standpoints. Given greater microbial activity (Doran et al., 1996) and a high proportion of weed seeds in surface layer of soil in no-till (Yenish et al., 1992), it is possible that soil microbial agents could be especially effective in no-till systems. Pitty et al. (1987), for example, found that fungal colonization of foxtail (*Setaria* spp.) caryopses was related to crop residue placement in soil. In no-till plots, colonization occurred predominately on seeds near the soil surface whereas moldboard plowing and deeper residue incorporation resulted in greater colonization of seeds at lower depths. Because weed control is one of the major obstacles to a more widespread use of no-till system, relationships among residue management and seed deterioration need to be determined and integrated into weed management strategies. The potential for enhancing weed seed decay may be much greater under no-till, because surface applications of microbial inoculants and soil amendments can be placed in close proximity to the weed seeds. The higher moisture level of the no-till soil surface (Cardina et al., 1991) is also more conducive to the survival of applied biological control microbes.

To test the hypothesis that changes in the soil surface environment associated with no-tillage will encourage greater microbial decay of resident weed seeds, we conducted field experiments in long term (15 year) replicated treatments of conservation and no-till in eastern Washington. The long-term no-till plots, compared to those managed by reduced tillage (chisel/disk), contained greater total carbon, nitrogen, and electrical conductivity. Wild oat (a highly dormant genotype, AN265) seeds were placed in the field in the fall, near the time when natural seed shed would be complete, and recovered the following year. These field studies provided limited evidence in the support of our original hypothesis. In the first year of our studies wild oat mortality was similar in both no-till and reduced-till systems (31% and 33%, respectively). However, the proportion of dead seed was greater following 7 months burial (mean depth of 2.5 cm) in no-till (12.2%) compared to reduced tillage treatments (6.8%) when the experiment was repeated (Gallandt et al., unpublished data). Fungicide seed treatment effects were likewise inconsistent. Fungicide seed treatment reduced overwinter mortality (30% with fungicide vs. 37% untreated control; $P = 0.001$) in the first field season, whereas mortality was similar with and without seed treatment when the experiment was repeated. Because wild oat decay was variably affected by the improved soil quality at the surface of long-term no-till treatments, it is unlikely that accelerated seed decay will contribute to management efforts targeting this species in conservation tillage systems. Further seed decay research including a range of weed species and soil environments, however, is essential for a more complete understanding the processes affecting seed decay in soil (Kremer, 1998).

Safe-Sites and Weed Establishment

To successfully establish, seeds must experience conditions and resources that promote germination and avoid hazards to establishment - i.e., safe-sites are required (Naylor, 1985). Grower and researcher observations suggest that the abundance of safe-sites can be manipulated. Ridge-till guidelines from a successful Iowa farmer are to "pack seed with a small press wheel and cover seed with loose soil," reasoning that "... the loose soil will dry out and weed seeds will not germinate" (Thompson et al., 1995). Noting the importance of seed-soil contact, Donald and Ogg (1991) cited observations that jointed goatgrass emergence in wheat rows was greater in dry than wet years, and greater in wheel-tracked areas than in uncompacted soil.

Jointed goatgrass is particularly well-adapted to no-till systems because its germination and establishment is optimal at shallow depths (0 to 5 cm) (Morrow et al., 1982). "Safe-sites" within this vertical profile provide conditions and resources that promote germination and offer protection from hazards to establishment (Harper, 1977). Safe-sites can be thought of as an "environmental sieve" that determines final seedling densities (Naylor, 1985). It is likely that quantity of surface residues and disturbance are key factors regulating the abundance of jointed goatgrass safe sites, and that both of these factors are significantly affected by the type of drill used for direct seeding. Using the appropriate drill (e.g., resulting in the desired level of soil disturbance) to manipulate the amount of soil disturbance and residue incorporation could be central to the success of reducing weed establishment in direct-seed cropping systems.

These hypotheses were recently tested by sowing jointed goatgrass spikelets at a range of densities prior to seeding winter wheat using contrasting no-till drills. The experiment was established at two field locations on the same experimental farm in eastern Washington. One site that had no prior history of no-till management, the other had a seven-year history of no-tillage. Residue treatments were established in the spring of the year at which time plots were seeded to either spring pea (low residue) or spring wheat (high residue). After harvesting these crops the jointed goatgrass spikelets were distributed and winter wheat was subsequently seeded using either a high- or low-disturbance no-till drill (Yielder, USDA III, and Cross-Slot, respectively; see Baker et al., 1996). Residue had a strong influence on the rate of jointed goatgrass emergence, total emergence, and fecundity during the 1998-1999 season (Gallandt, E.R. and S. Early, unpublished data). Contrary to the original hypothesis, drill-related disturbance had little effect on jointed goatgrass recruitment. Jointed goatgrass establishment was less ($P < 0.05$) in high residue following spring wheat compared to low residue treatments (Figure 1). Of greater interest, however, are the apparent effects of time under no-till management - apparent because the contrasting tillage histories were not included in the same experiment, and lack of homogeneity of variance precluded our combining the data sets. Nevertheless, it is notable that the long-term no-till site had an average of 50% less total jointed goatgrass emerge, compared with the conventional tillage (first year of no-till) site (Figure 1). This effect was also noted when the experiment was repeated in these no-till plots and at a nearby conventional site with conventional history. The decreased level of recruitment suggests that, over time, no-till systems may acquire weed suppressive attributes related, presumably, to modifications in the soil surface environment.

COVER CROPPING

In the broadest sense, cover crops are grown specifically for soil improvement. Depending on the context and the species used, cover crops protect the soil from erosion, add organic matter, capture and recycle nutrients, provide residues (and thus a mulch layer) for weed suppression and water conservation. They may also serve as an emergency forage source in animal-based systems, and, in the case of leguminous species, as an on-farm source of nitrogen typically referred to as "green manures" (Sarrantonio, 1994). Improvements in characteristics such as organic matter content, abundance of water stable aggregates and microbial biomass confirm the contribution of cover cropping to improving soil quality (Doran et al., 1996). The contribution of cover crops to weed management may be a result of disturbance events associated with establishment or termination of growth, competition during the growth, physical and/or allelopathic effects of the cover crop residues, or through general improvements in soil fertility (Samson, 1991). The design of cropping systems with cover crops at various points in time contributes often needed diversity of crop competition and disturbance needed to apply multiple stresses against weed populations (Liebman and Gallandt, 1997).

Competitive Ability of Cover Crops

Many cover crop species are chosen for their rapid canopy development and therefore their ability to compete with weeds. Cereals, for example, oat (*Avena sativa*), are recognized as highly competitive relative to other crops (van Heemst, 1985). Winter rye (*Secale cereale*), buckwheat (*Fagopyrum esculentum*), *Brassica spp.*, and sorghum-sudangrass (*Sorghum bicolor* X *S. bicolor* var. sudanense) are further examples of cover crops common to temperate cropping systems that are considered to be exceptionally competitive (Sarrantonio, 1994). When grown as short season, summer cover crops, e.g., oat and buckwheat, weed growth is not likely a concern because growth may be terminated prior to the production of viable weed seed. Competitive ability is of primary importance for full-season cover crops chosen for biomass and nitrogen accumulation, e.g., pea (*Pisum sativum*), oat, and hairy vetch (*Vicia villosa*) mixtures (Jannink et al., 1996) in which mowing to prevent weed seed production is not an option. Intercropping mixtures of cereals with legume green manures such as red clover (*Trifolium pratense*) resulted in less weed biomass at the time of cereal harvest, late fall, and following spring regrowth (Liebman and Davis, 2000).

Mixtures of cover crop species may be sown in an attempt to combine desirable attributes. Interspecific mixtures, e.g., intercrops or polycultures, may exhibit over-yielding relative to their component species grown in monoculture due to more complete capture of available resources (Vandermeer, 1990). Furthermore, with greater resource capture by desirable components of the mixture, fewer resources are available to support weed growth (Liebman and Dyck, 1993). By extension, this concept has been investigated in efforts to improve the competitive ability of a growing cover crop (Creamer et al., 1997; Jannink et al., 1996). In a study of thirteen cover crop polycultures (21 species) in Ohio, four species - hairy vetch, barley, rye, and crimson clover - were determined to possess adequate competitive ability, adequate biomass production, the ability to overwinter, rapid establishment, and they were easily killed (Creamer et al., 1997). Likewise, subsequent field studies also conducted in Ohio demonstrated that rye was the essential weed suppressive component in mixtures of rye and pea; weed suppression was greatest with mixtures including 50% or more rye (Akemo et al., 2000).

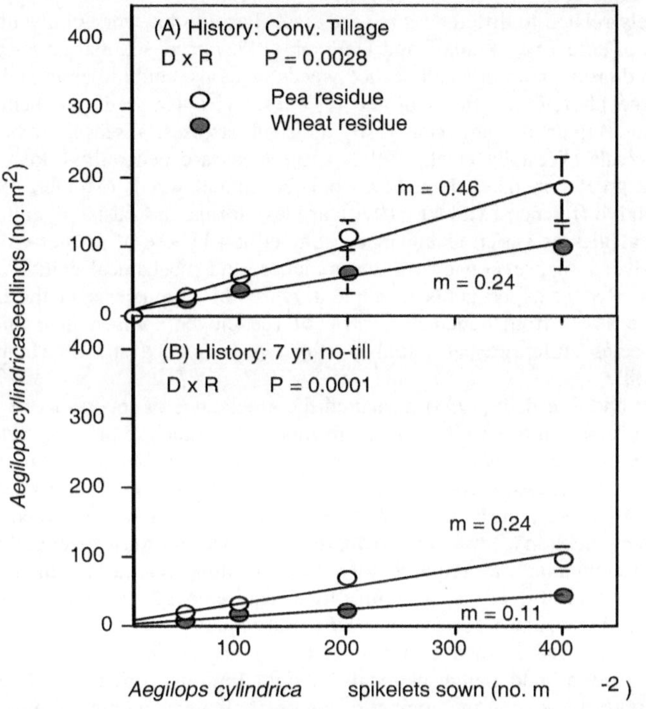

Figure 1. Probability (see slopes) of jointed goatgrass (*Aegilops cylindrica*) establishment in nearby fields with contrasting tillage histories. Pea (*Pisum sativum*) and spring wheat (*Triticum aestivum*) were sown in the spring of 1998. Following harvest of the spring crops jointed goatgrass spikelets were sown (randomly dispersed using a device which dropped them from a height of 1 meter) and winter wheat was seeded using high- and low-disturbance no-till drills. Spikelet density (D) by residue (R) effects are shown averaged over drill-related-soil-disturbance, which was not significant. Gallandt E.R., S.E. Early (unpublished data).

Weed Suppressive Residues

Surface Residues -Mulch Effects

Following termination of their growth, cover crop residues may be retained as mulch on the soil surface or incorporated into the soil surface. Compared to fallow treatments some level of weed suppression, generally measured as a reduction in establishment, is commonly observed in both systems although the former has received considerably more attention. In addition to reducing weed densities, cover crop residues may also, or alternatively, reduce the fitness of weeds relative to the crop (Williams et al., 1998). The selectivity between crops and weeds appears to be most closely related to differences in seed size. Early in his work on the allelopathic properties of cereal rye, Putnam and DeFrank (1983) proposed that large-seeded and transplanted crops, but not small-seeded weeds or crops, could tolerate and evidently outgrow the phytotoxic effects of the residues. Heavier seeds are better able to tolerate and repair damage caused by nutrient stresses, disease, herbivory, and allelochemicals (Westoby et al., 1996). Crop seeds are generally 1 to 3 orders of magnitude greater in mass than the associated annual weeds (Mohler, 1996). This fundamental difference, Mohler (1996) argues, forms the basis of cultural weed control strategies as varied as enhanced competition by use of greater crop seeding densities, weed suppression by residue mulches, and mechanical cultivation within crop rows. Recent experiments to explicitly test this hypothesis in the context of allelopathic stress from aqueous extracts of red clover confirm that small (light) seeded species suffer greater inhibition than larger-seeded species (Liebman and Davis, 2000).

Mohler and Teasdale (1993) measured the emergence of several weed species in response to increasing mass of surface residues of winter rye or hairy vetch. Both cover crops reduced densities of velvetleaf (*Abutilon theophrasti*), redroot pigweed (*Amaranthus retroflexus*), common lambsquarters (*Chenopodium album*), witchgrass (*Panicum capillare*), curly dock (*Rumex crispus*), common chickweed (*Stellaria media*), and dandelion (*Taraxacum officinale*), however, for most species, 2 to 4 times the natural field rate was required to reduce seedling emergence. In this type of system small-seeded weed species with a light requirement for germination were most sensitive to the residues (Teasdale, 1996). Subsequent work by these authors has shown that successful weed establishment through mulches is dependent on the ability to grow around mulch elements in light limiting conditions (Teasdale and Mohler, 2000). Increasing the number of layers of elements within the mulch and the proportion of mulch volume that was solid were features contributing to the reduction in weed establishment. Although winter rye and hairy vetch residues can reduce weed density early in the season, the level and duration of weed suppression is insufficient to eliminate the need for post emergence herbicides (Teasdale, 1996). Strategies are being tested to determine whether high residue cultivation equipment can replace herbicides in this system. Residues are being moved from the interrow space to the crop row thereby increasing the residue mass to 2 times the natural field rate in-row; rolling can then compact the residue and increase its density to maximize the weed suppressive effects of the mulch (Mohler and Teasdale, personal communication).

Incorporated Residues - Green Manures

The ubiquitous presence of glucosinolates in Brassicaceae has prompted considerable research related to the possible pest management contribution of cover crops within this family. Isothiocyanates derived from glucosinolates in the Brassicaceae are reported to have insecticidal, fungicidal, nematicidal, as well as phytotoxic properties (Brown and Morra, 1997). Boydston and Hang (1995) found weed density was reduced by 85 and 73%, and weed biomass reduced by 96 and 50% in two years, respectively, of potato grown following fall-sown rapeseed (*Brassica napus* 'Jupiter'). Although the level of weed control was commercially acceptable in only one of the two years, the practice could make an important contribution to a more comprehensive weed management program.

Because of their ability to contribute nitrogen to support the growth of subsequent crops, and their contribution to weed management, leguminous cover crops are particularly valuable in low-external-input cropping systems (Liebman and Davis, 2000; Liebman and Ohno, 1998). Moreover, rotations in which nitrogen fertility is supplied primarily by legume green manures accumulate organic nitrogen over time, suggesting that low carbon-to-nitrogen organic residues (i.e., legume cover crops) can contribute to the long-term improvement of soil quality (Drinkwater et al., 1998). In contrast to winter rye systems, in which the cover crops are grown to establish mulch to suppress weeds in a subsequent no-till crop, legume cover crops are often incorporated to facilitate decomposition and release of nitrogen.

Effects on Weed Establishment. Incorporated legume green manures have long been known to reduce establishment of subsequent crops (Fred, 1916); establishment of many small-seeded weed species is likewise frequently reduced. Residues and aqueous extracts of field-grown crimson clover and hairy vetch were phytotoxic to wild mustard (*Brassica kaber*), pitted morningglory (*Ipomoea lacunosa*), and Italian ryegrass (*Lolium multiflorum*) in laboratory and greenhouse bioassays (White et al., 1989). In the field, incorporated crimson clover (*Trifolium incarnatum*) increased the time to 50% emergence by 5 days and reduced common lambsquarters total emergence by 27% (Dyck and Liebman, 1994).

Red clover is inexpensive, grows well intercropped with spring cereals, and reliably over-winters in most northern U.S. cropping regions (Sarrantonio, 1994). Sampling field plots following monoculture spring wheat and a wheat/red clover intercrop, Ohno et al., (2000) bioassayed soil/water extracts and found that red clover amended soil reduced wild mustard radical elongation 20% at 8 days after residue incorporation. Bioassay results were most closely correlated with the concentration of soluble phenolic compounds in the extracts. In a separate but related series of experiments, Conklin et al., (2002) used a soil bioassay (see Dabney et al., 1996) instead of water extracts to study the effects of incorporated red clover residues on weeds. Consistent with the results of Ohno et al. (2000), soil amended with red clover residues reduced growth of wild mustard. Furthermore, this soil bioassay demonstrated that, during the first 40 days following incorporation, incidence of *Pythium* spp. infection was 10 to 70% greater in the amended soils compared to an unamended control (Figure 2). These results suggest that phenolic compounds, known to reduce membrane integrity (Balke, 1985), thereby increase the susceptibility of small seedlings to the flush of *Pythium* spp. growth that occurs following green manure incorporation (Rothrock and Kirkpatrick, 1995).

Field studies in which known quantities of several weed species were sown following incorporation of wheat stubble or wheat stubble + red clover residues, confirm that red clover residues may affect establishment of other weed species in addition to wild mustard. Red clover residue reduced establishment of redroot pigweed, velvetleaf, and wild buckwheat (*Polygonum convolvulus* L.) 34, 36, and 31%, respectively, compared to the control without red clover (Gallandt and Molloy, unpublished). The rate of emergence (i.e., time to 50% emergence) for each of these weed species was similar across residue and incorporation treatments. Contrasting methods of residue incorporation were compared in an attempt to modulate the residue-mediated effects of red clover. The typical technique used in our field studies, flail mowing followed by rototilling, was compared to incorporation of standing clover using a powered rotary spader (Imants Model 210 w/power harrow, Imants USA Inc., Perkiomenville, PA). The latter method being more consistent with goals related to reducing soil disturbance, but also likely to reduce the effects of red clover on weed establishment because residue decomposition would be delayed relative to the finely chopped and thoroughly incorporated treatments. Lack of significant interaction of residue type and method of incorporation affecting establishment of redroot pigweed, wild mustard, and wild buckwheat ($P = 0.200$, 0.252, and 0.182, respectively), failed to support the hypothesis that chopped, thoroughly incorporated residues (Rototilled) would be more suppressive than buried, coarse residues (Spader). Apparently the weed suppressive effects of legume green manures may, in part, be related to release of allelochemicals before incorporation. Blackshaw et al., (2001), for example, reported weed suppression by yellow sweetclover (*Melilotus officinalis*) was similar whether the cover crop residue was harvested for hay, left on the soil surface as a mulch, or incorporated.

It should be noted that cover crop mulch systems may not always suppress weed emergence (Moore et al., 1994). Likewise, incorporated residues may fail to suppress weed establishment. For example, results of Blum et al., (1997) from field studies in North Carolina are in direct contrast to generally consistent weed suppressive effects of red clover we have observed in Maine. Incorporated green residues of crimson clover, subterranean clover, rye, or wheat were variably found to stimulate emergence of pitted morningglory, redroot pigweed, and prickly sida (*Sida spinosa*) compared to a reference treatment without a cover crop. In rye-based mulch systems, inconsistent residue effects may be related to genotypic variation in allelochemical concentrations, which may differ by an order of magnitude (Burgos et al., 1999). Presumably such variation could occur in other commonly used cover crops as well.

Effects on Interspecific Interference. In addition to effects on establishment, green manuring may also affect the competitive ability of those weeds that successfully establish. This hypothesis was tested in a late-season sweet corn (*Zea mays*) crop grown following either an early spring-sown crop of crimson clover, or bare fallow (Dyck et al., 1995). Minimizing confounding effects of the green manure's nitrogen contribution, fallow treatments amended with 45 kg N ha^{-1} could be compared to the clover-amended plots determined to have a N equivalency value of 55 kg N ha^{-1}. Common lambsquarters biomass was 72% lower in the clover treatments 2 weeks after emergence, showing early season effects of the residue. At the final harvest, weed biomass remained 39% lower in the clover treatments compared to the fertilizer treatments. Although the sweet corn crop showed similar evidence of early season stress related to the clover amendments, it more fully recovered. Consequently, yield loss due to common lambsquarters interference was 8% in clover treatments compared to 28% in those receiving N fertilizer. These residue effects were more consistent and of a greater magnitude that in similar experiments in which a full season silage corn was grown following fall-sown and winter-killed crimson clover (Dyck and Liebman, 1995).

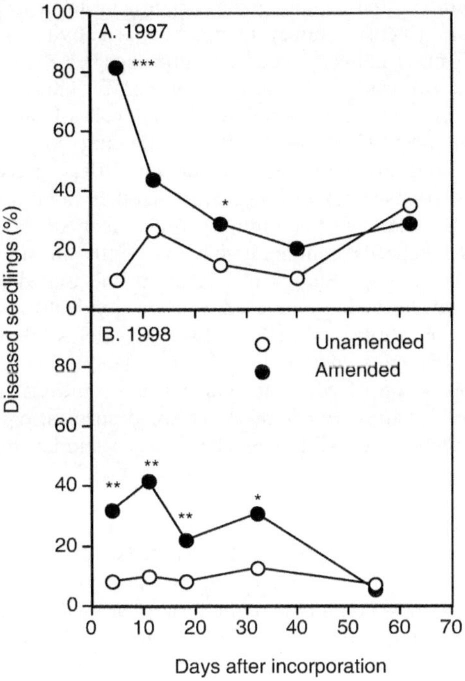

Figure 2. Incidence of disease, primarily *Pythium* spp., in wild mustard (*Brassica kaber*) seedlings in 1997 (A) and 1998 (B). Soil samples were collected prior to, and following, incorporation of wheat stubble (unamended) or wheat stubble plus red clover and compost (amended). At each sampling time, treatments designated *, **, or *** are significantly different (P = 0.05, 0.01, and 0.001 respectively). Adapted from Conklin et al. (2002).

In a series of experiments similar to those described above, dry bean (*Phaseolus vulgaris* L.) and wild mustard interference was studied following spring wheat either without amendment, with ammonium nitrate fertilizer, or incorporation of red clover green manure (Liebman and Gallandt, 2002). Wild mustard biomass was equivalent, 37% lower, and 53% lower than in the ammonium nitrate fertilized treatments in the three years of these experiments (Figure 3). Bean seed yield was similar in the red clover and fertilizer treatments in two of three years, and greater in the red clover treatment in one of three years (data not shown). Comparing organic sources of nitrogen (red clover green manure + compost) to ammonium nitrate fertilizer, Davis and Liebman (2001) likewise found wild mustard-related yield loss in sweet corn to be lower in the organic system or equal in the organic and fertilizer-based systems. Although leguminous green manures lack consistency in their effects on weed emergence and weed/crop interference, they offer sufficient additional benefits to be important components in sustainable cropping systems.

The ability to contribute at multiple points - nitrogen fertility, organic carbon and nitrogen cycling, reducing weed establishment, and reducing the competitive ability of weeds relative to crops - is likely an important consideration for growers, improving the likelihood that legume green manures be incorporated into the design of future cropping systems. However, because selectivity is largely based on differences in seed mass between weed and crop species (Mohler, 1996), residue-mediated weed control from cover crops will not fit into all cropping systems. Additional pitfalls include the potential for high levels of disease (Dabney et al., 1996) and/or insect pests (Hammond, 1990) following cover crop incorporation. Despite these potential problems, cover cropping offers greater opportunity for stressing weed populations at numerous points than other soil improving practices.

In situations where land is not limiting, full season cover crops are likely to contribute rotational benefits commonly observed only in animal-based cropping systems that include forages. Survey of farmers in the Canadian prairie with such systems indicated that, following forages, 83% of respondents observed weed control benefits for 2 or 3 years (Entz et al., 1995). Sampling fields following termination of a crop of alfalfa (*Medicago sativa*), Ominski et al., (1999), found, consistent with the survey results, suppression of wild oat, wild mustard, and Canada thistle (*Cirsium arvense*). However, densities of several other weed species, e.g., redroot pigweed, common lambsquarters, and wild buckwheat, were similar following cereals or alfalfa.

SOIL AMENDMENTS

Organic amendments, most commonly in the form of animal manures or compost, can lead to rapid and substantial improvements in soil quality. In a potato-based cropping system, for example, annual application of 22 metric tonnes ha^{-1} compost and 45 metric tonnes ha^{-1} *Bovine* manure increased soil organic matter by 0.4% following a single application; after four years the amended system's organic matter exceeded that of the unamended system by 1.4% (4.7% vs. 3.3%, respectively) (Gallandt et al., 1998b).

These soil management systems were part of a cropping system comparison established in 1991 to determine, among other questions, whether improvements in soil quality affected pest management. Improvements in soil quality did not benefit efforts to manage insects or diseases; rather, flea beetle populations were consistently greater in the amended system. However, in two of the first five years, weed biomass in plots managed without herbicides was lower in potato grown in the amended compared to the unamended soil management system (Gallandt et al., 1998a). Consistent with anecdotal reports from local organic farmers, the "healthy soil" apparently promoted a "healthy crop" better able to compete with weeds. Limited data available at the time supported this hypothesis. Potato leaf area was greater in the amended soil management system or equal in amended and unamended systems, and seed bank data suggested that the pool of potential recruits was similar in both systems (Gallandt et al., 1998b). However, because weed biomass was measured once, late in the growing season, the demographic distribution of competing weeds was not known.

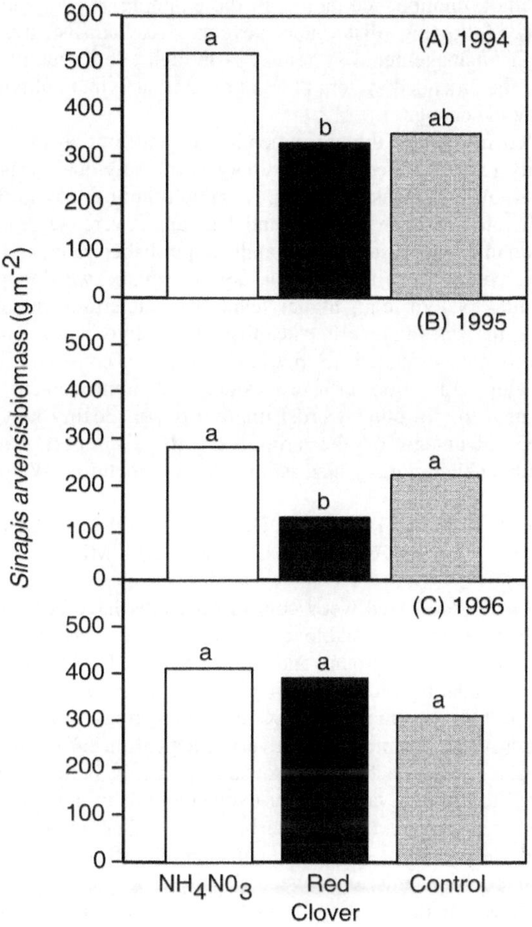

Figure 3. Wild mustard (*Sinapis arvensis)* biomass in 1994, 1995, and 1996 as affected by soil management. Within a year, columns not topped by the same letter are significantly different ($p < 0.05$). Adapted from Liebman and Gallandt (2002).

Growers have reported that improvements in soil quality dramatically improve the efficacy of cultivation (Anonymous, 1994). Thus, one explanation for reduced weed biomass in amended compared to unamended potato (Gallandt et al., 1998a) could be related to differential cultivation efficacy. In other words, soil management could affects the demographic profile of weeds competing with potato. Careful monitoring of common lambsquarters individuals (the dominant species in this experiment) within permanently marked quadrats, however, failed to support this hypothesis

(Figure 4; Gallandt, unpublished data). In the example shown, and in 3 subsequent years, mortality following disturbance events was consistently similar in the contrasting soil management systems. Although soil quality has improved considerably in the amended system there is no evidence that cultivation efficacy has been affected by the change.

Having failed to explain the soil amendment contribution to weed management based on density related effects, growth analysis of individuals was compared. The majority of parameters measured, e.g., common lambsquarters height, biomass, relative growth rate, nitrogen content, and leaf area, were generally similar in the amended and unamended systems (Gallandt, unpublished data). Although there are several further avenues of inquiry that could explain weed suppression in the amended cropping system, e.g., higher levels of deleterious rhizobacteria (Li and Kremer, 2000), the "healthy soil" "healthy plant" hypothesis may ultimately be supported. Mechanisms related to nutrient release (Davis and Liebman, 2001), rhizobacteria, and cultivation efficacy aside, soil improvement could generally improve plant growth. In other words, improving soil quality may simply enhance the initial size advantage of the crop, i.e., its disproportionate advantage in competition with smaller plants—size asymmetric competition (Weiner et al., 2001).

SOIL MANAGEMENT - THE FOUNDATION OF AN ECOLOGICAL APPROACH TO WEED MANAGEMENT

Alternatives to herbicide-based weed control programs have been inspired by rising costs of inputs coupled with stable commodity prices, and public concern over widespread contamination of ground and surface waters. Resulting legislation reflects these concerns, mandating reductions in pesticide usage and increase in integrated approaches to pest management. Lastly, in certain areas, herbicide-resistant weed populations are a further, immediate and very practical incentive for alternative weed management strategies. As a first step towards reducing herbicide reliance growers were encouraged to use cultivation either exclusively or in combination with banded and/or reduced rate application of herbicide. Further improving efficiency of herbicide-based weed control programs, considerable effort has been placed on the development of decision aids, e.g., weed density/crop yield loss relationships that could be combined with scouting information, costs of control, and projected revenues, to ensure efficient use of herbicides. Recognizing the limitations of these efforts focused on tweaking the efficiency of herbicide-based systems, an increasing number of researchers have been calling for strategies that modify the weed's environment placing stress at multiple points in it life cycle. Often referred to as ecologically-based weed management, efforts in this context are expanded beyond seedling mortality to reduce weed establishment, to increase the competitive ability of the crop, to minimize or attempt to eliminate weed seed production and dispersal, and to accelerate decline of the weed seed bank. While herbicides and cultivation are not necessarily eliminated from such a system, their preeminent role in weed management is diminished and they are used as a last rather than a first resort.

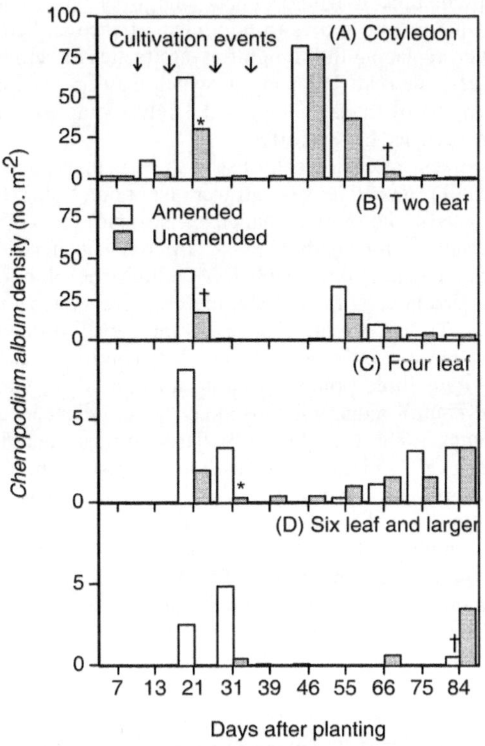

Figure 4. Common lambsquarters (*Chenopodium album*) density, by growth stage (A – D) recorded in permanently established quadrats within potato managed without herbicides, with fertility supplied by synthetic fertilizers (Unamended) or a combination of fertilizers, animal manure and compost (Amended). Gallandt E.R. (unpublished).

What tactics may be combined in an ecologically-based weed management system? Crop residues, soil disturbance, and timing of field operations can be managed to minimize weed establishment. Crop competition can be favored by choosing a competitive cultivar, optimizing the density and spatial arrangement of the crop, optimally timing delivery of resources based on the crop's demand, and, of course, by reducing weed density or delaying weed emergence relative to crop emergence using herbicides or cultivation. Enhancing the competitive ability of the crop will further contribute by reducing weed seed production. Depending on the coincidence of weed and crop seed maturity, post-harvest herbicide application, mowing, or tillage may also be used to minimize seed production. Although modeling results indicate that annual weed dynamics are greatly affected by seed bank mortality rates, other than disturbance to encourage germination, strategies to manage this key process are only recently being studied. Considering these many individual tactics, combined unpredictably over time and across the field environment in a diverse crop rotation, it would appear that we have a reasonable depth and breadth of theory and proven tactics to minimize reliance on herbicides (Liebman et al., 2001). However, replacing the straightforward tactics of herbicide application and cultivation with these alternatives, many of which may be accompanied with greater risk and greater inputs of management, will likely require incentives beyond what current data and experience have to offer.

Further development of ecologically-based weed management systems must keep in mind the need for incentives for grower adoption. Using the organization of MacRae et al., (1990) the initial changes - striving for greater efficiency and substituting cultivation for herbicides - offered financial benefits in most circumstances that are readily documented. At a higher level, however, ecologically-based weed management calls for redesigning cropping systems to make them more weed suppressive. To encourage change at this level ecologically based weed management must place a priority on tactics that contribute to the function of the cropping system at multiple points. Depreciating multiple, complex management efforts over weed control alone will not likely be sufficient incentive to encourage widespread adoption. Thus, ecologically based tactics should provide desirable ecosystem services, e.g., biodiversity, carbon sequestration, green spaces. They should enable growers to better pay themselves as high-level managers as they replace purchased inputs with intensified monitoring and decision making. The tactics must also provide multiple benefits to the cropping system, e.g., maintenance or improvement in soil quality, suppression of insect pests or pathogens, or contribution to fertility management by fixing atmospheric nitrogen or cycling excess nutrients in the field. The juncture of weed and soil management appears to offer the greatest opportunity to simultaneously address these goals.

REFERENCES

Akemo M.C., Regnier E.E., Bennett M.A. Weed suppression in spring-sown rye (*Secale cereale*) - pea (*Pisum sativum*) cover crop mixes. Weed Technol 2000; 14:545-49.

Anonymous. Healthy soil smooths early cultivation. The New Farm 1994; March/April:16-18.

Baker C.J., Saxton K.E., Ritchie W.R. *No-Tillage SeedingSe ience and practice*. Wallingford, UK: CAB International, 1996.

Balke N.E. "Effects of allelochemicals on mineral uptake and associated physiological processes." In *The Chemistry of Allelopathy*, A.C. Thompson, ed. Washington, D.C.:American Chemical Society, 1985; pp. 161-77.

Blackshaw R.E., Moyer J.R., Doram R.C., Boswell A.L. Yellow sweetclover, green manure, and its residues effectively suppress weeds during fallow. Weed Sci 2001; 49:406-13.

Blum U., King L.D., Gerig T.M., Lehman M.E., Worsham A.D. Effects of clover and small grain cover crops and tillage techniques on seedling emergence of some dicotyledonous weed species. Am J Alternative Agr 1997; 12:146-61.

Boydston R.A., Hang A. Rapeseed (*Brassica napus*) green manure crop suppresses weeds in potato (*Solanum tuberosum*). Weed Technol 1995; 9:669-75.

Brown P.D., Morra M.J. Control of soil-borne plant pests using glucosinolate-containing plants. Adv Agron 1997; 61:167-231.

Brust G.E., House G.J. Weed seed destruction by arthropods and rodents in low-input soybean agroecosystems. Am J Alternative Agr 1988; 3:19-25.

Burgos N.R., Talbert R.E., Mattice J.D. Cultivar and age differences in the production of allelochemicals by *Secale cereale*. Weed Sci 1999; 47:481-85.

Cardina J., Regnier E., Harrison K. Long-term tillage effects on seed banks in three Ohio soils. Weed Sci 1991; 39:186-94.

Conklin A.E., Erich M.S., Liebman M., Lambert D., Gallandt E.R., Halteman W.A. Effects of red clover (*Trifolium pratense*) green manure and compost soil amendments on wild mustard (*Brassica kaber*) growth and incidence of disease. Plant Soil 2002; 238:245-56.

Copeland L.O., McDonald M.B. *Seed Science and Technology*. New York: Chapman and Hall, 1995.

Creamer N.G., Bennett M.A., Stinner B.A. Evaluation of cover crop mixtures for use in vegetable production systems. Hortic Sci 1997; 32:866-70.

Cromar H.E., Murphy S.D., Swanton C.J. Influence of tillage and crop residue on postdispersal predation of weed seeds. Weed Sci 1999; 47:184-94.

Dabney S.M., Schreiber J.D., Rothrock C.S., Johnson J.R. Cover crops affect sorghum seedling growth. Agron J 1996; 88:961-70.

Davis A.S., Liebman M. Nitrogen source influences wild mustard growth and competitive effect on sweet corn. Weed Sci 2001; 49:558-66.

DeFelice M.S., Brown W.B., Aldrich R.J., Sims B.D., Judy D.T., Gluethle D.R. Weed control in soybeans (*Glycine max*) with reduced rates of postemergence herbicides. Weed Sci 1989; 37:365-74.

Donald W.W., Ogg Jr A.G. Biology and control of jointed goatgrass (*Aegilops cylindrica*), a review. Weed Technol 1991; 5:3-17.

Doran J.W., Sarrantonio M., Liebig M.A. Soil health and sustainability. Adv Agron 1996; 56:1-54.

Drinkwater L.E., Wagoner P., Sarrantonio M. Legume-based cropping systems have reduced carbon and nitrogen losses. Nature 1998; 396:262-65.

Dyck E., Liebman M. Soil fertility management as a factor in weed control: the effect of crimson clover residue, synthetic nitrogen fertilizer, and their interaction on emergence and early growth of lambsquarters and sweet corn. Plant Soil 1994; 167:227-37.

Dyck E., Liebman M. Crop-weed interference as influenced by a leguminous or synthetic fertilizer nitrogen source: II. Rotation experiments with crimson clover, field corn, and lambsquarters. Agric Ecosyst Environ 1995; 56:109-20.

Dyck E., Liebman M., Erich M.S. Crop-weed interference as influenced by a leguminous or synthetic fertilizer nitrogen source: I. Double cropping experiments with crimson clover, sweet corn, and lambsquarters. Agric Ecosyst Environ 1995; 56:93-108.

Entz M.H., Bullied W.J., Katepa-Mupondwa F. Rotational benefits of forage crops in Canadian prairie cropping systems. J Prod Agric 1995; 8:521-29.

Forcella F., Buhler D.D., McGiffin M.E. "Pest Management and Crop Residues." In *Crops Residue Management*, J.L. Hatfield, B.A. Stewart, ed. Boca Raton, Florida: Lewis Publishers, 1994; pp. 173-89.

Fred E.B. Relation of green manures to the failure of certain seedlings. J Agric Res 1916; 5:1161-76.

Gallandt E.R., Liebman M., Corson S., Porter G.A., Ullrich S.D. Effects of pest and soil management systems on weed dynamics in the Maine potato ecosystem project: 1991-1995. Weed Sci 1998; 46:238-48.

Gallandt E.R., Liebman M., Huggins D.R. Improving soil quality: implications for weed management. J Crop Prod 1999; 2:95-121.

Gallandt E.R., Mallory E.B., Alford A.R., Drummond F.A., Groden E., Liebman M., Marra M.C., McBurnie J.C., Porter G.A. Comparison of alternative pest and soil management strategies for Maine potato production systems. Am J Alternative Agr 1998; 13:146-61.

Hammond R.B. Influence of cover crops and tillage on sweet corn maggots (Diptera:Anthomyiidae) populations in soybeans. Environ Entomol 1990; 19:510-14.

Harper J.L. *Population Biology of Plants*. San Diego, CA: Academic Press, Inc., 1977.

Jannink J.L., Liebman M., Merrick L.C. Biomass production and nitrogen accumulation in pea, oat, and vetch green manure mixtures. Agron J 1996; 88:231-40.

Karlen D.L., Mausbach M.J., Doran J.W., Cline R.G., Harris R.F., Schuman G.E. Soil quality: a concept, definition, and framework for evaluation (a guest editorial). Soil Sci Soc Am J 1997; 61:4-10.

Kladivko E.J. Tillage systems and soil ecology. Soil Tillage Res 2001; 61:61-76.

Kremer R.J. Management of weed seed banks with microorganisms. Ecol Appl 1993; 3:42-52.

Kremer R.J. "Microbial interactions with weed seeds and seedlings and its potential for weed management." In *Integrated Weed and Soil Management*, J.L. Hatfield, D.D. Buhler, B.A. Stewart, eds. Chelsea, MI: Sleeping Bear Press, 1998; pp. 161-79.

Li J., Kremer R.J. Rhizobacteria associated with weed seedlings in different cropping systems. Weed Sci 2000; 48:734-41.

Liebman M., Davis A.S. Integration of soil, crop, and weed management in low-external-input farming systems. Weed Res 2000; 40:27-47.

Liebman M., Dyck E. Crop rotation and intercropping strategies for weed management. Ecol Appl 1993; 3:92-122.

Liebman M., Gallandt E.R. Differential responses to red clover residue and ammonium nitrate fertilizer by bean and wild mustard. Weed Sci 2002; 50:521-29.

Liebman M., Gallandt E.R. "Many little hammers: ecological approaches for management of crop-weed interactions." In *Ecology in Agriculture,* L.E. Jackson, ed. San Diego, CA: Academic Press, 1997; pp. 291-343.

Liebman M., Ohno T. "Crop rotation and legume residue effects on weed emergence and growth: implications for weed management." In *Integrated Weed and Soil Management*, J.L. Hatfield, D.D. Buhler, B.A. Stewart, ed. Chelsea, MI:Ann Arbor Press, 1998; pp. 181-221.

Liebman M., Mohler C.L., Staver C.P. *Ecological Management of Agricultural Weeds.* Cambridge, UK: Cambridge University Press, 2001.

Lonsdale W.M. Losses from the seed bank of *Mimosa pigra*: soil micro-organisms vs. temperature fluctuations. J Appl Ecol 1993; 30:654-60.

MacRae R.J., Hill S.B., Mehuys G.R., Henning J. Farm-scale agronomic and economic conversion from conventional to sustainable agriculture. Adv Agron 1990; 43:155-98.

Mohler C.L., Teasdale J.R. Response of weed emergence to rate of *Vicia villosa* Roth and *Secale cereale* L. residue. Weed Res 1993; 33:487-99.

Mohler C.L. Ecological basis for the cultural control of annual weeds. J Prod Agric 1996; 9:468-74.

Moore M.J., Gillespie T.J., Swanton C.J. Effect of cover crop mulches on weed emergence, weed biomass, and soybean (*Glycine max*) development. Weed Technol 1994; 8:512-18.

Morrow L.A., Young F.L., Flom D.G. Seed germination and seedling emergence of jointed goatgrass (*Aegilops cylindrica*). Weed Sci 1982; 30:395-98.

Mt. Pleasant J., Schlather K.J. Incidence of weed seed in cow (*Bos* sp.) manure and its importance as a weed source for cropland. Weed Technol 1994; 8:304-10.

Naylor R.E.L. "Establishment and peri-establishment mortality." In *Studies on Plant Demography: A Festschrift for John L Harper* , J. White, ed. London, UK:Academic Press, Inc., 1985; pp. 95-109.

Ohno T., Doolan K., Zibilske L.M., Liebman M., Gallandt E.R., Berube C. Phytotoxic effects of red clover amended soils on wild mustard seedling growth. Agric Ecosyst Environ 2000; 78:187-92.

Ominski P.D., Entz M.H., Kenkel N. Weed suppression by *Medicago sativa* in subsequent cereal crops: a comparative survey. Weed Sci 1999; 47:282-90.

Pitty A., Staniforth D.W., Tiffany L.H. Fungi associated with caryopses of *Setaria* species from field-harvested seeds and from soil under two tillage systems. Weed Sci 1987; 35:319-23.

Putnam A.R., DeFrank J. Use of phytotoxic plant residues for selective weed control. Crop Prot 1983; 2:173-81.

Rothrock C.S., Kirkpatrick T.L. The influence of winter legume cover crops on soil borne plant pathogens and cotton seedling diseases. Plant Dis 1995; 79:167-71.

Samson R.A., In *Fifth Annual REAP Conference* , Macdonald College, Ste-Anne-de-Bellevue, Quebec, 1991, pp. 11-22.

Sarrantonio M. *Northeast Cover Crop Handbook.* Emmaus, PA: Rodale Institute, 1994.

Teasdale J.R., Mohler C.L. The quantative relationship between weed emergence and the physical properties of mulches. Weed Sci 2000; 48:385-92.

Teasdale J.R. Contribution of cover crops to weed management in sustainable systems. J Prod Agric 1996; 9:475-79.

Thompson D., Thompson S., Thompson R. *Attributes in Agriculture.* Thompson On-Farm Research, Boone, IA, 1995.

Uri N.D. *Conservation Tillage in U.S. Agriculture - Environmental, economic, and policy issues*. New York, NY: Food Products Press, The Haworth Press, Inc., 1999.

van Heemst H.D.J. The influence of weed competition on crop yield. Agric Syst 1985; 18:81-93.

Vandermeer J.H. "Intercropping." In *Agroecology*, C.R. Carroll, J.H. Vandermeer, P. Rosset, eds. New York, NY:McGraw-Hill Publishing Co., 1990; pp. 481-516.

Weiner J., Griepentrog H., Kristensen L. Suppression of weeds by spring wheat *Triticum aestivum* increases with crop density and spatial uniformity. J Appl Ecol 2001; 38:784-90.

Westoby M., Leishman M., Lord J. Comparative ecology of seed size and dispersal. Phil Trans R Soc Lond 1996; 351:1309-18.

White R.H., Worsham A.D., Blum U. Allelopathic potential of legume debris and aqueous extracts. Weed Sci 1989; 37:674-79.

Williams M.M., Mortensen D.A., Doran J.W. Assessment of weed and crop fitness in cover crop residues for integrated weed management. Weed Sci 1998; 46:595-603.

Yenish J.P., Doll J.D., Buhler D.D. Effects of tillage on vertical distribution and viability of weed seed in soil. Weed Sci 1992; 40:429-33.

CHAPTER 14

WEED MANAGEMENT IN LOW-EXTERNAL-INPUT AND ORGANIC FARMING SYSTEMS

MATT LIEBMAN[1], LAMMERT BASTIAANS & DANIEL T. BAUMANN[3]

[1]Iowa State University, Department of Agronomy, Ames, IA, 50011-1010, USA; [2]Wageningen University, Department of Plant Sciences, Crop and Weed Ecology Group, P.O. Box 430, 6700 AK, Wageningen, The Netherlands; [3]Swiss Federal Research Station for Fruit-Growing, Viticulture, and Horticulture, CH-8820, Wädenswil, Switzerland

During the past quarter century, growing numbers of government policy makers, scientists, consumers, and farmers have expressed concerns over the impacts of conventional farming practices on environmental quality, human health, and the economic viability of farm families and rural communities. Particularly in western Europe, and to a certain extent in the USA, Canada, and other countries, these concerns have begun to translate into changes in public policy, research priorities, and market opportunities that favor the development of low-external-input (LEI) and organic farming systems. For example, during the 1980s in Sweden, a >50% reduction in agricultural pesticide use was mandated and achieved through coordinated sets of regulations, research and extension education activities, and economic incentives (Bellinder et al., 1994; Matteson, 1995). Similar approaches have been initiated in other European countries (Matteson, 1995). Concurrently, consumer demand for organic crop and livestock products has grown 20 to 25% per annum in the USA, many European countries, and Japan (Geier, 1998; Myers and Rorie, 2001). Sales of organic products in 2000 were estimated at $7.8 billion in the USA (Myers and Rorie, 2001) and at least $7.5 billion in Europe and the UK (Soil Association, 2001). Continued growth is expected in the amount of land managed with LEI and organic methods and in associated economic activity.

Weed management is a focal point for all farming systems, but it is especially critical in LEI and organic systems. The avoidance or exclusion of herbicides in these systems can lead to situations where weed density and weed competition against crops rise to unacceptable levels unless large amounts of mechanical cultivation or hand labor are used (Vereijken, 1994; Rasmussen and Ascard, 1995). Heavy reliance on both of these options is undesirable, however, since frequent cultivations can reduce soil quality and crop health, and hand labor for weeding generally has high cash and opportunity costs. If requirements for cultivation and hand weeding are to be minimized in LEI and organic farming systems, a better understanding of weed ecology and the full range of factors regulating weed density, growth, and competitive ability is needed (Liebman and Gallandt, 1997).

Inderjit (ed.), Weed Biology and Management, 285-315.

This chapter examines existing and emerging approaches for weed management in LEI and organic farming systems. After first discussing information requirements for weed management in such systems, tactics that can serve as components of multifaceted weed management strategies are described. Some use of herbicides may be included in LEI systems, but the emphasis in this chapter is on tillage options, soil fertility and moisture regimes, crop residue management practices, methods to improve crop competitive ability, cropping system diversification, and other tactics whose refinement is likely to make reductions in herbicide use more viable in LEI systems and to improve the performance of organic systems. Three case studies from Switzerland, the USA, and the Netherlands are presented that illustrate how combinations of tactics may be put to work against weeds in different LEI and organic farming systems. These case studies include a look at some of the shortcomings and remaining challenges of weed management. The chapter ends by identifying high-priority topics for future research in LEI and organic systems. The data and examples used in this chapter are drawn almost exclusively from temperate areas, with which we are most familiar, but we believe there are ample opportunities to develop LEI and organic systems in tropical regions as well.

INFORMATION REQUIREMENTS

Conventional weed management systems based on herbicides are designed for use in many different locations against broad spectra of weed species. In contrast, weed management systems that avoid or exclude the use of herbicides are more site-specific and require more careful consideration of the impacts of management tactics on individual weed species and weed community composition. Although general principles can be identified, actual weed management practices must be chosen, integrated, and adapted locally based on location-specific information.

Information requirements for the design of weed management strategies in LEI and organic farming systems occur at several organizational levels. First, information must be assembled concerning key components and functions of the agroecosystem that affect and are affected by weed management. Next, the abundance, distribution, and ecological characteristics of key weed species present within the agroecosystem must be determined. Analysis of this information should result in the identification of weed management priorities, after which different tactical options can be selected and tested with respect to their efficacy and suitability within the specific agroecosystem. After evaluation these strategies can be adjusted and refined.

Agroecosystem Characteristics

A large number of materials, organisms, and processes collectively define and determine the character of an agroecosystem. Within this complexity are certain key environmental factors, pieces of infrastructure, species, activities, and relationships that shape the context for developing and implementing weed management strategies. By studying the system-level context for weed management, key constraints and best opportunities for desirable changes can be identified by farmers and the researchers and extension professionals working with them. Conway (1985) and Conway and McCraken (1990) have described techniques for agroecosystem analysis in detail. The following factors are among those that could be included in agroecosystem analyses that have an explicit focus on weed management.

Soil Characteristics
Soil characteristics of different fields within a farm affect the timing of tillage and cultivation operations that kill, damage, or move weed plants and propagules. Soil characteristics also affect the availability of nutrients and water to crops and weeds, and the choice of crops within rotation sequences. *Seasonal climatic patterns* also determine when tillage and cultivation operations are possible, as well as the timing of crop and weed growth and resource use.

As discussed later, the *sequence of crops* grown on different fields and the associated crop management practices strongly influence opportunities for weed recruitment, growth, and reproduction. The types of crops that a farmer can consider producing are determined, in many cases, by the opportunities that exist for *marketing* them and by *government programs*, such as subsidy payments. LEI and organic products may be directed into special marketing channels and may receive price premiums in the marketplace or special government subsidies. Integration of crops with *livestock production* determines not only the types of crops grown on the farm and within rotation sequences, but also manure management practices that affect weed seed survival and dispersal within and among farms.

The *equipment* available for weed control and crop production affects the range of options for suppressing weeds and establishing and maintaining crop species. Equipment may be owned by the farm operator, or rented or borrowed within time constraints imposed by commercial firms and other farmers. *Labor* used for weed control and other farm operations may be provided by the farm family itself, by members of the surrounding community, or by contracted workers brought in from longer distances. Typically there are *labor bottlenecks* within the year during which labor requirements for weed management may conflict with labor needs for other on-farm and off-farm activities. Weed control options are also affected by the farmer's *financial situation*, access to *technical information*, interactions with *people with experience*, and short-term and long-term *goals*.

Often agroecosystem characteristics appear to be unchangeable and trivially obvious. However, closer analysis may reveal opportunities to implement desirable changes that were not immediately apparent. For example, analysis of yearly crop production patterns and climatic conditions may indicate when weed-suppressive cover crops could be introduced into rotation sequences. Successful management of a cover crop and its residue for weed suppression will depend on the availability of appropriate equipment, access to relevant information, impacts on additional pests, and other factors that may be identified by dialogue among farmers, scientists, extension personnel, agricultural consultants, and others.

Weed Abundance, Community Composition, and Ecological Characteristics

To develop and improve a weed management strategy for a given farm or set of farms, knowledge of key components and processes of the target agroecosystem must be complemented with information concerning weed abundance and community composition. Because the individual weed management tactics commonly used in LEI and organic farming systems typically are less effective and less persistent than herbicides, there is a particular need in these systems to match specific weed problems with sets of tactics applied at specific times.

Sampling to determine weed abundance and community composition serves additional purposes, which are not restricted to LEI and organic systems. Monitoring these parameters can indicate whether weed suppression tactics are working over short- and long-term horizons, whether changes in control tactics need to be initiated at particular locations on a farm or at particular times within a crop sequence, and whether "new," noxious species have arrived and need to be controlled before they become widespread. Accurate assessments of weed density also are needed for threshold-based management strategies, but the details and effectiveness of such strategies remain the subjects of considerable debate.

Methods to track weed abundance and community composition in farmers' fields range from hand-drawn maps produced by visually assessing the vegetation at one's boot-tip at several stops during a walk across a field, to computer-drawn maps produced by counting weeds within quadrats at numerous locations that are recognized by satellite-assisted global positioning systems. The choice of methods in a given situation will depend on the level of precision needed and the time, money, labor, and equipment that are available. Regardless of the actual method used, however, the dominant weed species within fields should be noted, the general patterns of their spatial distributions should be identified, and some repeatable estimate of their abundance (e.g., leaf area coverage of the soil, stem density, etc.) should be made.

After the major species of the weed community have been identified, key aspects of their ecology that affect their responses to management tactics need to be determined through discussions, library research, and experiments. Identification of temporal patterns of emergence, growth, and reproduction of dominant weed species is useful for assessing the potential impacts of control measures imposed at different times. Consideration of weed phenological patterns can indicate which species are likely to be suppressed by particular control methods and which species are likely to escape unless other control measures are used. Similarly, knowledge of species-specific responses to burial at different depths in the soil can indicate how long a given weed is likely to persist as a seed or vegetative structure on or in the soil, and whether it is likely to emerge as a seedling or vegetative sprout following a period of quiescence. Weed species also differ with regard to their responses to resource conditions and allelochemicals. Knowledge of these species-specific responses can suggest ways to reduce weed density and impair weed performance through management practices that affect soil fertility and moisture, light interception, and the quantity, quality, and location of crop residues and other organic matter amendments.

Biological and ecological data, in conjunction with information obtained from agroecosystem analyses and vegetation monitoring, create the knowledge base from which tactical options for weed management can be evaluated. Enlarging this knowledge base as tactics are tested under field conditions will encourage rapid development of weed management strategies that effectively attack vulnerable points in weed life cycles.

TACTICAL OPTIONS

As noted earlier, cultivation by machine or by hand is the primary curative approach for managing weeds in farming systems managed without herbicides. It is also the tactical option most often employed when herbicide use is reduced in LEI systems. Important innovations and improvements in cultivation equipment are now being made, particularly in Europe, and the effectiveness of mechanical tactics is increasing. Nevertheless, cultivation used alone as the major line of defense against weeds may still degrade soil, damage crops, and control weeds inconsistently. These shortcomings indicate the need for additional weed management tactics, particularly those that improve the prevention of future weed problems, rather than the short-term cure of existing problems (Jordan, 1996; Bastiaans et al., 2000).

Different tactical options affect weeds at different life stages: within the soil seed bank and during germination, seedling emergence, vegetative growth and resource consumption, reproductive growth, and dispersal. By combining tactics that suppress weeds at several life stages, long-term regulation of weed populations is likely to be more successful. Moreover, because of the diversity of physiologies, morphologies, and phenologies present in different weed species, combinations of diverse tactics are more likely than narrow strategies to be successful for managing multispecies weed communities. Here, we briefly summarize some of the important non-herbicide options available to LEI and organic farmers.

Tillage Practices

Tillage practices affect weed seed placement within the soil profile and physical characteristics of the soil (Cousens and Moss, 1990; Mohler, 1993). Weed seeds initially deposited on the soil surface are buried deepest by moldboard plowing, but this technique also brings back to the soil surface older, dormant seeds that previously had been buried deeply. In contrast, minimum- and zero-tillage practices retain weed seeds on or close to the soil surface. The temperature, atmospheric, and light conditions to which weed seeds are exposed are strongly altered when plowing disturbs soil; these conditions are more constant when tillage is minimized. In general, moderate to severe soil disturbances enhance the germination of most weed species (Mohler, 2001a).

Tillage effects on weed seedling density are determined by the combined impacts of weed seed placement within the soil profile, seed dormancy status at the time of burial, seed survival at different depths, germination responses to soil disturbance, and the ability of seedlings to emerge from different depths (Baskin and Baskin, 1989; Mohler, 1993). In general, weed seed survival decreases with proximity to the soil surface, whereas the probability that a newly germinated seedling will emerge from the soil increases with the seed's proximity to the soil surface, particularly for small-seeded species in disturbed soil (Mohler, 1993). Across species, the maximum seed burial depth from which a weed seedling can emerge successfully is correlated with seed mass: species with heavier seeds can emerge from greater depths in the soil (Roberts et al., 1982; Benvenuti et al., 2001).

Minimum-tillage systems can be either advantageous or disadvantageous for weed management, depending on whether soil conditions are adequate for weed germination and emergence, and whether new seeds are added to the soil seed bank (Mohler, 1993, 2001b). If weed seeds germinate, but replenishment of the seed bank is strongly restricted through management practices, annual weed populations are likely to decline more rapidly in minimum-tillage conditions than in plowed soil. Alternatively, if weed emergence and reproduction are relatively unfettered and seeds remain near the soil surface, weed populations will increase more rapidly under minimum-tillage conditions than in plowed soil.

In many LEI and organic farming systems, flexible tillage strategies may be the best choice for weed management (Mohler, 2001a). After seasons in which weed seed production is strongly suppressed, fewer weeds are likely to be present in the succeeding crop if minimum-tillage techniques are used. In contrast, after seasons in which weed seed production is not effectively suppressed, moldboard plowing can be useful for burying seeds that would otherwise germinate in the succeeding crop. A certain percentage of these deeply buried seeds will die before being returned to the soil surface by subsequent tillage.

Cussans and Moss (1982) modeled the effects of tillage on the population dynamics of blackgrass (*Alopecurus myosuroides* Huds.) and found that deep plowing at least every fifth year was required to counteract increases in blackgrass densities that occurred in years of zero-tillage. Their predictions closely match actual farmer practices for producing winter wheat in the UK (Cousens and Mortimer, 1995, p. 198). Similarly, Sakai et al. (1999) modeled the effects of different depths of tillage on densities of annual weeds and predicted that weed seedling density would be lowest when deep tillage (40 cm) was interspersed in alternate years with shallow tillage (15 or 20 cm depth). Alternation between tillage systems also varies selection pressures on perennial weeds, which are favored by minimum tillage systems and suppressed by plowing (Mohler, 2001a). Thus, over time, the interspersion of plowing with minimum- or zero-tillage is likely to prevent the build-up of weed species well-adapted to a particular tillage regime, as well as offer the farmer opportunities to address weed pressures that change due to variable weather conditions, conflicts with other farm activities, labor availability, and other factors.

Combinations of tillage practices are also possible in a single field in a single year through the use of "zone tillage" techniques in which soil is disturbed only in a narrow band where the crop is sown. Although the approach is not yet well developed, zone tillage may be useful for restricting soil disturbance and weed seedling emergence to only a small fraction of the field area, thus allowing cultivation and other direct weed management tactics to be concentrated spatially. For example, Forcella and Lindstrom (1988) and Exner and Thompson (1992) described a ridge tillage system in which early-season soil disturbance is confined to the truncation of soil ridges as crop seeds are sown into the ridges. Inter-row cultivation later in the season kills weeds between ridges and buries weeds emerging with the crop on ridges. Melander and Rasmussen (2000) experimented with a biennial cultivation system in which narrow bands of soil are repeatedly cultivated during and after the production of a cereal crop with the goal of depleting weed seed banks, and then, in a succeeding year, a crop such as sugarbeet is sown without further soil disturbance into the previously cultivated bands. Weeds between the bands can be controlled by harrowing. Approaches such as these offer important opportunities for wedding soil conservation with weed management and could be refined if they receive more attention from researchers and equipment manufacturers.

Soil Fertility and Moisture Management

Because of their fixed root habit, weeds (and crops) are highly responsive to soil chemical, physical, and biological conditions. Furthermore, because small advantages in early season seedling growth can translate into large differences in size and light interception ability later in the season, early responses to soil conditions can be critical for determining competitive interactions between weeds and crops.

Fertilizer application and irrigation are two common farming practices that affect soil resource conditions in ways that can be used for weed management. Using these practices to manipulate soil conditions for weed management depends on creating and exploiting differential responses between weeds and crops such that weeds are inhibited and crop performance is enhanced. However, because weed species differ in their adaptations to soil conditions, it should be expected that reliance on any one soil management regime would result in a change in weed community composition. Thus, a diversity of soil management techniques is needed to manage diverse weed communities.

Experiments conducted with synthetic fertilizers indicate that nutrients can be placed where they are available to crops, but less available to weeds (DiTomaso, 1995). For example, Rasmussen et al. (1996) found that banding nitrogen 5 cm below the depth at which barley was sown decreased weed biomass 55% and increased barley grain yield 28%, compared with broadcasting the same quantity of fertilizer on the soil surface. The investigators concluded that fertilizer placement close to the crop row favored the crop at the expense of the weeds. Similarly, injection of liquid pig or cow manure into soil at a depth of 8 to 10 cm reduced weed density and biomass and increased barley yield, relative to surface application of manure followed by shallow (3 to 4 cm depth) incorporation with a cultivator (Rasmussen, 2002).

The timing of fertilizer application can also be exploited for weed management. For some weed-crop combinations, crop growth and competitive ability are favored by delayed application of nutrients. Angonin et al. (1996) found that over a range of densities, biomass of ivyleaf speedwell (*Veronica hederifolia* L.) was reduced by half when N fertilizer was applied at the stem elongation stage of wheat development compared to when it was applied earlier, at the tillering stage. The weed had no effect on wheat N uptake and yield when N fertilizer was applied at stem elongation, whereas it reduced wheat N uptake and yield when fertilizer was applied at tillering. The investigators noted that growth, development, and potential competitive effect of ivyleaf speedwell were concentrated early in the growing season, and that delaying N application reduced the weed's competitive ability while benefiting the crop.

For other weed-crop combinations, early pulses of fertilizer may provide greater benefit to crops than later pulses. In experiments with winter wheat infested with downy brome (*Bromus tectorum* L.), Ball et al. (1996) found that delaying application of N fertilizer until spring had either no effect or a stimulatory effect on weed biomass production, and either no effect or a negative effect on wheat yield, compared with N application at planting. Nitrogen fertilizer increased wheat yield without increasing downy brome biomass only when it was applied during the fallow period preceding crop production. Anderson (1991) attributed this result to movement of fertilizer N into lower layers of the soil profile before growth of the crop and weed began, and to differences in their rooting habits: the shallow rooted weed was unable to extract fertilizer N at lower depths, whereas the deeper rooted crop gained access to that source of nutrients. Consideration of this case and that of wheat infested with ivyleaf speedwell leads to the conclusion that the impacts of different times of fertilizer application depend on the specific characteristics of the weed and crop species involved.

Animal manures offer important opportunities for recycling nutrients and improving soil physical, chemical, and biological properties, but they can also carry weed seeds into fields where they had not previously been present (Dastgheib, 1989; Mt. Pleasant and Schlather, 1994). Aerobic composting is one way to address this problem, although its effectiveness can be species-specific. Eghball and Lesoing (2000) reported that aerobic composting of dairy manure and beef cattle feedlot manure was entirely effective in killing seeds of common cocklebur (*Xanthium strumarium* L.), giant foxtail (*Setaria faberi* Herrm.), ivyleaf morningglory (*Ipomoea hederacea* (L.) Jacq.), redroot pigweed (*Amaranthus retroflexus* L.), and shattercane (*Sorghum bicolor* (L.) Moench), but killed only 83 to 96% of velvetleaf (*Abutilon theophrasti* Medicus) seeds. Both temperature elevation and exposure to chemical products of decomposition, such acetic acid, are responsible for weed seed mortality during the composting process (Shirilipour and McConnell, 1991; Grundy et al., 1998; Ozores-Hampton et al., 1999). Turning and mixing compost is necessary to prevent weed seeds on the outer surface of piles from escaping the thermal and chemical conditions responsible for seed mortality.

Irrigation water can serve as a vector for weed seeds (Kelley and Bruns, 1975; Wilson, 1980), but this problem can be reduced by grazing, mowing, or burning ditch banks to keep them free of weeds and by screening irrigation water to remove weed seeds before they enter crop fields (Mohler, 2001b). Certain irrigation techniques can deliver water to crops while minimizing its availability to weeds. Grattan et al. (1988) compared three irrigation methods for producing tomato under nearly rain-free conditions and found that subsurface drip irrigation greatly reduced weed growth and increased tomato production, compared to furrow and sprinkler irrigation. The former method concentrated moisture in the crop root zone, whereas the latter two methods dispersed water across the field surface, making moisture more accessible to weeds.

Irrigation timing can strongly affect the abundance and growth of weeds in rice fields. Moody and Drost (1983) reported that the density and biomass of Chinese sprangletop (*Leptochloa chinensis* (L.) Nees) were nil when rice was flood irrigated five days after seeding, but increased progressively to 72 plants m^{-2} and 46 g m^{-2}, respectively, as flooding was delayed until 20 days after planting. While these results suggest that early flood irrigation is desirable for rice production, continued reliance on this approach may create shifts in weed community composition toward adapted weed species. This has occurred in California, where water-seeding of rice and continuous flooding began in the 1920s as a method to control severe infestations of barnyardgrass (*Echinochloa crus-galli* (L.) Beauv.), which is well adapted to dry-seeding, but not early flooding (Seaman, 1983). Over time, however, early watergrass (*Echinochloa oryzoides* (Ard.) Fritsch) and late watergrass (*Echinochloa phyllopogon* (Stapf) Koss.) have become important weeds and have largely replaced barnyardgrass. The former two species have large seeds, which allow germination and emergence through flood water up to 30 cm deep; barnyardgrass has smaller seeds and is unable to emerge through deep water (Barrett, 1983; Seaman, 1983). As a consequence of this species shift, rice growers in California are now advised to consider alternating between water-seeding and dry-seeding practices (Williams et al., 1992).

Crop Residue Management and Mulching Practices

Crop residue management is critical for conserving soil and improving soil quality. It can also be used to regulate weed emergence and growth. When incorporated into soil, residues of many crop species, including cereals, crucifers, and legumes, release chemical compounds (allelochemicals) that can inhibit germination and early seedling growth of many plant species (Boydston and Hang, 1995; Weston, 1996). Liebman and Davis (2000) suggested that allelochemicals are likely to be more toxic to small-seeded species (e.g., redroot pigweed) than to large-seeded species (e.g., soybean) or to transplanted seedlings for three reasons. First, small-seeded species tend to germinate close to the soil surface, where concentrations of residue-derived phytotoxins may be highest. Second, small-seeded species tend to have greater amounts of root length per unit of root mass and thus proportionally greater amounts of absorptive surface area through which phytotoxins may enter. Third, compared with large-seeded species and transplants, small-seeded species have fewer reserves with which to tolerate or detoxify chemical stress agents.

Because the seeds and vegetative propagules of most agronomic crops are considerably larger than those of the weeds that infest them (Mohler, 1996), allelopathic crops might be exploited for their differential effects on crops and weeds in a range of cropping systems. For example, Dyck and Liebman (1994) reported that incorporation of crimson clover residue into soil reduced the emergence rate, density, and biomass production of common lambsquarters (*Chenopodium album* L.), a small-seeded weed, while having negligible effects on the emergence and growth of sweet corn, a large-seeded crop. Similarly, Boydston and Hang (1995) found that rapeseed residue incorporated into soil before planting potato reduced weed density 73-85% and reduced weed biomass 50-96%, while raising potato tuber yield 10-18%, compared with a fallow treatment. Boydston and Hang (1995) suggested that isothiocyanates were responsible for the weed suppression, and that the residue's positive effect on potato growth may have been due to improved crop nutrition, suppression of soil-borne pathogens, as well as reduced competition from weeds.

Crop residues used as surface mulches can suppress weed emergence and growth due to both allelopathic and physical effects (e.g., changes in soil thermal regime) (Teasdale, 1998). Crop residues on the soil surface can also reduce weed densities by physically impeding weed seedling emergence and intercepting light that cues weed germination (Teasdale and Mohler, 2000). Differential effects of residues on crops and weeds might be obtained with either zone tillage techniques that remove residue just over the crop row (Kaspar et al., 1990), or the use of crop seeds with large size, independence from light requirements for germination, and enough vigor to push through several centimeters of mulch material (Liebman and Mohler, 2001). The crop residue/mulch approach has been used successfully for weed management in soybean (Liebl et al., 1992), sweet corn (Mohler, 1991), and other crops. Special attention must be directed, however, to the management of surface germinating weeds, such as dandelion (*Taraxacum officinale* Weber in Wiggers), which can be well-adapted to minimum tillage systems (Liebman et al., 1995).

Weed seed density can be substantially reduced before planting a crop by covering soil with polyethylene tarps for several weeks, a process that has been called "solarization" (Stapleton and DeVay, 1995). Solarization kills weed seeds by raising soil temperatures, as well as by altering soil atmospheric, chemical, and microbiological conditions. Incorporation of certain cover crop residues into soil before solarization can increase control of soil-borne plant diseases (Gamliel and Stapleton, 1997), and this combination of practices might also increase weed suppression. The costs of solarization tarp materials, installation, removal, and disposal - are relatively high (US \$ 750 to \$ 1500 ha^{-1}; Stapleton and DeVay, 1995) and probably can be borne only for high value crops. The development of new thin-layer mulch materials that are temporarily effective for sealing the soil surface, but which are photodegradable or biodegradable, would reduce the consumption of petrochemicals in the manufacture of mulch materials and might reduce costs of solarization and mulch disposal (Stapleton and DeVay, 1995).

A variety of materials can be added to fields as mulches to suppress weed seedling recruitment during periods of crop production, as well as to conserve soil moisture (Monks et al. 1997; Schonbeck, 1998). Polyethylene is widely used for this purpose, but a number of lower-cost organic materials also are available. Hay and straw can be used, but may serve as sources of germinable seeds and ultimately plants that compete with crops. The use of composted yard wastes, newspaper, and other organic materials that do not contain viable seeds may avoid these problems. Areas between crop rows not covered with mulch require additional weed control measures.

Management Practices to Increase Crop Competitive Ability

The ability of crops to yield well in the presence of weeds and to suppress weed establishment, growth, and reproduction can be affected by a number of farming practices that increase the proportion of available resources captured by crops rather than weeds.

Plant size early in the growing season strongly affects interspecific interactions. Transplanting rather than direct seeding horticultural crops, such as tomato, provides an initial size advantage that can translate into reductions in the length of the critical period for weed control, in weed biomass production, and in crop yield loss to weed competition (Weaver, 1984; Weaver et al., 1992). For cereal crops, such as wheat and barley, seedling emergence rate and early-season height and mass have been shown to be correlated positively with seed size (Chastain et al., 1995a, 1995b). Kolbe (1980) separated barley seed populations into fractions with lighter or heavier seed weights and showed that stands grown from heavier seeds had higher yield and lower weed cover than stands grown from smaller seeds.

Substantial increases in crop competitive ability can be achieved by ensuring that weed emergence occurs after crop emergence (Cousens et al., 1987; Chikoye et al., 1995). Differences between crops and weeds with regard to phenology and response to temperature conditions can be exploited for this purpose through adjustments in tillage practices and planting date. For warm-season crops, such as maize and soybean, delaying planting for several weeks beyond the normal date can allow large numbers of weed seedlings to emerge and be killed before the crop is sown

(Gunsolus, 1990; Forcella et al., 1993). The stale seedbed approach, which is widely used in carrot and onion production, involves soil disturbance by tillage several weeks ahead of crop sowing to promote weed emergence; after weeds are killed by flaming or shallow cultivation, the crop is sown with minimal further soil disturbance (Mohler, 2001a). For crops that are better adapted to cool conditions than the weeds that infest them, early planting can give the crop a competitive advantage. Vezina (1992) reported that two warm-season grasses, yellow foxtail (*Setaria pumila*) and barnyardgrass, had no effect on spring wheat yield when the crop was sown in late April or early May, but reduced wheat yield when the crop was sown later.

Crop production often occurs in row widths designed to accommodate cultivation equipment, but where row widths can be narrowed and crops planted in more equidistant arrangements, earlier shading of interrow areas can be achieved and crop competitive ability sometimes can be increased (Teasdale and Frank, 1983; Murdock et al., 1986; Murphy et al., 1996). The effectiveness of narrowing row width for weed suppression appears to be linked to the relative heights of the crop and weed species involved. Weeds that are inherently shorter than the crop are likely to be more susceptible to a narrowing of crop rows, whereas weed species that have the potential for growing taller than the crop are much less likely to be susceptible to row narrowing (Schnieders, 1999; Mohler, 2001c).

Reductions in row width are often accompanied by increases in crop population density, but even without a reduction in row width, increases in density generally make crops more competitive against weeds. Species for which this has been demonstrated include maize, pea, peanut, rapeseed, rice, safflower, soybean, and wheat (Mohler, 1996). Increases in crop density raise the quantity of resource-intercepting surfaces (leaf and root surface area) that the crop has early in the season and result in a greater proportion of available resources being captured by the crop, rather than by weeds (Mohler, 2001c).

Crop genotypes that have high leaf area expansion and height growth rates early in the season often have superior ability to suppress weed growth (Mohler, 2001c). Nevertheless, on an absolute basis, yields of cultivars with high competitive ability against weeds may be inferior to yields of less competitive cultivars. Bastiaans et al. (1997) compared two rice varieties differing in growth characteristics and found that under weed-free conditions, the shorter variety with slower leaf growth (IR8) produced 57% more grain than a taller, leafier variety (Mahsuri). In mixtures with purple rice (a weed), IR8 still produced considerably more grain than Mahsuri (25 to 48%), despite the fact that the biomass of purple rice in plots sown with Mahsuri was 80% less than in plots sown with IR8. This trade-off between competitive ability and yield potential exists in other crops, although it is not universal (Mohler, 2001c). Where the trade-off is observed, breeding efforts are needed to determine whether high competitive ability and high yield potential can be expressed in a single genotype.

Intercropping combines two or more crops whose resource consumption characteristics are physiologically, temporally, or morphologically complementary. By combining crop species that differ in the way they use light, water, and nutrients, intercropping can prevent the crops from fully competing with one another (Vandermeer, 1989). Consequently, intercrops may use a greater share of available resources and produce more yield per unit land area than monocultures of the component species (Willey, 1979, 1990). Greater resource use by intercrops than monocultures can, in turn, provide improved opportunities for suppressing weeds through resource competition. For example, Dyke and Barnard (1976) reported that the biomass of quackgrass (*Elytrigia repens* (L.) Nevski) in barley intercropped with red clover was 29 to 88% less than in monocultured barley. Although Dyke and Barnard (1976) did not measure barley yield, results of experiments conducted by Stewart et al. (1980) and Kunelius et al. (1992) indicate that intercropping red clover with barley has little or no effect on barley yield, at least in humid climates, suggesting that increased weed suppression through barley/red clover intercropping need not come at the cost of a yield penalty. Abraham and Singh (1984) found that a grain sorghum/fodder cowpea intercrop intercepted more light, captured greater quantities of nutrients (N, P, and K), produced higher yields, and contained lower weed densities and less weed biomass compared with monocultured sorghum. Similar patterns of resource use, crop yield, and weed suppression have been observed for intercrops and monocultures of sorghum and pigeonpea (Natarajan and Willey, 1980a, 1980b; Shetty and Rao, 1981).

Results of experiments conducted in a range of cropping systems indicate that various management practices which increase crop competitive ability can be combined effectively. Malik et al. (1993) found that increased bean density, reduced interrow distance, and the use of a semi-vining, rather than a bush variety maximized bean seed yield and minimized weed biomass production. Teasdale (1995) reported that when maize density was increased and row spacing was decreased, herbicide use could be reduced 75% without any reduction in maize yield and weed control relative to results obtained from a normal density, normal row spacing treatment receiving standard herbicide rates. Bulson et al. (1997) observed that total crop yield rose and weed biomass decreased as the sowing density of wheat/faba bean intercrops was increased above levels used for normal density monocultures of the component crops.

Cropping System Diversification in Time

Temporal diversification of cropping systems through use of rotations and cover crops can improve nutrient accretion and retention, soil structure, soil conservation, and suppression of certain insect pests and crop diseases (Smith et al., 1987; Hargrove, 1991; Karlen et al., 1994). Consequently, these practices play key roles on low-external-input and organic farms. Crop rotation and cover cropping can also contribute to weed management due to effects on the availability of regeneration niches and growth resources used by weeds (Liebman and Dyck, 1993; Teasdale, 1998; Liebman and Staver, 2001). Several salient points and illustrative examples are noted here.

A regeneration niche comprises a species-specific set of environmental conditions required to ensure that a mature plant is replaced by an individual of the next generation (Grubb, 1977). By rotating crops with different planting dates and growth periods, contrasting competitive characteristics, and dissimilar management practices, the regeneration niches of different weed species can be disrupted and increases in densities of key weed species prevented. Blackshaw (1994) reported that downy brome density remained relatively stable when winter wheat was rotated with spring-sown rapeseed, whereas the weed's density increased rapidly when wheat was grown continuously. Suppression of downy brome was attributed to changes in the herbicide regime and in the timing of tillage practices that reduced seedling survival. Covarelli and Tei (1988) noted that when herbicides were withheld, seed and seedling densities of the summer annual weed barnyardgrass were lower in a cool season/warm season crop rotation (winter wheat/maize) than in a continuous summer crop sequence (maize/maize). Schreiber (1992) found that giant foxtail seedling density tended to be greatest in continuous maize, intermediate in a two-year maize/soybean rotation, and lowest in a three-year maize/soybean/winter wheat rotation. Suppression of giant foxtail in the winter wheat phase of the rotation was attributed to allelopathic effects of wheat straw, but also may have been influenced by phenological differences between the weed and the wheat crop.

As the intensity of chemical weed control is lessened, the importance of disrupting regeneration niches through rotation increases. Kegode et al. (1999), for example, found that weed seed production could be reduced to low levels in continuous maize and maize-soybean sequences when standard herbicide doses were used, whereas weed seed production increased greatly in these simple cropping systems when herbicide use was minimized. In contrast, weed seed production was low to moderate in rotations that included maize, soybean, wheat, and alfalfa, regardless of the intensity of herbicide inputs.

Rotation of perennial forage crops, such as alfalfa, with annual crops, such as wheat or maize, can reduce densities of annual weeds in the annual crops, though it may increase the density of perennial weeds and certain annual species adapted for survival and reproduction in forage crops (Entz et al., 1995; Clay and Aguilar, 1998; Ominski et al., 1999). Mowing and grazing regimes have a large impact on weed dynamics during forage phases of a rotation sequence and may have to be adjusted to optimize both forage quality and weed suppression. Pino et al. (1998) used field trials and a matrix model to study population dynamics of broadleaf dock (*Rumex obtusifolius* L.) in wheat/alfalfa rotations. Model projections indicated that early harvest (starting 1 April) of alfalfa reduced broadleaf dock density, whereas later harvest (starting 25 April) could allow density of the weed to increase.

Cover crops that occupy a field before or after periods of 'main crop' production suppress weed growth and the production of weed seeds and vegetative propagules that affect subsequent crops. McLenaghen et al. (1996) compared five different cover crops and a fallow control treatment and found that weed growth was inversely proportional to the ground cover produced by the cover crops. Teasdale and Daughtry (1993) reported that live hairy vetch reduced weed density 70-78% and reduced weed biomass 52-70% compared to a fallow treatment. Measurements made after vetch had completed the majority of its growth showed that an average of 87% of sites beneath the cover crop received <1% of unobstructed sunlight. Hairy vetch reduced the red (660 nm) to far-red (730 nm) ratio of transmitted light by 70% and reduced daily maximum soil temperature and diurnal soil temperature amplitude. Weed suppression was attributed to all of these changes in environmental conditions. As discussed earlier, chemical characteristics of cover crop residues also may be important for weed suppression.

Despite the general trend toward lower weed densities in temporally diversified cropping systems, such systems are not always better than simpler systems for suppressing weeds. For example, in a ten-year study, Doucet et al. (1999) observed that weed densities in maize rotated with soybean and winter wheat were sometimes lower than in continuous maize, but often did not differ between the two cropping systems. Moreover, weed densities in soybean were often higher in the three-crop rotation than in continuous soybean. If diversification of cropping systems is to be used effectively as a tool for improving weed suppression, factors leading to different weed management outcomes between different systems and between years and locations for a single system need to be identified.

Within cropping systems, weed densities can vary markedly between different crop phases. Buhler et al. (2001) measured densities of weed seeds and mature plants within a five-year rotation system (maize-soybean-maize-oat-hay) and found that weed populations grew rapidly in the oat and hay phases of the rotation before being reduced to lower levels in the maize and soybean phases. Increases during the oat and hay phases apparently occurred because weeds were able to produce seeds before the oat stubble and forage crops were mowed. Earlier mowing might have helped prevent the observed increases in weed density (Norris and Ayres, 1991). Determining how different phases within a crop sequence affect weed population dynamics and identifying how management strategies should be altered within particular phases are critical for the development and implementation of weed suppressive cropping systems.

Other Approaches

As interest in LEI and organic farming systems increases, additional options for reducing weed density, growth, and competition against crops are likely to become available. For example, new mowing machinery may offer an alternative to interrow cultivation for weed suppression. Donald (2000) investigated the use of a specially constructed mower in soybean and found that it provided excellent control of weeds between crop rows, although weeds within crop rows remained problematic unless herbicides were applied.

New opportunities may also exist for the use of biocontrol agents. The release of artificially cultured microbes and insects to suppress weeds has been the focus of a considerable amount of research but, with certain exceptions, these methods seem unlikely to be widely available to farmers through commercial sources in the near future. Moreover, biological control of weeds with microbial pathogens and insects poses a certain degree of risk with regard to potential impacts on non-target organisms (Simberloff and Stiling, 1996). Improved exploitation of resident microbes, insects, and vertebrates for weed suppression may overcome concerns regarding the ecological safety of introduced biocontrol agents.

A number of studies have demonstrated that resident microbes, invertebrates, and vertebrates can have substantial impacts on weed seed density (Menalled et al., *in press*). Fellows and Roeth (1992) found that indigenous soil-borne microbes reduced numbers of germinable *Sorghum bicolor* seeds by about 60% as compared to control treatments that were protected with fungicides. Cromar et al. (1999) reported that ground-dwelling invertebrates and other organisms (e.g., birds and rodents) consumed about 30% of the common lambsquarters and barnyardgrass seeds they placed in field plots. Cardina et al. (1996) noted that indigenous invertebrates and rodents consumed velvetleaf seeds placed in field plots at an average rate of 11% per day, leading to an almost 80% reduction in seed density over a four-week period.

Because the seed phase serves as a bottleneck within the life cycle of annual weeds, organisms that attack weed seeds may have large effects on weed population dynamics (Swanton et al., 1999; Menalled et al., *in press*). The abundance and impact of resident weed seed predators and seed pathogens can be affected by crop species (Zhang et al., 1998), tillage system (Pitty et al., 1987; Cromar et al., 1999), residue management (Brust and House, 1988), and other factors. In general, it appears that resident weed seed predators can be conserved by reducing the frequency and intensity of tillage, delaying post-harvest tillage, and protecting non-crop habitats in close spatial association with highly disturbed crop fields (Menalled et al., *in press*). Additional research is needed to determine more precisely which farming practices are most desirable for increasing the impacts of resident weed biocontrol agents.

CASE STUDIES

The following three case studies illustrate how diverse weed management tactics can be used to suppress target weed species in a range of agroecosystems. In the first two examples, we present data from field experiments that indicate the impacts of ecologically based, multitactic approaches on individual weed species common to the experiment areas. In the third example, we examine the impacts of various tactical options on a mixed community of weed species found on a set of commercial organic farms.

Case Study 1: Management of Common Groundsel in Leek and Celery

Common groundsel (*Senecio vulgaris* L.) is a short-lived annual species common throughout the world. In temperate climates, it is capable of multiple generations per year. The weed's short generation time, high fecundity, and rapid germination make it a troublesome pest in many cropping systems (Müller-Schärer and Frantzen, 1996). It can be especially problematic in vegetable production systems in which crops are slow to develop canopy cover and hence are poor competitors for light.

Leek (*Allium porrum* L.) is an economically important vegetable crop in much of northern Europe. Like many other *Allium* crop species, leek's slow growth rate and open canopy structure make it highly susceptible to weed infestation throughout the entire period when the crop is produced. Although common groundsel individuals that emerge late in the crop cycle do not affect crop yield and quality, they may still interfere with mechanical harvest and produce seeds that can cause infestations in subsequent crops (Baumann et al., 2000). Mechanical or chemical destruction of late-emerging common groundsel individuals is difficult because of mechanical damage to the crop or chemical residues on the harvested product.

Germination, growth, and seed production of common groundsel are reduced by shading (Hilton, 1983; Leishman et al., 1999; Baumann et al., 2001). Thus, one option for improving the management of common groundsel in leek is to grow the crop in mixture with a second cash crop capable of capturing light that would otherwise be available for weed growth. In a field experiment conducted in Wädenswil, Switzerland (47°13' N, 08°40' E), Baumann et al. (2000) investigated the weed suppression characteristics of a leek-celery intercropping system. Related experiments had previously shown that leek/celery intercrops intercepted more light earlier in the growing season than did leek monocultures (Baumann et al., 2000).

The treatment design of the experiment involved two factors: crop system and time of common groundsel seedling establishment. Celery (*Apium graveolens* L.) and leek seedling transplants were used to create monocultures of both crops and an intercrop. The leek/celery intercrop was formed as a row-by-row substitutive mixture of the two monocultures (i.e., the density of each crop in the mixture was half its density in pure stand). Pre-germinated seeds of common groundsel were sown between crop rows at 10-day intervals from 10 to 70 days after crop establishment.

Results of the experiment indicated that final biomass of common groundsel planted 10 days after crop establishment did not differ significantly among the three cropping systems. In contrast, when common groundsel was established 20, 30, and 40 days after the crops, the weed produced 58%, 76%, and 94% less biomass, respectively, in the intercrop than in the leek monoculture (Figure 1A). No significant differences in common groundsel biomass were detected among cropping systems when the weed was sown 50 or more days after the crops. Capitula production by common groundsel (Figure 1B), which serves as an indicator of the weed's reproductive output, and common groundsel seedling density (Figure 1C) were strongly reduced by intercropping relative to monoculture production of leek.

Analysis of the competitive relationships between the crops and the weed showed that crop yield and quality as well as the suppressive ability of the intercropping system can be optimized with simple management practices (Baumann, 2001). From the standpoint of weed management, the potential advantages of this system are clear: it reduces the period of time for which mechanical cultivation, hand-weeding, or herbicide is required to strongly suppress common groundsel biomass and seed production. Because of its impact on common groundsel reproduction, the leek/celery intercropping system is likely to reduce infestations of this weed in subsequent crops, although this has not been tested.

The potential complementarity between cultivation and intercropping is clearly indicated by results of the experiment and additional tactical options have been identified. These include adjustments in row spacing and transplant density to further increase competitive ability of the intercrop. The use of certain cultivars might also improve crop quality and weed suppression. A rust fungus, *Puccinia lagenophorae* Cooke, has been tested as a biological control agent for common groundsel (Müller-Schärer and Frantzen, 1996), and because effects of the fungus on the weed are intensified by crop competition (Paul et al., 1993), it may prove to be a useful tactical option.

Case Study 2: Management of Wild Mustard in Sweet Corn

Wild mustard (*Brassica kaber* (LC) DC. Wheeler) is an annual weed found in a wide range of crops grown in the northern tier states of the USA (Bridges and Baumann, 1992) and elsewhere. Nitrogen fertilizer can increase canopy height and area, photosynthetic rate, biomass, competitive ability, and seed production of *Brassica* spp. (Liebman, 1989; Liebman and Robichaux, 1990). Conversely, nutrient deficits during early development can reduce wild mustard's later growth and competitive ability. Alkämper et al. (1979) noted that relative to a single dose of NPK fertilizer applied at planting, delayed application of NPK fertilizer to a mixture of maize and wild mustard increased the ability of the crop to competitively suppress the weed. Thus, fertility management is likely to be useful for wild mustard management.

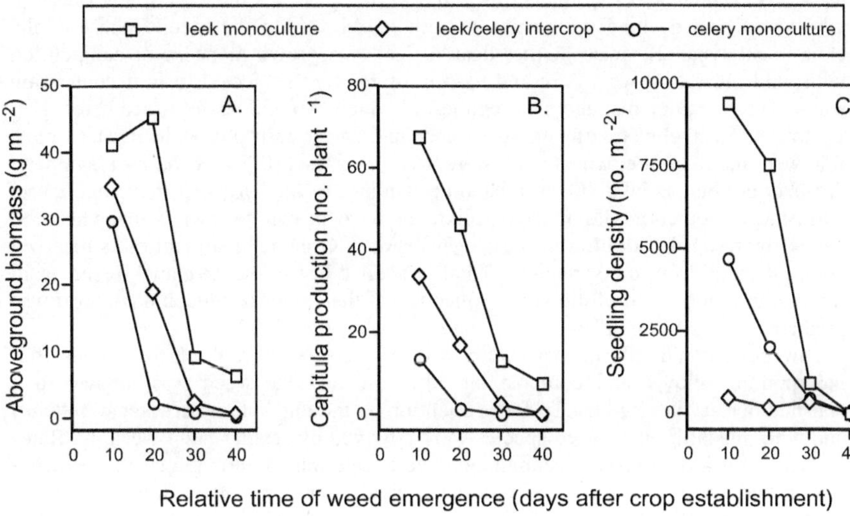

Figure 1. Biomass (A), capitula production (B), and seedling density (C) of common groundsel emerging at different times after the establishment of leek monocultures, celery monocultures, and a leek/celery intercrop. Standard errors for biomass, capitula and seedling density data are 7.3, 5.9, and 1058, respectively. Adapted from Baumann et al., 2000.

Wild mustard can be suppressed by allelochemicals contained in residues of red clover (Ohno et al., 2000), which is widely used as a companion crop for wheat, barley, and oat, and as a soil-improving green manure for subsequent crops. Red clover fixes a substantial amount of atmospheric N_2 and mineralization of organic N contained in red clover residues can satisfy much or all of the N requirement of a succeeding maize crop (Fox and Piekielek, 1988; Stute and Posner, 1995).

Relative to a single dose of N fertilizer applied at planting, decomposing legume residues and other organic materials, such as composted manure, may function as slow-release N sources (Dyck et al., 1995; DeLuca and DeLuca, 1997). For large-seeded crops, such as maize, which have substantial energy and nutrient reserves within their seeds, low rates of nutrient availability early in the growing season do not inhibit final yield if N becomes available before the start of the grand period of crop growth and nutrient uptake (Magdoff, 1991). Because wild mustard is sensitive to early season nutrient deficits whereas maize is not, the use of organic materials as slow-release N sources may be a useful tactic for manipulating interactions between these two species towards a desirable outcome. As noted earlier, large-seeded species are also likely to be less sensitive than small seeded species to allelochemicals (Liebman and Davis, 2000). Thus, use of red clover as a N source may have little negative effect on maize seedlings, but a much greater effect on wild mustard seedlings.

To compare the effects of synthetic and organic N sources on crop-weed interactions, a field experiment was conducted in Stillwater, Maine, USA (44°56' N,

68°42' W) (Davis and Liebman, 2001). In the first phase of the experiment (year 1), wheat was grown alone or in mixture with red clover. In the second phase of the experiment (year 2), sweet corn (sugar maize) was grown alone or in competition with wild mustard. Sweet corn and wild mustard were subjected to four contrasting soil and crop residue management regimes, of which two will be discussed here.

In the 'Synthetic' treatment, sweet corn and sweet corn plus wild mustard plots followed monoculture wheat and were treated with 110 kg N ha^{-1} of synthetic fertilizer applied as NH_4NO_3 in a band at planting. In the 'Organic' treatment, sweet corn and sweet corn/wild mustard mixtures were grown following the wheat/red clover intercrop. In addition to red clover residue, 'Organic' treatment plots received compost made from dairy manure. Total N availability in the 'Organic' treatment in the first growing season following application of the clover residue and compost was predicted to be 110 kg N ha^{-1}.

Sweet corn and wild mustard were sown on the same day, about two weeks after incorporating clover and compost into the soil, and the weed was thinned to a constant density in all plots. To focus attention on the interaction between sweet corn and wild mustard, other weed species were removed by regular hand weeding. Rates of wild mustard biomass accumulation were determined throughout the growing season and sweet corn ear yield was determined at crop maturity. Soil inorganic N concentration (NO_3-N plus NH_4-N) was measured several times during the second repetition of the experiment.

Results of the experiment indicated that soil and crop residue management could reduce the growth and competitive ability of wild mustard. Averaged over the two repetitions of the experiment, maximal relative growth rate of wild mustard was 12% lower in the 'Organic' treatment than in the 'Synthetic' treatment (Figure 2). The competitive impact of wild mustard on sweet corn ear yield was reduced by the organic N sources in one out of two years. In 1997, sweet corn ear yield and yield loss due to wild mustard competition were comparable between the 'Organic' and 'Synthetic' treatments (Figure 3A). In contrast, in 1998, sweet corn yield was unaffected by wild mustard in the 'Organic' treatment, whereas yield was reduced 20% by wild mustard in the 'Synthetic' treatment (Figure 3B).

To determine the impacts of the two soil and crop residue management treatments on wild mustard emergence, special test rows of wild mustard seeds were sown. Averaged over 1997 and 1998, wild mustard seedling emergence at seven days after planting (DAP) was 22% lower in the 'Organic' treatment than in the 'Synthetic' treatment (data not presented). Thus, if wild mustard density had not been held constant between treatments, differences in cropping system performance between 'Organic' and 'Synthetic' treatments might have been even greater than observed. In 1998, sweet corn seedling emergence at 7 DAP was lower in the 'Organic' than 'Synthetic' treatment, but final crop density was not affected.

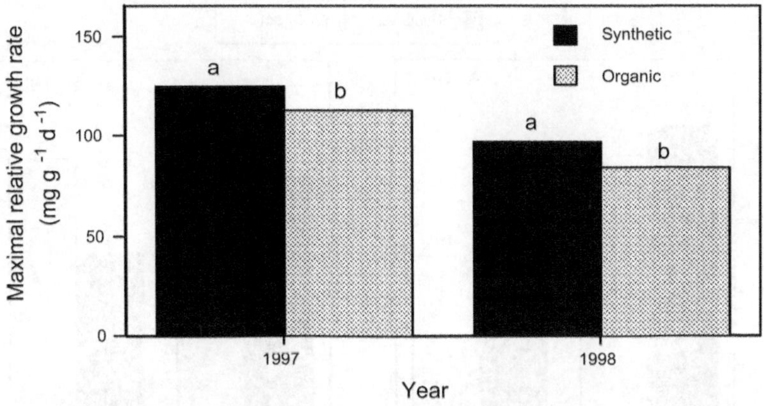

Figure 2. Maximal relative growth rate (RGR_{max}) of wild mustard grown with sweet corn. Plots were treated with 110 kg N ha^{-1} as ammonium nitrate fertilizer ('Synthetic') or with red clover residue and composted dairy manure ('Organic'). Within years, columns topped by different letters differ significantly ($p < 0.05$). Adapted from Davis and Liebman (2001). Note that the 'Synthetic' treatment corresponds to what was called 'Early' in the original publication.

Measurements of soil N status indicated that less inorganic N was present at 9 and 27 DAP in the 'Organic' treatment than in the 'Synthetic' treatment (Davis and Liebman, 2001). In laboratory bioassays, wild mustard seedling growth in soil drawn from the 'Organic' treatment was lower than in the 'Synthetic' treatment and this was coincident with significantly higher incidence of infection by *Pythium* spp. (Conklin et al., 2002), fungal pathogens known to be influenced by additions of organic materials (Cook and Baker, 1983, pp. 95-96). Bioassays detected no differences in sweet corn seedling growth and fungal infection between the two soil and crop residue treatments (Conklin et al., 2002). As noted previously, allelotoxins released by red clover residues also were likely to have suppressed wild mustard seedling emergence and growth (Ohno et al., 2000).

In addition to changing the timing of N availability, soil microbial dynamics, and soil biochemistry, use of composted manure and red clover residue in place of synthetic N fertilizer may contribute to weed suppression and crop protection through other mechanisms. Aerobic composting of manure kills weed seeds thermally and chemically (Shirilipour and McConnell, 1991; Grundy et al., 1998; Ozores-Hampton et al., 1999), thus reducing the capacity of manure to serve as a vector for weed seed dispersal. Red clover established with cereal crops will continue to grow after cereal harvest and can suppress the growth of perennial weeds until the field is prepared for planting the next "main" crop (Dyke and Barnard, 1976). Thus, multiple methods of weed suppression can be addressed through the use of organic matter amendments to soil. Combination of this approach with precision cultivation and other weed management tactics should be expected to improve its value to farmers.

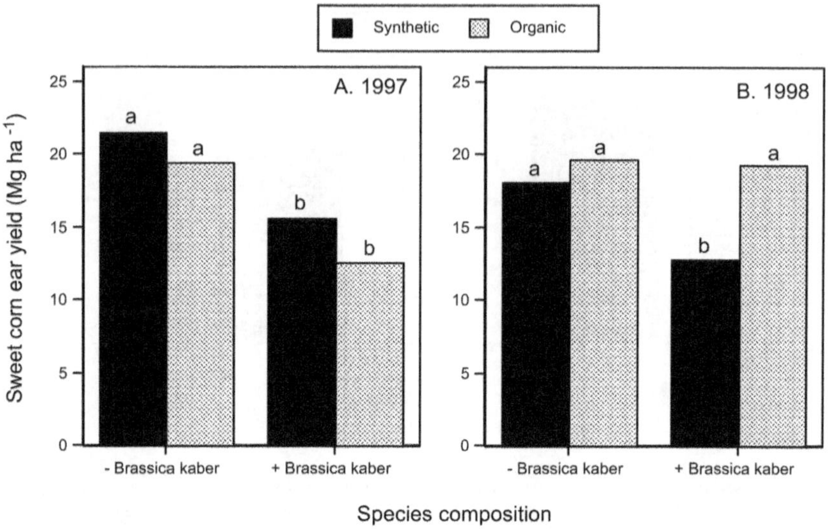

Figure 3. Ear yield of sweet corn grown in 1997 (A) and 1998 (B) with or without competition from wild mustard. Plots were treated with110 kg N ha^{-1} as ammonium nitrate fertilizer ('Synthetic') or with red clover residue and composted dairy manure ('Organic'). Within years, columns topped by different letters differ significantly ($p < 0.05$). Adapted from Davis and Liebman (2001). Note that the 'Synthetic' treatment corresponds to what was called 'Early' in the original publication.

Case Study 3: Management of Multiple Weed Species in Diverse Rotations, with Special Attention to Common Chickweed

Weed management at the level of whole fields and farms involves multiple weed species, although typically one or several species are particularly problematic. This situation is exemplified by weed communities on organic farms in the central clay area of the Netherlands. Most of these farms use long rotations (>6 years) containing various combinations of grains (barley, oat, spelt, triticale, and wheat), vegetables (bean, broccoli, cabbage, carrot, celeriac, endive, onion, pea, potato, and sweet corn), annual or perennial forage crops (grass-clover mixtures and silage maize), and, in some cases, sugar beet (Mertens, 1998; Vereijken, 1999).

"Multifunctional" rotation sequences were established on 10 of these farms in a collaborative project organized by the DLO Institute for Agrobiology and Soil Fertility, a government research unit (Vereijken, 1994). Although specific crops varied among farms, each rotation was intended to maintain soil fertility and crop health with a minimum of inputs and a minimum of environmental damage. With respect to weed management, the rotation sequences involved alternation between highly competitive mown crops (e.g., grains) and less competitive lifted row crops (e.g., carrot and onion). Alternation between these contrasting crop types was intended to diversify the selection pressures and stresses placed on the weed community, suppress weed reproduction, and reduce weed densities, particularly in the less competitive crops (Vereijken, 1999; Bastiaans et al., 2000).

Surveys of the organic farms showed that despite the intended outcome of the rotation sequences, weed management remained a serious problem. In 1997, only one of the 10 farms used <500 hours of hand-weeding; this was the only farm that had no need for off-farm sources of labor for weeding (Vereijken, 1999). The other farms required up to 2400 hours of hand-weeding, mostly directed at control of common chickweed (*Stellaria media* (L.) Vill.) in carrot and onion fields (Vereijken, 1999). Additional laborers for hand weeding were available only at a relatively high cost to the farmers.

Mertens (1998) conducted monthly surveys of the species composition, abundance, and distribution of weeds on six organic farms in the region, of which five were operated commercially and one was university-run. Her results indicated that although many weed species were present on the commercial farms, the weed communities were dominated by common chickweed (Table 1). On the university-run farm, common chickweed also was the most abundant species, but considerable densities of annual bluegrass (*Poa annua* L.) and shepherd's-purse (*Capsella bursa-pastoris* (l.) Medicus) were also present (Table 1). In addition to hand weeding, direct weed control tactics used on the farms included spring-tine harrowing, flaming, interrow cultivation, and ridging.

Mertens (1998) noted that the density of flowering individuals of common chickweed was three times greater in highly competitive crops, such as wheat, than in poorly competitive crops, such as carrot and onion, and she suggested that the high hand-labor requirements for controlling common chickweed in the poorly competitive crops were largely the result of weed seed being added to the soil during production of the competitive crops. That is, crop species such as wheat that did not themselves suffer from the presence of common chickweed nonetheless permitted the weed to reproduce abundantly, thus creating dense infestations of the weed in subsequent carrot and onion crops.

By providing information concerning farming system components relevant to weed management (e.g., crop sequences, hand-labor expenditures) and weed community composition, the work of Mertens (1998) and Vereijken (1999) provides a base for developing improved weed management systems on the targeted set of Dutch organic farms. The next step is to analyze ecophysiological, morphological, and phenological characteristics of the dominant, problematic weed species that may render them vulnerable to new control tactics.

Germination and growth of common chickweed can occur at temperatures close to freezing and because the species can make cleistagamous, self-pollinated flowers, viable seed can be produced well into the late fall (Roberts and Lockett, 1975; Holm et al., 1977; Storkey and Cussans, 2000). Germination of common chickweed can be increased by exposure to nitrate (Steinbauer et al., 1955; Roberts and Lockett, 1975) and this species' growth responses to fertilizer N can be proportionally greater than those of the crops it infests (Lawson and Wiseman, 1979). Common chickweed is capable of early reproduction (within 30 to 40 days after emergence - Miura et al., 1995) and can produce, under favorable circumstances, $>11 \times 10^6$ seeds per hectare (Salisbury, 1961). Seeds of common chickweed can persist for a number of years in the soil seed bank. However, over three- to four-year periods in which fields were plowed annually, cropped with wheat, and kept free of reproductive weeds, numbers of viable common chickweed seeds in the soil declined 30% per year (Lawson et al., 1993).

Table 1. Mean density of the five most prevalent weed species observed over a seven-month period, April through October 1997, on five commercial farms and one university-run farm in Flevoland, The Netherlands. Source: Mertens (1998).

Commercial farms		University-run farm	
Weed species	Mean density (plants m^{-2})	Weed species	Mean density (plants m^{-2})
Stellaria media	12.4	*Stellaria media*	19.4
Poa annua	0.7	*Poa annua*	12.7
Senecio vulgaris	0.6	*Capsella bursa-pastoris*	7.7
Urtica urens	0.5	*Taraxacum officinale*	2.0
Sonchus spp.	0.4	*Solanum nigrum*	1.8
Other species	2.0	Other species	3.3

Control tactics can be directed at seed, vegetative, and generative life stages of common chickweed. Aerobic composting that creates a pile temperature of at least 55° C for three days kills seeds of common chickweed (Grundy et al., 1998). Fall and spring harrowing of winter wheat can reduce common chickweed density and biomass significantly (Wilson et al., 1993). Common chickweed density also can be reduced by flaming before crop emergence (Ascard, 1995), and by mulching with straw (Koster et al., 1997) or a 3 cm layer of compost (Ligneau and Watt, 1995). Mertens and Jansen (2002) found that common chickweed biomass and seed production in spring wheat could be reduced by narrowing the distance between crop rows and increasing crop density.

Other tactical options for suppression of common chickweed are suggested by its ecological characteristics. Minimizing the availability of inorganic N in the soil, through the use of cover crops that immobilize and sequester N, may reduce common chickweed density, biomass, and seed production in leguminous crops (e.g., pea and bean) capable of satisfying all or part of their N requirement through fixation of atmospheric N_2. Shallow cultivation can promote common chickweed germination (Roberts and Feast, 1972; Holm et al., 1977) and could allow depletion of the soil seed bank between periods of crop production. Cover crops, such as *Trifolium* spp., sown either with a cereal crop or into its stubble, may compete with and limit reproduction by common chickweed plants growing after harvest of the cereal crop (Ross et al., 2001). Because most common chickweed seedlings emerge from seeds buried within 2-3 cm of the soil surface (Cussans et al., 1996; Grundy et al., 1996), depletion of the soil seed bank through cultivation and cover cropping in one season, followed by minimal tillage for crop seedbed preparation in the succeeding season may minimize common chickweed density in carrot, onion, and other poorly competitive crops.

FUTURE PRIORITIES FOR RESEARCH

The preceding discussion has emphasized practical measures for managing weeds in LEI and organic farming systems. Although it is clear that applying certain sets of these practices can be effective for managing weeds under some conditions, weed management remains a high priority for many LEI and organic farmers (Peacock and Norton, 1990; Walz, 1999), and much more knowledge must be developed and disseminated before weeds are managed effectively on the majority of farms operating with little or no reliance on herbicides. Based on our assessment of the current situation in North America and Europe, four paths seem especially promising for future improvements in weed management for LEI and organic farming systems. While their emphases differ, these paths are complementary and could be pursued simultaneously.

Prioritization of Weed Control Efforts in Time and Space

There are good reasons to believe that better prioritizing weed management activities in both temporal and spatial dimensions may provide substantial benefits. Using matrix models to study weed population trajectories in different crop rotation systems, Mertens et al. (2002) showed that crop sequence, not just crop diversity, has strong effects on weed population size. That is, for a given set of crops (say A and B) that differ in their ability to suppress weed growth and reproduction, weed population growth rate for the crop sequence ABAB can differ greatly from that in the sequence AABB, due to different combinations of natural mortality factors and management activities. Consequently, over the long term, weed density and costs for weed control could be lower in rotations that include the same crops at similar frequencies but in different sequences.

Norris (1999) argued that by intensifying weed control efforts to the point that weed seed production was virtually eliminated and immigration of weed seeds was prevented, large reductions in weed control costs could be realized over the long-term. Such a strategy suggests that it might be valuable to address the weed problems of entire farms by sequentially concentrating weed management activities and depleting the weed seed banks of individual fields. By limiting subsequent recolonization and invasion of fields, permanent reductions in weed density might be achieved across larger areas. The total effort expended in such an approach might be less than that expended in revisiting weed problems in a partially effective manner year after year (Norris, 1999).

Breeding to Increase the Deleterious Effects of Crops on Weeds

Given the large amount of effort devoted to and benefit accrued from breeding crops for tolerance and resistance to pathogens and insect pests, it is surprising how little effort has been devoted to developing weed-suppressive and weed-tolerant cultivars. In rice, one crop for which attention has been focused on weeds in screening and selection work, genotypes with increased ability to compete for resources and with significant allelopathic potential have been identified (Bastiaans et al., 1997; Olofsdotter, 2001a). Olofsdotter (2001b) described a multi-stage protocol in which (i) laboratory, greenhouse, and field studies demonstrated that certain rice cultivars have strongly suppressive allelopathic effects against monocot and dicot weed species; (ii) allelochemicals responsible for weed suppression were isolated, identified, characterized; (iii) correlations between inhibition of weed growth and allelochemical concentrations were established; (iv) genetic maps for quantitative trait loci associated with allelopathy were constructed; and (v) breeding was initiated and greenhouse and field testing were conducted to produce rice cultivars with improved allelopathic and competitive ability. Similar approaches should be undertaken for other crops with particular attention to the issue of whether weed-suppressiveness carries a yield penalty. If weed-suppressiveness is linked to lower yields in existing cultivars, research is needed to determine whether breeding can overcome this deficiency.

Improved Understanding of Ecological Factors Regulating Weed Emergence, Seed Survival, and Invasiveness

There remain a large number of basic ecological questions whose answers could greatly improve weed management in LEI and organic farming systems. For example, to what degree is weed density limited by numbers of appropriate sites for germination and establishment (see, for example, Evans and Young, 1970), rather than numbers of seeds and vegetative propagules (see, for example, Hartzler, 1996)? If the former is important, soil and crop residue management practices might be developed to maximize suppression of weed seedling recruitment. If seed numbers regulate weed population size, how might pre-dispersal and post-dispersal mortality of weed seeds and vegetative propagules be increased without damaging crop performance? Preventing future weed problems is a key objective for LEI and organic farming systems (Jordan, 1996), but is only achievable if the process of weed invasion is better understood. Which crop and soil management factors minimize the possibility of new weed species colonizing a crop field?

Technical Innovation

Although we do not advocate abandoning weed management tactics based on ecological principles for those built on improvements in farm equipment, it is clear that new field machinery and particularly computer-based informational technologies may have large impacts on weed management within the next two decades. Cultivator guidance systems that use electronic sensors and computers to rapidly adjust the location of cultivating tools are already beginning to give rise to a new generation of machines that can operate at high speeds close to crop rows with minimal chance of crop damage (Mohler, 2001b). The combination of electronic sensors and computers with new weed-cutting technologies, such as CO_2 lasers (Heisel et al., 2001) and high-pressure water jets (Fogelberg and Blom, 2002), may allow the development of machines that attack individual weeds within crop rows. Other machines are now being tested that kill weed seeds via steam injection in the row area where crops will be planted (Melander et al., 2002). Localization of steam application to narrow bands rather than to entire fields focuses weed control where it is needed most, while saving energy and increasing the speed at which fields are treated. Given the current emphasis on technical innovations for site-specific conventional agricultural systems, including localized herbicide application (e.g., Blackshaw et al., 1998), technical breakthroughs for weed management in LEI and organic farming systems seem quite possible. Care should be taken to integrate them with other ecologically-based tactics.

Progress in addressing these four pathways toward improved weed management is most likely to be obtained from a mixture of experimentation, modeling, and systems-level analysis of existing and prototypic farms. The process involved will be difficult, but is necessary if the full potential of LEI and organic farming systems is to be realized.

ACKNOWLEDGMENTS

This chapter was conceived and written when ML and DTB were working with the Crop and Weed Ecology Group of Wageningen University. ML's stay there was made possible by a grant from the C.T. de Wit Graduate School for Production Ecology and Resource Conservation and a leave from Iowa State University; these are gratefully acknowledged. DTB's stay in Wageningen was made possible by a leave from the Swiss Federal Research Station for Fruit-Growing, Viticulture and Horticulture, Wädenswil, and that also is gratefully acknowledged. Finally, we thank A.S. Davis, M.J. Kropff, S.K. Mertens, and P. Vereijken for generously sharing their insights and data with us.

REFERENCES

Abraham C.T., Singh S.P. Weed management in sorghum-legume intercropping systems. J Agric Sci, Cambridge 1984; 103:103-115.

Alkämper J., Pessios E., Long D.V. Einfluss der Düngung auf die Entwicklung und Nahrstoffaufnahme verschiedener Unkräuter in Mais. Proceedings of the 3rd European Weed Research Society Symposium, Mainz, Germany, 1979; pp. 181-192.

Anderson R.L. Timing of nitrogen application affects downy brome (*Bromus tectorum*) growth in winter wheat. Weed Tech 1991; 5:582-585.

Angonin C., Caussanel J.P., Meynard J.M. Competition between winter wheat and *Veronica hederifolia*: influence of weed density and the amount and timing of nitrogen application. Weed Res 1996; 36:175-187.

Ascard J. Effects of flame weeding on weed species at different developmental stages. Weed Res 1995; 35:397-411.

Ball D.A., Wysocki D.J., Chastain T.G. Nitrogen application timing effects on downy brome (*Bromus tectorum*) and winter wheat (*Triticum aestivum*) growth and yield. Weed Tech 1996; 10:305-310.

Barrett S.C.H. Crop mimicry in weeds. Econ Bot 1983; 37:255-282.

Baskin J.M., Baskin C.C. "Physiology of dormancy and germination in relation to seed bank ecology." In *Ecology of Soil Seed Banks*, M.A. Leck, V.T. Parker, R.L. Simpson, eds. San Diego, CA: Academic Press, 1989; pp. 53-66.

Bastiaans L., Kropff M.J., Goudriaan J., van Laar H.H. Design of weed management systems with a reduced reliance on herbicides poses new challenges and prerequisites for modeling crop-weed interactions. Field Crops Res 2000; 67:161-179.

Bastiaans L., Kropff M.J., Kempuchetty N., Rajan A., Migo T.R. Can simulation models help design rice cultivars that are more competitive against weeds? Field Crops Res 1997; 51:101-111.

Baumann D.T. Competitive suppression of weeds in a leek-celery intercropping system - an exploration of functional biodiversity. Ph.D. thesis, Crop and Weed Ecology Group, Wageningen University, The Netherlands, 2001.

Baumann D.T., Bastiaans L., Kropff M.J. Effects of intercropping on growth and reproductive capacity of late-emerging *Senecio vulgaris* L., with special reference to competition for light. Ann Bot 2001; 87:209-217.

Baumann D.T., Kropff M.J., Bastiaans L. Intercropping leeks to suppress weeds. Weed Res 2000; 40:359-374.

Bellinder R.R., Gummesson G., Karlson C. Percentage-driven government mandates for pesticide reduction: the Swedish model. Weed Tech 1994; 5:545-552.
Benvenuti S., Macchia M., Miele S. Quantitative analysis of emergence of seedlings from buried weed seeds with increasing soil depth. Weed Sci 2001; 49:528-535.
Blackshaw R.E. Rotation affects downy brome (*Bromus tectorum*) in winter wheat (*Triticum aestivum*). Weed Tech 1994; 8:728-732.
Blackshaw R.E., Molnar L.J., Chevalier D.F., Lindwall C.W. Factors affecting the operation of the weed-sensing Detectospray system. Weed Sci 1998; 46:127-131.
Boydston R., Hang A. Rapeseed (*Brassica napus*) green manure crop suppresses weeds in potato (*Solanum tuberosum*). Weed Tech 1995; 9:669-675.
Bridges D.C., Baumann P.A. "Weeds causing losses in the United States." In *Crop Losses Due to Weeds in the United States: 1992,* D.C. Bridges, P.A. Baumann, eds. Champaign, IL: Weed Science Society of America, 1992; pp. 75-147.
Brust G.E., House G.J. Weed seed destruction by arthropods and rodents in low-input soybean agroecosystems. Amer J Alt Agric 1988; 3:19-25.
Buhler D.D., Kohler K.A., Thompson R.L. Weed seed bank dynamics during a five-year crop rotation. Weed Tech 2001; 15:170-176.
Bulson H.A.J., Snaydon R.W., Stopes C.E. Effects of plant density on intercropped wheat and field beans in an organic farming system. J Agric Sci, Cambridge 1997; 128:59-71.
Cardina J., Norquay H.M., Stinner B.R., McCartney D.A. Postdispersal predation of velvetleaf (*Abutilon theophrasti*) seeds. Weed Sci 1996; 44:534-539.
Chastain T.G., Ward K.J., Wysocki D.J. Seedbed residue and seed size relationships in winter barley. Agron J 1995a; 87:517-520.
Chastain T.G., Ward K.J., Wysocki D.J. Stand establishment responses of soft white winter wheat to seedbed residue and seed size. Crop Sci 1995b; 35:213-218.
Chikoye D., Weise S.F., Swanton C.J. Influence of common ragweed (*Ambrosia artemisiifolia*) time of emergence and density on white bean (*Phaseolus vulgaris*). Weed Sci 1995; 43:375-380.
Clay S.A., Aguilar I. Weed seedbanks and corn growth following continuous corn or alfalfa. Agron J 1998; 90:813-818.
Conklin A.E., Erich M.S., Liebman M., Lambert D., Gallandt E.R., Halteman W.A. Effects of red clover (*Trifolium pratense*) green manure and compost soil amendments on wild mustard (*Brassica kaber*) growth and incidence of disease. Plant Soil 2002; 238:245-256.
Conway G.R. Agroecosystem analysis. Agric Admin 1985; 20:31-55.
Conway G.R., McCraken J.A. "Rapid rural appraisal and agroecosystem analysis." In *Agroecology and Small Farm Development*, M.A. Altieri, S.B. Hecht, eds. Boca Raton, FL: CRC Press, 1990; pp. 221-235.
Cook R.J., Baker K.F. *The Nature and Practice of Biological Control of Plant Pathogens.* St. Paul, MN: American Phytopathological Society, 1983.
Cousens R., Brain P., O'Donovan J.T., O'Sullivan P.A. The use of biologically realistic equations to describe the effects of weed density and relative time of emergence on crop yield. Weed Sci 1987; 354:720-725.
Cousens R., Mortimer M. *Dynamics of Weed Populations.* Cambridge, UK: Cambridge University Press, 1995.
Cousens R., Moss S.R. A model of the effects of cultivation on the vertical distribution of weed seeds within the soil. Weed Res 1990; 30:61-70.
Covarelli G., Tei F. Effet de la rotation culturale sur la flore adventice du mais. In VIIIeme Colloque International Sur la Biologie, l'Ecologie et la Systematique des Mauvaises Herbes, Vol. 2. Paris, France: Comite Francais de Lutte Contre les Mauvaises Herbes and Leverkusen, Germany: European Weed Research Society, 1988; pp. 477-484.
Cromar H.E., Murphy S.D., Swanton C.J. Influence of tillage and crop residue on postdispersal predation of weed seeds. Weed Sci 1999; 47:184-194.
Cussans G.W., Moss S.R. Population dynamics of annual grass weeds. Proceedings of the British Crop Protection Conference - Weeds, 22-25 November 1982, Brighton, England. Croydon, UK: British Crop Protection Council, 1982; pp. 91-98.
Cussans G.W., Raudonius S., Brain P., Cumberworth S. Effects of depth of seed burial and soil aggregate size on seedling emergence of *Alopecurus myosuroides, Gallium aparine, Stellaria media* and wheat. Weed Res 1996; 36:133-141.
Dastgheib F. Relative importance of crop seed, manure, and irrigation water as sources of weed infestation. Weed Res 1989; 29:113-116.
Davis A.S., Liebman M. Nitrogen source influences wild mustard growth and competitive effect on sweet corn. Weed Sci 2001; 49:558-566.

DeLuca T.H., DeLuca D.K. Composting for feedlot manure management and soil quality. J Prod Agric 1997; 10:235-241.
DiTomaso J.M. Approaches for improving crop competitiveness through manipulation of fertilization strategies. Weed Sci 1995; 43:491-497.
Donald W.W. Timing and frequency of between-row mowing and band-applied herbicide for annual weed control in soybean. Agron J 2000: 92:1013-1019.
Doucet C., Weaver S.E., Hamill A.S., Zhang J. Separating the effects of crop rotation from weed management on weed density and diversity. Weed Sci 1999; 47:729-735.
Dyck E., Liebman M. Soil fertility management as a factor in weed control: the effect of crimson clover residue, synthetic nitrogen fertilizer, and their interaction on emergence and early growth of lambsquarters and sweet corn. Plant Soil 1994; 167:227-237.
Dyck E., Liebman M., Erich M.S. Crop-weed interference as influenced by a leguminous or synthetic nitrogen source: 1. Doublecropping experiments with crimson clover, sweet corn, and lambsquarters. Agric Ecosyst Env 1995; 56:93-108.
Dyke G.V., Barnard A.J. Suppression of couchgrass by Italian ryegrass and broad red clover undersown in barley and field beans. J Agric Sci (Cambridge) 1976; 87:123-126.
Eghball B., Lesoing G.W. Viability of weed seeds following manure windrow composting. Compost Sci Util 2000; 8:46-53.
Entz M.H., Bullied W.J., Katepa-Mupondwa F. Rotational benefits of forage crops in Canadian prairie cropping systems. J Prod Agric 1995; 8:521-529.
Evans R.A., Young J.A. Plant litter and establishment of alien annual weed species in rangeland communities. Weed Sci 1970; 18:697-703.
Exner D.N., Thompson R. *Reducing Weed Pressure in Ridge-Till.* Bulletin SA-2. Ames, IA: Iowa State University Cooperative Extension, 1992.
Forcella F., Eradat-Oskoui K., Wagner S.W. Application of weed seedbank ecology to low-input weed management. Ecol Appl 1993; 3:74-88.
Forcella F., Lindstrom M.J. Movement and germination of weed seeds in ridge-till crop production systems. Weed Sci 1988; 36:56-59.
Fellows G.M., Roeth F.W. Factors influencing shattercane (*Sorghum bicolor*) seed survival. Weed Sci 1992; 40:434-440.
Fogelberg F., Blom A. Water jet cutting for weed control. Proceedings of the 5th Workshop of the European Weed Research Society Working Group on Physical and Cultural Weed Control, 11-13 March 2002, Pisa, Italy. D.C. Cloutier, ed. Ste. Anne de Bellevue, Quebec, Canada; Institut de Malherbologie, 2002; p. 237.
Fox R.H., Piekielek W.P. Fertilizer N equivalence of alfalfa, birdsfoot trefoil, and red clover for succeeding corn crops. J Prod Agric 1988; 1:313-317.
Gamliel A., Stapleton J.J. Improvement of soil solarization with volatile compounds generated from organic amendments. Phytoparasitica 1997; 25 (supplement):31S-38S.
Geier B. The organic market: opportunities and challenges. ILEIA Newletter 1998; 14(4):6-7.
Grattan S.R., Schwankl W., Lanini W.T. Weed control by subsurface drip irrigation. Calif Agric 1988; 42(3):22-24.
Grubb P.J. The maintenance of species richness in plant communities: the importance of the regeneration niche. Biol Rev 1977; 52:107-145.
Grundy A.C., Mead A., Bond W. Modeling the effect of weed-seed distribution in the soil profile on seedling emergence. Weed Res 1996; 36:375-384.
Grundy A.C., Green J.M., Lennartsson M. The effect of temperature on the viability of weed seeds in compost. Compost Sci Tech 1998; 6(3):26-33.
Gunsolus J.L. Mechanical and cultural weed control in corn and soybeans. Am J Alter Agric 1990; 5:114-119.
Hargrove W.L., ed. *Cover Crops for Clean Water.* Ankeny, IA: Soil and Water Conservation Society, 1991.
Hartzler R.G. Velvet leaf (*Abutilon theophrasti*) population dynamics following a single year's seed rain. Weed Tech 1996; 10:581-586.
Heisel T., Schou J., Christensen S., Andreasen C. Cutting weeds with a CO_2 laser. Weed Res 2001; 41:19-30.
Hill J.E., De Datta S.K., Real J.G. "*Echinochloa* competition in rice: a comparison of studies from direct-seeded and transplanted flooded rice." In Proceedings of the Symposium on Weed Management, B.A. Auld, R. C. Umaly, S. S. Tjitrosomo, eds. Special Publication No. 38, Bogor, Indonesia: Biotrop, 1990; pp. 115-129.
Hilton J.R. The influence of light on the germination of *Stellaria media*. New Phytol 1983; 94:29-37.
Holm L.G., Plucknett D.L., Pancho J.V., Herberger J.P. *The World's Worst Weeds: Distribution and Biology.* Honolulu, HI: East-West Center Books, 1977; pp. 450-455.

Jordan N. Weed prevention: priority research for alternative agriculture. J Prod Agric 1996; 9:485-490.

Karlen D.L., Varvel G.E., Bullock D.G., Cruse R.M. Crop rotations for the 21st century. Adv Agron 1994; 53:1-45.

Kaspar T.C., Erbach D.C., Cruse R.M. Corn response to seed-row residue removal. Soil Sci Soc Amer J 1990; 54:1112-1117.

Kegode G.O., Forcella F., Clay S. Influence of crop rotation, tillage, and management inputs on weed production. Weed Sci 1999; 47:175-183.

Kelley A.D., Bruns V.F. Dissemination of weed seeds by irrigation water. Weed Sci 1975; 23:486-493.

Kolbe W. Effect of weed control on grain yield of different winter barley cultivars with reference to sowing time, seed rate and seed size, in long-term trials at Höfchen and Laacherhof experimental stations (1968-1980). Pflanzenshutz-Nachrichten Bayer 1980; 33:203-219.

Koster A.T.J, van der Meer L.J., de Jong K.Y., van Haaster A.J.M., Kok B.J., van Aanholt J.T.M. Strategies for effective weed control in the future. Proceedings of the Seventh International Symposium on Flower Bulbs, Herzliya, Israel, 10-16 March 1996, H. Lilien-Kipnis, A. Borochov, A.H. Halevy, eds. Acta Horticulturae 1997; 430(II):669-675.

Kunelius H.T., Johnston H.W., MacLeod J.A. Effect of undersowing barley with Italian ryegrass or red clover on yield, crop composition and root biomass. Agric Ecosyst Env 1992; 38:127-137.

Lawson H.M., Wright G. M., Wilson B.J., Wright K.J. Seedbank persistence of five arable weed species in autumn-sown crops. Proceedings of the Brighton Crop Protection Conference — Weeds — 1993. Farnham, Surrey, UK: British Crop Protection Council, 1993; pp. 305-310.

Lawson H.M., Wiseman J.S. Competition between weeds and transplanted spring cabbage: effects of nitrogen top-dressing. Hort Res 1979; 19:25-34.

Leishman M.R., Sanbrooke K.J., Woodfin R.M. The effects of elevated CO_2 and light environment on growth and reproductive performance of four annual species. New Phytol 1999; 144:455-462.

Liebl R.A., Simmons F.W., Wax L.M., Stoller E.W. Effect of rye (*Secale cereale*) mulch on weed control and soil moisture in soybean (*Glycine max*). Weed Tech 1992; 6:838-846.

Liebman M. Effects of nitrogen fertilizer, irrigation, and crop genotype on canopy relations and yields of an intercrop/weed mixture. Field Crops Res 1989; 22:83-100.

Liebman M., Corson S., Rowe R.J., Halteman W.A. Dry bean responses to nitrogen fertilizer in two tillage and residue management systems. Agron J 1995; 87:538-546.

Liebman M., Dyck E. Crop rotation and intercropping strategies for weed management. Ecol Appl 1993; 3:92-122.

Liebman M., Davis A.S. Integration of soil, crop, and weed management in low-external-input farming systems. Weed Res 2000; 40:27-47.

Liebman M., Gallandt E.R. "Many little hammers: ecological management of crop-weed interactions." In *Ecology in Agriculture*. L.E. Jackson, ed. Orlando, FL: Academic Press, 1997; pp. 287-339.

Liebman M., Mohler C.L. "Weeds and the soil environment." In *Ecological Management of Agricultural Weeds*, M. Liebman, C.L. Mohler, C.P. Staver. Cambridge, UK: Cambridge University Press, 2001; pp. 210-268.

Liebman M., Robichaux R.H. Competition by barley and pea against mustard: effects on resource acquisition, photosynthesis and yield. Agric Ecosyst Env 1990; 31:155-172.

Liebman M., Staver C.P. "Crop diversification for weed management." In *Ecological Management of Agricultural Weeds*, M. Liebman, C.L. Mohler, C.P. Staver. Cambridge, UK: Cambridge University Press, 2001; pp. 322-374.

Ligneau L.A.M., Watt T.A. The effects of domestic compost upon the germination and emergence of barley and six arable weeds. Ann Appl Biol 1995; 126:153-162.

Magdoff F. Understanding the Magdoff pre-sidedress nitrate test for corn. J Prod Agric 1991; 4:297-305.

Malik V.S., Swanton C.J., Michaels T.E. Interaction of white bean (*Phaseolus vulgaris* L.) cultivars, row spacing, and seeding density with annual weeds. Weed Sci 1993; 41:62-68.

Matteson P. The "50% pesticide cuts" in Europe: a glimpse of our future? Amer Entom 1995; 41(4):210-220.

McLenaghen R.D., Cameron K.C., Lampkin N.H., Daly M.L., Deo B. Nitrate leaching from ploughed pasture and the effectiveness of winter catch crops in reducing leaching losses. New Zealand J Agric Res 1996; 39:413-420.

Melander B., Heisel T., Jorgensen M.H. Band-steaming for intra-row weed control. Proceedings of the 5th Workshop of the European Weed Research Society Working Group on Physical and Cultural Weed Control, 11-13 March 2002, Pisa, Italy. D.C. Cloutier, ed. Ste. Anne de Bellevue, Quebec, Canada: Institut de Malherbologie, 2002; pp. 216-219.

Melander B., Rasmussen K. Reducing intrarow weed numbers in row crops by means of a biennial cultivation system. Weed Res 2000; 40:205-218.

Menalled F.D., Liebman M., Renner K.A. "The ecology of weed seed predation in herbaceous cropping systems." In *Handbook of Sustainable Weed Management*, D.R. Batish, ed. Binghamton, NY: Haworth Press, *in press*.

Mertens S.K. Weed Communities on Six Ecological Farms in Flevoland. M.Sc. thesis, Departments of Theoretical Production Ecology and Ecological Agriculture, Wageningen University, Wageningen, The Netherlands, 1998.

Mertens S.K., van den Bosch F., Heesterbeek J.A.P. Weed populations and crop rotations: exploring dynamics of a structured periodic system. Ecol Appl 2002; 12:1125-1141.

Mertens S.K., Jansen J. Weed seed production, crop planting pattern, and mechanical weeding in wheat. Weed Sci 2002; 50:748–756.

Miura R., Kobayashi H., Kusanagi T. Effects of emergence time on survival and plant size in natural populations of *Stellaria media* and *S. neglecta*. Weed Res-Japan 1995; 40:179-186.

Mohler C.L. Effects of tillage and mulch on weed biomass and sweet corn yield. Weed Tech 1991; 5:545-552.

Mohler C.L. A model of the effects of tillage on emergence of weed seedlings. Ecol Appl 1993; 3:53-73.

Mohler C.L. Ecological bases for the cultural control of weeds. J Prod Agric 1996; 9:468-474.

Mohler C.L. "Mechanical management of weeds." In *Ecological Management of Agricultural Weeds*, M. Liebman, C.L. Mohler, C.P. Staver. Cambridge, UK: Cambridge University Press, 2001a; pp. 139-209.

Mohler C.L. "Weed life histories: identifying vulnerabilities." In *Ecological Management of Agricultural Weeds*, M. Liebman, C.L. Mohler, C.P. Staver. Cambridge, UK: Cambridge University Press, 2001b; pp. 40-98.

Mohler C.L. "Enhancing the competitive ability of crops." In *Ecological Management of Agricultural Weeds*, M. Liebman, C.L. Mohler, C.P. Staver. Cambridge, UK: Cambridge University Press, 2001c; pp. 269-321.

Monks C.D., Monks D.W., Basden T., Selders A., Poland S., Rayburn E. Soil temperature, soil moisture, weed control, and tomato (*Lycopersicon esculentum*) response to mulching. Weed Tech 1997; 11:561-566.

Moody K., Drost D.C. The role of cropping systems on weeds in rice. Proceedings of the Conference on Weed Control in Rice, 31 August-4 September 1981, Los Baños, The Philippines. Los Baños, The Philippines: International Rice Research Institute, 1983; pp. 73-86.

Mt. Pleasant J., Schlather K.J. Incidence of weed seed in cow (*Bos* sp.) manure and its importance as a weed source for cropland. Weed Tech 1994; 8:304-310.

Müller-Schärer H., Frantzen J. An emerging system management approach for biological weed control in crops: *Senecio vulgaris* as a research model. Weed Res 1996; 36:483-491.

Murdock E.C., Banks P.A., Toler J.E. Shade development effects on pitted morningglory (*Ipomoea lacunosa*) interference with soybeans (*Glycine max*). Weed Sci 1986; 34:711-717.

Murphy S.D., Yakubu Y., Weise S.F., Swanton C.J. Effect of planting patterns and inter-row cultivation on competition between corn (*Zea mays*) and late emerging weeds. Weed Sci 1996; 44:865-870.

Myers S., Rorie S. Facts and stats: the year in review. 2001. Available on-line at: http://www.organicandnaturalnews.com/articles/0c1feat1.html.

Natarajan M., Willey R.W. Sorghum-pigeonpea intercropping and the effects of plant population density. 1. Growth and yield. J Agric Sci, Cambridge 1980a; 95:51-58.

Natarajan M., Willey R.W. Sorghum-pigeonpea intercropping and the effects of plant population density. 2. Resource use. J Agric Sci, Cambridge 1980b; 95:59-65.

Norris R.F. Ecological implications of using thresholds for weed management. J Crop Prod 1999; 2:31-58.

Norris R.F., Ayres D. Cutting interval and irrigation timing in alfalfa: yellow foxtail invasion and economic analysis. Agron J 1991; 83:552-558.

Ohno T., Doolan K., Zibilske L.M., Liebman M., Gallandt E.R., Berube C. Phytotoxic effects of red clover amended soils on wild mustard seedling growth. Agric Ecosyst Env 2000; 78:187-192.

Olofsdotter M. Getting closer to breeding for competitive ability and the role of allelopathy–an example from rice (*Oryza sativa*). Weed Tech 2001a; 15:798-806.

Olofsdotter M. Rice–a step toward use of allelopathy. Agron J 2001b; 93:3-8.

Ominski P.D., Entz M.H., Kenkel N. Weed suppression by *Medicago sativa* in subsequent cereal crops: a comparative survey. Weed Sci 1999; 47:282-290.

Ozores-Hampton M., Stoffella P.J., Bewick T.A., Cantliffe D.J., Obreza T.A. Effect of age of composted MSW and biosolids on weed seed germination. Compost Sci Util1 999; 7:51-57.

Paul N.D., Ayres P.G., Hallett S.G. Mycoherbicides and other biocontrol agents for *Senecio* spp. Pesticide Sci 1993; 37:323-329.

Peacock L., Norton G.A. A critical analysis of organic vegetable crop protection in the U.K. Agric Ecosyst Env 1990; 31:187-197.

Pino J., Sans F.X., Masalles R.M. Population dynamics of *Rumex obtusifolius* under contrasting lucerne cropping systems. Weed Res 1998; 38:25-33.

Pitty A., Staniforth D.W., Tiffany L.H. Fungi associated with caryopses of *Setaria* species from field-harvested seeds and from soil under two tillage systems. Weed Sci 1987; 35:319-323.
Rasmussen J., Ascard J. Weed control in organic farming systems. In *Ecology and Integrated Farming Systems,* M. Glen, M.P. Greaves, H.M. Anderson, eds. Chichester, UK: Wiley Publishers, 1995; pp. 49-67.
Rasmussen K. Influence of liquid manure application method on weed control in spring cereals. Weed Res 2002; 42:287-298.
Rasmussen K., Rasmussen J., Petersen J. Effects of fertilizer placement on weeds in weed harrowed spring barley. Acta Agric Scandinavica, Section B, Soil Plant Sci 1996; 46:192-196.
Roberts H.A., Chancellor R.J., Hill T.A. "The biology of weeds." In *Weed Control Handbook – Principles (Volume 1, 7th edition).* H.A. Roberts, ed. Oxford, UK: Blackwell Scientific Publications, 1982; pp. 1-36.
Roberts H.A., Feast P.M. Emergence and longevity of seeds of annual weeds in cultivated and undisturbed soil. J Appl Ecol 1972; 10:133-143.
Roberts H.A., Lockett P.M. Germination of buried and dry-stored seeds of *Stellaria media.* Weed Res 1975; 15:199-204.
Ross S.M., King J.R., Izaurralde R.C., O'Donovan J.T. Weed suppression by seven clover species. Agron J 2001; 93:820-827.
Sakai K., Sasao A., Shubusawa S., Ohara M. Weed ecological system modeling for optimizing tillage process. Proceedings of the ASAE-CSAE-SCGR Annual International Meeting, Toronto, ON, 18-21 July 1999. ASAE Paper No. 995032, 1999.
Salisbury E.J. *Weeds and Aliens.* London, UK : Collins, 1961.
Schnieders B.J. *A Quantitative Analysis of Interspecific Competition in Crops with a Row Structure.* Ph.D. thesis, Department of Theoretical Production Ecology, Wageningen University, Wageningen, The Netherlands, 1999.
Schonbeck M.W. Weed suppression and labor costs associated with organic, plastic, and paper mulches in small-scale vegetable production. J Sust Agric 1998; 13(2):13-33.
Schreiber M.M. Influence of tillage, crop rotation, and weed management on giant foxtail (*Setaria faberi*) population dynamics and corn yield. Weed Sci 1992; 40:645-658.
Seaman, D.E. Farmers' weed control technology for water-seeded rice in North America. Proceedings of the Conference on Weed Control in Rice, 31 August-4 September 1981, Los Baños, The Philippines. Los Baños, The Philippines: International Rice Research Institute, 1983; pp. 167-179.
Shetty S.V.R., Rao A.N. Weed management studies in sorghum/pigeonpea and pearl millet/groundnut intercrop systems — some observations. Proceedings of the International Workshop on Intercropping, 10-13 January 1979, Hyderabad, India. R.W. Willey, ed. Patencheru, India: ICRISAT, 1981; pp. 238-248.
Shirilipour A., McConnell D.B. Effects of compost heat and phytotoxins on germination of certain Florida weed seeds. Soil Crop Sci Soc Florida Proceedings 1991; 50:154-157.
Simberloff D., Stiling P. How risky is biological control? Ecology 1996; 77:1965-1974.
Smith M.S., Frye W.W., Varco J.J. Legume winter cover crops. Adv Soil Sci 1987; 7:95-139.
Soil Association. Organic facts and figures, May 2001: European organic food production. 2001. Available on-line at: http://www.soilassociation.org/SA/SAWebDoc.nsf.
Stapleton J.J., DeVay J.E. "Soil solarization: a natural mechanism of integrated pest management." In *Novel Approaches to Integrated Pest Management*, R. Reuvani, ed. Boca Raton, FL: Lewis Publishers, 1995; pp. 309-322.
Steinbauer G.P., Grigsby B., Correa L., Frank P. A study of methods for obtaining laboratory germination of certain weed seeds. Assoc Official Seed Anal Proc 1955; 45:48-52.
Stewart R.H., Lynch K.W., White E.M. The effect of growing clover cultivars in association with barley cultivars upon grain yield of the barley crop in the year of sowing and the subsequent year. J Agric Sci, Cambridge 1980; 95:715-720.
Storkey J., Cussans J.W. Relationships between temperature and the early growth of *Triticum aestivum* and three weed species. Weed Sci 2000; 48:467-473.
Stute J.K., Posner J.L. Legume cover crops as a nitrogen source for corn in an oat-corn rotation. J Prod Agric 1995; 8:385-390.
Swanton C.J., Griffiths J.T., Cromar H.E., Booth B.D. Pre- and post-dispersal weed seed predation and its implications to agriculture. Proceedings of the 1999 Brighton Crop Protection Conference–Weeds. Farnham, Surrey, UK: British Crop Protection Council, 1999; pp. 829-834.
Teasdale J.R. Influence of narrow row/high population corn (*Zea mays*) on weed control and light transmittance. Weed Technol 1995; 9:113-118.
Teasdale J.R. "Cover Crops, Smother Plants, and Weed Management." In *Integrated Weed and Soil Management*, J.L. Hatfield, D.D. Buhler, B.A. Stewart, eds. Chelsea, MI: Ann Arbor Press, 1998; pp. 247-270.

Teasdale J.R., Daughtry C.S.T. Weed suppression by live and dessicated hairy vetch (*Vicia villosa*). Weed Sci 1993; 41:207-212.

Teasdale J.R., Frank J.R. Effect of row spacing on weed competition with snap beans (*Phaseolus vulgaris*). Weed Sci 1983; 31:81-85.

Teasdale J.R., Mohler C.L. The quantitative relationship between weed emergence and the physical properties of mulches. Weed Sci 2000; 48:385-392.

Vandermeer J. *The Ecology of Intercropping.* Cambridge, UK: Cambridge University Press, 1989.

Vereijken P. Designing prototypes. Progress Report 1 of the Research Network on Integrated and Ecological Arable Farming Systems for EU and Associated Countries. AB-DLO, Wageningen, The Netherlands, 1994.

Vereijken P. A methodical way of prototyping integrated and ecological arable farming systems (I/EAFS) in interaction with pilot farms. European J Agron 1999; 7:235-250.

Vezina L. Influence de la date de semis sur la compétition de peuplements de *Setaria pumila* et d'*Echinochloa crus-galli* avec l'orge et le blé de printemps. Weed Res 1992; 32:57-65.

Walz E. Third Biennial National Organic Farmers' Survey. Santa Cruz, CA: Organic Farming Research Foundation, 1999.

Weaver S.E. Critical period of weed competition in three vegetable crops in relation to management practices. Weed Res1984; 24:317-325.

Weaver S.E., Kropff M.J., Groeneveld R.M.W. Use of ecophysiological models for crop-weed interference: the critical period of weed interference. Weed Sci 1992; 40:302-307.

Weston L.A. Utilization of allelopathy for weed management in agroecosystems. Agron J 1996; 88:860-866.

Willey R.W. Intercropping: its importance and research needs. Part 1. Competition and yield advantages. Field Crops Abstr 1979; 32:1-10.

Willey R.W. Resource use in intercropping systems. Agric Water Manage 1990; 17:215-231.

Williams J., Wick C., Scadaci S., Klonsky K., Chaney D., Livingstone P., Tourte L. Sample costs to produce organic rice, water-seeded, in the Sacramento Valley. Davis, CA: University of California Cooperative Extension, 1992.

Wilson B.J., Wright K.J., Butler R.C. The effect of different frequencies of harrowing in the autumn or spring on winter wheat, and on control of *Stellaria media, Galium aparine* and *Brassica napus*. Weed Res 1993; 33:501-506.

Wilson R.G. Dissemination of weed seeds by surface irrigation water in western Nebraska. Weed Sci 1980; 28:87-92.

Zhang J., Drummond F.A., Liebman M. Effect of crop habitat and potato management practices on the population abundance of adult *Harpalus rufipes* (Coleoptera: Carabidae) in Maine. J Agric Entom 1998; 15:63-74

CHAPTER 15

IMPROVEMENT OF ALLELOPATHY IN CROPS FOR WEED MANAGEMENT

POSSIBILITIES, BREEDING STRATEGIES AND TOOLS

MARIA OLOFSDOTTER[1] & SVEN BODE ANDERSEN[2]

[1]*Forskningskoordinator, Øresund Food Network, Gl. Kongevej 1, 4, 1610 København V, Denmark;* [2]*Department of Agricultural Sciences, KVL, Thorvaldsens vej 40, 1871 Frederiksberg C, Denmark.*

Weeds are a major concern in all agricultural systems around the world. In contrast to the progress observed with respect to disease resistance, few attempts have been made to breed crops to be competitive against weeds. This is partly due to the success of herbicides as an effective solution against weeds. The apparent complex nature of competitive ability in crops and the widespread impression that competitive ability is negatively correlated with yield is another reason weed suppressive crops have not been selected for. However, recent research with rice shows that it is possible to increase competitive ability without affecting yielding ability (Jones et al., 1997; Olofsdotter et al., 1999). By increasing our understanding of plant competitive ability, we may potentially improve a crop's competitive ability, the key to successful breeding of weed-fighting cultivars (Mortimer et al., 1999). Theoretically, enhanced crop production, independent of weed growth, can be obtained in three different ways: (i) growing plants that will yield well in spite of weed infestations (weed tolerant crops or cultivars), (ii) growing plants that suppress the surrounding flora (weed-suppressing crops or cultivars) (Lemerle et al., 2001) or (iii) both strategies used at the same time. For long-term weed reduction, greater benefits exist with the later strategy of selecting for weed suppression and thereby increasing competitive ability. Increased crop competitive ability, will suppress weed impair their regenerative capacity, thus decreasing weed abundance in the field over time.

Competitive ability can be summarized by several general features, which together provide better fitness in a given environment. Some features are genetically regulated and include early vigor, fast and voluminous vegetative growth, high nutrient acquisition and use efficient, tolerance to pests, tolerance to abiotic stress and finally chemical defense mechanisms through mycorrhiza or allelopathy (Caton et al., 1999; Grace, 1990; Jordan and Jannink, 1997; Kirk et al., 1998; Kropff et al., 1995; Moolsri et al., 1999; Olofsdotter et al., 1999; Ranasinghe and Crabtree, 1999).

However, due to the complexity of the inheritance of many of these traits, limited research work has been performed to determine and understand the genetics of these features. This leads to a situation where breeders who might be willing to work towards more competitive cultivars do not have the knowledge or tools necessary for undertaking such a breeding program.

Inderjit (ed.), Weed Biology and Management, 317-328.

One of the genetically driven forces in competitive ability, mentioned above, is the allelopathic potential of the plant. Allelopathy can be defined as chemicals released from higher plants which influence development and growth of one or more other plant species in close proximity to the donor species. The allelochemicals causing the effect might be directly phytotoxic or indirectly phytotoxic through mediation in the soil environment (Inderjit, 2001). In rice, field allelopathy has been reported to account for 34 % of variation in weed suppressing ability of different rice cultivars when grown in association with Echinochloa crus-galli (L.) Beauv. (Olofsdotter et al., 1999). Similarly, cultivar screenings conducted with rice, wheat and barley have shown that allelopathic potential varies between cultivars indicating the possibility for manipulating the trait through breeding (Courtois and Olofsdotter, 1998; Wu et al., 1999; Wu et al., 2001; Liu and Lovett , 1993a). This paper reviews the present knowledge of allelopathy genetics, outlines possibilities for improving overall competitive ability in crops via allelopathy, and suggests suitable tools for genetic improvement of crop allelopathy.

ALLELOPATHY IN CROPS

Over time, many crops have been reported to possess allelopathic properties (Inderjit and Keating, 1999). However, few reports have been published on variability within crop species as far as allelopathic potential is concerned.

Cucumber

The first modern evidence of allelopathy in a crop was in cucumber (*Cucumis sativus*), in an experiment involving a germplams collection of 526 accessions of cucumber originating from 41 countries. The experiment showed that several accessions could inhibit growth of *Panicum miliaceum* L. and *Brassica hirta* Moench. One accession reduced growth of the test species by 87%, and 25 other accessions showed more than 50% growth inhibition (Putnam and Duke, 1974). A later field study showed that allelopathic activity, which was measured in the laboratory, could not be directly translated to field conditions. One allelopathic cucumber accession reduced the total weed population by 54% but the growth inhibition did not occur consistently across the field. It was also shown that rainfall suppressed the allelopathic effect (Lockerman and Putnam, 1979). This indicates, not surprisingly, that environmental conditions can influence the allelopathic effects of a crop.

Barley

For centuries, barley (*Hordeum vulgare* L.) has been known as a `smother' crop capable of suppressing weeds. Part of the possible explanation for this was found when Overland (1966) demonstrated that barley produced and released phytotoxic substances in concentrations that inhibit growth of several common weeds. In barley as well as in oats a group of chemicals, gramines, is believed to be responsable for this allelopathic potential. The amount of gramines varies with barley cultivar (Yoshida et al., 1993). However, their results also indicate that gramine content in the leaves does not necessarily correlate with the amount released from the cultivars into the soil. Looking closer into the released gramines from barley, revealed no cultivar differences in the release of the gramine hordenine as such, but when comparing *H. vulgare* with *H. spontaneum* Koch. a different pattern in the release was apparent (Lovett et al., 1994). *H. vulgare* cultivars released more hordenine over a longer time with a maximum release (2μg/plant/day) at 36 days after sowing (Liu and Lovett, 1993b). However, recent screening of barley accessions for allelopathic potential against weeds shows that there are cultivar differences in the weed suppressive ability (Kudsk et al., 1998). These results together indicate that content and release of hordenine might not be a reliable measure for allelopathic potential of barley.

Similarly, a large screening among 3000 *Avena* spp. accessions revealed biotype differences in content of another gramine, scopoletin. Scopoletin is a known phytotoxic compound that causes root growth inhibition in several plant species. Among the 3000 *Avena* spp., 25 accessions contained significantly higher scopoletine concentration when compared to a control cultivar 'Garry'. Among these, four cultivars contained up to three times more scopoletin and inhibited growth of wild mustard in sand culture. Plants in close association with allelopathic accessions were chlorotic, stunted and/or twisted, indicating that inhibition was caused by a chemical compound (Fay and Duke, 1997).

Rye

Many studies on allelopathy in cereals have reported weed suppression from rye mulch (Putnam and DeFrank, 1983; Shilling et al., 1987; Barnes and Putnam, 1986; Barnes et al., 1986). Rye generally provides good mulching material with strong weed suppressive ability. The resulting weed suppression, however, can be due to a combination of several factors such as shading and changes in water potential as well as chemicals released from the degrading plant material. Such chemicals released from degrading crop residues may never be released from the living plant itself. There are indications that the weed suppressive ability of rye mulch varies with cultivar (Barnes and Putnam, 1987) but the allelopathic effects of different rye cultivars growing in the field are still unknown.

All above-mentioned crops have shown indications of cultivar differences with respect to their allelopathic ability. However, little has been done to study the genetics behind such cultivar differences, leaving breeders short of tools necessary for improving crop cultivars for weed suppressive ability. Only very recent studies on wheat and rice have evaluated genetics of weed supression and these findings are discussed below.

Rice

Allelopathy in rice was first discovered accidentally in the late 1980s. In a seed increase plot for the Arkansas rice germplasm bank, a plant breeder observed that certain rice cultivars had a weed free zone around the base of the plant. (Bob Dilday, pers. comm.). The field had a natural mono-species weed infestation of ducksalad (*Heteranthera limosa* (Sw.) Willd.). At the time of the observation, the rice crop was not large enough to shade the area around the base of the plant and a chemical effect was indicated. Since then, 12,000 rice cultivars have been screened for allelopathy and 3.4% of the accessions have been found to be allelopathic against one of the weed species evaluated (Dilday et al., 1998; Hassan et al., 1998; Olofsdotter et al., 1999). Aside from allelopathy against ducksalad, the American group has also screened for rice allelopathy against redstem (*Ammannia coccinea* Bottb.). Four hundred and twelve rice accessions were identified as having allelopathic potential against ducksalad, 145 against redstem, but only 16 accessions showed effect on both weed species (Dilday et al., 1998). This indicates that rice allelopathy may be selective, which means that a rice cultivar allelopathic against one weed is not necessarily allelopathic against other weed species. This selective action indicates that several chemical compounds are involved in rice allelopathy and/or that weeds have different mechanisms to break down the allelochemicals. It also suggests that allelochemicals in rice are not likely to be common and widespread secondary metabolites such as simple phenolic acids. Since the initial findings in Arkansas, several research groups around the globe have been studying rice allelopathy. Besides the weed species mentioned above, rice with weed suppressive activity against broadleaf signalgrass (*Brachiaria platyphylla* (Griseb.) Nash), rice flatsedge (*Cyperus iria* L.), sprangletop (*Leptochloa* spp.), and barnyardgrass (*Echinochloa crus-galli*) have been reported (Hassan et al., 1998; Kim and Shin, 1998; Olofsdotter et al., 1999).

Allelopathy effects in rice are, as shown above, quite widespread and can suppress a wide range of weed species selectively. Accessions with allelopathic activity have been shown to originate from all different rice producing areas in the world suggesting that allelopathy occurs in rice accessions and cultivars which are genetically diverse. A small number of the rice cultivars show allelopathic effects against a range of weeds. Among these is the cultivar Taichung native 1 (TN 1) originating from Taiwan, showing allelopathic effect against at least five important weed species (*E. crus-galli, E. colona* L. Link, *Trianthema portulacastrum* L., *H. limosa* and *A. coccinea*) (Dilday et al., 1998; Olofsdotter and Navarez, 1996; Pheng et al., 2001). Because of this broad-spectrum effect, TN 1 has been selected as a model cultivar from which to identify the responsible allelochemicals. This identification work uses bioassay-directed isolation to ensure identification of both known and unknown compounds (Rimando et al., 1998). In bioassay-directed isolation every fraction in the fractionation procedure are tested in a bioassay to ensure that all phytotoxic compounds are identified. So far, four fractions of pure chemicals extracted from rice leaves with selective phytotoxicity against *E. crus-galli* have been characterized (Rimando et al., 2001).

Wheat

Inspired by the success in rice, a project was started searching for allelopathic differences in wheat germplasm in Australia. Laboratory screening of 453 wheat accessions showed continuous variation in wheat allelopathy against annual ryegrass (*Lolium rigidum* Gaudin.) similar to that found in rice (Wu et al., 2000a, b). Among wheat varieties studied, 63 were highly allelopathic, inhibiting root growth by more than 81%. These highly allelopathic accessions originated from 23 different countries, suggesting that allelopathy in wheat is widespread (Wu et al., 2000a). The responsible allelochemicals in wheat have not yet been published. However, many reports have suggested allelopathy in both wheat and rye to be related to the production of hydroxamic acids in the plant leaves. Wheat and rye produce, among others, the hydroxamic acids, DIBOA (2,4-dihydroxy-1,4(2H)-benzoxazin-3-one) and DIMBOA (2,4-dihydroxy-7-methoxy-1,4-benzoxazin-3-one).Both acids produce the degradation product BOA (2(3H)-benzoxazolinone) known to be phytotoxic to several plant species (Perez, 1990; Perez and Ormeno-Nunez, 1993; Barnes and Putnam, 1987). According to a screening of Swedish wheat cultivars for insect resistance (Escobar and Niemeyer, 1993), the quantities of acids that are produced and released vary according to genotype and environment. Besides cultivar differences, it was also found that old cultivars release more DIMBOA than modern high yielding ones. Although some of the hydroxamic acids are known to be phytotoxic, there is no conclusive evidence indicating that these chemicals are the only active allelochemicals in wheat and rye. Recently, a new hydroxamic acid, (2R)-2-beta-D-glucopyranosyloxy-4,7-dimethoxy-2H-1,4-benzoxazin-3(4H)-one (GHDIBOA) produced by wheat was found (Kluge et al., 1997). This indicates that unknown compounds are still to be found and the search for allelochemicals in wheat should continue, preferably without prior decision as to which compounds to search for using bioassay-guided isolation (Rimando et al., 1998).

TOOLS FOR IMPROVEMENT OF ALLELOPATHIC POTENTIAL

The ability of some rice and wheat plants to suppress competing weeds through the release of phytotoxic chemicals is apparently affected by multiple genes. Since the growth environment of the plant also affects the trait, culturing methods and treatments during growth may also be important for its ability to release phytotoxic chemicals. A general improvement of the allelopathic strength of cultivars may be obtained through breeding for the trait in the future. To assist in such future attempts, recent technological improvements in plant genetics may be useful.

Allelopathy in both rice and wheat is apparently quantitatively inherited. Improvement of such quantitative traits by traditional methods is slow and resource demanding if based upon selection of phenotypes in the field, because the phenotype is affected both by several genes and expression of these genes may be altered by the environment. With DNA marker technologies it is now possible to dissect such quantitative traits into their underlying genes, named quantitative trait loci (QTLs). This means that each gene locus for the trait can be localized to a specific chromosome area. Each locus can be assigned a name and different mutations (alleles) of the locus affecting the trait can be identified. Such dissection of a quantitative trait into its underlying QTLs therefore enables a much more systematic breeding for the trait. QTL mapping of quantitative traits like allelopathy needs specially constructed offspring from crosses between parents with different genes for the trait. Such offspring are subsequently evaluated carefully for their expression of the quantitative trait. In addition, the offspring are analyzed for segregation of large numbers of genetic markers, which are molecularly recognizable points distributed across the chromosomes. QTLs for the trait are found through statistical comparison of inheritance of both the trait and the markers among the offspring, using special statistical procedures. The main principle of such QTL analysis is that markers for chromosomal segments with a QTL are inherited together with the trait (Andersen and Torp, 2001; Jensen et al., 2001). Offspring (mapping populations) used for such QTL mapping can be of two major types. The easiest mapping populations to produce are those in which the plants are still partially heterozygous, like F_2 or backcross of F_1

to one parent. Heterozygosity of the material means that offspring from each plant will still segregate genetically and therefore in reality only one plant of each genotype is available for trait evaluation. Mapping populations consisting of homozygous plants like recombined inbreds (RIL) or chromosome doubled haploids (DH) requires more resources to prepare. However, because of the homozygosity of this material each genotype can be multiplied into homogenous lines, which enables evaluation of each line for the trait in replicated trials in different environments (Andersen and Torp, 2001).

In addition to suitable genetic marker technology, which is now well developed for several major plant species, reliable methods for measurement of the trait value is needed for each trait to be mapped. In the case of allelopathy, this means efficient methods are needed to measure allelopathic ability of the offspring towards different weed species under conditions, which will correlate well with field behavior.

Identification of major QTLs responsible for allelopathic ability in major crop plants will enable much more efficient concentration of such genes in cultivated material, because they can be manipulated by selection of genetically linked markers. In most cases, QTL mapping of a quantitative trait in plants leads to identification of a number of QTLs, while the actual gene responsible for the quantitative trait loci remains unknown. A much more efficient search for desirable alleles in such genes and more efficient manipulation of such mutations can be obtained if the sequences of the actual genes can be obtained (Daly and Day, 2001; Fukai et al., 2001), as exemplified by beta-amylase affecting malting quality in barley (Kaneko et al., 2001) and high molecular weight glutenines affecting baking quality in wheat (Rousset et al., 2001). Such detailed knowledge of the genes also enables approaches using genetic engineering to learn how the genes and their interacting molecular environment exert their efforts. In this way, one can find true molecular strategies for improvement of the traits. Unfortunately, for most true QTLs, such identification and isolation of the underlying candidate genes is still a rather difficult task. This is because QTLs cannot in general be mapped very accurately and if additional information about the mechanism of function is not known, it is difficult to know which of the many genes in a QTL chromosome segment is affecting the trait. For allelopathic QTLs, such information may be gained through studies of amounts and release of chemical compounds among offspring used for the QTL mapping procedure. Changes of potential allelochemical compounds, which will co-segregate with one or more of the mapped QTLs associated with allelopathic ability, will be good candidates for an explanation of the gene action. Detailed study of structure and metabolic pathways of such compounds may subsequently lead to identification of major regulatory enzymes for their control and thereby identification of candidate genes for mapped QTLs. Such identification of major genes and steps in biosynthetic pathways may form the basis of a strategy for the use of allelopathy in future agronomic production.

ALLELOPATHY GENETICS

The variability in allelopathic potential within crop species suggests that genetic differences for the trait may be utilized for breeding of weed-suppressive crop cultivars. However, except for the recent reports in rice and wheat, few attempts to understand the genetics of allelopathy have been reported. Since the active allelochemicals are not known, a good phenotyping system, to evaluate the allelopathic potential of a given genotype, must be used to account for all effects in a quantitative genetics approach. In both rice and wheat, such screening systems have been developed (Navarez and Olofsdotter, 1996; Wu et al., 2000b).

Rice

From laboratory experiments with rice, the calculated broad-sense heritability for reduction of barnyardgrass root growth was 0.85. Heritability measures the fraction of variation determined by genes and varies between 0 and 1. Following allelopathy screenings using the same evaluation techniques have produced comparable estimates of heritability ranging from 0.7 to 0.85, (Jensen et al., 2000, 2001). As expected, screening results revealed continuous variation between accessions for allelopathic strength. From these results, it is clear that allelopathic activity in rice is inherited in a quantitative manner, but with an apparently high heritability, thereby offering good prospects for future breeding-based manipulation of the trait (Courtois and Olofsdotter, 1998). The high heritability of allelopathic effects of rice, together with their selective action against different weed species, may eventually indicate the action of major genes affecting pathways for biosynthesis or release of allelochemicals.

To study the genetics of allelopathy in rice, several crosses were made between allelopathic and non-allelopathic parents (Olofsdotter and Courtois, unpublished). One of these offsprings has been developed into an F6 population of 252 recombinant inbreed lines suitable for QTL analysis.

Developing a population of recombinant inbreed lines takes several years. While waiting for the development of the designed population, work was initiated to screen parental material for other already available RIL populations, which had been developed for mapping of other characteristics. Among these, a population was found, which was derived from a cross between IAC 165, a japonica upland cultivar, highly allelopathic against *Echinochloa crus-galli*, and CO 39, an indica irrigated cultivar, which was only weakly allelopathic (Jensen et al., 2001a). This population had been developed previously to study blast resistance and had also been used to study root morphology (Mauleon, 1995; Shen et al., unplublished data). The population has 250 RILs produced by single seed descent. Using this population, four QTLs were identified to be located at 3 different chromosomes (Jensen et al., 2001a, b). The identified QTLs accounted for 35% of the variation found in the evaluation of the lines. This means that there are still more genes affecting the trait to be detected among these lines. Since the population also had been mapped for root morphology traits, it could also be used to establish any correlation between allelopathy in rice and root morphology. Correlation analysis showed that no significant correlation between the two traits. This is an important finding because such correlation could have impact on breeding efforts if allelopathy required special root morphology or root mass to be effective (Jensen et al., 2001a, b).

Laboratory work on mapping the rice population designed for mapping allelopathic traits was finished recently. This population consists of recombinant inbred lines from a cross between an AUS 196 (non-allelopathic) and AC 1423 an allelopathic parent, also developed through single seed decent. The map from this population is still under preparation, but phenotyping data show large variation in allelopathic potential in this population. Of the total variation for allelopathy in this mapping population, 71% could be explained by gene regulation, which is an acceptably high genetic determination for breeding purposes (Jensen et al., 2000). The phenotyping data suggest that the allelopathic parent, AC 1423, is the strongest allelopathic parent in the cross, but there are indications that the non-allelopathic parent, AUS 196, also contributes some genes adding to the allelopathic potential in this cross.

Other research groups have identified genes for allelopathy in rice. Five QTLs on five different chromosomes explaining 63.5% of the allelopathic potential against ducksalad (*Heteranthra limosa*) were found using an F2 population of a cross between the commercial variety Rexmont (non-allelopathic) and an allelopathic breeding line PI 312 777 (Ebana et al., 2001). These authors, however, used rice extracts for phenotyping the population and it remains to be seen, how and if similar chemical compounds are released to the soil from living rice plants.

Wheat

Similarly, the allelopathic strength in wheat against ryegrass (*Lolium rigidum*) has been reported to vary continuously indicating quantitative inheritance (Wu et al., 2000 a, b). When the origin of the Australian allelopathic germplasm was studied, it was found that 30 accessions shared a common allelopathic parent, Condor, while another 12 accessions shared a non-allelopathic parent, Pavon. Comparison of these accessions separately produced two distinct groups of activity against ryegrass. Lines with Condor background showed stronger allelopathic activity than lines with Pavon background (Wu et al., 2000a). These results with wheat indicate that breeding for allelopathy should be possible also in this species.

BREEDING STRATEGIES FOR INCREASING OVERALL COMPETITIVE ABILITY USING ALLELOPATHY

Knowledge of the genetic and molecular basis of allelopathy in plants will ultimately provide us with the possibility to breed more crop cultivars which are more competitive against weeds. Usefulness of such approaches in agriculture will depend on identification of allelopathic compounds or genes with strong weed suppressive effects, which can be released in an effective manner into the soil environment.

In rice, allelopathic activity appears to be selective (Olofsdotter et al., 1997). This selective action can be due to different allelochemicals or to different amounts of such chemicals released from different rice cultivars. Independently of the mode of selectivity the most beneficial type of allelopathic crop would be a non-selective one, capable of suppressing a wide spectrum of weeds in the crop. Such systems would be simple and efficient to use for the farmer. However, most literature suggests that this will not be immediately possible because only few plant accessions have shown a wide species range in their phytotoxicity. Non-selective action against weeds might be possible to obtain using a strategy maximizing several features contributing to overall competitive ability of the crop species. One successful attempt to do so was achieved at West African Rice Development Agency (WARDA). Researchers from different disciplines collaborated to design a rice "plant type" with favorable morphological competitive traits. A wide cross between *Oryza glaberrima* (African rice) and *Oryza sativa* was achieved to combine morphological competitive traits from *O. glaberrima* with yielding ability from *O. sativa* (Jones et al., 1997; Dingkuhn et al., 1997). Some of these "new African rice" cultivars have shown substantial productivity gains (24%) when compared to local traditional cultivars in farmer-controlled evaluations as well as increased overall competitive ability against weeds, measured by reduced weed biomass in the field, when compared to *O. sativa* (Anonymous, 1999).

Exploitation of selective types of allelopathy to improve crop competitiveness against many weed species will require a more complicated strategy to succeed. One strategy could include pyramiding of several genes with strong effects against most weed species in each crop cultivar. It must be expected that weeds that are tolerant to the allelopathic compounds or with some ability to escape their effects will occur. If the allelopathic cultivars are eventually used as the only weed management tool there will be changes in the weed flora towards a "harder to control" one, as known from all other weed management strategies when applied alone (Haas and Streibig, 1982). Additionally, weeds might develop resistance towards the allelochemicals released from the crop. The time frame for such development is dependent on the number of genes involved at the target site of the allelochemical, in principle similar to development of herbicide resistance. The general rule is that the larger the number of genes involved, the longer it will take for resistance to develop. Depending on durability of the trait, genes or cultivars with such allelopathic effects will have to be exchanged with new ones harboring new genes for the trait. Alternatively, durability of the allelopathic traits can be prolonged through crop rotations with different cultivars of the same or different crop species with alternative enhanced capability to effectively compete with weeds. This will reduce selection for tolerance among the weed population and has successfully been applied for reducing development of herbicide resistance in parts of the world. An alternative to pyramiding of many genes in one cultivar would be to grow mixtures of more or less genetically identical lines with different genes suppressing one or a few weed species each. An advantage of such a multi line approach is that relative amounts of each component line in the mixture can be adjusted each year according to changes in the weed composition and new allelopathic genes can be introduced easily to replace existing ones, against which tolerance has developed among the weed.

In any case, allelopathy alone will not solve all weed problems. A breeding strategy that tries to optimize the overall competitive ability of the plant will probably be most successful. Beside allelopathy, such a strategy should include both morphological traits and physiological traits adding to competitive ability. This can only be done with input of knowledge from a wider range of scientists from different disciplines like agronomy, weed science, plant physiology, chemistry, biotechnology as well as plant breeding.

REFERENCES

Andersen S.B., Torp A.M. "QTL mapping in crop plants." In *MolecularTtechniques in Crop Improvement,* S. M. Jain, D. S. Brar, B. S. Ahluwalia, eds. The Netherlands: Kluwer Academic Publishers, 2001.

Anonymous. Impact of varietal improvements in West African crop ecologies. West Africa Rice Research Brief 1999; pp. 1-2.

Barnes J.P., Putnam A.R. Evidence for allelopathy by residues and aqueous extracts of rye (*Secale cereale*). Weed Sci 1986; 34:384-390.

Barnes J.P., Putnam A.R. Role of benzoxazinones in allelopathy by rye (*Secale cereale* L:). J Chem Ecol 1987; 13:889-906.

Barnes J.P., Putnam A.R., Burke B.A. "Allelopathic activity of rye (*Secale cereale* L.)." In *The Science of Allelopathy,* A. R. Putnam, C. S. Tang, eds. New York, NY: John Wiley & Sons, 1986; pp. 271-286.

Caton B.P., Mortimer A.M., Foin T.C., Hill J.E., Gibson K.D., Fischer A.J. Weed morphology effects on competitiveness for light in direct-seeded rice. November 22-27, Bangkok, Asian-Pacific Weed Science Society. Proceedings of the 17th Asian-Pacific Weed Science Society conference, 1999; pp. 116-120.

Courtois B., Olofsdotter M. "Incorporating the allelopathy trait in upland rice breeding programs." In *Allelopathy in Rice,* M. Olofsdotter, ed. Los Banos, Philippines: International Rice Research Institute, 1998; pp. 57-68.

Daly A.K., Day C.P. Candidate gene case-control association studies: advantages and potential pitfalls. British J Clin Pharmac 2001; 52:489-499.

Dilday R.H., Yan W.G., Moldenhauer K.A., Gravois K.A. "Allelopathic activity in rice for controlling major aquatic weeds." In *Allelopathy in Rice,* M. Olofsdotter, ed. Los Banos, Philippines: International Rice Research Institute, 1998; pp. 7.26.

Dingkuhn M., Jones M.J., Johnson D.E., Fofana B., Sow A. "*Oryza sativa* and *O. glaberrima* gene pools for high-yielding, weed competitive rice plant types." In *Breeding Strategies for Rainfed Lowland Rice in Drought-Prone Environments,* S. Fukai, M. Cooper, J. Salisbury, eds. ACIAR Proceedings. 1997; pp. 144-155.

Ebana K., Yan W., Dilday R.H., Namai H., Okuno K. Analysis of QTL associated with the allelopathic effect of rice using water-soluble extracts. Breed Sci 2001; 51:47-51.

Escobar C.A., Niemeyer H.M. Potential of hydroxamic acids in breeding for aphid resistance in wheat. Acta Agric Scand Sec B Soil Plant Sci 1993; 43:163-167.

Fay P.K., Duke W.B. An assessment of allelopathic potential in *Avena* germplasm. Weed Sci 1997; 25:224-228

Fukai E., Nishio T., Nasrallah M.E. Molecular genetic analysis of the candidate gene for MOD, a locus required for self-incompatibility in *Brassica rapa.* Mole Gene Genom 2001; 265:519-525.

Grace J.B. "On the relationship between plant traits and competitive ability." In *Perspectives on Plant Competition,* J. B. Grace, D. Tilman, eds. New York, NY: Academic Press, 1990; pp. 51-65.

Haas H., Streibig J.C. "Changing patterns of weed distribution as a result of herbicide use and other agronomic factors." In *Herbicide Resistance in Plants,* J. Gressel, H. LeBaron, eds. New York, NY: John Wiley & Sons, 1982; pp. 57-79.

Hassan S.M., Aidy I.R., Bastawisi A.O., Draz A.E. "Weed management using allelopathic rice varieties in Egypt." In *Allelopathy in Rice*, M. Olofsdotter, ed. Los Banos, Philippines: International Rice Research Institute, 1998; pp. 27-38.

Inderjit. Soil: environmental effects on allelochemical activity. Agron J 2001; 93:79-84.

Inderjit, Keating K.I. Allelopathy: principles, procedures, processes and promises for biological control. Adv Agron 1999; 67:141-231.

Jensen L.B., Courtois B., Shen L., Li Z., Olofsdotter M., Mauleon R.P. Locating genes controlling allelopathic effects against barnyardgrass in upland rice. Agron J 2001a; 93:21-26.

Jensen L.B., Olofsdotter M., Courtois B. Genetic control of allelopathy in rice. Penang Malaysia, August 7-11. Second Asian-Pacific conference on chemical ecology- current scenario and trends in chemical ecology, 2001b.

Jensen L.B., Olofsdotter M., Courtois B. "Genetic control of allelopathy in rice (*Oryza sativa* L.)." In *Rice Allelopathy,* K. U. Kim, D. H. Shin, eds. Kyungpook National University, Taegu, Korea, 2000; pp. 27-40.

Jones M.J., Dingkuhn M., Aluko G.K., Semon M. Interspecific *Oryza sativa* L. x *O. glaberrima* Steud. progenies in upland rice improvement. Euphytica 1997; 92:237-246.

Jordan N.R., Jannink J.L. Assessing the practical importance of weed evolution: a research agenda. Weed Res 1997; 37:237-246.

Kaneko T., Zhang W.S., Ito K., Takeda K. QTL mapping for enzyme activity and thermostability of beta-amylase in barley (*Hordeum vulgare* L.). Breed Sci 2001; 51:99-105.

Kim K.U., Shin D.H. "Rice allelopathy research in Korea." In *Allelopathy in Rice,* M. Olofsdotter, ed. Los Banos, Philippines: International Rice Research Institute, 1998; pp. 39-44.

Kirk G.J.D., George T., Courtois B., Senadhira D. Opportunities to improve phosphorus efficiency and soil fertility in rainfed lowland and upland ecosystems. Field Crops Res 1998; 56:73-92.

Kluge M., Grambow H.J., Sicker D. (2R)-2-beta-D-glucopyranosyloxy-4,7-dimethoxy-2H-1,4-benzoxazin-3(4H)-one from *Triticum aestivum.* Phytochemistry 1997; 44:639-641.

Kropff M.J., Lotz L.A.P., Weaver S.E., Bos H.J., Wallinga J., Migo T. A two parameter model for prediction of crop loss by weed competition from early observations of relative leaf area of the weeds. Ann Appl Biol 1995; 126:329-346.

Kudsk P., Olsen J., Mathiassen S.K., Brandt K., Christensen L.P. Allelopathy in barley. 18. Danske Planteværnskoference, 43-45. Danmarks Jordbrugsforskning. DJF Rapport, 1998.

Lemerle D., Gill G.S., Murphy C.E., Walker S.R., Cousens R., Mokhtari S., Peltzer S.J., Coleman R., Luckett D.J. Genetic improvement and agronomy for enhanced wheat competitiveness with weeds. Aust J Agric Res 2001; 52:527-548.

Liu D.L., Lovett J.V. Biologically active secondary metabolites of barley. I. Developing techniques and assessing allelopathy in barley. J Chem Ecol 1993b; 19:2217-2230.

Liu D.L., Lovett J V. Biologically active secondary metabolites of barley. II. Phytotoxicity of barley allelochemicals. J Chem Ecol 1993a; 19:2231-2244.

Lockerman R.H., Putnam A.R. Evaluation of allelopathic cucumbers (*Cucumis sativus*) as an aid for weed control. Weed Science 1979; 27:54-57.

Lovett J.V., Hoult A.H.C., Christen O. Biologically active secondary metabolites of barley. IV. Hordenine production by different barley lines. J Chem Ecol 1994; 20:1945-1954.

Mauleon R.P. *RFLP mapping of genes conferring resistance to blast in the rice cultivars IAC 165 and CO 39 across environments.* University of the Philippines, Los Baños, Laguna, Philippines, 1995.

Moolsri S., Kow-in P., Chaitep W., Mortimer A.M. Rice-weed competitiveness. 1B, 721-725. November 22-27, Bangkok, Asian-Pacific Weed Science Society. Proceedings of the 17th Asian-Pacific Weed Science Society conference, 1999.

Mortimer A.M., Caton B.P., Hill J.E. On ecological issues in the development of sustainable weed management. 1 A, 45-50. Bangkok, Thailand, Asian Pacific Weed Science Society. Proceedings of the 17th Asian-Pacific Weed Science Society Conference. Weeds and Environmental Impact, 1999.

Navarez D., Olofsdotter M. Relay seeding technique for screening allelopathic rice (*Oryza sativa*). Proceedings of the second International Weed Control Congress, Department of Weed Control & Pesticide Ecology, Copenhagen, Denmark, 1996; pp. 1285-1290.

Olofsdotter M., Navarez D. Allelopathic rice in *Echinochloa crus-galli* control. Proceedings of the second International Weed Control Congress, Department of Weed Control & Pesticide Ecology, Copenhagen, Denmark, 1996; pp. 1175-1182.

Olofsdotter M., Navarez D., Rebulanan M. Rice allelopathy - where are we and how far can we get? Brighton, BCPC. Brighton Crop Protection Conference, 1997; pp. 99-104.

Olofsdotter M., Navarez D., Rebulanan M., Streibig J.C. Weed-suppressing rice cultivars - does allelopathy play a role. Weed Res 1999; 39:441-454.

Overland L. The role of allelopathic substances in the smother crop barley. Am J Bot 1966; 53:423-432.

Perez F.J. Allelopathic effect of hydroxamic acids from cereals on *Avena sativa* and *A. fatua*. Phytochemistry 1990; 29:773-776

Perez F.J., Ormeno-Nunez J. Weed growth interference from temperate cereals: the effect of hydroxamic-acid-exuding rye (*Secale cereale* L.) cultivars. Weed Res 1993; 33:115-119.

Pheng S., Adkins S.W., Jahn G.C., Olofsdotter M., Nesbitt H.J. Towards the production of allelopathic rice crops in Cambodia. Beijing, China. May 28- June 2. Proceedings of the 18th Asian-Pacific Weed Science Society conference, 2001; pp. 187-197.

Putnam A.R., DeFrank J. Use of phytotoxic plant residues for selective weed control. Crop Prot 1983; 2:173-181.

Putnam A.R., Duke W.B. Biological suppression of weeds: evidence for allelopathy in accessions of cucumber. Science 1974; 185:370-371.

Ranasinghe L.L., Crabtree G.D. Plant characteristics associated with rice (*Oryza sativa* L.) - barnyardsgrass (*Echinochloa crus-galli* L. Beauv.) competition. November 22-27, Bangkok, Asian-Pacific Weed Science Society. Proceedings of the 17th Asian-Pacific Weed Science Society conference, 1999; pp. 99-104.

Rimando A.M., Dayan F.E., Czarnota M.A., Weston L.A., Duke S.O. A new photosystem II electron transfer inhibitor from *Sorghum bicolor*. J Nat Prod 1998; 61:927-930.

Rimando A.M., Olofsdotter M., Dayan F.E., Duke S.O. Searching for rice allelochemicals: an example of bioassay-guided isolation. Agron J 2001; 93:16-20

Rousset M., Brabant P., Kota R.S., Dubcovsky J., Dvorak J. Use of recombinant substitution lines for gene mapping and QTL analysis of bread making quality in wheat. Euphytica 2001; 119:81-87.

Shilling D.G., Liebl R.A., Worsham A.D. "Rye (*Secale cereale* L.) and wheat (*Triticum aestivum* L.) mulch: the suppression of certain broadleaved weeds and the isolation and identification of phytotoxins." In *The Chemistry of Allelopathy: Biochemical Interactions Among Plants,* A. C. Thompson, ed. Washington DC: American Chemical Society, 1987; pp. 243-272.

Wu H., Pratley J., Lemerle D., Haig T. Crop cultivars with allelopathic capability. Weed Research 1999; 39:171-180.

Wu H., Pratley J., Lemerle D., Haig T. Evaluation of seedling allelopathy in 453 wheat (*Triticum aestivum*) accessions against annual ryegrass (*Lolium rigidum*) by the equal-compartment-agar method. Aust J Agric Res 2000a; 51:937-944.

Wu H., Pratley J., Lemerle D., Haig T. Laboratory screening for allelopathic potential of wheat (*Triticum aestivum*) accessions against annual ryegrass (*Lolium rigidum*). Aust J Agric Res 2000b; 51:259-266.

Wu H., Pratley J., Lemerle D., Haig T. Allelopathy in wheat (*Triticum aestivum*). Ann Appl Biol 2001; 139:1-9.

Yoshida H., Tsumuki H., Kanehisa K., Corcuera L.J. Release of gramine from the surface of barley leaves. Phytochemistry 1993; 34:1011-1013.

CHAPTER 16

WEED MANAGEMENT UNDER NO-TILLAGE SYSTEMS IN TROPICAL REGIONS

ITAMAR F. SOUZA[1] & LUIS WAGNER R. ALVES[2]

[1]*Lavras Federal University, Agronomy Department, C. P. 37, 37200-000, Lavras, MG, Brazil;* [2]*University of Alfenas, Agriculture School, Alfenas, MG, Brazil*

The no-tillage (NT) planting system is a complex technique involving no soil tillage associated with plant straw (plant residues) on soil surface and crop rotations, aiming better weed control, crop productivity and environmental safety. Planting into soil that has not been tilled recently, is a crop production system that breaks all traditional patterns and recommendations for agriculture and livestock. An understanding of water-soil-plant relationships is needed to accept these changes and take economical advantage of the system without causing adverse environmental impact (Gassen and Gassen, 1996). In the mid 70's Frank Dikstra and Manoel Pereira, farmers associated with ABC (Arapoti, Batavo, and Castrolandia counties) Foundation, found in NT a solution to soil erosion problems in Paraná State, Southern Brazil. They became the diffusers of NT in Brazil where the planted area under this system reached 11 millions hectares, which represents 30% of the total grain production area (FEBRAPDP, 2000). No tillage has experienced great expansion in the South, Northeast, and North of the country.

The NT system has brought changes in weed species. While annual species tend to increase, grassy species tended to decrease; and hard-to-control species, such as *Commelina benghalensis* and *Sida rhombifolia*, have been frequently noticed. The increasing number of herbicides in the market allowed farmers to adopt NT planting systems. This has altered both environmental safety and crop-weed ecological relationships. With the adoption of NT, Brazilian agriculture has experienced increased productivity at lower production costs.

Inderjit (ed.), Weed Biology and Management, 329-343.

WEED MANAGEMENT IN CONSERVATION TILLAGE

The expansion of agricultural frontiers in Brazil in the 80's required new technologies to replace hand work on expanded cropping areas. Handweeding in such areas was inadequate and chemical weed control has increased exponentially. This is not true in smaller farms, where all crop production relies on family work and is not dependent on external inputs. Chemical weed control is important for perennial species and for high infestations of annuals where uncontrolled weeds will affect yield (Almeida, 1991). Therefore, the absence of perennial species and low infestation by annuals are two important factors that should be considered before adopting no-till system for crop production.

Straw Deposition (Weed or Crop Residues)

Any plant growing where it is not desired is identified as a weed. Therefore, those plants that spontaneously grow during a fallow period may be useful in NT systems when well managed and when they produce enough residues to reduce the next flush of weed germination. Surface plant residues may alter soil moisture, light intensity, and temperature extremes. These factors are of importance for seed germination and dormancy. They also affect seedling development, functioning as a physical barrier for plant growth, alter soil C/N ratios, and some are allelopathic. Finally, plant residues on soil surface may shelter insects and fungi, which may be pests and pathogens of weed seeds and seedlings.

A mower or roller (mechanical management) can substitute for chemical control of existing plants (weed or crop) with the advantage of more uniform straw distribution on soil surface and better weed control in the next crop. Plant residues on the soil surface have four basic objectives:

1. To create high mulch density under some situations in tropical areas. Species used to produce mulch are grown during the summer season. For example: mucunas (*Mucuna aterrina* and *M. pruriens*) which are inter-cropped with corn and kept on the field to be killed by frost in the winter. Winter crops are seeded into mucuna residues. Under other situations cover crops are broadcast or row seeded in the winter and killed before summer crop seeding. For example, radish (*Raphanus sativus*), lupine (*Lupinus albus*).
2. For animal feeding in the winter and for mulching in the summer. Several crops produce enough biomass to feed animals in the winter and re-grow to cover the soil before summer crop planting. For example, black oat (*Avena strigosa*), ryegrass (*Lolium multiflorum*), rye (*Secale cereale*).
3. For grain yield and mulching. In this case growers must grow the crop, harvest the grain, and use the straw for mulching. For example, oat (*Avena sativa*), wheat (*Triticum aestivum*).
4. Fallow managed weeds for mulching. Cultivating crops in the winter fallow in tropical countries may not be desirable but there is a need to manage weeds and prevent seed production. One good example is the infestation of alexandergrass (*Brachiaria plantaginea*), which grows profusely in drybean (*Phaseolus vulgaris*) areas. The weed must be managed with a grass herbicide and the crop planted into

the weed residue. To maintain the undisturbed soil seedbed it should not be planted when excessively dried and the tractor should not run fast (Buzatti, 1999).

Plant residues left on the soil under NT system affect weed seed germination by the following three processes (Pitelli, 1997; Theisen and Vidal, 1999).

1. Physical effect - lower temperatures may have detrimental effects on photoblastic seeds by decreasing radiation on soil surface; also high straw density inhibits seedling emergence.
2. Chemical effect - allelochemicals released/exuded may influence different weed species.
3. Biological effect – increasing populations of soil microorganisms such as fungi and bacteria play an important role in weed seed deterioration in the soil.

The decomposition index of a plant tissue may determine the initial effect on organisms (weeds). Soft plant tissue decomposes and releases chemicals faster in the soil environment than hard plant tissue. For example, vetch (*Vicia sativa*) decomposes and releases chemicals faster than wheat (Almeida, 1991).

Cereals are reported to possess several allelochemicals. For example, barley (*Hordeum vulgare*) - phenolic, benzoic, phenylacetic, and phenylbutyric acids (Tousson, 1968); corn (*Zea mays*) - phenolic aldehydes, *p*-hydroxybenzylaldehyde, fluro-glucinol, resorcinol, and butyric, phenylacetic, benzoic, syringic, *p*-coumaric, *trans*-cynamic, and caffeic acids (Chou and Lin, 1976); wheat - acetic, propionic, and butyric acids (Tang and Waiss, 1978). More recently, other chemical groups have been studied such as sorgoleone, a quinone, in *Sorghum*, the non protein amino acids hordenine and guanine in barley, and the hydroxamic acids DIBOA (2,4,-dihydroxy-1,4-benzoxazin-3-one), DIMBOA (2,4-dihydroxy-7-methoxy-1,4-benzoxazin-3-one), MBOA (methyl-benzoxazolinone), and BOA (benzoxazolinone) in rye and corn.

The thickness of residue layers is of importance to maintain weed control for a certain period and to determine the control level. For example, thinner residue layer will have the higher the emergence percentage of alexandergrass, although a small percentage of soil seed bank (4.47%) germinated, indicating the ability of alexandergrass seeds to remain dormant (Theisen and Vidal, 1999).

Oats

It is used for animal feeding, green manure, and mainly as cover crop and its straw serves as a mulch in a no-till system. Decomposing plant residues secrete toxins such as scopoletin, sculin, and other allelochemicals having effects on weed seed germination and seedling growth. The effects on weeds are caused by a sum of allelopathy and competition for light and other growth factors. Souza (1992) found that oat straw decomposes quicker than rye and wheat straw but the larger amount of straw yielded by oat results in better weed control.

Rye

Rye is used as green manure and cover crop, due to its resistance to drought and high allelopathic capacity. Compounds such as the hydroxamic acids - BOA, MBOA, and DIBOA - are potent allelochemicals reported from rye. Souza (1996) showed a 30% reduction of beggartick (*Bidens pilosa*) and smallflower galinsoga (*Galinsoga parviflora*) infestation in soybean when rye was planted as a cover crop in winter.

Perez and Ormeno-Nunez (1993) indicated a 83% wheat biomass reduction by BOA and DIBOA.

Wheat

Among winter cereals, wheat is one that yields the least amount of straw but synthesizes and secretes large amount of secondary compounds. This is important for no till weed control system (Almeida, 1988). Phenolic acids such as ferulic, *p*-coumaric, protocatechuic, caffeic are among the compounds that might have joint action in a mixture (Inderjit et al., 2002).

Schreiber (1992) studied the effect of wheat - soybean rotation during 12 years on giant foxtail (*Setaria faberi*) population control and found that it was reduced due to allelopathic effects of wheat straw. The same author had a study case involving soybean - wheat - corn rotation under conventional and no tillage systems for giant foxtail control and concluded that the reduction of soil disturbance increased giant foxtail population but it can be decreased with a wheat and corn rotation. Economic analysis also indicated that the intensity of weed control below threshold for grain yield is not indicated due to high costs of herbicides.

Sorghum

Under certain conditions, sorghum (*Sorghum bicolor*) (BR 303 variety) straw has allelopathic effects on weeds and may interfere with soybean grain yield (Peixoto, 1999). Planting time play an important role on the amount of allelochemicals exudation by sorghum. This is true for tropical regions where straw decomposition is somewhat fast. Sorghum allelopathy can be associated with reduction of herbicide rates on the next crop. No tilled soil also reduces weed infestations with time, or at least changes weed population, and all of these three factors (sorghum allelopathy, reduced herbicide rates, and no tilled soil) together will improve weed management in agroecosystems.

The sorghum cultivars to be used as cover crop should be considered from weed management standpoint. Correia (2001) studied the effect of sorghum varities – Saara, DK 860, Ambar and Summer fallow – on the management of certain weeds. The species – lionsear (*Leonotis nepetifolia)*, alligatorweed (*Alternanthera philoxeroides)*, smooth pigweed (*Amaranthus hybridus*), redroot pigweed (*A. retroflexus*), spiny amaranth (*A. spinosus*), threelobe morninglory (*Ipomoea triloba*), tropical spiderwort (*Commelina benghalensis*), and apple of Peru (*Nicandra physalodes*) were better controlled by Saara and Ambar sorghum hybrid straw than by DK 860 when no herbicide was applied on soybeans. Only Ambar showed good control of such species when 15 $g.ha^{-1}$ of imazamox herbicide was applied. When a higher herbicide rate was used (30 $g.ha^{-1}$), DK 860 and Ambar gave better control (Correia, 2001). Sorghum variety behavior may be associated with sorgoleone concentration (Einhellig and Souza, 1992). Santos (1996) extracted more sorgoleone from the BR 304 (8.1 mg/100 roots) than CMXSX 376, AG 2002, and AG 2501 sorghum hybrids.

Weeds

Weeds do not constitute a good cover plant due to low mass volume and quick decomposition (Monegat, 1991). The key point for using weeds as cover plant in no tillage system is to control them before they set seeds.

Some growers in Brazil consider two important species for mulching in no tillage system. These are tropic ageratum (*Ageratum conyzoides*) and chickweed (*Stellaria media*). Tropic ageratum germinates in January and leaves high mass density on the soil but yields up to 40,000 seeds/plant. Chickweed grows profusely in February/March in a prostate habit with high capacity of seed production, therefore good weed management is required.

Soil cover by summer crops

Jack beans (Canavalia ensiformis)

One of the most important green manure and alternative crop for weed control in many countries. Fernandes et al. (1999) considered jack beans a very efficient weed control crop even at low planting densities. This fact is attributed to its dense canopy and the allelopathic compounds (canavaline and cananine) exuded by the plant (Vickery and Vickery, 1981). The crop has been recommended as alternative to herbicides for nutsedge (*Cyperus* spp) control in high infested areas. Herbicides such as glyphosate, sulfosate, sulfentrazone, EPTC, halosulfuron (Rodrigues and Almeida, 1998) have been used to control nutsedges under normal circumstances. For high infested or old infested areas with tubers located at certain depth in soil layer; jack beans is used for nutsedge control.

Mucunas (Stilozobium spp)

Several mucuna species are found in many countries and are used as animal and human food. They are also used to recover depleted soils. For weed management, mucunas have been recommended to control nutsedges and cogongrass (*Imperata cylindrica*). It is necessary to grow mucunas for two or more years before weed infestation is reduced (Narwal, 1994). Dihydroxyphenylalanine (DHPA) is in the plant at concentrations of 0.5 to 1.5% fresh weight and is one of the most important allelochemicals acting on weed growth. This compound alone has little effect on grass and leguminous species, but mucunas yield 20 to 30 tons of biomass/ha meaning 200 to 300 kg of DHPA per hectare.

Crotalarias (Crotalaria spp)

Crotalaria species are reported to control weed biomass in soybeans (Rodrigues, 1993). In terms of controlling weed biomass (g/m^2), *C. juncea* was found to be most effective followed by *C. mucronata, C. spectabilis, C. brevidens* and *C. paulinea.*

Sunflower (Helianthus annuus)

Sunflower is a potent allelopathic summer cover crop although few research results on sunflower varieties for crop rotation and succession have been reported. Narwal (1994) has demonstrated that crop rotation of sunflower and oats for five years reduced weed infestation. He showed that sunflower and the EPTC herbicide have similar abilities to control weeds. Guaianolide, a terpenoid allelochemical, has been found to be the compound responsible for broadleaved weeds growth reduction (Narwal, 1994).

Weeds

A species of goosegrass (*Eleusine coracana*) has been grown by "Agronorte Pesquisas" (2001) company in Brazil and recommended for animal feeding and cover crop. *E. coracana* is of African origin and yields up to 18 tons of biomass per hectare. Agronorte Pesquisas indicates *E. coracana* as weed growth inhibitor by mulching and allelopathy through root exudates during the plant development.

Fallow periods between winter and summer crops are important to decide the number and timing of herbicide applications (Almeida, 1991). In North of Paraná State, Brazil, the fallow period after winter crop harvesting (summer fallow) goes from August to the end of November, when summer species are planted. The winter fallow (after summer crop harvesting) is in March. In mid-South, on the contrary, winter fallow is longer when winter crops are planted in June and summer fallow is shorter (1-2 weeks).

WEED POPULATION DYNAMICS IN CONSERVATION TILLAGE

Shifting from conventional tillage to no-tillage causes an alteration of stress imposed by the systems on the environment and consequently on weed populations. In the past, plowing and harrowing had such impact that many weed species did not survive the soil disturbance, high light intensity, and temperature extremes. The resulting weed population was adapted to that disturbed environment through complex dormancy mechanisms, ability to germinate from deep soil layers, and tolerance to extreme temperatures.

Weed seed distribution in the soil layers became more uniform such that plowing and harrowing only moved seeds from a deeper layer to shallower layer, maintaining the same seed bank for germination. Those species that were not adapted had populations reduced and are considered today as species of secondary importance (Pitelli, 1997).

The introduction of NT in Southern Brazil brought changes in weed species dynamics. In this system annual weeds decreased, perennials increased, and more specifically, grass weeds increased, due to new impacts on the soil environment caused by the system. Reduction of soil disturbance promotes a temporary reduction in weed populations in the agroecosystem (Pitelli, 1997):

1. A great proportion of seeds in the soil will be kept at sufficient depth to not germinate.
2. Seeds will be laid and kept on top where birds and rodents will destroy them. This is true for seeds that require some time for physiological maturity or for breaking dormancy before germinating.
3. Higher number of seeds on soil surface will facilitate seedling emergence and control measures will be more efficient.
4. Some species that did not survive under conventional tillage may do so under NT and their populations will increase. These changes can be observed on old NT areas that have experienced different populations from conventional tillage.

Crop Rotation and Succession Affecting Weed Population Dynamics

Crop varieties of fast initial growth, narrow row spacing, and high seed density in the row are considered to have better competitive abilities (Pitelli, 1997). Crop rotation (winter and summer crop planted in the same year) is of importance to shift the crop environment and prevent high weed infestations. Rotating herbicides with different mechanism of actions will also keep tolerant weed species from building up. These practices help environmental equilibrium, reduce external inputs, and prevent development of a weed population resistant to herbicides. For example wild poinsettia (*Euphorbia heterophylla*) and balloonvine (*Cardiospermum halicacabum*) populations resistant to imazaquin have developed in Paraná.

Ruedell (1995) reported that early grass weed infestation was sufficiently reduced by a corn crop planted in winter to avoid corn grain yield loss but did not reduce late infestations to reduce soybean grain yield in the summer. Corn may also be used as soil cover in the summer to plant winter crops although herbicides are necessary during crop development. Species dynamics in the two systems are quite different. Ruedell (1995) showed a 10-fold reduction of broadleaved weeds when soil was under NT for 10 years and only 2.7-fold reduction under conventional tillage. Horseweed (*Conyza bonariensis*) and Asiatic dayflower (*Commelina communis)* are well adapted to NT and they establish during the fall and winter. It is easier to control them with herbicides at this time. In spring, weeds are well developed and higher rates are necessary (Gassen and Gassen, 1996).Densities of grass weeds tend to be higher under NT (Buzatti, 1999). Perennial weeds that do not resist soil disturbance rarely grow well under conventional tillage but grow profusely under NT. In all cases the potential growth of a weed population is indicated by the soil seed bank and the population growth may be reduced by reduction of seed set by mother plants (Figure 1).

Timing for Weed Management in Conservation Tillage

Winter Weed Management

It is done in April/June in tropical regions by herbicides, mowing, or rolling (stalk chopping) 10-20 days after summer crop harvesting on those areas where infestation is high or where weed populations resistant to herbicides occur (Buzatti, 1999).

Herbicide Combinations for Winter Weed Management

Post-Emergence and Pre-Emergence Tank Mix Herbicides

Some characteristics of pre-emergence compounds to be considered before application are listed below.

1. Water solubility – Pre-emergence herbicides should be quite water soluble since rainfall and dew are the two most important factors to carry the compound down into the soil.

2. Straw retention - The thickness, origin, and decomposition stage of the mulch on soil surface are three of many factors to be considered before choosing pre-emergence herbicides. For example, diclosulan effect on *Sida* species depends on thickness of the straw layer (Marochi, 1994). Clomazone efficacy was reduced as wheat straw thickness increased even with 22 mm of sprinkler irrigation after herbicide application (Rodrigues, 1993). Imazaquin was tested as pre-emergence over wheat straw and a 7 day period without rainfall followed application. The author suggested that this time was sufficient to cause low beggartick and alexandergrass control.
3. Photosensitivity - Herbicides are always exposed to sunlight to some extent before being incorporated into the soil. Less light sensitive compounds should be preferred.

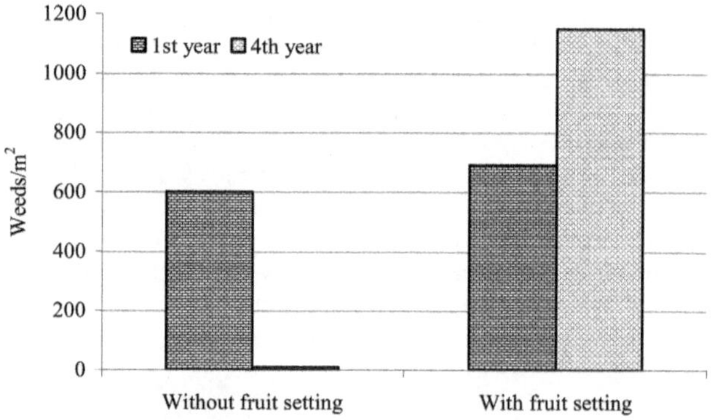

Figure 1. Weed density under oat plant residue managed before and after fruit setting. Paraná State, PR, Brazil.

Besides pre-emergence herbicide characteristics, weed density is also a factor that determines the performance of chemical weed management. Pre-emergence compounds should not be used in tank mix with post-emergence when 30% of soil surface is covered by weeds or cover crop straws. In this case the soil is so protected by plant residues that the herbicide will not reach the soil at sufficient amount to keep weed germination under control. Also, pre-emergence herbicide efficacy is reduced where organic matter content is over 3.5% and soil clay content is high. Rates should be increased or post-emergence compounds may be recommended (Gassen and Gassen, 1996). Marochi (1994) did some experiments in Paraná State in Brazil and found almost 100% control of beggartick, arrowleaf sida (*Sida rhombifolia*), and alexandergrass when glyphosate, 2,4-D (post-emergence herbicides), and flumetsulan (pre-emergence herbicide) were applied in tank mix or sequentially.

Summer Weed Management

Weed infestation after summer crop seeding may decrease yield and quality. Summer weed management is done before summer crop planting and after winter crop harvesting. Buzatti (1999) suggests the use of herbicides as chemical control, such as glyphosate, sulfosate, 2,4-D, paraquat, paraquat + diuron, cyanazine, and flumyoxazin.

Herbicide combinations and rates are dependent on weed species, weed growth stage, and next crop species to be planted. In Campos Gerais Region of Paraná State, Brazil, glyphosate + 2,4-D is often recommended but a 10 day period before soybean planting should be observed. Weed management in fallow periods is the basis for the success of NT system to prevent hard-to-control weeds growth.

Sequential Herbicide Applications Following Pre-Emergence

Residual herbicides are applied after winter vegetation present at planting is killed and summer crop seeding but before summer crop and weed emergence. Pre-emergence application with a water soluble herbicide is recommended just before rainfall, e.g. flumetsulan (water solubility of 5650 ppm), to facilitate quick herbicide penetration through plant residue.

Post-emergence herbicides are usually necessary at rates dependent on weed species and growth stage. For almost all crops in tropical countries selective herbicides are available.

Weed Management in Soybeans

A herbicide tank mix or sequential application is always recommended for soybean weed control. The components as well as doses to be applied depend on weed species and infestation level, growth stage, and cost of each combination. Table 1 lists some herbicide combinations used in Brazil for broadleaved and grass weed control. Herbicides listed are selected examples out of many compounds recommended for soybeans (Table 1).

Weed Management in Corn

For post-emergence weed control in corn the herbicides are to be applied at the initial growth stage of the weeds with few exceptions. Many of them have residual effects but plant residues will impede these compounds reaching the soil surface. A list of herbicides recommended for corn is shown by Table 2.

Fall Weed Management

After summer crop harvest, late season germinating weeds grow profusely and may set seeds to re-infest the field. Management of such species should be done 10-20 days after crop harvesting in tropical countries. Weed management may be done through chemical or mechanical methods depending on cost, equipment, and winter crop and weed species (Buzatti, 1999). Chemical weed management is done by herbicides such as glyphosate, sulfosate, 2,4-D, paraquat, paraquat+diuron, cyanazine, and flumioxazin. Species with capacity of re-growing after herbicide application should be controlled by systemic compounds (Gassen and Gassen, 1996).

Higher concentration (80%) of atrazine was detected in the 0-10 cm layer of uncovered soil before irrigation (Figure 2 A) in comparison with 60% under 9 cm layer thickness of black oat after 20 cm irrigation (Figure 2 B) (Fornarolli et al, 1998). These results suggest that 20 mm irrigation is not sufficient to leach atrazine down to the soil surface when covered with black oat straws.

Table 1. Herbicide and herbicide combinations for weed control in soybeans (Buzatti, 1999).

Herbicide	a.i./L	g a.i./ha	Species Controlled
Imazethapyr+lactofen+ Bentazon	100+240+600	50+72+600	Broadleaf
Imazethapyr+lactofen+ Chlorimuron	100+240+250	50+72+7.5	Broadleaf
Fomesafen+bentazon	250+600	125+600	Broadleaf
Lactofen+bentazon	240+600	96+600	Broadleaf
Fomesafen+imazethapyr	250+100	125+60	Broadleaf
Clethodin	240	96-120	Grass
Clethodin+fenoxaprop-p	50+50	40+40	Grass
Fenoxaprop	120	150-210	Grass
Fenoxaprop-p	110	695-935	Grass
Fluazifop-p	125	185-250	Grass
Haloxyfop-r	120	48-60	Grass
Propaquizafop	100	125	Grass
Quizalofop-p	50	75-100	Grass
Sethoxydim	184	230	Grass

Weed management with pre-emergence herbicides in conservation tillage is always indicated by industry and Brazilian research. In fact there are several options going from mechanical methods to contact and systemic post-emergence compounds that may be applied at a convenient time. A good choice of one of these alternatives allows the grower a good weed management and pre-emergence herbicides may not be necessary (Souza, 2002).

There are many limitations in using these chemicals under such situations (no tillage system) which always increase costs and decrease efficacy. For example, residue layer thickness and plant origin decrease pre-emergence herbicide movement down to soil surface; soil microflora in conservation tillage causes pre-emergence herbicide degradation; retention of the compound by straw exposes it to sunlight for degradation; planting furrows drain herbicides and increase concentrations such that crop injury may occur. For these and other reasons, we suggest alternate methods for weed management in tropical regions before new adapted herbicides can be developed.

Costs of Herbicides for Weed Management in Conservation Tillage

Land preparation for soybean planting under no till is US$ 2.86 per hectare less than in conventional tillage (Table 3, Sub-total A). This is not true for corn crop when the same author found US$ 1.62/ha higher than in conventional tillage. Herbicides applied at planting time and post emergence in NT are US$ 3.36.ha^{-1} less than in CT for soybeans, and herbicide costs for weed control in corn are US$ 5.50.ha^{-1} higher in NT than in CT. (Table 3, Sub-total B).

No-tillage does not depend on a high amount of herbicides for a long period because straw deposition on soil surface increases with time and suppress weeds. Therefore, herbicide costs will decrease and this is true when the understanding of crop-weed dynamics is considered.

Table 2. Herbicides and herbicide combinations for initial post-emergence weed control in corn (Silva, 1998, modified).

Herbicide	a.i./L	g a.i./ha	Species Controlled
Atrazine	500	2000-3000	Grass/broadleaf
Atrazine + oil	400	2000-2800	Grass/broadleaf
Bentazon	600	540-720	Broadleaf
Cyanazine	500	1500-2500	Grass/broadleaf
2,4-D amine	720	500-1000	Broadleaf
Nicosulfuron	40	40-60	Grass/broadleaf
Atrazine+metolachlor	200+300	(1000+1500)-(1400+2100)	Grass/broadleaf
Atrazine+metolachlor+nicosulfuron	200+300+40	1000+1500+40	Grass/broadleaf
Atrazine+simazine	250+250	(875+875)-(1750+1750)	Grass/broadleaf
Atrazine+nicosulfuron	500+40	1000+40	Grass/broadleaf
Atrazine+alachlor+nicosulfuron	300+180+40	1800+1080+40	Grass/broadleaf
Cyanazine+simazine	250+250	(1250+1250)-(2000+2000)	Grass/broadleaf

To reduce the use of herbicides under tropical conditions, an alternative to a 2-crop rotation is proposed by Kliewer et al (1998). Two sequences were tested: soybean-wheat-soybean, and corn-wheat-soybean. In the first sequence, a fallow period of 62 days before wheat planting was observed. In the corn-wheat-soybean sequence, sunflower and crotalaria were planted before wheat. The authors found in the first case that five herbicide applications were necessary with a cost of US\$ 107.66.ha^{-1} to control weeds all year round. In the second, sunflower and crotalaria grew up 6m height at 60 days after planting, yielding 5.7 ton of straw per hectare and no herbicide was used. Fast growing sunflower and crotalaria reduced weed populations and few of them grew up afterwards. The authors stated that this technique does not require herbicide applications to plant wheat in the winter and no post-emergence herbicide was necessary. It cost US\$ 37.39 and US\$ 36.62 for planting sunflower and crotalaria, respectively.

To plant the soil during the winter in Brazilian Savanah region a corn crop succession with pearl-millet (*Pennisetum glaucum*), corn, sunflower, sorghum, and pigeon pea (*Cajanus cajan*) is proposed by Pasqualetto (1999). Pearl-millet-corn allowed higher infestation of broadleaved weeds while sunflower-corn and sorghum-corn were quite efficient. Broadleaved weed control in the summer was achieved by sorghum-soybean. Soybeans grain yield was reduced by pearl millet-soybean combinations.

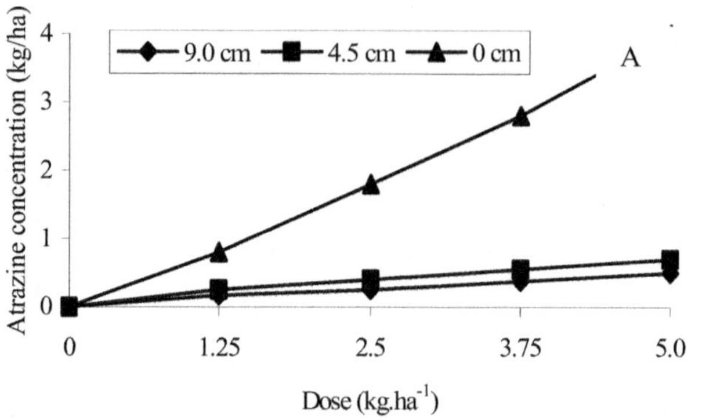

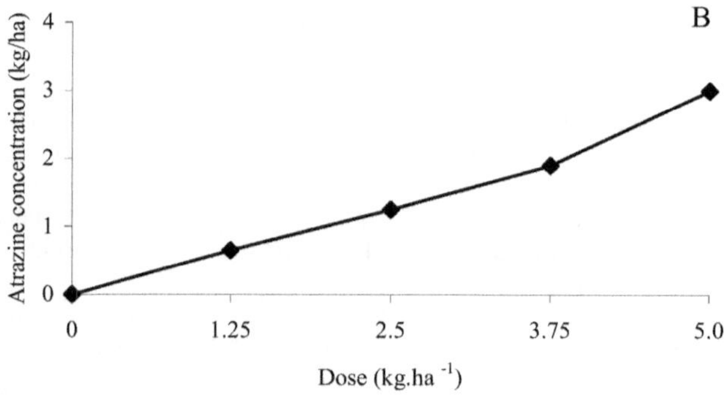

Figure 2. Effects of black oat straw thickness on atrazine concentration in 0-10 cm soil layer before (A) and after (B) 20 mm irrigation (Fornarolli, 1998).

SEED LONGEVITY IN CONSERVATION TILLAGE

The viability of crop seeds is shorter than weed seeds once in the soil. For examples, lotus (*Nelumbo nucifera*) seeds found on bottom of an old lake in Central Asia were 1,040 years old, determined by radiolabel carbon technique; curly dock (*Rumex crispus)* seeds 100 years old are mentioned by Lorenzi (1994). Seed longetivity of redroot pigweed (*Amaranthus retroflexus)*, Virginia pepperweed (*Lepidium virginicum)*, Common purslane (*Portulaca oleracea)* and jimsonweed (*Datura stramonium)* is 40, 40, 40 and 80 years, respectively. The mechanism of dormancy of weed seeds, which is a characteristic of wild seeds, is responsible for their longevity. Also, seed dormancy is broken over time so that seeds from the same mother plant keep germinating for years.

Table 3. Soil preparation and herbicide costs (US\$.ha^{-1}) for corn and soybean under NT and CT (conventional tillage) systems. Brazil, 1999. (Mello Filho and Mendes, 1999).

	Soybeans		Corn	
	NT	CT	NT	CT
Post-emergence herbicide for winter vegetation control	12.15	-	15.19	-
Herbicide before planting	2.93	-	2.93	-
Harrowing	-	17.94	-	16.50
Sub-total A	15.08	17.94	18.12	16.50
Herbicide during planting time	2.93	11.64	35.79	30.29
Herbicide during crop development	37.72	32.57	2.93	2.93
Sub-total B	40.65	44.01	38.72	33.22
Total	55.73	61.95	56.84	49.72

In some cases, germination percentage of weed seeds is low so that resulting weed densities do not compete severely with crop plants. As result, less effort is needed for weed control during crop development. Most of the time the NT system by itself is not sufficient to reduce the weed population to a level that does not compete at all with crops and a second weed control practice is necessary.

CONCLUSIONS

New technologies designed for conservation tillage have been applied with success in Brazil. This agricultural technology has expanded very quickly in South and Brazilian Savannah Region where soils are quite water and wind eroded. There is always a shift of weed population as one moves from conventional to conservation tillage. Decrease of weed populations under no-till are accompanied by modifications of many other agricultural factors such as plant disease and insect dynamics, plant nutrition, and stress resistance.

REFERENCES

Agronorte Pesquisas. *Considerações gerais sobre o cultivo do pé de galinha (Eleusine coracana) como fonte de biomassa, descompactação do solo e integração agropecuária no sistema plantio direto.* Agronorte Pesquisas, ed. Sinop, MT, Brazil, 2001.

Almeida, F.S. *Alelopatia e as plantas*. Parana Agricultural Institute, ed. Circular 53, October/88, Londrina, PR, Brazil, 1988

Almeida F.S. *Controle de plantas daninhas e plantio direto.* Paraná Agricultural Institute, ed. Londrina, PR, Brazil. 34 p. 1991.

Bianchi M.A. "Manejo integrado de plantas daninhas". In *A cultura do milho no plantio direto.* B.C.Campos, ed. FUNDACEP FECOTRIGO, Cruz Alta, R. S., Brazil, 1998.

Buzatti W.J.S. "Controle de plantas daninhas no sistema plantio direto na palha". In *Plantio direto: atualização tecnológica.* V. Pauletti and R. Segandfredo, eds. Fundação Cargil/Fundação ABC, Paraná, Brazil. p. 97-111, 1999.

Chou C.H., Lin J. Y. Autointoxication mechanism of *Oryza sativa*: phytotoxic effects of decomposing rice residues in soil. J Chem Ecol 1976; 3: 353-367.

Correia N.M. Palhadas de variedades de sorgo (*Sorghum bicolor*) no controle de plantas daninhas e nodesenvolvimento da soja (*Glycine max*). M.Sc. thesis, Agronomy Department, Lavras Federal University, Lavras MG, Brazil, 2001.

Einhellig F.A., Souza I.F. Sorgoleone found in grain sorghum root exudates. J Chem Ecol 1992; 18: 1-11.

Federação de plantio direto na palha FEBRAPDP, Brasília, DF, Brazil, 2000.

Fernandes M.F., Barreto A.C., Filho J. Fitomassa de adubos verdes e controle de plantas daninhas em diferentes densidades populacionais de leguminosas. PAB; 1999: 1593-1600.

Fornarolli, D.A., Rodrigues, B.N., Lima, J., Valerio, M.A.. Influência da cobertura morta no comportamento do herbicida atrazine. Planta Daninha 1998: 97-107.

Gassen D.N., Gassen F.R. Plantio direto: o caminho do futuro. Aldeira Sul, ed. Passo Fundo, RS, Brazil. 207 p. 1996.

Inderjit, Streibig J., Olofsdotter M. Joint action of phenolic acid mixtures and its significance in allelopathy research. Physiol Plant 2002; 114:422-428

Jacobi U.S., Fleck N.G. Avaliação do potencial alelopático de genótipos de aveia no início do ciclo. PAB 2000: 11-19.

Kliewer I., Casaccia J., Vallejos F. Viabilidade da redução do uso de herbicidas e custos no controle de plantas daninhas nas culturas de trigo e soja no sistema de plantio, através do emprego de adubos verdes de curto período. Proceedings of National seminar on weed management in no-till system sponsored by Brazilian Association of No-till Planting System; 1998, Passo Fundo, RS, Brazil. p. 120-123. 1998.

Lorenzi H. Manual de identificação e de controle de plantas daninhas: plantio direto e convencional. Instituto Plantarum, ed. Nova Odessa, SP., Brazil. 1994.

Lorenzi H. Manual de identificação e de controle de plantas daninhas: plantio direto e convencional. Instituto Plantarum, ed. Nova Odessa, SP., Brazil, 339p. 2000.

Marochi A.I. 1994. Eficiência de flumtsulam aplicado em mistura com glyphosate no controle de infestantes na cultura da soja em sistema de plantio direto. Dow Elanco, ed. São Paulo, SP, Brazil.

Mello Filho G.A., Mendes D.S. Estimativa de custo de produção de soja nos sistemas plantio direto e convencional, safra 1999/2000. EMBRAPA Agropecuaria Oeste, ed. Dourados, MS, Brazil. Technical communication n. 2. p. 1-3. 1999.

Monegat C. *Plantas de cobertura do solo: características e manejo em pequenas propriedades*. C. Manegat, ed. Chapeco, SC, Brazil. 337 p. 1991.

Narwal S.S. "Exploitation of allelopathy for crops production". In *Allelopathy in Crop Production.* S.S. Narwal, ed. Scientific Publishers, Jodhpur, India, 1994; pp. 162-217.

Pasqualetto, A. Sucessão de culturas como alternativas de produção em plantio direto no cerrado. Ph.D. thesis, Agronomy Department, Viçosa Federal University, Viçosa, MG, Brazil. 1999.

Peixoto M.F. Residuos de sorgo e doses de imazamox no controle de plantas daninhas na soja sob plantio direto. M.Sc. thesis, Agronomy Department, Lavras Agricultural University. 1999.

Perez F.J., Ormeño-Ñunez J. Weed growth interference from temperate cereals: the effect of a hydroxamic acids exuding rye cultivars. Weed Res 1993: 115-119..

Pitelli R. A. Dinâmica de plantas daninhas no sistema plantio direto sponsored by Brazilian Weed Science Society during the Chemical Weed Control Symposium; 1997; Dourados, MS, Brazil. 1997.

Rodrigues B.N. Influência da cobertura morta no comportamento dos herbicidas imazaquin e clomazone. Planta Daninha 1993: 21-28.

Rodrigues B.N., Almeida F.S. *Guia de herbicidas*. B.N.Rodrigues and F.S. Almeida, eds. Paraná Agricultural Institute, Londrina, PR, Brazil. 1998.

Ruedell J. Plantio direto na região de Cruz Alta. om FUNDACEP/Basf, ed. Cruz Alta, RS, Brazil. 1995.

Santos O. G. Alelopatia de genótipos de sorgo em culturas de hortaliças. MS thesis, University of Brasilia, Brasilia, DF, Brazil. 1996.

Schreiber M.M. Influence of tillage, crop rotation, and weed management on giant foxtail (*Setaria faberi*) population dynamics and corn yield. Weed Sci 1992:645-653.

SIDRA-IBGE. 2000. Censo agropecuário 1995. Silva, J.B. "Manejo de plantas daninhas na cultura da soja". In: *Manejo de plantas daninhas nas culturas da soja e do milho*. J.A. Carvalho and N.M. Correia, eds. Uberlândia Federal University, Uberlândia, MG, Brazil: 35-169. 1998.

Skora Neto F. Controle de plantas daninhas em plantio direto nas pequenas propriedades sponsored by Paraná Agricultural Institute during the Encontro latino americano sobre plantio direto na pequena propriedade; Londrina, PR, Brazil, 1993.

Souza I.F. Allelopathy in no-till system in Brazil. Proceedings of the International plant protection congress, Rio de Janeiro, RJ, Brazil, 1992.

Souza I.F. Efeitos alelopaticos do centeio na região do Alto Paranaiba, Minas Gerais. Ciência e Agrotecnologia 1996: 245-248.

Tang C.S., Waiss A.C. Short-chain fatty acids as growth inhibitors in decomposing wheat straw. J Chem Ecol 1978; 4: 225-232.

Theisen G., Vidal R.A. Efeito da cobertura do solo com resíduos de aveia preta nas etapas do ciclo de vida do capim marmelada. Planta Daninha 1999: 189-196.

Tousson T.A. Nature of phytotoxic substances during plant residue decomposition in soil. Phytopathology 1968: 41-45.

Vickery M.L., Vickery B. "Compounds derived from aminoacids". In *Secondary Plant Metabolism.* University Park Press, Baltimore, Maryland, USA. 1981.

CHAPTER 17

SOIL SOLARIZATION
AN ECO-FRIENDLY APPROACH FOR WEED MANAGEMET

N. T. YADURAJU & J. S. MISHRA
National Research Centre for Weed Science,
Jabalpur - 482 004 (M. P.) India

The search for new effective, economic and eco-friendly methods for controlling weeds is a continuous process, as none of the existing methods can be used in all situations. Manual weeding though effective and commonly used in many under developed and developing countries is expensive and time consuming, and hence economically restricted to certain crops and seasons. Use of herbicides for controlling weeds is very effective and economical but has been criticized by the environmentalists for polluting the ecosystem. Besides, there is inherent danger of resurgence of weeds resistant to herbicides. Use of solar energy as soil solarization for controlling soil-borne pests including weeds, pathogens and nematodes has potential for reducing the dependency on chemicals.

At present three basic approaches exist for the disinfestation of soil. These are heating (steaming), fumigation and solarization (Katan, 1981). Fumigation is the most common and widespread method of soil disinfestation. Various non-selective fumigants, including methyl bromide, chloropicrin, carbon disulfide, formaldehyde, metham sodium and more selective material, such as ethylene dibromide are used (Wilhelm, 1966; Krentzer, 1960). These affect various populations of soil microflora and fauna. Under certain conditions, some of the fumigants may acts as eradicants which can destroy most of the soil microflora. Fumigation frequently may suppress plant growth. Compared to steam and chemical fumigation, soil solarization can be regarded as a relatively mild treatment. Nevertheless, a broad spectrum of pathogenic fungi, weeds, nematodes and other pests are controlled by solarization in diverse agricultural systems. Further, in view of the increasing public demand for organic products in recent years, curtailment of pesticide use, threats to environmental pollution and the everlasting need to control pathogens, weeds and pests, the solarization technique offers hope for the future.

Soil mulching is a common method employed in extreme weather conditions to moderate soil temperature and to improve crop growth and yield. Although, plant residues are traditionally used, paper and plastics have been used lately for the purpose. Besides other benefits, mulching suppresses weeds primarily by cutting off light. Soil solarization is a special mulching technique developed in the seventies by Katan and associates in Israel, where moist soil is covered by thin transparent polyethylene film during summer months for several weeks to trap the heat and disinfest the soil. This hydro-thermal process results in elevation of temperatures to levels that are lethal to many soil-borne plant pathogens, insects and weeds, and causes other physical and biological changes in soil which are beneficial to crop

Inderjit (ed.), Weed Biology and Management, 345-362.

growth (Katan, 1981; Stapleton and Lopez, 1988). The method has been variously named as solar sterilization, solar heating or pasteurization, plastic or polyethylene mulching or tarping etc. This approach of killing weed seeds and propagules seems to have greater potential in tropical and sub-tropical regions where air temperature goes upto 45 °C during summer months.

PRINCIPLES OF SOLAR HEATING

Visible solar radiation and short-wave infrared radiation are responsible for heating of the soil through the mulch. The radiation from earth (terrestrial radiation) is normally in the form of long wave infrared radiation, which are not permeable through glass or plastic. Hence, heat is trapped beneath the transparent polyethylene mulch. Polyethylene also reduces heat convection and evaporation of water from the soil. The formation of water droplets on the inner surface of the polyethylene film also reduces the transmissivity to long wave infrared radiation substantially, resulting in better heating due to an increase in its greenhouse effect (Malik and Tran, 1973). The following are to be borne in mind for maximum benefit from soil solarization.

1. Transparent not black polyethylene should be used since the former transmits most of the solar radiation that heats the soil.
2. Soil mulching should be carried out during the period of high temperatures and intense solar radiation.
3. Soil should be kept wet during mulching to increase thermal sensitivity of resting structures and improve heat conduction.
4. The thinnest polyethylene tarp possible (25-30 μm) is recommended, since it is both cheaper and somewhat more effective in heating due to better radiation transmittance, than thicker films.
5. The mulching period should be sufficiently extended, usually four weeks or longer, to heat the soil at deeper layers.

The key factors involved in solarization effects on soil-borne organisms are temperature, soil moisture and possibly volatile compounds. The damaging effect of high temperatures has been attributed to metabolic and structural changes in cells that become irreversible with increasing temperature (Christiansen, 1978).

FACTORS AFFECTING SOIL SOLARIZATION

Soil Preparation

The heating of the soil is best when the polyethylene film is laid close to the soil with a minimum of airspace to reduce the insulating effect of an air layer. Hence, good land preparation is essential to provide a smooth even surface.

Soil Characteristics

Dark soils absorb more radiation than light soils; this may partly account for the higher maximum temperatures achieved in some soils. Small differences in soil characteristics or moisture content can translate into large differences in soil heat transfer characteristics (Smith, 1964).

Soil Moisture

Adequate soil moisture is required at the beginning of solarization for two different purposes: (a) to increase soil heat conductivity and (b) to sensitise the weed seeds and other resting structures to high temperature. Moist soil, either irrigated before mulching or irrigated under the plastic film increases the thermal heat transfer or conduction in the soil (Mahrer et al., 1984).

Polyethylene Film Types and Characteristics

Transparent polyethylene films are more efficient than black films in trapping solar radiation (Horowitz et al., 1983; Bhasker et al., 1998; Mudalagiriappa et al., 1999) and reducing weed emergence (Table 1). Mulching of the soil with low-density clear polyethylene sheets (0.1 mm thick) proved more effective than mulching with black plastic in reducing bromrape (*Orobanche* spp.) parasitism on the following crop of eggplant (Braun et al., 1984). Chase et al. (1998) found that soil solarization through transparent polyethylene reduced the emergence of purple nutsedge (*Cyperus rotundus* L.) (5 plants/m^2) as compared to black low-density polyethylene (35.7 plants/m^2) after 5 weeks. The lower emergence was attributed to lethal soil temperature resulting from solarization, which killed *Cyperus* tubers. In rows mulched with black films, rhizomes punctured the film and leaf expansion occurred above the film. Besides clear transparent polyethylene costs less and has high strength. It also allows maximum transmittance of radiation from 0.4 μm to 36 μm (Waggoner et al., 1960). Thinner films, 19-25 μm are more effective for solar heating than thicker films (50-100 μm) and are proportionately less expensive (Stapleton and DeVay, 1986). Higher soil temperature under thin films could be attributed to favourable properties such as higher transmission and lower reflection (Stapleton, 1997). Basavaraju and Nanjappa (1999) observed that soil solarization for 45 days with thin films (25μm) recorded significantly lower weed population in chilli and delayed the emergence of crabgrass (*Digitaria marginata* L.) to an extent of 40 days as compared to non-solarization. The transparent polyethylene of 0.05 mm thickness was superior to that of 0.10 mm thickness (Habeeburrahman and Hosmani, 1999b). However, Patel et al. (1995) found that thickness of transparent polyethylene did not significantly alter pest control as much as duration of solarization did. Clear mulch warms the soil more effectively, and plastic mulches have been developed that filters out photosynthetically active radiation (PAR), but let through infrared light to warm the soil. These infrared transmitting (IRT) mulches have been shown to be effective at controlling weeds (Majek and Neary, 1991).

Table 1. Effect of solarization with black and transparent plastic on weed emergence after plastic removal (Horowitz et al., 1983).

Species emerged	Weed emergence (No./0.5 m^2)		
	Control	Black	Transparent
Pigweed (*Amaranthus* spp.)	30[a]	12[b]	4[bc]
Common purslane (*Portulaca oleracea* L.)	78[a]	68[a]	4[b]
Henbit (*Lamium amplexicaule* L.)	146[a]	0[b]	4[b]
Bull mallow (*Malva niceaensis* All.)	330[a]	107[b]	16[c]
Compositae*	524[a]	0[b]	0[b]
Total annuals	1294[a]	194[b]	44[c]
Broomrape (*Orobanche crenata* Forsk.)	40[a]	9[b]	0[b]

*Mainly daisy (*Chrysanthemum segetum* L.) and annual sowthistle (*Sonchus oleraceus* L.). Figures in the same line followed by the same letter do not differ at 5% level of significance

Shoots of the field bindweed (*Convolvulus arvensis* L.) appeared in great numbers after solarization with black plastics but none emerged with the transparent plastic (Horowitz, 1980). At a depth of 5 cm, the maximum temperature was increased by 9.3°C for black and 17-19°C for transparent plastic. Soil temperature under plastic cover is a function of incoming solar radiation and thermal characteristics of the mulching material and the soil (Mahrer and Katan, 1981). Type of transparent plastic differs in their transmittance characteristics and resulting heating (Table 2). Thin transparent polyethylene has good mechanical and thermal properties and a relatively low cost (Horowitz et al., 1983). Soil solarization with thermal-infrared-retentive (TIR) film resulted in higher soil temperature than with a 30μm low-density polyethylene (LDPE) clear film. With TIR films, greater proportions of emerged purple nutsedge plants were killed by foliar scorching, and 6 weeks of soil solarization was more effective at reducing purple nutsedge density than with the LDPE film (Chase et al., 1999).

Beds – Width and Direction

For control of weeds where vegetables are planted on beds, the beds must be pre-formed and solarized. Studies on beds have shown excellent weed control, whether irrigated after bedding and before tarping, receiving natural rainfall before tarping, or with drip irrigation under the tarp. Mahrer and Katan (1981) found that there was usually a 2 to 4 °C lower temperature at the edge of the mulch than at the centre, at the same soil depth. Horowitz et al. (1983) measured the mulch width of polyethylene as related to maximum soil temperature and irrigation, prior to or during solarization and observed that there was no weed emergence from a 20- to 140- cm mulch width and that the temperatures at the 5 cm depth were comparable. Temperatures were generally as much as 5°C higher at the 15 cm depth with the widest mulch width prior to and during irrigation while solarization was in progress. Beds running in a north-south direction were preferable to an east-west direction to avoid a lower temperature on one side of the bed (Elmore, 1991).

Table 2: Effect of plastic type on soil temperature during solarization and weed control (Horowitz et al., 1983)

Type of plastic	Maximum soil temperature (°C) at depth		Weed emergence (No./m^2) after plastic removal			
			Pigweed		Common purslane	
	5 cm	15 cm	1 month	11 months	1 month	11 months
Control	30.4	28.2	92^a	70^a	24^a	12^a
Black polyethylene	39.7	31.4	8^b	1^b	29^a	2^b
Transparent P.V.C*	47.4	35.9	0^b	0^b	12^b	2^b
Transparent thin (0.03 mm) polyethylene	49.5	36.1	1^b	0^b	8^b	3^b
Transparent thick (0.10 mm) polyethylene	47.7	35.0	1^b	1^b	6^b	2^b

*P.V.C. = Polyvinyl chloride (0.09 mm)
Figures in the same column followed by the same letter do not differ at 5% level of significance.

Timing and Duration of Soil Solarization

Soil solarization is most effective when it is done during the warmest part of summer and plastic sheeting left in place for as long as practical. However, a duration of 4-6 weeks would be adequate under many situations (Yaduraju, 1993; Habeeburrahman and Hosmani, 1996; Singh et al., 2000). Shorter duration is often sufficient in tropical countries. In India, April-May in the south and May-June in northern areas experience intense radiation and are best suited for solarization (Yaduraju and Kamra, 1997). The land is also idle during this period, which is important, as many farmers do not like to lose a crop in favour of soil solarization. Soil solarization for 40 days was found more effective in controlling weeds as compared to 20 days (Bhasker et al., 1998). Solarization is most effective in the upper 10 cm of soil; however, pest control at greater depths may be achieved sometimes, with extended periods of polyethylene mulching. Egley (1983) found that just one week of solarization treatment significantly reduced the number of viable seeds of prickly sida (*Sida spinosa* L.), common cocklebur (*Xanthium pensylvanicum* Wallr.), velvetleaf (*Abutilon theophrasti* Medic.) and spurred anoda (*Anoda cristata* L. Schlecht.). When the duration was increased to 2 weeks, additional species were controlled. In India, Singh et al. (2000) observed that solarization for a period of 3 weeks during May to June was enough to control most of the annual weeds in soybean (Table 3).

Soil solarization for a period of 5 weeks effectively controlled most summer and winter annual weeds at Rehovot, Israel for a period of at least 5 months but particularly controlled perennial weeds such as purple nutsedge, Johnsongrass (*Sorghum halepense* (L.) Pers.) and bermudagrass *(Cynodon dactylon* (L.) Pers.) (Rubin and Benjamin, 1983). Solarization for a period of 10 weeks effectively controlled perennial weeds (Rubin and Benjamin, 1983). Horowitz et al. (1983) obtained excellent weed suppression after 2 weeks of solarization (Table 4). The effect was very clear on common purslane and appreciable on the field bindweed.

CAUSES OF SEED DEATH

According to available information, following mechanisms may be involved in the weed control process using soil solarization (Rubin and Benjamin, 1984):

- direct thermal killing of germinating or even dormant seeds;
- thermal breaking of seed dormancy followed by thermal killing;
- thermally induced changes in CO_2/O_2, ethylene and other volatiles which are involved in seed dormancy release followed by thermal killing;
- direct effect of high temperature interacting with toxic volatiles released from decomposing organic matter or seed metabolism;
- indirect effects via microbial attack of seeds weakened by sub-lethal temperature.

Table 3: Effect of period of solarization on population and dry weight of weeds and seed yield of soybean (Singh et al., 2000).

Period of solarization	Weed density/m^2		Weed dry weight (g/m^2)		Seed yield (kg/ha)
	30 DAS	60 DAS	30 DAS	60 DAS	
Control	96.9^a	122.9^a	43.9^a	69.6^a	831^a
Solarization for 3 weeks	0.0^b	43.6^b	0.0^b	46.1^b	1391^b
Solarization for 4 weeks	0.0^b	40.1^b	0.0^b	32.2^b	1403^b
Solarization for 5 weeks	0.0^b	38.8^b	0.0^b	32.1^b	1473^b
Summer ploughing	83.6^a	79.4^c	20.8^c	76.4^a	910^a

Figures in the same column followed by the same letter do not differ at 5% level of significance.
DAS = Days after sowing

Table 4. Effect of duration of soil solarization in early summer in Newe-Yaar (Israel) on soil temperature and weed emergence after plastic removal (Horowitz et al., 1983).

Duration (weeks)	Maximum soil temperature (^{0}C) at depth		Weed emergence after plastic removal			
			Common purslane		Field bindweed	
	5 cm	15 cm	3 months	12 months	3 months	12 months
0	32.5	29.5	1322^a	137^a	17^a	18^a
2	46.5	37.7	1^b	16^b	11^a	2^b
4	44.2	37.6	0^b	16^b	1^b	2^b
6	45.3	39.5	0^b	10^b	0^b	5^b

Figures in the same column followed by the same letter do not differ at 5% level of significance.

EFFECT ON WEED EMERGENCE

Elevated temperature in soil following soil solarization results in reduction in the population of soil borne pathogens, nematodes and weeds. The response to solarization in weeds varies with weed species. Soil solarization was most effective at controlling broad-leaved weeds than sedges and grasses (Reddy et al., 1998). Abdullah (1998) also found that seed bed solarization gave 100, 80 and 16 % weed reduction for annual broad leaf weeds, annual grasses and perennial weeds, respectively in onion. Winter weeds are generally more susceptible, whereas summer weeds especially *Cyperus* spp. and field bindweed are generally more resistant. Nasr-Esfahani et al. (2000) reported that soil solarization effectively reduced the population of almost all weeds by around 100 % except for purple nutsedge and spiny sowthistle (*Sonchus asper* (L.) Hill), which was reduced by 59 and 44 %, respectively. At the International Crop Research Institute for Semi-Arid Tropics, Hyderabad (India), Chauhan et al., (1988) observed a marked decrease in annual weeds in chickpea and pigeonpea due to solarization. The perennials such as purple nutsedge and bermuda grass however recovered gradually. A large number of annuals and many perennial weeds are sensitive to soil solarization. However, a few weed species *viz*., pigweed, bermudagrass, three flower beggarweed (*Desmodium* spp.), *Astragalus* spp., *Cyperus* spp., purple nutsedge, yellow nutsedge (*Cyperus esculentus*), sweet clover (*Melilotus* sp.), yellow sweet clover (*Melilotus indica* (L.) All.*), M. sulcatus*, horseweed (*Conyza canadensis* (L.) Cronq.), crabgrass, mallow (*Malva* sp.) defy solarization treatment (Yaduraju, 1993). Common purslane was also reported to be resistant to solarization (Tamietti and Valentino, 2000). Several workers (Egley, 1983; Rubin and Benjamin, 1984; Yaduraju, 1993) have also observed the tolerance of sweet clover to soil solarization. While a solarization for just 10 days gave complete control of littleseed canarygrass (*Phalaris minor* Retz.) and winter wild oat (*Avena ludoviciana* Durieu.) – the most dominant grass weeds in winter season crops in India-there was an increase in the population of sweet clover with 40 day of solarization. The heat tolerance of yellow sweet clover has been attributed to its thick seed coat (Yaduraju, 1993).

In perennials with an established underground system of deep roots, rhizomes or tubers, the failure of solarization is probably due to the limited penetration of heat in soil beyond 10 cm depth (Horowitz et al., 1983). Most perennials are capable of regenerating rapidly from partially damaged underground organs. The survival of purple nutsedge tubers in the soil has been reported by several workers (Kumar et al., 1993; Horowitz et al., 1983; Rubin and Benjamin, 1984) and it has been attributed to heat resistance of the tubers. As the solarization effect diminishes with soil depth, weeds that are capable of emergence from deeper layers survive the treatment. Although, solarization did not control purple nutsedge directly, it weakened the plants to the extent it was more susceptible to applications of glyphosate (Gonzalez et al., 1992).

Parasitic weed broomrape (*Orobanche* spp.) has been effectively controlled with solarization (Mauromicale et al., 2001; Meti, 1993; Krishnamurthy et al., 1993; Abu-Iramaileh, 1991; Jacobsohn et al., 1980). Mulching with clear plastic for 36 days resulted in almost complete eradication of *O. aegyptiaca* from carrot and eggplant fields soon afterward, and in partial control of other weeds (Jacobsohn et al., 1980; Katan et al., 1976). At Jerusalem, Israel, no carrot plant was parasitised with broomrape upto 110 days after planting in solarized plots whereas 90% of the plants got parasitized at 75 days after planting in non-solarized plots (Jacobsohn et al., 1980). In Sudan, soil solarization controlled broomrape by 72-100 % (Braun et al., 1988). Soil solarization controlled broomrape for two successive years in faba bean and tomato in Egypt (Satour et al., 1991), and in lentil and faba bean by over 90 %. (Sauerborn et al., 1989). In cabbage, solarization for 2-6 weeks alone killed broomrape seeds at soil surface, but had no significant effect on seeds below the surface. Solarization with chicken manure, however, killed broomrape seeds at all depths. (Haidar and Sidahmed, 2000).

Soil solarization for 4-6 weeks in summer not only gave excellent weed control in soybean but also in the succeeding wheat crop (Yaduraju and Ahuja, 1996). The weed control in soybean was better with soil solarization alone than with manual weeding (1 or 2) and pre-emergence application of pendimethalin (0.75-1.5 kg/ha). Soybean yields were significantly higher with soil solarization than with pendimethalin. This was attributed to the poor control of dominant weed tropical spiderwort (*Commelina benghalensis* L.) and *Digera arvensis* Forsk. with the herbicide. Soil solarization, however, gave excellent control of these weeds. Application of crop residues and soil solarization with 0.05 and 0.075 mm transparent polyethylene resulted in maximum weed reduction and the highest yield in groundnut-frenchbean cropping system (Mudalagiriappa et al., 2001).

EFFECT OF SOIL SOLARIZATION ON BURIED WEED SEEDS

The reserves of dormant weed seeds in agricultural soils act as a potential source of persistent weed problems that often require repeated control measures (Egley, 1983). A reduction in the number of dormant weed seeds in the soil should also correspondingly reduce weed persistence and weed control requirements. Hence, soil solarization would be desirable as a means of reducing the dormant weed seed reserves in the soil. However, solarization was not found to eliminate dormant weed seeds from the germination zone (Egley, 1983). The treatments killed non-dormant seeds and greatly reduced the number of weed seedlings that otherwise would have emerged. Soil solarization for one week, significantly reduced the percentage of buried seeds (64-98%) of prickly sida, common cocklebur, velvetleaf and spurred anoda remaining in the soil (Table 5). The emergence of grasses and pigweed was also significantly reduced due to solarization for just one week. However, emergence of purple nutsedge was increased in some instances (Table 6). Five weeks of solarization with clear polyethylene films at Waimanalo, Hawaii raised the mean soil temperature at 15 cm depth by 5.8 ^{0}C in spring and 7.2 ^{0}C in summer and increased the final sprouting percentage of *C. rotundus* in the field from 74 to 97 % in spring and 97 to 100 % in summer (Miles et al., 2002). Solarization reduced annual blue grass (*Poa annua* L.) seed survival from 89 to 100 % in the upper 5 cm of soil, but did not reduce survival below 5 cm (Peachey et al., 2001).

Although weed seeds in the surface 5 cm of soil are killed to a great extent, soil solarization has rather a poor effect on weed seeds at deeper layers because of lower temperatures (Kumar et al., 1993). Chandrakumar et al. (2002) also reported higher soil temperature in top 0-5 cm soil depth as compared to 5-10 cm soil depth. Solarization for 40 days killed common dayflower in the top 11 cm of soil but seeds of annual *Cyperus* spp. and barnyardgrass (*Echinochloa crus-galli* (L.) Beauv.) only within the upper 3 to 4 cm (Standifer et al., 1984). It is possible that all weed seeds are unaffected at deeper layers but weeds with bold seeds are capable of emerging from lower depths while small-seeded weeds lie dormant.

Effect of Soil Solarization on Crop Growth and Yield

Soil solarization can have a tremendous impact on crop growth and yield by eliminating or reducing the infestation of soil-borne pathogens, nematodes and weeds. The magnitude of increase depends upon the type of pest problem and the degree of control. Control of weeds by solarization increased the yield of onion by 100-125% (Katan et al., 1980), groundnut by 52% (Grinstein et al., 1979), sesame by 72% (Stapleton and Lopez, 1988), carrot by 28 %, cabbage by 34 %, beet by 37 %, greengram by 32% (Ricci et al., 2000) and soybean by 77% (Kumar et al., 1993; Singh et al., 2000). Solarization is very effective in controlling the parasitic weed broomrape and a yield of 78 t/ha of carrots was reported from solarized plots, while no yield was obtained from the non-solarized plots (Jacobsohn et al., 1980).

Solarization has been found to enhance the growth and yield of crop plants even in places where there is no infestation of either soil-borne pathogens, nematodes or weeds. This could be attributed to several chemical and biological changes in soil caused by solar radiation when covered by clear plastic films especially when the soil has a high moisture content (Yaduraju and Kamra, 1997). With adequate control of weeds and nematodes through chemical or mechanical methods, solarization still enhanced the yield of soybean (Yaduraju, 1993).

Effect on Soil Properties and Mineral Nutrients

Solarized soils contained higher levels of soluble mineral nutrients than untreated soils (Baker and Cook, 1974; Stapleton et al., 1985). Significant increases in ammonium-nitrogen, nitrate-nitrogen, Ca^{+2}, Mg^{+2} and electrical conductivity were consistently found. Phosphorus, K^+ and Cl^- were also increased in some soils. However, the availability of other micro-nutrients *viz.*, Fe^{+2}, Mn^{+2} and Cu^{+2} were not increased (Stapleton and DeVay, 1986). An increase in nitrate-N and ammonical-N was observed in New Delhi soils but organic carbon content was not altered (Kumar and Yaduraju, 1992).

Table 5. Effect of soil solarization through transparent polyethylene cover upon survival of buried weed seeds (Egley, 1983).

Weeds	Viable seeds remaining 4 weeks after removal of cover (%)	
	Not covered	Covered
Prickly sida (*Sida spinosa L.*)	76	5[a]
Common cocklebur (*Xanthium pensylvanicum* Wallr.)	62	9[a]
Velvetleaf (*Abutilon theophrasti* Medic.)	73	47[a]
Spurred anoda (*Anoda cristata* (L.) Schlecht.	66	45[a]
Pitted morningglory (*Ipomoea lacunosa* L.)	46	34
Common purslane (*Portulaca oleracea* L.)	96	94
Johnsongrass (*Sorghum halepense* (L.) Pers.)	44	57

Means followed by "a" are significantly different from corresponding non covered (control) at the 5% level of significance as determined by the standard T-test.

Table 6. Effects of soil solarization on weed seedling emergence (Egley, 1983).

Weed species	Not covered	Weeds emerged (no./m^2)			
		Period covered (weeks)			
		1	2	3	4
Grasses	841[a]	306[b]	41[c]	15[c]	9[c]
Pigweed (*Amaranthus* species)	300[a]	65[b]	13[b]	0[b]	0[b]
Morningglory (*Ipomoea* spp.)	76[a]	54[ab]	17[b]	0[b]	2[b]
Horse purslane (*Trianthema portulacastrum* L.)	74[a]	22[a]	26[a]	4[a]	0[a]
Purple nut sedge (*Cyperus rotundus* L.)	24[a]	43[a]	43[a]	9[a]	22[a]
Total[b]	1385[a]	495[bc]	140[c]	28[c]	33[c]

[a] Within each weed class, means followed by the same letter do not differ at the 5% level according to Duncan's multiple range test.
[b] Total includes weeds in addition to those identified in the table.

However, with extensive studies in different soils types and nutrient sources, it has been shown that the increases in the levels of soil nutrients are transient and do not persist long (Yaduraju and Kamra, 1997).

Effect on Soil Microorganisms

Effects of solarization are more selective on soil microorganisms in comparison with most other methods of soil disinfestation (Krentzer, 1965; Baker and Cook, 1974). Solarization causes changes in soil biota and substrate that provide a favourable environment for colonization by microorganisms with greater competitive ability. These organisms are usually saprophytes, rather than phytopathogens, which may subsequently inactivate surviving phytopathogenic fungi, bacteria, nematodes, and weed seeds that are damaged or weakened by solarization. These effects may persist for several seasons. Several investigations have been conducted on the effect of solarization on native as well as inoculated *Rhizobium*. The native populations in pigeonpea and chickpea were decreased with solarization

(Linke et al., 1990; Rupela and Sudershana, 1990). A population of *Rhizobium*, sufficient to cause heavy nodulation of bean roots, survived solarization in Israel (Katan, 1981). Improved yields of mungbean, soybean and groundnut in response to seed inoculation of *Rhizobium* (Table 7) were observed in New Delhi (Yaduraju, 1993).

Effect of Soil-Borne Fungi

Verticillium and *Fusarium* wilts of several crops have been successfully controlled by solarization, as well as diseases caused by *Bipolaris sorokiniana, Didymella lycopersicii, Phytophthora cinnamomi, Plasmodiophora brassicae, Pysenochaeta lycopersici, P. tervestris, Pythium myrothecium, P. ultimum, Rhizoctonia solani, Sclerotium oryzae, S. rolfsii* and *Thielaviopsis basicola*. Pathogenic fungi including *Pythium irregulare, S. cepivorum* and *Sclerotinia minor* were reduced in artificially inoculated soils (Stapleton and DeVay, 1986). However, the destruction of beneficial organisms such as arbuscular mycorrhizal (AM) may also occur, there by reducing the positive effect of solarization. Schreiner et al. (2001) found that solarization did not reduce the infectivity of AM fungi immediately after the solarization period. But infectivity was greatly reduced 8 months after solarization. Solarization apparently reduced AM fungi in soil indirectly by reducing weed populations that maintained ineffective propagules.

Table 7. Effect of solarization and *Rhizobium* inoculation on seed yield (kg/ha) of some legumes (Yaduraju, 1993).

Crop	Solarized		Non-Solarized	
	- Inoculation	+ Inoculation	- Inoculation	+ Inoculation
Soybean	1833	2683	717	1383
Groundnut	1489	1756	844	1112
Mungbean	1191	1315	547	556

Effect on Soil-Borne Bacteria

The thermal sensitivity of soil-borne bacteria depends upon the nature of the individual taxa. *Agrobacterium* spp., fluoresent pseudomonads, pectolytic pseudomonads and some Gram – positive bacteria have been reduced by 69-98% due to solarization. However, pseudomonads rapidly recolonized treated soil and no significant difference between treatments was apparent 3-6 months later (Stapleton and DeVay, 1986). Actinomycetes and *Bacillus* spp., many of which are thermo-tolerant, were sometimes reduced to a much lesser extent (45-58%) or were even increased (26-158%) following solarization (Stapleton and DeVay, 1982). Increased colonization (183-631%) of plant roots by plant growth-promoting fluorescent pseudomonads from inoculated seed also occurred following soil solarization (Stapleton and DeVay, 1984).

Effect on Soil-Borne Nematodes and Mites

Solarization reduced the population of plant parasitic nematodes including *Tylenchus, Heterodera, Xiphinema, Hoplolaimus, Pratylenchus* by about 42 to 100% (Kumar et al., 1993; Katan, 1984; Barbercheck and Von Broembsen, 1986; Stapleton and DeVay, 1984), but the nematode levels had largely recovered after 70 days (Kumar et al., 1993). With regards to soil-borne mites, solarization has been used to control the plant-parasitic mite, *Rhizoglyplus robini*, in Israel (Katan, 1984). Soil solarization significantly reduced ant and earthworm numbers but had no effect on millipede population (Ricci et al., 1999).

Applicability and Potential of Soil Solarization

Soil solarization is an unique method of pest control. It is (i) non-hazardous, (ii) user-friendly, (iii) environmentally benign, (iv) non-dependent on fossil fuel, (v) effective on a variety of pests including soil-borne fungi, bacteria, nematodes and weeds, (vi) often effective for more than one season or a year, and (vii) stimulatory to crop growth. It has great applicability in organic agriculture and high value crops.

Cost Effectiveness and Long-Term Benefits

Solarization falls into the medium price range of soil disinfestation treatments. As solarization technology advances, the overall cost of application should decrease. The major constraint in developing countries is the high cost of the treatment. Thinner films, which are more efficient, would be more economical. Control of different types of pests, in successive crops, lower fertilizer requirements and increased yield of crops, all will have to be taken into consideration while computing the economics of soil solarization treatment. In areas, where only 2-4 weeks treatment is sufficient and where summers are long, the same polyethylene films can be used in more than one place in the same season. Further, in a long-term study, it was seen that soybean grown without solarization on a field otherwise solarized for two successive summers gave yield of soybean similar to 3-year solarization, provided the soil was not disturbed (Yaduraju and Ahuja, 1996). These measures will help to cut costs substantially.

Application of Soil Solarization

The technology for applying plastic films to large acreages already exists (Pullman et al., 1984) and is similar to use of soil fumigation. Solarization is also ideal for treating nurseries. Duxbury (2002) found that the combination of vitavax-200 and solarization of soil in the rice nursery controlled the root-knot nematode *Meloidogyne graminicola* and increased the rice yield upto 45%. He further noted that around 30-40% rice yield was increased just by solarizing the soil in the rice nursery. It is widely used in raising nurseries of tobacco and vegetables in India (Patel et al., 1995; Sudha et al., 1998; Reddy et al., 1998). Vizantinopoulos and Katrains (1993) reported that solarization gave better weed control than pre-emergence pendimethalin+atrazine, imazaquin+metolachlor, metribuzin+alachlor or acetachlor + atrazine at their recommended rates in maize in Greece. It can reduce the use of toxic soil fumigants. This technology may be more promising in high value crops, such as in vegetable growing, floriculture, etc. In India, pre-plant-solarization films may be left in place, after plant emergence, as post-plant mulch. Soil solarization has been shown to enhance degradation of pesticide residues in soil (Gopal et al., 1997), hence could be employed for decontamination of polluted soils.

Soil solarization is a fully integrated pest control method, based on easily manipulated physical factors, but its use is restricted to regions with intense sunshine and field is free of crops for about one month or more at the time of mulching. Solarization can be viewed as part of soil preparation before planting. To date, there are no contra-indications against use of solarization in any crop, except economic considerations. In estimating its economic value, it should be considered not only as a possible replacement for herbicide and other pesticide treatments but also as an eventual solution for situations in which no other safe and efficient method is available.

LIMITATIONS OF SOIL SOLARIZATION

Despite huge benefits, soil solarization has not found large-scale adoption yet, possibly because of some of the following limitations:

- Solarization is in fact indirectly dependent on fossil fuel, since they are used in the production of the plastics.
- Disposal of plastic film is by for the greatest problem. Disposal by burning pollutes the environment and disposal in a landfill takes up considerable space. The answer to this problem could be use of biodegradable plastic films.
- Retention of solarization films as production mulches usually requires additional inputs of paint and labour to paint the mulch white so that cooler soils prevail for the production of the crop.
- Removal and reuse of the film is not feasible in large-scale operations that utilize machines to lay plastic films.
- The technique is more useful for crops grown immediately following the solarization period. Thus, in the northern hemisphere, it is more useful for fall-grown crops than spring-grown crops.
- Lack of persistence of nematode control and poor control of some weeds mean that supplemental control methods are often needed, particularly in crops that occupy a field for a long duration.
- In some climates cloudiness and rainfall during the hottest part of the year can limit effectiveness.
- Difficult to retain the films intact during period of heavy winds.

CONCLUSION

Soil solarization is a novel technique for the control of soil borne diseases, nematodes and weeds. Covering the moist soil for a period of 2-6 weeks during hot summer months can elevate the surface soil temperature by about 6-16 ^{0}C depending upon the soil type, moisture content, thickness of polyethylene films and the intensity of solar radiation. Besides pest control, the polyethylene mulching also alters the soil chemical and biological properties. These changes usually bring about enhancement in crop growth and yield. The technique is successful in controlling a wide range of annual grasses, broadleaved weeds, sedges and some perennials in a variety of crops.

Despite several advantages, the large-scale adoption of this technology has not taken place primarily due to economic reasons. Disposal and non-biodegradation of plastic in the environment are the major concerns, which need to be seriously addressed. Development of biodegradable polyethylene films and safer alternative methods of re-use could allay such fears by environmentalists. However, it has definite advantage in nurseries, where solarization can reduce or eliminate use of soil disinfectants such as methyl bromide, chloropicrin, carbon disulfide, formaldehyde, metham sodium, etc, which are hazardous or poisonous. It can also find application in high value crops and in organic agriculture.

REFERENCES

Abdullah M.M.F. Improving vegetable transplant using soil solarization. II. Onion, "*Allium cepa*" Proceedings of the Seventh Conference of Agricultural Development Research, Cario, Egypt, 15-17 December, 1998.

Abu-Iramaileh B.E. Soil solarization controls broomrapes (*Orobanche* spp.) in host vegetable crops in Jordan Valley. Weed Technol 1991, 5:575-581.

Baker K.F., Cook R.J. *Biological control of plant pathogens*. Freeman, San Francisco, C.A., 1974.

Barbercheck M.E., Von Broembsen S.L. Effect of soil solarization on plant-parasitic nematodes and *Phytophthora cinnamomi* in South Africa. Plant Dis 1986; 70:945-450.

Basavaraju H.K., Nanjappa H.V. Weed dynamics in chilli-maize cropping sequences as influenced by solarization. Indian J Weed Sci 1999, 31:183-186.

Bhasker K.V., Nanjappa H.V., Ramachandrappa B.K. Solarization for weed control in sunflower (*Helianthus annuus* L.). Mysore J Agric Sci 1998; 32:142-147.

Braun M., Burgstaller H., Walker H. Critical evaluation of control methods for *Orobanche ramosa* L. occurring in smallholder vegetables farms of the Khartoum province, Sudan. Proceedings of third International Symposium on Parasitic weeds, Aleppo, Syria. pp. 245-249, 1984.

Braun M., Koch W., Mussa H.H., Steifvaler M. "Solarization for weed and pest control, possibilities and limitations." In *Weed Control in Vegetable Production*. Proceedings of meeting of the EC experts group, Stuttgard.28-31 October, 1986, Rottendrum, Balkome, 1988; pp.169-178.

Chandrakumar S.S, Nanjappa H.V., Ramachandrappa B.K., Baig M.K. Weed control in sunflower (*Helianthus annuus* L.) through soil solarization. Crop Res (Hisar) 2002, 23:287-289.

Chase C.A., Sinclair T.R., Locascio S.J. Effect of soil temperature and tuber depth on *Cyperus* spp. control. Weed Sci 1999; 47:467-472.

Chase C.A., Sinclair T.R., Shilling D.G., Gilreath J.P., Locascio S.J. Light effect on rhizome morphogenesis in nut sedges (*Cyperus* spp.): implications for control by soil solarization. Weed Sci 1998; 46:575-580.

Chauhan Y.S., Nene Y.L., Johansen C., Haware M.P., Sayene N.P., Singh S., Sharma S.B., Sahrawat K.L., Burford J.R., Rupela O.P., Kumar Rao J.V.D.K., Sithanantham S. Effect of soil solarization on pigeonpea and chickpea. Res Bull No. 11, ICRISAT, Patancheru, Andhra Pradesh, India. 1988.

Christiansen M.N. "The physiology of plant tolerance to temperature extremes." Pages 173-191 In *Crop Tolerance to Suboptimal Land Conditions*.G.A. Jung, ed. American Society of Agronomy, Madison, Wisconsin, 1978; Special Publication No. 32.

Duxbury J.M.Improving crop productivity and sustainability in rice (*Oryza sativa*)-wheat (*Triticum aestivum*) cropping system. Extended Summaries Vol. 1. 2nd International Agronomy Congress, November, 26-30 2002, New Delhi, India :31-32.

Egley G.H. Weed seed and seedling reduction by solarization with transparent polyethylene sheets. Weed Sci 1983; 31:404-409.

Elmore C.L. "Weed control by solarization." In *Soil Solarization,*. J. Katan, J.E. DeVay, eds. Boca Raton: CRC Press Inc., 1991; pp. 62-72.

Gonzalez T.R, Agudo B.M., Lopez G.M.C., Zaragoza L.C. Integrated use of solarization and glyphosate in the control of *Cyperus rotundus* L. Proceedings of the 1992 congress of the Spanish Weed Science Society. 1992, 337-340.

Gopal M., Mukherjee I., Yaduraju N.T. Soil solarization: A tool for decontamination of soil pesticides. Proceedings of National Seminar on Plant Protection in India. Plant Protection Association of India, December 22-24, 1997, Hyderabad, India. R-25.

Grinstein A., Orion D., Greenberger A., Katan J. "Solar heating of the soil for the control of *Verticillium dahliae* and *Pratylenchus thornei* in potatoes." In *Soil-Borne Plant Pathogens*. B. Schippers, W. Gams, eds. London: Academic Press, 1979; pp.431-438.

Habeeburrahman P.V., Hosmani M.M. Effect of soil solarization on weed growth and yield of succeeding rainy season sorghum (*Sorghum bicolor*). Indian J Agron 1996; 41:54-57.

Haidar M.A., Sidahmed M.M. Soil solarization and chicken manure for control of *Orobanche crenata* and other weeds in Lebanon. Crop Prot 2000; 19: 169-173.

Horowitz M. Weed research in Israel. Weed Sci 1980; 28: 457.

Horowitz M.; Regve Y., Herzlinger G. Solarization for weed control. Weed Sci 1983; 31 : 170-179.

Jacobsohn R; Greenberger A; Katan J.; Levi M., Alon H. Control of Egyptian broomrape (*Orobanche aegyptiaca*) and other weeds by means of solar heating of the soil by polyethylene mulching. Weed Sci 1980; 28: 312-318.

Katan J. "Soil solarization." In *Second International Symposium on Soil Disinfestation. Acta Hortic No. 152,* C. Van Assche, ed. Leuven, Belgium, International Society for Horticultural Science, 1984,Wageningen, Netherlands, p. 227.

Katan J. Solar heating (solarization) of soil for control of soil-borne pests. Annu Rev Phytopathol 1981; 19: 211-236.

Katan J., Greenberger A.; Alon H.; Grinstein A. Solar heating by polyethylene mulching for the control of diseases caused by soil-borne pathogens. Phytopathology 1976; 66:683-688.

Katan J., Rotem I., Finkel Y., Daniel J. Solar heating of the soil for the control of pink root and other soil borne diseases in onions. Phytoparasitica 1980; 8: 39-50.

Krentzer W.A. "Soil treatment." In *Plant Pathology: An Advanced Treatise* Vol.3 II. *The Diseased Population, Epidemics and Control,* J.G. Horsfall, A.E. Dimond, eds. New York: Academic Press, 1960.

Krentzer W.A. "The reinfestation of treated soil." In *Ecology of soil-borne pathogens: Prelude to Biological Control*, K.F. Baker, W.C. Snyder, eds. Berkeley and Los Angeles. University of California Press, 1965. p. 495.

Krishnamurthy G.V.G., Raju C.A., Murty G.V.G.K., Pieterse A.H. Effect of solarization on the germinability of broomrape seeds (micro-plot study). Proceedings of the 3rd International Workshop on Orobanche and related Striga Research. Amsterdaqm, Netherlands, pp. 493-495.1993.

Kumar B., Yaduraju N.T. Effect of soil solarization on the temperature and physico-chemical properties of soil. Plasticulture, 1992; 94: 13-20.

Kumar B., Yaduraju N.T., Ahuja K.N., Prasad D. Effect of soil solarization on weeds and nematodes under tropical Indian conditions. Weed Res 1993; 33: 423-429.

Linke K.H., Saxena M.C., Sauerborn J., Marsi H. Effect of soil solarization on the yield of food legumes and on pest control – Proceedings of the first International Conference on Soil Solarization, Amman, 139-147, 1990.

Mahrer Y., Katan J. Spatial soil temperature regime under transparent polyethylene mulch: numerical and experimental studies. Soil Sci 1981; 131: 82-87.

Mahrer Y., Naot O., Rawitz E., Katan J. Temperature and moisture regimes in soils mulched with transparent polyethylene. Soil Sci Soc Am J 1984; 48 : 362-367.

Majek B.A., Neary P.E. Selective wavelength transmitting mulches for yellow nutsedge control. Proceedings Brighton Crop Protection Conference-Weeds, Brighton, U.K., 1991; pp. 263-268.

Malik M.A.S., Tran U.V. A simplified mathematical- model for predicting the nocturnal output of solar stile. Solar Energy 1973; 14: 371-385.

Mauromicale G., Restuccia G., Marchese M. Soil-solarization, a non-chemical technique for controlling *Orobanche crenata* and improving yield of faba bean. Agronomie 2001, 21: 757-765.

Meti S.S. Control of *Orobanche cernua* in bidi tobacco by soil solarization. M.Sc. Thesis. UAS, Dharwad India 1993.

Miles J.E., Kawabata O., Nishimoto R.K. Modeling in purple nutsedge sprouting under soil solarization. Weed Sci 2002, 50: 64-71.

Mudalagiriappa N.H.V., Ramachandrappa, B.K. Effect of crop residue and solarization on weeds and yield of groundnut-frenchbean cropping system. Karnataka J Agric Sci 2001; 14: 462-464.

Mudalagiriappa N.H.V., Ramachandrappa B.K. Effect of soil solarization on weed growth and yield of *kharif* groundnut (*Arachis hypogaea*). Indian J Agron 1999; 44: 396-399.

Nasr-Esfahani M., Akhiani A., Fatemi R., Hassanpore H. Soil solarization effect on soil-borne fungal diseases, nematodes and weeds in autumn cucumber fields. J Sci Technol Agric Nat Res 2000. 4:111-123.

Patel H.R., Makwana M.G., Patel B.N. The control of nematodes and weeds by soil solarization in tobacco nurseries. Plasticulture 1995; 107:21-27

Peachey R.E., Pimkerton J.N., Ivors K.L., Miller M.L., Moore L.W. Effect of soil solarization, cover crops and metham on field emergence and survival of buried annual blue grass (*Poa annua*) seeds. Weed Technol 2001; 15: 81-88.

Pullman G.S., De Vay J.E., Elmore C.L., Hart W.H. Soil solarization – a non chemical method for controlling diseases and pests. Cooperative Extension, Division of Agriculture and Natural Resources, Leaflet 21377. Davis, California, University of California, 1984.

Reddy C.N., Reddy M.D., Devi M.P. Solarization for weed control in vegetable nurseries. Indian J Weed Sci 1998; 30:88-89.

Ricci M.D.S.F., Almedia D.L. De, Fernandes M. Do C.A., Ribeiro R. De L.D., Cantanheide M.C. Dos S. Effect of soil solarization on purple nutsedge population density and on productivity of vegetable crops under organic cultivation. Pesquisa Agropecuaria Brasileira 2000; 35: 2175-2179.

Ricci M.S.F., Almedia D.L. De, Ribeiro R. De L.D., Aquino A.M. De, Pereira D.C., De-Polli H., Reis, V.M., Eklund C.R. *Cyperus rotundus* control by solarization. Biol Agric Hort 1999; 17:151-157.

Rubin B., Benjamin A. Solar heating of the soil: effect on weed control and soil incorporated herbicides. Weed Sci 1983; 31:810-825.

Rubin B., Benjamin A. Solar heating of the soils: involvement of environmental factors in the weed control process. Weed Sci 1984; 32:138-142.

Rupela O.P., Sudarshana M.R. Displacement of native *rhizobia* nodulating chickpea (*Cicer arietinum* L.) by an inoculant strain through soil solarization. Biol. Fertil. Soils 1990; 10: 207-212.

Satour M.M., EL-Sherif E.M., El-Ghareeb-L., El-Hadad S.A., El-Wakil H.R. Achievements of soil solarization in Egypt. FAO Plant production and protection papers (1991). No. 109:200-212

Sauerborn J., Linke K.H., Saxena M.C., Koch W. Solarization: a physical control method for weeds and parasitic plants (*Orobanche* spp.) in Mediterranean agriculture, Weed Res 1989; 29: 391-397.

Schreiner R.P., Ivors K.L., Pinkerton J.N. Soil solarization reduces arbuscular mycorrhizal fungi as a consequence of weed suppression. Mycorrhiza 2001; 11:273-277.

Singh V.P.; Dixit A.; Prasadbabu M.B.B., Mishra J.S. Effect of period of soil solarization and weed control measures on weed growth and yield of soybean (*Glycine max* L.). Extended Summaries, International Conference on Managing Natural Resources for Sustainable Agricultural Production in the 21st Century Feb. 14-18, 2000; New Delhi, India. Vol. 3: 1457-1459. 2000.

Smith E.M. Potential field for heat transfer in soil covered by different plastic mulches. Proceedings of the National Agricultural Plastic Conference 1964, 5: 80-92.

Standifer L.C., Wilson P.W., Gorbel R.S. Effect of solarization on weed seed populations. Weed Sci 1984; 32: 569-573.

Stapleton J.J. Mode of action of solarization and bio-fumigation. Proceedings of the 2nd International Conference on Soil Solarization and Integrated management of soil-borne pests, 16-21 March, 1997, Aleppo, Syria.

Stapleton J.J., DeVay J.E. Effect of soil solarization on populations of selected soil-borne micro-organisms and growth of deciduous fruit tree seedlings. Phytopathology 1982; 72: 323-326.

Stapleton J.J., DeVay J.E. Soil solarization: a non-chemical approach for management of plant pathogens and pests. Crop Prot 1986; 5:190-198.

Stapleton J.J., DeVay J.E. Thermal components of soil solarization as related to changes in soil and root microflora and increased plant growth response. Phytopathology 1984; 74:255-259.

Stapleton J.J., Lopez J.G.G. Mulching of soils with transparent (solarization) and black polyethylene films to increase growth of annual and perennial crops in south-western Mexico. Trop Agric (Trinidad) 1988; 65(1) : 29-32.

Stapleton J.J.; Quick J., DeVay J.E. Soil solarization: effect on soil properties, crop fertilization and plant growth. Soil Biol and Biochem 1985; 17: 369-373.

Sudha T., Nanjappa H.V., Udikeri S.S. Influence of solarization on weed management in tomato and tobacco nursery. Mysore J Agric Sci 1998; 32: 269-274.

Tamietti G., Valentino D. Effectiveness of soil solarization against soil-borne plant pathogens and weeds in Piedmont (northern Italy). Proceedings of the Fifth International Symposium on Chemical and Non-chemical soil and Substrate Disinfestation. Torino, Italy, 11-15 September, 2000.

Vizantinopoulos S., Katranis N. Solarization in Greece. Weed Res 1993;33:225-230.

Waggoner P.E., Miller P.M., De Roo H.C. Plastic mulching: principles and benefits. Conn Agric Exp Stn Bull. 1960; 634. 44 pp.

Wilhelm S. Chemical treatment and inoculum potential of soil. Annu Rev Phytopathol 1966; 4, 53.

Yaduraju N.T. The role of soil-solarization in weed management. Proceedings of International Symposium. Indian Society of Weed Science, Hisar, November 18-20, 1993; Vol. I : 343-349.

Yaduraju N.T., Ahuja K.N. Effect of solarization with or without weed control on weeds and productivity in soybean-wheat system. Proceedings of Second International Weed Control Congress. Copenhagen. 1996; 721-727.

Yaduraju N.T., Kamra A. "Soil solarization-a novel non-chemical method of pest control". In *Ecological Agriculture and Sustainable Development*, Vol.2, Proceedings of the International Conference on

Ecological Agriculture Towards Sustainable Development, Chandigarh, India, 15-17 November, 1997. (Edited by Dhaliwal, G.S., Arora, R.,Randhawa, N.S., Dhawan, A.K.). Indian Ecological Society, Chandigarh, 2:189-201.

CHAPTER 18

RATIONALE, APPROACH AND ADOPTION OF INTEGRATED WEED MANAGEMENT

PRASANTA C. BHOWMIK[1] & INDERJIT[2]

[1]Department of Plant and Soil Sciences, University of Massachusetts, Amherst, MA 01003-7245 USA and [2]Department of Botany, University of Delhi, Delhi 110 007, India

The management of pests in crop production is a key component among all farming practices. Weed management is one of the components of pest management. The practice of integrated pest management (IPM) is old in its use of cultural strategies or other control methods that served as the basis for pest control before the advent of pesticides. However, the current focus of IPM for weeds has been expanded to include weed biology, bio-economics, modeling of crop-weed interference, and cropping system effects on weed dynamics.

The term IPM has been defined in many different ways. Shaw (1982) suggested that there is no widely accepted definition from the classic, academic point of view. IPM has been defined as a control strategy in which a variety of cultural, biological and chemical control methods are combined to provide stable and long-term pest control (Shaw, 1982). IPM has also been defined as a biologically oriented pest control strategy that deemphasizes pest control chemicals (Goldstein, 1978). In general, the goals of IPM range from maximizing profit margins to safeguarding natural resources and minimizing the impact on the environment (Burn et al., 1987). According to the U.S. National IPM Coordinating Committee (1988), the primary goals of IPM programs are to reduce pesticide use and subsequent environmental impact and to rely more on other strategies to control pests.

The concept of integrated weed management (IWM) is not new, and was defined by Buchanon (1976) as the deliberate selection, integration and implementation of effective weed control measures with due consideration of economic, ecological and sociological consequences. Thus, IWM is undoubtedly broader and more descriptive than just weed control. IWM emphasizes integration of control tactics with all other practices that influence the ecosystem, and links weed control back to the larger picture of ecosystem management.

Inderjit (ed.), Weed Biology and Management, 363-373.

RATIONALE

Environmental awareness and concerns are growing not only in the developed countries, but also in developing countries. Various aspects of current production systems in agriculture have been under serious review. The emphasis has been to change the focus of production practices from maximum yield potential to sustainable yield. So, our challenge is to have a production system that will produce economic yield, while preserving environmental sustainability. For the challenges of producing enough food for the growing population in the next 25 years, we must continue to develop new technologies for IWM.

APPROACH

Integrated weed management combines direct and indirect weed control strategies. Direct weed control strategies include cultural and chemical means. Indirect control implies agronomic practices, including weed management strategies designed to promote crop yields and suppress weeds. These include choice of crop variety, time of sowing, seeding rate, nutrient management, and crop sequence. Weed management strategies include cultural, biological and chemical control methods. The impact of one weed management component must be analyzed in relation to yield loss and weed population growth, subsequent spread, and possible future weed shifts. If the component practices in IWM are to be considered viable, sustainable and environmentally safe, they must be evaluated in relation to guidelines as suggested in Table 1.

Table 1. Components of an integrated weed management system

What does the IWM component contribute to:

- Development of weed species abundance to reduce interference with crop.
- Dominance of a species in the composition of the existing weed flora.
- Substitution of a species in the composition of the existing weed flora.
- Change in succession of species.
- Change in genetic structure of the targeted species.
- Acceptance of weed management system in relation to environment.
- Enhanced profitability.
- Improvement of overall pest management at the farm level.
- Social acceptance of practices those are appropriate to stake holders.

Weed Biology

Current studies in weed biology are in transition, largely due to the effects of agricultural practices on weeds, cropping systems, and the environment. The research emphasis has changed based on the need to understand basic weed biology (Bhowmik, 1993a). It is our role to predict how weed species, populations, and biotypes evolve in response to selection pressure primarily due to agricultural practices. This knowledge will help develop weed management practices that are sustainable to our environment.

Seed Bank Dynamics

The vast majority of seeds entering the seed bank in arable land come from annual weeds. The species composition and density of weed seeds in soil is likely to influence their population dynamics. Research emphasis must be made regarding weed seed bank dynamics and its implications on predicting weed species emergence (Bhowmik, 1997; Buhler et al., 1997). Prediction of weed seedling emergence will help in discovering weed management options that can be optimized based on resources. Although we have various models (Bridges et al., 1989; Buhler et al., 1997; Forcella et al., 1996; Harvey and Forcella, 1993; Lybecker et al., 1991; Mohler, 1993) to predict weed seed germination or seedling emergence, we need to redirect our efforts to use these models in designing weed management options that result in optimization of production inputs. Our research efforts must focus on economic analysis of weed control systems if we are to make decisions on optimizing farm inputs for sustainable agriculture.

Dynamics of Reproductive Plant Parts of Perennial Weed Species

Common milkweed (*Asclepias syriaca* L.) can disperse from roadsides into fertile cultivated lands. The subsequent spreading into cultivated lands may be related to the removal of annual weeds by the increased use of herbicides that do not control common milkweed (Bhowmik, 1993b). Thus, after seedling establishment either from seed or root fragment, patches expand in cultivated land and spreading occurs into adjacent areas.

Changing crop production practices have encouraged the spread and density of common milkweed infestations during the past two decades (Evetts and Burnside, 1975). Farmers now use shorter crop rotations and cultivate less, which provides more favorable conditions for seedling establishment, plant development, and subsequent spread. Also, the elimination of timely tillage allows common milkweed seedlings to become established permanently. Plant density varied from 1 to 3 plants/m^2 in east central Missouri sites (Sauer and Feir, 1974), while the density varied from 11,819 to 88,226 stalks/ha in cultivated fields in southern Ontario (Bhowmik and Bandeen, 1976).

Critical Period of Weed Interference

The concept of using critical period of weed interference in optimizing herbicide use within a given tillage system is a valid component for IWM program. There are two components to the critical period. The first is the length of time when weed control efforts must be maintained to prevent crop yield loss, while the second component is the length of the time when weeds can remain in the crop before they interfere with crop growth, resulting in yield loss.

The information on critical period can potentially be used to reduce the need for long-term residual weed control with herbicides (Hall et al., 1992; Weaver, 1984; Weaver et al., 1987). It can also be used to optimize the rate and timeliness of postemergence herbicides. This information can be used to determine the timing of cover crop planting as well as mechanical cultivation. The detailed information on the critical period is essential and research must continue to refine the needed information.

Cropping Systems

Competitive crops can influence the establishment of weed species. In general, narrow-row spaced crops are more desirable than wide-row spaced crops because early shading suppresses weed growth. Row spacing of 45 to 60 cm for corn and 19 to 38 cm for soybeans resulted in superior control of purple (*Cyperus rotundus* L.) or yellow (*Cyperus esculentus* L.) nutsedge at harvest when compared to spacings of 90 and 76 cm, respectively (Chapel and Leasure, 1980; Chaudhury, 1981).

Research has indicated that nutsedge species lack a competitive edge with crops once the crops form a dense canopy (Keeley and Thullen, 1978; Kelley et al., 1983). Light requirements, dry matter and tuber production of yellow nutsedge increased in direct proportion to increased amounts of light (Keeley and Thullen, 1978). A 30% shade reduced dry matter and tuber production by 32%, while 80% shade reduced dry matter and tuber production about 94%. Both purple and yellow nutsedges were sensitive to shading (Keeley and Thullen, 1978). It was also noted that shading decreased the partitioning of plant biomass into tubers and rhizomes, resulting in increased partitioning into leaves. Growing competitive crops in rotation with cotton has reduced populations of yellow nutsedge tubers by 97 to 99% within 3 years (Keeley and Thullen, 1983).

In general, management practices that influence the interactions of nutsedge with crops can be adopted in a weed management program. Date of planting, row spacing, crop cultivars, competitive crop species, crop rotation, fertilization, and method of planting (transplanting vs seeding) are some of the management practices that can be implemented.

Tillage Systems

In recent years, one of the most important changes occurring in agriculture is the increased acceptance of conservation tillage. Conservation tillage advocates the maintenance of crop residue cover on 30% of the soil surface. This practice is within the realm of sustainable agriculture. Reduced tillage is often an integral part of the conservation tillage to reduce soil erosion and surface runoff (Hairston et al., 1984).

In conservation tillage, weed communities vary compared to the conventional tillage. The presence of crop residue increased weed suppression and may result in reduced herbicide use. The potential for reduction in total herbicide use is greater in conservation tillage, particularly in no-till, than can be achieved in conventional tillage systems. This reduction in total herbicide use will likely be a factor in population dynamics of weed seed banks (Bhowmik and Bekech, 1993; Froud-Williams, 1988; Mulugeta and Stoltenberg, 1997) and of rhizomes and tubers (Bhowmik, 1995; Chandler and Swanton, 1990). However, conservation tillage typically relies heavily on pesticides, particularly herbicides (Hinkle, 1983; Johnson, 1994; Yoo and Touchton, 1989).

Alternative Methods of Control

We continue to search for alternative weed control methods. One of the important question arises is how effective are the alternative methods? The alternative methods of weed control must be better or equal to the current system. Gillmeister et al. (1989)

reported that decisions about controlling weeds in cabbage production should be made with farm profitability as a chief consideration. They also pointed out that the expected profitability of the alternative weed control strategies depends to some extent on the size of the weed population. Gillmeister et al. (1990) concluded that because of concerns about groundwater quality and food safety, reducing the routine use of preemergence herbicides may become an increasingly important component of weed management strategies in the years ahead. Some of the alternative methods include integration of cultivation and herbicide treatment, use of cover crops and others.

Integration of Cultivation and Herbicide Banding

To reduce herbicide use in a given row crop, band application of herbicides in combination with cultivation is an option. Initial flushes of annual weeds in row crops can be controlled by shallow cultivation of the soil, requiring two to four passes. The banding of herbicides over the crop rows and the removal of weeds between the rows with one to two passes of inter-tillage cultivators constitute an alternative weed control method. Our own data indicated that the application of bensulide, DCPA, or trifluralin at 4.5, 5.6, or 0.56 kg/ha followed by one cultivation 4 weeks later provided excellent weed control in early transplanted cabbage and resulted in acceptable yields with quality produce (Bhowmik, 1983). The treatments resulted in reduced herbicide use and this approach would be a part of IWM.

Use of Cover Crops

The use of cover crops and their mulches during the crop-free periods in row crops is a way of suppressing weeds. This would be of real importance in conservation tillage systems. Cover crops can be planted into a standing crop or into a stubble area after crop harvest. The timing of planting cover crops is very important in relation to its establishment.

Use of Natural Products through Allelopathic Concepts

There is a need to design a total system approach for crop production to resolve economic and environmental consequences of agricultural pests (Lewis et al., 1997). Allelopathy is an emerging strategy to control agricultural pests or at least lessen the use of herbicides. Two main applications of allelopathy include: crops and their residues with allelopathic potential, and allelochemicals as natural herbicides (Inderjit and Bhowmik, 2002; Bhowmik and Inderjit, 2003). The aim of this section is not to write an in-depth review of the role of allelopathy in natural weed control but to highlight allelopathy as an important strategy of IWM. Certain cover crops (Table 2) and their residues are known to release chemicals into the environment that can suppress weed populations (Bhowmik and Inderjit, 2003; Fujii, 1992; Inderjit and Keating, 1999; Weston, 1996).

Although there is not much success in commercialization of natural products from higher plants as herbicides, natural products have some benefits over synthetic herbicides, e.g., their novel structure, short half-life and multiple mode of action (Duke et al., 2000). Some of the natural products from higher plants widely studied are ailanthone from *Ailanthus altissima* (mill.) Swingle (Heisey, 1996), artemisinin from *Artemisia annua* L. (Duke et al., 1987; Lydon and Duke, 1988), and sorgoleone from *Sorghum bicolor* (L.) Moench (Einhellig and Souza, 1992; Einhellig et al., 1993; Weston et al., 1999). Natural products have another potential advantage over synthetic herbicides, in that they can have multiple modes of action whereas synthetic herbicides may have only a single mode of action. For example, sorgoleone inhibits photosystem II (PS II), hydroxyphenylpyruvate dioxygenase (HPPD) (Czarnota et al., 2001; Gonzalez et al., 1997; Meazza et al., 2002; Rimando et al., 1998) and respiratory electron flow (Einhellig et al., 1993; Rasmussen et al., 1992). Tellimagrandin II, an algicidal and cyanobactericidal compound from a freshwater microphyte, *Myriophyllum spicatum,* has two modes of action, inhibition of PS II and inhibition of microalgal exoenzyme formation (Gross, 1999; Leu et al., 2002). There are, however, certain constraints in implementing natural products as commercial herbicides, e.g., *in vivo* activity may not match in vitro activity, synthesis may be difficult, physico-chemical properties may restrict use as herbicides, and soil degradation may be rapid (Inderjit and Bhowmik, 2002). In spite of these limitations, natural plant products have a promising future in IWM (Duke et al., 2002; Bhowmik and Inderjit, 2003; Inderjit and Duke, 2003).

Table 2. Cover crops with allelopathic potential against weed species.

Crops	Chemicals
Black mustard (*Brassica nigra)*	Allyl isothiocyanate
Buckwheat (*Fagopyrum esculentum*)	Fatty acids
Oat (*Avena fatua)*	Scopoletin, phenolics acids
Rye (*Secale cereale)*	Benzoxaziones, phenolics acids
Sorghum (*Sorghum bicolor)*	Phenolic acids, dhurrin, sorgoleone
Sudangrass (*Sorghum sudanense)*	Phenolic acids, dhurrin, sorgoleone
Sunflower (*Helianthus annuus)*	Heliannuols
Velvetbean (*Mucuna pruriens)*	L-3,4-dihydroxyphenylalanine
Wheat (*Triticum aestivum)*	Hydroxamic acids

New Herbicide Chemistry

The introduction of compounds with new chemistry is one of the major ways to reduce the total herbicide use. The new sulfonylurea herbicides are characterized by low dosage rates and low mammalian toxicity. Some of these herbicides include triasulfuron for use in small grains, and halosulfuron, nicosulfuron and primisulfuron for use in corn. The recommended rate is between 5 to 45 g ai/ha. The rates of such new chemistry herbicides are about 20 to 300 times lower than those of commonly used herbicides such as triazines or acetanilides. As a result of the recent introduction of several low-rate herbicides, there was a declining trend in herbicide consumption in the United States in 1990. This decline represents the IWM mandated pesticide use reduction. Of course, new lower rate products may be more potent, so evaluation of pesticide load on a mass per unit area basis is not entirely appropriate.

In addition to the low-rate characteristics, other characteristics such as physical and chemical properties of the new compounds must be taken into consideration. For example, a given low-rate compound may have a high leaching potential, an unfavorable degradation pattern in the soil, or a high vapor pressure, all of which could partially or fully offset the benefits from a low use rate. The use of low-rate compounds must be integrated into the practices of weed management in a cropping system. New discoveries of such desirable compounds must be continued and efforts must be made to incorporate these new tools for our IWM system.

Modeling of Crop Weed Interference

The development of IWM system is a complex task and must be supported by a thorough understanding of crop-weed interference. Computer models must be developed to integrate available information on interference, critical period, seedling emergence, and other parameters and to predict the outcome of crop-weed interference (Ghersa and Holt, 1995; Kropff, 1988). Weed density-yield model are the basis for development of economic weed threshold models that predict the weed density at which specific post-emergence weed control measures are warranted. However, population dynamics models can be used for assessing the long-term implications of control measures on weed populations (Cousens et al., 1986, 1987; Doyle et al., 1986; Kropff and Spitters, 1991). A mechanistic approach to modeling can be used to predict crop-weed interference over a wide range of conditions (Doyle et al., 1986; Orwick et al., 1978).

System Approach

Research information from large-scale IWM projects is very limited. Little is known about the long-term effects of different management practices on a given environment. This type of project must be designed with the participation of weed scientists, entomologists, plant pathologists, ecologists, agronomists, soil scientists, economists, extension personnel, computer modelers, and others. Projects must be designed to compare different high- and low-input production systems over a period of several years. Projects may also include crop rotation, fertilization programs, and weed management options under various tillage systems. Information on soil fertility, weed density, seed dynamics, insect and disease incidence, crop yield and the economic return to the farmer must be available to the user group for any adoption programs. Farm profitability is one of the factors to be considered for adoption of any new technology (Gillmeister et al., 1989). Therefore, economic value of the system approach to IWM must be warranted to the farmers in relation to profitability so that the farmers remain in the farming business.

ADOPTION

Formal extension programs to promote IPM adoption in the northeastern United States were begun in the late 1970s (Bajwa and Kogan, 1996). As more farmers adopt IPM practices, it has become difficult to identify "non-IPM" farms to use in side-by-side comparisons of cost, pesticide use, related practices (Hollingsworth and Coli, 2001). Let us examine the IPM program adoption as well as adoption level in the northeastern U. S. (Table 3).

Table 3. IPM adoption by northeastern U.S. sweet corn growers (Hollingsworth and Coli, 2001).

Level of IPM Adoption[1,2]	State[3] CT	DE	MA	MD	ME	NJ	PA	All states column %
Low	34.5	22.2	38.1	58.2	59	38.8	76.8	58
	10	4	24	32	23	45	73	
	14.7	9.2	32	28	19.8	59	48.3	
Moderate	55.2	61.1	47.6	40.0	38.5	56.9	23.2	44
	16	11	30	22	15	66	22	
	12.7	7.9	27.6	24.1	17.1	50.9	44.7	
High	10.3	16.7	14.3	1.8	2.6	4.3	0	5
	3	3	9	1	1	5	0	
	1.5	1	3.3	2.9	2.1	6.1	5	
Total number of growers	29	18	63	55	39	116	95	n = 415

[1]For each level of adoption, the first number in the column under each state is the percentage of growers scoring for that level; the second number is the cell frequency (number of growers); and the third number is the expected cell frequency from X^2 analysis.
[2] X^2 is significant at $p \leq 0.001$ (12 df). Caution is advised in interpreting results where cells have expected counts <5.
[3] CT = Connecticut, DE = Delaware, MA = Massachusetts, MD = Maryland, ME = Maine, NJ = New Jersey, PA = Pennsylvania.

Adoption practices of IPM vary from state to state within a region. More than half of northeastern U.S. sweet corn growers were classified as low-level IPM adopters and only 5% of growers were classified as high-level IPM adopters. This type of adoption may be true with IWM programs. Limited studies have been done to identify the nature and adoption of weed management practices.

Education and Extension

The information on new technologies on IWM must be communicated to the user group, farmers, ranchers, and other interested groups. The successful implementation of an IWM system is dependent upon the thorough transfer of information and technology. Progressive and innovative farmers must be identified and they must be informed. Workshops, tours, seminars, and on-site demonstration trials need to be carefully organized to highlight new changes or any alternative methods at hand to be adopted. This core group of farmers will then serve as the main vehicle for extension of new information and implementation of new practices within the farm community. The extension personnel must be involved in this education process. The latest multi-media approach can help in designing many outlets that will provide information to the farmers.

CONCLUSION

Weed scientists must re-evaluate and redirect their weed research programs by developing long-term weed management alternatives based on a holistic approach. Emphasis must be made to understand the cropping systems rather than short-term solutions of weed control by chemical methods. Societies must address the appropriateness of particular technologies for weed control because of legitimate concerns over environmental quality and resource use (Blacklow, 1997). We must be willing to implement new IWM strategies for our sustainable weed management. Hollingsworth and Coli (2001) stated that future IPM adoption will be enhanced by developing a suite of IPM systems that recognize consumer and grower interest in protecting human health and the environment, as well as the diversified, direct-sale nature of much of the region's agriculture. Similarly, IWM systems must identify their problems and address potential barriers to adoption and in turn take advantage of incentives to adoption whenever possible.

REFERENCES

Bajwa W. I., Kogan M. Compendium of IPM definitions. Oregon State University, Integrated Plant Protection Center, Corvallis. Web site http://www.ippc.orst.edu/ipmdefinitions/, 1996.

Bhowmik P. C. Weed biology: importance to weed management. Weed Sci 1997; 45:349-356.

Bhowmik P. C. Integrated technique for controlling *Elytrigia repens* populations. 9th European Weed Research Society Symposium, July 10-12, 1995, Budapest, Hungary. 2:611-618.

Bhowmik, P. C. Weed biology: its importance to integrated weed management systems. Proceedings of the International Symposium, Indian Society of Weed Science, Vol. 1:57-65; 1993a November 18-20; Hisar, India.

Bhowmik P. C. Biology and control of common milkweed (*Asclepias syriaca*). Rev Weed Sci 1993b; 6:227-250.

Bhowmik P. C. Evaluation of herbicides for weed control in transplanted cabbage. Weed Science Society of America, Abstract no. 85A; 1983; p. 35.

Bhowmik P. C., Bandeen J. D. The biology of Canadian weeds. 19. *Asclepias syriaca* L. Can J Plant Sci 1976; 56:579-589.

Bhowmik P. C., Bekech M. M. Horseweed (*Conyza canadensis*) seed production, emergence, and distribution in no-tillage and conventional-tillage corn (*Zea mays*). Agronomy (Trends Agric Sci) 1993; 1:67-71.

Bhowmik P. C., Inderjit. Challenges and opportunities in implementing allelopathy for natural weed management. Crop Prot 2003; 22:661-671.

Blacklow M. Sustainable weed management-fact or fallacy. In *Proceedings 16th. Asian-Pacific Weed Science Society Conference*, A.Rajan, ed. Kuala Lumpur, Malaysia, 1997; pp. 1-13.

Bridges D. C., Sharpe P. J. H., Chandler M. Modeling distribution of crop and weed seed germination time. Weed Sci 1989; 37:724-729.

Buchanan, G. A. , Weeds and weed management in cotton. Proceedings of the Beltside Cotton Production Research Conf. 1976; 166-168.

Buhler D. D., Hartzler R. G., Forcella F. Implications of weed seedbank dynamics to weed management. Weed Sci 1997; 329-336.

Burn A. J., Coaker T. H., Jepson P. C. *Integrated Pest Management.* Academic Press, San Diego, CA, 1987.

Chandler K., Swanton C. J. Effect of tillage on control of quackgrass [*Agropyron repens* (L.) Beauv.]. Proceedings of Quackgrass Symposium, October 24-25, London, Ontario, 1990; pp. 135-150.

Chapel B. C., Leasure J. K. The effect of row spacing on yellow nutsedge control in soybeans. Proceedings of the North Central Weed Control Conference. 1980; 35:14-15.

Chaudhary A. H. Effect of population and inter-row spacing on yield of maize and control of weeds with herbicides in the irrigated Savanna. Exp Agric 1981; 17:389-397.

Cousens R., Brain P., O'Dpnpvan J. T., O'Sullivan P. A. The use of biologically realistic equations to describe the effects of weed density and relative time of emergence on crop yield. Weed Sci 1987: 35:720-725.

Cousens R., Doyle C. J., Wilson B. J., Cussans, G. W. Modelling the economics of controlling *Avena fatua* in winter wheat. Pestic Sci 1986; 17:1-12.

Czarnota M. A., Paul R. N., Dayan F. E., Nimbal C. I., Weston L. A. Mode of action, localization of production, chemical nature, and activity of sorgoleone: a potent PS II inhibitor in *Sorghum* spp. root exudates. Weed Technol 2001; 15:813-825.

Doyle C. J., Cousens R., Moss S. R. A model of the economics of controlling *Alopecurus myosuroides* Huds. In winter wheat. Crop Prot 1986; 5:143-150.

Duke S. O., Vaughn K. C., Croom E. M., Elsohly H. N. Artemisinin, a constituent of annual wormwood (*Artemisia annua*), is a selective phytotoxin. Weed Sci 1987; 35:499-505.

Duke S.O., Dayan F.E., Ramagni J.G., Rimando A.M. Natural products as sources of herbicides: current status and future. Weed Res 2000; 40:90-111.

Duke S.O., Dayan F.E., Rimando R.M., Schrader K.K., Aliotta G., Anna O., Romagni J.G. Chemicals from nature for weed management. Weed Sci 2002; 50:138-151.

Einhellig F. A., Souza I. F. Phytotoxicity of sorgoleone found in grain sorghum root exudates. J Chem Ecol 1992; 18:1-11.

Einhellig F.A., Rasmussen J.A., Hejl A.H., Souza I.F. Effects of root exudate sorgoleone on photosynthesis. J Chem Ecol 1993; 19:369-375

Evetts L. L., Burnside O. C. Effects of early competition on growth of common milkweed. Weed Sci 1975; 23:1-3.

Forcella F., King R. P., Swinton S. M, Buhler D. D., Gunsolus J. L. Multiyear validation of a decision aid for integrated weed management. Weed Sci 1996; 44:650-661.

Froud-Williams R.J. "Changes in weed flora with different tillage and agronomic management systems." In *Weed Management in Agroecosystem: Ecological Approaches,* M. A. Altieri, M. Liebman, eds. Boca Raton, FL: CRC Press, 1988, pp. 213-236.

Fujii Y., Shibuya T., Yasuda T. Allelopathy of velvetbean: its discrimination and identification of L-DOPA as a candidate of allelopathic substances. Jpn Agric Res Quat 25:238-247.

Ghersa C. M., Holt J. S. Using phenology prediction in weed management: a review. Weed Res 1995; 35:461-470.

Gillmeister W.L., Moffitt L.J., Bhowmik P.C., Geoffrey P. Information value in weed management. Northeast J Agric Resour Econ 1990; 19:24-27.

Gillmeister W.L., Moffitt L.J., Bhowmik P.C., Geoffrey P. Profitability of Cabbage Weed Control Alternatives. University of Massachusetts Bull no. C-199, 1989, 10 p.

Goldstein J. *The Least is Best Pesticide Strategy.* The JG Press, Emmaus, PA, 1978, 205 p.

Gonzalez V.M., Kazimir J., Nimbal C., Weston L. A., Cheniae G. M. Inhibition of a photosystem II electron transfer reaction by the natural product sorgoleone. J Agric Food Chem 1997; 45:1415-1421.

Gross E.M. "Allelopathy in benthic and littoral areas: case studies on allelochemicals from benthic cyanobacteria and submerged macrophytes." In *Principles and Practices in Plant Ecology: Allelochemical Interactions*, Inderjit, K. M. M. Dakshini, C. L. Foy, eds. Boca Raton, FL: CRC Press, 1999; pp. 179-199.

Hairston J. E., Sanford J. O., Hayes J.C., Reinschmiedt L.L. Crop yield, soil erosion, and net returns from five tillage systems in the Mississippi Blackland prairie. J Soil Water Conser 1984; 39:391-395.

Hall M.R., Swanton C.J., Anderson G.W. The critical period of weed control in grain corn (*Zea mays*). Weed Sci 1992: 40:441-447.

Harvey S. J., Forcella F. Vernal seedling emergence model for common lambsquarters (*Chenopodium album*). Weed Sci 1993; 41:309-316.

Heisey R.M. Identification of an allelopathic compound from *Ailanthus altissima* (Simaroubaceae) and characterization of its herbicidal activity. Am J Bot 1996; 83:192-200.

Hinkle M.K. Problems with conservation tillage. J Soil Water Conserv 1983; 38:201-206.

Hollingsworth C.S., Coli W.M. IPM adoption in northeastern U.S.: an examination of the IPM continuum. Am J Altern Agric 2001; 16:177-183.

Inderjit, Bhowmik P.C. "Allelochemicals phytotoxicity in explaining weed invasiveness and their function as herbicide analogues." In *Chemical Ecology of Plants: Allelopathy in Aquatic and Terrestrial Ecosystems,* Inderjit, A. U. Mallik, eds. Basal, Switzerland: Birkhauser-Verlag AG, 2002; pp 187-197.

Inderjit, Duke S.O. Ecophysiological aspects of allelopathy. Planta, 2003; in press.

Inderjit, Keating K.I. 1999 Allelopathy: principles, procedures, processes, and promises for biological control. Adv Agro 1999; 67:141-231

Integrated Pest Management: A National Plan for Future Direction. Prepared by the National IPM Coordinating Committee, 1988; p. 12.

Johnson, R. P. Influence of no-till on soybean cultural practices. J Prod Agric 1994; 7(1):43-49.

Keeley P. E., Thullen R. J. Light requirements of yellow nutsedge (*Cyperus esculentus*) and light interception by crops. Weed Sci 1978; 26:10-16.

Keeley P. E., Thullen R. J. Influence of yellow nutsedge (*Cyperus esculentus*) - free periods on yield of cotton (*Gossypium hirsutum*). Weed Sci 1983; 31:803-807.

Keeley P. E., Thullen R. J., Miller J. H., Carter C. H. Comparison of six cropping systems for yellow nutsedge (*Cyperus esculentus*) control. Weed Sci 1983; 31:63-67.

Kropff M. J. Modelling the effects of weeds on crop production. Weed Res 1988; 28:465-471.

Kropff M.J., Spitters C.J.T. A simple model of crop loss by weed competition from early observations on relative leaf area of the weeds. Weed Res 1991; 31:97-105.

Leu E., Krieger-Liszkay A., Goussias C., Gross E.M. Polyphenolic allelochemicals from the aquatic angiosperm *Myriophyllum spicatum* inhibit photosystem II. Plant Physiol 2002; 130:2011-2018

Lewis W. J., van Lenteren J. C., Phatak, S. C., Tumlinson, J. H. 1997. A total system approach to sustainaable pest management. Proc Natl Acad Sci USA 94:12243-12248

Lybecker D.W., Schweizer D.W., King R.P. Weed management decisions in corn based on bioeconomic modeling. Weed Sci 1991; 39:124-129.

Lydon J, Duke S.O. Glyphosate induction of elevated levels of hydroxybenzoic acids in higher plants. J Agric Food Chem 1988; 36:813-818

Meazza G., Scheffler B. E., Tellez M. R. , Rimando A. M., Romagni J. G., Duke S. O., Nanayakkara D., Khan I. A., Abourashed E. A., Dayan F. E. The inhibitory activity of natural products on plant *p*-hydroxyphenylpyruvate dioxygenase. Phytochemistry 2002; 60:281-288.

Mohler C.L. A model of the effects of tillage on emergence of weed seedlings. Ecol Appl 1993; 3:53-73.

Mortimer M., Caton B.P., Hill J.E. On ecological issues in the development of sustainable weed management. Proc. 17th Asian-Pacific Weed Science society Conference. 1993; I(A) 45-50.

Mulugeta D., Stoltenberg D. E. Weed and seedbank management with integrated methods as influenced by tillage. Weed Sci 1997; 45:706-715.

Orwick P. L., Schreiber M. M., Holt J. A. Simulation of foxtail (*Setaria viridis* var. *robusta-alba*, *Setaria viridis* var. *robusta-purpurea*) growth: the development of SETSIM. Weed Sci 1978; 26:691-699.

Rasmussen J. A., Hejl A. M., Einhellig F. A., Thomas J. A. Sorgoleone from root exudate inhibits mitochondrial functions. J Chem Ecol 1992; 18:197-207.

Rimando A.M., Dayan F. E., Czarnota M.A., Weston L.A., Duke S.O. A new photosystem II electron transport inhibitor from *Sorghum bicolor*. J Nat Prod 1998; 61:927-930.

Sauer D., Feir D. Population and maturation characteristics of the common milkweed. Weed Sci 1974; 22:293-297.

Shaw W.C. Integrated weed management systems technology for pest management. Weed Sci 1982; 30:2-12.

Weaver S.E. Critical period of weed competition in three vegetable crops in relation to management practices. Weed Res 1984; 24:317-325.

Weaver S.E., Smits N., Tan C. S. Estimating yield losses of tomatoes (*Lycopersicon esculentum*) caused by nightshade (*Solanum* spp.) interference. Weed Sci 1987; 35:163-168.

Weston L.A., Nimbal C.I., Jeandet P. "Allelopathic potential of grain sorghum (*Sorghum bicolor* [L.] Moench) and related species." In *Principles and Practices in Plant Ecology: Allelochemical Interactions*, Inderjit, K. M. M. Dakshini, C. L. Foy, eds. Boca Raton, FL: CRC Press, 1999; pp 467-477.

Weston L.A. Utilization of allelopathy for weed management in agroecosystems. Agron J 1996; 88:860-866.

Yoo K.H., Touchton J.T. Runoff and soil loss by crop growth stage under three cotton tillage systems. J Soil Water Conserv 1989; 44:225-228.

CHAPTER 19

ADJUVANTS

TOOLS FOR ENHANCING HERBICIDE PERFORMANCE

JERRY M. GREEN[1] & CHESTER L. FOY[2]

[1]*DuPont Crop Protection, Stine-Haskell Research Center, Bldg. 210, Newark, DE 19714-0030, USA;* [2]*Virginia Polytechnic Institute and State University, Department of Plant Pathology, Physiology, and Weed Science, Blacksburg, VA 24061-0331, USA*

INTRODUCTION

Adjuvants have been used as tools to enhance pesticide performance for a long time. More than 200 years ago, growers used adjuvants such as tar, sugar, or tree sap to stick Bordeaux mixture fungicide to grapes. In the late 1880's, soap was used with kerosene to destroy insect eggs, and later to increase insecticide activity of arsenic (Gillette, 1890). Fish and whale oil were also commonly used adjuvants. In 1942, surfactant enhancement was reported in some of the first experiments with 2,4-D (Zimmerman and Hitchcock 1942). Later, triazine herbicide activity was greatly enhanced with paraffinic oils (McWhorter, 1956) and then with oil-surfactant emulsions (crop oil concentrates) (Holstun and Bingham, 1960). Adjuvants transformed triazine soil herbicides into foliar ones. Today, adjuvants are universal components of herbicide sprays.

Adjuvants influence many aspects of the application process (Figure 1). Determining the best adjuvant must be done systematically as there are too many adjuvants and potential mixtures to evaluate all possible combinations empirically. Applying the right herbicide with the right concentration and type of adjuvant is a major task for the user. The right adjuvant depends on weed characteristics (e.g., species, growth stage, and cuticle), environmental conditions (e.g., temperature, light, humidity, and rainfall), application method (e.g., spray tank environment, spray volume, and droplet size), and physicochemical properties of the herbicide. In practice, the user depends on the herbicide label for directions. These directions range from mandatory requirements, optional recommendations, prohibitions, to no mention at all.

Inderjit (ed.), Weed Biology and Management, 375-401.

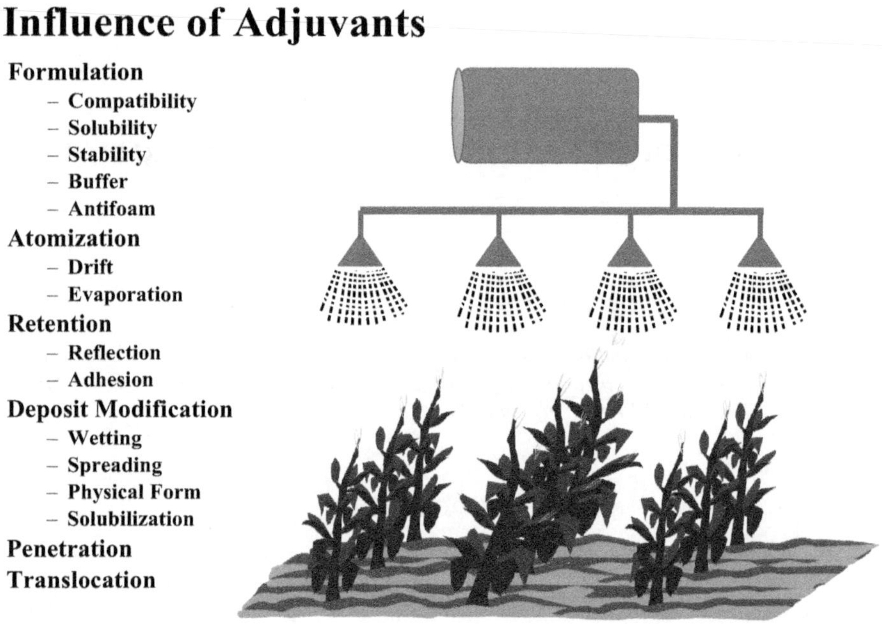

Figure 1. Adjuvants influence the physical and chemical properties of spray solutions to enhance herbicide performance.

Terminology

Even experts do not agree on some of the basic definitions. In North America, adjuvants are defined as any materials added to a spray solution or formulation to modify the action of an agrochemical or the physical properties of the spray solution. However, Europeans usually restrict adjuvants to spray tank additives (Hazen, 2000). When in the formulation, non-pesticidal components are called formulants in Canada and inerts in the USA. Most governmental regulations restrict active ingredient to pesticides, but many adjuvant labels still refer to their components as active and inactive. Large differences in opinions still exist on how to define surfactant and the use of oxymorons like 'spreader sticker' must be resolved. The lack of standardised terminology has created a wide array of confusing terms that are used on labels (Table 1). The good news is that many organisations are cooperating to get universally accepted definitions.

Adjuvants can be broadly grouped into two functional categories: utility adjuvants that improve application characteristics and activator adjuvants that enhance biological activity, usually increasing absorption. Adjuvants can also be classified by chemistry.

Table 1. Terminology used to describe adjuvants on adjuvant and pesticide labels.

Activator	Acidifier/Buffer	Additive
Adherent	Antiflocculant	Antifoam/Defoamer
Antifreeze	Attractant	Basic Blend
Chelating Agent	Cleaner	Colorant/Dye
Compatibility Agent	Cosolvent	Coupler
Crop Oil Concentrate	Deposition Agent	Detergent
Dispersant	Drift Control Agent	Emulsifier
Evaporation Reducer	Extender	Fertiliser
Foam Marker	Formulant	Inert
Humectant	Methylated Seed Oil	High load COC
Polymers	Modified Vegetable Oil	Penetrator
Repellant	Petroleum Oil Concentrate	Preservative
Rainfast Agent	Retention Aid	Solubiliser
Surfactant	Spreader Sticker	Synergist
Thickener	Translocation Aid	*UV* Protectant
Vegetable Oil	Water Conditioner	Wetting Agent

Both chemical and functional classification systems have limitations because adjuvants are usually chemical mixtures with multiple functions.

Regulation

Regulatory agencies in many countries pay little attention to adjuvants. Likely reasons for this inattention include small markets, less perceived risk, and their favourable role in reducing pesticide risks. Adjuvants have historically been perceived as inactive or inert and generally regarded as safe (GRAS). However the adjuvant system can have worse toxicological characteristics than the herbicide, e.g., the warning label for some glyphosate formulations is principally due to the tallow amine ethoxylate (TAE) adjuvant system and not the herbicide. The 1996 Food Quality Protection Act (FPQA) changed how inerts are classified and regulated in the USA. FPQA requires Environmental Protection Agency (EPA) to review all adjuvants to ensure 'reasonable certainty of no harm'. Some countries regulate adjuvants similarly to pesticides and require chemical disclosure, efficacy trials, and environmental and toxicological studies. Such regulations discourage tank-mix adjuvants and encourage these functions be put into the herbicide formulation. The lack of federal regulations in the USA has encouraged nine states to regulate tank-mix adjuvants and the adjuvant industry through the Chemical Producers and Distributors Association (CPDA) to begin to self-regulate. Still herbicide labels are the primary source for tank-mix adjuvant recommendations and the most common and circuitous way that governments regulate adjuvant use.

Marketing

The agricultural tank-mix adjuvant market is currently about $1 billion or 3% of the pesticide market (Underwood, 2000). With minimal patent protection or governmental regulations, manufacturers do not disclose chemical compositions to prevent competitors from copying formulations without any investment or monetary compensation. This non-proprietary and unregulated nature encourages secrecy in a business that has been slow to rid itself of a 'snake oil' image. Adjuvants can be very profitable, but marketing adjuvants is a two-edged sword: to develop specialised products tailored for specific applications and to develop all-purpose products made for all herbicides and conditions. These contradictory goals confuse users trying to align technical label requirements with the label descriptions of their locally available adjuvants. Ironically, growers may use as many as five adjuvants in an attempt to get 'all the good things' into a single spray application, but they rarely blame the adjuvants for performance failures.

In 1996, Monsanto put the adjuvant industry in turmoil when it introduced glyphosate with a complete adjuvant system and restricted tank-mix adjuvants. Although the same amount of adjuvant was still being used, chemical dealers lost a major portion of their tank-mixture business. Their business recovered when growers increased the use of utility adjuvants to combat glyphosate drift, foaming, and inactivation by hard water. The glyphosate adjuvant system is still evolving as indicated by ever increasing 200 plus patent publications. The original 15 to 17% tallow amine ethoxylate system used before 1991 was reduced to 7 to 8% TAE to avoid a warning label (Berger and Jimenez, 1997; Franz et al., 1997). The surfactant load recently increased back to higher level with safer surfactant system. Monsanto currently sells more than eight glyphosate formulations with many others being introduced by generic manufacturers. The current commercial situation has been described as high load, medium load, and no load glyphosate or by analogy to the car industry as Chevrolet, Ford, and Dodge glyphosate (Ohmes et al., 2000). With so many formulations and over 200 tank-mixture recommendations, getting the right adjuvants with glyphosate is very difficult.

Physicochemical Properties

Adjuvants control herbicidal performance by controlling the physicochemical properties of the spray solution (Green and Hazen, 1998). Electrostatic charge may be the most important characteristic for ionic surfactant while other hydrophilic interactions are more important for nonionic surfactants. These important physicochemical properties include effects on solubility, hydrophilic-lipophilic balance (HLB), surface tension, molecular size, spreading, and type of spray deposit (Table 2).

The most visible physicochemical property is the wetting of a leaf surface. It involves a fundamental process, the displacement of air from the solid surface with an aqueous mixture. Adjuvants vary greatly in their ability to influence the wetting. For example, petroleum oil concentrates spread greatly, but slowly while silicone surfactants quickly spread droplets up to 100 times more than water alone. A spray liquid will only wet a leaf surface if the attractive forces between the liquid and surface are greater than the cohesive forces of the liquid. For spreading to occur spontaneously, the surface free energy must decrease during the process and the droplet must maintain a surface tension gradient. The concept can be described mathematically where the spreading coefficient (S):

$$S = \gamma_{SA} - (\gamma_{SL} + \gamma_{LA})$$

where γ_{SA} = critical surface tension, γ_{SL} = interfacial tension between the solid and liquid, and γ_{LA} = surface tension of the liquid. If S is positive, the liquid will wet the surface spontaneously. If S is negative, wetting will not occur. Spreading varies with surfactant type and the nature of the surface.

Table 2. Adjuvant physicochemical properties that influence herbicide performance.

Contact Angle	Chemical Structure	Critical Micelle Concentration
Film Formation	Dilational Modulus	Dynamic Surface Tension
Melting Point	Ethylene Oxide Content	Hydrophilic-Lipophilic Balance
Molecular Weight	Humectancy	Ion Exchange Capacity
Partition Coefficient	Ionic Class	Static Surface Tension
pH	Deposit Physical Form	Size of Hydrophobe
Solubility	Spreading / Wetting	UV Absorption
Viscoelasticity	Viscosity	Volatilisation

Many adjuvant properties are interrelated. For example, surface tension or energy influences contact angle, critical micelle concentration (CMC), and spreading and wetting. Surface energy is the result of intermolecular forces such as hydrogen bonding and van der Waals. Adjuvants change surface energy and the amount of work needed to move a molecule away from one molecule and bring it into contact with another.

After spreading and surface tension, contact angle is probably the best known adjuvant property (Figure 2). Contact angle is calculated from droplet dimensions (Foy and Smith, 1969). The type and amount of adjuvant as well as the structure and composition of the leaf surface determine the contact angle. Surfactants generally reduce contact angle as concentrations increase from 0.00001 to 0.1% of the spray

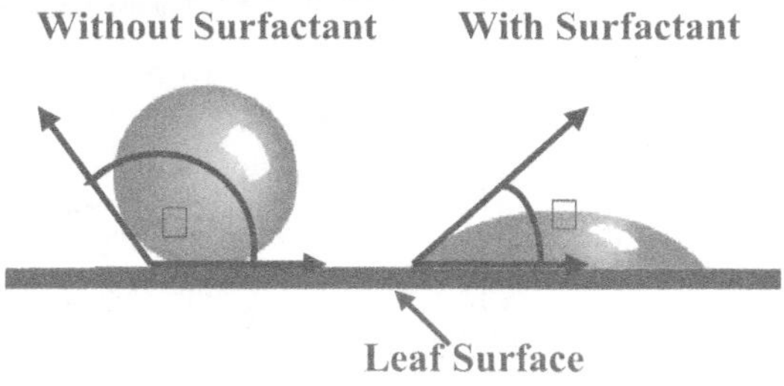

Figure 2. Contact angle θ measured as the tangent from the point of droplet contact with the leaf surface.

volume. The leaf surface can greatly change the contract angle. Harr et al. (1991) illustrated distinct differences in contact angles with the same surfactant solution on the leaf surfaces of thirty weeds and later in an analogous publication on fourteen crops (Harr and Guggenheim, 1995). Spray solution droplets with low contact angles form better contact with the leaf surface and can be more potent (Matysiak and Nalewaja, 1999; Chachalis et al., 2001).

Understanding the important adjuvant physicochemical properties and building those properties into the formulation or prescribing tank-mix adjuvants with the right type and amount of adjuvant for the spray mixture are essential research activities during the development of new herbicide products. Additional information on these properties is given when each adjuvant type is discussed.

Biology

Plant response is a function of both the herbicide and adjuvant dose. Figure 3 shows an example where herbicide alone was ineffective at twice its recommended use rate (70 g/ha), but was effective at 10-fold lower rate with an adjuvant. Environmental conditions can also greatly affect herbicide activity and even eliminate the need for adjuvant under some conditions. For example, herbicide alone can control dicotyledonous weeds with thin cuticles that were grown under low light and in high humidity (Kudsk and Streibig, 1993; Green, 2001a). Some view adjuvants as insurance for environmental variations.

There are essentially two ways to compare the biological effects of adjuvants (Figure 4). Herbicide and adjuvant rates can be fixed and the biological results

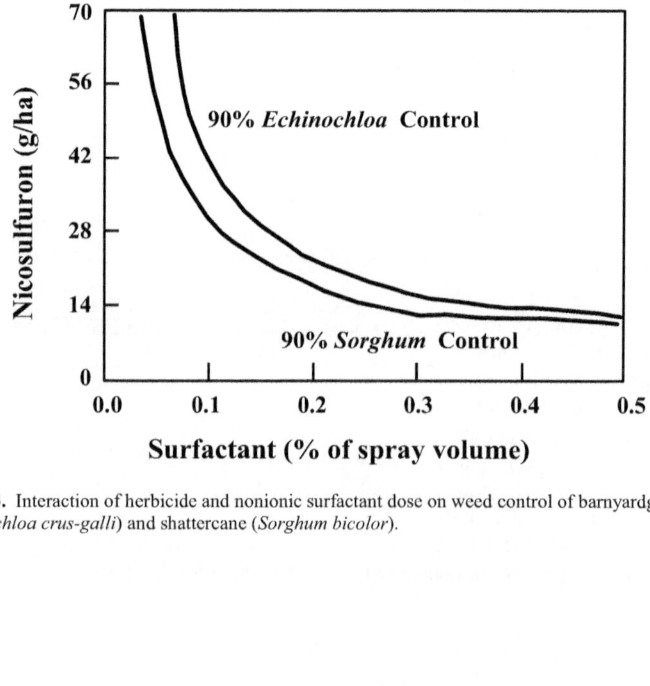

Figure 3. Interaction of herbicide and nonionic surfactant dose on weed control of barnyardgrass (*Echinochloa crus-galli*) and shattercane (*Sorghum bicolor*).

compared with analysis of variance or herbicide or adjuvant dose can be varied and potency compared with dose response analysis (Streibig et al., 1993; Green, 1999a). The first technique is common, but vulnerable to the shape of the response curve. For example, many weeds are totally controlled at recommended use rates and thus will not show treatment differences. Weeds not controlled may show a big adjuvant effect because of the slope of the response curve. With analysis of variance, researchers may mistakenly conclude the effect is species specific. Response curve analysis calculates the change in potency with a factor that is independent of the shape of the response curve.

A widely held belief is that different surfactants interact differently with different herbicides on different species (Stock and Holloway, 1993). The structure-activity response surface for six weed species supports this (Figure 5). For example, increasing surfactant molecular weight, the number of CH_2 or ethylene oxides (EO) groups, increased rimsulfuron activity on velvetleaf (*Abutilon theophrasti* Medicus), but lost activity with the largest molecular weight surfactant on giant foxtail (*Setaria faberi* Herrm.). Increasing surfactant molecular weight increased common cocklebur (*Xanthium strumarium* L.) control if the right balance between the hydrophobic and hydrophilic groups was maintained. The most effective surfactants across species had 17 to 24 carbons in the hydrophobic and a hydrophilic group with 13 to 30 EO (Green, 2001b).

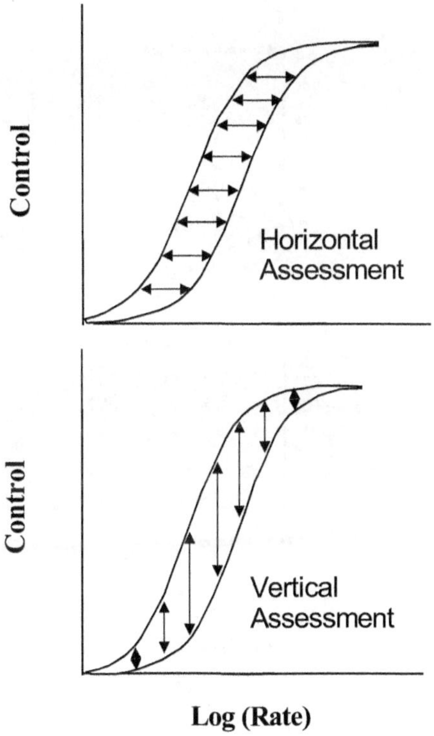

Figure 4. Comparison of horizontal (dose) and vertical (effect) biological assessments.

ACTIVATOR ADJUVANTS

Activator adjuvants are used universally with postemergence herbicides to enhance biological activity. The most common adjuvants are surfactants or surface active agents. Surfactants enhance biological activity and facilitate or enhance the emulsifying, dispersing, spreading, sticking, or wetting properties of liquids. Oil adjuvants are also commonly used to enhance activity and as carriers for herbicides.

Surfactants

Surfactants is a clever contraction of SURFace ACTive AgeNTS for substances that reduce surface tension at interfaces and improve spreading and contact with the leaf surface. Surface tension (γ) is a force that acts perpendicular and inward from the boundaries of the surface, decreasing the area of the interface. Static or equilibrium surface tension (EST) refers to the measurement at equilibrium while dynamic surface tension (DST) refers to the measurement before equilibrium while the created surface is very new, usually a fraction of a second. Surface tension influences droplet formation, transport, evaporation, impact, and retention. Increasing surfactant concentration generally decreases DST and increases droplet retention and cuticle penetration.

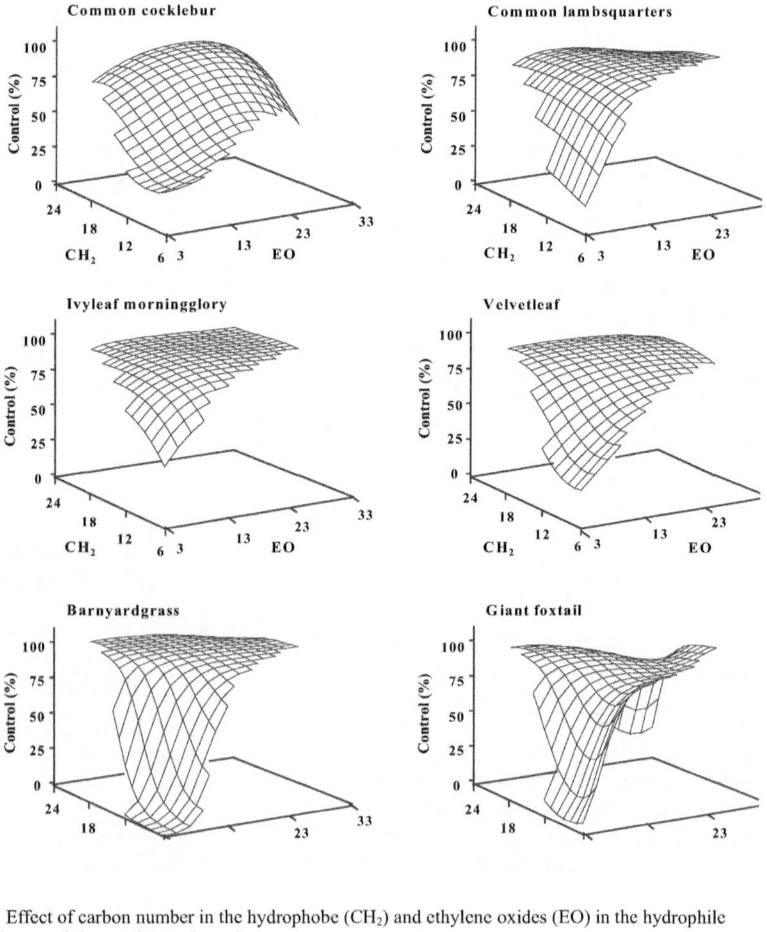

Figure 5. Effect of carbon number in the hydrophobe (CH_2) and ethylene oxides (EO) in the hydrophile on rimsulfuron activity on six weed species.

Surfactants are amphiphilic with both a lipophilic hydrocarbon portion and a hydrophilic polar portion. Characteristic features of surfactants are adsorption to surfaces and self-assembly (Holmberg et al., 2003). Most surfactant molecules aggregate into micelles, but some do not form micelles and assemble into lamellar structures. Silicone surfactants are an example of surfactants being classified by the nature of the hydrophobe, but most surfactants are classified by the ionic nature of the polar group. The polar group can be nonionic or ionic [cationic (positive charge), anionic (negative charge), or amphoteric (both positive and negative charges)]. Most nonionic surfactants have ethylene oxide chains (also called oxyethylenes or ethoxylates) as the polar groups, but using sugar or polyols derived from sugar as the polar group is increasing because of improved safety and the perceived link to a natural product.

The molecular balance of the hydrophilic and lipophilic groups is the HLB. One method to calculate HLB is to determine the percent of the molecule on a molecular weight basis that is hydrophilic divided by 5, e.g., a surfactant that is 60% hydrophilic has an HLB of 12 (Griffin, 1954). Other methods are also used to determine HLB. For ionic surfactants, HLB is often estimated experimentally by observing dispersibility in water (no dispersion = 1 to 3, poor dispersion = 3 to 6, unstable milky dispersion = 6 to 8, stable milky dispersion = 8 to 10, translucent to clear dispersion = 10 to 13, and clear solution > 13). Purists struggle with this impreciseness, but HLB is often the most important surfactant property in both agricultural and non-agricultural applications.

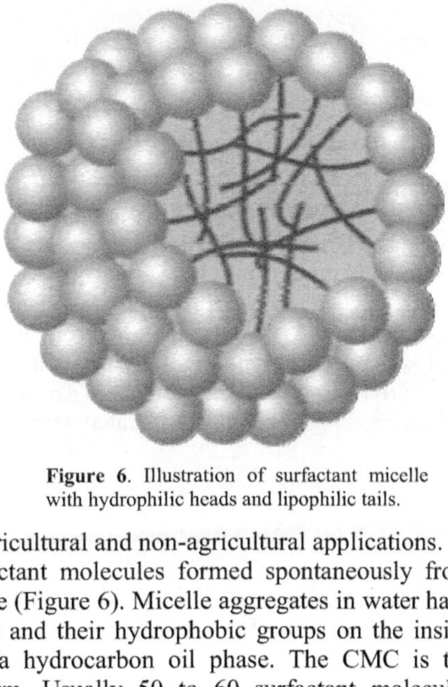

Figure 6. Illustration of surfactant micelle with hydrophilic heads and lipophilic tails.

A micelle is an assembly of surfactant molecules formed spontaneously from unassociated molecules in the bulk phase (Figure 6). Micelle aggregates in water have their hydrophilic groups on the surface and their hydrophobic groups on the inside creating an internal zone resembling a hydrocarbon oil phase. The CMC is the concentration where micelles first form. Usually 50 to 60 surfactant molecules aggregate to form a micelle and the radius of the micelle is close to the maximum length of the hydrophobe, usually in the range of 1.5 to 3 nm. Surfactants with big hydrophobes have lower CMC values than surfactants with small hydrophobes. For nonionic surfactants, each additional carbon in the hydrophobe usually lowers the CMC by a factor of three. For ionic surfactants, each additional hydrocarbon lowers the CMC by a factor of two. Increasing the size of the hydrophilic group increases the CMC, e.g., there is a slight increase from 4 to 12 EO groups and above 12 the effect is greater as the large hydrophilic groups strongly resist micelle formation. Surfactant structure, temperature, and electrolyte concentration all influence the how and when micelles form.

Some surfactant effects occur only above the CMC (Mysels, 1969). If the concentration is below the CMC, the surfactant unimers migrate to the wetting front and are adsorbed to surfaces. At the CMC, surfactant has been maximally absorbed onto all surfaces. Above the CMC, the individual surfactant molecules or unimers aggregate into micelles. The micelles are in a dynamic equilibrium with the unimers, spontaneously releasing surfactant molecules to maintain equilibrium. An adequate supply of unimers is essential to maintain the wetting front. Micelles can emulsify lipophilic herbicides and reduce the interaction with the leaf surface. Surfactants improve contact between the spray deposits and plant surfaces by:

1. Uniformly depositing the spray solution on the plant;
2. Increasing retention or sticking of spray droplets to the plant;
3. Increasing penetration through hairs, scales, or other leaf structures;
4. Preventing crystallisation of spray deposits; and
5. Slowing drying and increasing water retention in the spray deposits.

Nonionic Surfactants

Nonionic surfactants do not have any ionic charge and are the 'workhorse' adjuvants for postemergence herbicides. Nonionic surfactants usually get their hydrophilicity from a chain of ethylene oxides and lipophilicity from a hydrocarbon chain (Figure 7). When glucose or another sugar serves as the hydrophilic group, they are called alkyl polysaccharides or alkyl polyglycosides (Lauchet, 2001). The "saccharide" name comes from the sugar component and the "glycoside" comes from the covalent bonding of two sugars. These surfactants are biodegradable, compatible with fertiliser solutions, and used commercially in some glyphosate formulations. Block co-polymers are a special type of surfactant with more than one hydrophilic or lipophilic part. Usually, lipophilic propylene oxide $[-(CH_2)_3\text{-}0\text{-}]$ or butylene oxide $[-(CH_2)_4\text{-}0\text{-}]$ are alternated with hydrophilic ethylene oxide chains. The hydroxy end-cap (-OH) of ethylene oxide chains can be methylated to increase lipophilicity and reduce chemical reactivity associated with hydroxyl groups.

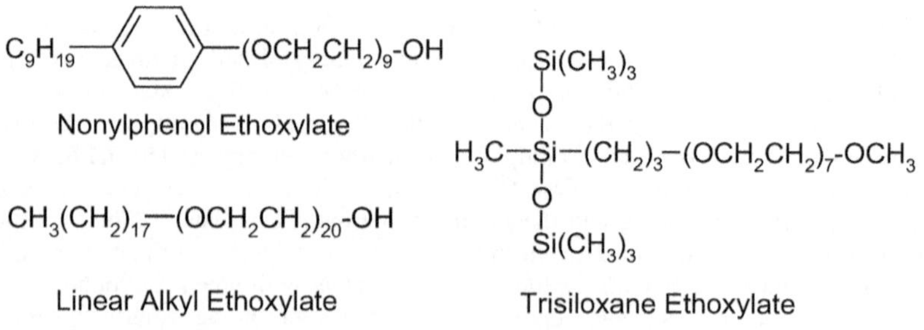

Figure 7. Examples of nonionic surfactants.

The plant cuticle is a complex lipophilic membrane that forms the outer surface of the aerial portion of plants. Penetration through the cuticle is the most important barrier to herbicide uptake. How do surfactants increase absorption through the waxy plant cuticle? Transport through the cuticle is a diffusion process consisting of sorption into, diffusion through, desorption from the cuticle (Kirkwood, 1999). Adjuvants act by changing the activation energy for these processes by solubilising the herbicide, hydrating or solubilising the cuticle, and changing the viscosity and crystalline structure of epicuticular waxes. Hydrophilic herbicides probably penetrate through polar pathways and lipophilic herbicides through the non-polar waxes. Maximum herbicide uptake occurs when the surfactant and herbicide penetrate together (Schönherr et al., 1999). The amount of a surfactant that absorbs into the cuticle is usually dependent on EO content with more absorption occurring with more lipophilic surfactants (Coret et al., 1993).

Most evidence shows that surfactants do not solubilise the cuticle. Riederer and Schönherr (1990) found no detectable amount of cuticular wax was lost when the cuticles were treated with excessive surfactant because much of the wax is in an ordered crystalline state and contains very long hydrophobic alkyl chains. Coret and Chamel (1995) used differential scanning calorimetry to show that wax melted at lower temperature after surfactant absorption and suggested that increased absorption of oil soluble herbicides was due to increased fluidity and thus increased diffusion across the cuticle. Surfactants can induce unspecific "plasticizing" of cuticular wax, but they do not actually solubilise or alter crystalline structure.

The mechanism for increasing herbicide absorption is probably different for high EO (hydrophilic) and low EO (lipophilic) nonionic surfactants. Increased water permeability of cuticles can be measured after high EO surfactants are applied (Riederer and Schönherr, 1990). For a given relative humidity, surfactants with high EO contents increase hydration of the cuticle. For example, nonylphenol surfactant with 17 EO increased water absorption into isolated tomato cuticles (Coret and Chamel, 1995). An increased hydration state may well increase the diffusion of water-soluble herbicides across the cuticle.

The optimum surfactant HLB for a particular herbicide can generally be predicted based on the water solubility of the herbicide, *e.g.,* low HLB surfactants work best with water insoluble herbicides and high HLB surfactants work best for water-soluble herbicides (Figure 8). For example, surfactants with a high HLB are more effective with the hydrophilic herbicide glyphosate while more lipophilic surfactants are more effective with the lipophilic quizalofop-P ethyl. Surfactants with intermediate HLB values are the most effective with the intermediately soluble nicosulfuron. Surfactants with the wrong HLB can reduce or deactivate glyphosate activity (Gaskin and Holloway, 1992). The optimum HLB can even change for the same herbicide. For example, when pH of the spray solution increases and changes the ionic state of nicosulfuron, the optimum surfactant HLB also changes (Green and Cahill, 2003). Unfortunately, matching the HLB with the herbicide is not currently possible, as most tank-mix adjuvant manufacturers do not disclose this information to users.

Stomatal penetration (infiltration) of spray droplets from a water-based solution is a complex process and does not occur in field spray applications of herbicides without surfactants. Stomatal infiltration is not possible unless the surface tension of the spray solution is significantly reduced by the use of surfactants in the formulation or spray tank. Most surfactants are not able to reduce the surface tension adequately to allow stomatal infiltration. However, the organosilicone surfactants (described below) reduce surface tension to the point where stomatal infiltration occurs.

Silicone surfactants are nonionic surfactants that are especially effective in reducing surface tension and increasing spread of spray droplets. This increased spreading can be dramatic and termed "super spreading". It was originally proposed to be due to the compact size of the lipophilic trisiloxane moiety allowing it to transfer readily from the liquid/air interface of the advancing solution, termed "molecular zippering." Less compact hydrophobes spread less. More recent reports attribute "super spreading" to a more traditional "surface tension gradient" mechanism where the greater the gradient, the greater the spreading (Nikolov et al., 1998). The special characteristic of silicones to form stable films and adsorb as the contact region is stretched is important to the spreading mechanism (Figure 9).

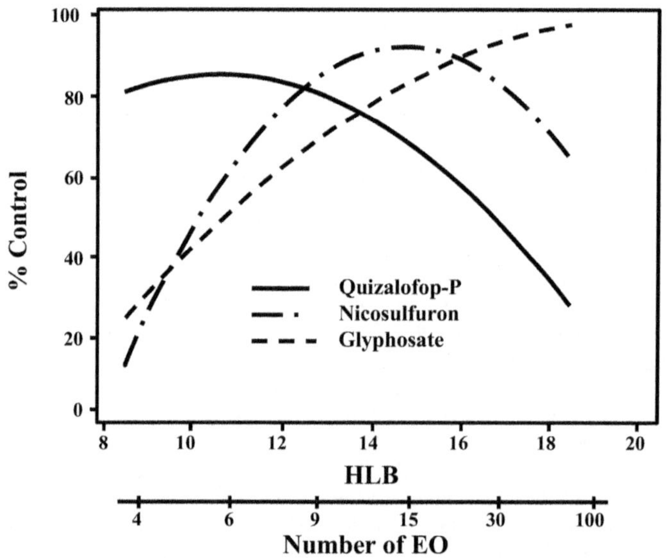

Figure 8. Effect of HLB on quizalofop-P ethyl, nicosulfuron and glyphosate activity on giant foxtail (Green 1999b).

Silicone surfactants quickly protect the herbicide from washing off by rain (Roggenbuck et al., 1993). This protection can be from rapid absorption into the cuticle or stomatal infiltration. For stomatal infiltration to occur, surface tension must be very low. For example, in field bean *(Vicia faba* L.) the critical surface tension for stomatal infiltration was 19 to 23 mN/m (dynes/cm) (Buick et al., 1993). The surface tension of a 0.25% Silwet® L-77® silicone solution is about 22 mN/m. Silicone surfactants also influence absorption of herbicides through the cuticle. Gaskin and Stevens (1993) reported absorption of glyphosate with silicones after stomatal infiltration was completed.

Some silicone surfactants are unstable when the pH is not between 6 and 8 (Murphy et al., 1991). Hydrolysis of silicon-oxygen bonds occurs under acidic and basic conditions. A silicone surfactant containing only one silicon atom, *i.e.,* no silicon-oxygen bonds, is claimed to be as effective as a trisiloxane and not sensitive to pH degradation (Klein et al., 1996). 'Super spreading' may also be lost when mixed with non-silicone adjuvants (Murphy et al., 1991). The incompatibility is attributed to competition between the compact hydrophobe of the organosilicone and the bulky hydrophobe of the conventional surfactants for the interface between the spreading drop and the leaf (jamming the 'zipper'). Some small conventional surfactants can be mixed with silicones to maintain super spreading (Policello and Murphy, 1993). Silicone surfactants can reduce activity when application volume is high and spray runs off, evaporation occurs too rapidly, or when small concentrated deposits are needed for maximum activity (Franz et al., 1997). Silicone surfactants do not work well for all herbicides (Figure 10).

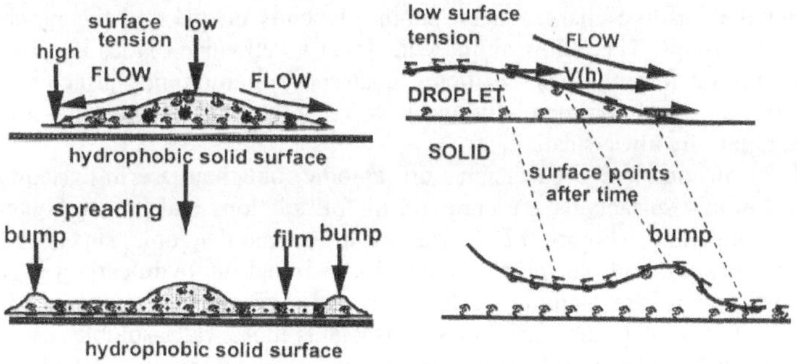

Figure 9. Silicone surfactant spreading process (Nikolov et al., 1998).

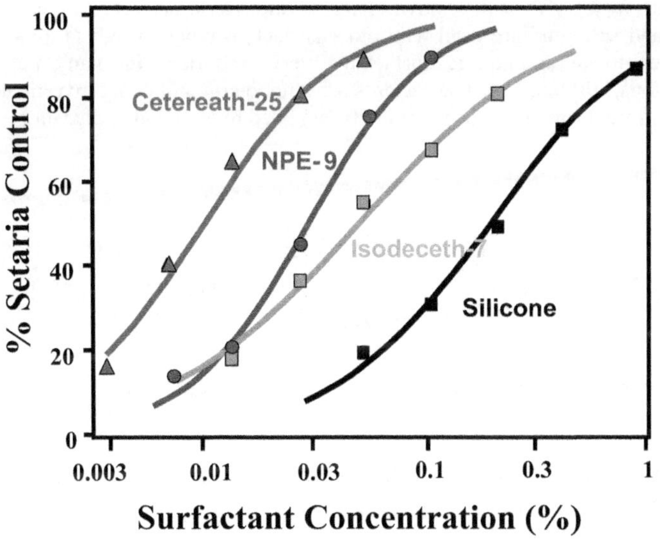

Figure 10. Effect of surfactant dose with nicosulfuron with Cetereath-25 (17 carbon alcohol, 25 EO), NPE-9 (nonylphenol , 9 EO), Isodeceth-7 (10 carbon alcohol, 7 EO), and a silicone surfactant (Green 1999a).

Ionic Surfactants

Surfactants with ionic charges are also commonly used in agriculture. Anionic and cationic groups raise surfactant HLB, increase water solubility, and may ionically pair with oppositely charged pesticide molecules to aid uptake. Other possible modes of action include neutralisation of cuticular pathways and creating hydrophilic pathways in the cuticle. Unexpected ion complexing is a potential problem for ionic surfactants.

Cationic surfactants are surface active agents that have a lipophilic segment and a cationic segment (positive ion) when placed in aqueous solution (Hazen, 2000). The most used surfactant type in agriculture may be the tallow amine ethoxylates as they are used with glyphosate (Figure 11). These surfactants are often referred to as cationic, but the positive charge on the amine depends on pH and the number of ethylene oxide groups. The tallow amine with 15 or so ethylene oxides is commonly used in glyphosate formulations. Cationic quaternary ammonium surfactants, also called "quats", are used commercially and have recently been patented for use with glyphosate. Quats are always cationic.

About 60% of commercial surfactants are anionic, but their use in agriculture is much less. Anionic surfactants are common in formulations and as components of tank-mixture adjuvants (Figure 12). The most common anionic surfactants are sulfates, carboxylates, and phosphates attached to a lipophilic hydrocarbon segment. Salts of fatty acids are anionic surfactants and common in many surfactant formulations. The polar groups make these surfactants more water-soluble.

Amphoteric surfactants are sold commercially to 'act like nonionic' surfactants, but technically have both a positive and negative charge (Figure 13). The ionic charges generally include an amino group and a carboxylic or sulfonic acid. At low pH, the amino group is protonated and the surfactant is cationic. At high pH, the carboxyl group is ionized and the surfactant is anionic. At intermediate pH, the surfactant can have both a negative and positive charge. Amphoteric surfactants tend to be mild, gentle, and safe, and are used with other surfactant types to reduce toxicity. These properties make amphoteric surfactants very important in many personal care applications. Protein-based surfactants are amphoteric and being investigated for use in agriculture. Proteins usually need to be acylated to get good surfactant properties.

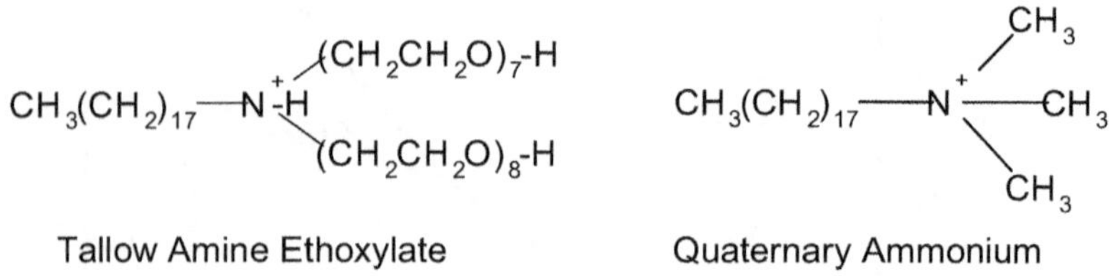

Figure 11. Examples of two cationic surfactants. The tallow amine ethoxylate is shown under acidic conditions.

Petroleum Oil Adjuvants

Oils used in agriculture are of two types: refined mineral oils (petroleum based) and seed oils (vegetable oils) (Hamilton, 1993). These oils and water do not mix and must be combined with surfactant emulsifiers for uniform distribution in spray mixtures. When the surfactant emulsifier load is high (15 to 20%), these mixtures are called oil concentrates. Both the oil and the surfactant contribute to the biological efficacy (Manthey et al., 1992).

Oils are usually more phytotoxic than surfactants and petroleum oils are usually more phytotoxic than vegetable oils. To minimise phytotoxicity, petroleum oils should be long chain paraffins with a minimum amount of naphthalene (Figure 14). This requires distillation and extensive refining such as solvent extraction to remove aromatics, dewaxing, and a final fractionation to recover a narrow boiling range oil. Nonphytotoxic mineral oil adjuvants usually have a high proportion of straight or branched hydrocarbon chains with lengths from 18 to 30 carbons.

Nonphytotoxic oils can also be used as carriers for oil soluble herbicides, but application volumes must be very low to be practical. Research on delivery systems such as the air-assist spray nozzle and positive displacement pump have used oil as a carrier at ultra low volumes (1 to 4 litres/ha) (Bohannan and Jordan, 1995). Using oil carriers can reduce surface tension, increase wetting, spreading, absorption (McWhorter et al., 1993), improve rainfastness, and reduce loss of carrier during and after application.

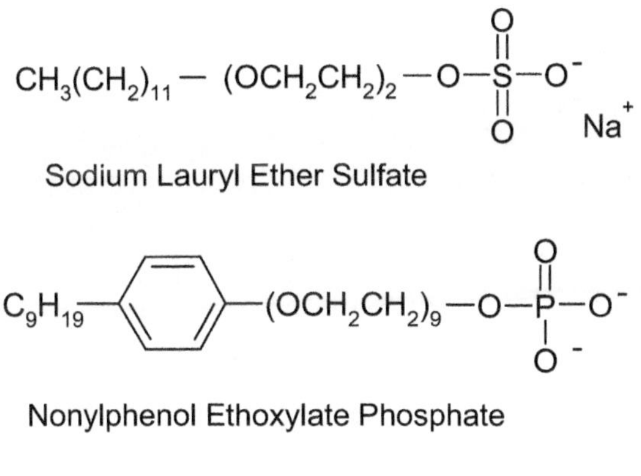

Figure 12. Examples of anionic surfactants.

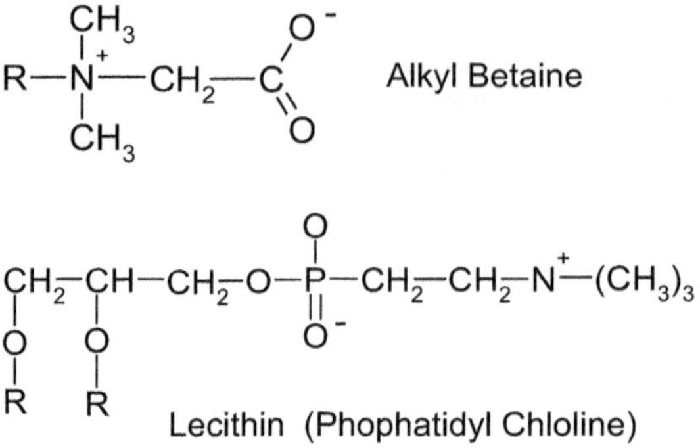

Figure 13. Examples of amphoteric surfactants.

Oil adjuvants are commonly used when conditions are hot and dry and cuticles are thick. Paraffinic oil can smooth the epicuticular wax or cause cracks in the cuticle. These cracks allow entry of hydrophilic materials that can cause swelling of the cutin layer and further enlarge surface faults that ultimately increase herbicide penetration (Foy and Smith, 1969). Some interpreted this as dissolving wax, but paraffin oils are poor solvents and only soften wax. Vegetable oils, particularly modified seed oils, are better solvents than paraffinic oils. Nalewaja and Skrzypczak (1986) found a correlation between wax solubilisation and herbicide penetration with methylated seed oils.

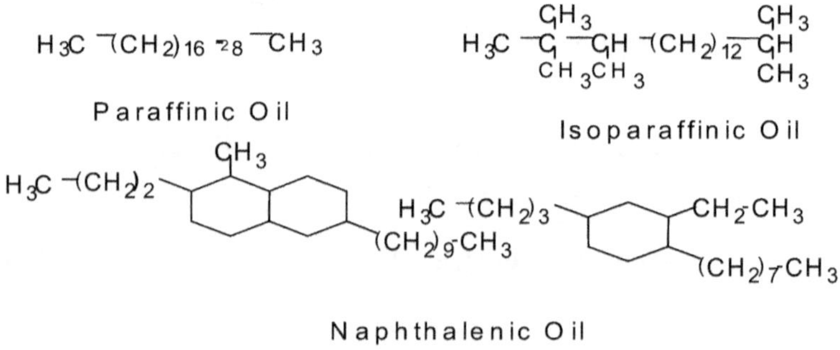

Figure 14. Examples of petroleum oils.

Emulsifiable oils are typically used at 0.5 to 1% of the spray volume or at 1 to 2 L/ha. These adjuvants contain nonphytotoxic oil (80 to 99%) and a surfactant (1 to 20%).

The purpose of the surfactant is to emulsify the oil in the water spray solution and lower the surface tension of the spray solution, but they can also interact with the cuticle and assist herbicide absorption. The exact mode of action of oils is unknown, but oils enhance droplet spreading on the plant surface and increase the absorption of the herbicide into the plant (Gauvrit and Cabanne, 1993). Santier and Chamel (1996) proposed that oils partition into the cuticle and increase the fluidity of cuticular components, thereby increasing herbicide diffusion. Several oil adjuvants have been introduced recently with high 40 to 50% surfactant emulsifier content. These adjuvants are hybrids of the traditional oil concentrate and surfactant adjuvants. The so-called 'high load COC' adjuvants are used at intermediate doses.

Vegetable Oils

There are two main types of seed oils used in agriculture: 'vegetable oils' and 'modified seed oils' (Figure 15). Vegetable oils are triglycerides or three fatty acids esterified to the carbons of glycerol. Modified seed oils are methyl or ethyl esters of fatty acids. Both oils need to be liquid so the fatty acids are often unsaturated such as oleic and linoleic acid. Triglyercide oils have a higher viscosity than methylated oils. The chemistry of these two oils is remarkably similar considering triglycerides are often perceived as much less potent than modified seed oils (Nalewaja, 1994; Urvoy and Gauvrit, 1991). One key difference is that vegetable oils usually have only 5 to 7% surfactant emulsifier while modified vegetable oil have 10 to 20% surfactant. Methylated seed oils are better solvents than other oil types. Manthey and Nalewaja (1992) reported methylated sunflower oil completely dissolved wax isolated from green foxtail, but others have not had similar effects with methylated canola oil on other species (Gauvrit and Cabanne, 1993).

Seed oils are extracted from plants by pressing or solvent extraction. The hydrocarbon chain lengths are even numbered with 16 or 18 carbons dominating. Seed oils need further purification to remove gums, mucilages, and phospholipids (e.g., precipitated with phosphoric acid then neutralised with sodium hydroxide), color (e.g., heating in the presence of Fuller's earth), and short-chain aldehydes, ketones, acids, and hydrocarbons (e.g., injecting steam). Citric acid can be added to chelate metals that cause oxidation and rancidity. Methyl esters can be prepared when fatty acids are removed from the glycerol backbone by hydrolysing the triglyceride or by inducing a trans-esterification where the triglyceride is reacted with methanol in the presence of sodium methoxide. Such methyl esters are termed methylated oils. Ethyl and butyl esters have also been prepared, but did not show any advantage over methyl esters. The composition of oils varies depending on the seed source and can influence activity (Nalewaja, 1994).

Ammonium Fertilisers

Ammonium fertilisers such as ammonium sulphate, ammonium nitrate, and ammonium polyphosphate are used as activator and utility adjuvants. Ammonium can activate the postemergence activity of weakly acidic herbicides on key weeds, but the exact mode of action is not certain and the effect is herbicide and species specific. The biological effect can be large as shown by Fielding and Stoller (1990) when absorption of thifensulfuron into velvetleaf increased more than 10-fold, from 4% to 45%.

The mode of action probably involves several mechanisms. Inorganic salts alter tissue permeability and interactions between water and macromolecules. For example, $(NH_4)_2SO_4$ increases betacyanin leakage and $CaCl_2$ counteracts this effect. Inorganic salts also change osmotic effects and surfactant CMC and micelle size. Salts affect surfactant hydration and change the optimum HLB. Some ammonium fertilisers are alkaline and may simply increase spray water pH and increase herbicide solubility. Fertilisers are often prepared with excess ammonia to minimise metal corrosion. Fertilisers can also act as humectants and prevent precipitation or crystallisation of the herbicide on the leaf surface (Nalewaja and Matysiak, 2000). Another mechanism is ion pairing. When ammonium forms salts with weak acid herbicides, it neutralises ionic charge and increases penetration especially through high pH leaf surfaces like velvetleaf (*Abutilon theophrasti*) (Wanamarta et al., 1993). Ammonium salts are known to diffuse more freely across the cuticle than other inorganic salts (e.g.,

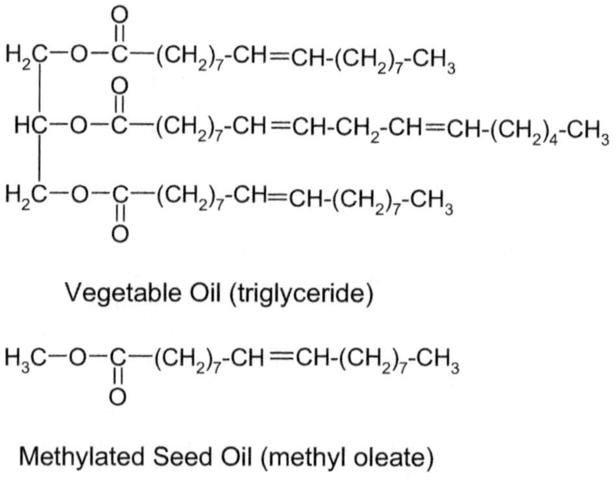

Figure 15. Examples of vegetable oils.

sodium , potassium, or calcium). Ammonium can also decrease apoplastic pH when it penetrates plant cells through induction of an ATPase hydrogen ion efflux. The more alkaline cell shifts the equilibrium of weak acid herbicide toward neutral forms that readily diffuse into cells and are trapped. This mechanism is known as "ion trapping". The best understood mechanism of ammonium sulphate is to reduce herbicide antagonism by hard water ions (see water conditioner discussion).

UTILITY ADJUVANTS

Utility adjuvants improve the spray application process. They do not directly enhance biological activity, but still have very high value by making the application easier and minimising spray tank interaction and unwanted side effects during application.

Water Conditioners and Compatibility Agents

Water conditioners minimise or prevent ions in the spray solution from reacting with herbicides to form precipitates or salts that are less readily absorbed by weeds. For example, cations in hard water can react with glyphosate and water conditioners are commonly used to prevent reduced activity. Nalewaja and Matysiak (1991) determined that the severity of glyphosate antagonism was iron > zinc > calcium = magnesium > sodium > potassium. Recent work has shown that the source of these ions may be the plant itself (Hall et al., 2000). For example, velvetleaf responds to ammonium sulphate and has 3.1% calcium whereas johnsongrass (*Sorghum halepense* (L.) Pers.) does not respond and only has 1% calcium.

Ammonium sulphate prevents the precipitation of the calcium salt of glyphosate and induces the formation of the ammonium salt that is more readily absorbed (Thelen et al., 1995). The metal chelator EDTA, and phosphoric, sulphuric, citric, tartaric, lactic, and oxalic acids all counteract hard water effects (Franz et al., 1997). Compatibility agents perform a similar function preventing chemical or physical interaction between materials in a spray mixture that could lead to a non-homogeneous or unsprayable mixture. For example, the exchange of cations between the sodium salt of bentazon and the ammonium salt of sethoxydim reduced sethoxydim penetration (Wanamarta et al., 1993). The chemistry and function of these agents is diverse (Figure 16).

pH Adjusters and Buffers

pH is the negative logarithm of the hydrogen ion concentration, pH = -log [H+]. Buffers are "chemical systems capable of resisting changes in pH." Buffers have conjugate acid-base pairs that act as proton donors and acceptors. Buffers can be acidic, neutral, or alkaline and pH adjusters may not have any buffering capacity. pH is often very important and adjuvant and pesticide manufacturers commonly try to control the pH of the spray mixture. When acidic or basic conditions can degrade herbicides, manufacturers put buffers into formulations or require tank mixtures to control pH. Adjuvants that adjust pH often improve the stability, compatibility, dispersion, solubilisation, and biological activity of herbicides. The most commonly used chemicals to reduce pH are phosphoric acid and fatty acids. The most commonly used chemical to increase pH is sodium carbonate (soda ash).

Liu (2002) showed that when a weak acid herbicide was used at a very low concentration, solubility was not a limiting factor and more uptake was found at low pH. When the concentration of the acid was increased, herbicide solubility was limiting and higher pH increased uptake. Although diffusion through the cuticle would be expected to be higher for the neutral, undissociated acid, solubility in the spray deposit was more important for bentazon under certain conditions. This relationship may explain why the relative water solubility of phenoxy acid salts correlates with their biological activity, i.e., the amine salt > ammonium salt > sodium and potassium (Su, 1991).

Herbicide solubilisation is probably required, but not sufficient for high foliar activity (Hess and Foy, 2000). A new type of adjuvant called 'Basic Blends' is used to solubilise herbicides by combining a surfactant or oil adjuvant with a component to raise the pH of the spray solution. The premise is that dry particulate formulations of some herbicides such as sulfonylureas are poorly soluble in acidic spray solutions but if the pH is increased, the particles readily dissolve and activity is enhanced (Woznica et al., 1996). Results vary depending on the herbicide, rate, particle size, surfactant type, and weed species (Woznica et al., 2001; Green and Cahill, 2003).

Humectants

Adjuvants influence the type of spray deposit formed and thus the amount of herbicide available for uptake. The amount of water decreases with time, but how it dries varies greatly. When a deposit forms a crystalline residue, there is no further herbicide uptake into the leaf unless the herbicide is re-solubilised. Humectants keep the spray deposit moist and extend the duration for uptake. Common humectants are glycerol, propylene glycol, diethylene glycol, polyethylene glycol, urea, and ammonium sulphate (Figure 17). Glucose and molasses have been used as humectants since 1894.

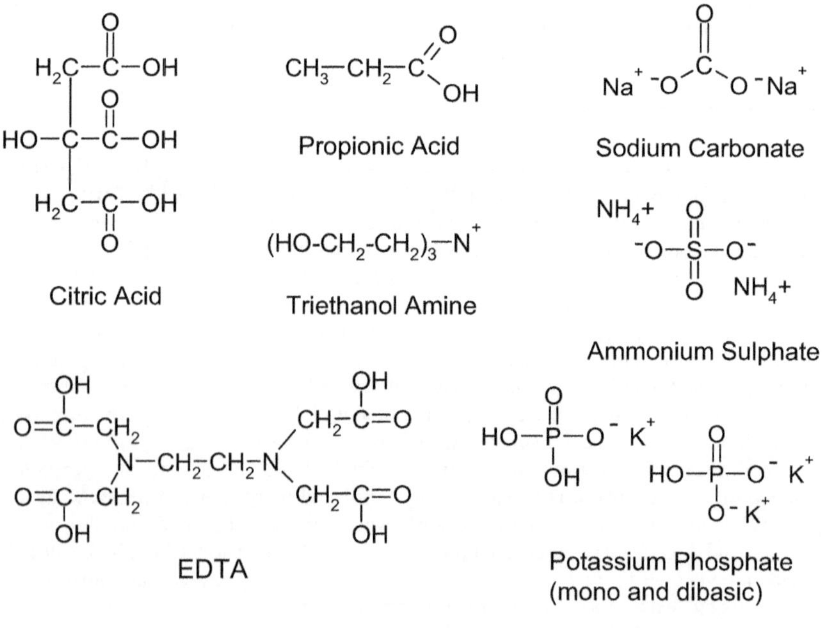

Figure 16. Examples of pH buffers and water conditioning agents.

Humectants extend duration of uptake by keeping the herbicide in solution. Once the herbicide deposit becomes dry and crystalline, absorption decreases. Humectants increase the water content of spray deposits by slowing the drying time but can also draw moisture from the environment. Silicone surfactants increase glyphosate rainfastness, but also reduce activity on some weeds because they dry too quickly. Adding a humectant overcomes this reduction without losing rainfastness (Gaskin and Stevens, 1993).

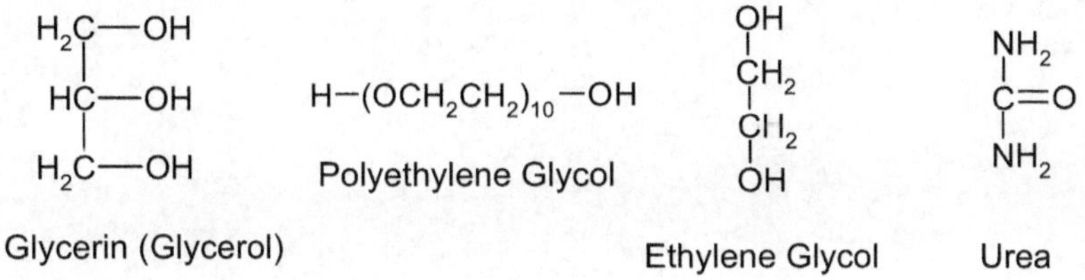

Figure 17. Examples of humectants.

Antifoam and Defoaming Agents

Antifoam agents suppress foam formation in the spray tank. Defoamers break foam after it forms. Many spray mixtures foam excessively and cause problems in rinsing the sprayer and may even lead to foam overfilling and running down the sides of the tank. Most of these adjuvants are dimethopolysiloxane-based, but silica, alcohol, and oils are also useful. These agents reduce surface tension, physically burst the air bubbles, or weaken the foam structure. It is much easier to prevent foam formation than to eliminate it after it forms.

Drift Control and Deposition Agents

Drift control agents reduce spray drift. These agents were usually based on large polymers such as polyacrylamides, polysaccharides, and gums. The most driftable part of the spray is the fine spray droplets, less than 150 μm in diameter (Downer et al., 1998). Drift control agents typically increase droplet size and sometimes larger droplets increase herbicide uptake (Feng et al., 2003). However, large droplets are more difficult to retain on the leaf as their kinetic energy increases as the seventh power of droplet diameter. Spray deposition is controlled by two factors: surface tension and energy dissipation during the collision process. Some drift control adjuvants improve deposition and are referred to as deposition aids. Guar gum is one and acts by reducing surface tension and increasing the viscoelasticity of the droplets (Bergeron et al., 2000). Marketers prefer the positive connotations associated with improving deposition to the negative connotations of spray drift.

Figure 18. Leaf angle of sicklepod (*Senna obtusifolia*) in sunlight (left) and 1 hour darkness (right).

Diurnal leaf movement is a significant factor and an interesting complication for the control of some weeds (Anderson and Koukkari, 1979). Dicotyledonous leaves are often horizontal during daylight and thus easily able to retain spray droplets, but change their orientation to vertical at night or under heavy cloud cover and shading (Figure 18). This leaf movement is called nyctinasty. Weed scientists have recently found significant differences in the activity of glyphosate because of leaf angle during diurnal leaf movements (Norsworthy et al., 1999). This leaf movement often occurs on difficult to control weeds where any reduced activity is obvious and has led to the humorous label that glyphosate is a 'nine to five' herbicide.

Stickers

Wind, physical contact, and rain can dislodge herbicide deposits. Stickers help keep deposits, usually solid, on the target leaf after droplet dry-down (Figure 19). Stickers do not evaporate, are viscous, and usually not water soluble so they resist rain wash-off. Stickers are most useful with dry wettable powder and granule formulations.

Fatty acids can be considered natural surfactants and are commonly used as stickers in many adjuvant formulations. The alkaline salts of fatty acids are soaps. Soaps are no longer used as adjuvants even though some still refer to adjuvants as soaps. Fatty acids also weakly enhance herbicide activity, reduce pH, and slow evaporation from spray deposits. Their primary use is as coupling agents to aid stability of oil concentrates. Fatty acids are inexpensive and difficult to remove completely from vegetable-based products. Fatty acids are 'naturally derived' and considered safe, but short chain fatty acids have considerable 'contact phytotoxicity' and are even used as bio-herbicides.

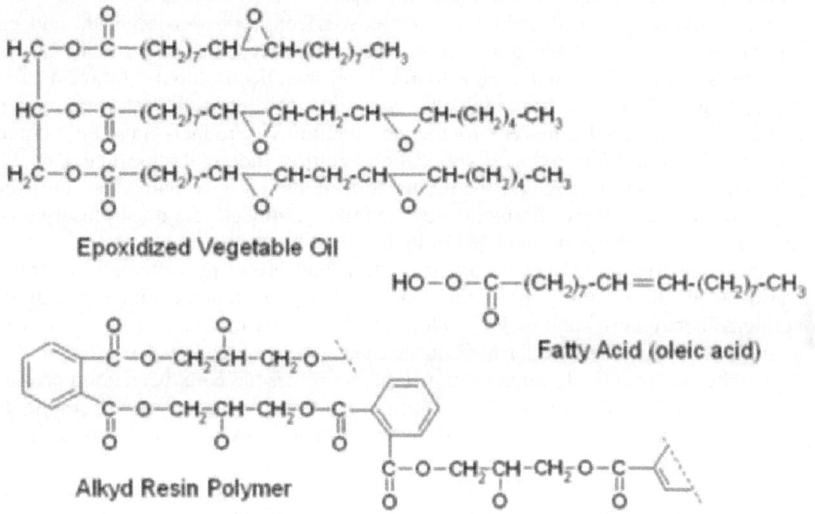

Figure 19. Examples of stickers.

UV Absorbents

Sunlight, particularly ultraviolet light, can degrade some herbicides. Photodegradation for some such as the cyclohexanedione herbicides can be as high as 50% in less than 10 minutes (Hazen and Krebs, 1992). Some adjuvants protect herbicides from sunlight. The protective effect can be physical or chemical; e.g., by increasing the rate of uptake into the cuticle or by absorbing the ultraviolet light. Adjuvants that increase the rate of uptake also make the herbicide rainfast.

Dyes, Foams, and Other Utility Adjuvants

Adjuvants have other functions to help herbicide application. Dyes are commonly used in formulations to visually spot spraying. Foams are used to mark the edge of spray applications to ensure complete coverage while not over-overlapping sprays. An emerging trend is to develop adjuvants with new functions such as translocation enhancement, seed treatment aids and protectants, and soil conditioners. Surfactants are well known to increase soil wetting and herbicide movement through soil (Foy, 1992). Singh (2002) recently showed that specific surfactants could increase herbicide binding to soil particles and thus decrease soil leaching.

FUTURE DIRECTIONS

Adjuvants have helped revolutionise the use of herbicides as tools for users to improve application and achieve more cost-effective, better-targeted, and more environmentally acceptable pest control. Adjuvants have shifted herbicide use from preventative, high dose applications to low dose, specifically foliarly targeted curative applications. There are no universal adjuvants to improve performance across herbicide types, weeds, and environmental conditions. Products must be tailored to the specific conditions of each application. Putting the right type and amount of adjuvant in the spray tank and the herbicide formulation is essential for consistently successful postemergence herbicide applications. Manufacturers must prescribe tank-mix adjuvants on the pesticide label or include in the formulation.

Adjuvant research is evolving from 'trial and error' to systematic approaches based on physicochemical properties and fundamental understandings of adjuvant-herbicide behaviour (Zabkiewicz, 2000). However, approaches that have focused on a single adjuvant attribute such as surface tension or spreading have failed. These approaches are intuitively obvious, but single attributes have not been good predictors of performance. Researchers must study multiple processes to determine how adjuvants influence the physical and chemical aspects of the spray application and herbicide absorption.

Discussion about adjuvants has rapidly evolved from disbelief of fantastic claims to technical disputes over composition, terminology, and cost-benefit. Usually the cost-benefit argument is easy as adjuvants are generally much less expensive than herbicides and can often be justified as 'insurance'. Despite many issues, adjuvants are successful and will continue to be the most cost effective and easiest way to reduce herbicide rates, minimise unwanted side effects, and improve herbicide performance. The future of adjuvants depends on the resolution of several apparent dichotomies.

1. Where the adjuvant system will be located? Many technical reasons support adjuvants and herbicides in separate packages, but governmental regulations, customer demands for convenience, and market demands for 'new and improved' products may force everything into one package. Users want all the good things in one package without a price premium.
2. Will adjuvant products evolve to specifically prescribed applications or more generally to fit a wide range of conditions and herbicides?
3. The key question may be how adjuvants will be regulated in the future. The adjuvant business may need to restructure to extract the money to pay regulatory costs or important products will disappear.

REFERENCES

Anderson R.N., Koukkari W.L. Rhythmic leaf movements of some common weeds. Weed Sci 1979; 26:401-415.

Berger P.D., Jimenez A.M. Pesticidal compositions of polyoxyalkylene alkylamine surfactants having reduced eye irritation. US Patent 1997; 5,703,015.

Bergeron V., Bonn D., Martin J-Y., Vovelle L. Controlling droplet deposition with polymer additives. Nature 2000; 405:772-775.

Bohannan D.R., Jordan T.N. Effects of ultra-low volume application on herbicide efficacy using oil diluents as carrier. Weed Technol 1995: 9:682-688.

Buick R. D., Buchan G.D., Field R.J. The role of surface tension of spreading droplets in absorption of a herbicide formulation via leaf stomata. Pestic Sci 1993; 38:227-235.

Chachalis D., Reddy K.N., Elmore C.D., Steele M.L. Herbicide efficacy, leaf structure, and spray droplet contact angle among *Ipomoea* species and smallflower morningglory. Weed Sci 2001; 49:628-634.

Coret J., Chamel A. Effects and possible mode of action of nonionic surfactants on the diffusion of glyphosate and chlorotoluron across isolated plant cuticles. Pestic Sci 1995; 43:163-166.

Coret J., Gambonnet B., Brabet F., Chamel A. Diffusion of three ethoxylated octylphenols across isolated plant cuticles. Pestic Sci 1993; 38:201-209.

Downer R.A., Mack R.E., Hall F.R., Underwood A.K. Roundup Ultra with drift management adjuvants. Proceedings of the Fifth International Symposium on Adjuvants for Agrochemicals; 1998 August 17 – 21; Memphis, TN. Memphis: ISAA, 1998.

Feng P.C.C., Chiu T., Sammons R.D., Ryerse J.S. Droplet size affects glyphosate retention, absorption, and translocation in corn. Weed Sci 2003; 51:443-448.

Foy C.L. "Influence of Certain Surfactants on the Mobility of Selected Herbicides in Soil." In *Adjuvants for Agrichemicals*, Chester L. Foy, ed. Boca Raton, FL: CRC Press, 1992.

Foy C.L., Smith L.W. "The Role of Surfactants in Modifying the Activity of Herbicide Sprays." In *Pesticidal Formulations Research,* Washington, DC: American Chemical Society, 1969.

Franz R.E., Mao M.K., Sikorski J.A., *Glyphosate: A Unique Global Pesticide.* Washington, DC: American Chemical Society, 1997.

Fielding R.J., Stoller E.W. Effects of additives on the efficacy, uptake and translocation of the methyl ester of thifensulfuron. Weed Sci 1990; 8:172-178.

Gaskin R.E., Holloway P. Some physicochemical factors influencing foliar uptake enhancement of glyphosate-mono(isopropylaminonium) by polyoxyethylene surfactants. Pestic Sci 1992; 34:195-206.

Gaskin R.E., Stevens P.J.G. Antagonism of the foliar uptake of glyphosate into grasses by organosilicone surfactants. Pestic Sci 1993; 38:185-200.

Gauvrit C., Cabanne F. Oils for weed control: uses and mode of action. Pestic Sci 1993; 37:147-153.

Gillette C. P. Experiments with arsenites. Iowa Agric Exp Stn Bull 1890; 10:401-420.

Green J.M. Optimizing alcohol ethoxylate surfactant activity at low doses. Weed Technol 1999a; 13:737-740.

Green J.M. Effect of nonylphenol ethoxylation on the biological activity of three herbicides with different water solubilities. Weed Technol 1999b; 13:840-842.

Green J.M. Factors that influence adjuvant performance. Proceedings of the Sixth International Symposium on Adjuvants for Agrochemicals; 2001 August 13 –17; Amsterdam. Amsterdam, ISAA, 2001a.

Green J.M. Weed specificity of alcohol ethoxylate surfactants applied with rimsulfuron. Weed Technol 2001b; 15:79-83.

Green J.M., Cahill W.R. Enhancing the biological activity of nicosulfuron with pH adjusters. Weed Technol 2003; 17: 338-345.

Green J.M., Hazen J.L. Understanding and using adjuvant properties to enhance pesticide activity. Proceedings of the Fifth International Symposium on Adjuvants for Agrochemicals; 1998 August 17 – 21; Memphis, TN. Memphis, ISAA, 1998.

Griffin W.C. Calculation of HLB of nonionic surfactants. J Soc Cosmet Chem 1954; 5:249-259.

Hall G.J., Hart C.A., Jones C.A. Plants as sources of cations antagonistic to glyphosate activity. Pest Manag Sci 2000; 56:351-358.

Hamilton R.J. Structure and general properties of mineral and vegetable oils used as spray adjuvants. Pestic Sci 1993; 37:141-146.

Harr J., Guggenheim, R. *The Leaf Surface of Major Crops.* Basil, Switzerland: Friedrich Reinhardt Verlag Ltd., 1995.

Harr, J., Guggenheim, R., Schulke, G., Falk, R. H. *The Leaf Surface of Major Weeds.* Waterswil, Switzerland: Fricker Ltd. Press, 1991.

Hazen J. L. Adjuvants – terminology, classification, and chemistry. Weed Technol 2000; 14:773-784.

Hazen J. L., Krebs P. J. "Photodegradation and Absorption of Sethoxydim as Adjuvant-Influenced Surface Effects." In *Adjuvants for Agrichemicals.* Chester L. Foy, ed. Boca Raton, FL: CRC Press, 1992.

Hartmann F., Green J.M. "Low Dose Urea-Surfactant Adjuvant for Sulfonylurea Herbicides." In *Pesticide Formulations and Application Systems: A New Century for Agricultural Formulations*, Jane C. Mueninghoff, Alan K. Viets, and Roger A. Downer, eds. West Conshohocken, PA: American Society for Testing Materials, 2001.

Hess F.D., Foy C.L. Interaction of surfactants with plant cuticles. Weed Technol 2000; 14:807-813.

Holstun, J.T., Bingham, S.W. Several triazines as selective postemergence herbicides in cotton. Weeds 1960; 8:187-197.

Holmberg K., Jönsson B., Kronberg B., Lindman B., *Surfactant and Polymers in Aqueous Solution.* Chichester, England: John Wiley & Sons, Ltd., 2003.

Kirkwood R.C. Recent developments in our understanding of the plant cuticle as a barrier to the foliar uptake of pesticides. Pestic Sci 1999; 55:69-77.

Klein D.K., Wilkowski S., Selby J. Silane surfactants - novel adjuvants for agricultural applications. Proc Southern Weed Sci Soc 1996; 162-167.

Kudsk P., Streibig J.C. "Formulation and Adjuvants." In *Herbicide Bioassays*, Jens C. Streibig, Per Kudsk, eds. Boca Raton, FL: CRC Press, 1993.

Lauchet F.J. "Foam Control Strategies for Formulating with Alkylpolyglycosides Surfactants." In *Pesticide Formulations and Application Systems: A New Century for Agricultural Formulations*, Jane C. Mueninghoff, Alan K. Viets, and Roger A. Downer, eds. West Conshohocken, PA: American Society for Testing Materials, 2001.

Liu Z.Q. Lower formulation pH does not enhance bentazone uptake into plant foliage. New Zealand Plant Protect 2002: 55:163-167.

Manthey F.A., Matysiak R., Nalewaja J.D. Petroleum oil and emulsifier affect the phytotoxicity of imazethapyr. Weed Technol 1992; 6:81-84.

Manthey F.A., Nalewaja J.D. "Relative Wax Solubility and Phytotoxicity of Oil to Green Foxtail *(Setaria viridis* (L) Beauv)." In *Adjuvants for Agrichemicals*. Chester L. Foy, ed., Boca Raton, FL: CRC Press, 1992.

Matysiak R., Nalewaja J. D. Temperature, adjuvants, and UV light affect sethoxydim phytotoxicity. Weed Technol 1999: 13:94-99.

McWhorter C.G. The comparative tolerance of crops to sprays of new herbicides directed to avoid contact with the foliage of crops. Proc. 9th South Weed Conf; 1956; 295-296, 1956.

McWhorter C.G., Ouzts C., Hanks J.E. Spread of water and oil droplets on johnsongrass *(Sorghum halepense)* leaves. Weed Sci 1993; 41:460-467.

Murphy G.J., Policello G.A., Ruckle R.E. Formulation considerations for trisiloxane based organosilicone adjuvants. Proc Brighton Crop Prot Conf - Weeds 1991; 355-362, 1991.

Mysels K.J. "Contribution of Micelles to the Transport of Water-Insoluble Substance through a Membrane." In *Pesticide Formulation Research,* W. Van Valkenburg, ed. Washington, DC: American Chemical Society, 1969.

Nalewaja J.D. Esterified seed oil adjuvants. Proc North Central Weed Sci Soc 1994; 49:149-156.

Nalewaja J.D., Matysiak, R. Salt antagonism of glyphosate. Weed Sci 1991; 39:622-628.

Nalewaja, J. D., Matysiak, R. Spray deposits from nicosulfuron with salts that affect efficacy. Weed Technol 2000; 14:740-749.

Nalewaja J.D., Skrzypczak G.A. Absorption and translocation of sethoxydim with additives. Weed Sci 1986; 34:657-663.

Nikolov A. D., Wasan D.T., Koczo K., Policello G.A. Mechanisms for "Superspreading": Role of Surface Tension Gradient and Surfactant Absorption. Proceedings of the Fifth International Symposium on Adjuvants for Agrochemicals; 1998 August 17 – 21; Memphis, TN. Memphis, ISAA, 1998.

Norsworthy J.K., Oliver L.R., Purcell L.C. Diurnal leaf movement effects on spray interception and glyphosate efficacy. Weed Technol 1999; 13:466-470.

Ohmes A., Kendig A., Barham R., Exell P. Chevrolet, Ford, and Dodge glyphosate. Proc of North Central Weed Science Soc 2000; 55:12.

Policello G.A., Murphy G.J. The influence of co-surfactants on the spreading ability of organosilicone wetting agents. Pestic Sci 1993; 37:228-230.

Riederer M., Schönherr J. Effects of surfactants on water permeability of isolated plant cuticles and on the composition of their cuticular waxes. Pestic Sci 1990; 29:85-94.

Roggenbuck F.C., Penner D., Burow R. F., Thomas B. Study of the enhancement of herbicide activity and rainfastness by an organosilicone adjuvant utilizing radiolabelled herbicide and adjuvant. Pestic Sci 1993; 37:121-125.

Rogiers L.M. New trends in the formulation of adjuvants. Proceedings of the Fourth International Symposium on Adjuvants for Agrochemicals; 1995 October 3-6; Melbourne, Rotura: New Zealand Forest Research Institute 1995.

Santier S., Chamel A. Penetration of triolein and methyloleate through isolated plant cuticles and their effect on penetration of quizalofop-ethyl and fenoxaprop-ethyl. Weed Res 1996; 36:167-174.

Schönherr J., Baur P. Modelling penetration of plant cuticles by crop protection agents and effects of adjuvants on their rates of penetration. Pestic Sci 1994; 42:185-208.

Schönherr J., Baur P., Buchhloz A. "Modelling Foliar Penetration: its Role in Optimizing Pesticide Delivery." In *Pesticide Chemistry and Bioscience: the Food-Environment Challenge*, Gerald T. Brooks and Terry R. Roberts, eds. Cambridge: Royal Society of Chemistry, 1999.

Singh M. Adjuvants and agrochemicals. Proceedings of the 2nd Ibero American Agricultural Adjuvant Symposium, 2002 October 3 to 4; Mexico City. Mexico City: Columbus Group, 2002.

Stock D., Holloway, P.J. Possible mechanisms for surfactant-induced foliar uptake of agrochemicals. Pestic Sci 1993; 38:165-177.

Streibig J. C., Rudemo, M., Jensen, J. E. "Dose Response Curves and Statistical Models." In *Herbicide Bioassays,* Jens C. Streibig and Per Kudsk, eds. Boca Raton, FL: CRC Press, 1993.

Su, S. Q. "Phenoxy Acids." In Guidelines for Herbicides, Sheng Quiang Su, ed. Beijing: Scientific Press, 1991.

Thelen K.D., Jackson E. P., Penner D. 2,4-D Interactions with glyphosate and sodium bicarbonate. Weed Technol 1995; 9:301-305.

Underwood A. K. Adjuvant trends for the new millennium. Weed Technol 2000; 14:765-772.

Urvoy C., Gauvrit C. Seed oils as adjuvants: penetration of glycerol trioleate, methanol oleate and diclofop-methyl in maize leaves. Proc Brighton Crop Protection Conf - Weeds 1991: 337-342, 1991.

Wanamarta G.; Kells J.J.; Penner D. Overcoming antagonistic effects of sodium bentazon on sethoxydim absorption. Weed Technol 1993; 7:322-325.

Woznica Z., de Villiers B.L., Messersmith C.G., Nalewaja J.D. Sulfosulfuron efficacy is affected by surfactants, pH, and salts. Proc Weed Sci Soc Am 2001: 41:18-19.

Woznica Z., Nalewaja J.D., Matysiak R. Inexpensive, low cost adjuvant composition for herbicidal sprays comprising amine to give alkaline pH, ammonium salt, nonionic surfactant and water, increasing herbicide effect. U.S. Patent 1996; 5,658,855.

Zabkiewicz J.A. Adjuvants and herbicide efficacy – present status and future prospects. Weed Res 2000; 40:139-149.

Zimmerman P.W., Hitchcock A.E. Substituted phenoxy and benzoic acid growth substances and the relation of structure to physiological activity. Contrib Boyce Thompson Inst 1942; 12:321-343.

CHAPTER 20

ALTERNATIVE WEED MANAGEMENT STRATEGIES FOR LANDSCAPE AND TURF SETTINGS

CÉCILE BERTIN & LESLIE A. WESTON

*Cornell University, Department of Horticulture*Ithaca NY, 14853 USA

Weeds consistently appear in less maintained landscape and turf settings, particularly in weak or stressed turf or unattended landscape beds. To achieve consistent control, it is necessary to better understand the reason why weeds are often persistent in these settings. Understanding the plant's interaction with its environment, including rhizosphere ecology, and the impact of stress on plant growth are important considerations in landscape and turf weed management. In landscapes, the most critical time for effective weed management is before planting. Effective landscape planting and design will help prevent future weed problems.

Even though chemical herbicides remain the predominant means for commercial weed management, there is an increased emphasis on developing alternative pest management strategies for turf and ornamentals. These alternatives include the use of organic products, cultural practices, biological control, and weed suppressive ground covers for long-term weed management.

INCREASED GREENSCAPE

The green industry has experienced a significant expansion over the last decade in terms of private and commercial landscape projects, golf courses, parks, and athletic fields. Currently in the U.S., there are nearly 46 million acres of maintained turfgrass including lawns, parks, golf courses, and highway right-of ways (Monaco, 2002). The sports turf market includes golf courses (public and private), athletic fields (football, baseball, and soccer), and other turf areas used for less common sports including polo, lacrosse, field hockey, and rugby. There are currently about 13,000 golf courses in the U.S. which average 25 acres each in fairways alone (Stueteville, 1994). New golf courses continue to be constructed around the world as the popularity of the game increases. Additionally, the popularity of many other field-played sports (such as soccer, football, and baseball) has fostered the construction of new fields. The landscape industry is also growing at a record pace, as new homes and commercial buildings are established and landscaped (Monaco, 2002). New plant materials have increased the popularity of landscaping and the public is spending more each year on landscape maintenance and plant establishment. In some states, the landscape industry has doubled in size over the past 10 years (Mc Niel and Weston, 1992).

Inderjit (ed.), Weed Biology and Management, 403-422.

In addition to private and public agencies and stakeholders who maintain turf and landscapes, State Departments of Transportation (SDOT) across the U.S. are managing turf and vegetation growing along rights-of-way in order to provide adequate site distances, visibility of signs and guiderails, prevent trees from establishing within the right-of-way and maintain roadway drainage so that pavement breakup is prevented. Because these agencies have obligations and the responsibilities to enhance, protect, and improve the environment, many SDOTs are seeking alternate strategies for vegetation management including prevention, monitoring, and control methods. Currently, the management program developed on most rights-of-way consists of periodic mowing in turf areas, whereas periodic herbicide treatments are used for vegetation beneath the guiderail and around signs where mowing is less feasible.

Concerned by environmental issues and public health fears associated with the use of herbicides, the Department of Transportation (DOT) has initiated a program in New York State to investigate numerous alternatives to herbicides for vegetation management, including the use of groundcovers or turf species possessing weed suppressive ability, organic products, advancements in mowing technology and mulches for weed suppression.

Landscape professionals and turfgrass managers often strive to maintain a perfect plant, defined as one that is healthy, vigorously growing, and free of all pest problems. Inevitably, when pests are encountered, pesticides are generally considered an important means for achieving control. Herbicides are valuable components of a landscape cultural program, but pest management includes more than simply selecting and applying the appropriate chemical to control specific pests. Therefore, we will discuss an integrated approach to weed management in these settings.

Both landscape and turfgrass weeds have successfully adapted to the growing conditions specific to the landscape itself. Large or hairy crabgrass (*Digitaria sanguinalis* (L.) Scop.) and small or smooth crabgrass (*Digitaria ischaemum* (Schreb. ex Schweig.) Schreb. ex Muhl.) are two closely related species which rank among the most troublesome in lawns throughout the U.S. Dandelion (*Taraxacum officinale* Weber in Wiggers), plantain species, thistles and other broadleaf weeds are also prevalent in lawns and landscape beds. Whatever the weed problem, it is most important to quickly identify the key weeds that are located in turf and landscape beds. The use of weed maps by professional landscape managers can be helpful in locating areas of intense infestation and prioritizing management efforts over time on a larger scale (Weston et al., 2001). Turfgrass managers specifically responsible for growing and maintaining turfgrass on golf courses, in residential and commercial landscapes, athletic fields, and school grounds generally have three goals in minds: (i) to protect the quality of turf, (ii) to safeguard public health and the environment, and (iii) to reduce management costs. For approximately 40 years, the dominant strategy for weed management in these settings has been the use of herbicides coupled with frequent mowing. Herbicides can be relatively inexpensive and are often used as a quick fix to solve a weed problem. Herbicides have now replaced many of the cultural weed management tactics used for centuries.

Unfortunately, the misuse of herbicides and other pesticides has created significant ecological and social problems. Off-site movement of herbicides caused by soil erosion, water movement, and wind make pesticides less effective and a source of ground water pollution. Chronic exposure of humans to pesticides is of key concern associated with long-term usage or exposure. Recent studies have suggested that the incidence of lymphoma in the U.S. may be greater in areas associated with greatest usage or exposure. The pesticides most often associated with an increase in lymphoma were 2,4-D ((2, 4-dichlorophenoxy)acetic acid) and triazine herbicides, products frequently used in home and agricultural settings (Faustini et al., 1996; Ibrahim et al., 1991). The American Lymphoma Foundation has recently published the results of human and mammalian exposure studies which correlate incidence of lymphoma with long-term exposure to certain pesticides including 2,4-D. In light of this, it is important to maintain the perspective that pesticides are just one tool of the many that can be used to effectively manage pests. In order to achieve safe, long-term pest management with minimal adverse effect on the surrounding environment, one should consider an integrated approach to weed management in the landscape.

In developing a landscape care program, one should first clearly determine the expectations in terms of quality desired and use patterns. Some settings require near perfect visual quality, and will be subjected to low traffic, whereas other settings need a high level of aesthetics but will also require the ability to recuperate from damage due to high levels of traffic. As in crop production, the principles of ecological lawn and landscape care are based on an understanding of the relationships between the biology of an organism, in this case the turf or ornamental plant, and its environment. A comprehensive program influences not only the landscape itself, but also its environment, i.e., soil, light, temperature, moisture, and human usage patterns (Dreistadt, 1994).

An overview of the important ecological concepts in a landscape system includes evaluation of plant interactions and available space within the system. For example: (1) adaptation - selecting plants that are adapted to a particular environment, (2) competition with other species and within a species, and (3) open space. If properly managed, healthy plants, including turfgrasses, will develop a dense canopy and resist damage from pests and traffic. The following sections explore the alternatives for weed management that are currently available or under investigation for use in landscape settings.

CHEMICAL CONTROL

During the 20th century, the development of chemicals for pest control was emphasized. The herbicide 2,4-D was first discovered during World War II and represents the beginning of modern day herbicide technology (Monaco, 2002). Herbicides in the landscape are widely used especially in commercial settings and managed home lawns and in fact, represent up to 12% of the herbicides used globally for weed management in all production systems. Herbicides are a convenient and generally inexpensive way to control weeds in some landscape situations, especially in larger and institutional landscapes which must minimize hand labor costs. Proper selection and application of herbicides can reduce competition and enhance desirable plant growth (Table 1). Furthermore, proper selection and application, and record keeping of herbicide usage are important when choosing a selective herbicide product for use in the landscape. Accurate information about the herbicide to be applied and about the environmental conditions at the time of application (soil and air temperature, soil moisture and pH, and wind speed and direction for instance) will help landscape and turf managers to avoid errors and achieve effective control.

Table 1. Checklist for selecting the appropriate herbicide (adapted from Dreistadt, 1994).

1. *Alternatives:* are practical alternatives to herbicides available?
2. *Situation*: is the weed problem at a planting site, new planting, or in an established landscape or turf setting?
3. *Timing:* are weeds at the right growth stage? Most herbicides are efficient within a specific time frame. For example, pre-emergence herbicides are only effective before weed seed germination.
4. *Weed species*: from previous monitoring, what weed species are causing or will cause the greatest problem?
5. *Registration:* are there any use restrictions to avoid injuring surrounding plants, people, or the environment? Check the label.
6. *Residual control:* what is the desired length of herbicide activity? What are the possibilities for off-site movement? Soil erosion, water, and wind can influence herbicides off-site movement making them less effective and a source of pollution.
7. *Application equipment*: do you have the proper equipment? Some herbicides are sold as either granular or powder formulations and require special equipment. Has the equipment been calibrated correctly to deliver the right amount of product?
8. *Expense*: what will the cost be? How many applications will be necessary? Compare with other weed management strategies.

Non-Selective Control

Herbicides have been effectively used in many landscape plantings for weed control. Classified by several criteria which include their selectivity and translocation in plants, chemistry, application method, application timing, persistence, and mode of action, herbicides are estimated to be used on nearly 50% of green space maintained in the U.S. today. Non-selective weed control includes the use of soil fumigants (volatile liquids or gases), which are effective against a wide range of soil-borne pests. Limited to small acreages and high-value cropping or production systems, fumigation is expensive because of the price of the product itself and the use of a plastic cover to trap fumigant vapors. Fumigation is often used in site preparation of golf course tees, greens or sand traps (Smeda and Weston, 1995).

The most common fumigants used in turf and ornamental beds in the past included methyl-bromide and metham (methylcarbamodithioic acid) or metham-sodium (sodium methyl-dithiocarbamate). Most recently, dazomet (Basamid) fumigant has been labeled for use in turf to control weeds, fungi, and nematodes (Park and Landschoot, 2002), and is considered a potential replacement for methyl-bromide.

Many non-selective herbicides such as glyphosate and glufosinate can be used for preplant control of problematic perennial weeds or turfgrasses. For instance, the more difficult to control bermudagrass can be controlled with a minimum of three applications (Johnson, 1988), and fescue, perennial ryegrass or Kentucky bluegrass can be usually controlled with one or two applications at a higher rate.

Spot treatment with a non-selective herbicide can be useful for control of perennial weeds in landscape beds. Wick applicators that sponge on an herbicide concentrate are often useful with non-selective products to prevent off-site movement. These products are also used to eliminate all vegetation before landscape bed preparation and installation. Due to the extensive number of herbicides available on the market and labeled for use in the landscape, we will restrict our discussion to an overview of these products.

Selective Control

Selective control can often be achieved with the use of a preemergence herbicide that is applied prior to weed seed germination. Many of these products actually act as inhibitors of cell division or cell elongation. Some of the more popular preemergence herbicides used in the landscape have been on the market for many years and include trifluralin (Treflan), oryzalin (Surflan), pendimethalin (PreM or Pendulum), and metolachlor (Pennant). Yellow nutsedge (*Cyperus esculentus* L.) and most of the annual grasses are generally controlled by metolachlor. Trifularin and oryzalin are effective on annual grasses and a few broadleaf weeds (Neal, 1994). The half-life of preemergence herbicides varies with the product, but most are effective for a relatively short period (six weeks or less) in the landscape (Table 2).

In contrast, postemergence herbicides are effective against existing weeds, and most have limited soil activity. Postemergent herbicide application should be done when the weeds are not too large and generally less than 12 inches in height. In some cases, both pre and postemergent products have a tendency to injure certain turfgrass, especially bentgrass (*Agrostis* spp.) at higher rates. Therefore, these products should be selected carefully and product labels consulted.

CULTURAL PRACTICES

A major reason for weed infestation in the landscape is poor turf and landscape management or generally poor growing conditions (Table 3). Any effort to control weeds in the landscape should begin with good cultural practices to optimize turf and ornamental growth and density and to maintain a healthy, insect and disease-free turf and landscape.

Choice of Cultivars

To achieve the goal of a high quality landscape, it is essential to identify cultivars or mixtures of species that are well adapted to the prevailing environmental conditions. Grasses are usually classified based on their ability to adapt to a particular environmental setting; for example, their adaptation to climatic conditions, moisture, light availability, and daytime temperatures. Cool-season grasses produce optimal shoot growth between 18 and 24 °C, and optimal root growth at 13 to 18 °C. Therefore, cool-season grasses grow best under spring and fall conditions and are less adapted to hotter, drier conditions. Winter and spring are the most important growth seasons for turf roots. As temperatures increase, cool-season grass root growth becomes less prolific. Knowledge of this seasonal growth cycle must be considered when developing appropriate lawn care strategies for fertilization and weed management.

Table 2. Approximate persistence of some herbicides used in landscape and turf settings (Vencill, 2002).

Herbicide	Approximate persistence in soil Half-Life (days)
Dazomet	7
Glufosinate	7
Glyphosate	47
Metolachlor	90 to 150
Oryzalin	20 to 128
Pendimethalin	44
Trifluralin	45

The proper selection of turfgrass species that meet quality expectations is also more likely to result in turf that successfully competes with undesirable species throughout the growing season. However, the selection of lawn grasses should not be exclusively based on climatic conditions. Other factors such as soil fertility, moisture, pH, and mowing requirements along with traffic, and pest tolerance should also be considered. As turf breeders develop more well-adapted turfgrass species, one's choice in lawn grasses has dramatically expanded.

For landscape ornamentals, the choice of species and cultivars has also greatly expanded in recent years due to new plant introductions. Careful attention must be paid to existing soil conditions, pH, available sunlight and exposure, and moisture availability. The use of pest resistant cultivars is important when considering a low maintenance landscape. Species adapted to local conditions usually thrive and are more competitive with weeds due to rapid establishment, resistance to pests or rapid rate of growth.

Mowing of Turfgrass

Cultural practices, when properly timed and executed with the goal of reducing disruption of a system's ecological balance, are essential to maximize landscape health. The primary cultural practices for turf include mowing, fertilization, and irrigation and for ornamentals include fertilization, irrigation, and timing of weed management.

Compared to any other cultural practice, mowing is the most critical for turf weed infestation. If mowed regularly at appropriate heights, turf can reach maximum density and consequently will compete favorably with weed species for light, available moisture and nutrients. Furthermore, mowing should be scheduled periodically and not too frequently, to allow a dense turf to establish and efficiently utilize available soil nutrients. Mowing will also help to prevent weed seed germination, and maturation and reduce accumulation of propagules in the seed bank (Turgeon, 1996).

Many broadleaf weeds are common in newly established turf and are easily eliminated with regular mowing. However, other weeds including goosegrass (*Eleusine indica* (L.) Gaertn.), annual bluegrass (*Poa annua* L.), common purslane (*Portulaca oleracea* L.), spotted and prostate spurge (*Euphorbia maculata* L. and *E. humistrata* Engeln. ex Gray), prostate knotweed (*Polygonum aviculare* L.), dandelion, broadleaf (*Plantago major* L.) and buckhorn (*Plantago lanceolata*) plantain are tolerant to mowing. Weeds with prostate or rosette-type growth habits maintain their growing points, or crowns, at or just below the soil surface. Thus, mowing frequently removes their older leaves but unfortunately does not affect the growing point and, consequently, new growth resumes (McCarthy and Murphy, 1994).

When turf is mowed too closely to the soil surface, its competitive ability is reduced and greater light penetration of the canopy occurs, which encourages seed germination at the soil surface. Mowing at low heights results in "scalping" of the turf and reduces its ability to regrow and tolerate higher light intensity and lower moisture. Additionally, other cultural practices such as aerification and vertical mowing can stimulate the germination of certain weed species as the soil surface is disturbed and weed propagules move to the soil surface (Weston et al., 2000).

Table 3. Weeds can be an indicator of different growing conditions (adapted from McCarthy and Murphy, 1994).

Soil Conditions	**Weeds**
Low soil pH	Sorrels, moss
High soil pH	Plantains
Soil compaction	Goosegrass, knotweed, annual bluegrass, plantains
Low soil nitrogen levels	Clovers, chickweeds, trefoil
Poor sandy soils	Sandbur
Poor drainage / overwatering	Sedges, rushes
Excessive surface moisture and mowing height too low	Spurges, pusley, knotweed, moss, and algae
Excessive nematode population	Algae, spurges

Cultivation and Thatch

Cultivation and dethatching have important effects in terms of weed control in turfgrass settings. Cultivation activities include core aerification, slicing, and vertical mowing. These activities, along with dethatching, can reduce thatch build-up and help prepare the turf for overseeding. Core aerification is very useful for reducing soil compaction. Soil compaction from traffic and tight soils will often decrease turf competition and favor some weeds, especially goosegrass and knotweed. It is important to conduct cultivation activities such as coring, slicing, and aerification during periods of active growth when turfgrasses are best able to recover from these practices. In areas where annual weeds are a particular problem, cultivation activities should be considered in the autumn months to reduce weed germination and competition.

Since light is often necessary to stimulate weed seed germination, weeds that germinate in the thatch layer usually die because of the environmental conditions encountered. In this respect, thatch provides some degree of weed suppression. On the other hand, too much thatch can cause turf decline by harboring disease-causing pathogens, reducing water infiltration and tying up pesticides and fertilizers. Power-raking and verticutting are typically used to remove thatch. Recent thought by turf specialists on this subject would suggest that dethatching in temperate climates is often a waste of time, and few benefits result (Rossi, pers. comm..).

Irrigation

Even though most lawns do not require regular watering, irrigation also has an important impact on weed encroachment. Infrequent and deep watering favors deeper and stronger turf root systems, which in turn allows turfgrass to successfully compete with weeds because of the presence of a stronger and denser turfgrass (Weston et al., 2000). Even though total quantities of water from rainfall may be large, the seasonal distribution of rainfall may not be optimal, thus the goal is to provide water during periods when water is deficient.

Irrigation is needed to compensate for water loss due to evapotranspiration (McCarty and Weinbrecht, 1992; Turgeon, 1996). Additionally, modified soils including golf greens and athletic fields demand higher water rates because the amount of water loss is increased by internal drainage in sandy soils. Over-irrigation favors weed seed germination whereas lack of moisture can result in decreased turf density. In the Northeastern and mid central U.S., homeowners are encouraged not to irrigate their lawns because very little long-term benefit is evident from irrigation (Rossi, pers. comm..). For other commercial maintenance personnel, several components of watering are important to keep in mind, especially timing and rate. Time of application should be scheduled when the evaporation rates are low, such as in the early morning or evening. The application rate also should not exceed the infiltration rate, which can be defined as the rate at which the soil can absorb the water applied. Again, it is important to apply fewer thorough waterings than many shallow irrigations.

The above discussion has dealt mainly with plant growth issues; however, irrigation is also important for activity of preemergence herbicides. Labels commonly specify the irrigation requirement after application if rainfall is inadequate. Many preemergent herbicides, especially granular products, require timely irrigation after application to move the herbicide into the soil when weed seeds germinate (Derr et al., 1997).

Fertilization

Appropriate fertilization, based on soil tests, that maintains a dense turfgrass canopy will assist in achieving optimal weed management. The amount and the timing are equally important. Soil tests are typically used to determine the level of application of phosphorus, potassium, and other necessary nutrients, except nitrogen which varies seasonally and is particularly mobile. In a perennial turf setting, a low level of nitrate-nitrogen is often observed because the perennial crop is continuously growing, and limited nitrate-nitrogen accumulation occurs, except when excessive applications are performed. Studies in the 1930's noted that weed control in turfgrass was influenced by soil fertility (Levy and Madden, 1931). After 8 years of adequate fertilization, the density of dandelion and large crabgrass was significantly reduced in fertilized turfgrass which was more densely established (Murray et al., 1983), compared to unfertilized turfgrass.

Fertilization can also be important for activity of preemergence herbicides. Competition between mature Kentucky bluegrass (*Poa pratensis* L.) and large crabgrass seedlings was an essential factor in achieving large crabgrass control with preemergence herbicides (Roberts et al., 1966). In these experiments, nitrogen and soil moisture were varied to determine the influence of these factors on preemergence herbicide performance. This study demonstrated that when soil was moistened and treated with a preemergent herbicide, Kentucky bluegrass competition accounted for at least 12 to 18% of large crabgrass control. Additionally, when turfgrass is adequately fertilized in the fall to encourage root growth, the resulting turf canopy in the spring will prove highly competitive with annual weeds (Weston et al., 2000).

Similarly, Murray et al. (1983) demonstrated that a dense turfgrass with adequate nitrogen fertility was infested with fewer large crabgrass plants than a thin turfgrass with low nitrogen fertility. In this study, when herbicides were not applied, Kentucky bluegrass competition alone accounted for 70% of the large crabgrass control. In Georgia, similar results were reported, and fertilizer could be substituted for herbicide application to maintain turfgrass density over time (Johnson and Bower, 1982). Many researchers have attributed the decrease in both dicot and monocot numbers in fertilized turf to better nitrogen fertility (Adams, 1980; Beard, 1973; Johnson, 1982; Mehnert and Madel, 1982; Murray et al., 1983). Numerous studies have shown that a combination of preemergence herbicides and adequate nitrogen fertility resulted in better control of large crabgrass and other weeds than either used alone.

A combination of cultural practices, including mowing height and fertilization, also influenced weed infestation, especially with respect to smooth crabgrass invasion in a red fescue (*Festuca rubra* L.) lawn. According to Emmons (1984), if red fescue was mowed at 5.6 cm, the turf density was improved and the occurrence of smooth crabgrass was decreased (4 to 14%) during a 4-year period in comparison with a turf mowed at 3.2 cm where the smooth crabgrass invasion reached 18 to 39%. Furthermore, the yearly addition of nitrogen fertilization at 49 or 88 kg/ha improved the turf stand; thus, smooth crabgrass population was reduced. Under low fertility rates and excessively low mowing height, heavy weed infestation and poor turf quality were consistently observed (Emmons, 1984).

For perennial ornamental species, adequate fertilization and irrigation are especially important for successful establishment. The first year of growth can be most difficult when maintaining a newly transplanted groundcover or ornamental. Therefore, adequate and timely irrigation, early in the day or in the evening, is preferable to minimize stress due to drought. Appropriate fall and spring fertilization, especially after recent establishment, will also ensure adequate growth and successful stands, if other environmental conditions are favorable for establishment of that species (Derr et al., 1997; Dreistadt, 1994).

Mulching

Other cultural means for weed management in ornamentals include the use of mulching materials in the landscape. Various types of mulches and their properties are described in the Table 4. Mulches act in a physical manner to exclude light, preserve soil moisture and suppress weed seedling germination. Organic mulches including bark, various hulls (pecan, cocoa, buckwheat), and municipal composts are effective in suppressing annual weeds growth. Their efficiency will be maximized if mulches are fairly coarse textured with a low water holding capacity (Neal, 1994). However, mulches do not provide complete weed control over the long-term; hence, additional management such as hand weeding or spot spraying is usually required. Because of the cost of supplemental weeding, certain landscapers favor the use of geotextiles and/or preemergent herbicides placed on the top of a thin layer of decorative mulch. Synthetic fabrics and geotextiles allow water and air to pass, but prevent weed seedling emergence. Due to their high initial expense in installation, fabrics in landscape settings are generally cost effective only if they remain in place for 4 or more years. Many types of synthetic mulches are available, but they have rarely shown complete long-term weed suppression in the landscape, especially when organic mulches are placed over the synthetic mulch to harbor blown in weed propagules. The labor involved and expense incurred during installation have relegated these materials for use on small areas or installations requiring special treatments, such as sand traps or sloped sites (Mc Niel and Weston, 1992).

Table 4. A description of various types of available mulches and their properties (adapted from Dreistadt, 1994).

Organic Mulches	Properties
Bark chips and ground bark	- Attractive, improves soil texture with gradual decomposition. - Can harbor pests such as earwigs, termites, and sowbugs. - Can be placed over plastics as decorative material
Wood chips	- Can be inexpensive - May contain some weeds - May not stay in place - May tie up soil nitrogen with decomposition - Compost well and add nitrogen to the soil first
Compost	- Excellent source of organic matter - Readily available or can be created - May harbor weed seeds if not properly composted - May promote crown disease if applied to contact trunk
Grass clippings and leaves	- Readily available, can be applied often - May contain weed seeds - May mat or reduce water penetration if not dried first - Better if composted before use
Peat moss	- Increases water-holding capacity if mixed into soil - Adds acidity to alkaline soils - Few or no weeds - Expensive

Hay and Straw	- Good water penetration - Less attractive - May contain weed seeds - May use soil nitrogen as it decomposes - Compost well before using or add nitrogen to soil
Rice hulls	- Benefit soil tilth - Slow to degrade - If not composted, may contain weed seed - May use soil nitrogen as it decomposes - Compost well before using or add nitrogen to soil
Leaf mold	- May add needed acidity to alkaline soils - Attractive
Newspaper (shredded)	- Readily available - Contains no weeds - Inexpensive - Attract earwigs and sowbugs - Ink may be toxic, do not use nearby edible plants - Not stable under windy conditions - Unattractive
Pine needles	- Adds acidity - Readily available - Leachates may be toxic to weed and young plants - Only for acidic soil-adapted plant species - Slow to break down
Sawdust	- Improves soil organic matter - Inexpensive or free - May mat or inhibit water penetration - May utilize available soil nitrogen with decomposition - Blows away and decomposes rapidly - Compost well before using or add nitrogen to soil - May contain organic matter that can harm young plants
Pressed heavy fibrous paper for mulching	- Good water and air penetration - Tend to break or tear after transplanting

Inorganic Mulches	Properties
Black plastic (polyethylene)	- Very effective - Easy to handle - Not permeable to air and water - Requires drip irrigation - Warms soil - Breaks down over time - Unattractive unless a top mulch is applied - Weeds can grow rapidly through tears and holes
Clear plastic	- Performs like black plastic - Encourages weed growth if no solarization is performed - Weeds can grow rapidly through tears and holes

Woven polypropylene and nonwoven polyester	- Very effective - Long lasting - Good air and water penetration - Expensive - Unattractive without a top mulch
Photodegradable plastic	- May not need to be removed - Degradation may not be complete - Must be exposed to light to degrade - If exposed to light, may not last as long as other materials
Crushed stone, gravel	- Attractive as a top mulch - May be weed infested if used alone - May get too hot - Expensive - May be time consuming to remove
Roofing/ Building paper	- Long lasting and durable - Unattractive unless a top mulch is applied - Expensive
Used carpet	- Long lasting - Unattractive unless a top mulch is applied - May produce an unpleasant odor when wet

Fire or Flame

Fire can be used to remove undesirable brush and weeds and it also reduces thatch and mat accumulation in turf. Additionally, controlled burning in the landscape results in control of some insects and diseases. However, the heat generated by burning will break seed dormancy of certain weeds. In most areas, open burning is regulated or restricted by governmental agencies. However, on a smaller scale, the use of flame burners has recently gained attention for weed management. Flame burners can be expensive and experience has shown that weed management is short-term. Frequent use of flame is required to maintain control. Specific concerns associated with the use of flame burners include uncontrolled fires that are generated by escaped sparks and possible human injuries resulting from exposure to burners. In our own experience in using flame in turf settings, turfgrass is damaged so extensively that flame cannot be considered for use in established lawns. They are, however, useful for treatment along sidewalks or driveways.

Solarization

Solarization is a simple, five-step process that kills insects, soil borne plant diseases, nematodes, harmful fungi, and weed seeds (Table 5). Soil that has been solarized allows plants to draw on the nutrients, especially nitrogen, calcium, and magnesium more readily, and seeds germinate more quickly. Plants grow faster and stronger, often maturing earlier with substantially higher yields than in unsolarized soil. The concept of solarization is based on the principle of a greenhouse, where a transparent covering, 3 or 6 mils plastic sheeting, traps the sun's heat. After several days of sunshine, soil temperatures rise to as high as 140 degrees at the surface and well over 100 degrees within the top 12 inches. Six weeks of sunny weather are necessary to pasteurize the soil. Most annual and certain perennial weeds are controlled by solarization for 6 months to one year. The susceptibility to soil solarization depends on the weed species; annual bluegrass and barnyardgrass (*Echinochloa crus-galli* (L.) Beauv.) are sensitive to solarization, whereas white sweetclover (*Melilotus alba* Medicus) is resistant (Dreistadt, 1994).

Table 5. Five-step process for solarization.

1. *Preparation of the soil:* Weeds or old crops are pulled. Stones or clumps that might raise the plastic and create air pockets that could cause uneven heating are removed.
2. *Moisten soil thoroughly:* For several hours soil should be thoroughly moistened or soaked to create a high humidity under the plastic, which acts as a vapor to kill unwanted pests.
3. *Dig a trench:* trench around the bed or plot 6-8 inches deep.
4. Lay a clear plastic sheet: use plastic 3-6 mils thick over the entire area, overlapping the trench on all sides.
5. *Wait*: Although cloudy weather will slow things down by cooling the soil, a few weeks of sunshine will improve the soil dramatically and inexpensively, and inexpensively. In areas with cool or cloudy summers, or to speed up the process a second sheet of plastic can be added. The second sheet of plastic over the ground-level sheet will be raised creating an airspace between which acts as a temperature buffer zone during cloudy weather. The combination of the two sheets of plastic serves to raise the soil temperature an additional 6 degrees.

BIOLOGICAL CONTROL

Biological methods are based on the use of natural antagonists of weeds as control agents. The intent of biological control is generally not to eradicate weed infestation, but to reduce the weed population below a level of economic or aesthetic threshold. Biological agents have not generally been considered for turf because they are often pathogenic to the turfgrass itself. A successful biological weed control agent should injure only the weeds species to be controlled, reproduce faster than the weed species themselves, and be adapted to the local environment.

Even though examples of biological control in the landscape are rare, the use of leaf spot pathogens *Bipolaris setariae* and *Piricularia grisea* to control goosegrass (Figliola et al., 1988) was moderately successful. Additionally, in Kentucky bluegrass, creeping bentgrass (*Agrostis stolonifera* L.), or annual bluegrass turfs, selected strains of *Sclerotinia sclerotiorum* were effective for control of dandelion (Riddle et al., 1991).

Bacteria are also candidates for biological control. *Xanthomonas campestris,* for instance, has been isolated and screened for the control of annual bluegrass (McCarty and Weinbrecht, 1992). *X. campestris* has shown considerable potential for controlling annual bluegrass without harming desirable turfgrass (Nishino and Tateno, 2000). Perennial ryegrass (*Lolium perenne* L.) can also be successfully controlled with *Xanthomonas* (Johnson, 1994). Dandelion has been the subject of biological research control with *Phoma herbarum* (Neumann and Boland, 1999; Teshler et al., 2000). In order to facilitate more complete control, a better understanding of the organism and its life cycle is required. Several studies in the landscape are underway to evaluate the efficacy of various rust (*Puccinea* spp.) organisms for yellow nutsedge and ground ivy (*Glechoma hederacea* L.) control. Several rust fungi have been reported to infect ground ivy in its native Eurasian range. In 1998, one of these rusts, *Puccinia glechomatis*, was found in North America on ground ivy in the mid Central U.S. and also in N.Y. State. Research has demonstrated that this rust fungus infects only plant species within the genus *Glechoma* (Boellmann and Schoeller, 2001; Schoeller, 2000). Recent studies in N.Y. show that this organism has potential to control dense infestations of ground ivy in a landscape setting (DiTommaso et al., 2002).

At McGill University in 1997, a fungal isolate was discovered that was highly effective and provided environmentally safe control for several broadleaf weeds, including dandelions. This isolate, MAC1, is a well-characterized plant pathogenic fungus with a narrow agronomic range, which can be cultured in liquid or solid fermentation. A single application of a granular formulation of MAC1 to dandelions in turf results in very rapid control with no injury to the grass. Lesions develop quickly and the plant can be killed within 7 days when conditions are suitable. Overall activity is often better than 2,4-D based herbicides and the fungus controls dandelion twice as rapidly as the herbicide.

The product does not perform well in hot and dry weather, but efficacy under these conditions can be improved by following some simple recommendations. Other broadleaf weeds besides dandelions can also be controlled, and the product is compatible with normal lawn maintenance practices including mowing, fertilization, and irrigation (Watson and Dupont, 1999). Owners of this technology are seeking partners to bring this product to market. Since relatively few examples of biocontrol in the landscape have actually been developed or even studied, this area offers ample opportunity for future product development. As homeowners and concerned applicators seek alternatives to herbicides, biocontrol products and other strategies are receiving increased attention.

Examples where insects have been used for pasture or landscape weed management also exist. Thistles in the genus *Carduus* and *Cirsium* rank as some of the most common and most troublesome weeds in fence rows, borders, pastures and haycrops across the Eastern U.S. Introducing natural enemies can aid in regulating the reproduction and spread of such thistles, especially the bull thistle. One such natural enemy is *Trichosirocalus horridus* (Panzer), the rosette weevil. This weevil feeds on the crown or center of the rosette. Native to Europe, this weevil was extensively studied by the USDA to insure that it would not damage other desirable plants in the U.S. Biological control of bull thistle is a good option since (1) the control agent is non-toxic, (2) it is non-polluting, (3) it can reduce the threat of groundwater contamination or off-site movement from herbicides, (4) it is self-dispersing, and (5) it is sustainable. Little additional effort is required once the weevils are established, while other control methods must be applied periodically to achieve long-term control (McDonald and Robbins, 1994).

OTHER ALTERNATIVE STRATEGIES

To address the goal of reducing herbicide applications and environmental impact over the long term, recent studies in managed turf settings and along roadsides have investigated the use of alternatives to herbicides for vegetation and weed management. The New York State Department of Transportation (NYSDOT), among others, has evaluated the use of mulch mats made from recycled tires, weed fabric mulches, and of the use of wildflower mixes to provide low-growing sustainable ground covers. Many of these projects included demonstrations designed along roadsides by DOT agencies. After several years of evaluation, the DOT found that none of these strategies was particularly effective for weed suppression, and most options were very expensive, thereby limiting their utility along roadsides and rights of ways (Weston, 2002a).

Organic Products

Organic herbicides including clove oil derivatives, corn gluten meal-based products, and acetic acid mixtures were evaluated for turf and vegetation control in the landscape by both the NYSDOT and other researchers (Table 6). A study was designed to observe and document on the long-term control of turfgrass species and lawn weed species by two new acetic acid type herbicides, one herbicidal soap (containing pelargonic acid), three commercially-available vinegars, and a traditional herbicide (RoundUp or glyphosate). The weed population included quackgrass (*Elytrigia repens* (L.) Nevski), large crabgrass, ground ivy, dandelion, violet (*Viola* spp.), plantain and Kentucky bluegrass.

Acetic acid was the most effective of the products evaluated. A 5% or greater concentration provided only short-term control of most perennial weeds. However, plantain and crabgrass were moderately controlled at the 5% rate, with repeated applications and higher concentrations providing greater overall control (Chinery and Weston, 2002). Acetic acid was somewhat more effective than pelargonic acid, but it was also more costly. This study was conducted during droughty conditions in N. Y. and is worth repeating under more typical summer conditions. Plant membranes are typically destroyed in a contact manner by application of these materials. Regrowth often occurs and repeat applications are needed.

Clove oil concentrates exhibited considerable postemergent activity, but they are quite costly in comparison to glyphosate. Glyphosate remains the more economic and effective product, but pesticide applicators may opt to use acetic acid despite higher costs, especially if local legislation encourages the use of non-traditional pesticides. Further studies are underway to evaluate the performance of other organic products, and hopefully to reduce their cost of application (Chinery and Weston, 2002).

Corn gluten meal was developed in the mid-1980s as a preemergent herbicide product for use in turf and home garden settings (Christians, 1983). Extracted from corn meal, corn gluten products contain high levels of organic nitrogen and were observed to inhibit root formation of germinating seeds. Weed suppression is thought to be due to these high levels of nitrogen which result in stimulation of turf growth and may suppress weed growth. The meal is also reported to contain weed suppressive allelochemicals. Overall, Cornell University studies suggested that turf growth stimulation by provision of nitrogen through corn gluten meal application suppressed weed growth only moderately over a 3 year period, and suppression of large crabgrass (15-20%) was similar to that when inorganic nitrogen was applied to stimulate turf growth (Senesac, 2000).

Table 6. Organic products currently available for weed management in turf.

Product	Activity	Weed Controlled	Effectiveness	Mode of Action
Corn Gluten Meal	Pre emergent	Small seeded annuals	Not particularly effective	Presence of nitrogen stimulates turf growth
Cramby Meal	Pre emergent	Small seeded annuals	100% control but not easy to find	Natural products with inhibitory activity are present, inhibit seed germination
Acetic Acid	Post emergent	Small seeded annuals	Up to 75% but need high rate	Destroys cuticle and plant membranes
Clove Oil	Post emergent	Annuals and perennials	30%	Unknown
Pelargonic Acid	Post emergent	Annuals and perennials	10%	Disrupts normal membrane permeability and cellular physiology leading to cell leakage of contacted tissue

Alternative or Weed Suppressive Plant Materials

Alternative plant materials that provide long-term weed suppression may be useful in the landscape. Groundcovers that are easily maintained, stress tolerant, and have dense low-growing foliage that prevents light penetration at the soil surface could provide useful suppression of weeds in the landscape. These groundcovers effectively compete with weed seedlings for space, light, fertility, and water. Certain groundcovers and turfgrass species also release natural herbicides or allelochemicals from their foliage or from their living roots.

In the past, weed suppressive groundcovers have been effective for annual weed control in orchards, vineyards, nurseries, and vegetable fields. Many of these groundcovers also exhibit resistance to insects, diseases, and other predators. The allelochemicals released by groundcovers can be chemically diverse, and may have an important impact on plant-plant interactions and plant adaptation to the local environment (Figure 2) (Weston, 2002a). For example, *Pachysandra* groundcovers produce a group of saponins that are potent plant growth and germination inhibitors. The common bearberry, *Arctostaphylos*, releases several complex terpenoids with inhibitory activity towards weeds and insects (Brielmann, 1999). Certain *Sedum spp.* appear to exhibit allelopathic suppression of weeds which is not based on density of biomass but rather, secondary product production (Weston, 2002b). In addition to their aesthetically appealing aspects and their production of allelochemicals, many groundcovers are effective competitors with weeds due to dense biomass formation. They can also reduce soil erosion on sloped sites. The formation of dense canopy quickly is important in suppression of annual and perennial landscape weeds. Our laboratory is currently evaluating the long-term establishment, aesthetic appeal, maintenance requirements and weed suppression provided by numerous newly developed ornamental groundcovers species as well as industry standards. We are seeking plant materials that are dense, low growing, easy to establish and maintain, and are also aesthetically appealing.

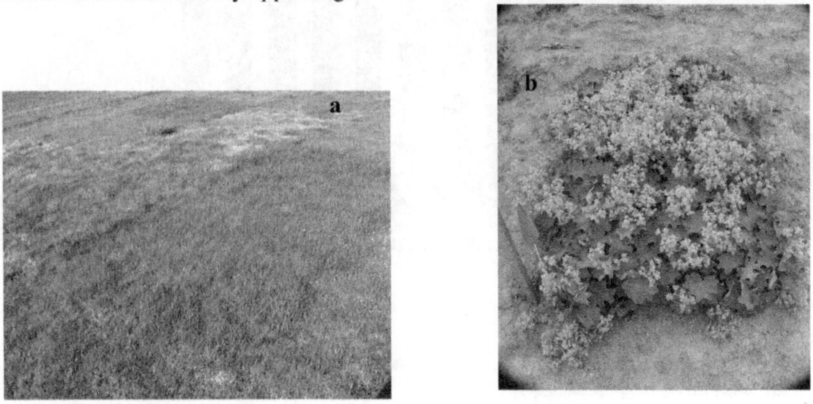

Figure 2. Weed suppressive fine fescue cv. Intrigue (a) and groundcover species Lady's Mantle (*Alchemilla mollis* L. 'Auslese').

Grasses such as perennial ryegrass, buffalograss (*Buchloe dactyloides*), and the tall and fine leaf fescues (*Festuca arundinacea* and *Festuca* spp.) also are known to be weed suppressive. Many are thought to be allelopathic as well (Peters and Luu, 1985; Peters, 1968). In past studies with fescue species at the Turfgrass Research Center in Ithaca, N.Y. and in Riverhead, LI, we have seen that weed suppressive characteristics exhibited by fine fescues are cultivar dependent (Figure 1a, b). Certain cultivars are consistently more suppressive than others, in several years of repeated field evaluation.

Weed suppression is associated with a dense growth habit and also the production of root exudates from the living roots of fine leaf fescue seedlings (Bertin et al., 2002). Fine fescue root exudates contain numerous compounds, which are potent inhibitors of root growth and seed germination. Structural elucidation of these inhibitors is now underway. The purified root exudates cause growth inhibition of roots and shoots of developing weeds which exhibit bleaching, chlorosis, and eventual necrosis (Bertin et al., 2002). At relatively low concentrations, seed germination is completely inhibited in certain sensitive species such as large crabgrass.

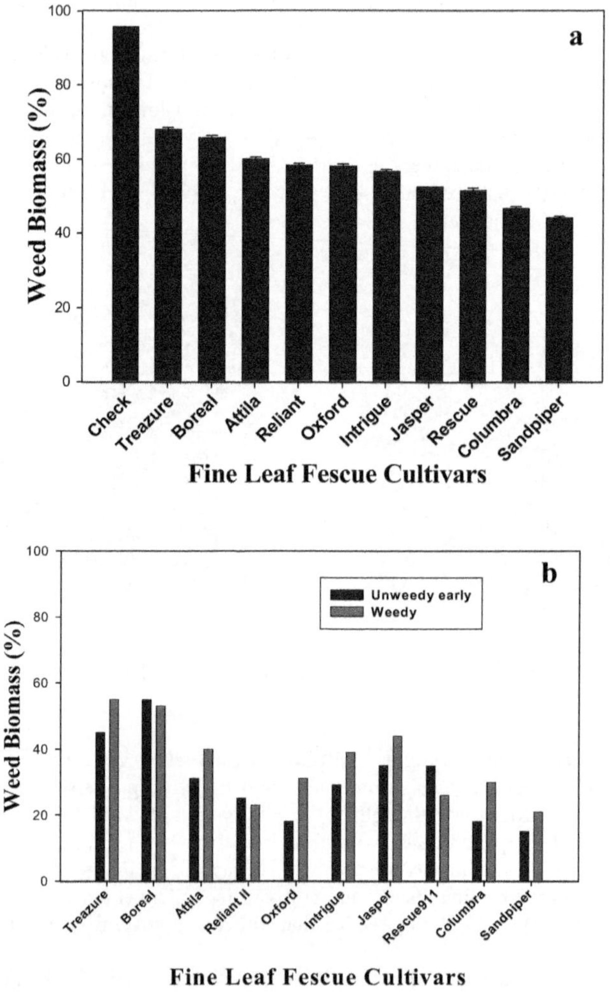

Figure 1. Evaluation of the weed biomass (%) after one year of establishment in fine leaf fescue cultivar trials in Ithaca, NY (a) and Riverhead, LI (b). Reliant, Sandpiper, Oxford, and Intrigue cultivars proved to be consistently suppressive in both locations, whereas Treazure and Boreal were not particularly suppressive in either location.

Very little effort has been expended towards the development of weed suppressive turf and ornamental crops through selection and breeding. Studies have shown that certain cultivars and species of both turf and ornamentals have great potential to be used for long-term weed suppression in the landscape. Use of appropriate cultivars is a highly effective means to suppress weeds over time in an environmentally sound and aesthetically appealing manner. Further studies are now underway with over 120 unique ornamental and turfgrass species under evaluation.

SUMMARY

Successful turf and landscape weed management is based on an understanding of the ecological environment that surrounds turf or landscape settings. Weed management before planting is essential, especially in landscape plantings, to prevent future weed infestation. Herbicides are still commonly used for weed management in turf and ornamentals, as they are cost effective and labor saving. However, turf and landscape managers are seeking alternatives for weed control including use of appropriate cultural practices such as mowing and mulching, biological control with pathogenic organisms, and use of allelopathic grass and groundcovers for enhanced suppression and aesthetical appeal.

RFERENCES

Adams W.A. "Effects of nitrogen and cutting height on the shoot growth, nutrient removal and turfgrass composition of an initially perennial ryegrass dominant sports turf." In *Proceedings of the 2nd International Turfgrass Research Conference*, J. B. Beard, ed., Munich, Germany. 1980. p. 343-350,

Beard J.B. *Turfgrass: Science and Culture,* Prentice Hall, Englewood Cliffs, NJ. 1973.

Bertin C., Rossi F., Weston L.A. Further evaluation of the allelopathic potential of fine leaf fescue. Proceedings of Northeastern Weed Science Society; 2002; 116.

Boellmann J., Schoeller M. Preliminary studies on the spread of the rust fungus *Puccinia glechomatis* in North America. Phytopathology 2001; 91:104.

Brielmann H.L. "Phytochemicals: the chemical components of plants". In *Natural Products from Plants.* P.B. Faufman et al., eds. Washington, D.C.: CRC Press, 1999; pp. 1-36.

Chinery D., Weston L.A. How well does organic weed control work?. Cornell University Turfgrass Times. 2002; 13:12-15.

Christians N.E. The use of corn gluten meal as a natural preemergence weed control in turf. Int. Turfgrass Soc Res J 1983; 7:284-290.

Derr J.E., Neal J.C., Kuhns L.J., Smeda R.J., Weston L.A., Elmore C., Wilen C.A., Ahrens J., Senesac A., Mervosh T. (eds.). Weed Management in Landscape and Nursury Plantings, ASHS Press. 1997; pp. 1-139.

Dreistadt A.H. *Pests of Landscape Trees and Shrubs: An Integrated Pest Management Guide.* ANR Publications, University of California, Oakland, CA. 1994.

Emmons R.D. *Turfgrass: Science and Management.* Delmar Publications, Albany, NY. 1984.

Faustini A., Settimi L., Pacifici R., Fano V., Zuccaro P., Forastiere F. Immunological changes among farmers exposed to phenoxy herbicides: preliminary observations. Occup Environ Med 1996; 53:583-585.

Figliola S.S., Camper N.D., Ridings W.H. Potential biological control agents for goosegrass (*Eleusine indica*). Weed Sci 1988; 36:830-835.

Ibrahim M.A., Bond G.C., Burke C.P., Cole P., Dost F.N., Enterline P.E., Gough M., Greenberg R.S., Halperin W.E., McConnell E., Munro I.C., Swenberg J.A., Zahm S.H., Graham J.D. Weight of the evidence on the human carcinogenicity of 2,4-D. Environ Health Perspect 1991; 96:213-222.

Johnson B.J. Oxadiazon treatments on overseeded putting-green turf. Weed Sci 1982; 30.

Johnson B.J. Glyphosate and SC-0224 for bermudagrass (*Cynodon* spp.) cultivar control. Weed Technol 1988; 2:20-23.

Johnson B.J. Biological control of annual bluegrass with *Xanthomonas campestris* pv. *poannua* in bermudagrass. HortScience 1994; 29:659-662.

Johnson B.J., Bower H.G. Management of herbicide and fertility levels on weeds and Kentucky bluegrass turf. Agron J 1982; 74:845-850.

Levy E.B., Madden E.A. Weeds in lawns and greens-competition effects and control by treatment with chemical sprays. New Zealand J Agric 1931; 42:406-421.

McDonald R.C., Robbins A.O. Biological control of bull thistle using a beneficial weevil [Online]. North Carolina Department of Agriculture and Consumer Sciences. http://www.ncagr.com/plantind/plant/ bioctrl /biopub/bullthis.htm (posted Nov 1994).

Mc Niel R.E., Weston L.A. Landscape fabrics as weed blockers: differences in weed cover suppression. Proc SNA 1992; 37:301-304.

McCarthy L.B., Murphy T.R. "Control of turfgrass weeds." In *Turf Weeds and their Control.* A. J. Turgeon, ed. ASA-CSSA, Madison. 1994; pp. 209-248.

McCarty L.B., Weinbrecht J.S. "Annual bluegrass control with *Xanthomomas campestris".* In *Turfgrass Research in Florida: A Technical Report.* T. E. Freeman, ed. Inst. of Food and Agric. Sci., Univ of Florida, Gainesville. 1992; pp. 327-328.

Mehnert C., Madel F. The influence of reduced nitrogen application on plant population, grass clipping yield and mineral deficiencies in ordinary turf. Rasen Turf Gazen 1982; 13:28-33.

Monaco, T.J.. "Lawn, turf, and ornamentals." In *Weed Science: Principles and Practices.* F.M. Ashton, ed. Wiley, New York. 2002. pp. 500-521.

Murray J.J., Klingman D.L, Nash R.G., Woolson E.A. Eight years of herbicide and nitrogen fertilizer treatments on Kentucky bluegrass (*Poa pratensis*) turf. Weed Sci 1983; 31:825-831.

Neal P.H. Plan before you plant. A 5 step process of developing a landscape weed management plan. Weed Facts 5. Cornell Cooperative Extension. 1994.

Neumann S., Boland G.J. Influence of selected adjuvants on disease severity by *Phoma herbarum* on dandelion (*Taraxacum officinale*). Weed Technol 1999; 13:675-679.

Nishino J., Tateno A. Camperico(R): a new bioherbicide for annual bluegrass in turf. Agrochem Japan 2000; 13-14.

Peters E., Luu K.T. "Allelopathy in tall fescue." In *The Chemistry of Allelopathy.* A.C. Thompson, ed. Washington, DC: American Chemical Society. 1985; pp. 273-283.

Peters J.E. Toxicity of tall fescue to rape and birdsfoot trefoil seeds and seedlings. Crop Sci 1968; 8:650-653.

Riddle G.E., Burpee L.E., Boland G.J. Virulence of *Sclerotinia sclerotiorum* and *S. minor* on dandelion (*Taraxacum officinale*). Weed Sci 1991; 39:109-118.

Roberts E.C., Markland F.E., Pellet H.M. Effects of bluegrass stand and watering regime on control of crabgrass with preemergence herbicides. Weeds 1966; 14:157-161.

Schoeller M. Rust on ground-ivy found for the first time in North America. Plant Dis Rep 2000; 84:371.

Senesac, A. Personnal Communication. 2000.

Teshler M., Teshler I., Stewart W.S., Green S., Watson A., DiTommaso A., Boland G., Sampson G. Development of a mycoherbicide for control of dandelion (*Taraxacum officinale*) in turf. Isreal J Plant Prot Sci 2000; 28:179.

Turgeon A.J. *Turfgrass Management.* Prentice Hall ed. Charles Stewart, Upper Saddle River, N.J., 1996.

Vencill W.K. (ed.) *Herbicide Handbook.* Lawrence, KS: Weed Science Society of America, 2002; pp. 1-493.

Watson A., Dupont S. Mycoherbicide products for broadleaf weed control in turf. J Pest Manage News 1999; 10(4):6.

Weston L.A. Weed suppressive ground covers - a more attractive way to manage weeds. Cornell University Turfgrass Times (CUTT); 2002a.; 12:4-7.

Weston L.A., Lamboy J., Rossi F. Weeds and Your Lawn. IPM Brochure. Cornell Univeristy, Ithaca, N.Y., 2001.

CHAPTER 21

BIOLOGICAL CONTROL OF ROOT PARASITIC WEEDS WITH PLANT PATHOGENS

JÜRGEN KROSCHEL & DORETTE MÜLLER-STÖVER

Institute of Plant Production and Agroecology in the Tropics and Subtropics, University of Hohenheim (380), D-70593 Stuttgart, Germany

Root parasitic weeds are nutrition specialists. Using special organs, the haustoria, they penetrate into the vessels of roots of host plants in order to supply themselves with water and nutrients. They can be either facultative or obligate, obligate parasitic weeds being considered as one of the most serious constraints to food production in many parts of the world. Obligate parasites are either hemi-parasites with chlorophyll, like species of the genus *Striga* (Scrophulariaceae), witchweeds, or holo-parasites without chlorophyll such as species of the genus *Orobanche* (Orobanchaceae), broomrapes. Forty *Striga* species are reported world-wide: 33 in Africa, 7 in Asia and 4 endemic in Australia. Eleven species are parasites on agricultural crops. The genus *Orobanche* has more than 100 species but only 7 are considered as economically significant parasitic weeds (Parker and Riches, 1993; Raynal-Roques, 1996).

ECONOMIC IMPORTANT *STRIGA* AND *OROBANCHE* SPECIES AND THEIR MAIN HOST PLANTS

Striga spp. hamper cereal and legume production in Sub-Saharan Africa. *Striga hermonthica* (Del.) Benth. and *S. asiatica* (L.) Kuntze are specific to grasses and cause severe production losses in sorghum (*Sorghum bicolor* (L.) Moench), millet (*Pennisetum americanum* L.), maize (*Zea mays* L.) and rice (*Oryza sativa* L.) while *S. gesnerioides* (Willd.) Vatke prefers dicot hosts and affects mainly cowpea (*Vigna unguiculata* (L.) Walp.). *Orobanche* spp. occur primarily in the Mediterranean region, West Asia and North Africa. *Orobanche crenata* Forsk., *O. ramosa* L., *O. aegyptiaca* Pers., *O. cernua* Loeffl., *O. cumana* Wallr., *O. minor* Sm. and *O. foetida* Poir. are major biotic limiting factors to the production of legumes like faba bean (*Vicia faba* L.), chickpea (*Cicer arietinum* L.), lentil (*Lens culinaris* Medick.), and to crops of the family Solanaceae (tomato, *Lycopersicon esculentum* Mill.; potato, (*Solanum tuberosum* L.; tobacco, *Nicotiana tabacum* L.) and Asteraceae (such as sunflower, *Helianthus annuus* L.) (Parker and Riches, 1993).

Inderjit (ed.), Weed Biology and Management, 423-438.

GROWTH AND DEVELOPMENT

After shedding, seeds of *Striga* and *Orobanche* are dormant and require a period of after-ripening before they will respond to a germination stimulant, which is released by roots of hosts and some non-host species (Figure 1).

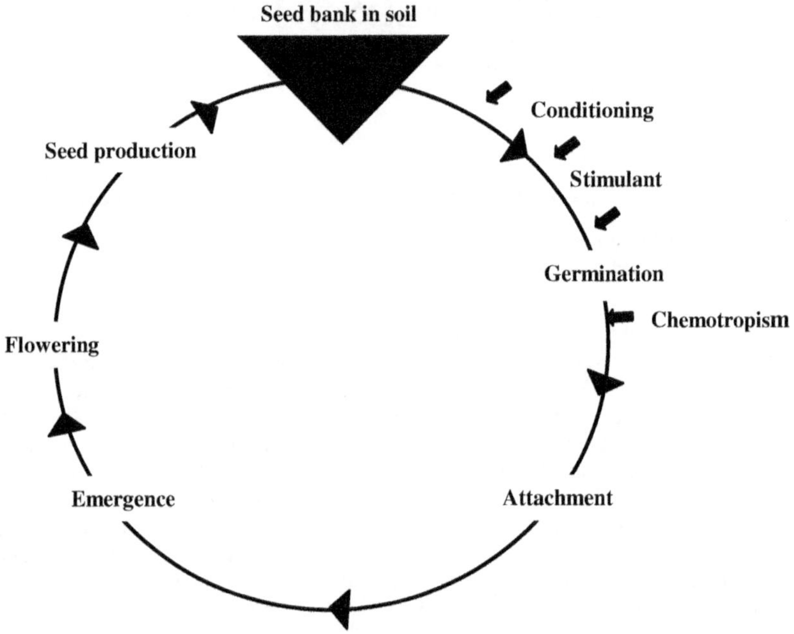

Figure 1. The life cycle of *Striga* and *Orobanche* species

In the dry state, seeds can remain dormant and viable for many years. A water-conditioning period of several days is needed for seeds to germinate. The necessity for seeds to receive a stimulant, which is released by the root tip of host plants, before they can germinate ensures that a host is available and near enough to be reached by a germ tube, which elongates and grows in direction of the origin of the signal. Most of the seeds in the soil will not be reached by the stimulant, but will remain viable until the following cropping periods. After attachment to the host root, the parasite starts to penetrate the host root epidermis and to develop the haustorium, which supplies the parasite with water, mineral nutrients and organic materials from its host. The penetration of haustorial cells into host tissue (xylem and/or phloem system) is achieved both mechanically, by pressure on the host endodermal cells, and by hydrolytic enzymes. After several weeks of underground development the parasite emerges above the soil surface and flowers and produces seeds. Numerous seeds (up to 100,000 or even more) are produced by a single plant and re-infest the field. Thus, if host plants are grown often, the seed population in the soil increases tremendously and cropping of host plants becomes more and more un-economical (Kroschel, 2001).

YIELD LOSSES AND CONTROL

Nowadays, the actual *Striga* infested area is estimated at 21 million hectares, while 16 million hectares of arable land are endangered by *Orobanche* (Sauerborn, 1991). Parasites of both genera cause losses of up to 100% on farmers' fields, which then often have to be abandoned due to their unproductivity. The management of parasitic weeds is still unsatisfactory as, with the exception of the use of glyphosate in faba bean to control *O. crenata,* present methods are not effective at controlling the underground development stages of the parasites. Presently, successful restoration of infested fields can only be achieved by the improvement of existing farming systems based on sound analysis of the parasitic weed problem and the application of an integrated control approach. Biological control using insects and fungal antagonists could be an important component of such an approach.

ANTAGONISTS OF PARASITIC WEEDS

Insects associated with Striga *and* Orobanche *species*

The use of insects in the control of *Striga* and *Orobanche* has been reviewed by Kroschel et al. (1999) and Klein and Kroschel (2002). Although many phytophagous insects have been collected on *Striga* and *Orobanche* species, most are polyphagous and the target weed species often are not their principle host plants. Oligo- and monophagous herbivores with potential in biocontrol of *Striga* spp. are larvae of the butterfly *Eulocastra argentisparsa* Hampson (Lepidoptera: Noctuiidae) in India; *Smicronyx* spp. (Coleoptera: Curculionidae) in India and Africa; and the shoot miner *Ophiomyia strigalis* Spencer (Diptera: Agromyzidae) in East Africa (Kroschel et al., 1999; Traoré et al., 1999). The fly *Phytomyza orobanchia* Kalt. (Diptera: Agromyzidae) and *Smicronyx cyaneus* Gyll. are of great interest in biocontrol of *Orobanche* spp. As a single method, biological control with herbivores will hardly be successful in tackling the parasitic weed problem. But in combination with other methods in an integrated control package, herbivores could play a role in lowering the parasitic weed population and reducing its reproductive capacity and spread. Most promising are inundative mass releases of *P. orobanchia*, which can reduce the *Orobanche* seed production by 95% (Klein et al., 1999). However, also classical biological control might be an option by introducing *P. orobanchia* into countries where *Orobanche* spp. are not endemic and have been unintentionally introduced, such as Chile (Norambuena et al., 1999, 2001).

Microorganisms Associated with Striga *and* Orobanche *Species*

The occurrence of pathogens on *Striga* species has been reported by several authors. Wager (1931) obtained *Pythium ultimum* in culture from *Striga* seeds in South Africa. Nag Raj (1966) reported the occurrence of *Alternaria* sp., *Cercospora* sp., and *Phoma* sp. on *S. asiatica* and *S. densiflora* (Benth.) Benth. in India. Meister and Eplee (1971) isolated *Curvularia geniculata, Fusarium roseum, F. solani, Rhizoctonia solani,* and

Sclerotium rolfsii on *S. asiatica* in North Carolina. Zummo (1977) reported three diseases of *S. hermonthica* in West Africa, including a leaf spot, caused by an undescribed species of *Cercospora*, a vascular wilt caused by *F. equiseti*, and a stem lesion caused by *Phoma* sp. However, none of these fungi was evaluated for its potential use as a biocontrol agent.

In recent years, in-depth surveys have been conducted to identify pathogens attacking *Striga* spp. in West Africa (Abbasher et al., 1995; Ciotola et al., 1995; Kroschel et al., 1996; Abbasher et al., 1999). *Fusarium* spp., soil-borne pathogens, were the most commonly isolated fungi from infected *S. hermonthica* (Table 1) (Abbasher et al., 1995). Among these, *F. oxysporum* was the predominant species (Abbasher et al.,1998).

Table 1. Fungi isolated from infected *S. hermonthica* plants in northern Ghana, their incidence and type of damage.

Fungi	Incidence	Type of damage
Alternaria alternata (Fr.) Kreissler	+	foliar
Bipolaris specifera Nelson	+	foliar
Cladosporium oxysporum Berk. and M.A. Curtis	+	foliar
Curvularia fallax Boed	+	foliar
Fusarium equiseti (Corda) Sacc.	+++	soil-borne
Fusarium equiseti var. *bullatum (*Sherb.) Wollenw.	+++	soil-borne
Fusarium oxysporum Schlecht	++	soil-borne
Fusarium proliferatum (Matsushima) Nirenberg	+	soil-borne
Fusarium solani (Mart.) Sacc.	++	soil-borne
Macrophomina phaseolina(Tassi) Goidan	++	soil-borne
Nodulisporium gregarium (Berk. and M.A. Curtis) J.A. Meyer	+	foliar
Phoma sorghina (Sacc.) Boer.,Dorenb. and Van Kesteren	++	foliar
Sclerotium rolfsii Sacc.	+	soil-borne

*+ = low, ++ = moderate, +++ = high (Abbasher et al., 1995)

Investigations on the use of fungal pathogens in biocontrol of *Orobanche* spp. started about 50 years ago in the former Soviet Union (Panchenko, 1974). Target species were mainly *O. aegyptiaca, O. cernua* and *O. ramosa;* and *Fusarium* spp. was the most frequently isolated pathogenic fungal species (Table 2). In the 1990s, detailed studies were carried out to identify potential antagonists of *O. crenata.* However, no pathogenic strains of *Fusarium* spp. could be isolated. Instead, *Ulocladium atrum* and *U. botrytis* were found (Linke et al., 1992; Müller-Stöver, 2001).

Limited work has been done with bacteria. *Azospirillum brasilense,* isolated from a sorghum field in Mali, significantly inhibited *in vitro* germination of *S. hermonthica* seed (Bouillant et al., 1996). Oswald et al. (2000) found significantly reduced germination of *Striga* seeds after the inoculation of maize seedlings (Kenyan land race "Local White") with plant growth promoting rhizobacteria (PGPR). In contrast, this effect was not observed with the commercial maize hybrid (Pioneer 3152), suggesting specificity of Kenyan PGPR-isolates to the local maize variety. A laboratory test was developed to screen ethylene-producing bacteria like *Pseudomonas syringae* pv. *glycinea* for ability to promote suicidal germination of *Striga* spp. seeds (Berner et al., 1999). Bacterial isolates that are potentially pathogenic to an *Orobanche* sp. that parasitizes tobacco plants have been recently isolated by Barghouthi et al. (2000).

Table 2. Fungi associated with different *Orobanche* species

Fungus	*Orobanche* species	Country	Reference
Colletotrichum lageniarum	*O. aegyptiaca*	Former Soviet Union	Stankevich (1971)
Fusarium oxysporum var. *orobanches*	*O. aegyptiaca*	Former Soviet Union	Kott (1971); Kabulov and Khalimov (1977); Wang et al. (1985)
F. oxysporum var. *orthoceras*	*O. aegyptiaca*	Former Soviet Union	Panchenko (1974)
F. arthrosporioides *F. oxysporum*	*O. aegyptiaca, O. cernua, O. ramosa*	Israel	Amsellem et al. (1999)
F. oxysporum	*O. cernua*	Iran	Mazaheri et al. (1991)
Sclerotium rolfsii	*O. cernua*	India	Raju et al. (1995);
Sclerotinia sclerotiorum	*O. cernua*	India	Narasimhan and Thirumalachar (1954)
F. oxysporum f. sp. *orthoceras*	*O. cumana*	Bulgaria	Bedi and Donchev (1991, 1995)
Ulocladium atrum	*O. crenata*	Syria	Linke et al. (1992)
Ulocladium botrytis	*O. crenata*	Syria, Egypt	Linke et al. (1992); Müller-Stöver (2001)
F. oxysporum f. sp. *orobanches*	*O. ramosa, O. mutellii*	Former Soviet Union	Timchenko and Dovgal (1973)
F. equiseti *F. solani*	*O. ramosa*	Hungary	Hodosy (1981)
F. lateritium	*O. ramosa, O. mutellii*	Bulgaria	Bozoukov and Kouzmanova (1994)
Rhizoctonia solani	*O. ramosa*	USA	Duafala et al. (1975, 1976)

BIOLOGICAL CONTROL WITH PLANT PATHOGENS

Biological Control Strategies

Natural enemies can be important factors in limiting the number and distribution of most living organisms. The basis of biological control is the calculated use of natural enemies to suppress the growth or reduce the population of a target pest species.

Two approaches of biological control of weeds are: the classical or inoculative approach, which consists of the release of host-specific natural enemies for a durable establishment into areas in which they did not previously occur. This approach could be suited to the use of insects for biocontrol of *Striga* spp. (Kroschel et al., 1999). The aim of the inundative or mycoherbicidal approach is to periodically overwhelm the target weed with large numbers of mass-reared endemic or exotic natural enemies that are non-persistent in the ecosystem. It is well suited to phytopathogenic fungi and applicable when biocontrol agents can be mass-produced under controlled conditions. Under favorable climatic conditions, application of massive doses of inoculum to a weed population will create a rapid and severe epidemic. Mycoherbicides can control weeds in annual crops on a comparable or even better level than chemical herbicides since they can be highly specific. Examples for mycoherbicides for commercial use are DeVine (*Phytophthora palmivora*) against milkweed vine (*Morrenia odorata* Lindl.) in citrus orchards and Collego (*Colletotrichum gloeosporioides* f.sp. *aeschynomene*) against northern jointvetch (*Aeschynomene virginica* (L.) B.S.P.), a leguminous weed in rice and soybean, which have been introduced in the USA in the early 1980s (Kenney and DeVine, 1986; Bowers, 1986). However, mycoherbicides have remained largely experimental due to numerous constraints to their production, application and commercialization which have been extensively reviewed by Auld and Morin (1995).

Pathogenicity of Fungi to Parasitic Weeds

Pathogenicity of the potential biocontrol agents has been tested in various ways (Müller-Stöver et al., 2001). Usually, fungal inoculum was applied in pot experiments pre-planting into the soil in different concentrations and was observed to reduce the emergence of parasite shoots (Abbasher and Sauerborn, 1992; Ciotola et al., 1995; Kroschel et al., 1996) and/or to cause disease symptoms on emerged and underground shoots (Bedi and Donchev, 1991) of various parasitic weeds tested. In some cases, the ability of the fungal antagonist to attack the target weed post-emergence was also investigated. Thomas et al. (1998) reported that 99% of *O. cumana* shoots sprayed with a suspension of *Fusarium oxysporum* f. sp. *orthoceras* conidia were killed within two weeks after inoculation. Abbasher and Sauerborn (1992) tested the pathogenicity of *F. nygamai* to emerged *Striga* plants by placing 2 g of solid fungal inoculum around the base of the emerged shoots which killed all plants within 28 days.

The impact of the fungal antagonists on the underground stages of parasitic weeds was studied in Plexiglass root chambers or polyethylene bags (Linke et al., 2001). One approach was to cultivate the host plants until underground parasite stages were formed and these were then inoculated with fungal propagules. Browning of *Orobanche* tubercle tissue could be observed as a result of the inoculation (Thomas et al., 1998; Amsellem et al., 2001) In a second approach, conidia suspensions were applied in different concentrations together with the parasite seeds onto the filter paper strip, which resulted in an inhibition of *Striga* or *Orobanche* seed germination, necrosis of the germ tubes and a decrease of the number of attachments on host roots (Abbasher and Sauerborn, 1992; Ciotola et al., 1995; Thomas et al., 1998 a, b; Elzein et al., 2000).

In vitro experiments in Petri dishes confirmed the impact of the fungal antagonists on the germination of the parasites. Ciotola et al. (1996) showed that *S. hermonthica* germination was reduced by 71% after the seeds had been exposed to a suspension of *Fusarium oxysporum* isolate *M12-4A* conidia during the preconditioning phase. Similarly, Abbasher et al. (1996) and Kroschel et al. (1996) reported the inhibition of the germination of different *Striga* species by various *Fusarium* strains. Sauerborn et al. (1996) and Thomas et al. (1999) further investigated the infection of seeds using cytological methods and found that mycelium of biocontrol fungi penetrated and colonized dormant seeds of both *Striga* and *Orobanche*. The fungal attack on the early developmental stages is of special interest with respect to the management of parasitic weeds, as most of the damage to the crop plants occurs before the parasite shoots emerge.

The fungal isolates currently under investigation for the biocontrol of *Orobanche* and *Striga* have been found to have different host ranges (Table 3). *Fusarium oxysporum* f. sp. *orthoceras* exclusively attacks *O. cumana*, the very closely related *O. cernua* and some susceptible biotypes of *O. aegyptiaca* (Thomas et al., 1998). Müller-Stöver (2001) observed a reduced germination rate of *O. crenata* seeds caused by *Fusarium oxysporum* f. sp. *orthoceras,* although no pathogenicity was observed towards later developmental stages of the parasitic weeds. FOXY and FARTH, isolated in Israel, attack *O. aegyptiaca, O. cernua* and *O. ramosa*, but are avirulent against *O. cumana* (Amsellem et al., 1999). *Ulocladium botrytis* is pathogenic to *O. crenata* and even more virulent to *O. cumana*, but does not attack *O. aegyptiaca* (Müller-Stöver, 2001).

Recent observations showed that Foxy2 significantly reduced the emergence of *S. hermonthica* and *S. asiatica,* although disease symptoms could only be observed on *S. hermonthica.* Furthermore, Foxy2 is non-pathogenic to *S. gesnerioides,* a result that clearly indicates its host specificity within the genus *Striga* (Elzein, 2003).

Pathogenicity Towards Non-Target Plants and Safety Aspects

Neither *Ulocladium atrum* and *U. botrytis* nor *O. crenata* were pathogenic on any of the tested crop plants. However, the results observed with *U. botrytis* were not always consistent (Linke et al., 1992; Müller-Stöver, 2001). *Fusarium* species are generally specific at the family or generic level (Boyette et al., 1984). Accordingly, none of the *Fusarium* isolates pathogenic to either *Striga* or *Orobanche* was found to attack any of the crop plants tested. Ciotola et al. (1995) showed that the sorghum, millet, maize, cowpea, cotton (*Gossypium* spp.) or okra (*Abelmoschus esculentus* (L.) Moench) cultivars tested were not affected by *F. oxysporum* M12-4A. Elzein and Kroschel (2003) tested 25 crop plants regarding their susceptibility to *Foxy2,* none of which showed any signs of infection. Similarly, the strains of *Fusarium* pathogenic to

Orobanche did not attack any of tested crop plants (Bedi, 1994; Amsellem et al., 2001; Müller-Stöver, unpublished data). As soil-borne fungi, *Fusarium* spp. are distributed worldwide and are known not only as plant pathogens, but because of their ability to produce mycotoxins, they can also affect humans and animals (Nelson et al., 1992). Important mycotoxins produced are moniliformin, zerealenone, trichotecene derivates and fumonisin B_1 which possess a broad range of biological activities. *F. moniliforme* (Sheldon), *F. proliferatum* (Matsushima) Nirenberg, *F. nygamai*, and *F. napiforme* Marasas are the most important producers of fumonisin B_1 (Nelson et al., 1992; Nelson et al., 1994; Thiel et al., 1992) which showed cancer promoting activity to rats (Gelderblom et al., 1988). *Fusarium* spp. also produce phytotoxins which are effective against several plants (Duke, 1986) and, therefore, their bioherbicidal effects have been tested against various weeds and crops (Hoagland, 1990; Abbas et al., 1991; Abbas and Boyette, 1992). Our investigations showed that fumonisin B_1 has a herbicidal effect on *Striga* spp., but it is still not clear if fumonisin B_1 plays an active role, either in the infection process or in the course of the disease, when the biocontrol agent, *F. nygamai*, is applied to control *S. hermonthica* (Kroschel and Elzein, 2003).

Table 3. *Fusarium* isolates currently under investigation for the biocontrol of *Striga* and *Orobanche* spp.

Fungus	Abbreviation	Target weed	Reference
Fusarium oxysporum	*F. oxysporum M12-4A*	*Striga hermonthica*	Ciotola et al. (1995)
F. oxysporum	*Foxy2*	*Striga hermonthica, Striga asiatica*	Kroschel et al. (1996), Elzein and Kroschel (2003)
F. oxysporum f. sp.*orthoceras*	*FOO*	*Orobanche cumana, O. cernua, O. aegyptiaca (?)*	Bedi and Donchev (1991)
F. oxysporum	*FOXY*	*Orobanche aegyptiaca, Orobanche ramosa, Orobanche cernua*	Amsellem et al. (1999)
F. arthrosporioides	*FARTH*	*Orobanche aegyptiaca, Orobanche ramosa, Orobanche cernua*	Amsellem et al. (1999)

F. oxysporum does not belong to the group of *Fusarium* spp. producing fumonisin B_1 which makes this fungal species a more interesting candidate for the biological control of parasitic weeds. However, *F. oxysporum* isolates under investigation for biological control should also be tested for their toxin production. *F. oxysporum* M12-4A produced small amounts of fusaric acid (which is not generally regarded as a mycotoxin) and dehydrofusaric acid, but none of the other toxins reported to be produced by some strains of *F. oxysporum* (Savard et al., 1997). *Fusarium oxysporum* f. sp. *orthoceras* was found to produce fusaric acid as well and did not produce some of the trichotecene derivates, however, the testing of mycotoxin production has not yet been completed (Thomas, 1998 a, b).

Application

A very simple form of applying a fungal biocontrol agent is to collect infected *Orobanche* shoots and distribute them in fields with non-infected *Orobanche* populations in order to enhance the development of diseases. Such a procedure was reported to be applied by farmers in northwestern China (Wang et al., 1985).

Another simple method is the use of a spore or mycelial suspension. Fungi of the genus *Fusarium* easily grow in liquid cultures and the resulting suspensions can be used as a soil drench or a post-emergence spray application. Thomas et al. (1998b) applied a conidia suspension of *Fusarium oxysporum* f. sp. *orthoceras* pre-planting into the soil and observed an increasing level of *O. cumana* control with increasing dosages of inoculum (Figure 2).

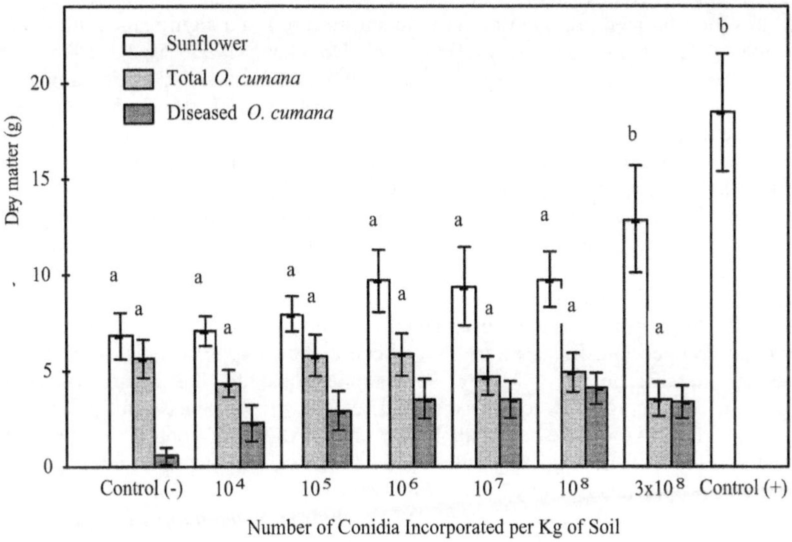

Figure 2. Effect of unformulated conidia of *Fusarium oxysporum* f.sp. *orthoceras* incorporated pre-planting into the soil on the dry-matter production of sunflower and *O. cumana*, and disease incidence on the parasite.

Mazaheri et al. (1991) applied an aqueous suspension of conidia of *F. oxysporum* grown in Potato Dextrose Broth into the planting holes of tobacco in the field. This treatment reduced number and dry weight of *Orobanche* plants by 80-90% and increased the tobacco yield by approximately 80%. Linke et al., (1992) sprayed conidia suspensions of *Ulocladium atrum* on emerged *O. crenata* shoots in the field and in the plastic house and obtained a complete reduction of the parasite, provided temperature was around 20°C and the relative humidity between 50 and 80%. Amsellem et al. (2001) found that a suspension of freshly prepared, chopped mycelia of FOXY and FARTH infected *O. aegyptiaca* incubated in polyethylene bags more quickly and reached a higher maximum level of control, than conidia. This suspension was also used for the dipping of tomato roots before transplanting which led to a significantly reduced number and biomass of broomrape tubercles.

Pathogens that attack their target weed at or below the soil line can also be delivered in a solid formulation applied to the soil. A simple possibility is the application of dried mycelium and spores. Iliev et al. (1998) developed a method to dry biomass of *Fusarium oxysporum* f. sp. *orthoceras* obtaining a product that was efficient in pot trials in doses of about 1 g per 10 kg of soil. However, a granular formulation offers several advantages to the application of mere mycelium or spores. It provides the biocontrol agent with a substrate necessary for proliferation and growth that can also be chosen to favour the desired fungal propagule. Thus, granules can serve as a site for continuous spore production whereas spores or mycelium applied alone may require repeated applications. A simple form of a granular formulation is the use of fungal propagules together with their growth substrate, e.g. infested cereal grains. Fungus-inoculated organic substrates or soil/maize feed mixtures incorporated pre-planting into the soil resulted in a significant control of *O. cumana* under field conditions (Bedi and Donchev, 1991; Bedi, 1994). The disadvantage of the application of solid substrates is that, usually, a large amount of inoculum is needed to obtain effective weed control. Bedi (1994) applied 800 kg of solid inoculum per ha. Also, solid substrates may have an inadequate shelf-life. To reduce the amount of inoculum, the application can be restricted to the furrow or the planting hole. Kott (1971) reported the reduction in the *O. aegyptiaca* infestation of tobacco by 100% after applying *Fusarium* sp., grown on a mixture of straw and maize flour, to the planting hole at a rate of 0.5 to 1 g per hole. However, as inoculated organic material is hardly suitable for commercial use, several techniques have been developed to encapsulate fungal propagules in a solid matrix which can also help to buffer the organism from environmental constraints, like rapid desiccation or microbial competition. Some efforts have been made to apply simple encapsulation methods, suitable for pre-planting soil applications, to the biocontrol agents pathogenic to parasitic weeds. Amsellem et al. (1999) successfully used the 'Stabileze' formulation method with FOXY and FARTH to control *O. aegyptiaca* grown in polyethylene bags. Granules made from wheat-flour and kaolin ('Pesta') containing propagules of Foxy2 and *Fusarium oxysporum* f. sp. *orthoceras* showed high efficacy in controlling *S. hermonthica* and *O. cumana* in the greenhouse (Kroschel et al., 2000; Müller-Stöver et al., 2002; Figures 3 and 4). The inoculum amount needed for a successful *O. cumana* control was calculated to be approximately 60% lower compared to the amount of solid inoculum Bedi (1994) applied in a field experiment in Bulgaria (Müller-Stöver, 2001). The application of 'Pesta' granules of Foxy2 in greenhouse experiments even allowed the reduction of the applied inoculum per kg soil by 95% compared to the wheat grain inoculum used before (Elzein, 2003), i.e., an amount between 76 and 125 kg per ha would be needed for the control of *S. hermonthica* under field conditions, depending on the cultivation system of sorghum or maize in semi-arid regions.

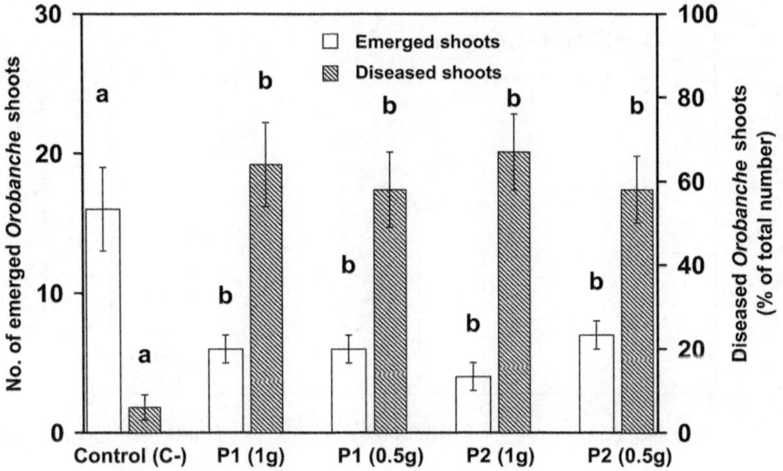

Figure 3. Effect of wheat-flour kaolin ('Pesta') granules containing different types *Fusarium oxysporum* f. sp. *orthoceras* inocula and applied in different dosages (0.5 g or 1 g per pot) on the total number and the proportion of diseased *Orobanche cumana* shoots. Bars within each treatment followed by the same letter are not significantly different according to Tukey's HSD test (emerged shoots) or the Nemenyi test (diseased shoots) at the $\alpha = 0.05$ level of significance. Vertical lines indicate SE. (C-) = sunflower plus *Orobanche*; P1 = granules containing microconidia; P2 = granules containing chlamydospore-rich biomass. One pot contained approximately 2 kg of soil. (Data from Müller-Stöver et al., 2002).

The type of inoculum used in the formulations greatly influenced their shelf-life. Amsellem et al. (1999) found that formulated chopped mycelia stored better than conidia, and Elzein et al. (2000) and Müller-Stöver (2001) reported a longer shelf-life for formulated chlamydospore-rich biomass than microconidia of the biocontrol agents.

Another approach, which aims to deliver the biocontrol agents directly into the rhizosphere of the host plants is the use of seed dressings. Coating the host plant seeds with the pathogen would allow a dramatic reduction of the inoculum needed for parasitic weed control. Sorghum seeds coated with air-dried *F. oxysporum* M12-4A inoculum completely prevented S. *hermonthica* emergence in a pot experiment (Ciotola et al., 2000). Foxy2 was shown to quickly colonize the root system of sorghum and significantly reduced the number of emerged *Striga* shoots when sorghum seeds had been coated with dried chlamydospore inoculum homogenized into 20% arabic gum (Elzein, 2003). However, this approach possibly does not work with all pathogens and additional problems may arise with incompatible fungicides or other biocontrol agents that are intended to be applied as seed dressings as well (Amsellem et al., 2001).

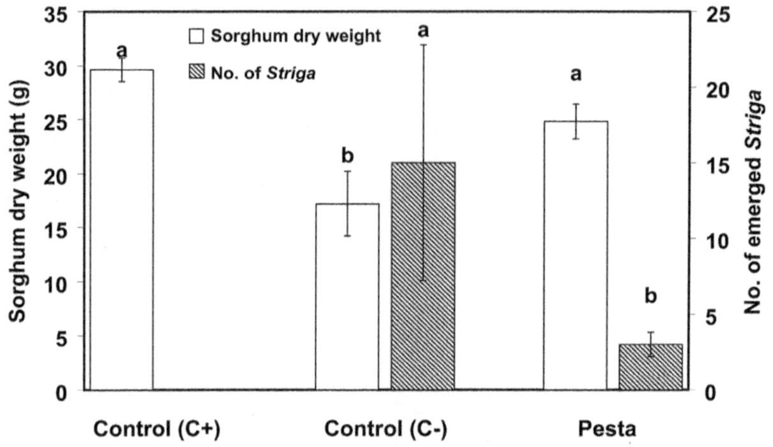

Figure 4. Effect of wheat-flour and kaolin ('Pesta') granules containing *Fusarium oxysporum* (Foxy 2) on the number of emerged *Striga* shoots and sorghum dry weight. Bars within each treatment followed by the same letter are not significantly different according to LSD test at the α= 0.05 level of significance. Vertical lines indicate SE. (C+) = sorghum only, (C-) = sorghum plus *Striga.* Pesta granules were applied at a dosage of 0.4 g per pot (approximately 2 kg of soil). (Data from Kroschel et al., 2000).

Efficacy

Most of the recent data on efficacy of the biocontrol agents is provided by pot experiments and sometimes it is difficult to compare them, due to the use of different application methods.

Inoculum of *F. nygamai* at a rate of 10 g per kg soil reduced the emergence of *S. hermonthica* by 100% (Abbasher and Sauerborn 1992). Ciotola et al. (1995) achieved the same efficacy and a 400% sorghum yield increase with 3 g of *F. oxysporum* M12-4A inoculum per pot (10 cm diam.), *Orobanche* was generally not completely controlled by the application of the biocontrol agents. Amsellem et al. (2001) observed a reduction of the number of tubercles to less than 20% of the control after a root dip and a drench of suspensions containing mycelium and spores of FOXY and FARTH. The application of 3 x 10^8 conidia of *Fusarium oxysporum* f. sp. *orthoceras* per kg soil led to a reduction of *O. cumana* emergence of 80% (Thomas et al., 1998a, b). However, 90% *Orobanche* shoots showed disease symptoms at this inoculum concentration, thus contributing to weed control through a further reduction of *Orobanche* seed production. In contrast, *Ulocladium botrytis* applied to *O. crenata* or *O. cumana* only resulted in an increased number of diseased shoots but not in a decrease of the number of *Orobanche* shoots (Müller-Stöver, 2001).

Data of efficacy of the currently investigated biocontrol agents under field conditions are still rare. In small-scale field experiments in Bulgaria, Bedi (1994) achieved successful control of *O. cumana* by pre-planting inculation of *Fusarium oxysporum* f. sp. *orthoceras* grown on barley grains or a wheat straw/corn feed mixture. After soil-incorporation of 800 kg/ha in a depth of 10-15 cm, *O. cumana* emergence was reduced by 63% and the sunflower yield was increased by 18 fold. However, this efficacy could not be confirmed under Israeli field conditions, when the fungus had been applied in a granular formulation (Müller-Stöver, unpublished data).

OUTLOOK

Although numerous microorganisms have been described to be pathogenic to parasitic weeds, no commercial bioherbicide has been developed so far. This may be due to limited research resources and/or a lack of interest of industry to investigate in the development of biocontrol methods for parasitic weeds because of the relatively small market size. Another reason may be unsatisfactory control efficacy of the potential biocontrol agents, especially under field conditions where many biotic and abiotic factors can be detrimental to disease development. Possibilities to increase control efficacy may be to (i) keep searching for new biocontrol candidates and to screen them for their control potential and (ii) try to enhance the virulence of fungi already under investigation by means of fermentation, formulation and application techniques. In this regard it would be advantageous if all research activities could be concentrated on one or two pathogens (instead of many working groups working on many different fungi in small research projects) and to put more effort into devising reliable, practical formulations which could show that biocontrol can be a relevant tool for parasitic weed management.

REFERENCES

Abbas H.K., Boyette C.D., Hoagland R.E., Vesonder R.F. Bioherbicidal potential of *Fusarium moniliforme* and its phytotoxin, fumonisin. Weed Sci 1991; 39:673-677.

Abbas H.K., Boyette C.D. Phytotoxicity of fumonisin B_1 on weed and crop species. Weed Technol 1992; 6:548-552.

Abbasher A.A., Sauerborn J. *Fusarium nygamai*, a potential bioherbicide for *Striga hermonthica* control in sorghum. Biocontrol 1992; 2:291-296.

Abbasher A.A., Kroschel J., Sauerborn J. Microorganisms of *Striga hermonthica* in Northern Ghana with potential as biocontrol agents. Biocont Sci Technol 1995; 5:157-161.

Abbasher A.A., Hess D.E., Sauerborn J., Kroschel J. Effect of different *Fusarium* spp. on seed germination of *Striga hermonthica* (sorghum and millet strains), *S. asiatica* and *S. gesnerioides*. Proceedings of the 6th Int. Symp. on Parasitic Weeds; 1996; Cordoba.

Abbasher A.A., Hess D.E., Sauerborn J. Fungal pathogens for biological control of *Striga hermonthica* on sorghum and pearl millet in West Africa. Afr Crop Sci J 1998; 6:179-188.

Abbasher A.A., Hess D.E., Sauerborn J., Kroschel J. "Biological control of *Striga*." In *Report on the ICRISAT Sector Review for Striga Control in Sorghum and Millet,* D.E. Hess, J.M. Lenné eds. Bamako: International Crops Research Institute for the Semi-Arid Tropics, 1999; pp. 138.

Amsellem Z., Zidack N.K., Quimby P.C. Jr, Gressel J. Long term dry preservation of viable mycelia of two mycoherbicidal organisms. Crop Protect 1999; 18:643-649.

Auld B.A., Morin L. Constraints in the development of bioherbicides. Weed Technol 1995; 9:638-652.

Barghouthi S., Faqieh F., Badrieh A. Effect of bacteria on broomrape germination. 3rd International Weed Science Congress; 2000; Foz do Iguassu.

Bedi J.S. Further studies on control of sunflower broomrape with *Fusarium oxysporum* f. sp. *orthoceras*- a potential mycoherbicide. Proceedings of the Third International Workshop on *Orobanche* and related *Striga* research; Amsterdam, The Netherlands. Royal Tropical Institute, 1994.

Bedi J.S., Donchev N. Results on mycoherbicides control of sunflower broomrape (*Orobanche cumana* Wall.) under glasshouse and field conditions. Proceedings of the 5th Int. Symp. on Parasitic Weeds; 1991; CIMMYT, Nairobi.

Bedi J.S., Donchev N.A technique for macroconidia production by *Fusarium oxysporum* f.sp. *orthoceras*, a biocontrol agent for *Orobanche cumana*. Plant Dis Res 1995; 10:62-63.

Berner D.K., Schaad N.W., Völksch B. Use of ethylene-producing bacteria for stimulation of *Striga* spp. seed germination. Biol Cont 1999; 15:274-282.

Bouillant M.L., Kabir M., Alexandre G., Jacoud C., Miche L., Bally R., Sallé G., Ouédraogo O. Inhibition of *Striga* seed germination by bacteria isolated from soil under sorghum field. Proceedings of the 6th Int Symp on Parasitic Weeds; 1996; Cordoba.

Bowers R.C. Commercialization of Collego – an industrialist's view. Weed Sci 1986; 34: 24-25.

Boyette C.D., Templeton G.E., Oliver L.R. Texas gourd (*Cucurbita texana*) control with *Fusarium solani* f. sp. *cucurbita*. Weed Sci 1984; 32:649-355.

Bozoukov H., Kouzmanova I. Biological control of tobacco broomrape (*Orobanche* spp.) by means of some fungi of the genus *Fusarium*. Proceedings of the 3rd International Workshop on *Orobanche* and related *Striga* Research. 1994; Royal Tropical Institute; Amsterdam.

Ciotola M., Watson A.K., Hallett S.G. Discovery of an isolate of *Fusarium oxysporum* with potential to control *Striga hermonthica* in Africa. Weed Res 1995; 35:303-309.

Ciotola M., Hallett S.G., Watson A.K. Impact of *Fusarium oxysporum* isolate M12-4A upon seed germination of *Striga hermonthica* in vitro. Proceedings of the 6th International Symposium on Parasitic Weeds; 1996; Cordoba.

Ciotola M., DiTommaso A., Watson A.K. Chlamydospore production, inoculation methods and pathogenicity of *Fusarium oxysporum* M12-4A, a biocontrol for *Striga hermonthica*. Biocon Sci Technol 2000, 10:129-145.

Duafala T., Gold A.H., Sagen J., Wilhelm S. Apparent biological control of branched broomrape, *Orobanche* ramosa. Proc Am Phytopathol Soc 1975; 3:113.

Duafala T., Wilhelm S., Gold A.H., Sagen J.E. Rhizoctonia disease of broomrape, a possible biological control. Proc Am Phytopathol Soc 1976; 3:272.

Duke S.O. Microbially produced phytotoxins as herbicides - a perspective. Rev Weed Sci 1986; 2:15-44.

Elzein A., Kroschel J., Adamasu A., Fetene M., Müller-Stöver D. Fungi isolated from *Orobanche* spp. in Ethiopia and their potential use for biological control. Proceeedings of the X International Symposium on Biological Control of Weeds; 2000; Bozeman, Montana.

Elzein A., Sauerborn J., Müller-Stöver D., Kroschel J. Effect of storage temperature, granule size, and inoculum type on the viability of *Fusarium oxysporum*, a pathogen of *Striga hermonthica*, encapsulated in wheat-kaolin ("Pesta") granules. 3rd International Weed Science Congress; 2000; Foz do Iguassu.

Elzein A., Kroschel J., Sauerborn J. Host specificity assessments of *Fusarium oxysporum* (FOXY2), a potential antagonist of *Striga hermonthica*: Implications for its use as an acceptable mycoherbicide. Proceedings of the 7th Int. Symp. on Parasitic Weeds; 2001; Nantes.

Elzein A., Kroschel J. Host range studies of *Fusarium oxysporum* "Foxy 2": an evidence for a new forma specialis and its implications for *Striga* control. BioControl 2003 (accepted).

Elzein A. Development of a granular mycoherbicidal formulation of *Fusarium oxysporum* "Foxy2" for the biological control of *Striga hermonthica* (Del.) Benth. Tropical Agriculture – Advances in Crop Research (2), Weikersheim: Margraf Verlag, 2003; (in press).

Gelderblom W.C.A., Jaskiewicz K., Marasas W.F.O., Thiel P.G., Horak R.M., Vleggaar R., Kriek N.P.J. Fumonisins - Novel mycotoxins with cancer-promoting activity produced by *Fusarium moniliforme*. Appl Environ Microbio 1988; 54: 1806-1811.

Hoagland R.E. "Microbes and microbial products as herbicides, an overview". In *Microbes and Microbial Products as Herbicides,* R.E. Hoagland, ed. Washington DC: American Chemical Society, 1990. pp. 2-52.

Hodosy Z. Biological control of broomrape, *Orobanche ramosa*, a tomato parasite. I. Occurrence and adaptability of *Fusarium* species to control broomrape in Hungary. Zöldsegtermesztesi Kutato Intezet Bulletinje 1981; 14:21-29.

Iliev I., Kiryakov I., Shindrova P., Entcheva V. A possibility for industrial production of a bioherbicide for control of *Orobanche cumana*. Proceedings of the 4th International *Orobanche* Workshop; 1998; Albena.

Kabulov D.T., Khalimov M.K.-H. Some biological characteristics of Egyptian broomrape and ways of controlling it by means of the fungus *Fusarium*. Weed Abstracts 1977; 26:228.

Kenney D.S. DeVine – the way it was developed – a industrialist's view. Weed Sci 1986; 34:15-16.

Klein O., Kroschel J., Sauerborn J. "Efficacité de lachers supplémentaires de *Phytomyza orobanchia* Kalt. [Diptera: Agromyzidae] pour la lutte biologique contre l'*Orobanche* au Maroc". In *Advances in Parasitic Weed Control at On-farm Level. Vol. II. Joint Action to Control Orobanche in the WANA Region,* J. Kroschel, M. Abderabihi, H. Betz, eds. Weikersheim: Margraf Verlag, 1999. pp. 161-171.

Klein O., Kroschel J. Biological control of *Orobanche* spp. with *Phytomyza orobanchia*, a review. BioControl 2002; 47:245-277.

Kott S.A. Biological control of broomrape. Weed Abstracts 1971; 20:187.

Kroschel J., Hundt A., Abbasher A.A., Sauerborn J. Pathogenicity of fungi collected in northern Ghana to *Striga hermonthica.* Weed Res 1996; 36:515-520.

Kroschel J., Müller-Stöver D., Elzein G., Sauerborn, J. The development of mycoherbicdes for the management of parasitic weeds of the genus *Orobanche* and *Striga-* a review and recent results. Proceedings of the X International Symposium on Biological Control of Weeds; 2000; Bozeman, Montana.

Kroschel J., Jost A., Sauerborn J. "Insects for *Striga* control - Possibilities and constraints". In *Advances in Parasitic Weed Control at On-farm Level. Vol. 1. Joint action to Control Striga in Africa,* J. Kroschel, H. Mercer-Quarshie, J. Sauerborn, eds. Weikersheim: Margraf Verlag, 1999. pp. 117-132.

Kroschel J. *A Technical Manual for Parasitic Weed Research and Extension.* Dordrecht: Kluwer Academic Publisher, 2001.

Klein O., Kroschel J. Biological control of *Orobanche* spp. with *Phytomyza orobanchia,* a review. BioControl 2002; 47:245-277.

Kroschel J., Elzein A. Bioherbicidal effect of fumonisin B1 on parasitic weeds of the genus *Striga.* Biocont Sci Technol 2003 (in press).

Linke K.H., Scheibel C., Saxena M.C., Sauerborn, J. Fungi occurring on *Orobanche* spp. and their preliminary evaluation for *Orobanche* control. Trop Pest Manage 1992; 38:127-130.

Linke K.H., Joel D.M., Kroschel, J. "Observations of the underground development". In *A Technical Manual for Parasitic Weed Research and Extension,* J. Kroschel, ed. Dordrecht: Kluwer Academic Publishers, 2001. pp. 53-59.

Mazaheri A., Vaziri M., Moayed-Zadeh N. Studies on the chemical control of broomrape (*Orobanche aegyptiaca* Pers.) in tobacco (*Nicotiana tabacum* L.) fields. Proceedings of the 5th Int. Symp. on Parasitic Weeds; 1991; Nairobi.

Meister C.W., Eplee R.E. Five new fungal pathogens of witchweed (*Striga lutea*). Plant Disease Reporter 1971; 55: 861-863.

Müller-Stöver D. *Possibilities of biological control of* Orobanche crenata *and* O. cumana *with* Ulocladium botrytis *and* Fusarium oxysporum *f. sp.* orthoceras. Agroecology (3), Laubach: APIA-Verlag, 2001.

Müller-Stöver D., Thomas H., Abbasher A.A. "Application of control methods. Biological – Pathogens". In *A technical manual for parasitic weed research and extension,* J. Kroschel, ed. Dordrecht: Kluwer Academic Publishers, 2001. pp. 144-151.

Müller-Stöver D., Kroschel J., Thomas H., Sauerborn J. Chlamydospores of *Fusarium oxysporum* Schlecht f.sp. *orthoceras* (Appel and Wollenw.) Bilai as inoculum for wheat flour-kaolin granules to be used for the biological control of *Orobanche cumana* Wallr. Europ J Plant Protect 2002; 108:221-228.

Nag Raj T.R. Fungi occurring on witchweed in India. Commonwealth Inst Biol Cont Tech Comm 1966; 7:75-80.

Narasimhan M.J., Thirumalachar M.J. A sclerotinia disease of *Orobanche ceruna* in Bihar (India). Phytopathologische Zeitschrift 1954; 22:421-428.

Nelson P.E, Plattner R.D, Shackelford D.D., Desjardins A.E. Fumonisin B1 production by *Fusarium* species other than *F. moniliforme* in section Liseola and by some related species. Appl Environ Microbiol 1992; 58:984-989.

Nelson P.E., Dignani M.C., Anaissie E.J. Taxonomy, biology and clinical aspects of *Fusarium* species. Clinical Microbial Rev 1994; 7:479-504.

Norambuena H., Kroschel J., Klein, O. "Introduction of *Phytomyza orobanchia* for *Orobanche* spp. biocontrol in Chile". In *Advances in Parasitic Weed Control at on-farm level. Volume II: Joint Action to Control Orobanche in the WANA region,* J. Kroschel, A. Abderabihi, H. Betz, eds. Weikersheim: Margraf Verlag, 1999. pp. 197-204.

Norambuena H., Diaz J., Kroschel J., Klein O., Escobar S. Rearing and field release of *Phytomyza orobanchia* on *Orobanche ramosa* in Chile. Proceedings of the 7th Int. Symp. on Parasitic Weeds; 2001; Nantes.

Oswald M., Sauerborn J., Oswald A., Mbogo Okoth P. Interaction of Plant Growth Promoting Rhizobacteria (PGPR) with maize and *Striga hermonthica* (Del.) Benth. seeds. 3rd International Weed Science Congress; 2000; Foz do Iguassu.

Panchenko V.P. Use of Microorganisms in control of Egyptian broomrape parasitizing watermelons. Mikologiya i Fitopatologiya 1974; 8:122-125.

Parker C., Riches C.R. *Parasitic Weeds of the World - Biology and Control.* Wallingford: CAB International, 1993.

Raju C.A., Krishna Murty G.V.G., Nagarajan K., Chari M.S. A new disease of *O. cernua* parasitizing tobacco, caused by *Sclerotium rolfsii.* Phytoparasitica 1995; 23:235-237.

Raynal-Roques A. A hypothetic history of *Striga* – a preliminary draft. Proceedings of the 6th Int Symp on Parasitic Weeds; 1996; Cordoba.

Sauerborn J. The economic importance of the phytoparasites *Orobanche* and *Striga.* Proceedings of the 5th Int. Symp. on Parasitic Weeds; 1991; Nairobi.

Sauerborn J., Dörr I., Abbasher A.A., Thomas H., Kroschel J. Electron microscopic analysis of the penetration process of *Fusarium nygamai*, a hyperparasite of *Striga hermonthica.* Biol Cont 1996; 7:53-59.

Savard M.E., Miller J.D., Ciotola M., Watson, A.K. Secondary metabolites produced by a strain of *Fusarium oxysporum* used for *Striga* control in West Africa. Biocont Sci Technol 1997; 7: 61-64.

Stankevich G.L. On specialized form of *Colletotrichum lagenarium* Halst. et El. on *Orobanche aegyptiaca.* Mikologiya i Fitopatologiya 1971; 5: 209-211.

Thiel P.G., Marasas W.F.O., Sydenham E.W., Shephard G.S., Gelderblom W.C.A. The implications of naturally occurring levels of fumonisins in corn for human and animal health. Mycopathologie 1992; 117: 3-9.

Thomas H. *Das Potential von Pilzen zur Kontrolle von* Orobanche *spp. unter Berücksichtigung von Anbausystemen im Terai, Nepal.* PLITS 16 (4), Weikersheim: J Margraf Verlag, 1998.

Thomas H, Sauerborn J., Müller-Stöver D., Ziegler A., Bedi J.S., Kroschel J. The potential of *Fusarium oxysporum* f.sp. *orthoceras* as a biological control agent for *Orobanche cumana* in sunflower. Biol Cont 1998; 13:41-48.

Thomas H., Heller A., Sauerborn J., Müller-Stöver D. *Fusarium oxysporum* f.sp. *orthoceras*, a potential mycoherbicide, parasitizes seeds of *Orobanche cumana* (sunflower broomrape): a cytological study. Ann Bot 1999; 83:453-458.

Timchenko V.I., Dovgal E.S. Microbiological control method for broomrape on vegetable crops. Weed Abstracts 1973; 22:172.

Traoré D., Vincent C., Stewart R.K. *Smicronyx guineanus* Voss and *S. umbrinus* Hustache (Coleoptera: Curculionidae): potential biocontrol agents of *Striga hermonthica* (Del.) Benth. (Scrophulariaceae). In *Report on the ICRISAT Sector Review for* Striga *Control in Sorghum and Millet,* D.E. Hess, J.M. Lenné, eds. Bamako, International Crops Research Institute for the Semi-Arid Tropics, 1999; pp. 138.

Wang Z., Zhu G., Ma T. Control of Egyptian broomrape (*Orobanche aegyptiaca* Pers.) with *Fusarium orobanches* Jacz. Chinese J Biol Cont 1985; 1:27-31.

Zummo N. Diseases of giant witchweed, *Striga hermonthica* in West Africa. Plant Disease Reporter 1977; 61:428-430.

CHAPTER 22

RECENT APPROACHES TO *OROBANCHE* MANAGEMENT

A REVIEW

YAAKOV GOLDWASSER[1] & YESHAIAHU KLEIFELD[2]

[1]*Institute of Plant Science & Genetics, Faculty of Agriculture, Food & Environmental Sciences, The Hebrew University of Jerusalem, Rehovot 76100, Israel.* [2] *10 Yehuda Hanassi, Kiryat Tivon 36047, Israel.*

Parasitic plants account for approximately 1% of angiosperm species and are present in 22 botanical families. Several of the parasitic species are important agricultural weeds, infest a wide range of crops around the globe, and pose a major threat to the food security of numerous communities. The agriculturally important parasitic weeds fall into four main groups:

1. Viscaceae and Loranthaceae (mistletoes): green hemiparasites that parasitize aerial parts of shrubs and trees;
2. Cuscutaceae (dodders): holoparasitic twining stem parasites that attack dicot plants;
3. Scrophulariaceae (figworts and witchweeds): contains hemi-parasitic root parasites that invade roots of host plants;
4. Orobanchaceae (broomrapes): holoparasitic root parasites that invade dicot plants.

The Orobanchaceae consists of 14 genera of chlorophyll-lacking root holoparasites of which the genus *Orobanche* (common name broomrape), includes more than 100 species that subsist on broad-leaf plants, depleting them of nutrients, minerals and water. Several review papers and books on *Orobanche* have been published (Kasasian, 1973a; Pieterse, 1979; Musselman, 1980; Ramaiah, 1987; Foy et al., 1989; Parker and Riches, 1993; Press and Graves, 1995; Dhanapal et al., 1996; Alonso, 1998). The start of the new millennium is an appropriate time to summarize the past and update methods for *Orobanche* control, and point out the prospects for new tools to combat the devastating effects of this parasitic plant in the future. Even though significant advances have been made in the study of parasitic plant biology and control measures, most of this information has not percolated to the farmers, especially in developing countries because of setbacks in technology transfer and lack of resources.

Inderjit (ed.), Weed Biology and Management, 439-466.

Numerous worldwide surveys, evaluations and studies describing the parasitic plant problem have been conducted, but research on practical and feasible control measures has been minimal. Most of the devastating cases of parasitic weeds occur in developing countries of Africa and Asia, thus the big seed and chemical companies show little interest in this market and have failed to develop and register resistant crop varieties and specific herbicides for combating parasitic plants. Furthermore, ongoing inadvertent introduction of *Orobanche* species to the USA, South America, Australia and Germany have not to date stimulated research on control. The number of these introduced infestations will increase in future years due to increasing global travel and trade.

More emphasis should be given to the development of tools for parasitic plant control and to ensure the utilization of these tools by farmers. Potential control tools exist (resistant varieties, selective herbicides, and herbicide resistant crop plants) but lack of funds and interest impede fighting the plague and insuring the food stability of millions of people. This review summarizes research regarding the actual and potential control of *Orobanche* in the field. Most of the data presented in this chapter is based on research conducted by the authors in Israel.

DESCRIPTION OF THE *OROBANCHE* PROBLEM

Biology

The *Orobanche* genus includes six species that are of agricultural importance and cause devastating yield and quality losses to many agricultural crops around the globe (Parker and Riches, 1993; Riches and Parker, 1995). The *Orobanche* life cycle is extremely specialized for parasitism (Graves, 1995; Stewart and Press, 1990). Small *Orobanche* seeds (0.2-0.4 mm) germinate in the soil only after a preconditioning period of moist and suitable temperatures, and only in response to germination stimulants released by specific host plant roots. These conditions ensure that only seeds within the rhizosphere of an appropriate host root will germinate. The parasite seedling radicle can grow only a few millimeters and must contact a host root a few days after germination before exhausting its limited energy resources. As a result of contact with the host root, the radicle forms a specialized parasitic-plant organ, the haustorium, which attaches to the root, penetrates the epidermis and cortex tissues, and ultimately fuses into the root vascular system (Dörr, 1996; Joel et al., 1988; Kuijt, 1977). The parasite then draws water, minerals and assimilates out of the host plant. Following the accumulation of metabolites, a tubercle is formed and finally a floral shoot emerges above the soil surface. Flowers develop on the floral shoots, leading to the production of capsules capable of yielding hundreds of thousands of seeds per plant (Foy et al., 1989; Holm et al., 1997; Pieterse, 1979).

Distribution

Orobanche species parasitize the roots of numerous important broadleaf crop plants and weeds. These parasites are widespread throughout the Middle East and the Mediterranean regions with some incursions into Southern, Northern and Eastern Europe, Africa, the Far East, New Zealand, Australia, North, Central and

South America. Most introductions of *Orobanche* species have been to regions with Mediterranean-like climate. Although each *Orobanche* species has a discrete and in some cases narrow host range, the spectrum of crops parasitized by these parasitic weeds is very large and includes vegetable, field, fodder, flower, and herb crops from various botanical families. The main distribution areas of the agriculturally significant *Orobanche* species are:

1. *O. aegyptiaca* Pers.- Mediterranean, Middle East, North Africa, Far East, Southern Europe;
2. *O. ramosa* L.- Mediterranean, Middle East, North Africa, Southern and Eastern Europe, Australia, Central and South America;
3. *O. crenata* Forsk.- Mediterranean, Middle East, North Africa;
4. *O. cernua* Loeffl.- Mediterranean, Middle East, North Africa, Far East, Southern Europe;
5. *O. cumana* Wallr.- Mediterranean, Middle East, Southern and Eastern Europe.
6. *O. minor* Sm.- Northern and Southern Europe, Australia, New Zealand, Africa, North America.

Host Range and Damage

The major agriculturally important *Orobanche* species and their major host crops are:

- *O. aegyptiaca* (Egyptian broomrape) and *O. ramosa* (branched broomrape): Solanaceae (tomato, potato, eggplant, pepper, tobacco); Fabaceae (chickpea, lentil, vetch, faba bean, clover, peanut); Brassicaceae (cabbage, cauliflower, rapeseed, horseradish, mustard); Cucurbitaceae (melon, watermelon, cucumber); Asteraceae (sunflower, lettuce, safflower); Umbelliferae (carrot, parsley, celery, dill); Cannabiaceae (hemp); Linaceae (flax);
- *O. crenata* (crenate broomrape): Apiaceae (carrot, parsley, celery, dill); Fabaceae (chickpea, lentil, vetch, faba bean, clover, pea); Geraniaceae (*Pelargonium graveolens*);
- *O. cernua* (nodding broomrape): Solanaceae (tomato, eggplant, tobacco); Lamiaceae (clary sage);
- *O. cumana* (sunflower broomrape): Asteraceae (sunflower); *O. minor* (small broomrape): Fabaceae (clover, alfalfa);
- Other minor species: *O. lutea* (yellow broomrape): alfalfa and clover in central and southern Europe; *O. foetida* (fetid broomrape): faba bean, clover and alfalfa in Spain, northern Morocco and Tunisia.

As obligate parasites, *Orobanche* spp. are totally dependent on their host plants. The attached parasite functions as a strong metabolic sink, often named "super-sink", strongly competing with other parts of the host plant for water, mineral and assimilate absorption and unloading. The diversion of these substances to the parasite causes moisture and assimilate starvation, host plant stress and growth inhibition that result in crop yield and quality reduction. In case of severe infestations in sensitive crops, *Orobanche* can cause total collapse and death of crop plants. Furthermore, the parasite causes severe reduction in yield quality as a result of foliage wilting and sudden exposure to radiation (tomato), decrease of tuber size (potato, carrot), decrease of leaf size, color and quality (tobacco and herbs) and decrease of sugar levels and size (carrot). The contamination of crop seeds, fodder or herbs by *Orobanche* plant material or seeds reduces produce quality. Since *Orobanche* is classified as a "noxious weed", the presence of one parasite seed in a batch of crop seed disqualifies the marketing of the entire seed lot (Parker and Riches, 1993; Foy et al., 1989).

Researchers tried to evaluate and predict crop damage based on *Orobanche* seed counts in soil samples. These predictions turned out to be problematic because samples and evaluation require many soil samples to achieve significant correlation. Furthermore, *Orobanche* damage is influenced not only by infestation level but also by environmental factors, field fertility, and response of crop variety (Dhanapal et al., 1996; Foy et al., 1989).

Why is Orobanche Difficult to Control

Orobanche is extremely difficult to control in agricultural crops due to its underground location, close association with host roots, and complex mechanisms of seed dispersal, germination and longevity. Because the parasite germinates only in response to host root exudates and then attaches and develops underground on the host plant for the major part of its life, it is inaccessible to conventional control methods such as tillage and herbicide treatments. Furthermore, when the plant becomes visible above ground, much of the damage has already been done and control would be futile. The late appearance of parasite shoots above the soil and the lack of a photosynthetic system as a potential herbicide target make *Orobanche* species difficult for conventional herbicide control.

The characteristics of *Orobanche* seeds account for much of the difficulty in controlling this parasitic weed. The extremely small seeds are produced in vast numbers and can remain viable in the soil or in storage for many years. Reports of seed longevity in fields range from 13 years (Parker and Riches, 1993) and in Israel up to 35 years (Kleifeld, unpublished). The tiny seeds are easily dispersed to near and far by wind, water and livestock. Human practices are significantly responsible for distributing *Orobanche* seeds by transporting and using contaminated agricultural vehicles, farm implements and produce containers (by direct seed contamination or through clinging of contaminated soil). Further parasite seed distribution is caused by transportation of contaminated plant material (such as crop seeds and hay) and contaminated soil and manure movement. The use of organic manure from livestock fed with contaminated hay is a cause of further seed dispersal, since the parasite seeds do not lose their viability while passing through animal's digestive systems (Jacobsohn et al., 1987). In Egypt, non-infested irrigated fields in newly reclaimed desert are being contaminated with broomrape seeds carried in manure or topsoil from other locations in the country, resulting in large scale infestations (ME Borollosy, personal communication). In Israel, non-infested tomato fields became infested by the transport and use of *Orobanche*-contaminated tomato containers.

HISTORICAL MEASURES FOR *OROBANCHE* CONTROL

The first scientific paper on *Orobanche* spp. was probably published in Germany in 1829 (Wegmann, 1998). Published information for *Orobanche* management started in the mid 19th century, but most of the literature appeared after World War II, increasing towards the end of the second millennium (Wegmann, 1998;

Parker and Riches, 1993). Most control measures in the past have been of limited success because they were inefficient or expensive or hard to achieve.

Prevention

Methods aimed at preventing *Orobanche* infestation are directed at avoiding seed production and distribution. Since massive amounts of tiny seeds are produced continuously for many weeks and are easily dispersed in air, water and soil, it is almost impossible to prevent seed transfer from a heavily infested site to its close surroundings. In such cases the only prevention method for the farmer is the discontinuation of growing *Orobanche* host crops, but this measure does not prevent this site from being a source of infestation by seed transfer or by the parasitism of *Orobanche* on susceptible weeds. Prevention methods could be more effective if the initial specific infestation is sparse and measures are taken to prevent long distance seed transfer. The individual farmer could be responsible for some of the prevention measures, but in most cases cooperation with neighbors and intervention by local, regional or national authorities is required. However, in all cases it is important to document and monitor every initial spot of infestation as well as infested fields in the neighborhood.

Jacobsohn (1986) lists the phytosanitary measures for avoiding the dispersal of *Orobanche* seeds from infested locations as follows:

1. Prevent movement of infested soil by vehicles, farm machinery and planting material;
2. Clean farm machinery and equipment i.e., tractors, combines, containers, etc;
3. Prevent spreading seeds by farm animals. Since *Orobanche* seeds may pass animal digestive systems without losing viability, grazing on or feeding hay from broomrape infested fields is prohibited, as well as using animal manure from unknown sources;
4. Do not use infested seeds or planting material, only officially certificated plant material;
5. Do not use *Orobanche*-contaminated water from ponds or reservoirs since they may be a source of seed infestation;
6. Prevent *Orobanche* seed dispersal by wind or water soil erosion.

Cultural Control

The scientific literature documents very successful cultural methods for *Orobanche* management with excellent results of preventing parasite damage and significantly increasing crop yields, but in many cases successful management by similar means could not be repeated. There is considerable variation in *Orobanche* populations and their virulence, and different responses to various environments such as soil conditions, humidity, and temperature. Linke (1999) listed some of the cultural measures for *Orobanche* management. These are: crop rotation, altering of sowing dates, tillage systems, soil fertility, flooding, and hand weeding.

Crop Rotation, Trap and Catch Crops

A crop rotation system includes *Orobanche* host crops, trap crops and catch crops and non-host crops. Most publications and reviews dealing with *Orobanche* control and management describe crop rotation as a strategy for reducing parasite infestation, but only few suggest concrete guidelines. One exception is the proposal of rice cropping in which flooding throughout the growing season destroys *Orobanche* seeds (Parker and Riches, 1993; Sauerborn and Saxena, 1986). Theoretically, repeated planting with non-host crops for many seasons should deplete the parasite seed bank in the field. However, we have evidence of very heavy *O. aegyptiaca* infestations of fields after 30-35 years of repeated non-host cultivation and cases of *O. crenata* infestations following more than 20 years of farming various non-hosts. There is an agreement that monoculture with the same *Orobanche* host crop, or with other hosts of the same *Orobanche* species, rapidly increases *Orobanche* infestation. We have documented evidence that a small spot of infestation could develop into a large-scale heavily infested field as a result of 2-3 years of mono cropping (Kleifeld, unpublished).

The major factor for *Orobanche* infestation in a crop rotation system is the long viability of its seeds. Thus, heavy infestations may remain in a field despite absence of host crops for several years. Parasite seeds buried in the soil beneath the crop root zone by deep plowing can be brought up to the crop host root zone a few years later as a result of additional plowing. Infestation can also be sustained by susceptible host weeds growing in the field and on field borders (Braun et al., 1984).

The best rotational crops for reducing *Orobanche* seed bank in infested fields are "trap crops"- crops which exude stimulants that induce *Orobanche* seed germination but no viable attachment to the host roots is established, i.e., the stimulation yields suicidal parasite germination. Resistant varieties that induce parasite seed germination, but do not support the young parasite after attachment, may serve as excellent trap crops as well (Goldwasser et al., 1997; Eizenberg, 2002). In the scientific literature many efficient trap crops are mentioned, but most of them were examined *in vitro* or in small pots. Our experience with *O. aegyptiaca*, and *O. cumana* in heavily infested *Orobanche* fields is that a 20-30% reduction in seed germination does not affect infestation and crop yield reduction.

- Trap Crops for *O. crenata* are: Sorghum (*Sorghum vulgare* Pers.), barley (*Hordeum vulgare* L.), vetch (*Vicia villosa* var. *dasycarpa)* and purple vetch. (*V. atropurpurea*), clover (*Trifolium alexandrinum*), flax (*Linum usitatissimum*), and coriander (*Coriandrum sativum*).
- Trap Crops for *O. cernua, O. aegyptiaca* and *O. ramosa* are: Pepper (*Capsicum annuum*), sorghum (*Sorghum bicolor* (L.) Moench.), cowpea (*Vigna unguiculata* (L.) Walp.), hemp (*Hibiscus subdariffa*), mungbeans, (*Phaseolus aureus*) flax, alfalfa (lucerne) (*Medicago sativa* L.), soybean (*Glycine max* (L.) Merr.), vetches (*Vicia* spp.) and chickpea (*Cicer arietinum* L.) (Krisnamurthy and Rao 1976; Krishnamurthy et al., 1977; Abu-Irmaileh, 1984; Sauerborn and Saxena, 1986; Al-Menoufy, 1991; Saxena et al., 1994; Kleifeld et al., 1994a).

There is contradicting information about the efficacy of trap crops: the same crop species is described in one case as reducing 50–70% of the parasite seed bank and in another case it achieves very poor results (Kleifeld et al., 1994a; Parker and Riches, 1993). The reasons for these discrepancies could be the influence of the tested media, aqueous solution versus soil and the close connection between trap crop root systems with the parasitic seeds, which occur in small pots, compared to the more sparse relations in a naturally infested field. In addition, many researchers do not specify the trap crop variety. In some cases there are extremely different responses among varieties of parsley, pepper, cabbage and vetch to the same *Orobanche* species. Moreover, different populations of the same *Orobanche* species, collected from various locations, responded differently to the same flax variety used as a trap crop (Goldwasser et al., 1991; Kleifeld et al., 1994a).

An additional cultural means for reducing *Orobanche* seed bank in the soil is use of "catch crops"- i.e., planting an *Orobanche* host crop, which will be destroyed after inducing parasite seed germination and attachment. In this case vigorous and densely planted host crops are preferred, such as faba bean (*Vicia faba*), for *O. crenata, O. aegyptiaca* and *O. ramosa* and white mustard (*Sinapis alba*) and lentil (*Lens culinaris*) for *O. aegyptiaca* and *O. ramosa*. Most of the reviewers (Parker and Riches, 1993; Sauerborn and Saxena, 1986; Dhanapal et al., 1996) point out that in infested fields several repeated cropping seasons with catch crops were needed for significant reduction of the parasite seed bank. This cropping system is not economical since most of the catch crops are used as green manure. Even an efficient catch or trap crop stimulates only a low germination percentage of the parasitic seed bank in a single season. These values are very disappointing for a heavy infestation, but could be important as components of an integrated *Orobanche* management strategy (Kleifeld et al., 1994b).

Planting Dates

There are reports that altering of planting dates influences *Orobanche* attack and damage to host crops. In most cases this phenomenon is a result of parasite and host plant differential response to low temperatures. The best documented cases of decreased parasitism are for delayed planting of *O. crenata* on faba bean (van Hezewijk, 1994; Mesa-Garcia and Garcia-Torres, 1984), on lentil and chickpea (Linke and Saxena, 1991), and on carrot (Eizenberg et al., 2001a). Similar cases were reported for *O. aegyptiaca* attacking carrots, tobacco, pepper and *Brassica campestris* L. (Chaudhary, 1995), *O. cumana* attacking sunflowers (Castejon-Munoz et al., 1993), and *O. cernua* attacking pepper. The same delay in planting dates in non-infested fields results in yield reduction, thus researchers suggested overcoming this problem by using early maturing varieties (Linke and Saxena, 1991).

Tillage System

There are many reports on the use of tillage systems as cultural means for *Orobanche* management (Pieterse, 1979; Parker and Riches, 1993; Linke, 1999). Deep plowing seems to reduce *Orobanche* infestation and crop damage. Following build-up of a heavy parasite seed-bank in a field which was not deeply cultivated, plowing to a depth of 30-40 cm could bury most of *Orobanche* seeds to a depth that is hardly reached by host roots, at least at the critical early growth period. Repeated experiments with this method did not always yield similar results. Additional deep plowing in the next season brings the seeds up again and renews infestation. In light soils, deep plowing mixes soil layers and does not turn them over. Our experience has proven that shallow cultivations following solarization or soil fumigation with methyl bromide on heavy clay soils prevented re-infestation with *Orobanche* by avoiding the raising of parasite seeds from deep soil layers where control was incomplete.

Soil Fertility

Increasing soil fertility by adding fertilizers or manure was suggested as a tool for *Orobanche* suppression. Improving soil fertility affects the ability of the host plant to cope with the parasitism, primarily on *Orobanche*-infested sites that are located on poor soils (ter Borg, 1986). Adding nitrogen fertilizers is reported to have a positive effect on reducing *Orobanche* infestation and damage while no effect was achieved by adding phosphorus and potassium (Linke, 1999). *In vitro* and pot experiments have proved that adding nitrogen compounds significantly decreased *Orobanche* seed germination (Parker and Riches, 1993). Furthermore, researchers demonstrated that nitrogen in the ammonium form inhibits radicle elongation of *Orobanche* spp. seedlings (Westwood and Foy, 1999). In moist soils, ammonium is rapidly converted into nitrate, therefore its effect on *Orobanche* seeds is short and limited (Parker and Riches, 1993).

Flooding and Irrigation

Flooding is mentioned as a cultural mean for managing *Orobanche*. A significant effect was achieved when infested fields were flooded for a long period, i.e., as with rice growing, but when flooding was for shorter periods, the effect was not consistent (Parker and Riches, 1993; Linke, 1999). There are reports that irrigation reduces *Orobanche* infestation (Linke, 1999), but it is not clear if the effect is due to an increase in host crop vigor, or to a direct effect on the parasite. At least in the case of *O. cumana* attacking sunflower in Israel, total yield loss is common under dry land cropping; while irrigating at the time of sunflower flowering and seed filling stage ensures some commercial yield.

Hand Weeding

Hand weeding is extremely important in stopping the spread of infestation from the foci of new developing infestations. Hand weeding of emerged parasite shoots in infested fields has only minor affect on the crop yield since damage has already occurred. Careful and continuous hand weeding of shoots prior to seed distribution might decrease the parasite seed bank for future crops, but it will take many years to achieve significant seed bank reduction in a heavily infested field. It is important to destroy the collected *Orobanche* flowering shoots since seed maturing and dispersal continues from detached shoots (Parker and Riches, 1993).

Chemical Control

Foy et al. (1989) thoroughly reviewed the attempts of chemical control of *Orobanche*. The authors list thirty-six pesticides from various groups, which showed activity on the parasite at different growth stages. Most of the presented results came from pot experiments and could not have been repeated under field conditions because most of the mentioned herbicides were not selective to the host crops and their use would be limited. Another practical problem is that most of the selective herbicides were limited in their movement in the soil, so despite being selective to the host crop they would not reach parasite seedlings that germinate adjacent to host crop roots in deep soil layers.

In conclusion, an effective chemical for controlling *Orobanche* seed-bank has to be selective for crops grown in the following season, or have single season residual effect. For the control of root-attached *Orobanche*, host crop selectivity is the most important characteristic, but the movement of the active chemical in the host plant towards the new roots, or in the soil into the host crop root zone is also very essential. Three groups of chemicals have been shown to possess potential to control *Orobanche*. These are: soil fumigants, residual soil applied herbicides and post emergence applied herbicides.

Soil Fumigants

Methyl bromide. Wilhelm et al. (1958) suggested injecting methyl bromide as a preplant soil application. The fumigant showed good activity in controlling some dangerous soil born diseases, insects, soil nematodes and weeds. Since methyl bromide is not active on pests at dry dormant stages, it is important to keep the treated soil layer moist for a few days prior to application. To avoid evaporation the application is performed as "hot gas" under plastic sheets or "cold gas"-injected under plastic mulch. The plastic cover has to be kept for at least 24 hours and the treated soil must be ventilated to prevent toxic effects on the following planted crop (Wilhelm, 1962).

Methyl bromide application succeeded as a large-scale national project in eradication of *O. ramosa* in introduced infestations in California (Wilhelm et al., 1959), but failed to give long-term control when applied in small private farms in Israel (Kleifeld at al., 1987b). Californian farmers signed a contract to stop growing *Orobanche* host crops after the fumigation, and hence prevented the development of parasite seeds that had escaped treatment from becoming a new source of infestation. Israeli farmers used methyl bromide for renewing the fertility of small fields for intensive cropping and repeated planting of *Orobanche* host crops. This resulted in rapid *Orobanche* re-infestation originating from seeds that escaped control or were easily transferred from neighboring infested fields.

An important reason for some failures of methyl bromide applications was the medium to heavy clay soils, which were difficult to fumigate properly. It is very difficult in these soils to reach a uniform optimal soil water content (about 40-60 % of field capacity) needed for preconditioning of *Orobanche* seeds and enabling fumigant mobility and distribution. Soil cultivation of such wet clay soil causes soil compaction, leading to soil structure damage and decreased pest control. However, in Israel, methyl bromide is currently used by individual farmers in greenhouses and open fields prior to planting high value, vegetable and field crops every 5-6 years in a highly intensive crop rotation system. Finally, world health and agricultural authorities are gradually reducing and ultimately banning the use of methyl bromide by 2005 because of its negative environmental effects (United Nation Environmental Protection Service, 1992).

Methylisothiocyanate. Fumigation by compounds that release methylisothiocyanate was suggested for *Orobanche* eradication. Metham sodium, applied directly by injection or by chemigation via irrigation systems into the soil, or dazomet incorporated mechanically into the soil, followed by irrigation that releases the toxic ingredient, were found to be very effective for *Orobanche* control. Methylisothiocyanate was effective in deeper soil layers, but very ineffective on the surface, because of its rapid evaporation (Goldwasser et al., 1995).

1,3-dichloropropen. This fumigant was applied by injection into the soil followed by sealing of the surface by rolling or frequent sprinkler irrigation (Jacobsohn et al., 1991). This treatment was very effective against *O. crenata*, but less potent against *O. aegyptiaca.* The difficulties in application of 1,3-dichloropropen and the narrow pest control range limit its utilization to small-scale intensive farming only. Other tested soil fumigants, which had shown some activity in *Orobanche* control, failed due to difficulties in field application or were eventually banned because of danger to the user and/or the environment.

Residual Soil Applied Herbicides

Most soil applied herbicides, especially those showing host crop selectivity were tested for efficacy in *Orobanche* control. The results of these tests were generally negative.

Propyzamide. This herbicide that is mainly used as a selective herbicide in Fabaceae and Asteraceae crops, efficiently controlled *Orobanche* spp. in infested pot experiments. When propyzamide was applied to *O. cumana*-inoculated sunflower in the field, only delayed parasite infestation was achieved (Kleifeld et al., 1987a). The low field efficacy of propyzamide in *Orobanche* control is a result of a short soil residual effect in warm and moist soils, which is common in the sunflower growing season. Preplant soil incorporation of higher propyzamide doses improved *Orobanche* control, but did not prevent *Orobanche* parasitism in heavily infested fields.

Dinitroanilines. Several reports were published on the effect of mechanically incorporated herbicides belonging to the dinitroaniline group (trifluralin, dinitralin, pendimethalin etc.) on *Orobanche* (Parker and Riches, 1993), but none were efficient in field trials.

Post Emergence Applied Herbicides

The first and most important innovation in this area was the report by Kasasian (1973a,b) regarding the tolerance of faba bean to glyphosate applied in repeated treatments at low rates. Low rates of glyphosate were sufficient to achieve *O. crenata* control. Other researchers adopted this information and were able to show effective parasite control by glyphosate mainly in various legumes. Other crops belonging to the Solanaceae and Asteraceae were heavily damaged by the same glyphosate doses. In practice, glyphosate treatment was rarely adopted by farmers suffering from *Orobanche* infestation because of difficulties in deciding about the necessity of treating an invisible enemy and the determination of accurate timing and rate for the split application. Moreover, mixed cropping of various species in the same field could damage the glyphosate sensitive crops. The use of glyphosate in recent control approaches and as a part of an integrated *Orobanche* management system is discussed later.

RECENT *OROBANCHE* CONTROL APPROACHES

Chemical Control

Glyphosate

Glyphosate is a nonselective broad spectrum, post-emergence, foliar applied herbicide that is absorbed through the leaves and is translocated primarily in the symplast from the point of application to all metabolic active parts of the plant ("metabolic sinks"). The mode of action of glyphosate is inhibition of the enzyme EPSP synthase of the shikimic acid pathway, inhibiting aromatic amino acid synthesis and thus protein synthesis and growth (Amrhein et al., 1980). The attached root parasite acts as a strong metabolic active part of the host plant, rapidly absorbing and accumulating the herbicide through its direct connection with the host root vascular system. Despite the fact that glyphosate is defined as a non-selective herbicide, some plants belonging to the Apiaceae and Fabaceae families tolerate low rates of glyphosate. Tolerance of plants to glyphosate has been mainly attributed to degradation of the herbicide to nontoxic metabolites. This phenomenon has been exploited for the selective control of *Orobanche* with glyphosate in tolerant crops. To increase efficacy and selectivity, herbicides should be split-applied at low rates at precise timings during early *Orobanche* development stages (Table 1). It should be mentioned that glyphosate inhibits flavonoid phytoalexin biosynthesis and might cause leaf diseases on treated crops (Sharon et al., 1992).

As mentioned earlier, glyphosate was first suggested by Kasasian (1973a,b) and later by Arjona-Berral et al (1988) for *O. crenata* control in post emergence application in faba bean, pea and lentil. Effective *O. crenata* control in faba bean was achieved by post application of 60-80 g/ha in Morocco (Schmitt et al., 1979), Egypt, Syria and Spain (Kukula and Masri, 1984; Mesa Garcia et al., 1984). The small parasitic tubercles attached to the host plant roots were sensitive to the herbicide in comparison with the relatively high tolerance of the host plant. The herbicide might be applied repeatedly, with no significant damage to the crop. The same phenomenon was found later in vetch, celery, carrot and lately in parsley (Jacobsohn and Kelman, 1980; Jacobsohn and Levy, 1986; Kleifeld, 1999). In vitro studies showed that glyphosate at 0.2-0.4 kg/ha provided excellent suppression of *O. aegyptiaca* in *Vicia sativa* L. While absorption and translocation was similar between parasitized and non-parasitized plants, up to 32% of ^{14}C-glyphosate was metabolized by *V. sativa* and 27% of the translocated labeled herbicide was accumulated in *O. aegyptiaca* tubercles attached to *V. sativa* roots (Nandula et al., 1999). Glyphosate effectively controlled *O. aegyptiaca* in naturally infested parsley fields. In recent field trials, the application of 36 g/ha glyphosate two weeks after each parsley cutting completely controlled *Orobanche* with no crop injury (Goldwasser et al., 2003).

The recent development of genetically engineered glyphosate-resistant crop varieties creates new possibilities for *Orobanche* control and will be discussed later in this chapter.

Sulfonylureas

The sulfonylureas are systemic herbicides absorbed through roots, shoots and foliage of plants with rapid acropetal and basipetal translocation through the xylem and phloem. The mode of action of these herbicides is inhibition of acetolactate synthase (ALS) also called acetohydroxyacid synthase (AHAS), a key unique plant enzyme in the biosynthesis of the branched-chain amino acids isoleucine, leucine and valine (Schloss, 1995). Blockage of this pathway interferes with synthesis and cell growth. These herbicides accumulate in meristematic growing points of the plant such as apical buds at the tips of roots, shoots, leaves and reproductive structures. These herbicides are effective in low rates when applied pre-emergence and post-emergence on annual and perennial grass and broadleaf weeds. Sulfonylurea herbicides can persist in the soil from several weeks to several years depending on the herbicide and application rate. Soil degradation can be by microbial action and chemical hydrolysis.

Plant selectivity to the ALS inhibiting herbicides is primarily through rapid degradation to nontoxic metabolites while susceptible species are unable to detoxify the herbicide before plant death occurs. Plants can metabolize sulfonylurea herbicides through hydroxylation of the phenyl ring by cytochrome P450 (Frear et al., 1991; Hinz et al., 1997). Selectivity of wheat and other crop and weed grasses to chlorsulfuron triasulfuron and sulfosulfuron, and of corn to rimsulfuron is due to plant rapid metabolism of the herbicide (Ahrens, 1994; Hatzios, 1998; Olson et al., 2000). The mode of resistance of specific Solanaceae crop species to sulfonylurea herbicides has not been studied, but is probably due to metabolism of the herbicide to non-phytotoxic products, as reported for *Solanum nigrum* L. tolerance to chlorsulfuron (Ahrens, 1994). Resistance could also be due to an altered ALS/AHAS binding site as in selected mutations in transgenic crops. Sulfonylurea–resistant weed biotypes selected in the field by heavy use of one of the herbicides usually have cross-resistance to other sulfonylurea herbicides. The mechanism of resistance in this case is an altered herbicide-binding site in the ALS target enzyme (Beyer et al., 1988; Ahrens, 1994).

Chlorsulfuron. This herbicide is one of the first and extensively used sulfonylurea herbicides developed for control of a wide range of weeds in wheat, barely and oats. Chlorsulfuron applied pre-emergence and post-emergence as a soil or foliar treatment is highly potent and highly persistent in the soil (Ahrens, 1994).
At first we conducted in vitro studies to test the herbicides activity on *O. aegyptiaca* germination and radical elongation in Petri dish without host plants. Ten μM chlorsulfuron reduced radical length by 60%. Next we performed laboratory studies with *Orobanche* parasitizing tomatoes in nutrient solutions in plastic bags and found that 0.25 μM chlorsulfuron applied at very early parasitism stages (up to tubercle formation), killed all parasites on tomato roots (Hershenhorn et al., 1998a). These promising results led to greenhouse studies in which tomatoes were grown in *Orobanche* contaminated soil in pots. In these studies we found that 3.75 g/ha of chlorsulfuron applied directly to the soil completely controlled *O. aegyptiaca* (Hershenhorn et al., 1998b). This herbicide, which has a longer soil activity than most other sulfonylureas, was applied directly into the soil by "chemigation" (delivering of the herbicide through irrigation water) to tomato transplants after establishment in the field. The phytotoxic contact of the herbicides with the host foliage was avoided by using very dilute solutions, and by washing of the herbicide into the soil by additional sprinkler irrigation. Three split applications of 2.5 g/ha chlorsulfuron through sprinkler irrigation, starting at 14 days after tomato planting and at intervals of 10-14 days followed at each application by 300 m^3/ha irrigation, controlled 80-90% *Orobanche* without phytotoxicity to the tomato crop (Table 1). Single or double split application, application through a drip system or at high volume spray did not sufficiently control *Orobanche* throughout the growing season (Hershenhorn et al., 1998c). Because of the soil persistence of chlorsulfuron, we applied the herbicide through drip-irrigation systems to control late season *Orobanche* emergence around drip emitters in tomato. This treatment effectively controlled *Orobanche* emergence (Kleifeld et al., 1999), but its efficacy was inconsistent in other trials. According to our experience, the success of this application depends on specific irrigation, soil and field conditions.

Triasulfuron. This sulfonylurea herbicide was developed for pre and post-emergence control of broad leaf weeds in wheat and barley and categorized as somewhat persistent (Ahrens, 1994). Triasulfuron was tested for *O. aegyptiaca* control concurrently with chlorsulfuron in laboratory, greenhouse and field experiments as described above. Triasulfuron at 10 μM reduced parasite radicle growth by 56% in Petri dish, efficiently controlled *O. aegyptiaca* on tomato at 2.5 and 25 μM in plastic bag studies and at 3.75 g/ha in greenhouse experiments. In tomato field studies, three chemigation applications of 7.5 g/ha efficiently and selectively controlled *O. aegyptiaca* (Table 1). This herbicide was less potent and with shorter residual activity on *O. aegyptiaca* than chlorsulfuron. A single application of 7.5 g/ha triasulfuron sprayed on potato foliage severely damaged the crop and further experimentation with the herbicide in potato was discontinued (Goldwasser et al., 2001).

Rimsulfuron. This sulfonylurea herbicide was developed originally for early post emergence control of broad leaf and grass weeds in corn (Hatzios, 1998). The selectivity of this herbicide to the Solanaceae family led to its registration for weed control in tomato and potato (Reinke et al., 1991). Rimsulfuron degrades rapidly in soil and water mainly via chemical pathways while microbial degradation plays a minor role. Because of its rapid degradation, rimsulfuron is categorized as non-persistent.

Rimsulfuron was tested concurrently with chlorsulfuron and triasulfuron as mentioned above. Rimsulfuron at 10 μM reduced parasite radicle growth by 48% in Petri dish, efficiently controlled *O. aegyptiaca* on tomato at 5 μM in plastic bag and pot studies. However, in drip-irrigated tomatoes in the field *Orobanche* control was poor (Hershenhorn et al., 1998a,b,c). Since the residual activity of rimsulfuron in soil is short, repeated applications were necessary for *O. aegyptiaca* control throughout the growth season (Kleifeld et al., 1994b).

Three repeated doses of rimsulfuron at 12.5 g/ha each followed by irrigation, sprayed on potato foliage two weeks after crop emergence and re-applied at two week intervals effectively and selectively controlled *O. aegyptiaca* with no damage to potato yield or tuber quality (Goldwasser et al., 2001). Rimsulfuron achieved efficient *Orobanche* control in potato fields but not in tomato fields because potato fields are sprinkler irrigated while tomato fields are drip-irrigated and the herbicide is rapidly leached around the drip emitters.

Sulfosulfuron. This recently developed sulfonylurea herbicide is registered worldwide for pre and post-emergence of grass and broad leaf weeds in wheat. This systemic and somewhat persistent herbicide is absorbed through foliage and roots of plants with rapid acropetal and basipetal translocation. Though sulfosulfuron was initially developed and registered for controlling an array of grass and broad leaf weeds in wheat, its selectiveness to some broadleaf crop species has recently led to its registration for weed control in potato in Poland (Anonymous, 1995; Hatzios, 1998).

The mode of resistance of specific Solanaceae species to sulfosulfuron has not been studied, but is probably due to metabolism of the herbicide to non-phytotoxic products. In extensive research conducted in Israel during the last two years, sulfosulfuron has proven to be highly efficient and selective for *O. aegyptiaca* control in processing tomatoes (Eizenberg et al., 2001b). While chlorsulfuron and triasulfuron were most effective when applied by chemigation, sulfosulfuron can be sprayed on tomato foliage followed by sprinkler irrigation to wash the herbicide into the soil where it is absorbed directly by the young parasites or via the tomato host roots. Greenhouse experiments with activated charcoal suggested that the herbicide acts mainly through the soil and not by translocation through the host tomato plant. To achieve good parasite control, high herbicide rates at early developmental stages of the parasite are needed: two or three applications of 37.5 g/ha starting two weeks after tomato planting and repeated at two week intervals (Table 1). The recommended application rates for weed control in wheat and potato are a single treatment of 10-40 g/ha only.

Imidazolinones

The imidazolinones are ALS-inhibiting herbicides with the same mode of action and similar characteristics as the sulfonylurea herbicides. These herbicides are used pre-emergence and post-emergence for control of annual and perennial grass and broadleaf weeds. Most of these herbicides have medium to long soil persistence. Residual weed control by imidazolinones varies from a week up to two years or more depending on the specific herbicide and application rate. Degradation in the soil is primarily through microbial breakdown. Dry weather and cool temperatures are responsible for longer persistence in the soil due to low microbial activity under these conditions (Shaner, 1991).

Various legumes are resistant to some of the imidazolinone herbicides and this resistance has led to selective use of these herbicides in certain legume crops. Legumes are tolerant to imazapyr because they can metabolize it to an inactive form (Shaner, 1989). Some weed biotypes have altered ALS binding sites conferring resistance to imidazolinone herbicides. Garcia Torres et al. (1998) reported selective *O. crenata* control in faba bean by pre-emergence and post-emergence applications of imazethapyr, imazapyr and imazaquin. In our studies we have found that crops belonging to other botanical families are imidazolinone-tolerant: split application with various imidazolinone herbicides on potato, sunflower and parsley foliage selectively controlled *O. ramosa, O. cumana,* and *O. crenata* respectively. In these cases the herbicides were extensively translocated to the attached root parasite directly through the host plant, in contrast to the mode of control with sulfonylurea herbicides that act on the parasite directly through the soil. This method eliminates the need for irrigation following application.

Imazethapyr. This herbicide was developed for the control of many broadleaf and grass weeds by pre-plant, pre-plant incorporated and post-emergence applications in soybean, peanuts and edible legumes (Ahrens, 1994). This herbicide was the first of the imidazolinone group to be registered for *Orobanche* control. A post emergence application of 20 g/ha on garden and field pea (*Pisum sativum* and *Pisum arvense*, respectively) one month after planting, and an additional treatment of 20-40 g/ha two weeks later, was selective to pea and efficient in *Orobanche* control (Jacobsohn et al., 1998) (Table 1). Imazethapyr has been registered at these rates for post-emergence *O. crenata* control in peas in Israel and at 75-100 g/ha pre-emergence applications *for O. crenata* control in faba bean in Spain (Garcia-Torres et al., 1998). There are reports of some promising results of *O. crenata* control by faba bean and pea seed treatments with imazethapyr (Jurado-Exposito et al., 1996, 1997, 1999).

Imazapic. A recently developed imidazolinone herbicide, imazapic is registered for pre and post-emergence control of grass and broad leaf weeds in peanut and in soybean (Thomson, 1997).

In field screening of imidazolinone herbicides, imazapic was the most selective to parsley and extremely toxic to young *Orobanche*. In large-scale commercial field experiments we completely controlled *O. crenata* and *O. aegyptiaca* by split application of imazapic at 2.5-5.0 g/ha (Goldwasser et al., 2003). We applied the herbicide by spraying on parsley foliage at the rosette stage with 5-7 leaves before the first cutting, and on the new sprouts after each cutting (Table 1). Imazapic may cause slight crop retardation when crop growth is slowing down, in the winter and at very high temperatures in the summer (Kleifeld et al., 1998). In carrots, three applications of 5 g/ha each, controlled *O. aegyptiaca and O. crenata* (Jacobsohn et al., 1996; Kleifeld et al., 1998).

In peanuts, two applications of 2.4 g/ha imazapic each, sprayed 60 DAP and 80 DAP on crop foliage, completely controlled *O. aegyptiaca* with no crop damage (Birati and Benyamini, 1997).

Table 1. Chemical control of *Orobanche* in field trials.

Herbicide	Formulation	Rate (g ai/ha)	Application Method	Crop	*Orobanche* controlled	Reference
Glyphosate	360 g/l SL	80 + 80 + 80	POST[a]	Faba bean	*O. crenata ; O. aegyptiaca*	Kukula and Masri (1984) Messa Garcia et al. (1984)
		90 + 90	POST	Vetch	*O. crenata ; O. aegyptiaca*	Kleifeld (1999)
		36 + 36 + 36	POST	Parsley	*O. crenata ; O. aegyptiaca*	Kleifeld (1999)
		36 + 36 + 36	POST	Carrot	*O. crenata ; O. aegyptiaca*	Jacobsohn and Kelman
		54 + 54	POST	Celery	*O. crenata*	(1980); Kleifeld (1999)
Sulfonylureas						
Chlorsulfuron	75% WG	2.5 + 2.5 + 2.5	CHEM+IRRI[b]	Tomato	*O. aegyptiaca ; O. cernua*	Hershenhorn et al. (1998c)
Rimsulfuron	25% WG	25 + 25 + 25	CHEM+IRRI	Tomato	*O. aegyptiaca*	Kleifeld et al. (1994a)
		12.5 +12.5+ 12.5	CHEM+IRRI	Potato	*O. aegyptiaca,*	Goldwasser et al. (2001)
Triasulfuron	75% WG	7.5 + 7.5 + 7.5	CHEM+IRRI	Tomato	*O. aegyptiaca; O. ramosa*	Hershenhorn et al. (1998c)
Sulfosulfuron	75% WG	37.5 + 37.5 + 37.5	POST+IRRI[c]	Tomato	*O. aegyptiaca*	Eizenberg et al. (2001b)
Imidazolinones						
Imazapic	240 g/l SL	2.4 + 4.8 + 4.8	POST	Parsley	*O. aegyptiaca; O. crenata*	Kleifeld et al. (1998)
		2.4 + 2.4 + 2.4	POST	Carrot	*O. aegyptiaca; O. crenata*	Kleifeld et al. (1998)
		2.4 + 2.4	POST	Peanut	*O. aegyptiaca*	Birati and Benyamini (1997)
		2.4 + 2.4 + 2.4	POST	Sunflower	*O. cumana*	Aly et al. (2001)
Imazethapyr	100 g/l SL	20 + 20+ 20	POST	Pea	*O. crenata*	Jacobsohn et al. (1998)

POST= Sprayed on host foliage at 200 l/ha spray volume.
[b]CHEM+IRRI= Chemigated by sprinkler irrigation and washed into the root zone by additional sprinkler irrigation.
POST+IRRI= Sprayed on host foliage at 200 l/ha spray volume and followed by 300 m^3/ha sprinkler irrigation.

Two to three sequential applications of 2.4 g/ha imazapic, starting at the 4-6 leaf growth stage and continuing at two week intervals, efficiently and selectively controlled *O. cumana* infestation in dry land confectionary sunflowers. Only a few degenerated inflorescences emerged above the soil in the treated plots. This dramatic control doubled the yield of the imazapic-treated plots compared to the untreated control. In irrigated sunflower, a new wave of *O. cumana* emerges following the irrigation that is applied at the flower bud stage. At this stage imazapic treatments have to be discontinued due to their damage to sunflower inflorescence (Aly et al., 2001).

In *O. ramosa* and *O. aegyptiaca*-infested potato fields, three doses of imazapic at 4.5 g/ha each, sprayed two weeks after crop emergence and re-applied at two week intervals controlled *Orobanche* infestation. Although imazapic treatments increased crop vigor and potato yield, in light sandy loam fields imazapic treatments caused deformation of tubers (Goldwasser et al., 2001). In field studies with *O. aegyptiaca*-infested processing tomatoes, imazapic at three applications of 5 g/ha provided excellent *Orobanche* control, but caused dropping of tomato flower buds, vigorous vegetative growth and delay in fruit ripening (Kleifeld et al., 1998).

Herbicide Resistant Crops

The recent development of transgenic herbicide-resistance crops, and especially those resistant to amino-acid inhibiting herbicides, has opened up new opportunities for *Orobanche* control. The use of these transgenic crops for parasitic plant control will intensify in the future with the identification and utilization of additional herbicide-resistant genes.

It was proposed that target site herbicide resistance would allow the systemic translocation of the un-metabolized herbicide to the attached parasite (Gressel, 1992). Metabolism-based herbicide resistance can be used when applying the herbicide to the parasites directly through the soil or in cases in which the attached parasite rapidly absorbs the herbicide from the treated host before it has been detoxified or when multiple applications are made to compensate for the herbicide metabolism.

The use of herbicide-resistant crops as an *Orobanche* control measure was suggested by Foy et al. (1989) and later examined by Joel et al. (1995). Chlorsulfuron–resistant tobacco, glyphosate–resistant rape and asulam-resistant tobacco planted in *Orobanche*-infested pots were treated with their respective herbicides. Later, asulam-resistant potato plants were specifically engineered for *Orobanche* control (Surov et al., 1997). In all the above cases the herbicides were applied to the host crop foliage, translocated to the roots and effectively reduced the number of attached *O. aegyptiaca* parasites. The application of glufosinate to glufosinate-resistant tomato based on metabolism resistance, and not target site resistance, did not control *O. aegyptiaca* (Joel et al., 1995). This phenomenon is explained by the metabolic degradation of the herbicide but could have been caused because this herbicide is non-mobile in xylem and phloem. All the above-reported cases were conducted in the lab but no follow-up research has been conducted in commercial fields.

An additional approach for utilizing herbicide resistant crops for parasitic plants control is seed coating of herbicide resistant susceptible crops. ALS-resistant corn seed dressing with the magnesium salt of imazapyr was effective in *Striga hermonthica* control (Abayo et al., 1997; Kanampiu et al., 2001; Kanampiu et al., 2002). Similar results are reported with asulam dressing of asulam-resistant tobacco seeds (Joel et al., 2000).

We have found that amino acid-inhibiting herbicides are very effective in *Orobanche* control. As a general rule, glyphosate and imidazolinone herbicides were found to be selective and thus efficient mainly on Fabaceae, Asteraceae and Apiaceae while sulfonylurea herbicides were selective mainly to Solanaceae crops. In our experiments imidazolinones were more damaging to host plant flowering and thus have to be applied early in the season or to crops of which the inflorescences are of no economic importance. Imidazolinones could have adverse affects on underground storage organs such as potato tubers and carrots. Efficient control was achieved when herbicides were delivered at early stages of parasitism, either by translocation through the host plant (glyphosate, imidazolinones) or directly through the soil solution (sulfonylureas). In all cases repeated applications, usually at up to 2 week intervals were needed to insure a prolonged parasite-lethal concentration of the herbicide on one hand and a host-safe concentration on the other.

The frequent application of the same herbicide or herbicides from the same group on crops including genetically engineered herbicide resistant crops contains the danger of the development of herbicide-resistant weeds. Many ALS-resistant weed populations have evolved after 3-5 years of ALS-inhibiting herbicide use. These herbicides will exert a stronger selection pressure on parasitic plants (Gressel et al., 1996).

Solarization

Katan et al. (1976) were the first to use solar radiation as a control method for soil-borne plant pathogens and Jacobsohn et al (1980) described its effect on *O. aegyptiaca* control. The solarization method is based on the trapping of radiation in moist soil (achieved by irrigation prior to mulching or during the solarization process by drip irrigation) by clear polyethylene mulch for at least 5-6 weeks in the summer. *Orobanche* seed and soil-borne disease control is achieved in the upper 15-20 cm, where the temperature exceeds 50-55 ^{0}C. Effective solarization is achieved when the plastic mulch is maintained intact throughout the solarization period, and for large area applications, effective gluing techniques and relatively thick plastic films are needed. Although solarization applied under favorable conditions was effective in controlling many weed species, *Orobanche* control and rhizomes of perennial weeds from soil layers deeper than 15-20 cm escaped control especially on heavy clay and clay-loam soils (Parker and Riches, 1993; Linke, 1999). Solarization is used on limited areas, mainly integrated with lower rates of soil fumigants for soil-borne pest control including *O. crenata,* especially in very hot regions for shallow rooted cash crops, carrots and winter legumes (Kleifeld, 1999).

Biological Control

The young attached parasite acts as a strong metabolic sink on the host plant, rapidly accumulating water and nutrients for its latter growth stages. Such a nutrient-rich plant organ is a good target for attacking pests, some of them could be species-specific and could be used for biological control. The collection of biocontrol candidates from adult *Orobanche* shoots may yield secondary saprophytes only.

Fungi

Using fungi as *Orobanche* biocontrol agents could solve the difficulty of reaching the underground young parasitic organs at early parasitism stages. This will require careful host-specificity tests, since some host crop diseases could be found on the parasite organs too. Major problems have to be worked out following the isolation and identification of a very specific and useful fungus for *Orobanche* control. Techniques of mass production, formulation and shelf life have to be carefully considered, since these factors could determine the bioherbicide efficacy. Using a fungus as a biocontrol agent for early developmental stages of *Orobanche* must include an effective method of inoculation i.e., incorporation of inoculum into the host growth media. This could be preformed by watering, or by host seed dressing. In conventional farming many fungicides are applied to seeds, soil and plants, which could easily damage beneficial fungi used for biocontrol.

Attempts to isolate fungi from *Orobanche* spp. for biological control were described by several researchers (Amsellem et al., 2001a). Recently, two promising cases should be mentioned:

1. *Fusarium oxysporum* f. sp. *orthoceras,* which was isolated from sunflower broomrape (*O. cumana*) by Bedi and Donchev (1991). An organic substrate containing the fungus was incorporated mechanically into the upper soil layer (15 cm) of *O. cumana*-inoculated small-scale sunflower field trials. This treatment reduced *O. cumana* infestation in relation to the applied inoculum rates. The fungus attacks parasite seeds and all underground stages and is species specific (Bedi, 1994; Thomas et al., 1998, 1999);
2. *Fusarium arthrosporioides*, strain E4a and *Fusarium oxysporum* strain E1d, which were isolated from juvenile Egyptian broomrape (*O. aegyptiaca*) parasitizing Melon (*Cucumis melo*) (Amsellem et al., 2001a). Both strains were found very specific to *O. aegyptiaca, O. ramosa* and *O. cernua,* which parasitize pepper and tomato in the warm season, but not to *O. cumana* which parasitizes Fabaceae and Umbelliferae host species in the cool season. The two fungi were not pathogenic to the *Orobanche* host crops melon, potato, tomato, pepper, carrot, celery, chickpea and sunflower. Comparison of the pathogenic efficacy and shelf life showed that inoculation with chopped mycelia and conidia of both fungi was significantly better than inoculation with conidia only. An experimental formulation of chopped mycelia harvested from cultures effectively controlled young *Orobanche in vitro* after a year of storage at 5^0C (Amsellem et al., 2001b). Despite the promising results obtained in-vitro and in soil filled pots, in field trials *Orobanche* control was partial and not efficient. So far no commercial application of *Fusarium* spp. has been attempted in *Orobanche*-contaminated fields.

In addition to research aimed at improving formulation and application techniques, researchers have suggested improving biocontrol of pathogenic fungi by introducing genes for hypervirulence into the fungi (Amsellem, 2001b). A safety procedure suggested for preventing virulent lines from further development and spreading is the genetic engineering of the fungi with genes for suicide after completing their host control (Gressel, 2001). Another approach suggests improving the efficacy of the biocontrol fungi by adding low concentrations of fungal toxins to the mycoherbicide suspension. Such low amounts of toxins do not cause toxic effects, but suppress host defense mechanisms (Vurro and Ellis, 1997).

Bacteria

Published studies on effective bacterial agents for *Orobanche* control are scarce. Barghouthi et al (2000) reports that bacteria were isolated from diseased *Orobanche* shoots and various bacterial candidates showing *Orobanche* biocontrol activity *in vitro* were found.

Insects

Many insects feed on *Orobanche* spp., but few have been identified as monophagous. Most attention has been paid to the insect *Phytomyza orobanchia*, a *Diptera*, sub-order *Brachycera* (flies) of the family *Agromyzidae*. This small fly (2.6-2.8 mm long for the female) is restricted to *Orobanche* spp., including *O. crenata*, *O. aegyptiaca*, *O. ramosa*, *O. cernua*, *O. cumana* and *O minor*. Occurrence of *P. orobanchia* has been reported from most countries in which *Orobanche* spp. occur (Kroschel and Klein, 1998).

This insect and its life cycle have been well known for a long time and reported by Kaltenbach as early as 1872 (Kroschel and Klein, 1998). The adults are attracted by *Orobanche* flowers and feed on its nectar. Eggs are laid under the epidermis of shoots and flowers and the larvae penetrate into the plant tissue and feed on the shoots and capsules. The damage to the parasite is to seed production, which is reduced by 30-80% as a function of population density and environmental conditions. Temperature and humidity, as well as available *Orobanche* plants, dictate the number of the fly generations, which vary from one per year in Germany (Hendel, 1920) to six per year in the Kirghiz Republic (Moiseeva and Mamrailiev, 1969). In many temperate countries *Orobanche* (and especially *O. crenata*) flowering in early spring escape from heavy *Phytomyza* damage, since the low soil winter temperatures kill high amounts of pupa (Kurbanov, 1970). Deep soil tillage of infested *Orobanche* shoots destroys pupae and inhibits adults from hatching (Trenchev, 1981). There are no reports on the effect of high temperatures, but during the summer there is an increase in the natural enemies (bacteria, fungi and parasitoids) of *P. orobanchia* (Horvath, 1987). Rains or irrigation in summer *Orobanche*-host crops lead to heavy insect infestation resulting in intensive insecticide application, which controls *Phytomyza* as well (Okazova, 1973).

Mass release of *P. orobanchia* adults to increase their efficacy in reducing *Orobanche* seed production, was successful in the former Soviet Union and other East European countries (Trenchev, 1981; Horvath, 1987). Since no artificial feed was suitable, mass production was obtained by collecting and storing infected *Orobanche* shoots during the winter season (Klyueva and Pamukchi, 1982) and various methods were suggested for adult release (Kroschel and Klein, 1998). A reduction of up to 5% of viable *O. crenata* seed production was achieved by *P. orobanchia* release (650 flies per 10,000 *Orobanche* shoots), compared to 50% viable seeds from plots with natural fly infestation (Amsellem et al., 2001b).

Since *Orobanche* seed bank in heavily infested fields is very high and seed viability is extremely long, this biological control method is primarily recommended for newly or low infested locations. A more suitable prospect is to include this method as part of IOM programs (Amsellem et al., 2001a, b).

Resistant Cultivars

Resistant crops offer a practical, effective, and environmentally friendly strategy for *Orobanche* control. However, cases of susceptible crops in which resistant varieties have been identified are scarce and there are difficulties maintaining resistance for long periods. The vast number of *Orobanche* seeds and variability of *Orobanche* populations lead to rapid selection and build-up of virulent races that overcome resistance. An overview of crop resistance to *Orobanche* and the mechanisms and genetics involved was published by Parker and Riches (1993) and recently a thorough up-to-date review was presented by Alonso (1998). Substantial and field-useful levels of resistance have been found in several host/parasite systems:

Sunflower

Selection of *Orobanche*-resistant oil sunflower in Russia is reported by Pustovoit as early as 1916 (Parker and Riches, 1993). Following a few years of use of the resistant variety, a new virulent *O. cumana* race developed, requiring the breeding of new resistant material. The breaking of resistance varieties reoccurred 5 times and presently there are no resistant cultivars to Race 5 of *O. cumana* (Cubero, 1986; Antanova, 1994). A similar phenomenon is reported for Bulgarian and Romanian oil sunflower cultivars in which 5 subsequent virulent races of *O. cumana* evolved on resistant varieties (Encheva and Shindrova, 1994). In Spain, breaking of confectionery sunflower resistance has been extremely rapid and there are reports of 6 subsequent virulent races and the corresponding identification of 7 resistance genes (Rodriguez et al., 2001).

In Israel, several *O. cumana* resistant confectionery sunflower varieties have been developed through breeding, all from the same source of 'D.Y.3' resistant plants identified in a heavily infested field. The proposed mechanism of resistance is phytoalexin production that causes degeneration of *O. cumana* and *O. aegyptiaca* attachments and tubercles approximately 3 weeks after parasite penetration. Resistance is temperature dependent: plants are resistant only at high temperatures (Eizenberg, 2002). It is important to mention that sunflower is grown at high summer temperatures. In the past year there have been reports of resistance inconsistency in these varieties for reasons not yet studied.

In all breeding programs, resistance sources were within land races and wild sunflower species. In most cases resistance has been attributed to histo-mechanical barriers such as host root lignification and to phytoalexin induction in sunflower roots (Alonso, 1998; Eizenberg, 2002).

Faba bean

Resistance of faba bean to *O. crenata* was first identified in line 'F 402' in Egypt in the 1970's (Nassib et al., 1982). This line led to the breeding of the commercial variety 'Giza 402' which was the basis for further resistant varieties (Khalil et al., 1994). Breeders in Spain used 'Giza 402' to breed 'Baraca', a high yielding resistant local variety (Cubero et al., 1992). The mechanism of resistance is reported to be a localized formation of corky host root tissue, accumulation of a red-staining substance and necrosis that inhibit development of the parasite (Zaitun et al., 1994; Perez-de-Luque et al., 2001).

Vetch

We have discerned distinct polymorphisms among vetch genotypes in response to *O. aegyptiaca* and *O. crenata*. Common vetch (*Vicia sativa*) varieties are susceptible to the parasite while purple vetch (*V. atropurpurea*) and hairy vetch (*V. villosa*) are resistant. The resistance of *V. atropupurea* varieties 'Sadot' (from Israel) and 'Popany' (from Australia) to *Orobanche* is evident in spite of the fact that both induce a high rate of *Orobanche* seed germination. The resistance results from inability of the parasite haustorium to penetrate into the vascular cylinder of the host root and connect with the host vascular system. A chemical and/or a mechanical barrier active at the endodermis cell layer seem to obstruct the intruding haustorium (Goldwasser et al., 1999, 2000). Though there has been a single report of a case of breaking of resistance (Joel and Portnoy, 1997), *V. atropurpurea* varieties are widely grown without parasitism, probably because vetch is grown only once in a few years in a crop rotation system. Resistance to *O. crenata* has been reported in local-cultivars of common vetch in Spain and Syria (Gil et al., 1987; Linke et al., 1993). However, the reported resistant '473-A' Spanish variety failed to show resistance to *Orobanche* populations in Israel.

Pepper

We screened hot, sweet and paprika *Capsicum annuum* varieties for *O. aegyptiaca* resistance and found that the sweet pepper varieties were the most resistant to *O. aegyptiaca*. Among the tested sweet pepper varieties, 'Maor' and 'Odem' had the highest levels of resistance (Hershenhorn et al., 1996).

Parsley

In greenhouse and field screening we found that the commercial curled parsley (*Petroselinum crispum*) varieties 'Garbo' and 'Garland' supported less *O. crenata and O. aegyptiaca* parasitism than susceptible varieties. These varieties suffered minimal yield and quality damage in heavily *Orobanche*-infested fields (Goldwasser and Kleifeld, 2002). Initial observations suggest that this partial resistance is characterized by dying-off of root-attached *Orobanche* tubercles (Goldwasser et al., 2003).

Integrated Orobanche Management (IOM)

There is not much practical experience and reports regarding IOM. Researchers dealing with *Orobanche* control have suggested many ideas but only few examined this approach in the field.

The first vital element in an integrated *Orobanche* management approach is knowledge. It is important to have a local, regional or national system with information that can recognize *Orobanche* presence, identify its species and strain, and estimate the means and rate of infestation in neighboring fields and in the surrounding environment. Such a system should warn the farmers of the potential hazards of infestation. It should also generate research, advise and coordinate local means of prevention to stop the dispersal of the parasite. The precise taxonomic identification will dictate the designing of specific crop rotation programs with preference to non-host crops and minimizing the presence of susceptible host crops and weeds. In the case of economic necessity to grow an *Orobanche*-susceptible crop, a tolerant variety and an appropriate planting date in which parasitism is reduced should be considered. Trap and catch crops should be important constituents in the proposed crop rotation plan.

In the case of a new infestation which is localized in spots or when *Orobanche* appears in limited distribution along crop rows, (indicating contamination of the planting material), intensive hand pulling or cutting, or herbicide spot treatment before seed set, followed by totally destroying the collected *Orobanche* shoots, should delay or reduce future infestation.

Heavy *Orobanche* infestation dictates the employment of control measures, chemical and biological and if available, resistant varieties. Herbicide rotation in coordination with crop rotation, restricted herbicide use and tank mixes of herbicides from different chemical groups can ensure the efficacy and long term effect of chemical control. In many cases these very potent herbicides effectively control other weeds, some of them extremely troublesome, or that serve as *Orobanche* host plants- such as wild carrot control by glyphosate in legumes and nightshade control by sulfosulfuron in tomato. These factors have to be considered in designing weed management programs. In the future it may be possible to use genetically engineered crop material, resistant to translocated *Orobanche* herbicides or to the parasite itself.

There is no chance to succeed in entirely solving an *Orobanche* infestation problem by employing any one of these means individually. The effect of the most aggressive and efficient treatment, such as methyl bromide soil fumigation, does not persist for a long period if cropping with *Orobanche* host-crops is repeated. The only way to improve the control and to maintain a long-term effect is to use additional management methods. Since there will always be some individuals in the *Orobanche* population that will evade or overcome any chemical, biological or mechanical control method, they will soon become the source for a new tolerant/resistant population. Therefore, shallow tillage should follow soil fumigation to prevent the raising of viable seeds from non-treated soil layers, and parasites that escaped treatment should be hand-pulled to prevent reproduction of tolerant/resistant biotypes.

When using biocontrol methods, host-crop seeds or transplant roots treated with a biocontrol agent should enable the crop to reach strong vigor and become more tolerant to late application of additional control means. Early weakening of the parasite is essential for succeeding in biological control. A tolerant crop or variety might encourage development and increase aggressiveness of biocontrol agents (fungi, bacteria). Parasite weakening can also be accomplished by prior herbicide chemical seed dressing. Biological control using insects should come after reduction of parasite population by other control means.

CONCLUSIONS AND FUTURE PROSPECTS

The most promising results involving chemical control of *Orobanche* have been mainly with the amino-acid inhibiting herbicides. Efficient and selective *Orobanche* control requires accurate application rates, timing and techniques, while carefully assessing crop and parasite development, soil and environmental conditions and irrigation methods. In future *Orobanche* management that will utilize transgenic crops there is a possibility to use high herbicide doses because of the high magnitude of resistance. Constant efforts should continue in the search of efficient soil fumigants. No effective methyl bromide substitutes have been identified yet, and the suspension of this potent soil fumigant could leave a big gap in our ability to eradicate *Orobanche* infestations. New herbicide formulations and application methods including seed dressing and slow release should increase our arsenal of *Orobanche* control tools.

Promising biological control studies have been reported for *Fusarium* spp. and *Phytomyza orobanchia*. These biological agents still have to be adapted to mass production, formulation and distribution in field conditions. Ongoing efforts should continue to identify additional biological agents and new inhibitors, inducers and phytotoxins involved in parasitic plant-plant interactions. These molecules could be utilized in disrupting the normal association between parasite and plant. The identification of new resistance and tolerance sources and the breeding of new varieties will continue to be an ongoing effort.

The new era of molecular biology and genomics has the potential to significantly enhance our understanding of plant-parasitic plant interactions. Utilization of this knowledge together with new molecular techniques will allow the developing of novel strategies for parasitic plant control. Genetic engineering of additional herbicide-resistant *Orobanche*-susceptible crops will expand the possibilities of *Orobanche* chemical control. In the future the direct genetic engineering of *Orobanche*-resistant crop varieties will be possible as well as genetic manipulation of biological agents to enhance their performance in *Orobanche* control. A setback to the genetic engineering strategy is the recent objection to "genetically modified organisms" that has slowed down the expected rapid development and release of transgenic crops. Most of the recent chemical control research reported for *Orobanche* has been performed with experimental herbicides. Agrochemical companies and institutions should be encouraged to develop and register herbicides and crop varieties for *Orobanche* control.

The integrated approach is crucial to avoid, or at least delay, the breaking of *Orobanche*-resistant crop varieties and herbicide-resistant varieties by virulent *Orobanche* races. IOM is not only important in improving *Orobanche* control, it is essential for good agriculture practice and the sustainability of our agricultural systems for future generations.

ACKNOWLEDGMENT

We thank Professor J. Gressel for his comments and suggestions regarding this manuscript.

REFERENCES

Abayo G.O., English T., Eplee R.E., Kanampiu F.K., Ransom J.K., Gressel J. Control of parasitic witchweeds (*Striga* spp.) on corn (*Zea mays*) resistant to acetolactate synthase inhibitors. Weed Sci 1997; 46:459-466.

Abu-Irmaileh B.E. Effect of planting flax on subsequent infestation of tomato by *Orobanche ramosa*. Proceedings of the Third International Symposium on Parasitic Weeds; Aleppo, Syria. ICARDA, 1984.

Ahrens W.H. ed. *Herbicide Handbook of the Weed Science Society of America. 7th edition*. Champaign, IL: WSSA, 1994.

Al-Menoufy O.A. Crop rotation as a control measure of *Orobanche crenata* in *Vicia faba* fields. Proceedings, of the International Workshop in *Orobanche* Research; Tubingen, FRG. Eberhard-Karls-Univesitat, 1991.

Alonso C.A. Resistance to *Orobanche* and resistance breeding: a review. Proceedings of the Fourth International Workshop on *Orobanche*; Albena, Bulgaria. Institute for Wheat and Sunflower, 1998.

Aly R., Eizenberg H., Goldwasser Y., Hershenhorn J., Golan S., Kleifeld Y. Broomrape (*Orobanche cumana)* control in sunflower (*Helianthus annuus*) in fields. Weed Technol 2001; 15:306-309.

Amerhein N., Deus B., Gehrke P., Steinrüken H.C. The site of the inhibition of the shikimate pathway by glyphosate. Plant Physiol 1980; 66:830-834.

Amsellem Z., Kleifeld Y., Kerenyi Z., Hornok L., Goldwasser Y., Gressel J. Isolation, identification, and activity of mycoherbicidal pathogens from juvenile broomrape plants. Biol Cont 2001a; 21:274-284.

Amsellem Z., Barghouthi S., Cohen B., Goldwasser Y., Gressel J., Hornok L., Kerenyi Z., Kleifeld Y., Klein O., Kroschel J., Sauerborn J., Muller-Stover D., Thomas H., Vurro M., Zonno M.C. Recent advances in the biological control of *Orobanche* (broomrape) species. BioControl 2001b; 46:211-228.

Anonymous. Mon 37500 Technical data sheet. Monsanto, St. Louis, MO, 1995.

Antanova T.S. Biochemical aspects of the development of new virulent forms in the Moldavian population (race C) of *Orobanche cumana* Wallr. against the background of resistant sunflower cultivars. Proceeding of the Third International Workshop on *Orobanche* and related *Striga* research; Amsterdam, The Netherlands. Royal Tropical Institute, 1994.

Arjona-Berral A., Mesa-Garcia J., Garcia-Torres L. Herbicide control of broomrape in peas and lentils. *FAO Plant Protect Bull* 1988; 36:175-178.

Barghouthi S., Faqieh F., Badrieh A. Effect of bacteria on broomrape germination. Third International Weed Science Congress; Foz do Iguassu, Brazil. IWSC, 2000.

Bedi J.S. Further studies on control of sunflower broomrape with *Fusarium oxysporum* f. sp. *Orthoceras*- a potential mycoherbicide. Proceedings of the Third International Workshop on *Orobanche* and related *Striga* research; Amsterdam, The Netherlands. Royal Tropical Institute, 1994.

Bedi J.S., Donchev, N. Results on mycoherbicide control of sunflower broomrape (*Orobanche cumana* Wallr.) under glasshouse and field conditions. Proceedings of the Fifth International Symposium of Parasitic Weeds; Nairobi, Kenya. CIMMYT, 1991.

Beyer E.M., Duffy M.J., Hays J.V., Schlueter D.D. "Sulfonylureas herbicides". In *Chemistry, Degradation and Mode of Action,* Kearney P.C., Kaufman D.D., eds. New York: Marcel Dekker, 1988.

Birati Y., Benyamini Y. Control of *Orobanche* in peanuts with imazapic. Field report. Tel-Aviv, Israel. Luxemburg Industries (Pamol) Ltd, 1997.

Braun M., Burgstaller H., Walter H. Critical evaluation of control methods for *Orobanche ramosa* L. occurring in smallholder vegetable farms of Khartoum Province, Sudan. Proceedings of the Third International Symposium on Parasitic Weeds; Aleppo, Syria. ICARDA, 1984.

Castejon-Muñoz M., Romero-Muñoz F., Garcia-Torres L. Effect of planting date on broomrape (*Orobanche cumana* Loefl.) infection on sunflower (*Helianthus annuus* L.). Weed Res 1993; 33:171-176.

Chaudhary N.K. *Orobanche* damage in winter crops. J Inst Agric Ani Sci (Rampur, Nepal) 1995; 16:61-65.

Cubero J.I. Breeding for resistance to *Orobanche* and *Striga*: a review. Proceedings of the Workshop on Biology and Control of *Orobanche*; Wageningen, The Netherlands. LH/VPO, 1986.

Cubero J.I, Moreno M.T., Hernandez, L. A faba bean (*Vicia faba* L.) cultivar resistant to broomrape (*Orobanche crenata* Forsk.). Proceedings of the First European Conference on Grain Legumes; Angers, France. Association Europeenne des Proteagineux, 1992.

Dhanapal G.N., Struik P.C., Udayakumer M., Timmermans P.C.J.M. Management of broomrape (*Orobanche* spp.): a review. J Agron Crop Sci 1996; 175:335-359.

Dörr I. New results on interspecific bridges between parasites and their hosts. Proceedings of the Sixth International Parasitic Weed Symposium; Cordoba, Spain. Junta de Andalusia, 1996.

Eizenberg H. The resistance mechanism of sunflower (*Helianthus annuus* L.) to *Orobanche* spp. Ph.D thesis, The Hebrew University of Jerusalem, Israel, 2002.

Eizenberg H., Tanaami Z., Jacobsohn R., Rubin B. Effect of temperature on the relationship between *Orobanche* spp. and carrot (*Daucus carota* L.). Crop Protect 2001a; 20:415-420.

Eizenberg H., Goldwasser Y., Hershenhorn J., Plakhine D., Kleifeld Y. *Orobanche aegyptiaca* control in tomato with MON 37500; Weed Science Society of America Abstracts 2001b; 41:176.

Encheva V., Shindrova P. Broomrape (*Orobanche cumana* Wallr.): a hindrance to sunflower production in Bulgaria. Proceedings of the Third International Workshop on *Orobanche* and related *Striga* research; Amsterdam, The Netherlands. Royal Tropical Institute, 1994.

Foy C.L., Jain R., Jacobsohn R.. Recent approaches for chemical control of broomrape (*Orobanche* spp.). Reviews of Weed Sci 1989; 4:123-152.

Frear D.S., Swanson H.R., Thalacker F.W. Induced microsomal oxidation of diclofop, triasulfuron, chlorsulfuron, and linuron in wheat. Pestic Biochem Physiol 1991; 41:274-287.

Garcia-Torres L., Lopez-Granados F., Jurado-Exposito M., Diaz-Sanchez J. Chemical control of *Orobanche* in legumes: achievements and constraints. Proceedings of the Fourth International *Orobanche* Workshop; Albena, Bulgaria. Institute for Wheat and Sunflower, 1998.

Gil J., Martin L.M. Cubero J.I. Genetics of resistance in *Vicia sativa* to *Orobanche crenata* Forsk. Plant Breed 1987; 99:134-143.

Goldwasser Y., Kleifeld Y., Golan S., Bargutti A. The use of flax as a bioassay for soil infestation by Egyptian broomrape. Proceedings of the Fifth International Symposium of Parasitic Weeds; Nairobi, Kenya. CIMMYT, 1991.

Goldwasser Y., Kleifeld Y., Golan S., Bargutti A., Rubin B. Dissipation of metham-sodium from soil and its effect on the control of *Orobanche aegyptiaca*. Weed Res 1995; 35:445-452.

Goldwasser Y., Kleifeld Y., Plakhine D., Rubin B. Variation in vetch (*Vicia* spp.) response to *Orobanche aegyptiaca*. Weed Sci 1997; 45:756-762.

Goldwasser Y., Hershenhorn J., Plakhine D., Kleifeld Y., Rubin B. Biochemical factors involved in vetch resistance to *Orobanche aegyptiaca*. Physiol Mol Plant Pathol 1999; 54:87-96.

Goldwasser Y., Plakhine D., Kleifeld Y., Zamski E., Rubin B. The differential susceptibility of vetch *(Vicia* spp.) to *Orobanche aegyptiaca*: anatomical studies. Ann Bot 2000; 85:257-262.

Goldwasser Y., Eizenberg H., Hershenhorn J., Plakhine D., Blumenfeld T., Buxbaum H., Golan S., Kleifeld Y. Control of *Orobanche aegyptiaca* and *O. ramosa* in potato. Crop Protect 2001; 20:403-410.

Goldwasser Y., Kleifeld Y. Tolerance of parsley varieties to *Orobanche*. Crop Protect 2002; 21:1101-1107.

Goldwasser Y., Eizenberg H., Golan S., Kleifeld Y. Control of *Orobanche crenata* and *O. aegyptiaca* in parsley. Crop Protect 2003; 22:295-305.

Graves J.D. "Host–plant responses to parasitism". In *Parasitic Plants,* Press M.C., Graves J.D., eds. London, UK: Chapman and Hall, 1995.

Gressel J. Potential failsafe mechanisms against the spread and introgression of transgenic hypervirulent biocontrol fungi. Trends in Biotechnology 2001; 19:149-154.

Gressel J. "The need for new herbicide-resistant crops". In *Achievements and Developments in Combating Pesticide Resistance*, Denholm I., Devonshire A.L., Hollomon D.W., eds. London, UK: Elsevier, 1992.

Gressel J., Segel L.E., Ransom J.K. Managing the delay of evolution of herbicide resistance in parasitic weeds. Internat J Pest Manage 1996; 42:113-129.

Hatzios K.K., ed. *Herbicide handbook- Supplement to Seventh Edition*. Lawrence, KS: WSSA, 1998.

Hendel F. Die pläaktischen Agronomyziden (Diptera) Prodomus einer monographie. Archiv für Naturgeschichte 1920; 84:109-174.

Hershenhorn J., Goldwasser Y., Plakhine D., Herzlinger G., Golan S., Russo R., Kleifeld Y. Role of pepper (*Capsicum annuum*) as a trap and catch crop for control of *Orobanche aegyptiaca* and *O. crenata.* Weed Sci 1996; 44:948-951.

Hershenhorn J., Plakhine D., Goldwasser Y., Westwood J.H., Foy C.L., Kleifeld Y. Effect of sulfonylurea herbicides on early development of Egyptian broomrape (*Orobanche aegyptiaca*) in tomato (*Lycopersicon esculentum*). Weed Technol 1998a; 12:108-114.

Hershenhorn J., Goldwasser Y., Plakhine D., Lavan Y., Herzlinger G., Golan S., Chilf T., Kleifeld Y. Effect of sulfonylurea herbicides on Egyptian broomrape (*Orobanche aegyptiaca*) in tomato (*Lycopersicon esculentum)* under greenhouse conditions. Weed Technol 1998b; 12:115-120.

Hershenhorn J., Goldwasser Y., Plakhine D., Aly R., Blumenfeld T., Bucsbaum H., Herzlinger G., Golan S., Chilf T., Eizenberg H., Dor E., Kleifeld Y. *Orobanche aegyptiaca* control in tomato fields with sulfonylurea herbicides. Weed Res 1998c; 38:343-349.

Hinz J.R.R., Owen M.D.K., Barrett M. Nicosulfuron, primisulfuron, and bentazon hydroxylation by corn (*Zea mays*), wooly cupgrass (*Eriochloa villosa*) and shattercane (*Sorghum bicolor*) cytochrome P-450. Weed Sci 1997; 45:474-480.

Holm L., Doll J., Holm E., Pancho J., Herberger J. *Worlds Weeds: Natural Histories and Distribution.* New York: Wiley, 1997.

Horvath Z. Investigations on *Phytomyza orobanche* Kalt. (Dipt.: Agromyzideae), a possible biocontrol agent of *Orobanche* spp. (Orobanchaceae) in Hungary. Proceedings of the 4th International Symposium on Parasitic Flowering Plants. Marburg, F.R.G. Philipps-Universität, 1987.
Jacobsohn R. Broomrape avoidance and control: agronomic problems and available methods Proceedings, Workshop on Biology and Control of *Orobanche*; Wageningen, The Netherlands. LH/VPO, 1986.
Jacobsohn R., Kelman Y. Effectiveness of glyphosate in broomrape (*Orobanche* spp.) control in four crops. Weed Sci 1980; 28:692-699.
Jacobsohn R., Greenberger A., Katan J., Levi M., Alon H. Control of Egyptian broomrape (*Orobanche aegyptiaca*) and other weeds by means of solar heating of soil by polyethylene mulching. Weed Sci 1980; 28:312-316.
Jacobsohn R., Levy D. Glyphosate for *Orobanche* in various crops: problems and promises. Proceedings, Workshop on Biology and Control of *Orobanche*. Wageningen, The Netherlands. LH/VPO, 1986.
Jacobsohn R., Ben-Ghedalia D., Marton K. Effect of animal's digestive system on the infectivity of *Orobanche* seeds. Weed Res1987; 27:87-90.
Jacobsohn R., Kleifeld Y., Agrawal V.P., Jha P., Marton K. Soil fumigation with Telone II for broomrape (*Orobanche* spp.) control. Proceedings of the International Workshop on *Orobanche* Research; Obermarchtal, Tuibingen. Eberhard-Karls-Universitat, 1991.
Jacobsohn R., Tanaami Z., Eizenberg H. Selective control of broomrape in carrot and vetch with foliar applied imidazolinone herbicides. Phytoparasitica 1996; 24:207.
Jacobsohn R., Dor Z., Tanaaami Z., Ovdat N. Crenate broomrape control in garden and field peas with foliarly applied imazethapyr. Proceedings of the Fourth International *Orobanche* Workshop; Albena, Bulgaria. Institute for Wheat and Sunflower, 1998.
Joel D.M., Kleifeld Y., Losner-Goshen D., Gressel J. Transgenic crops against parasites. Nature 1995; 374:220-221.
Joel D.M., Portnoy V.H. A virulent race of *Orobanche crenata*. Hassadeh, September:27-28, 1997.
Joel D.M., Losner-Goshen D., Goldman-Guez T., Portnoy V.H. The haustorium of *Orobanche*. Proceedings of the Fourth International *Orobanche* Workshop; Albena, Bulgaria. Institute for Wheat and Sunflower, 1988.
Joel D.M., Plakhine D., Creange P., Dupuis J.M., Kamoda J., Gressel J. Broomrape (*Orobanche aegyptiaca*) control using asulam-resistant seeds coated with asulam. WSSA abstracts 2000; 40:137.
Jurado-Exposito M., Castejon- Muñoz M., Garcia-Torres L. Broomrape (*Orobanche crenata*) control with imazethapyr applied to pea (*Pisum sativum*) seed. Weed Technol 1996; 10:774-780.
Jurado-Exposito M., Garcia-Torres L., Castejon- Muñoz M. Broad bean and lentil seed treatment with imidazolinones for the control of broomrape (*Orobanche crenata*). J Agric Sci 997; 129:307-314.
Jurado-Exposito M., Castejon- Muñoz M., Garcia-Torres L. Uptake and translocation of imazethapyr in peas as affected by parasitism of *Orobanche crenata* and herbicide application methods. Weed Res 1999; 39:129-136.
Kanampiu F.K., Ransom J.K., Gressel J. Imazapyr seed dressings for *Striga* control on acetolactate synthase target-site resistant maize. Crop Protect 2001; 20:885-895.
Kanampiu F.K., Ransom J.K., Gressel J., Jewell D., Friesen D., Grimanelli D., Hoisington D. Appropriateness of biotechnology to African agriculture: *Striga* and maize as paradigms. Plant Cell Tissue Organ Cult 2002; 69:105-110.
Kasasian L. Control of *Orobanche*. PANS 1973a; 19: 368-371.
Kasasian L. The chemical control of *Orobanche crenata* in *Vicia faba* and the susceptibility of 53 cultivars of *V. faba* to *O. crenata*. Proceedings, Symposium on Parasitic Weeds; Malta. European Weed Research Council, 1973b.
Katan J., Greenberger A., Alon H., Grinstein A. Solar heating by polyethylene mulching for the control of diseases caused by soil-borne pathogens. Phytopathology 1976; 66:683-688.
Khalil S.H., Saber H.A., El-Sherbeeny M.H., El-Hady M.M., Saleed S.R. Present state of *Orobanche* resistance breeding in faba bean in Egypt. Proceedings of the Third International Workshop on *Orobanche* and related S*triga* research, Amsterdam, The Netherlands. Royal Tropical Institute, 1994.
Kleifeld Y. *Orobanche* management and control in Israel. Proceedings, Regional Workshop on Joint Action to Control *Orobanche* in the WANA-region: Experiences from Morocco; Rabat, Morocco. GTZ GmbH, 1999.
Kleifeld Y., Graph S., Herzlinger G. 1987a. The effect of propyzamide on broomrape attacking sunflowers. Proceedings of the 4th International Symposium on Parasitic Flowering Plants. Marburg, F.R.G. Philipps-Universität, 1987.
Kleifeld Y., Golan S., Graph S., Joel D.M. Application of methyl bromide in heavy soils for broomrape eradication. Phytoparasitica 1987b; 16:378.
Kleifeld Y., Goldwasser Y., Herzlinger G., Joel D.M., Golan S., Kahana D. The effects of flax (*Linum usitatissimum*) and other crops as trap and catch crops for control of Egyptian broomrape (*Orobanche aegyptiaca* Pers.) Weed Res 1994a; 34:37-44.

Kleifeld Y., Goldwasser Y., Herzlinger G., Golan S., Blumenfeld T., Buxbaum H. Selective control of broomrape in tomatoes with rimsulfuron. Proceedings of the Third International Workshop on *Orobanche* and related *Striga* research; Amsterdam, The Netherlands. Royal Tropical Institute, 1994b.

Kleifeld Y., Goldwasser Y., Plakhine D., Herzlinger G., Golan S., Hershenhorn J. Selective control of *Orobanche* spp. with imazamethapyr. Proceedings of the Fourth International *Orobanche* Workshop; Albena, Bulgaria. Institute for Wheat and Sunflower, 1998.

Kleifeld Y., Goldwasser Y., Plakhine D., Eizenberg H., Herzlinger G., Golan S. Selective control of *Orobanche* spp. in various crops with sulfonylurea and imidazolinone herbicides. Proceedings, Regional Workshop on Joint Action to Control *Orobanche* in the WANA-region: Experiences from Morocco; Rabat, Morocco. GTZ GmbH, 1999.

Klyueva M.P., Pamukchi G.V. Technology of the use of *Phytomyza.* Zascita Rastenij 1982; 27:33-34.

Krishnamurty S., Rao U.M. Control of *Orobanche* through crop rotation. Indian Farming 1976; 25:23.

Krishnamurthy S., Lai R., Nagarajan K. Further studies on the effect of various crops on the germination of *Orobanche* seeds. PANS 1977; 23:206-208.

Kroschel J., Klein O. "Biological control of *Orobanche* spp. with *Phytomyza orobanchia* Kalt.: a Review." In *Joint Action to Control Orobanche in the WANA-region: Experiences from Morocco.* Proceedings, Regional Workshop on Joint Action to Control *Orobanche* in the WANA-region: Experiences from Morocco; Rabat, Morocco. GTZ GmbH, 1998.

Kuijt J. Haustoria of phanerogamic parasites. Ann Rev Phytopathol 1977; 17:91-118.

Kukula S., Masri H. Integrated cultural practices and chemical control of *Orobanche crenata* in faba bean. Proceeding of the Third International Symposium on Parasitic Weeds; Aleppo, Syria. ICARDA, 1984.

Kurbanov T. The use of *Phytomyza* on the fields of Solchos. Biological control of the Egyptian broomrape in vegetables and melons. Sel'skoe genus Chozjajstvo Uzbekistan 1970; 9:50.

Linke K.H. Status quo of *Orobanche* management: preventive, cultural and physical control. Proceedings, Regional Workshop on Joint Action to Control *Orobanche* in the WANA-region: Experiences from Morocco. Rabat, Morocco. GTZ GmbH, 1999.

Linke K.H., Sauerborn J., Saxena M.C. Orobanche *Field Guide.* FR Germany: University of Hohenheim, 1989.

Linke K.H., Saxena M.C. Toward an integrated control of *Orobanche* spp. in some legume crops. Proceedings, International Workshop in *Orobanche* Research; Tubingen, FR Germany. Eberhard-Karls-Univesitat, 1991.

Linke K.H., Abd El-Moneim A.M. Saxena M.C. Variation in resistance of some forage legumes species to *Orobanche crenata* Forsk. Field Crops Res 1993; 32:277-285.

Mesa-Garcia J., Garcia-Torres L. Effect of bean (*Vicia faba* L.) planting dates on broomrape (*Orobanche crenata* Forsk.) phenology and competition. Proceeding of the British Crop Protection Conference, Weeds 1984; 2:757-764.

Mesa-Garcia J., de Haro A., Garcia-Torres L. Phytotoxicity and yield response of broad bean (*Vicia faba*) to glyphosate. Weed Sci 1984; 32:445-450.

Moiseeva N., Mamraliev J. Broomrape (*Orobanche* sp.) can be eaten by *Phytomyza* sp. Sel'skoe Khozyiaistvo Kirgizii 1969; 7:24.

Musselman L.J. The biology of *Striga, Orobanche* and other parasitic weeds. Ann Rev Phytopathol 1980; 30:369-389.

Nandula V.K., Foy C.L., Orcutt D.M. Glyphosate for *Orobanche aegyptiaca* control in *Vicia sativa* and *Brassica napus*. Weed Sci 1999; 47:486-491.

Nassib A.M., Ibrahim A.A., Khalil S.A. "Breeding for resistance to *Orobanche*". In *Faba Bean Improvement*, Hawtin G., Webb C., eds, The Hague, The Netherlands: Martinus Nijhoff, 1982.

Okazova A.G. *Phytomyza* in tobacco of the Krim. Zascita Rastenij 1973; 18:21-22.

Olson B.L.S., Al-Khatib K., Stahlman P., Isakson P.J. Efficacy and metabolism of MON 37500 in *Triticum aestivum* and weedy grass species as affected by temperature and soil moisture. Weed Sci 2000; 48:541-548.

Parker C., Riches C. *Parasitic weeds of the world: Biology and Control.* Wallingford, UK: CAB International, 1993

Perez-de-Luque A., Cubero J.I., Rubiales D., Joel D.M. Histology of incompatible interactions between *Orobanche crenata* and some host legumes. Proceedings of the Seventh International Parasitic Weed Symposium; Nantes, France, 2001.

Pieterse A.H. The broomrapes (Orobanchaceae)- a review. Abst Trop Agric 1979; 5:9-35.

Press M.C., Graves J.D. eds. *Parasitic Plants*. London, UK: Chapman and Hall, 1995.

Ramaiah K.V. Control of *Striga* and *Orobanche* species- a review. Proceedings of the 4th International Symposium on Parasitic Flowering Plants. Marburg, F.R.G. Philipps-Universität, 1987.

Reinke H., Rosenzweig A., Clausm K.M., Chisholm C., Jensen P. DPX-E 9636, experimental sulfonylurea herbicide for potatoes. Proceedings of the Brighton Crop Protection Conference; Brighton, UK, 1991.

Riches C.R, Parker C. "Parasitic plants as weeds". In *Parasitic Plants,* Press M.C., Graves J.D., eds. London, UK: Chapman and Hall, 1995.

Rodriguez-Ojeda M.I., Fernandez-Escobar J., Alonso L.C. Sunflower inbred line (KI-374), carrying two recessive genes for resistance against a highly virulent Spanish population of *Orobanche cernua* Loelf./*O. cumana* Wallr. Race "F". Proceedings of the Seventh International Parasitic Weed Symposium; Nantes, France, 2001.

Sauerborn J., Saxena M.C. A review on agronomy in relation to *Orobanche* problems in faba bean (*Vicia faba* L.). Proceeding of a workshop on biology and control of *Orobanche.* Wageningen, The Netherlands. LH/VPO, 1986.

Saxena M.C., Linke K.H., Sauerborn J. Integrated control of *Orobanche* in cool-season food legumes. Proceeding of the Third International Workshop on *Orobanche* and Related *Striga* Research. Amsterdam, The Netherlands. Royal Tropical Institute, 1994.

Schloss J.V. "Recent advances in understanding the mechanism and inhibition of acetolactate synthase". In *Herbicides Inhibiting Branch Chain Amino Acid Biosynthesis,* Setter J, ed. New York: Springer Verlag, 1995.

Schmitt U., Schülter K, Boorsma P.A. Chemical control of *Orobanche crenata* in broadbean. FAO Plant Protection Bulletin 1979; 27:89-91.

Shaner D.L. *Factors Affecting Soil and Foliar Bioavailability of the Imidazolinone Herbicides*. Princeton, NJ: American Cyanamide Company, 1989.

Shaner D.L., O'Conner S.L., eds. *The Imidazolinone Herbicides*. Boca Raton, FL: CRC Press, 1991.

Sharon A., Amsellem Z., Gressel J. Glyphosate suppressing of an elicited defense response; increased susceptibility of *Cassia obtusifolia* to a mycoherbicide. Plant Physiol 1992; 98:654-659.

Stewart G.R., Press M.C. The physiology and biochemistry of parasitic angiosperms. Ann Rev Plant Physiol Plant Mole Biol 1990; 41:127-151.

Surov T., Aviv D., Aly R., Joel D.M., Goldman-Guez T., Gressel J. Generation of transgenic asulam-resistant potatoes to facilitate eradication of parasitic broomrapes (*Orobanche* spp.), with the *sul* gene as the selectable marker. Theo Appl Genet 1997; 96:132-137.

ter Borg SJ. Present and future of *Orobanche* research; summary and conclusions. Proceedings of a workshop on Biology and control of *Orobanche*; Wageningen, The Netherlands. LH/VPO,1986.

Thomas H., Sauerborn J., Muller-Stover D., Ziegler A., Bedi J., Kroschel J. Potential of *Fusarium oxysporum* f. sp. *orthoceras* as a biological control agent for *Orobanche cumana* in sunflower. Biol Cont 1998; 13:41-48.

Thomas H., Heller A., Sauerborn J., Muller-Stover D. *Fusarium oxysporum* f. sp. *orthoceras* a potential mycoherbicide, parasites seeds of *Orobanche cumana* (sunflower broomrape): a cytological study. Ann Bot 1999; 83:453-458.

Thomson WT. ed. *Agricultural Chemicals. Book II. Herbicides.* Fresno, CA: Thomson Publications, 1997.

Trenchev G. The possibility of using *Phytomyza orobanchia* Kalt. for the control of broomrape. Rasteniev"dni Nauki 1981; 18:112-119.

United Nations Environmental Protection Service. Report of the Fourth Meeting of the Parties to the Montreal Protocol on Substances that Deplete the Ozone Layer. Copenhagen. UNEP/OzL. Pro. 4/15,23-25. 11.92, 1992.

Vurro M., Ellis B.E. Effect of fungal toxins on induction of phenylalanine ammonia lyase activity in elicited cultures of hybrid poplar. Plant Sci 1997; 126:29-38.

van Hazewijk M.J. Germination ecology of *Orobanche crenata*-implication for cultural control measures. Ph.D. Thesis. The Netherlands: Amsterdam University, 1994.

Wegmann K. Progress in *Orobanche* research during the past decades in: current problems of *Orobanche* Researches. Proceedings of the Fourth International Workshop on *Orobanche*; Albena, Bulgaria. Institute for Wheat and Sunflower, 1998.

Westwood J.H., Foy C.L. Influence of nitrogen on germination and early development of broomrape (*Orobanche* spp.). Weed Sci 1999; 47:2-7.

Wilhelm S., Benson R.C., Sagen J. Studies on the control of broomrape on tomatoes: soil fumigation by methyl bromide is promising control. Plant Dis Rep 1958; 42:645-651.

Wilhelm S., Srokan R.C.E., Benson R.C., Sagen J.E., Carpenter T. Large scale soil fumigation against broomrape. Phytopathology 1959; 49:530-532.

Wilhelm S. History of broomrapes *Orobanche ramosa* and *O. ludoviciana* and their control by preplant soil injection with methyl bromide solutions. Proceedings of the 16th International Horticulture Congress; Brussels, 1962.

Zaitun F.M.F., ter Borg S.J. Resistance against *Orobanche crenata* in Egyptian and Spanish faba beans.Proceedings of the Third International Workshop on *Orobanche* and related *Striga* research; Amsterdam, The Netherlands. Royal Tropical Institute, 1994.

CHAPTER 23

A REVIEW ON WEED CONTROL IN SUGARBEET

FROM TOLERANCE ZERO TO PERIOD THRESHOLD

JAN PETERSEN

Institute of Sugar Beet Research (IfZ), Goettingen, Germany

Sugarbeet (*Beta vulgaris* spp. *rapacea* (Koch) Döll) are grown in many countries, mainly in temperate and Mediteranean climates (Table 1). In most of the regions the sugarbeet are sown in early spring. In some countries winter beet production (sowing in autumn) is quite common (e.g. Spain) and under very favourable conditions sugarbeet are drilled nearly throughout the year (e.g. California with the exception of July and August). In only a few cases sugarbeet are transplanted (e.g. Japan). Under dry conditions sugarbeet are irrigated (e.g. Western USA, Spain, Turkey). Therefore, due to these circumstances, weed infestation in sugarbeet and weed control strategies are quite different in each area. The weed flora is also influenced by crop rotation, soil type and other factors.

Sugarbeet are planted in wide rows from 45 to 75 cm apart (depending on machinery and irrigation system, etc.). This, and the fact that young plants grow relatively slowly, are some of the reasons that sugarbeet are characterized as weak competitors. Without weed control yield will be reduced significantly, up to 95%. Depending on harvesting and beet cleaning systems, weeds can also hinder harvest and processing of the beets. In addition, weeds can be sources for pests and diseases in sugarbeet (Gleißl et al., 1989; Schlösser, 1989) or following crops (Dewar et al., 2000a). These are the reasons why sugarbeet growers pay particular attention to weed control. In most of the growing areas the tolerance level of weeds in sugarbeet by the growers is virtually zero. Intensive mechanical or chemical weed control (or combinations of both methods) is the common standard. However, for economical and ecological reasons the focus of current research is to investigate how many and at which periods of the growing season weeds can be tolerated, without causing a yield reduction.

IMPORTANT WEEDS IN SUGARBEET

Sugarbeet are grown in various countries under different climatic, soil and agronomic conditions, and in each region a more or less typical weed flora occurs (Table 2). The use of a limited range of herbicides in the past caused a shift in the weed flora. For example, *Polygonum convolvulus* L. or *Aethusa cynapium* L. were not mentioned as important sugarbeet weeds in Germany 1959/64 (Hanf, 1959; Rademacher, 1964). Today these two species are very common (Merkes et al., 2001). Another reason which causes changes in the weed flora are modifications to crop rotations.

Inderjit (ed.), Weed Biology and Management, 467-483.

In the present economic climate farmers must change their rotations to include more high value crops and thus the diversity of crops on farms is decreasing. In some cases the result is that volunteer crops have become major weeds in sugarbeet. In some regions volunteer potatoes, oilseed rape and sunflowers cause very serious weed control problems (Merkes et al., 2001). In reduced tillage systems, other volunteers like cereals and maize can occur (Sullivan and Fischer, 1971).

Table 1. Sugarbeet growing areas (FAO, 2001)

	Country	Area [ha]*
Africa	Egypt	55,000
	Marocco	31,000
Asia	Azerbaijan	82,000
	China	256,000
	Iran	186,000
	Japan	70,000
	Kazakhstan	18,000
	Kyrgyzstan	24,000
	Syria	31,000
	Turkey	420,000
	Uzbekistan	22,000
North America	Canada	20,000
	USA	560,000
South America	Chile	48,000
Europe	Belarus	57,000
	Croatia	20,000
	Czech Republic	61,000
	EU	1,900,000
	Hungary	58,000
	Latvia	13,000
	Lithuania	28,000
	Moldova	65,000
	Poland	333,000
	Romania	56,000
	Russia	810,000
	Slovakia	32,000
	Slovenia	11,000
	Switzerland	18,000
	Ukraine	776,000
	Yugoslavia	45,000
World**		6,131,000

* harvested area in 2000
** incl. other countries < 10,000 ha (Afghanistan, Albania, Bulgaria, Ecuador, Iraq, Lebanon, Macedonia, Pakistan, Portugal, Tunisia: together: 25,000 ha)

In many sugarbeet growing areas dicot weeds of the families Chenopodiaceae, Asteraceae and Polygonaceae are of major importance. This can be explained by ecological conditions (soil, climate etc.) and because sugarbeet belongs to the Chenopodiaceae family as well. Also crop management strategies have a strong influence on the weed flora. Many selective sugarbeet herbicides are weak in controlling weeds from the Polygonaceae and Asteraceae families.

In general, the monocots are less important compared to dicot weeds. However, some species of the Poaceae and Cyperaceae families can cause serious weed control problems. For example, herbicide resistant biotypes (*Alopecurus myosuroides* Huds.) (Moss et al., 1999), late emerging species (*Echinocloa crus-galli* (L.) Beauv., *Setaria* sp., etc.) and perennial species (*Elymus repens, Cyperus esculentus* L., etc.). Perennial broadleaf weeds can cause serious weed control problems. Most of the herbicides will not give adequately control and mechanical methods are not completely effective due to regrowth from vegetative propagules. Important perennial dicot weeds in sugarbeet field are: *Calystegia sepium* (L.) R. Br., *Cirsium arvense* (L.) Scop., *Convolvulus arvensis* L., *Mentha arvensis* L., *Polygonum amphibium* L., *Sonchus arvensis* L., *Rumex* sp. *Tusilago farfara* L., volunteer potatoes and *Equisetum arvense* L. (spore plant). *Cuscuta campestris* Yuncker, a parasitic weed, can also appear on sugarbeet, particularly in the Mediteranian area. All perennial and parasitic weeds are very difficult to control. In addition to mechanical methods, in sugarbeet crop rotations soil tillage and weed control in previous crops can have a big influence (positive or negative) on population dynamics of these tough weeds.

Weed beet are a particular problem in sugarbeet. Normally, sugarbeet are biennial plants. In the first year the root grows and in the second year the inflorescence and finally the seeds are produced. For sugar production the plants are harvested at the end of the first year. Under unfavourable conditions (e.g. cold spring) some plants are vernalized and flower in the first year. This process, termed bolting, results in viable seed production. If seeding plants (bolters) are not removed, seeds with long dormancy builds up in the soil seed bank. Consequently, annual weed beet can germinate and compete with next sugarbeet crop in the rotation. Another way of introducing weed beet into fields is the introgression of the wild beet (e.g. *Beta vulgaris* ssp. *maritima* (L.) Arcang.) genome into sugarbeet during seed production. The results of the crosses between wild and sugarbeet are plants that are annual in habit and appear as bolters in sugarbeet (Boutry et al., 1993). Because weed and sugarbeet are very close relatives, only systems that use some form of mechanical control mechanisms are successful. Bolters and weed beet should be removed before seeds have been produced.

SUGARBEET - WEED INTERACTIONS

Weed-crop interactions are based on competition for water, nutrients and light and allelopathic effects may also play a small role. In sugarbeet weed interference, all these factors are important too, but light is of prime importance (Kropff and Spitters, 1992). Due to the fact that a lot of weeds can grow above the sugarbeet canopy (e.g. *Amaranthus* sp. or *Chenopodium* sp.) and reduce the amount of photosynthetic radiation reaching the crop, these weeds are stronger competitors compared to smaller weeds (e.g. *Lamium* sp. or *Stellaria media* (L.) Vill.) (Schäufele, 1986; Schäufele, 1991; Mittler et al., 2002). This effect of reduced light interception by the crop is also the reason for observations that broadleaf weeds are more competitive in sugarbeet than annual grasses (Zimdahl and Fertig, 1967). On the other hand, there are some differences between sugarbeet varieties that affect their sensitivity to weeds. Some varieties have more leaves that are horizontal in habit so that the canopy closes earlier and late emerging or smaller weeds will be suppressed more effectively than in varieties with a more erect habit (Lotz et al., 1991). However, some varieties with lower leaf areas but with a vertical leaf habit have higher yield potential. For these varieties, a higher intensity of weed control may be required.

N-fertilizer can also affect competition between sugarbeet and weeds (Scott and Moisey, 1972; Bräutigam and Schäufele, 1994). Leaf area of sugarbeet increased at high levels of N-fertilizers, while the weeds did not benefit from higher N-levels as measured in dry matter yield or N-uptake. However, higher levels of nitrogen can have a negative influence on beet quality due to a decrease of extractable sugar content.

Beside competition, weeds in sugarbeet can have negative as well as positive effects on the crop. Weeds can be sources or host plants for pests (e.g. *Heterodera schachtii;* Geißl et al., 1989) and diseases (e.g. Beet Necrotic Yellow Vein Virus; Schlösser, 1989). In the early growth stages weeds can have a stimulation effect on sugarbeet. Results from El Titi (1986) showed that more sugarbeet emerged and survived in weedy plots (weeds during first 30 days after planting) compared to weed free plots. This effect was explained by the distraction of pests like *Onychiurus fimatus* away from the crop (Ulber, 1980). Furthermore, weeds can be alternative host plants for sugarbeet pests or sources for pest antagonists. Delayed weed control can reduce aphid populations on beets (Dewar et al., 2000a). Aphids are vectors for virus yellows disease in sugarbeet (Beet Yellow Virus and Beet Mosaic Virus). When aphid populations on the beets are reduced by the presence of weeds, virus yellows infections also decreased (Dubois et al., 1993). Because of the ecological and phytopathological positive balance of weed effects in sugarbeet, differentiated and selective weed control is recommended, instead of trying to eliminate the weeds completely (Bosch, 1987).

Thresholds

Using post-emergence herbicides it is possible to take thresholds into consideration before applying treatments. Some work has been done in sugarbeet to define thresholds based on weed density (Neururer, 1975; Weatherspoon and Schweizer, 1971; Schweizer, 1981; Schweizer and Bridge, 1982; Schweizer, 1983; Schweizer, 1985). However, until today weed control in sugarbeet is not based on density thresholds. Various reasons make it difficult to determine critical weed densities in sugarbeet. First of all, sugarbeet grow very slowly during early stages of crop development. Weeds are able to germinate over long periods and time of emergence (crop and weeds); weed species and growing conditions change yearly. As a result in some cases a specific weed density can cause a significant yield loss whilst in others it will not (Figure 1). Furthermore, some weeds survive a herbicide treatment, but are weaker competitors than untreated plants. All these observations lead to the conclusion that sugarbeet yield losses cannot be accurately predicted from weed density alone (Dexter, 1998).

Another idea is to predict yield losses by determining weed coverage. This concept showed success to some extent, but the problem is that reliable prediction of yield loss is only possible after BBCH 19 of the sugarbeet (Mittler et al., 2002). This is too late to avoid yield losses due to early competition and too late for effective weed control, because herbicide efficacy strongly depend on weed size.

The concept of period thresholds, which means the period in which weeds should not compete with the sugarbeet therefore not cause yield loss (Figure 2) (Nieto et al., 1968) appears to be a helpful concept for integrated crop protection in less competitive crops (Dawson, 1996).

Table 2: Important weeds in sugarbeet (an extract)

Family	Species	Asia[1]	N-America	E-Europe	S-Europe	W-Europe
Amaranthaceae	*Amaranthus retroflexus*	x	x	x	x	x
	Amaranthus lividus	x		x	x	x
Caryophyllaceae	*Stellaria media*	x	x	x	x	x
Chenopodiaceae	*Atriplex patula*			x		x
	Chenopodium album	x	x	x	x	x
	Chenopodium hybridum			x		x
	Chenopodium polyspermum	x		x		x
	Chenopodium glaucum; C. leptphyllum		x			
	Chenopodium aristatum; C. serotinum	x[2]				
	Kochia scoparia		x			
	Salsola kali; S. collina[2]	x	x		x	
Commelinaceae	*Commelina communis*	x				
Asteraceae	*Ambrosia artemisiifolia*[3]	x	x	x	x	
	Bidens tripartita					x
	Cirsium arvense	x	x	x	x	x
	Galinsoga parviflora			x	x	x
	Galinsoga ciliata				x	x
	Ixeris chinensis; I. polycephala	x[2]				
	Matricaria chamomilla			x	x	x
	Senecio vulgaris	x				x
	Sonchus oleraceus	x	x		x	x
	Taraxacum mongolicum	x[2]				
	Xanthium strumarium; X. spinosum; X. sibiricum[2]	x				
Convolvulaceae	*Convolvulus arvensis*	x	x		x	x
Brassicaceae	*Brassica napus*					x
	Brassica nigra		x		x	
	Capsella bursa-pastoris	x	x			x
	Descurainia sophia	x	x	x		
	Erysimum cheiranthoides	x				
	Thlaspi arvense		x			x
	Sinapis arvensis	x	x	x		x
Cuscutaceae	*Cuscuta campestris*	x			x	
Cyperaceae	*Cyperus esculentus*		x		x	
	Cyperus rotundus		x			
Equisetaceae	*Equisetum arvense; E. diffusum; E. ramosissimum*	x[2]				
Euphorbiaceae	*Euphorbia esula*		x			
	Mercuriales annua	x				x
Poaceae	*Alopecurus myosuroides*	x				x
	Agropyron repens	x	x	x	x	x
	Agrostis alba	x				
	Avena fatua	x	x		x	x
	Cynodon dactylon		x		x	
	Digitaria sanguinalis	x	x		x	
	Echinochloa crus-galli	x	x	x	x	x

	Echinochloa oryzicola	x				
	Phalaris canariensis	x				
	Setaria sp.	x	x	x	x	
	Sorghum halepense	x	x		x	
Malvaceae	*Abutilon theothrasti*	x	x	x	x	x
	Malva neglecta		x			
Polygonaceae	*Polygonum aviculare; P. lapathifolium*	x		x	x	x
	Polygonum barbatum; P. bungeanum	x[2]				
	Polygonum convolvulus	x	x		x	x
	Polygonum persicaria		x	x	x	x
	Polygonum pensylvanicum		x			
	Polygonum nepalense	x				
Portulacaceae	*Portulaca oleracea*	x	x			
Rubiaceae	*Galium aparine*	x		x	x	
Scrophulariaceae	*Veronica persica*	x				x
Solanaceae	*Datura stramonium*	x	x		x	x
	Physalis heterophylla		x			
	Solanum nigrum	x	x	x	x	x
Apiaceae	*Aethusa cynapium*					x
Urticaceae	*Urtica urens*	x	x			x
Violaceae	*Viola arvensis*	x				x

After Ivashchenko, 1998; Petersen and Hurle, 1998; Sullivan and Fischer, 1971; Okuno, 2001 (pc. - personal communication); Özgür, 2000 (pc); Zhou Jianchao 2001 (pc.); [1)] China, Japan, Turkey; [2)] only in China; [3)] in China*: A. annua; A. lavandulaefolia; A. apiacea*

Quite a lot of work has been done in the past to define the critical period for weed control in sugarbeet (Scott and Moisey, 1972; Hack, 1981; Ammon and Kunz, 1982; Meyer et al., 1986; Montemurro et al., 1999; Kobusch, 2003). Depending on the conditions (weed flora and climate) the different authors found the critical periods of weed competition in sugarbeet range from 20 (14-28) to 50 (38-85) days after emergence of the crop. Weeds can be allowed to grow before and to germinate after this period. However, there was a large variation in the results of the experiments caused by changes in environmental conditions. Primarily, this was due to the fact that weeds and sugarbeet do not behave in the same way under different climatic conditions. To overcome this problem and find a more or less fixed threshold under all conditions, the critical period should be based up on crop growth stages. The experiments of Bräutigam (1998) and Wellmann (1999) demonstrated clearly that the critical period of competition in sugarbeet never started before the 4-leaf stage (BBCH 14) and ended normally with the 10- to 12-leaf stage (BBCH 19-32) of the beets. Consequently, the period when weeds in sugarbeet should be controlled is very short (21-28 days under normal growing conditions). If the sugarbeet grower wants to control the weeds in his fields according to the period threshold concept, he needs the tools to do so. Due to the lack of efficacy of most of the selective sugarbeet herbicides on larger weeds, other tools are needed. There are two possible solutions that would allow the period threshold concept to be practiced. The first solution is the mechanical way, but for economic reasons this is not acceptable for most growers. The second possible solution is the use of herbicide resistant varieties.

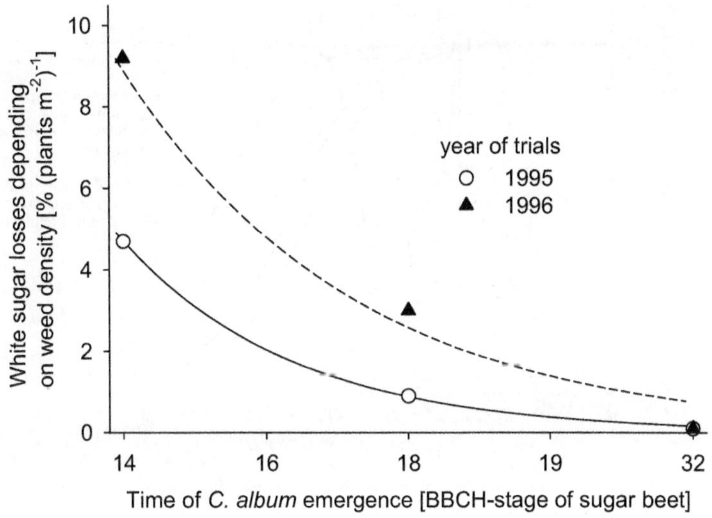

Figure 1. Competition effect of different densities of *Chenopodium album* on sugarbeet, depending on time of weed emergence in relation to crop development (BBCH-stage), Wellmann, 1999.

MECHANICAL WEED CONTROL

Wide distances between sugarbeet rows allows the use of cultivation equipment. Depending on the development stage of the crops, different kinds of cultivation tools must be used to avoid crop damage. Timing of use of mechanical cultivation is very important for high efficacy and crop safety. Sometimes weather conditions are not optimal when hoeing is needed. In particular, wet soil conditions do not allow the use of hoeing machines at all, or efficacy is decreased due to regrowth of the weeds under such conditions. Weeds in the rows are another problem if weed control is to depend only on mechanical methods. Harrows can be used also in the rows. However, the efficacy of harrows is limited to smaller weeds and harrows should not be used between emergence of the coleoptile from the seed and 2-leaf stage of the sugarbeet (Lüdecke, 1961). There are some tools for use on hoeing machines such as finger weeders or a torsion weeder, which are able to remove some weeds from the sugarbeet rows.

The efficacy of these tools is not as good as required. The consequence is that hand weeding is still necessary. Depending on the weed infestation level and weather conditions 70 to 300 hours per hactre are needed to remove the weeds in the row by hand. In industrial countries hand weeding is very expensive. Therefore nowadays hand weeding is limited to organic farms or very specific weed problems (e.g., weed beets). In countries where labour is cheaper than the use of herbicides, weeding by hand is quite common (e.g., Turkey and the countries of the former Soviet Union).

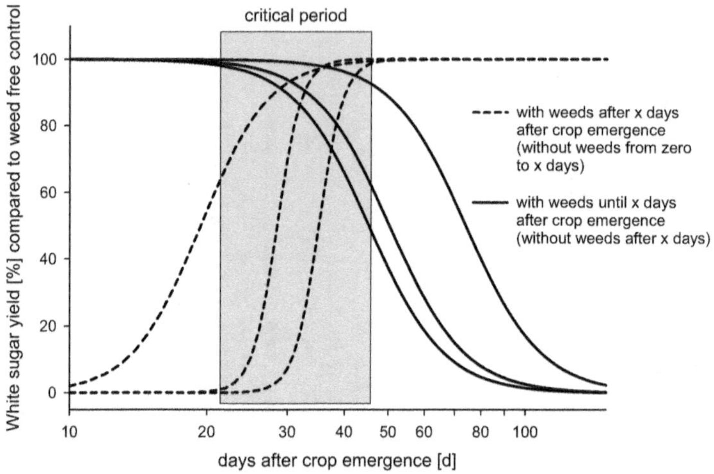

Figure 2. Critical period for weed control in sugarbeet – relative white sugar yield at 3 different sites or years depending on duration of weed competition (modified after Kobusch, 2003)

A possible option for reducing the use of herbicides is the integration of mechanical methods into chemical weed control systems. To overcome problems with weeds in the row, band spraying of herbicides (in the row) is often combined with the use of mechincal cultivation.

SELECTIVE HERBICIDES

In the early days of sugarbeet cultivation (in the 19th century), some inorganic chemicals were used for weed control in sugarbeet (e.g., NaCl or H_2SO_4). Before 1945, the use of calcium cyanamide incorporated into the soil before planting was a common method for weed control. After the development of 2,4-D as a herbicide in cereals, other organic compounds for weed control in sugarbeet were discovered (1946 propham, 1951 endothal). These herbicides were used as pre-emergence treatments primarily against dicot and some monocot weeds. Pre-planting applications with incorporation into the soil were carried out with TCA (1947), dalapon (1954), di-allate and tri-allate (1961) or cycloate (1966). All of these herbicides were used primarily against monocot weeds. High dosages and the persistence of compounds with side effects to non-target organisms are the reason that most of these herbicides have disappeared from the market. In the 60's and 70's new ingredients like ethofumesate, chloridazon, metamitron and phenmedipham were developed. Pre-emergence applications or post-emergence treatments with high dosages were common strategies for weed control. For ecological and economical reasons, herbicide use in sugarbeet changed completely in the late 80's. Concepts of reduced dose post-emergence treatments became the new standard. Therefore, mixtures of different active ingredients were applied on average 3 times. Due to the fact that weeds in sugarbeet can germinate over a long period, sequential applications at intervals of 8 to 14 days between two sprays is needed. For high efficacy of this method, the timing of application is very important. Weeds have to be small (cotelydon stage to 2 leaf stage) to ensure successful weed control. However, sometimes climatic conditions (especially soil moisture and relative air humidity) have a greater influence on herbicide performance than the growth stage of the weeds (May et al., 1997). The right dose, mixture of herbicides and timing have to be met according to the weed flora, soil and climatic conditions. Farmers need experience and a good knowledge in order to meet these requirements. In Europe and the US, the extension services (grower associations, sugar industry and governmental organisations) were very successful in developing the new low dose practice. Today post-emergence treatments of reduced dosages is common practice on nearly 80 % of the growing area in the west european countries. Table 3 gives an overview of the selective sugarbeet herbicides used.

Herbicide Resistant Sugarbeet

In the last decade, glyphosate and glufosinate resistant sugarbeet have been developed (Table 4). Resistance genes were transferred in both cases from different soil bacteria into the sugarbeet genome using biotechnology. In contrast to conventional sugarbeet herbicides, glyphosate and glufosinate are able to control larger weeds also. Consequently, growing herbicide resistant varieties can improve the flexibility and efficacy of chemical weed control (May, 2000). Furthermore, there is a high probability that the amount of active ingredient used will be reduced compared to conventional systems (Madsen and Jensen, 1995) and this reduction will continue long-term (Madsen et al., 1996). Herbicide resistant varieties will provide a useful solution for the control of specific weed problems such as volunteers (e.g., potatoes, weed beets, oilseed rape) or perennial and parasitic weeds, which are difficult to control with current beet herbicides, (Schäufele and Pfleiderer, 2000).

Using herbicide resistant sugarbeet varieties and the corresponding herbicides would allow a delay in weed control compared to common conventional application strategies. For the first time, weed control according to the concept of period threshold can be realized, because also larger weed can be managed by these new systems. As a consequence, fewer pests and diseases may appear on sugarbeet due to the presence of weeds during early growth stages of the beets (Ammon et al., 1996; Dewar et al., 2000a).

Most of the selective sugarbeet herbicides have some effect on sugarbeet growth (Figure 3); temporary symptoms will often appear on the leaves after applications. In some cases, this lack of selectivity will result in a yield reduction. Growing herbicide resistant varieties has the potential to increase yield due to the higher degree of selectivity present in the systems (Figure 4).

The high efficacy of glyphosate and glufosinate on larger weeds of most species will make it much easier to handle weed control in reduced tillage systems and the control of living mulches. Additionally, the new future-oriented development of precision weed control, which takes into account the spatial variability of weeds via patch treatment or sensor-driven systems, could be promoted by using herbicide resistant varieties (Hurle and Petersen, 2000).

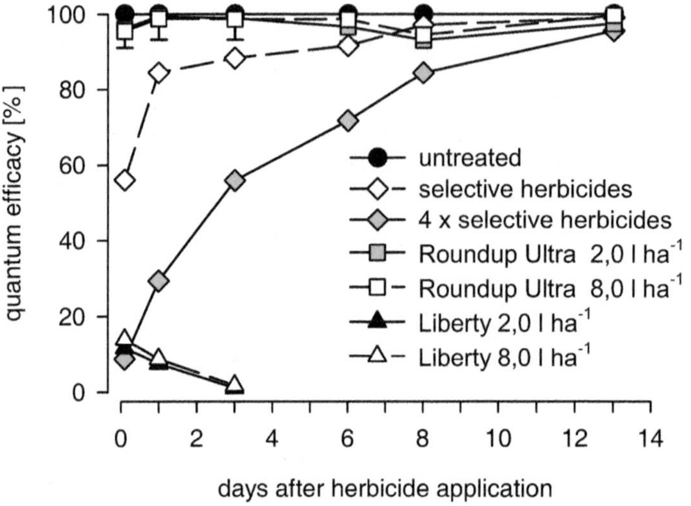

Figure 3. Chlorophyll fluorescence in glyphosate resistant sugarbeet variety treated with different herbicides (selective herbicides – mixture of six conventional sugarbeet herbicides applied in common dosage) after Beißner and Büttner, 2000.

Intensive use of more than one herbicide resistant crop (resistant to the same herbicide) in a region or in a crop rotation on one farm might cause agronomic or/and environmental problems such as outcrossing of resistance, selection of herbicide resistant weeds and volunteer crops, a shift in weed flora, surface water contamination, and damage to non target plants due to herbicide drift (Petersen and Hurle, 1998; Hurle and Petersen, 2000). To avoid these problems, management strategies will need to be put into place before herbicide resistant varieties are widely cultivated.

Mulch Systems

Normally sugarbeet are planted in spring and develop relatively slowly. Therefore the soil is sensitive to erosion (water and wind) and leaching of nutrients. Reduced tillage systems and use of cover crops can help to avoid these problems. In some countries (e.g., Germany, Merkes et al., 2001) integration of cover crops into a sugarbeet rotation is quite common (25 % of sugarbeet area). Normally, non winter hardy cover crops are planted after harvesting of cereals in late summer. Common species used as cover crops before growing sugarbeet are mustard (*Sinapis alba* L.), phacelia (*Phacelia tanacetifolia* Benth.) and radish (*Raphanus sativus* var. *oleiformis* (Stokes) Metzg.). Due to frost during winter the plants are killed and sugarbeet can be planted into the dead mulch in spring. In addition to control of erosion and leaching of nutrients, cover crops can also reduce nematode populations. Some mustard and radish varieties are nematode (*Heterodera schachtii*) resistant and will reduce their population. This is important, because *H. schachtii* is one the major pests in sugarbeet.

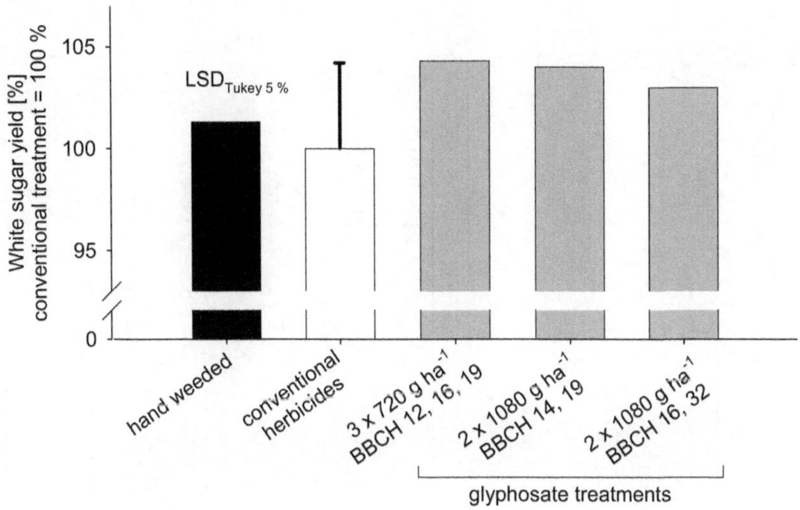

Figure 4. Effect of different weed control systems on white sugar yield in glyphosate resistant sugarbeet (IfZ field trials 1998-2000, n = 25; unpublished)

Weed suppression by these fast growing crops is effective in fall and is due to a dramatic reduction of light under the cover crop canopy (Auler, 1998). In cover crops from the Brassicaceae family allelopathy may also be an important factor for weed suppression. Glucosinolates are the characteristic secundary metabolites of this family (Fenwick et al., 1983). Degredation compounds from glucosinolates like isothiocyanates are biological active and can hinder weed germination (Al-Khatib et al., 1997). However, this factor is only important when cover crops are used as green manure and fresh plant material is incorporated into the soil (Petersen et al., 2001).

Mustard, phacelia and radish are not winter hardy. As a consequence, weeds that survive under the cover crop can re-grow and also new weeds can germinate in spring. Some weeds are able to survive under cover crops (e.g. *Galium aparine* L., *Viola arvensis* Murr.) (Garbe and Heitefuss, 1988), and other weeds like volunteer barley (*Hordeum vulgare* L.) are more sensitive to competition (Isselstein and Rauber, 1988). To avoid weed control problems after planting sugarbeet into cover crop mulch, a non selective herbicide is normally used before beets are planted.

Some cover crops are planted in spring before sugarbeet are sown to prevent wind erosion. For this, cereals like oats (*Avena sativa* L.) or barley are used (Fornstorm and Miller, 1996). However, these cover crops also compete with beets and have to be removed before the period threshold starts. This is normally done by using a graminicide, which provides good efficacy and high selectivity. Sowing of clover species (*Trifolium repens* L. or *Medicago lupulina* L.) between the rows after planting the sugarbeet is possible when used in combination with band spraying (in the rows), but due to competition from weeds and clover, sugar yield can be decreased (Haberland, 1996).

Table 3. Selective sugarbeet herbicides

Mode of action	Chemical Family	Active ingredient	HRAC* Group	Application
Inhibition of acetyl CoA carboxylase (ACCase)	Aryloxyphenoxy-propionates ('FOPs')	diclofop-methyl fluazifop-P-butyl haloxyfop-R-methyl propaquizafop quizalofop-P-ethyl	A	post
	Cyclohexanediones ('DIMs')	clethodim cycloxydim sethoxydim	A	post
Inhibition of acetolactate synthase (ALS)	Sulfonylureas	triflusulfuron-methyl	B	post
Inhibition of photosynthesis at photosystem II	Triazinones	metamitron	C1	pre/post
	Uracils	lenacil	C1	pre/post
	Pyridazinones	chloridazon	C1	pre/post
	Phenylcarbamates	desmedipham phenmedipham	C1	post post
Inhibition of carotenoid biosynthesis (unknown target)	Isoazolidinones	clomazone**	F3	pre
Microtubule assembly inhibition	Dinitroanilines	trifluralin	K1	post (i)
Inhibition of cell division	Chloroacetamides	metolachlor dimethenamid	K3	post post
Inhibition of lipid synthesis – not ACCase inhibition	Thiocarbamates	cycloate EPTC tri-allate	N	ppt (i) ppt (i) ppt

Table 3. Continued..

	Benzofuranes	ethofumesate	N	post
Action like indole acetic acid	Pyridine carboxylic acids	clopyralid	O	post
	Quinoline carboxylic acids	quinmerac	O	pre/post
Only in herbicide resistant varieties				
Inhibition of EPSP synthase	Glycines	glyphosate**	G	post
Inhibition of glutamine synthetase	Phosphinic acids	glufosinate**	H	post
Inhibition of acetolactate synthase (ALS)	Imidazolinones	imazamox**	B	post
Inhibitor of protoporphrinogen oxidase	Pyrimidindione	butafenacil**	E	post

* HRAC – Herbicide Resistance Action Committee (www.plantprotetion.org/HRAC/MOA.htm)
ppt – preplanting, pre –, post –emergence application, (i) - soil incorporation
** under development for use in sugarbeet

Table 4. Mode of action of sugarbeet relevant herbicide resistance systems.

System	Active ingredient (a.i.)	Target enzyme	Tolerance mechanism	Reference
Liberty Link	glufosinate 200 g a.i. L^{-1}	glutamine synthetase	detoxification	Donn & Eckes, 1992
Roundup Ready	glyphosate 360 g a.i. L^{-1}	5-enolpyruvylshikimate-3-phosphate synthase	non sensitive enzyme	Mannerlöf et al., 1997

Other living mulch systems (e.g. planting winter hardy cover crops) are possible in sugarbeet, but some research still needs to be done to integrate these cover crops into vegetation management (Ammon and Serafin, 1996). However, the introduction of herbicide resistant sugarbeet varieties will increase the use of reduced tillage systems in combination with different cover crops. Due to more easier and more cost effective weed control this is expected.

ACKNOWLEDGEMENT

I thank Michael John May IACR-Broom's Barn, UK for valuable discussions and carefully reading the manuscript.

REFERENCES

Al-Khatib K., Libbey C., Boydston R. Weed suppression with *Brassica* green manure crops in green pea. Weed Sci 1997; 45: 439-445.

Ammon H.U., Kunz P. Einfluss der Unkrautkonkurrenz zu bestimmten Entwicklungsstadien auf den Ertag der Zuckerrübe. Mitteilungen schw. Landwirtschaft 1982; 30: 29-34.

Ammon H.U., Dubois D., Bohren C. Späte Unkrautbekämpfung in Rüben mit Glufosinat – ein Anbauverfahren zur Nützlingsförderung. Zeit Pflanzenkrank Pflanzenschutz, Sonderheft 1996; XV: 593-598.

Ammon H.U., Serafin F. Streifenfrässaat von Rüben in Zwischenfruchtbestände und Regulation der Grünbedeckung mit Glyphosat und Glufosinat. Zeitschrift für Pflanzenkrankheiten und Pflanzenschutz, Sonderheft 1996; XV: 281-289.

Auler T. *Unkrautunterdrückung mit* Sinapis alba *L. als Bodendecker*. Dissertation Universität Hohenheim, 1998.

Beißner L., Büttner G. Herbicide stress in sugarbeet: physiology, symptomatology and economical relevance. Proceedings of the 63th IIRB Congress, Interlaken, 149-161, 2000.

Bosch J. Der Einfluß einiger dominater Ackerunkräuter auf Nutz- und Schadarthropoden in einem Zuckerrübenfeld. Zeit Pflanzenkrank Pflanzenschutz 1987; 94: 398-408.

Boutry P., Mörchen M., Saumitou-Lapprate P., Vernet P., van Dijk H. The origin and evolution of weed beets: consequences for the breeding and release of herbicide-resistant transgenic sugarbeet. Theor Appl Genet 1993; 87: 471-478.

Bräutigam H. *Untersuchungen zur Konkurrenz zwischen Unkraut und Zuckerrüben – Auftreten, Ursachen und Konsequenzen für die Unkrautregulierung*. Dissertation Universität Göttingen, Cuvillier Verlag, Göttingen, 1998.

Bräutigam H., Schäufele W.R. Einfluß der Stickstoffdüngung auf die Verunkrautung in Zuckerrüben. Zeitschrift für Pflanzenkrankheiten und Pflanzenschutz, Sonderheft 1994; XIV: 467-476.

Dawson J.H. The concept of period thresholds. In Proceedings of European Weed Research Society 5th Symposium, Ecomonic Weed Control, Stuttgart-Hohenheim, 327-331, 1996.

Dewar A.M., Haylock L.A., May M.J., Beane J., Perry R.N. Glyphosate applied to genetically modified herbicide-tolerant sugarbeet and volunteer potatoes reduced populations of potato cyst nematodes and the number and size of daughter tubers. Ann Appl Biol 2000a; 136: 179-187.

Dewar A.M., Haylock L.A., Bean K.M., May M.J. Delayed control of weeds in glyphosate tolerant sugarbeet and the consequences on aphid infestation and yield. Pest Manage Sci 2000b; 56: 345-350.

Dexter A.G. Weed control guide for sugarbeet. www.sbreb.org/98/weed/98p3.htm, 1998.

Donn G., Eckes P. Basta-verträgliche Kulturpflanzen durch Übertragung eines synthetischen Phosphinothricin-Acetyltransferase-Gens. Zeitschrift für Pflanzenkrankheiten und Pflanzenschutz, Sonderheft 1992; XIII: 499-504.

Dubois D., Häni A., Ammon H.U., Keller S. Einfluss der Unkrautbekämpfung auf den Ertrag und Virusbefall von Zuckerrüben. Landwirtschaft Schweiz 1993; 6: 133-138.

El Titi A. Unkrautkonkurrenz im Zuckerrübenanbau und ihre praktische Ausnutzung. Zeitschrift für Pflanzenkrankheiten und Pflanzenschutz 1986; 93: 136-145.

FAO. www.fao.org, 2001.

Fenwick G.R., Heaney R.K., Mullin W.J. Glucosinolates and their breakdown products in food and food plants. Crit Rev Food Sci Nutr 1983; 18: 123-301.

Fornstrom K.J., Miller S.D. Sugarbeet establishment in living mulches. J Sugarbeet Res 1996; 33: 15-29.

Garbe V. Heitefuss R. Beeinflussung der Verunkrautung durch Senf-Zwischenfrüchte in Systemen der Mulchsaat zu Zuckerrüben. Zeitschrift für Pflanzenkrankheiten und Pflanzenschutz, Sonderheft 1988; XI: 323-327.

Geißl W., Bachthaler G., Hoffmann G.M. Untersuchungen zur Wirtseignung verschiedener geographischer Herkünfte von Ackerunkräutern gegenüber dem Rübenzystenälchen *Heterodera schachtii* Schmidt. Weed Res 1989; 29: 221-228.

Haberland R. Unkrautregulierung in Zuckerrüben durch Klee-Einsaat. Zeitschrift für Pflanzenkrankheiten und Pflanzenschutz, Sonderheft 1996; XV: 267-274.

Hack H. Vergleich der chemischen Unkrautbekämpfung in Zuckerrüben mit einer mechanischen Sauberhaltung zu unterschiedlichen Zeitabständen. Zeitschrift für Pflanzenkrankheiten und Pflanzenschutz, Sonderheft 1981; IX: 355-363.

Hanf M. Unkräuter des Rübenackers. Praktische Blätter für Pflanzenbau und Pflanzenschutz 1959; 54: 162-174.

Hurle K., Petersen J. Cultivation of herbicide resistant crops: weed management and environmental aspects. Proceedings: The Biosafety Results of Field Tests of Genetically Modified Plants and Microorganisms – 5th Int. Symposium, Braunschweig 1998, 122-126, 2000.

Isselstein J., Rauber R. Zwischenfrüchte zur Kontrolle von Ausfallgerste und Unkräutern vor Zuckerrüben-Mulchsaat. Zeitschrift für Pflanzenkrankheiten und Pflanzenschutz, Sonderheft 1988; XI: 329-338.

Ivashchenko O. Besonderheiten der Verunkrautung im Zuckerrübenanbau der Ukraine. Zeitschrift für Pflanzenkrankheiten und Pflanzenschutz, Sonderheft 1998; XVI: 615-617.

Kobusch H. *Zeitbezogene Schadensschwelle in Zuckerrüben und Untersuchungen zu ihrer Anwendbarkeit in einer glufosinatresistenten Transformante.* Dissertation Universität Hohenheim, 2003; in preperation.

Kropff M.J., Spitters C.J.T. An eco-physiological model for interspecific competition, applied to the influence of *Chenopodium album* L. on sugarbeet. I: Model description and parameterization. Weed Res 1992; 32: 437-450.

Lotz L.A.P., Groeneveld R.M.W., de Groot N.A.M.A. Potential for reducing herbicide inputs in sugarbeet by selecting early closing cultivars. Proceedings of the Bristish Crop Protection Conference, Weeds, Brighton 1991, 1241-1248, 1991.

Lüdecke H. *Zuckerrübenanbau.* Paul Parey Verlag, Berlin, 1961.

Madsen K.H., Jensen J.E. Weed control in glyphosate-tolerant sugarbeet (*Beta vulgaris* L.). Weed Res 1995; 35: 105-111.

Madsen K.H., Blacklow W.M., Jensen J.E. Simulation of herbicide-use in crop rotation with transgenic herbicide resistant sugarbeet. Proceedings of the Second International Weed Control Congress, Copenhagen 1996, 1387-1391, 1996.

Mannerlöf M., Tuvesson S., Stehen P., Tenning P. Transgenic sugarbeet tolerant to glyphosate. Euphytica 1997; 94: 83-91.

May M.J., Lainsbury M.A., Hilton J.G. Effect of weather on efficacy of herbicide in sugarbeet. Proceedings of the British Crop Protection Conference – Weeds, Brighton 1997, 857-862, 1997.

May M.J. Efficacy and selectivity of RR and LL weed control techniques compared to classical weed control systems. Proceedings of the 63th IIRB Congress, Interlaken 2000, 163-170, 2000.

Merkes R., Muggele H., Sauer M., Kröhl M. Produktionstechnik zu Zuckerrüben – Kostensenkung, Umweltschonung, Nachhaltigkeit. Zuckerindustrie 2001; 128: 804-811.

Meyer H., Widmer U., Ammon H.U. Konkurrenz der Unkräuter und Einfluss auf die Unkrautbekämpfung im Zuckerrübenbau. Proceedings of the IIRB-Conference 1986, 263-275, 1986.

Mittler S., Petersen J., Koch H.J. Bekämpfungsschwellen bei der Unkrautbekämpfung in Zuckerrüben. Zeitschrift für Pflanzenkrankheiten und Pflanzenschutz, Sonderheft 2002; XVIII: 499-509.

Montemurro P., Castrignano A., Fracchiolla M., Lasorella C. The critical period for weed control in sugarbeet. Proceedings of the 11th European Weed Research Society Symposium, Basel, 67, 1999.

Moss S.R., Clarke J.H., Blair A.M., Culley T.N., Read M.A., Ryan P.J., Turner M. The occurrence of herbicide-resistant grass-weeds in the United Kingdom and a new system for designating resistance in screening assays. British Crop Protection Conference – Weeds, Brighton 1999, 179-184, 1999.

Neururer H. Weitere Erfahrungen in der Beurteilung der tolerierbaren Verunkrautungsstärke. Zeitschrift für Pflanzenkrankheiten und Pflanzenschutz, Sonderheft 1969; VII: 63-69.

Nieto J.H., Brando M.A., Gonzalez J.T. Critical periods of the crop growth cycle for competition of weeds. PANS 1968; 14: 159-166.

Petersen J., Hurle K. Einführung von herbizidtoleranten Sorten: Konsequenzen für die Unkrautbekämpfung. Zeitschrift für Pflanzenkrankheiten und Pflanzenschutz, Sonderheft 1998; XVI: 365-372.

Petersen J., Belz R., Walker F., Hurle K. Weed suppression by release of isothiocyanates from turn-rape mulch. Agron J 2001; 93: 37-43.

Rademacher B. Die Bedeutung der Verunkrautung für den Zuckerrübenanbau. Wiss. Aussprache Chem. Unkrautbekämpfung in Zuckerrüben mit Pyramin (BASF, Lu/Rhein), 1964; 5-10.

Schäufele W.R. Einfluß hochwachsenden Unkrautes (*Chenopodium album*) auf Entwicklung und Ertrag der Zuckerrübe. Zeit Pflanzenkrank Pflanzenschutz 1986; 93: 125-135.

Schäufele W.R. Einfluß niedrigwachsender Unkräuter zwischen den Reihen auf den Ertrag von Zuckerrüben. Gesunde Pflanze 1991; 43: 175-179.

Schäufele W.R., Pfleiderer U.E. Ansätze zur Bekämpfung von Ausfallraps, Kartoffeldurchwuchs und Unkrautrüben in herbizidresistenten Zuckerrüben – erste Ergebnisse. Zeitschrift für Pflanzenkrankheiten und Pflanzenschutz, Sonderheft 2000; XVII: 403-409.

Schlösser E. Rizomania. XI. Weeds as alternate hosts for virus and vector? Med. Fac. Landbouww. Rijksuniv. Gent 1989; 54/2b: 479-484.

Schweizer E.E. Broad leaf weed interference in sugarbeet (*Beta vulgaris*). Weed Sci 1981; 29: 128-133.

Schweizer E.E., Bridge L.D. Sunflower (*Helianthus annuus*) and Velverleaf (*Abutilon theophrasti*) interference in sugarbeets (*Beta vulgaris*). Weed Sci 1982; 30: 514-519.

Schweizer E.E. Common lambsquarters (*Chenopodium album*) interference in sugarbeets (*Beta vulgaris*). Weed Sci 1983; 31; 5-8.

Schweizer E.E. Powell Amaranth (*Amaranthus powellii*) interference in sugarbeet (*Beta vulgaris*). Weed Sci 1985; 33: 518-520.

Scott R.K., Moisey F.R. The effect of weeds on sugarbeet crop. Proceedings of the 11th British Weed Control Conference, Brighton 1972, 491-498, 1972.

Sullivan E.F., Fischer B.B. "Weed Control". In *Advances in Sugarbeet Production: Principles and Practices*. R.T. Johnson, Alexander J.T., Rush G.E., Hawkes G.R. eds. The Iowa State University Press, Ames, 1971; pp. 71-106.

Ulber B. Untersuchungen zur Nahrungswahl von *Onychiurus fimatus* Gisin (Onychiuridae, Collembola), einem Aufgangsschädling der Zuckerrübe. Zeitschrift für angewandte Entomologie 1980; 90: 333-346.

Weatherspoon D.M., Schweizer E.E. Competition between sugarbeets and five densities of Kochia. Weed Sci 1971; 19:125-128.

Wellmann A. *Konkurrenzbeziehungen und Schadensprognose in Zuckerrüben bei variiertem zeitlichen Auftreten von* Chenopodium album *(L.) und* Chamomilla recutita *(L.) Rauschert*. Dissertation Universität Göttingen, Cuvillier Verlag Göttingen, 1999.

Zimdahl R.L., Fertig S.N. Influence of weed competition on sugar beet. Weeds 1967; 15: 336-339.

CHAPTER 24

CONTROL OF PROBLEM WEEDS AND NET RETURNS WITH HERBICIDE PROGRAMS IN PEANUT (*ARACHIS HYPOGAEA* L.)

W.J. GRICHAR[1], R.G. LEMON[2], P.A. DOTRAY[3] & B.A. BESLER[4]

[1]Research Scientist, Texas Agric. Expt. Stat., Beeville, TX 78102; [2]Associate Professor, Texas Cooperative Extension, College Station, TX 77843; [3]Associate Professor, Texas Agric. Expt. Stat., Lubbock, TX 79401; [4]Research Associate, Texas Agric. Expt. Stat., Beeville, TX

Peanuts (*Arachis hypogaea* L.) are the twelfth most valuable cash crop in the United States and the fourth most valuable oil crop worldwide (Bridges et al., 1994a). India (5.4 million metric tons), China (5.0 million metric tons) and the United States (1.8 metric tons) are the largest producers in the world and grow about 70% of the world crop. While peanuts produced in the United States are primarily used for food, other countries produce peanuts primarily for oil. It is the third major oilseed of the world next to soybean (*Glycine max* (L.) Merr.) and cotton (*Gossypium hirsutum* L.) (Bridges et al., 1994a).

The cultivated peanut or groundnut originated in South America (Bolivia and adjoining countries) and is now grown throughout the tropical and warm temperature regions of the world. Remnants of peanut have been found at archaeological sites dating from 1500 B.C. on the northern coast of Peru. This crop was grown widely by native peoples of the New World at the time of European expansion in the sixteenth century and was subsequently taken to Europe, Africa, Asia, and the Pacific Islands. Peanut was introduced to the southeastern United States during 1700's (Bridges et al., 1994a).

Peanuts have several unique features that contribute to challenging weed management. First, most peanut cultivars grown in the United States require a fairly long growing season of 140 to 160 days depending on cultivar and geographical region (Henning et al., 1982; Wilcut et al., 1995a). Because of this long growing season, soil-applied herbicides may not provide season-long control and mid-to-late season weed problems are common. Secondly, peanut has a prostrate growth habit, a relatively shallow canopy, and is slow to shade inter-rows allowing weeds to be more competitive with the peanut plant (Walker et al., 1989; Wilcut et al., 1995a). Additionally, peanut fruit develops underground on pegs originating from branches that grow along the soil surface. This prostrate growth habit and pattern of fruit development restricts cultivation to an early season control option (Brecke and Colvin, 1991; Wilcut et al., 1995a). With conventional row spacing (91 to 102 cm), complete ground cover may not be attained until 8 to 10 weeks after planting. In some areas of the U.S. peanut growing region, complete canopy closure may never be attained.

Inderjit (ed.), Weed Biology and Management, 485-515.

Because of the low growing nature of peanuts, weeds that germinate early and are not controlled "escape" relatively late in the growing season (Buchanan et al., 1982). Covering peanuts and weeds with soil during cultivation is not practical and in fact can lead to more problems. Peanuts covered with soil create conditions conducive to foliage, stem, and pod diseases caused by soil-borne fungi (Buchanan et al., 1982; Porter et al., 1982).

Weed removal is extremely difficult once weeds become established in the peanut row. After both peanut and weeds achieve some growth, mechanical removal with tractor-mounted cultivators is impossible. Hand weeding is difficult, costly, and unrealistic under modern day conditions (Buchanan et al., 1982). Consequently, peanut growers have readily accepted chemical weed control practices.

Weeds compete with the peanut plant for sunlight, moisture, and nutrients throughout the growing season. However, weeds can also reduce harvesting efficiency. Weeds are particularly troublesome during digging and inverting procedures (Young et al., 1982). Weed biomass slows field-drying of peanut vines and pods and increases the likelihood of exposure to rainfall, which can also increase harvesting losses (Young et al., 1982; Wilcut et al., 1995a). The fibrous root system of annual grasses is extremely difficult to separate from the peanut (Wilcut et al., 1994a).

During digging and combining operations, under weedy conditions, the peanut pods can become detached from the peanut vines and are left on the soil surface. These detached peanut pods cannot be recovered with current mechanized harvesting equipment. Peanut growers apply preplant incorporated (PPI) and preemergence (PRE) herbicides to control annual grasses and broadleaf weeds. Postemergence (POST) herbicides are applied to control hard-to-kill or escaped weeds. Due to major differences in environmental and edaphic characteristics, problem weeds can vary from region to region.

This chapter covers various aspects of peanut weed management control in the United States. The main objective of this chapter is to present weed control strategies and practices for growing and managing a weed-free peanut crop. Although the discussion deals mainly with weed control in the United States, these practices can also be adapted in other countries where peanut is grown.

PROBLEM WEEDS

Weeds are discussed based on importance and availability of control measures (Table 1). Although broadleaf weeds are a serious problem, annual and perennial grasses are more competitive and detrimental to peanut yield than broadleaf weeds (Wilcut et al., 1995a).

Table 1. Common and Latin names of weeds occurring in peanut fields (Wilcut et al., 1995a).

Common name[a]	Latin name	Bayer code
Amaranth palmer	*Amaranthus palmeri* S. Wats.	AMAPA
Anoda, spurred	*Anoda cristata* (L.) Schlecht.	ANVCR
Beggarweed, Florida	*Desmodium tortuosum* (Sw.) D.C.	ACVCR
Bermudagrass, common	*Cynodon dactylon* (L.) Pers.	CYNDA
Burgherkin	*Cucumis anguria* L.	CUMAN
Carpetweed	*Mollugo verticillata* L.	MOLVE
Cocklebur, common	*Xanthium strumarium* L.	XANST
Copperleaf, hophornbeam	*Acalypha ostryifolia* Riddell	ACCOS
Crabgrass, large	*Digitaria sanguinalis* (L.) Scop.	DIGSA
Crabgrass, species	*Digitaria* spp.	DIGZZ
Crabgrass, southern	*Digitaria ciliaris* (Retz.) Koel.	DIGSP
Croton, tropic	*Croton glandulosus* Muell. Arg.	CVNGS
Crowfootgrass	*Dactyloctenium aegyptium* (L.) Willd.	DTTAE
Crownbeard, golden	*Verbesina encelioides* (Cav.) Benth & Hook. F. ex Gray	VEREN
Devil's-claw	*Proboscidea louisianica* (Mill.) Thellung	PROLO
Eclipta	*Eclipta alba* (L.) Hassk.	ECLAL
Foxtail species	*Setaria* spp.	SETZZ
Goosegrass	*Eleusine indica* (L.) Gaertn.	ELEIN
Horsenettle, Carolina	*Solanum carolinense* L.	SOLCA
Horsenettle, robust	*Solanum dimidiatum* Raf.	SOLDM
Jimsonweed	*Datura stramonium* L.	DATST
Johnsongrass	*Sorghum halepense* (L.) Pers.	SORHA
Lambsquarters, common	*Chenopodium album* L.	CHEAL
Morningglory, entireleaf	*Ipomoea hederacea* var. *integriuscula Gray*	IPOHG
Morningglory, ivyleaf	*Ipomoea hederacea* (L.) Jacq.	IPOHE
Morningglory, pitted	*Ipomoea lacunosa* L.	IPOLA
Morningglory, smallflower	*Jacquemontia tamnifolia* (L.) Griseb.	IAQTA
Morningglory, tall	*Ipomoea purperea* (L.) Roth	PHBPU
Nightshade, silverleaf	*Solanum elaeagnifolium* Cav.	SOLEL
Nutsedge, purple	*Cyperus rotundus* L.	CYPRO
Nutsedge, yellow	*Cyperus esculentus* L.	CYPES
Panicum, fall	*Panicum dichotomiflorum* Michx.	PAQIN
Panicum, Texas	*Panicum texanum* Buckl.	PANTE
Purslane, common	*Portulaca oleracea* L.	POROL
Pusley, Florida	*Richardia scabra* L.	RCHSC
Ragweed, common	*Ambrosia artemisiifolia* L.	AMBEL

Table 1. Continued

Sandbur	*Cenchrus* spp.	CENZZ
Sage, lanceleaf	*Salvia reflexa* Hornem.	SALRE
Sedges	*Cyperus* spp.	CYNZZ
Senna, coffee	*Cassia occidentalis* L.	CASOC
Sicklepod	*Senna obtusifolia* (L.) Irwin & Barneby	CASOB
Sida, prickly	*Sida spinosa* L.	SIDSP
Signalgrass, broadleaf	*Brachiaria platyphylla* (Griseb.) Nash	BRAPP
Spurge, prostrate	*Euphorbia humistrata* Engelm. ex Gray	EPHHT
Spurge species	*Euphorbia* spp.	EPHZZ
Smartweed, Pennsylvania	*Polygonum pensylvanicum* L.	POLPY
Starbur, bristly	*Acanthospermum hispidum* DC.	ACNHI
Sunflower, prairie	*Helianthus petiolaris Nutt.*	HELPE
Trumpetcreeper	*Campsis radicans* (L.) Seem. ex Bureau	CMTRA
Velvetleaf	*Abutilon theophrasti* Medicus	ABUTH

[a]Common names of weeds and Bayer codes are a WSAA-approved computer code from Composite List of Weeds, Rev. 1989. Available from WSAA, 1508 W. University, Champaign, IL 61821.

Broadleaf Weeds

Pigweed Species

Broadleaf weeds can be a problem in many peanut producing areas of the United States. Palmer amaranth (*Amaranthus palmeri* S. Wats) is generally not ranked as a troublesome weed in most major crops; however it is a common weed in many cultivated crops around the world (Grichar, 1997a). Pigweed (*Amaranthus* spp.) is listed as one of the 10 most common weeds in all major peanut-growing states in the United States and Palmer amaranth the fourth most common weed in South Carolina (Tables 2 and 3) (Dowler, 1998).

The prolific seed production potential of *Amaranthus* spp. allows for rapid spread (Klingaman and Oliver, 1994). Palmer amaranth competes effectively for light, water nutrients, and space due to its rapid erect growth and allelopathic potential (Menges, 1987, 1988). Palmer amaranth biomass residues affect seedling growth of grain sorghum (*Sorghum bicolor* L. Moench), cabbage (*Brassica oleracea* var. *capitata* L.), carrot (*Daucus carota* L.), and onion (*Allium cepa* L.) (Menges, 1988).

Pigweed species can be controlled with dinitroaniline or chloroacetamide herbicides (Wilcut et al., 1995a). Metolachlor applied PPI or PRE can control pigweed, although not as well or consistently as the dinitroanilines (Wilcut, 1991a). Dimethenamid has been shown to control pigweed species in other crops (Gaeddert et al., 1997; Owen et al., 1998; Tonks et al., 1999). Tonks et al. (1999) found that dimethenamid controlled redroot pigweed (*Amaranthus retroflexus* L.) better than metolachlor or pendimethalin.

Diclosulam, a new triazolopyrimidine sulfonanilide herbicide, has controlled pigweed species in soybean and peanut (Braxton et al., 1997; Dotray et al., 1998; Grichar et al., 1999). Diclosulam in combination with ethalfluralin applied PPI controlled Palmer amaranth 99% while ethalfluralin alone controlled Palmer amaranth 77% (Grichar et al., 1999).

Palmer amaranth can be effectively controlled with several POST herbicides. Imazapic at rates as low as 0.04 kg/ha controlled greater than 95% Palmer amaranth while imazethapyr at 0.05 to 0.07 kg/ha, acifluorfen alone or in combination with bentazon or 2,4-DB, and lactofen alone controlled greater than 90% Palmer amaranth in 2 of 3 years (Grichar, 1997b). Tank mixes of 2,4-DB with many POST herbicides, such as acifluorfen and bentazon, help control many broadleaf species, particularly if the broadleaf weeds are larger than the recommended size for treatment (Buchanan et al., 1982).

Lactofen provided greater than 80% control of Palmar amaranth (Grichar, 1997b). Grichar (1994) noted that lactofen controlled 92% spiny amaranth (*Amaranthus spinosus* L.) when applied early POST but less than 80% when applied late POST.

A study was conducted in Texas during the 1997 and 1998 growing season to evaluate the use of soil-applied herbicides alone or in combination with POST herbicides. Results indicated that the use of POST only herbicides provided erratic control (Table 4). Acifluorfen alone or in combination with bentazon controlled greater than 70% Palmer amaranth while imazethapyr or imazapic controlled greater than 80% Palmer amaranth. Bentazon, pyridate, or 2,4-DB alone failed to control Palmer amaranth. Bentazon has been shown to provide minimal control of pigweed species (Buchanan et al., 1982; Grichar, 1994; Wilcut et al., 1994a).

Use of a soil applied herbicide alone controlled less than 80% Palmer amaranth (Table 4). However, when soil applied herbicides were used in combination with POST herbicides, control of Palmer amaranth was over 90% in most instances. Grower acceptance of dinitroaniline herbicides has been a major factor contributing to increased of peanut yields and quality. These herbicides combined with other improved practices (improved varieties, disease management, agronomic practices, etc.) have contributed to a threefold increase in yield over the past fifty years (Warren, 1998).

Morningglory Species

Morningglories (*Ipomoea* spp.) are a problem in most peanut growing areas of the United States. Morningglories have been identified as one of the ten most common weeds in peanut and they also rank as one of the most troublesome weeds in Florida, Georgia, Mississippi, and North Carolina (Table 3) (Dowler, 1998).

Dinitroaniline and chloroacetamide herbicides do not effectively control morningglories (Grichar, 1997a; Richburg et al., 1995a). Trifluralin provided only 30% control of morningglory in Oklahoma (Chandler and Santelmann, 1970) while Jordan et al. (1993) noted that metolachlor demonstrated less than 44% control of *Ipomoea* spp. Peanut herbicide systems including treatments at peanut emergence, early POST, or late POST consistently controlled Morningglories better than metolachlor alone (Jordan et al., 1993)

Table 2. Most common weed species in peanut in the eight major U.S. peanut-producing states, listed in order of prevalence (Dowler, 1998).

Alabama	**Florida**	**Georgia**	**North Carolina**
Florida beggarweed	Crabgrass spp.	Florida beggarweed	Large crabgrass
Nutsedge spp.	Florida beggarweed	Texas panicum	Goosegrass
Morningglory spp.	Sicklepod	Sicklepod	Broadleaf signalgrass
Cocklebur	Florida pusley	Cocklebur	Morningglory spp.
Sicklepod	Pigweed spp.	Nutsedge spp.	Prickly sida
Bristly starbur	Morningglory spp.	Morningglory spp.	Nutsedge spp.
Crabgrass spp.	Nutsedge spp.	Bristly starbur	Pigweed spp.
Ragweed	Hairy indigo	Crabgrass	Common ragweed
Broadleaf signalgrass	Sandbur spp.	Pigweed spp.	Common
lambsquarters			
	Texas panicum	Prickly sida	Cocklebur
Oklahoma	**South Carolina**	**Texas**	**Virginia**
Pigweed spp.	Southern crabgrass	Pigweed spp.	Large crabgrass
Yellow nutsedge	Sicklepod	Yellow nutsedge	Lambsquarters
Prickly sida	Nutsedge spp.	Texas panicum	Morningglory spp.
Crownbeard	Palmer amaranth	Morningglory spp.	Fall panicum
Eclipta	Goosegrass	Broadleaf signalgrass	Pigweed spp.
Johnsongrass	Morningglory spp.	Large crabgrass	Common ragweed
Crabgrass spp.	Florida beggarweed	Bermudagrass	Nutsedge spp.
Texas panicum	Common ragweed	Purslane	Prickly sida
Morningglory spp.	Broadleaf signalgrass	Purple nutsedge	Cocklebur
Spurge spp.	Texas panicum	Eclipta	Spurred anoda

Table 3. Most troublesome weed species in peanut in the eight major U.S. peanut producing states, listed in order cf severity (Dowler, 1998).

Alabama	**Florida**	**Georgia**	**North Carolina**
Florida beggarweed	Florida beggarweed	Florida beggarweed	Nutsedge spp.
Bristly starbur	Hairy indigo	Nutsedge spp.	Eclipta
Nutsedge spp.	Sicklepod	Bristly starbur	Common ragweed
Sicklepod	Nutsedge spp.	Morningglory spp.	Prickly sida
Horsenettle spp.	Bristly starbur	Sicklepod	Sicklepod
Tropic croton	Morningglory spp.	Tropic croton	Carolina horsenettle
Ragweed	Common dayflower	Burgherkin	Florida beggarweed
Maypop passionflower	Cocklebur	Prickly sida	Common bermudagrass
Wild poinsettia	Texas panicum	Wild poinsettia	Texas panicum
Goosegrass	Goosegrass	Eclipta	Morningglory spp.
Oklahoma	**South Carolina**	**Texas**	**Virginia**
Yellow nutsedge	Sicklepod	Yellow nutsedge	Bermudagrass
Eclipta	Florida beggarweed	Morningglory spp.	Purple nutsedge
Crownbeard	Nutsedge spp.	Sandhills amaranth	Eclipta
Horsenettle	Tropic croton	Tumble pigweed	Common ragweed
Pigweed spp.	Broadleaf signalgrass	Purple nutsedge	Spurred anoda
Hophornbeam copperleaf	Morningglory spp.	Crownbeard	Texas panicum
Spurge spp.	Palmer amaranth	Texas panicum	Morningglory spp.
Johnsongrass	Goosegrass	Hophornbeam copperleaf	Yellow nutsedge
Texas panicum	Texas panicum	Protrate spurge	Horsenettle
Prickly sida	Johnsongrass	Eclipta	Prickly sida

Ethalfluralin plus diclosulam applied PPI showed at least 88% control of entireleaf morningglory *(Ipomoea hederacea* var. *integriuscula* Gray) (Bailey et al., 1999a). Acifluorfen plus 2,4-DB and lactofen plus 2,4-DB applied POST provided at least 98% control of entireleaf, ivyleaf [*Ipomoea hederacea* (L.) Jacq.], and pitted (*Ipomoea lacunosa* L.) morningglory regardless of PRE herbicide treatment (Bailey et al., 1999a). Although entireleaf and ivyleaf morningglory are more tolerant to acifluorfen and lactofen than pitted morningglory (Higgins et al., 1988), both species are controlled by 2,4-DB (Wilcut et al., 1994a, 1995a,b). Conversely, pitted morningglory has become tolerant to 2,4-DB and is more readily controlled by acifluorfen and lactofen (Barker et al., 1984). However, late POST applications of 2,4-DB have been shown to control greater than 85% pitted morningglory (Grichar, 1997a).

Preemergence applications of flumioxazin have controlled *Ipomoea* spp. (Grichar and Colburn, 1996b; Wilcut and Cranmer, 1997). In Texas, pitted and ivyleaf morningglory were controlled greater than 75% with flumioxazin PRE (Grichar and Colburn, 1996b). In North Carolina, weed management systems which included flumioxazin controlled greater than 75% entireleaf, ivyleaf, and pitted morningglory. A rate response was observed with flumioxazin PRE at 0.07 and 0.11 kg/ha.

Imazapic and imazethapyr have provided excellent morningglory control (Grichar, 1997a; Richburg et al. 1995 a, b, 1996; Webster et al., 1997). Grichar (1997a) found that imazapic at 0.04 and 0.07 kg/ha provided at least 80% pitted morningglory control with early POST and POST applications. Imazapic at 9 g/ha has provided pitted morningglory control which varied from 33 to 85% (Newsom and Shaw, 1994). Research by Wilcut et al. (1994b) found that imazapic was more effective than imazethapyr applied either PPI, PRE, or early POST.

Control of pitted morningglory with imazethapyr was not consistent with PPI applications (Grichar, 1997a). Shaw et al. (1990) reported greater than 90% pitted morningglory control with imazethapyr POST when applied to weeds at the three-leaf stage of development. They noted that imazethapyr applied at the six-and nine-leaf stages controlled morningglory less than 85%.

Smallflower morningglory [*Jacquemontia tamnifolia* (L.) Griseb.] has been controlled at least 88% with PPI imazethapyr applications (Richburg et al., 1995a,b). Rates of imazapic or imazethapyr as low as 36 g/ha have given complete control of smallflower morningglory (Richburg et al., 1995b). Webster et al. (1997) noted no difference in morningglory control with imazapic at 36 and 72 g/ha applied PPI or PRE; however, imazapic applied early POST was more effective than those same rates applied POST. They attributed the improved control at early POST to the smaller weed size present at early POST relative to those at POST. Imazapic, applied at the labeled rate (72 g/ha) early POST or POST, controlled *Ipomoea* morningglory complex at least 96%. Webster et al. (1997) concluded that imazapic has both foliar and soil activity on the most common *Ipomoea* morningglory species found in the southeastern U.S.

Paraquat does not control smallflower morningglory; however, bentazon does (Wilcut et al., 1995a). Many growers in Virginia and the southeastern U.S. use sequential applications of bentazon plus paraquat for morningglory control. However, paraquat is not used in the southwest U.S. because of the potential for peanut yield reduction (Grichar, 1998). Peanuts are tolerant to paraquat if applications are made prior to pegging and fruit development (Wehtje et al., 1986; Wilcut and Swann, 1990). Tolerance at this stage has been attributed to the size and nutrient reserves of the seed (Schroeder and Warren, 1971). Peanut tolerance to paraquat is not cultivar dependent (Knauft et al., 1990; Wehtje et al., 1991) nor is it influenced by seed size (Wehtje et al., 1991). Paraquat can be applied from emergence until 28 days after emergence (Anonymous, 2001). Paraquat applications after this 28 day period increase the chance of injury and yield reductions (Wehtje et al., 1986; Brecke and Colvin, 1988).

Florida Beggarweed

Florida beggarweed [*Desmodium tortuosum* (Sw.) DC] is among the most troublesome weeds in peanut in the southeastern U.S. (Cardina and Brecke, 1991). A weed-free period of 4 to 6 week was required to eliminate peanut yield reductions from Florida beggarweed (Hauser and Buchanan, 1981). Florida beggarweed germinating within a few weeks of peanut often forms a dense canopy over the crop by harvest (Cardina and Brecke, 1991). Peanut yields were reduced 20 to 40% by one Florida beggarweed plant per meter of row (Buchanan et al., 1982; Hauser et al., 1982).

There has been some confusion in the description of Florida beggarweed. Taxonomists have described it as a trifoliate, although seedlings are completely unifoliate (Cardina and Brecke, 1991). Some authors have described the weed as non-nodulating, when in fact it nodulates freely in peanut fields (Cardina and Brecke, 1991). Florida beggarweed has been described as 0.5 m tall at maturity but elsewhere up to 3.5 m tall (Duncan and Foote, 1975; Radford et al., 1968). Although germination and emergence were reported to be as early as April in Georgia (Cardina and Hook, 1989), Florida beggarweed has been considered a late-season weed that escapes PRE control measures and becomes apparent too late for effective POST control (Buchanan et al., 1982).

Few soil-applied herbicides provide acceptable control of Florida beggarweed. Norflurazon applied PRE at 1.3 kg/ha showed 85% control of Florida beggarweed (McLean et al., 1995). Dimethenamid applied PPI controlled less than 60% Florida beggarweed when used alone but control was improved when dimethenamid was followed by POST herbicide systems (Edenfield et al., 2001). Flumioxazin at 0.07 to 0.1 kg/ha controlled Florida beggarweed 89 to 100% over eight trials (Wilcut and Crammer, 1997).

The current commercial POST Florida beggarweed control program consists of paraquat plus bentazon applied once or twice within 28 days of peanut emergence followed by chlorimuron applied at 60 days after peanut emergence (Webster et al., 1997; Wilcut et al., 1995a). However, by 60 days after emergence, Florida beggarweed plants are generally taller than the 25-cm height specified on the chlorimuron label (Anonymous, 2001).

Florida beggarweed control with imazapic can be variable. Brecke et al. (1995) noted that imazapic controlled from 40 to 80% Florida beggarweed. Similarly, Richburg et al. (1995b) concluded that Florida beggarweed control with imazapic at 72 g/ha ranged from 23 to 90%. Imazapic control of Florida beggarweed may be affected by the amount and frequency of rainfall early in the growing season (Wilcut et al., 1994c). Drought-stressed Florida beggarweed may be more tolerant to imazapic than plants that are actively growing (Richburg et al., 1995b).

Wehtje et al. (2000a) reported that Florida beggarweed control with imazapic was independent of application method (foliage only, soil only, or soil plus foliage) when Florida beggarweed seedlings were less than 20 days old. However, a soil-only application of imazapic did not adequately control older seedlings, but foliage-only and foliage plus soil applications were equally effective at controlling Florida beggarweed. They concluded that reduced translocation by trifoliate foliage contributed to imazapic tolerance of older Florida beggarweed seedlings.

Imazapic has controlled Florida beggarweed better than imazethapyr (Wilcut et al., 1994b). Imazethapyr applied PPI or early POST provided inconsistent beggarweed control. Imazethapyr tank-mixed with either alachlor or metolachlor applied early POST improved Florida beggarweed control over imazethapyr alone (Wilcut et al., 1994b).

POST applications of chlorimuron and pyridate have been found to control Florida beggarweed (Hicks et al., 1990; Wehtje et al., 2000b; Wilcut et al., 1989). Hicks et al. (1990) concluded that control of Florida beggarweed with pyridate increased with the rate. Sequential pyridate applications or adding 2,4-DB to pyridate did not improve Florida beggarweed control. However, Florida beggarweed control with pyridate was enhanced when a chloroacetamide herbicide was applied PRE.

Wehtje et al. (2000a) reported that Florida beggarweed control was 86% with chlorimuron compared to 73% without chlorimuron when averaged across herbicide systems which included PRE, early POST, and mid POST herbicide systems. Chlorimuron cannot be applied to peanut until 60 days after peanut emergence due to crop tolerance (Johnson et al., 1992; Wilcut et al., 1989). Once the peanut plant has reached this size, chlorimuron absorption is relatively minimal, and the herbicide which is absorbed is readily metabolized (Wilcut et al., 1989). However, peanut injury can occur even when the correct application timing is observed. Chlorimuron was implicated in yield suppression in four out of 15 trials conducted across the peanut-production region (Brown et al., 1993). It was concluded that the risks of yield loss from Florida beggarweed competition was greater than that from chlorimuron-induced crop injury.

Eclipta

Eclipta (*Eclipta prostrata* L.) is an annual herbaceous plant native to Asia (Holm et al., 1977). It is a member of the Asteraceae family and is reported to be a weed in 17 crops in 35 countries around the world (Holm et al., 1977). Eclipta distribution in North America is primarily in the southern U.S., lower Midwest and along the East Coast (Steyermark, 1981). Eclipta can be an annual or perennial plant, depending on location. In South Africa and India, eclipta grows as an annual and as a perennial. In the Philippines, it flowers year-round and has been shown to produce more than 17,000 seeds per plant (Holm et al., 1977).

Eclipta can be a troublesome weed in cotton, peanut, rice (*Oryza sativa* L.), soybean, and various ornamentals (Altom et al., 1995; Berchielli-Robertson et al., 1989; Grichar and Colburn, 1996b; Sharma and Amritphale, 1988; Wilcut et al.,

1995a,b). It is usually found in poorly drained areas, along streams and ditches, in marshes, and on the dikes of rice paddies. However, it is common in lawns and in upland conditions where rainfall is over 120 cm or more.

Once introduced into a field, eclipta spreads quickly and can become a severe problem. Melouk et al. (1992) reported that eclipta serves as a host for Sclerotinia blight (*Sclerotinia minor* Jagger). Sclerotinia blight infects approximately 25% of Oklahoma peanut fields and can reduce yields 25 to 50% (Jackson et al., 1993). Therefore, eclipta can be important as a weed and a disease host.

Eclipta is not controlled by the dinitroaniline herbicides (Bailey et al., 1999a,b; Grichar and Colburn, 1996a; Wilcut et al., 1995a). Wilcut et al. (1991a) found that imazethapyr at 0.105 kg/ha applied PRE gave 95% eclipta control. However, the use rate for imazethapyr in peanut in the U. S. cannot exceed 0.071 kg/ha per year (Anonymous, 2001). Imazethapyr at the normal use rate (0.071 kg/ha) applied PPI, PRE, or EPOST controlled eclipta 67 to 75% (Wilcut et al., 1991a). They also reported that POST applications of imazethapyr controlled less than 50% eclipta and no rate response was observed.

Excellent eclipta control has been reported from a herbicide system using pendimethalin PPI, followed by metolachlor PRE, followed by acifluorfen plus bentazon applied POST (Wilcut et al., 1991a,b). Metolachlor has been shown to suppress early season eclipta germination (Anonymous, 1992); however, inconsistent eclipta control has been reported with metolachlor. Jordan et al. (1993) concluded that metolachlor alone controlled eclipta only 59% while Grichar and Colburn (1996b) noted greater than 88% eclipta control at 3 of 4 locations. In the same study, dimethenamid controlled eclipta greater than 90% when applied PPI or PRE. Grichar and Colburn (1996b) attributed the inconsistent control of eclipta with chloroacetamide herbicides due to certain rainfall conditions. Shrefler et al. (1994) noted that for soils high in organic matter, alachlor and dimethenamid produced better annual grass control than with metolachlor when less than 1.0 mm of rainfall was received during the week following herbicide application.

Metolachlor applied in combination with imazapic or imazethapyr PRE has provided greater than 80% early season eclipta control but poor (50%) late season control (Altom et al., 1995). At other locations, metolachlor plus imazethapyr have controlled eclipta 47 to 98% (Grichar and Colburn, 1996b).

Flumioxazin alone or in combination with metolachlor has provided excellent early-season eclipta control (Grichar and Colburn, 1996a). However, late season control of eclipta with flumioxazin alone was rate dependent while metolachlor plus flumioxazin controlled eclipta 65 to 70%.

Diclosulam has provided excellent eclipta control (Bailey et al, 1999a, b; Dotray et al., 1998). Ethalfluralin applied PPI followed by diclosulam applied PRE at 17 to 35 g/ha controlled 100% eclipta (Bailey et al., 1999b). Similar results were reported when diclosulam was applied PPI (Bailey et al., 1999a).

Eclipta germinates over a broad range of environmental conditions. Optimum eclipta germination occurs in warm soils, generally prevailing after application of soil-applied herbicides (Altom and Murray, 1996). Therefore, it is necessary to use POST herbicides for improved control during the growing season. Grichar (1997a) and Altom et al. (1995) found that eclipta was more effectively controlled with POST herbicides when applied to eclipta that was less than 5 cm in height. Excellent POST eclipta control has been reported with acifluorfen plus bentazon (Wilcut et al., 1991a).

Grichar (1997a) reported greater than 90% eclipta control when acifluorfen plus bentazon was applied early POST to eclipta less than 5 cm in height, but less than 50% control was observed when applications were made to eclipta which was 7 to 10 cm tall.

Imazapic POST does not provide consistent control of eclipta (Bailey et al., 1999a,b; Grichar and Colburn, 1996a); however, imazapic controlled eclipta greater than 85% season-long when applied to eclipta less than 1.5 cm in height (Grichar 1997a). Lactofen has provided excellent (99%) eclipta control (Jordan et al., 1993). Lactofen applied early POST at 0.22 kg/ha provided 100% eclipta control in 2 of 3 years (Grichar, 1997a) and greater than 90% control in another study (Altom et al., 1995). Lactofen applied at peanut emergence or late POST has provided erratic eclipta control (Altom et al., 1995; Grichar, 1997a). When lactofen has been applied in combination with metolachlor at peanut emergence followed by lactofen or lactofen plus 2,4-DB applied POST, eclipta control has been greater than 90% (Grichar and Colburn, 1996b). However, lactofen is not registered for peanut use in the United States.

Pyridate alone or in combination with 2,4-DB also has provided excellent eclipta control (Grichar and Colburn, 1996a). Grichar (1997a) reported that pyridate applied to eclipta seedlings less than 5 cm tall controlled greater than 95% eclipta but when applied to eclipta seedings at least 7 cm in height, control was inconsistent and varied from 20 to 100%. Altom et al. (1995) also found good to excellent control with pyridate when applied to small sized eclipta.

Other Broadleaf Weeds

The following broadleaf weeds are problems only in certain peanut growing areas of the U.S. and are usually not identified as problematic throughout the entire Peanut Belt.

Prickly sida

Prickly sida (*Sida spinosa* L.) is mainly a problem in the southeastern peanut growing region. Prickly sida is listed as a troublesome weed in five peanut-producing states of the U.S. (Dowler, 1998). Bentazon and imazethapyr POST will control prickly sida when applied to small sized plants (Wilcut et al., 1994b; York et al., 1995), but if larger than two-leaf, imazethapyr POST is ineffective (Wilcut et al., 1991a, 1994b). Wilcut et al. (1991a) noted that imazapic controlled at least 95% five-leaf prickly sida.

The dinitroaniline herbicides do not control prickly sida (Wilcut et al., 1995a). However, when diclosulam was added to ethalfluralin PPI, prickly sida was controlled at least 95% (Bailey et al., 1999a,b). Bailey et al. (1999b) reported that POST herbicides did not improve prickly sida control over ethalfluralin applied PPI followed by diclosulam applied PRE. Prickly sida was controlled less than 60% with POST herbicides when ethalfluralin was the only soil applied herbicide in the system.

Metolachlor alone does not control prickly sida (Askew et al., 1999; Jordan et al., 1993). However, when flumioxazin was added to metolachlor PRE, prickly sida control was greater than 95%. Askew et al. (1999) reported control of prickly sida did not increase with the rate with flumioxazin. Flumioxazin effectively controlled prickly sida in other studies (Wilcut and Cranmer, 1997; Wilcut et al., 1995b).

Paraquat and acifluorfen has not been reported to provide adequate prickly sida control (Wilcut et al., 1991a,b, 1994c), while bentazon controlled small prickly sida greater than 85% (Jordan et al., 1993). Wesley and Shaw (1992) noted better prickly sida control in soybean with lactofen as compared to acifluorfen.

Sicklepod

Sicklepod [*Senna obtusifolia* (L.) Irwin and Barneby] is listed among the ten most troublesome weeds in Alabama, Florida, Georgia, North Carolina, and South Carolina (Dowler, 1998). In early work, Bridges et al. (1984) found that neither herbicides nor a two cultivation program provided consistent and acceptable weed control and peanut yield in an area infested with sicklepod.

Dinitroaniline and chloroacetamide herbicides only suppress the early-season sicklepod growth (Cardina and Swann, 1988; Wehtje et al., 1988; Wilcut et al., 1995a). Imazethapyr soil applied or POST provides poor sicklepod control (Richburg et al., 1995a; Wilcut et al., 1994b).

Herbicides applied POST have shown good activity against sicklepod. Both paraquat and 2,4-DB are effective at controlling sicklepod (Wilcut et al., 1995a). Imazapic at 36 g/ha or more controlled at least 80% sicklepod (Wilcut et al., 1996; Webster et al., 1997). Hicks et al. (1990) found that as the pyridate rate increased, sicklepod control also increased up to 70%. However, when 2,4-DB was added to pyridate, sicklepod control increased from 42 to 80%. The improved control of sicklepod obtained with this tank mixture is a synergistic response that can be partially attributed to enhanced pyridate absorption (Hicks et al., 1998).

Golden Crownbeard

Golden crownbeard [*Verbesina encelioides* (Cav.) Benth. & Hook f. ex Gray] is generally found in all Texas peanut growing regions and portions of Oklahoma (Grichar and Sestak, 1998, 2000). Golden crownbeard grows up to 1 m tall and is multi-branched with grayish green leaves. In south Texas, golden crownbeard can be seen growing along roadsides and in fallow fields throughout the year. Since golden crownbeard can survive almost year-round, it has been implicated as serving as a host for tomato spotted wilt virus and thrip vectors (Grichar and Sestak, 1998, 2000).

Dinitroaniline and chloroacetamide herbicides alone controlled less than 80% golden crownbeard (Grichar and Sestak, 2000). When a PPI application of ethalfluralin was followed by imazapic or lactofen applied POST, control was 100%. Ethalfluralin plus imazethapyr controlled 80 to 88% golden crownbeard while imazethapyr alone applied PPI controlled golden crownbeard 86%.

In other research, imazapic and imazethapyr alone applied POST provided inconsistent golden crownbeard control (Grichar and Sestak, 1998). They attributed the poor control results with imazapic and imazethapyr to the amount and frequency of rainfall soon after application and weed size at the time of herbicide application. Richburg et al. (1993, 1994) noted that under low rainfall conditions less yellow nutsedge (*Cyperus esculentus* L.) root absorption of imazapic and imazethapyr occurred, resulting in reduced yellow nutsedge control. Richburg et al. (1995a) also reported that applications of imazapic and imazethapyr to bristly starbur (*Acanthospermum hispidum* DC.) plants taller than 4 cm killed only the terminal stem and plants recovered rapidly.

Numerous POST herbicides (acifluorfen alone, acifluorfen plus bentazon or 2,4-DB, bentazon alone, pyridate alone, or pyridate plus 2,4-DB) have provided excellent golden crownbeard control (Grichar and Sestak, 1998). However, the addition of 2,4-DB to either acifluorfen or pyridate did not improve golden crownbeard control over those herbicides alone.

Bristly starbur

Bristly starbur (*Acanthospermum hispidum* DC.) is a herbaceous annual native to Central and South America that has emerged as a major broadleaf weed problem in the southeast (Walker et al., 1989). Left uncontrolled, bristly starbur can be up to 0.5 m taller than the peanut at harvest, producing a dense canopy which may interfere with fungicide applications and harvesting, thereby reducing peanut yield (Walker et al., 1989).

While the dinitroanoline herbicides provide poor bristly starbur control, the chloroacetamide herbicides suppress bristly starbur (Colvin et al., 1990). Therefore, POST herbicides are required to provide adequate season-long control. Bentazon controls bristly starbur but paraquat does not (Wilcut et al., 1995a). Bristly starbur was controlled less than 70% by paraquat plus bentazon (Wilcut et al., 1994 a,c). Paraquat plus bentazon controlled bristly starbur 100% when evaluated within 10 days of early POST application but due to lack of residual activity, later germinating bristly starbur escaped. Imazethapyr will control bristly starbur as a soil or POST application, although POST applications are generally more consistent and effective when applied to plants less than 5 cm tall (Wilcut et al., 1994b). Bristly starbur control with imazethapyr was not increased by the addition of bentazon, but control decreased when imazethapyr was tank-mixed with paraquat (Wilcut et al., 1994a). They attributed the decrease in control to rapid foliar necrosis from paraquat, which reduced foliar absorption of imazethapyr.

Imazapic provides excellent bristly starbur control (Webster et al., 1997; Wilcut et al., 1996). Webster et al. (1997) found that imazapic soil-applied controlled bristly starbur at least 92% and there were no differences between PPI and PRE application methods. Imazapic POST controlled up to five-leaf bristly starbur. The fact that imazapic has both foliar and soil-residual activity was important for controlling bristly starbur because of the season-long emergence pattern of this weed (Webster et al., 1997).

Annual Grasses

Annual grasses are a common occurrence in most peanut fields across the United States. Texas panicum (*Panicum texanum* Buckl.), southern crabgrass [*Digitaria ciliaris* (Retz.) Koel.], broadleaf signalgrass [*Brachiaria platyphylla* (Griseb.) Nash], and goosegrass [*Eleusine indica* (L.) Gaertn.] are the most prevalent. Of these, Texas panicum is ranked as one of the ten most common weeds in five of the eight major peanut growing states while crabgrass spp. are ranked as some of the most common in all eight states (Dowler, 1998). Competition between annual grasses and peanut not only reduces yield but can also reduce harvesting efficiency because peanut pods become embedded in the annual grasses extensive root system and stripped from the vine during digging (Buchanan et al., 1982; Grichar, 1991a).

Generally, control of annual grasses can be achieved with PPI applications of dinitroaniline herbicides (Buchanan et al., 1982; Grichar, 1991a; Wilcut et al., 1995a). Wilcut et al. (1987a) reported that the minimum input necessary to achieve consistent Texas panicum control was a dinitroaniline herbicide combined with at least one cultivation. The dinitroaniline herbicides can be applied PRE but they must be incorporated with mixing wheels, rainfall, or irrigation for effective control (Prostko et al., 2001). Initial studies in peanut indicated that ethalfluralin provided 85 to 100% control of goosegrass, southern crabgrass, and Florida pusley (*Richardia scabra* L.) when applied PRE (Brecke and Currey, 1980). However, later research showed that PRE applications of ethalfluralin or pendimethalin were not effective for weed management in peanut (Grichar et al., 1994; Wilcut et al., 1990). Prostko et al. (2001) reported that PRE applications of ethalfluralin or pendimethalin followed by irrigation were as effective as PPI applications in controlling Texas panicum, southern crabgrass, and crowfootgrass [*Dactyloctenium aegyptium* (L.) Willd.]. The effectiveness of soil-applied herbicides is dependent upon several factors, including movement of the herbicide into the soil either through rainfall or irrigation or by mechanical incorporation (Ross and Lembi, 1999).

An important advantage of using irrigation rather than tillage to incorporate herbicides is the potential to reduce production inputs and for use in reduced-tillage production systems. The estimated cost of applying a herbicide and incorporating it with 19 mm of water using a center-pivot irrigation system is $16/ha while the cost of using mechanical incorporation methods range from $21 to $32/ha, depending on the implement used (Prostko et al., 2001).

In field studies conducted in 1997 and 1998 ethalfluralin controlled 94% Texas panicum while dimethenamid and S-metolachlor controlled less than 70% (Table 5). The dinitroanilines are the only soil-applied herbicides registered for use in peanut that will provide full-season control of Texas panicum (Wilcut et al., 1995a).

Of the POST herbicides, only imazapic controlled greater than 70% Texas panicum when used without a soil-applied herbicide (Table 4). The imidazolinone herbicides will partially control annual grasses (personal observation). Imazethapyr soil applied will provide virtually no control; however, imazapic applied POST will control 60 to 70% Texas panicum and greater than 90% southern crabgrass and broadleaf signalgrass when applied to grasses less than 3 cm tall. The use of a dinitroaniline herbicide with either imazapic or imazethapyr improved Texas panicum control over the dinitroaniline herbicide alone (Grichar et al., 1999). However, imazapic alone will control rhizome and seedling johnsongrass [*Sorghum halepense*

(L.) Pers.], large crabgrass [*Digitaria sanguinalis* (L.) Scop], southern crabgrass, and broadleaf signalgrass (Wilcut et al., 1995a). Chloroacetamide herbicides do not adequately control large seeded annual grasses (Grichar, 1991; Wehtje et al., 1986); however, paraquat has provided excellent Texas panicum control. Increasing the paraquat rate resulted in a linear increase in Texas panicum control (Wehtje et al., 1986). Wehtje et al. (1986) obtained better control using paraquat with a late planting date. They attributed the increased control to the majority of the Texas panicum having germinated and emerged at time of application compared to a normal planting date when Texas panicum had not all germinated.

Ethal and dimethenamid in combination with POST herbicides provided excellent (86 to 97%) control of Texas panicum while dimethenamid in combination with POST herbicides controlled 86 to 97% (Table 4). S-metolachlor in combination with POST herbicides showed variable control of Texas panicum ranging from 45 to 94%.

Diclosulam or flumioxazin alone will not control annual grasses (Grichar and Colburn, 1996b; Grichar et al., 1999). When ethalfluralin plus diclosulam was applied PPI, Texas panicum control was at least 97% compared with 87% control for ethalfluralin alone (Grichar et al., 1999). Flumioxazin alone at 0.1 kg/ha PPI provided good Texas panicum control (greater than 80%) for up to 4 weeks after PPI treatment; however at 8 weeks after treatment, control was less than 60% (Grichar and Colburn, 1996a).

Control of annual grasses POST can be obtained with clethodim or sethoxydim (Grichar, 1991 a,b; Prostko et al., 2001). Grichar (1991b) found that sethoxydim applied early POST provided more effective annual grass control than late POST applications. However, grass size did not affect control with clethodim. Sethoxydim was found to provide poor Texas panicum control under less than ideal moisture conditions (Grichar, 1991b). Grichar (1991b) speculated that reduced moisture conditions resulted in less translocation of the herbicide into the plant (Chernicky et al., 1984; Fawcett et al., 1987). Clethodim controls annual grasses at lower use rates than sethoxydim. Prostko et al. (2001) noted 90% Texas panicum, southern crabgrass, or crowfootgrass control occurred with clethodim when following an application of a soil applied dinitroaniline herbicide.

Perennial Grasses

Johnsongrass (*Sorghum halepense* (L.) Pers.) is not a problem in most peanut producing states and is ranked as one of the ten most common weeds in Oklahoma only (Dowler, 1998). However, common bermudagrass [*Cynodon dactylon* (L.) Pers.] can be a very troublesome weed for peanut producers (Grichar and Boswell, 1989; Grichar, 1995; Wilcut, 1991a). No soil applied herbicides available for use in peanut provide control of bermudagrass. Consequently, POST applied graminicides are the only control option available.

Table 4. Late season Palmer amaranth and Texas panicum control using soil-applied and POST herbicides.

POST herbicide	None		Soil-applied herbicides (%) Ethalfluralin		Dimethenamid		S-metolachlor	
	AMAPA[a]	PANTE	AMAPA	PANTE	AMAPA	PANTE	AMAPA	PANTE
Acifluorfen	77	10	99	89	98	91	96	58
Acifluorfen + bentazon	75	0	93	88	98	87	96	64
Bentazon	37	7	91	96	93	88	88	45
Imazapic	93	80	100	97	99	97	100	94
Imazethapyr	82	67	98	97	97	90	97	84
Pyridate	40	13	95	90	94	91	84	51
2,4-DB	61	0	91	90	95	86	90	69
None	0	0	79	94	59	67	69	54
LSD (0.05) AMAPA	.... 19							
PANTE	.... 25							

[a] Bayer Code for Palmer amaranth is AMAPA; for Texas panicum, PANTE.

Common bermudagrass can reproduce vegetatively, from rhizomes and stolons, or by producing seeds (Mitich, 1989). The extensive network of rhizomes and stolons makes bermudagrass difficult to eradicate. Cultivation tends to spread the rhizomes and stolons throughout the field (Mitich, 1989). Sethoxydin and clethodim are registered in the U.S. for use in peanut to control common bermudagrass. Wilcut (1991a) concluded that sequential applications of sethoxydim or clethodim provided better control than a single application. Grichar (1995) noted that early POST applications of sethoxydim failed to control bermudgrass (less than 60%) while sequential applications controlled bermudagrass greater than 80%. Grichar (1995) also found that sequential clethodim applications gave consistently better control than a single application.

Nutsedge Species

Yellow and Purple Nutsedge

Yellow nutsedge is a light green perennial sedge having three-sided culms and growing to 1 m in height. A basal bulb is formed by a swelling of the culm below the soil surface and rhizomes grow out from this bulb to terminate in single underground tubers. Both purple nutsedge (*Cyperus rotundus L.)* and yellow nutsedge may grow in mixed stands and are difficult to distinguish before they have flowered. Purple nutsedge produces rhizomes with tubers in a chain compared with a single tuber produced by yellow nutsedge. Yellow nutsedge propagates at a very fast rate. It has been shown that one tuber placed in a field produced 1,900 plants, almost 7,000 tubers, and covered an area about 2 m in diameter within one year (Jansen, 1971). It is considered a weed in 21 crops in more than 30 countries (Holm et al., 1977). Yellow nutsedge has been found to reduce crop yield up to 89% in both agronomic and horticultural crops (Keeley, 1987).

Nutsedge tubers can contaminate peanut and confectionary products, increasing food industry liability (Brown, 1990). Yellow nutsedge is difficult to control due to its perennial growth habit and tubers, its primary means of propagation (Banks, 1983; Johnson and Mullinix, 1997). Yellow nutsedge seeds remain viable for many years but are highly susceptible to drought and desiccation, which lessens their role in the distribution of the weed (Horak and Holt, 1986; Lapham and Drennan, 1990). Therefore, to successfully manage nutsedge, programs must focus on tuber mortality.

Yellow nutsedge control and management depends on integrated systems of chemical, mechanical, and cultural control tactics (Beste et al., 1992; Grichar, 1992; Johnson and Mullinix, 1997). Several herbicides have been found to reduce tuber production and viability (Banks, 1983). However, control with herbicides often is variable due to depth of tubers, dormancy, and environmental factors that directly affect herbicide efficacy (Stoller et al., 1972, 1979; Thullen and Keeley, 1980). Even if yellow nutsedge is successfully controlled with herbicides, the cost of control is substantial in all crops, adding another dimension to economic losses from weeds (Johnson and Mullinix, 1997).

Control of yellow and purple nutsedge with shallow tillage is effective if used repeatedly in fallow areas or as part of a stale seedbed weed control system (Glaze,

1987). When properly timed, shallow tillage depletes tuber carbohydrate reserves (Stoller et al., 1972; Thullen and Keeley, 1980), exposes tubers to dry conditions (Thomas, 1969), or preconditions tubers to freezing in northern climates (Stoller et al., 1972). Mechanical control using cultivation is commonly used in agronomic crops but is effective only in inter-rows and is not repeated often enough to deplete tuber carbohydrate reserves (Johnson and Mullinix, 1997).

Most agronomic crops are effective competitors with yellow nutsedge. Cotton, corn (*Zea mays* L.), and other crops suppress growth of yellow nutsedge by shading (Stoller et al., 1979; Thullen and Keeley, 1980). Since peanut is a slow growing crop, possessing a prostrate growth habit, both yellow and purple nutsedge can easily compete with the crop (personal observation).

Nutsedge continues to increase in severity in many peanut growing regions due to: 1) use of dinitroaniline herbicides which do an effective job of controlling grass and small-seeded broadleaf weeds, thus reducing competition with yellow nutsedge, 2) spread of tubers by improper cleaning of equipment after use in an infested field, and 3) its high reproductive capacity (Grichar, 1992).

Several herbicides are registered for use in peanuts for nutsedge control. However, control can be extremely variable when conditions at application are less than ideal. Reasons for failure include marginal translocation of herbicides to sites of action, temporary inhibition of tuber sprouting, or inconsistent control when applied at different stages of growth and under various environmental conditions (Pereira et al., 1987).

The use of alachlor and metolachlor can delay tuber sprouting and kill yellow nutsedge (Armstrong et al., 1973; Grichar, 1992). Grichar et al. (1981) obtained slightly better control of yellow nutsedge with metolachlor than with alachlor. This increased metolachlor activity may be due to longer soil persistence (Skipper et al., 1976) or the greater acropetal movement of the herbicide in the plant (Obrigawitch et al., 1980). Grichar et al. (1981) found that metolachlor controlled weeds better when incorporated 5 to 7 cm deep with a power tiller than when applied to the soil surface.

Grichar et al. (1996) also found that metolachlor applied POST can control yellow nutsedge. They reported that POST applications of metolachlor followed by irrigation within 24 hours could be an effective method for yellow nutsedge control, and thereby reduce the chance of peanut injury from soil applications of metolachlor. However, Grichar et al. (1996) concluded that late POST applications provided inconsistent control due to the large size (30 cm tall) of yellow nutsedge plants at the time of herbicide application.

Cardina and Swann (1988) noted that peanut yields were not reduced when metolachlor was applied PPI at labeled rates, followed by irrigation or rainfall. However, metolachlor delayed and reduced peanut emergence and growth. Grichar et al. (2000) found that metolachlor controlled yellow nutsedge better than dimethenamid when applied at 0.6x to 2x the suggested label rates. They attributed the lack of season-long yellow nutsedge control with dimethenamid to poor soil persistence. Mueller et al. (1999) reported that soil concentrations of metolachlor were higher later in the season compared to dimethenamid and therefore metolachlor provided consistent control of annual grasses longer into the growing season. S-metolachlor was found to provide similar yellow nutsedge control as metolachlor (Grichar et al., 2001).

Yellow nutsedge can also be controlled with vernolate (Grichar et al., 1981; Hauser, 1965; Wilkinson, 1988). Vernolate controls both yellow and purple nutsedge (Hauser, 1965); however, control is short-lived. Because this herbicide has a high vapor pressure, it must be incorporated soon after application for maximum effectiveness (Hauser, 1965; Wilkinson, 1988). In addition, vernolate can be subjected to enhanced degradation by soil microorganisms (Wilkinson, 1988), which can reduce the duration of weed control to only 4 or 5 weeks (Wilkinson, 1988; Wilcut et al., 1995a).

Imazethapyr control of purple and yellow nutsedge has been variable (Grichar et al., 1992; Richburg et al., 1993, 1995b). Grichar et al. (1992) noted that imazethapyr PPI did provide more consistent control than PRE or POST applications. Grichar et al. (1992) speculated that environmental factors may account for the relatively poor control (less than 80%) following PRE applications of imazethapyr. Exposure to sunlight has been shown to degrade or alter the structure of imazethapyr (Basham and Lavy, 1987) and rainfall soon after application is essential for imazethapyr activation Also, imazethapyr control of nutsedge is obtained primarily through root absorption (Richburg et al., 1993).

Purple nutsedge control with imazethapyr has been more consistent. Purple nutsedge control with imazethapyr at 0.07 kg/ha applied PPI, PRE, or POST was 85% when rated early season and greater than 70% when rated late season (Grichar et al., 1992). It also has been reported that the addition of metolachlor to imazethapyr PRE provided more consistent purple nutsedge control than imazethapyr alone (Ray and Wilcox, 1969). Metolachlor is more effective against yellow nutsedge than purple nutsedge because roots arising from the yellow nutsedge basal bulb provide a more direct route for herbicide movement to the shoot than those of purple nutsedge (Obrigawitch et al., 1980). Variable control of purple nutsedge with imazethapyr POST has been attributed to weed size. Better control of purple nutsedge with imazethapyr was observed when the herbicide was applied to 2-cm to 6-cm weeds rather than 8-cm to 12-cm weeds (Wilcut et al., 1995a).

Imazapic POST has been more consistent and provided better control of purple and yellow nutsedge in field experiments than imazethapyr (Dotray and Keeling, 1997; Grichar and Nester, 1997; Richburg et al., 1995b; Wilcut et al., 1995a). In greenhouse experiments, imazapic was found to have foliar and soil activity on purple and yellow nutsedge (Richburg et al., 1994).

Bentazon and pyridate POST controlled yellow nutsedge satisfactorily when soil moisture is adequate for optimum plant growth and herbicide movement into and within the nutsedge plant (Grichar et al., 1981; Grichar, 1992; Wilcut et al., 1995a). In a study conducted in south Texas, two applications of bentazon provided control equivalent to three applications of pyridate (Grichar, 1992). Grichar (1992) found that nutsedge regrowth appeared three to four weeks after herbicide application and resulted in delayed nutsedge growth and maturity. He speculated that this may have allowed the peanut plant to effectively compete with the nutsedge throughout the growing season. In the southeast, pyridate following a soil application of dinithenamid improved yellow nutsedge control over pyridate alone (Edenfield et al., 2001). Neither bentazon nor pyridate provide control of purple nutsedge (personal observation).

In fields with a mixed infestation of yellow and purple nutsedge, the combination of bentazon and imazethapyr has been studied (York and Wilcut, 1995). They reported that all combinations of bentazon and imazethapyr were antagonistic, resulting in reduced purple nutsedge control. These results are in contrast to those from a previously reported study (Wilcut et al., 1994c) when similar control of yellow and purple nutsedge was obtained with imazethapyr applied alone and mixed with bentazon.

Diclosulam, soil applied at rates of 0.03 kg/ha or greater, controlled yellow nutsedge more than 90% and provided variable purple nutsedge control (Grichar et al., 1999). When diclosulam was added to metolachlor PRE, yellow nutsedge control was greater than 95% (Scott et al., 2001).

ECONOMIC ASSESSMENT OF WEED CONTROL

Economic Loss

Weed problems contribute to economic losses in several ways. Herbicide costs range from \$37 to \$124/ha, with a net cost to U.S. peanut producers in excess of \$70 million annually (Wilcut et al., 1995a). Weeds also increase the need for additional primary and secondary tillage operations adding another \$7 to \$20/ha in cost. Weeds that escape control cost producers \$49 to \$124/ha due to yield reductions and \$7 to \$62/ha due to reduced quality (Bridges et al., 1994a,b). Also, harvest efficiency is decreased resulting in yield quality reductions. Harvesting losses are estimated to range from \$7/ha in Alabama to \$17/ha in Oklahoma and South Carolina (Wilcut et al., 1995a). Estimated total income losses from control procedures for weeds, yield and quality reductions, increased cultural inputs, and reduced harvesting efficiency range from \$132/ha in Texas to \$391/ha in Florida. Total income loss for U.S. peanut producers can be in excess of \$100 million (Bridges et al., 1994 a, b).

Bridges et al. (1984) examined the efficiency and economics of herbicides and mechanical cultivation in weed control systems. They found that herbicides alone, or two cultivations alone, generally were inadequate for consistent and acceptable weed control, peanut yield, and net returns in an area infested with sicklepod and Florida beggarweed and containing a light and variable population of Texas panicum. However, minimum herbicide use along with two timely cultivations provided adequate control, good peanut yields, and net returns compared to results from two cultivations plus two hand hoeings. They further concluded that systems involving hand hoeing were disadvantageous due to the lack of a dependable labor supply at a reasonable price.

Wilcut et al. (1987a) reported that highest net returns followed the same pattern as yields in that the maximum net return was achieved with a dinitroaniline herbicide applied PPI followed by alachlor at ground cracking, and two in-season cultivations. Wilcut et al. (1987b) also reported that in areas with heavy Texas panicum infestations, highest net returns were obtained when a dinitroaniline herbicide was followed by two in-season cultivations.

In work with paraquat in the Virginia-Carolina peanut growing area, Wilcut and Swann (1990) reported that delaying the initial paraquat application until 2 week after ground cracking reduced yield by about 1200 kg/ha and resulted in a net loss of $400/ha compared with paraquat applied at cracking. Similar results were reported with bentazon plus acifluorfen (Wilcut, 1991b).

The residual control provided by imazethapyr and imazapic has allowed for the development of herbicide systems that have reduced the need for cultivation. Net returns have increased when imazethapyr has been included in a weed management system (Wilcut et al., 1991b). All herbicide systems which included imazethapyr POST provided greater economic returns than the hand weeded check.

Wehtje et al. (2000a) found that with Florida beggarweed, at least two weed control inputs were required for maximum yield and net economic return. They reported that a herbicide system which included paraquat plus bentazon early POST followed by pyridate plus 2,4-DB mid POST was less effective in terms of weed control but resulted in maximum net return. They attributed this to the relatively low cost of the paraquat plus bentazon system compared to the imazapic-based approach.

When diclosulam and flumioxazin systems were evaluated for morningglory spp., yellow nutsedge, or large crabgrass control, the net returns for those systems were seldom improved with additional inputs of POST herbicides (Scott et al., 2001). However, they did note that net returns were greatest with systems that used diclosulam or flumioxazin PRE.

In a study on common bermudagrass, Wilcut (1991a) concluded that a sequential treatment of fluazifop-P provided the maximum net return. Sethoxydim at 0.31 kg/ha applied once or twice provided the least net return.

Peanut Yields as a Result of Herbicide Inputs

In field studies conducted during the 1997 and 1998 growing seasons to evaluate herbicide inputs, all herbicide systems which included ethalfluralin and dimethenamid increased yield over the untreated check (Table 5). Peanut yields were low in 1997 because peanuts were not harvested for 3 weeks after digging due to extremely wet conditions. However, peanut yields from plots without any herbicide were less than 700 kg/ha in either year. In 1997, ethalfluralin alone applied PPI or imazethapyr alone applied POST increased yield over the untreated check. Dimethenamid or S-metolachlor alone did not result in a yield increase over the untreated check (Table 5). In a systems approach, only S-metolachlor followed by acifluorfen POST did not result in a yield increase over the untreated check.

In 1998 when used alone, acifluorfen and imazapic, applied POST increased peanut yield over the untreated check as did ethalfluralin, dimethenamid, and s-metolachlor, and all POST herbicide combinations (Table 5). Other studies have reported that competition from Texas panicum and Palmer amaranth can severely reduce peanut yield (Grichar, 1997b; Wilcut et al., 1995a).

Net Returns

Imazapic and imazethapyr were the most expensive herbicides used in weed control systems studied while 2,4-DB had the lowest cost (Table 6). Net returns closely followed trends in yield. Although imazapic and imazethapyr were the most expensive herbicide inputs, they provided the greatest economic return (Table 7).

When comparing POST herbicides alone in 1997, only imazethapyr increased returns over the untreated check (Table 7). Ethalfluralin alone increased net returns while dimethenamid and s-metolachlor alone did not. All herbicide combinations that included ethalfluralin PPI increased net returns over the untreated check. Dimethenamid followed by 2,4-DB applied POST and s-metolachlor followed by acifluorfen applied POST failed to increase net returns when compared to the untreated check. In 1998, acifluorfen, imazapic, and imazethapyr applied POST increased net returns compared to the untreated check (Table 7). Ethalfluralin, dimethenamid, and S-metolachlor alone or in combination with a POST herbicide increased returns over the untreated check.

Table 6. Herbicide costs for 1997 and 1998 averaged over three distributors.

		Soil-applied herbicides ($/ha)		
POST herbicide	None	Ethalfluralin	Dimethenamid	S-metolachlor
Acifluorfen	24.50	43.77	71.65	56.54
Acifluorfen + bentazon[a]	39.92	74.00	87.06	71.95
Bentazon	49.99	69.26	97.15	83.68
Imazapic	81.31	100.58	128.46	113.35
Imazethapyr	58.37	77.63	105.52	90.40
Pyridate	17.07	51.75	79.64	64.52
2,4-DB	15.26	34.53	62.42	63.16
None	0	19.27	47.15	32.04

[a] A premix marketed as Storm.

Although imazethapyr and imazapic were components of the most expensive systems, they showed the greatest net returns (Table 7). Less costly systems resulted in reduced weed control and a lower dollar return per hectare. These data are consistent with other research demonstrating the necessity of a soil-applied herbicide to provide a base for a peanut herbicide weed program. Many growers feel that the use of a soil-applied herbicide is not necessary when imazapic or imazethapyr are available to provide broad spectrum of weed control (personal observation).

Table 5. Peanut yields from herbicide treatments.

	Soil-applied herbicides (kg/ha)							
	None		Ethalfluralin		Dimethenamid		S-metolachlor	
POST herbicide	1997	1998	1997	1998	1997	1998	1997	1998
Acifluorfen	960	2360	1930	4380	1970	3830	1440	4770
Acifluorfen + bentazon[a]	660	1540	1740	2640	1900	3910	2190	2690
Bentazon	1270	1770	3750	2110	2970	1770	2990	
Imazapic	1230	3090	2430	4490	2280	3510	2570	4400
Imazethapyr	1700	1570	2060	3400	2210	2730	2340	3640
Pyridate	1200	1570	2060	3550	2200	2730	2050	2870
2,4-DB	1230	1160	1690	3020	2280	3550	2000	2500
None	540	680	1630	3950	1470	2080	1330	2250
LSD (0.05) 1997	....	....	....	1030	....	....	....	....
LSD (0.05) 1998	....	....	....	1230	....	....	....	....

[a] A premix marketed as Storm.

Table 7. Return per hectare for each herbicide treatment[a].

	Soil-applied herbicides ($/ha)							
	None		Ethalfluralin		Dimethenamid		S-metolachlor	
POST herbicide	1997	1998	1997	1998	1997	1998	1997	1998
Acifluorfen	699	1753	1410	3255	1412	2813	1028	3537
Acifluorfen + bentazon[b]	457	1120	1237	1915	1344	2858	1578	1955
Bentazon	907	1283	125	2756	1492	2140	1250	2169
Imazapic	845	2729	1753	3282	1589	2915	1823	3201
Imazethapyr	122	2269	175	2484	559	2539	1672	2652
Pyridate	887	1166	1500	2623	1578	1977	1480	2098
2,4-DB	911	859	1239	2241	1045	2612	1444	1820
None	407	512	1209	2956	1060	1520	970	1663
LSD (0.05) 1997	.... 692							
LSD (0.05) 1998	.... 822							

[a] Return per hectare was determined by subtracting herbicide costs from gross $ value/ha.
[b] A premix marketed as Storm

CONCLUSION

A major problem of the modern peanut farmer is the integration of effective cultural and chemical control methods into the most logical procedure for controlling specific weeds. A "prescription approach" which implies fitting control measures to specific weeds and environmental conditions may be used (Buchanan et al., 1982). A systems approach to weed control may also be incorporated into the peanut farming operation. Such systems could involve herbicides only, cultivation only, or a logical combination of both herbicides and cultivation. The timeliness of either the herbicide application or of cultivation is critical in a weed control system. Under some conditions, only one timely cultivation contributes substantially to controlling late-season weeds where herbicides were previously applied.

The use of a herbicide-crop rotation to reduce weed pressure is becoming widely used to reduce weed populations. Weeds difficult to control in one crop may be easily controlled in another. For example, Florida beggarweed, hard to control in peanuts, is relatively easy to control in corn or cotton. Over a period of years, controlling each weed at its most susceptible area in a rotation can reduce weed seed numbers considerably.

More research is needed to fully understand the biology and population dynamics of weeds in peanut production. This type of research could improve the understanding of the ability of certain plant species to become a problem in peanut production. As a result, perhaps a cure for weed management in peanut could be developed instead of the annual remedial approach that man has utilized since the beginning of intensive agriculture (Zimdahl, 1991).

REFERENCES

Altom J.V., Westerman R.B., Murray D.S. Eclipta (*Eclipta prostrata*) control in peanuts (*Arachis hypogaea* L.). Peanut Sci 1995; 22:114-120.

Altom J.V., Murray D.S. Factors affecting eclipta (*Eclipta prostrata*) seed germination. Weed Technol 1996; 10:727-731.

Ambasht R. Ecology of the underground parts of *Cyperus rotundus* L. Trop Ecol 1964; 5:67-74.

Anonymous. Peanuts: New weed. Prog Farmer 1992;107 (1):60.

Anonymous. *Crops Protection Chemical Reference*. New York: Chem. Pharmaceutical Publ. Co. and John Wiley & Sons, Inc., 2001.

Armstrong T.F., Meggitt W.F., Penner D. Yellow nutsedge control with alachlor. Weed Sci 1973;21:354-357.

Askew S.D., Wilcut J.W., Cranmer J.R. Weed management in peanut (*Arachis hypogaea*) with flumioxazin preemergence. Weed Technol 1999; 13:594-598.

Bailey W.A., Wilcut J.W., Jordan D.L., Swann C.W., Langston V.B. Response of peanut (*Arachis hypogaea*) and selected weeds to diclosulam. Weed Technol 1999a; 13:771-776.

Bailey W.A., Wilcut J.W., Jordan D.L., Swann C.W., Langston V.B. Weed management in peanut (*Arachis hypogaea*) with diclosulam preemergence. Weed Technol 1999b;13:450-456.

Banks P.A. Yellow nutsedge (*Cyperus esculentus*) control, regrowth, and tuber production as affected by herbicides. Weed Sci 1983; 31:419-422.

Barker M.A., Thompson Jr. L., Godley F..M. Control of annual morningglories (*Ipomoea* spp.) in soybeans (*Glycine max*). Weed Sci 1984; 32:813-818.

Basham G.W., Lavy T.L. Microbial and photolytic dissipation of imazaquin in soil. Weed Sci 1987; 35:865-870.

Berchielli-Robertson D.L., Gilliam C.H., Fare D.C. Preemergence herbicide control of *Eclipta alba* in container-grown plants. Hort Sci 1989; 24:779-781.

Beste C.E., Frank J.R., Bruckart W.L., Johnson D.R., Potts W. E. Yellow nutsedge (*Cyperus esculentus*) control in tomato with *Puccinia canaliculata* and pebulate. Weed Technol 1992; 6:980-984.

Braxton L.B., Barrentine J.L., Geselius T.C., Grant D.L., Langston V.B., Nolting S.P., Redding K.D., Richburg, III J.S., Sheppard B.R. Efficacy and crop tolerance of diclosulam soil-applied in peanuts. Proc South Weed Sci Soc 1997; 50:162.

Brecke B.J., Currey W.L. Weed control in peanuts with ethalfluralin. Peanut Sci 1980;7:124-127.

Brecke B.J., Colvin D.L. Influence of paraquat timing of application on peanut tolerance and weed control. Proc. South. Weed Sci Soc 1988; 41:62.

Brecke B.J., Colvin D.L. "Weed management in peanuts." In *CRC Handbook of Pest Management in Agriculture,* D. Pimentel, ed, Boca Raton, FL: CRC Press, 1991; pp. 239-251.

Brecke B.J., Colvin D.L., Muzyk K.R. Broadleaf weed control in peanut with AC 263,222. Proc South Weed Sci Soc 1995; 48:11.

Bridges D.C., Walker R.H., McGuire J.A., Martin N.R. Efficiency of chemical and mechanical methods for controlling weeds in peanuts (*Arachis hypogaea*). Weed Sci 1984; 32:584-591.

Bridges D.C., Kvien C.K., Hook J.E., Stark, Jr. C.R. "Weeds and herbicides of the southwestern peanut market area." In *An Analysis of the Use and Benefits of Pesticides in U. S. - Grown Peanuts: II Southwestern Production Region,* Tifton, GA: National Environmentally Sound Production Agriculture Laboratory, 1994a; pp. 1-39.

Bridges D.C., Kvien C.K., Hook J.E., Stark, Jr. C.R. "Weeds and herbicides of the Virginia-Carolina peanut market area." In *An Analysis of the Use and Benefits of Pesticides in U. S. - Grown Peanuts: III Virginia-Carolina Production Region.* Tifton, GA: National Environmentally Sound Production Agriculture Laboratory, 1994b; pp. 1-39.

Brown S. M. *Weed Facts: Yellow and Purple Nutsedge*. Tifton, GA. Georgia Cooperative Extension Service Bulletin 1043, 1990.

Brown S.M., Brecke B.J., Colvin D.L., Everest J.W., Gooden D.T., Grichar W.J., Johnson, III W.C., Swann C.W., Wehtje G.R., Wilcut J.W., York A.C. Peanut yield response to Classic (chlorimuron): Results from a beltwide evaluation. Proc South Weed Sci Soc1993; 46:41.

Buchanan G.A., Murray D.S., Hauser E.W. "Weeds and Their Control in Peanut." In *Peanut Science and Technology,* H.E. Pattee, C. T. Young , eds. Amer Peanut Res Educ Soc, Inc, 1982; pp. 206-249.

Cardina J., Swann C.W. Metolachlor effects on peanut growth and development. Peanut Sci 1988; 15:57-60.

Cardina J., Hook J.E. Factors influencing germination and emergence of Florida beggarweed (*Desmidium tortuosum*). Weed Technol 1989; 3:402-407.

Cardina J., Brecke B.J. Florida beggarweed (*Desmodium tortuosum*) growth and development in peanuts (*Arachis hypogaea*). Weed Technol 1991; 5:147-153.

Chandler J.M., Santelmann P.W. *Chemical Control of Several Problem Annual Weeds of Cropland.* Stillwater, OK: Oklahoma State Univ. Agric. Exp. Stn. Bull. B-674, 1970

Chernicky J.P., Gossett B.J., Murphy T.R. Factors influencing control of annual grasses with sethoxydim or Ro-13-8895. Weed Sci 1984; 32:174-177.

Colvin D.L., Brecke B.J., Hewitt T.D. Impacts of reduced alachlor usage on the efficacy and economics of peanut weed control. Weed Sci Soc Amer Abstr 1990; 30:10.

Dotray P.A., Keeling J.W. Purple nutsedge control in peanuts as affected by imazameth and imazethapyr application timing. Peanut Sci 1997; 24:113-116.

Dotray P.A., Keeling J.W., Grichar W.J., Prostko E.P., Lemon R.L., Osborne T.S., Brewer K.D. Performance of diclosulam in Texas peanut. Proc Amer Peanut Res Educ Soc 1998; 30:27.

Dowler C.C. Weed survey - Southern states. Proc South Weed Sci Soc 1998; 51:299-313.

Duncan W.H., Foote L.E. *Wildflowers of the Southeastern United States*. Athens: Univ. of Georgia Press, 1975; pp. 78-79.

Edenfield M.W., Colvin D.L., Brecke B.J., Shilling D.G., McLean H.H. Weed management in peanut (*Arachis hypogaea*) with pyridate and SAN 582 systems. Weed Technol 2001; 15:13-18.

Fawcett J.A., Harvey R.G., Arnold W.E., Bauman T.T., Eberlein C.V., Kells J.J., Moshier L.J., Slife F.W., Wilson R.G. Influence of environment on corn (*Zea mays*) tolerance to sethoxydim. Weed Sci 1987; 35:568-575.

Gaeddert J.W., Peterson D.E., Horak M.J. Control and cross-resistance of an acetolactate synthase inhibitor-resistant Palmer amaranth (*Amaranthus palmeri*) biotype. Weed Technol 1997; 6:108-112.

Glaze N.C. Cultural and mechanical manipulation of *Cyperus* spp. Weed Technol 1987; 1:82-83.

Grichar W.J., Boswell T.E., Merkle M.G.. *Control of Yellow Nutsedge with Selected Herbicides.* Texas Agric Exp Stn Bull 3858, 1981.

Grichar W.J., Boswell T.E. Bermudagrass (*Cynodon dactylon*) control with postemergence herbicides in peanut (*Arachis hypogaea*). Weed Technol 1989; 3:267-271.
Grichar W.J. Control of Texas panicum (*Panicum texanum*) and southern crabgrass (*Digitaria ciliaris*) in peanuts (*Arachis hypogaea*) with postemergence herbicides. Peanut Sci 1991a; 18:6-9.
Grichar W.J. Sethoxydim and broadleaf herbicide interaction effects on annual grass control in peanuts (*Arachis hypogaea*). Weed Technol 1991b; 5:321-324.
Grichar W.J. Yellow nutsedge (*Cyperus esculentus*) control in peanuts (*Arachis hypogaea*). Weed Technol 1992; 6:108-112.
Grichar W.J., Nester P.R., Colburn A.E. Nutsedge (*Cyperus* spp.) control in peanuts (*Arachis hypogaea*) with imazethapyr. Weed Technol 1992; 6:396-400.
Grichar W.J. Spiny amaranth (*Amaranthus spinosus* L.) control in peanut (*Arachis hypogaea* L.). Weed Technol 1994; 8:199-202.
Grichar W.J., Colburn A.E., Kearney N.S. Herbicides for reduced tillage production in peanut (*Arachis hypogaea*) in the southwest. Weed Technol 1994; 8:212-216.
Grichar W.J. Comparison of postemergence herbicides for common bermudagrass (*Cynodon dactylon*) control in peanut (*Arachis hypogaea*). Weed Technol 1995; 9:825-828.
Grichar W.J., Colburn A.E., Baumann P.A. Yellow nutsedge (*Cyperus esculentus*) control in peanut (*Arachis hypogaea*) as influenced by method of metolachlor application. Weed Technol 1996; 10:278-281.
Grichar W.J., Colburn A.E. Flumioxazin for weed control in Texas peanuts (*Arachis hypogaea* L.). Peanut Sci 1996a; 23:30-36.
Grichar W.J., Colburn A.E. Eclipta (*Eclipta prostrata* L.) control in peanuts (*Arachis hypogaea* L.) with soil-applied herbicides. Texas J Agric Natl Resour 1996b; 9:97-104.
Grichar W.J., Nester P.R. Nutsedge (*Cyperus* spp.) control in peanut (*Arachis hypogaea*) with AC 263, 222 and imazethapyr. Weed Technol 1997; 11:714-719.
Grichar W.J. Influence of herbicides and timing of application on broadleaf weed control in peanut (*Arachis hypogaea*). Weed Technol 1997a; 11:708-713.
Grichar W.J. Control of Palmer amaranth (*Amaranthus palmeri)* in peanut (*Arachis hypogaea*) with postemergence herbicides. Weed Technol 1997b; 11:739-743.
Grichar W.J. Effects of paraquat application and timing on peanut (*Arachis hypogaea* L.) growth, yield and grade. Texas J Agric Natl Resour 1998; 11:41-47.
Grichar W.J., Sestak D.C. Control of golden crownbeard (*Verbesina encelioides*) in peanut (*Arachis hypogaea*) with postemergence herbicides. Peanut Sci 1998; 25:39-43.
Grichar W.J., Dotray P.A., Sestak D.C. Diclosulam for weed control in Texas peanut. Peanut Sci 1999; 26:23-28.
Grichar W.J., Sestak D.C. Herbicide systems for golden crownbeard *(Verbesina encelioides*) control in peanut. Peanut Sci 2000; 27:23-26.
Grichar W.J., Lemon R.G., Sestak D.C., Brewer K.D. Comparison of metolachlor and dimethenamid for yellow nutsedge (*Cyperus esculentus* L.) control and peanut (*Arachis hypogaea* L.) injury. Peanut Sci 2000; 27:26-30.
Grichar W.J., Lemon R.G., Brewer K.D., Minton B.W. S-metolachlor compared with metolachlor on yellow nutsedge (*Cyperus esculentus*) and peanut (*Arachis hypogaea*). Weed Technol 2001; 15:107-111.
Hauser E.W. Preemergence activity of three thiocarbamate herbicides in relation to depth of placement in the soil. Weeds 1965; 13:255-257.
Hauser E.W., Buchanan G.A. Influence of row spacing, seeding rates, and herbicide systems on the competitiveness and yield of peanuts. Peanut Sci 1981; 8:74-81.
Hauser E.W., Buchanan G.A., Nichols R.L., Patterson R.M. Effects of Florida beggarweed (*Desmodium tortuosum*) and sicklepod (*Cassia obtusifolia*) on peanut (*Arachis hypogaea*) yield. Weed Sci 1982; 30:602-604.
Henning R.J., Allison A.H., Tripp L.D. "Cultural Practices." In *Peanut Science and Technology*, H.E. Pattee, C.T. Young , eds. Yoakum, TX: Amer. Peanut Res. Educ. Soc., Inc., 1982; pp. 123-138.
Hicks T.V., Wehtje G.R., Wilcut J.W. Weed control in peanuts with pyridate. Weed Technol 1990; 4:493-495.
Hicks T.V., Wehtje G.R., Grey T.L. The interaction of pyridate and 2,4-DB in peanut (*Arachis hypogaea*), Florida beggarweed (*Desmodium tortuosum*), and sicklepod (*Senna obtusifolia*). Weed Sci 1998; 46:284-288.

Higgins J.M., Whitwell T., Murdock E.C., Toler J.E. Recovery of pitted morningglory (*Ipomoea lacunosa*) and ivyleaf morningglory (*Ipomoea hederacea*) following applications of acifluorfen, fomesafen, and lactofen. Weed Sci 1988; 36:345-353.

Holm L.G., Plucknett D.L., Pancho J.U., Herberger J.P. *The World's Worst Weeds, Distribution and Biology*. Honolulu: Univ. Press of Hawaii, 1977.

Horak M.J., Holt J.S. Isozyme variability and breeding systems in populations of yellow nutsedge. Weed Sci 1986; 34:538-543.

Jackson K.E., Damicone J.P., Williams Jr E., Melouk H.A., Pratt P.W., Sholar J.R., Wu L. *Results of 1992 Plant Disease Control Field Studies*. Okla. Agric. Exp. Stn., Okla. State Univ. Res. Rept. P-930, 1993.

Jansen L. Morphology and photoperiodic responses of yellow nutsedge. Weed Sci 1971; 19:210-219.

Johnson III W.C., Holbrook C.C., Mullinix, Jr. B.G., Cardina J. Response of eight genetically diverse peanut genotypes to chlorimuron. Peanut Sci 1992; 19:111-115.

Johnson III W. C., Mullinix, Jr. B.G. Population dynamics of yellow nutsedge (*Cyperus esculentus*) in cropping systems in the southeastern coastal plain. Weed Sci 1997; 45:166-171.

Jordan D. L., Wilcut J.W., Swann C.W. Application timing of lactofen for broadleaf weed control in peanut (*Arachis hypogaea*). Peanut Sci 1993; 20:129-131.

Keeley P.E. Interference and interaction of purple and yellow nutsedge (*Cyperus rotundus* and *Cyperus esculentus*) with crops. Weed Technol 1987; 1:74-81.

Klingaman T.E., Oliver L.R. Palmer amaranth (*Amaranthus palmeri*) interference in soybeans (*Glycine max*). Weed Sci 1994; 42:523-527.

Knauft D. A., Colvin D.L., Gorbet D.W. Effect of paraquat on yield and market grade of peanut (*Arachis hypogaea*) genotype. Weed Technol 1990; 4:866-870.

Lapham J., Drennan D.S.H. The fate of yellow nutsedge (*Cyperus esculentus*) seed and seedlings in soil. Weed Sci 1990; 38:125-128.

McLean H.S., Wilcut J.W., Richburg III J.S., Brown S.M. Influence of norflurazon on Florida beggarweed control in herbicide systems for peanuts. Proc. South. Weed Sci. Soc 1995; 48:15.

Melouk H.A., Damicone J.P., Jackson K.E. *Eclipta prostrata*, a new weed host for *Sclerotinia* minor. Plant Dis 1992; 76:101.

Melouk H.A., Shokes F..M. "Peanut health management in peanut production." In *Peanut Health Management,* H.A. Melouk, F.M. Shokes, eds. St. Paul, MN: APS Press, 1995; pp. 1-6.

Menges R.M. Allelopathic effects of Palmer amaranth (*Amaranthus palmeri*) and other plant residues in soil. Weed Sci 1987; 35:339-347.

Menges R.M. Allelopathic effects of Palmer amaranth (*Amaranthus palmeri*) on seedling growth. Weed Sci 1988; 36:325-328.

Mitich L. W. Bermudagrass. Weed Technol 1989; 3:433-435.

Mueller T.C., Shaw D.R., Witt W.W. Relative dissipation of acetochlor, alachlor, metolachlor, and SAN 582 from three surface soils. Weed Technol 1999; 13:341-346.

Newsom L.J., Shaw D.R. Influence of cultivation timing on weed control in soybean (*Glycine max*) with AC 263,222. Weed Technol 1994; 8:760-765.

Obrigawitch T., Abernathy J.R., Gipson J.R. Response of yellow (*Cyperus esculentus*) and purple (*Cyperus rotundus*) nutsedge to metolachlor. Weed Sci 1980; 28:708-715.

Owen C.K., Arnold R.N., Gregory E.J. Annual grass and broadleaf weed control in dry beans with dimethenamid. Proc. West. Soc. Weed Sci 1998; 51:20-21.

Pereira W., Crabtree G., Williams R.C. Herbicide action on purple and yellow nutsedge (*Cyperus rotundus* and *C. esculentus*). Weed Technol 1987; 1:92-98.

Porter D. M., Smith D.H., Rodriguez-Kabanana R. "Peanut plant diseases." In *Advances in Peanut Science,* H. E. Pattee, H. T. Stalker, eds. Yoakum, TX: Amer. Peanut Res. Educ. Soc., 1982; pp. 326-410.

Prostko E.P., Johnson III W.C., Mullinix Jr. B.G. Annual grass control with preplant incorporated and preemergence applications of ethalfluralin and pendimethalin in peanut. Weed Technol 2001; 15:36-41.

Radford A.E., Ahles H.E., Bell C.R. *Manual of the Vascular Flora of the Carolinas*. Chapel Hill: Univ. North Carolina Press, p. 608. 1968

Ray B. R., Wilcox M. Chemical fallow control of nutsedge. Weed Res 1969; 9:86-94.

Richburg III J.S., Wilcut J.W., Wehtje G.R. Toxicity of imazethapyr to purple (*Cyperus rotundus*) and yellow nutsedges (*C. esculentus*). Weed Technol 1993; 7:900-905.

Richburg III J.S., Wilcut J.W., Wehtje G.R. Toxicity of AC 263,222 to purple (*Cyperus rotundus*) and yellow nutsedge (*C. esculentus*) . Weed Sci 1994; 42:398-402.

Richburg III J. S., Wilcut J.W., Wiley G.L. AC 263,222 and imazethapyr rates and mixtures for weed management in peanut (*Arachis hypogaea*). Weed Technol 1995a; 9:801-806.

Richburg III J. S., Wilcut J.W., Eastin E.F. Weed management in peanut (*Arachis hypogaea*) with imazethapyr and metolachlor. Weed Technol 1995b; 9:807-812.

Richburg III J. S., Wilcut J.W., Colvin D.L., Wiley G.R. Weed management in southeastern peanut (*Arachis hypogaea*) with AC 263,222. Weed Technol 1996; 10:145-152.

Ross M. A., Lembi C.A. "Herbicide incorporation techniques and equipment." In *Applied Weed Science,* C. Stewart, J. Stagman, M. Garnis, eds. 2nd ed. Upper Saddle River, NJ: Prentice-Hall, 1999; pp. 371-375.

Schroeder M., Warren.G.F. Relative sensitivity of several plants to dinoseb. Weed Sci 1971; 19:671-674.

Scott G. H., Askew S.D., Wilcut J.W. Economic evaluation of diclosulam and flumioxazin systems in peanut (*Arachis hypogaea*). Weed Technol 2001; 15:360-364.

Sharma N.K., Amritphale D. Effect of moisture stress and baslin on germination and root length of three weeds of soybean. Indian J Plant Physiol 1988; 31:440-443.

Shaw D.R., Ratnayake S., Smith C.A. Effects of herbicide application timing on johnsongrass (*Sorghum halepense*) and pitted morningglory (*Ipomoea lacunosa*) control. Weed Technol 1990; 4:900-903.

Shrefler J.W., Chandler J.M., Smith K.L. Performance consistency of some chloroacetamide herbicides as affected by rainfall patterns. Proc South Weed Sci Soc 1994; 47:41.

Skipper H. D., Gossett B.J., Smith G.W. Field evaluation and soil residual characteristics of CGA-24705 and alachlor. Proc South Weed Sci Soc 1976; 29:418-422.

Steyermark, J. A. *Flora of Missouri.* 6th ed. Ames. IA: Iowa State Univ. Press, 1981.

Stoller E.W., Nema D.P., Bhan V.M. Yellow nutsedge tuber germination and seedling development. Weed Sci 1972; 20:93-97.

Stoller E.W., Wax L.M., Slife F.W. Yellow nutsedge (*Cyperus esculentus*) competition and control in corn (*Zea mays*). Weed Sci 1979; 27:32-37.

Thomas P.E.L. Effects of desiccation and temperature on survival of *Cyperus esculentus* tubers and *Cynodon dactylon* rhizomes. Weed Res 1969; 9:1-8.

Thullen R.J., Keeley P.E. Competition between yellow nutsedge (*Cyperus esculentus*) and Japanese millet (*Echinochloa crus-* galli var. *frumentacea*). Weed Sci 1980; 28:24-26.

Tonks D..J., Eberlein C.V., Gutlieri M.J., Brinkman B.A. SAN 582 efficacy and tolerance in potato (*Solanum tuberosum*). Weed Technol 1999; 13:71-76.

Walker R.H., Wells L.W., McGuire J.A. Bristly starbur (*Acanthospermum hispidum*) interference in peanuts (*Arachis hypogaea*). Weed Sci 1989; 37:196-2000.

Warren G.F. Spectacular increases in crop yields in the United States in the twentieth century. Weed Technol 1998; 12:752-760.

Webster T. M., Wilcut J.W., Coble H.D. Influence of AC 263,222 rate and application method on weed management in peanut (*Arachis hypogaea*). Weed Technol 1997; 11:520-526.

Wehtje G. R., McGuire J.A., Walker R.H., Patterson M.C. Texas panicum (*Panicum texanum*) control in peanuts (*Arachis hypogaea*) with paraquat. Weed Sci 1986; 34:308-311.

Wehtje G., Wilcut J.W., Hicks T.V., McGuire J.A. Relative tolerance of peanut to alachlor and metolachlor. Peanut Sci 1988; 15:53-56.

Wehtje G., Wilcut J.W., McGuire J.A. Foliar penetration and phytotoxicity of paraquat as influenced by peanut cultivars. Peanut Sci 1991; 18:67-71.

Wehtje G., Padgett D., Martin Jr. N.R., Imazapic-based herbicide systems for peanut and factors affecting activity on Florida beggarweed. Peanut Sci 2000a; 27:17-22.

Wehtje G., Brecke B.J., Martin N. R. Performance and economic benefit of herbicides used for broadleaf weed control in peanut. Peanut Sci 2000b; 27:11-16.

Wesley M. T., Shaw D.R. Interactions of diphenylether herbicides with chlorimuron and imazaquin. Weed Technol 1992; 6:345-351.

White A.D., Coble H.D. Validation of HERB for use in peanut (*Arachis hypogaea*). Weed Technol 1997; 11:573-579.

Wilcut J.W., Wehtje G.R., Patterson M.G. Economic assessment of weed control systems for peanuts (*Arachis* hypogaea). Weed Sci 1987a; 35:433-437.

Wilcut J.W., Wehtje G.R., Patterson M.G., Cole T.A., Hicks T.V. Absorption, translocation and metabolism of foliar-applied chlorimuron in soybeans, peanuts, and selected weeds. Weed Sci 1989; 37:175-180.

Wilcut J.W., Swann C.W. Timing of paraquat applications for weed control in Virginia-type peanuts (*Arachis hypogaea*). Weed Sci 1990; 38:558-562.

Wilcut J.W., Wehtje G.R., Hicks T.V. Evaluation of herbicide systems in minimum and conventional - tillage peanuts (*Arachis hypogaea*). Weed Sci 1990; 38:243-248.

Wilcut J.W. Efficacy and economics of common bermudagrass (*Cynodon dactylon*) control in peanut (Arachis hypogaea). Peanut Sci 1991a; 18:106-109.

Wilcut J.W. Economic yield response to peanut (*Arachis hypogaea*) to postemergence herbicides. Weed Technol 1991b; 5:416- 420.

Wilcut J.W., Walls Jr. F.R., Horton D.N. Imazethapyr for broadleaf weed control in peanuts (*Arachis hypogaea*). Peanut Sci 1991a; 18:26-30.

Wilcut J.W., Walls Jr. F.R., Horton D.N. Weed control, yield, and net returns using imazethapyr in peanuts (*Arachis hypogaea*). Weed Sci 1991b; 39:238-242.

Wilcut J.W., York A.C., Wehtje G.R. The control and interaction of weeds in peanut (*Arachis hypogaea*). Rev .Weed Sci 1994a; 6:177-205.

Wilcut J.W., Richburg III J.S., Wiley G., Walls Jr. F. R., Jones S. R., Iverson M.J. Imidazolinone herbicide systems for peanut (*Arachis hypogaea* L.). Peanut Sci 1994b; 21:23-28.

Wilcut J.W., Richburg III J. S., Eastin E.F., Wiley G.R., Walls, Jr. F.R., Newell S. Imazethapyr and paraquat systems for weed management in peanut (*Arachis hypogaea*). Weed Sci 1994c; 42:601-607.

Wilcut J.W., York A.C., Grichar W.J., Wehtje G.R. "The Biology and Management of Weeds in Peanut (*Arachis hypogaea*). In

Advances in Peanut Science, H.E. Pattee, H.T. Stalker, eds. Stillwater, OK: Amer. Peanut Res. Educ. Soc., 1995a; pp. 207-244.

Wilcut J.W., York A.C., Jordan D.L. "Weed management systems for oil seed crops." In *Handbook of Weed Management Systems,* A. E. Smith, ed. New York, NY: Marcel-Dekker. 1995b; pp. 343-400.

Wilcut J.W., Richburg III J.S., Wiley G.L., Walls. Jr. F.R. Postemergence AC 263,222 systems for weed control in peanut (*Arachis hypogaea*). Weed Sci 1996; 44:615-621.

Wilcut J.W., Crammer J. Summary of flumioxazin performance in southeastern peanuts. Proc South Weed Sci Soc 1997; 50:7.

Wilkinson, R.E. " Carbamathioates." In *Herbicides: Chemistry, Degradation, and Mode of Action,* P. C. Kerney, D. D. Kaufman, eds. Vol. 3., New York, NY: Marcel Dekker, 1988; pp. 245-300.

York A.C., Wilcut J.W., Swann C. W., Jordan D.L., Walls Jr F.R. Efficacy of imazethapyr in peanut (*Arachis hypogaea*) as affected by time of application. Weed Sci 1995; 43:107-116.

York A.C., Wilcut J.W. Interaction of bentazon and imazethapyr applied postemergence to nutsedge (*Cyperus* spp.). Weed Technol 1995; 22:150-154.

Young J.H., Person N.K., Donald J.O., Mayfield W.H. "Harvesting, curing, and energy utilization." In *Peanut Science and Technology,* H.E. Pattee, C. T. Young, eds. Yoakum, TX: Amer. Peanut Res. Educ. Soc., 1982; pp. 458-487.

Zimdahl R.L. "The concept and application of the critical weed-free period." In *Weed Management in Agroecosystems,* M. A. Altieri, M. Liebman, eds. Boca Raton, FL: CRC Press, 1988; pp. 145

CHAPTER 25

COMPETITIVENESS OF RICE CULTIVARS AS A TOOL FOR CROP-BASED WEED MANAGEMENT

KEVIN D. GIBSON[1] & ALBERT J. FISCHER[2]

[1]Department of Botany & Plant Pathology, Purdue University, West Lafayette, IN 47906, USA; [2] Department of Vegetable Crops, University of California, Davis, CA 95616, USA

Rice (*Oryza sativa* L.) is one of the most important crops in the world both in terms of acreage planted and human consumption (Evans, 1998). Millions of people derive a substantial portion of their caloric intake from rice and rice is grown on every continent except Antarctica (IRRI, 1993). Rice is unique among crops in the variety of environments in which it can be grown which range from drought-prone sites to tropical rainforest and from sea level to the slopes of the Himalayas (IRRI, 1993). Cropping systems range from highly mechanized, high yielding, irrigated systems in which fields are continually flooded to lower-yielding rainfed upland rice where prolonged periods of moisture stress may occur (IRRI, 1993). Weed control is essential across rice cropping systems and the economic costs of weed management practices and yield losses can be substantial (Moody, 1996). Weed management practices are also diverse and range from hand-weeding to the near-exclusive reliance on herbicides for control. Despite the diversity of environments and weed management practices, these cropping systems all rely to some extent on vigorous crop growth to interfere with weed growth. Enhanced crop competitive ability has been proposed for reducing the costs of weed control and as an environmentally-compatible tool for rice farmers (Moody, 1996). The rationale for adopting more competitive cultivars varies with cropping systems. In the Ivory Coast, approximately half of the rice yield decline was attributed to increased weed pressure (Becker and Johnson, 1995). Johnson et al. (1998) suggested that more competitive upland rice cultivars might reduce the need for hand-weeding in West Africa where only a small proportion of farmers rely on herbicides for weed control. In contrast, rice production in California is highly dependent on herbicide use and work on competitive cultivars (Gibson et al., 2001a, b) was driven by concerns related to reducing pesticide loads in the environment and to the evolution of cross-resistant watergrass (*Echinochloa* spp.) populations (Fischer et al., 2000).

Inderjit (ed.), Weed Biology and Management, 517-537.

Competitive rice cultivars have several properties that position them for inclusion in weed management strategies: they can be readily integrated into existing agronomic practices, are environmentally benign, and impose no additional costs on farmers. Although there appears to be a role for competitive cultivars in integrated weed management practices, questions have been raised regarding the feasibility of relying on competitive cultivars for weed control while maintaining high yields (Cousens and Mokhtari, 1998; Jennings and Aquino, 1968). The purpose of this paper is to review the literature on varietal differences in competitive ability among rice cultivars, the plant traits that confer competitive ability to rice, the potential for inclusion of competitive rice cultivars in weed management, and to suggest areas for further research. We recognize that some rice cultivars may interfere with weed growth through allelopathic chemicals (Olofsdotter, 2001), however, we have limited our discussion to rice traits related to competition for resources.

Varietal Differences

Differences among cultivars in their abilities to compete for resources have been reported for many crops, including rice (Callaway and Forcella, 1992; Pester et al., 1999). Differences in competitive ability among rice (*O. sativa*) cultivars have been detected in a number of rice cropping systems and geographical regions (Table 1). Researchers have worked with mixed infestations of weed species (DeDatta et al., 1969; Kawano et al., 1974; Akobundu and Ahissou, 1985; Johnson et al., 1998; Ni et al., 2000; Gibson et al., 2001a, b; Gibson et al., 2003) and with single species stands of barnyardgrass (*Echinochloa crus-galli* (L.) Beauv.), junglerice (*E. colona* (L.) Link) (Smith, 1974; Fischer et al., 1997; Ntanos and Koutroubas, 2000), palisadegrass (*Brachiaria brizantha*) (Hochst. ex A. Rich) Stapf and signalgrass (*B. decumbens*) Stapf (Fischer et al., 1995; Fischer et al., 2001; McGregor et al., 1988), red rice (*Oryza sativa* L.) (Kwon et al., 1991a, b), and bearded sprangletop (*Leptochloa fascicularis* (Lam.) Gray) (Carey et al., 1994). Some studies relied on rice cultivars to serve as model weeds and extrapolated from intraspecific competition to interspecific competition (Jennings and Aquino, 1968; Kawano et al., 1974). Researchers have identified differences in competitive ability among tall, traditional cultivars and modern semi-dwarfs (Jennings and Aquino, 1968), among rice and the indigenous wild species African rice (*O. glaberrima* Steud.) varieties and their crosses (Dingkuhn et al., 1999), between semi-dwarf temperate water-seeded cultivars (Gibson et al., 2001b), among tropical semidwarfs (Fischer, 1997) and between *japonica* and *indica* types (Ntanos and Koutroubas, 2000).

Rice has tremendous genetic and phenotypic variability and, as evidenced by these studies, variability among cultivars in competitive ability is more commonly the rule rather than the exception across rice cropping systems.

Table 1. Studies selected to represent the range of locations, cropping systems and weed species in which differences in competitive ability among cultivars has been detected.

Study Location	Rice cropping system	Weed species studied	Reference
Asia			
Japan	Transplanted paddy rice	late watergrass (*Echinochloa oryzicola*)	Suzuki et al. (2002)
Philippines	Lowland rainfed	Rice as model weed	Jennings and Aquino (1968)
Philippines	Upland rainfed	Multiple species infestation	Garrity et al. (1992)
Philippines	Lowland rainfed (wet and dry seasons)	Multiple species infestation	Ni et al. (2000)
Africa			
Nigeria	Lowland rainfed	Multiple species infestation	Akobundu and Ahissou (1985)
Ivory Coast	Upland rainfed	Multiple species infestation	Johnson et al. (1998)
Latin America			
Colombia	Upland rainfed rice-pasture Intercropping	Palisade- and signalgrass	Fischer (2001)
Colombia	Lowland irrigated	Junglerice	Fischer et al. (1997)
Peru	Lowland	Rice as model weed	Kawano et al. (1974)
United States			
California	Temperate, irrigated, Water-seeded	Early watergrass (*E. oryzoides)* and late watergrass (*E. phyllopogon* = *E. oryzicola*)	Gibson et al. (2001b)
Southern United States	Temperate, irrigated, dry-seeded rice,	Barnyardgrass, red rice Broadleaf signalgrass, (*Brachiaria platyphylla,* Bearded sprangletop	Smith (1974); McGregor et al., (1988); Carey et al. (1994)

Yield and Competitive Ability

Although there is considerable evidence of varietal differences in competitive ability, relatively little has been done to capitalize on these differences in a rice breeding program. We believe two factors have particularly limited progress in this area. First, the effectiveness of chemical control has led researchers in rice and other crops to focus on herbicide use with less emphasis on other control methods (Norris, 1992; Jordan, 1993). Second, researchers who compared newly introduced semidwarf lines with their taller predecessors concluded that yields in rice were inversely correlated with competitive ability (Jennings and Aquino, 1968; Jennings and De Jesus, 1968; Jennings and Herrera, 1968; Kawano et al., 1974). Given the apparent tradeoff between yield and competitive ability, Jennings and Aquino (1968) suggested that there would be little reason to develop competitive cultivars as long as alternative control methods were available. Setting aside the argument that, in some rice cropping systems, herbicides are not readily available due to cost, restrictions in use, or problems with herbicide resistant weeds, there is a growing body of evidence that improved competitive ability may be possible in rice varieties without a substantial yield penalty. Garrity et al., (1992) evaluated 25 upland rice cultivars (improved semi-dwarf, intermediate height, and traditional tall cultivars) in field experiments in the Philippines. The authors found several intermediate (1.0 to 1.15 m) cultivars that were as competitive as the tall (1.4 m) traditional cultivars. The intermediate height cultivars were ranked among the top cultivars for yield in the International Upland Nursery. The authors suggested that weed suppression ability could be combined with high yield potential. Fischer (1997) grew 14 semidwarf cultivars weed-free and with junglerice at Palmira, Colombia. There were significant differences among varieties in competitive ability but a correlation between competitive ability and weed-free yields was not detected (Table 2). This lack of correlation was also reported for a similar experiment involving rice and signal- and palisadegrass (Fischer et al., 2001). Dingkuhn et al. (1999) characterized the growth and competitive ability of rice, African rice and their interspecific progenies at Mbe, Ivory Coast. The correlations detected between competitiveness and yield potential were negative but not statistically significant. High SLA, which contributed to rice competitiveness in that study, reduces the amount of assimilate needed to produce a given leaf area but can result in low leaf-level photosynthetic rates which may limit yields (Dingkuhn et al., 1999). However, the authors concluded that this effect might be small and suggested that competitiveness was compatible with high yield potential if cultivars had large specific leaf areas (SLA) early in the season and small specific leaf area (SLA) during the reproductive growth period. Fofana and Rauber (2000) conducted similar experiments with African rice and rice varieties and also concluded that high yielding weed-suppressive cultivars could be developed. Ni et al. (2000) reported that selection for competitiveness among semi-dwarf varieties was compatible with high yields. Finally, 6 to 8 water-seeded temperate rice cultivars were grown with and without the competition of a mixture of early and late watergrass in the Central Valley of California (Gibson et al., 2003). No significant correlation was detected between traits related to weed suppression and rice yields in monoculture.

Table 2. Rice weed-free yields and percent yield reduction resulting from junglerice competition (Palmira, Colombia). Modified from Fischer et al., 1997.

Rice variety	Weed-free yield	Yield reduction
	kg ha^{-1}	%
Inti	7443	44
Eloni	6979	48
Irga 409	8326	58
CICA 4	6227	44
CICA 9	7891	41
CICA 8	8025	28
Oryzica 3	10832	54
Ceysvoni	7411	47
IR 8	6578	55
Bluebelle	6208	48
Bluebonnet 50	6620	56
Oryzica Caribe 8	7380	50
Oryzica Yacu 9	8643	53
Oryzica Turipana 7	6618	49
LSD (0.05)	1151	13

The contradiction between more recent studies that suggest high yielding competitive rice cultivars can be developed and earlier studies that found an inverse correlation between rice yield and competitive ability may be related to the relative contribution of plant height to competitive ability among these studies. In earlier studies (Tanaka et al., 1966; Jennings and Aquino, 1968; Kawano et al., 1974), the competitive abilities and yields of tall traditional varieties were compared with that of dwarf varieties. The tall varieties were leafy and had strong early season growth while the shorter varieties were small leaved and had weak early season growth (Tanaka et al., 1966). Not surprisingly, the tall cultivars were superior competitors but relatively low yielding. It was concluded that respiration costs, mutual shading of rice leaves, leaf deterioration and reduced leaf area index (LAI) and photosynthetic capacity later in the season limited yields in the tall cultivars (Tanaka et al., 1966). In contrast, modern, high-yielding, short-statured rice ideotypes with erect leaves and few tillers have been selected to maximize light interception while reducing mutual shading, to limit respiration costs associated with large biomass and to optimize partitioning to grain instead of vegetative growth (Dingkuhn et al., 1991; Peng et al., 1994). As we have already discussed, differences in competitive ability have been found among cultivars of similar height (Garrity et al., 1992; Ni et al., 2000; Lanning et al., 1997). The lack of correlation between competitive ability and yield in these studies suggests that tradeoffs between yield potential and competitive ability may not be as severe when plants of similar height are compared. By selecting for traits other than plant height in a breeding program, it may be possible to avoid or minimize a tradeoff between competitiveness and rice yield potential. Certainly more detailed studies on the effect of vegetative traits on rice yields among cultivars of similar height should be conducted. However, even if some tradeoff between yield and competitive ability is unavoidable, it should be noted that the inherent yield capacities of varieties is only one component of rice productivity in realistic weedy fields and some tradeoffs may be acceptable in order to meet other management goals (e.g. weed suppression, reduced herbicide use, etc.).

Traits Related to Competitive Ability

Weed Tolerance and Weed Suppressive Ability
Two factors contribute to crop competitiveness with weeds: weed tolerance (WT), the ability to maintain high yields despite weed competition and weed suppressive ability (WSA), the ability of the crop to reduce weed growth through competition (Jannink et al., 2000; Goldberg and Landa, 1991). Differences in WSA among cultivars can be readily determined by assessing variation in weed seed or vegetative biomass production when grown with the crop. Assessing the WT of cultivars is more problematic because researchers must determine first that varietal yield differences reflect weed interference and not genotypic differences that would be present in the absence of weeds and second that differences in yields are not due to reduced weed pressure arising from cultivar differences in WSA (Jordan, 1992). Jordan (1992) used path analysis to independently estimate the effects of soybean cultivars and density on common cocklebur (*Xanthium strumarium* L.) (WSA) and the effects of the weed on soybean density and yield (WT). The semidwarf soybean cultivars differed in WSA but had similar tolerances to weed interference.

Competition for Light
Competitive ability in rice has been primarily associated with traits related to light capture (Khush, 1996), such as plant height, tillering, leaf morphology and leaf area (Table 3). These traits have been identified in many studies by relating the relative yields (yields in weedy plots/ yields in monoculture) of cultivars to rice growth parameters. Researchers in rice and other crops who use relative yield as an index for competitive ability rarely separate the effects of WT from WSA. In West Africa, researchers focused on relative yield as a measure of rice competitiveness (Dingkuhn et al., 1999) and concluded that SLA, tillering and duration of growth until crop maturity contributed to variation in competitive ability among rice, African rice and their crosses. No information was provided on variation in weed suppression among the cultivars. However, Johnson et al. (1998) evaluated the suppression of a mixed weed infestation and the relative yields of three upland rice cultivars. IG10, an African rice type, had higher yields in competition with weeds and reduced weed biomass more than the traditional and improved rice varieties suggesting that at least some of the variation in relative yield seen elsewhere (Dingkuhn et al., 1999) might be due to WSA. IG10 accumulated more biomass, had higher LAI and SLA, produced more tillers and partitioned more biomass to leaves early in the season than the rice cultivars. Several experiments were conducted in drill-seeded rice in the southern United States comparing the relative yields of cultivars grown with red rice (Kwon et al., 1991a,b) or with bearded sprangletop (Carey et al., 1994). Although these experiments were not specifically designed to relate rice traits to differences in competitive ability among cultivars, greater relative yields were attributed to superior plant height.

Table 3. Traits related to rice competitive ability from studies selected to represent a range of cropping systems.

Cropping system	Measure of competitive ability	Traits related to competitive ability	Study
Upland	Relative yield of rice	Specific leaf area, tillering, duration of growth until crop maturity	Dingkuhn et al. (1999)
	Reduction in weed biomass	Plant height, LAI	Garrity et al. (1992)
	Reduction in weed biomass	Root length density, root weight density, tillering, LAI	Fofana and Rauber (2000)
	Reduction in weed biomass	LAI, tillering	Fischer et al. (2001)
Lowland rainfed	Relative yield of rice	Height, leaf area, plant weight, N absorption	Kawano et al. (1974)
	Reduction in weed biomass	LAI, seedling initial biomass, growth rate*	Ni et al. (2000)
Irrigated a) puddled (tropical)	Reduction in weed biomass	LAI, tillering	Fischer (1997)
b) dry-seeded (temperate)	Relative yield of rice	Duration of growth until crop maturity	Smith (1974)
c) water-seeded (temperate)	Reduction in weed biomass	LAI, root dry weight	Gibson et al. (2003)

*The authors used path analysis to directly relate crop biomass to weed biomass. The traits listed were indirectly related to weed biomass through crop biomass.

The studies described above associated yield reductions due to weed competition with the competitive ability of rice. In other studies, rice competitive ability was evaluated and competitive traits identified based on intraspecific competition (Jennings and de Jesus, 1968; Jennnings and Herrera, 1968; Jennings and Aquino, 1968). Jennings and Aquino (1968) grew monogenically differentiated tall and dwarf plants in mixture and in monoculture. The authors also grew two tall, leafy, profusely tillering cultivars BJ and MTU in mixture and monoculture. They concluded that plant size and early vigor as measured by more tillers, greater leaf length, leaf number and LAI; superior height and dry weight were related to superior competitive ability. Similarly, Kawano et al. (1974) grew 25 rice cultivars in Peru in monoculture, with a mixed infestation of weed species, or in competition with rice cultivar IR8. There were differences in relative yield (yield in monoculture vs. yield when grown with IR8) among cultivars due to competition with IR8. These differences were negatively correlated with weed weight. The authors interpreted this as evidence that competition of rice with weeds was controlled by the same factors as competition with IR8. Plant height, leaf area, plant weight and N absorption in early growth stages were then identified as important competitive traits based on correlations between these traits and differences in yield between monoculture. Plant traits were not correlated directly with weed weight in this study.

Relatively few studies have been conducted to correlate weed biomass directly with rice plant traits. Garrity et al. (1992) reported that plant height was the character most strongly related to low weed weight in their evaluation of 25 upland rice cultivars in the Philippines. Leaf area was also negatively correlated with weed weight while tiller number was not significantly related to weed weight. Although height was important, the authors found that cultivars of the same height differed by as much as 2.0 Mg ha^{-1} in weed weight and they inferred that characteristics other than height can play a significant role in WSA. Ni et al. (2000) used path analysis to identify traits related to WSA in semi-dwarf cultivars in the Philippines. Crop biomass in weed-free plots at tillering was the best predictor of competitiveness against weeds. Initial rice seedling biomass, crop growth rate, and LAI indirectly affected weed biomass through their effect on rice biomass. The authors deliberately excluded tall traditional cultivars from their study in order to concentrate on high-yielding modern cultivars. Lanning et al. (1997) adopted a similar approach when they examined the ability of wheat and barley lines to intercept light reaching inter-row spaces. The authors identified cultivars of similar height but different shading abilities and found positive correlations between light penetration and wild oat (*Avena fatua* L.) growth. Fischer (1997) and Fischer et al. (2001), working with modern semi-dwarf cultivars, reported that rice LAI, tiller production, and canopy light interception, but not height, were negatively associated with weed biomass in irrigated and upland rice in Latin America (Table 4). Plant height was not an important component of competitive ability in their study. Estorninos et al. (2002) evaluated the interactions between two rice cultivars and two red rice cultivars. The authors also concluded that short-statured cultivars with high tillering capacity could compete effectively with red rice. It appears that while height may play an important role in crop:weed competition, taller rice plants are not necessarily required to increase competitiveness in rice.

Table 4. Correlations between irrigated rice traits and junglerice biomass, and between upland rice traits and palisadegrass biomass, when rice and these species were grown in competition (Colombia). Modified from Fischer (1997) and Fischer et al. (2001).

Rice trait	junglerice biomass	palisadegrass biomass
	------------------- r	-------------------
Leaf area index	-0.71*	-0.86**
Number of tillers	-0.80*	-0.72*
Plant height	0.1 ns	0.08 ns
Aboveground biomass	-0.72*	-0.81**
PPFD interception[a]	-0.81**	-0.82**

*, ** $P < 0.05$ and $P < 0.01$, respectively.
[a] PPFD, photosynthetic photon flux density.

Competition for Nutrients
While researchers have associated traits related to light capture with rice competitiveness, relatively little attention has been given to the role that roots and competition for nutrients might play in rice-weed interactions. Researchers have often assumed that belowground dynamics could be safely ignored because crop systems are so heavily fertilized that nutrients should not limit growth. However, weed and crop growth may be limited by nutrient availability even when N is added at rates high enough that monoculture growth is not limited (Exley and Snaydon, 1992). For example, even when higher rates of N were applied, soil resources were still more limiting than light when wild oat competed with three spring cereals (wheat, barley, and oats) (Satorre and Snaydon, 1992). The authors attributed this to increased demand for N under conditions of interspecific competition. Weeds can have a higher mineral content requirement than crops (Malicki and Berbeciowa, 1986) and reported that weeds compete strongly for nutrients even when relatively high rates of fertilizer were applied (Vengris et al., 1953, 1955). Also, high rates of fertilizer application may favor weed growth to a greater degree than the crop, resulting in the ability of the weed to compete for other factors, such as light, at subsequent stages of development (Zimdahl, 1993). LeStrange (1981) grew barnyardgrass in competition with rice at a range of N fertilizer rates. Higher N rates increased barnyardgrass growth and ability to shade rice in competition. Similar results were obtained for early and late watergrass species (J. E. Hill, unpublished data). Carlson and Hill (1985) reported that wild oat was able to utilize N fertilizer better than wheat and wheat yields declined in weedy plots as N fertilizer increased. Competition for nutrients between crops and weeds is important for several crops (Aspinall, 1960; Irons and Burnside, 1982; Gamboa and Vandermeer, 1988; Satorre and Snaydon, 1992). In rice, several studies designed to determine the relative importance of root competition suggest that belowground dynamics are more important than generally believed (Assemat et al., 1981; Perera et al., 1992; Gibson et al., 1999). Gibson et al., (1999) found that shoot-only interference by rice had relatively little effect on late watergrass, although substantial reductions in watergrass growth were observed when plants competed for light and nutrients (Table 5). The authors also reported that root and shoot growth

were highly correlated for both species. This is generally true for rice and other crops (Assemat et al., 1981; Yoshida, 1981; Klepper et al., 1984) indicating that researchers should not rely solely on correlations between aboveground traits and competitive ability as proof that belowground dynamics can be ignored. It should be noted, however, that Suzuki et al. (2002) reported that shoot competition between transplanted rice and early watergrass had a greater effect on rice growth than root growth. The authors suggested that the relative importance of root and shoot competition could vary with the traits of the weed species and with soil nutrient availability. For example, the authors noted that the strain of barnyardgrass grown by Perera et al. (1992) grew more vigorously than rice throughout the growing season but late watergrass grown under similar conditions grew more vigorously than rice only in the latter half of the growing season (Yamasue et al., 1997). Suzuki et al. (2002) also suggested that nitrogen accumulation patterns, which can vary substantially among and within weed species, might also affect the importance of root and shoot competition.

Table 5. Growth reduction (%) of late watergrass, relative to plants grown without competition, due to full, shoot, and root competition with rice. Late watergrass plants were grown in flooded plants alone or surrounded by 8 rice plants and harvested at 100 days after seeding. The negative value indicates late watergrass plants were taller grown with shoot competition relative to plants grown without competition. Modified from Gibson et al., 1999.

Type of competition	Tillers	Height (cm)	Aboveground dry weight (g)
Shoot and root competition	85	18	84
Shoot competition	37	44	17
Root competition	76	43	80

Rice cultivars can differ substantially in root growth and morphology (Slaton et al., 1990; Suzuki et al., 2002) and in nutrient uptake rates (Teo et al., 1995). Fofana and Rauber (2000) examined WSA among African rice, traditional and improved rice lines grown with and without a mixed-species weed infestation in the Ivory Coast. Weed biomass was negatively correlated with rice root growth at early growth stages and with shoot and root growth at later stages. Fischer et al. (1995) evaluated the early components of competition between upland rice and palisadegrass (Hochst. Ex A. Rich) and suggested that competition for soil resources occurred before rice and weed canopies overlapped. In a study suggesting the importance of root competition, Fofana and Rauber (2000) concluded that screening cultivars for root development could increase rice competitiveness with weeds, but cautioned that differences in WSA might also be related to variation in allelopathic ability among upland rice cultivars.

Crop Duration

Crop duration is another characteristic that might be selected to improve rice competitiveness. Cultivars that mature late in the season could conserve yields by allowing the crop to recover from competition with weeds that senesce early in the season. Alternatively, early maturity cultivars might conserve yields by avoiding competition with late-emerging weeds. Smith (1974) associated variation in relative yields among cultivars to the time required for cultivars (Starbonnet, Bluebelle) to reach maturity. Starbonnet was of similar height to but matured 20 days after Bluebelle; the greater WT of Starbonnet was attributed to its ability to recover from competition with barnyardgrass McGregor et al. (1988) examined the effect on rice cultivar growth and yields of the duration of interference by broadleaf signalgrass. McGregor et al. (1988) suggested that crop duration might contribute to competitive ability but warned that other traits might also play a role. The later-maturing rice cultivar Mars was taller, produced more dry weight and had smaller yield reductions following competition with the weed than c.v. Bond. Mars also reduced broadleaf signalgrass dry weight more than Bond. Dingkuhn et al. (1999) reported that late-maturing African rice cultivars in their study had higher relative yields. They suggested that crop duration could be an important component of competitive ability when weed interference was of limited duration. African rice can tiller strongly following weeding at 30 and 60 days but rice lacks this ability (Johnson et al., 1998). Fischer et al. (2000) examined the potential for competitive cultivars in rice–pasture cropping system in Latin America. In this system, upland rice can be sown for the profitable production of grain while grazing land is renovated. Perennial pasture species are undersown in rice and rice cultivars must be competitive enough to cover the cost of pasture renovation. The most WT cultivar in their study had the shortest time to maturity and completed grain-filling before pasture growth and thus interspecific competition peaked. The authors suggested that early maturity rice cultivars might be particularly well-suited for rice-pasture intercropping.

Simulation Modeling

Although most of the work on competitive rice cultivars has focused on field or greenhouse experiments, some effort has been made to assess competitive traits through simulation modeling. Simulation models have been used to describe the growth of rice under various environments (Kropff and van Laar, 1993; Dingkuhn, 1996; Caton et al., 1999a,b), to suggest new plant types for irrigated rice (Dingkuhn et al., 1991), and to explore rice:weed interactions (Graf et al., 1990a,b; Graf and Hill, 1992; Miller et al., 1993; Kropff and van Laar, 1993; Caton et al., 1999c). We are aware of only two attempts to expressly quantify the importance of various rice traits to competitive ability (Bastiaans et al., 1997; Lindquist and Kropff, 1996). Bastiaans et al. (1997) used sensitivity analysis of the model INTERCOM (Kropff and van Laar, 1993) to identify early relative leaf area growth rate, early relative height growth rate and maximum height as the most important morphological characteristics in the competition for light. Lindquist and Kropff (1996) reported that increasing the rate of early rice leaf area expansion in INTERCOM significantly reduced barnyardgrass seed production. Bastiaans et al., (1997) concluded that simulation models could help design rice cultivars that are more competitive with weeds. However, Bastiaans et al., (1997) used a calibration factor to account for differences among selected cultivars in crop growth. Since crop growth rate was therefore simulated in a descriptive rather than mechanistic fashion the authors cautioned that the model might not account for side effects (interactions among traits) that occur when parameter values were changed. It seems premature to us to consider a model adequate to identify a trait for selection in a breeding program before we can reliably account for possible interactions among traits. Furthermore, competition for nutrients and the interaction between roots and shoots that arise from such competition were not considered in INTERCOM by either study. In the only published work modeling root and shoot competition between rice and weeds, Graf et al., (1990b) estimated that competition for N accounted for 35% of the reduction in rice growth due to weed competition. If belowground dynamics are important, as suggested by Graf et al. (1990b) and other studies, then the structure of the INTERCOM and other models that exclude root competition is incomplete and cannot be corrected by increasing the complexity of light dynamics. Current simulation models in rice and other crops have limitations in complexity and structure arising from our insufficient understanding of key plant processes (Passioura, 1996) and in the availability of adequate data sets for rigorous calibration and validation. Although simulation models have great value as an aid for structuring research, constructing hypotheses regarding competitive traits (Boote et al., 2001), and for identifying gaps in our knowledge of weed:crop dynamics, additional insight into the primary processes governing plant physiology and morphological development may be needed before these models can play a strong applied role in selecting competitive traits for breeding programs.

RELEVANT AREAS FOR FUTURE RESEARCH

Trait Selection

Compelling evidence now exists that competitive high-yielding rice cultivars can be developed; current efforts should therefore focus on developing selection criteria for breeding programs. Competitive lines can be identified through either direct or indirect selection (Lemerle et al., 1996). In direct selection, each genotype would be grown with weeds and either weed biomass or crop relative yield used as a selection criterion. This approach is labor intensive, prone to measurement error, and requires a quantity of seed for each genotype that may not be available during the early stages of a breeding program (Lemerle et al., 1996; Jannink et al., 2000). In contrast, indirect selection is focused on identifying particular plant attributes associated with competitive ability. Indirect selection allows breeders to concentrate their efforts on particular traits so that selection can occur early in the breeding program without the need for screening large numbers of genotypes under weedy conditions. However, these traits must be identified prior to selection.

Jannink et al. (2000) assessed the value of indirect selection for WSA in soybean. The authors focused on four areas: heritability of traits and genetic correlation between traits and WSA, Genotype x Environment (GxE) interaction, the utility of a selection index when applied to new populations, and the relationship between WSA and other desirable agronomic traits. The authors reported that soybean early height heritability and genetic correlation with WSA made it an ideal selection criterion. GxE interaction was not significant for early height suggesting WSA would be stable across environments. Indirect selection increased predicted selection efficiency relative to direct selection on weed dry weight and their analysis suggested undesirable correlated responses to early height and WSA could be eliminated using restricted selection indices (Jannink et al., 2000). However, there were correlations between soybean competitive traits and flowering such that early maturity cultivars had greater initial growth and weed suppression than later maturity cultivars. The authors inferred that since selection for high-yielding late maturity cultivars would select against WSA, it might prove difficult to make rapid progress in developing more competitive soybean lines. However, in subsequent work, Jannink et al. (2001) determined that greater seed size in early maturity lines, an environmentally dependent characteristic, explained the perceived tradeoff between late maturity and WSA. The authors suggested that late maturing soybean cultivars with strong early season weed suppression could be developed. Such cultivars would provide season-long control thus reducing the likelihood of weed community shifts resulting from community enrichment in species released from competition following soybean senescence. We are aware of no reports on competitive rice cultivars in which all of these areas have been addressed in similar detail. Fischer (1997) and Fischer et al. (2001) reported that rice genotypes performed differently relative to each other in weedy and weed-free conditions with respect to the expression of traits associated with competitive ability. The authors suggested that cultivar selection would therefore have to be conducted directly in weedy conditions. If rice genotypes express differences in growth only in weedy environments then indirect selection for WSA would not be possible. Clearly additional work on this topic is required.

Regardless of whether direct or indirect selection is used, researchers will have to decide whether the criteria for selection will be based on weed tolerance, weed suppressive ability or both. There is currently no clear consensus for the use of WT or WSA as selection indices (Pester et al., 1999). Callaway and Forcella (1992) argued that, similarly to insect and disease resistant cultivars, WT cultivars could reduce yield losses and the need for weed control practices thereby saving millions of dollars in production costs. They suggested that weed suppression is of secondary importance because "the development of crops tolerant to weeds will be paralleled by other improvements in weed management techniques." However, Jordan (1993) argued that WSA should be emphasized in breeding programs rather than WT because weed-suppressive cultivars have the potential for reducing both weed pressure within-season, thereby minimizing yield losses, and for limiting weed seed production, which should lower weed-control costs and yield losses in subsequent years. WSA may be particularly important in managing herbicide resistance since the suppression of seed production by resistant weeds is a critical factor in limiting the spread of resistance.

As noted earlier, it is difficult to partition the contribution of weed tolerance and weed suppression to yields under weedy conditions. Such distinctions may be important if traits related to WT and WSA differ in heritability, correlation with undesired agronomic traits, or have high G x E interactions. For example, based on their review of studies involving competitive cultivars, Cousens and Mokhtari (1998) suggested that WSA might be more stable across environments than WT. We lean toward the argument developed by Jordan (1993) but note that WT and WSA share a common set of traits so it may be difficult to select for one without also selecting for the other. In any case, researchers should be explicit about their approach.

One approach to improve competitive ability is the introgression of competitive traits from wild rice species into domesticated rice. Hybrids of African rice and rice were previously sterile beyond the F1 generation. Recently, the sterility barrier was overcome by breeders (Jones et al., 1997) thereby allowing the superior weed competitiveness of African rice to be combined with the higher yield potential of improved rice races (Dingkuhn et al., 1998). Other rice species (wild perennial red rice, *O. rufipogon* Griff. and wild annual rice, *O. nivara*) are also potential sources for introgressing useful genes to elite rice cultivars. Moncada et al. (2001) used advanced backcross quantitative trait loci (AB-QTL) analysis to identify QTLs associated with agronomic traits in a population derived from interspecific crosses between wild perennial red rice and Caiapo, an upland rice variety from Brazil. Fifty-six percent of the trait-enhancing QTLs were derived from wild perennial red rice and the authors concluded that AB-QTL analysis was a useful tool for the genetic improvement of rice cultivars.

Crop-Based Management Systems

More competitive rice cultivars have the potential to fit readily into existing management practices. However, the optimal use of the crop to suppress weeds will require a consideration of how agronomic practices interact with crop growth to affect weed establishment and survival. Growers have a number of tools available for weed management including crop planting date and seed rate (Sarkar and Moody, 1981), fertility management (DiTomaso, 1995), water management (Williams et al., 1990), herbicide rate and application timing (Christensen, 1994, 1995; Lemerle et al., 1996; Gibson et al., 2002) and, in some rice systems, hand-weeding (Garrity et al., 1992). The successful integration of these tools into a crop-based weed management system will depend in part on our understanding of weed-crop interactions. We believe that research in the following three areas would be particularly helpful.

Mechanisms of Competition Between Rice and Weeds

We need to improve our understanding of how rice interferes with the growth of weeds and how weed species avoid or lessen the effects of resource competition with the crop. Most studies on competitive rice cultivars have linked traits related to light capture with weed suppression. However, weeds have several adaptations that allow them to avoid or lessen the effects of shading by the crop. Early watergrass (Gibson and Fischer, 2001) and purple ammania (*Ammannia coccinea* Rottb.) (Gibson et al., 2001) responded to shade with a combination of morphological and physiological changes that are typical of many weeds (Patterson, 1995), i.e. increased allocation to leaves, greater SLA, and lower respiration rates. The authors suggested that for both species, reductions in photosynthetic photon flux density (PPFD) due to shading by the crop would have to occur early in the season and be of sufficient magnitude to substantially reduce weed growth. Both species lessened the effects of shade in part by reducing biomass partitioning to roots and Gibson and Fischer (2001) speculated that if Early watergrass had been simultaneously subjected to light and nutrient limitations the ability of the weed to partition biomass from roots to leaves might have been compromised. Additional work on root-shoot interaction and fertility management would be useful.

Researchers have focused on light competition in rice but N availability can have a strong effect on rice:weed interactions (Ampong-Nyarko and DeDatta, 1993). Kim and Moody (1980) determined that pickerel weed (*Monochoria vaginalis* (Burm. f.) Kunth) was more competitive with rice for light at higher N levels. LeStrange (1981) reported that barnyardgrass growth increased and rice yields decreased in competition as N rates increased. Researchers working in water-seeded rice grew the crop with varying densities of watergrass and N rates (J.E. Hill, unpublished data). Watergrass density and N fertilizer rate interacted to affect rice yields and high N rates compensated to some degree for low weed densities. Gibson and Fischer (unpublished data) grew late watergrass at various light and N fertilizer rates and found that low N fertilizer rates increased the effect of intermediate shading on late watergrass. Late watergrass plasticity in leaf morphology and biomass partitioning was limited when the weed was grown at low N and reduced light. Optimizing crop competition in water-seeded rice may therefore include developing cultivars that compete strongly for light and N and/or the manipulation of fertility to favor rice uptake of N relative to weeds.

Critical Periods of Competition

The duration of competition between rice and weed can be of critical importance in determining the potential for competitive cultivars. Critical periods of competition between rice and weeds have been established for a number of cropping systems (Moody, 1978; Smith, 1988; Fischer et al., 1993). Weed species differ in emergence and growth relative to the rice and in the period during the rice growth cycle in which they affect yields. Smith (1968) determined that while barnyardgrass, ducksalad (*Heteranthera limosa* (Sw.) Willd.), and smallflower umbrellasedge (*Cyperus difformis* L.) interfered with rice early in the season, hemp sesbania (*Sesbania exaltata* (Raf.) Rybd. ex A. W. Hill) and northern jointvetch (*Aeschynomene virginica* (L.) B.S.P.) had little effect on drill-seeded rice until late in the season. Caton et al. (1997) found that purple ammania interference with rice occurred relatively late in the season. If cultivars are selected for WT then the timing of weed interference could be of great importance in identifying relevant traits. For example, growers might want to use early or late maturity cultivars depending on whether they wanted rice to avoid competition or have sufficient time to recover from competition with a particular weed species. However, if the focus is on WSA, then the timing of weed emergence relative to the crop, which can affect the ability of the crop to suppress weed growth (Chikoye and Swanton, 1995; Liebmann and Galllandt, 1997; Florez et al., 1999), might affect trait selection. Delaying weed emergence relative to the crop generally confers a competitive advantage on the crop and limits the effect of the weed on crop yields (Liebmann and Gallandt, 1997). Rapid canopy closure and shading by the crop can reduce the growth of weeds but it might also serve to truncate the period of weed germination and establishment. Gibson et al. (2002) found that short delays in watergrass emergence relative to rice greatly reduced watergrass seedling survival. When plots were kept weed-free for 30 days or longer, watergrass establishment was completely suppressed. Delaying watergrass emergence for only 15 days after seeding rice reduced watergrass biomass by 74%.

Weed emergence and establishment might be affected directly by low PPFD. However, changes in the ratio of red to far-red light that accompany canopy closure can inhibit germination or elicit changes in stem elongation and plant morphology that may affect weed:crop competition (Holt, 1995; Ballare and Casal, 2000). The selection of competitive traits as well as crop planting density and arrangement (Fischer and Miles, 1973; Weiner et al., 2001) depend in part on identifying when the weed is susceptible to interference. The mechanism by which rapid rice canopy closure affects weed establishment and growth will therefore require further investigation.

Interactions Between Control Practices and Crop Competition

Although competitive cultivars have the potential to improve weed control systems, it seems unlikely that even strongly competitive cultivars could suppress weed growth to such an extent that additional weed control measures become unnecessary. A more likely scenario is one in which the crop is used to augment control achieved by herbicides or cultural practices. Such an approach could improve control of problematic weeds, allow for reduced inputs (lower herbicide rates, less frequent hand-weeding), and provide a degree of suppression when herbicides or cultivation are impractical. Competitive cultivars might be particularly useful in reducing the growth and seed production of herbicide-resistant weeds without imposing additional selection pressure from herbicides. If the potential of a crop-based management strategy is to be realized, then the ability of the crop to replace or augment other control measures must be quantified. There is good evidence that competitive cultivars can reduce the need for other control measures. Fischer (1997) grew junglerice with irrigated rice cultivars at Palmira Colombia and concluded that competitive semidwarf cultivars could substantially reduce the number of herbicide applications needed when flooding did not provide sufficient weed suppression. Gibson et al. (2001) reported that the less competitive cultivar in their study required higher rates of herbicide to achieve the same level of weed suppression as their more competitive cultivar. At standard herbicide rates, weed biomass was substantially lower for the more competitive cultivar. The authors concluded that more competitive cultivars could improve weed control at current herbicide rates and possibly allow lower rates to be used. Dingkuhn et al. (1999) and Fofana and Rauber (2000) reported that fewer hand-weedings were necessary with more competitive rice cultivars in West Africa. Christensen (1994) used a negative exponential model with barley, winter wheat and winter rye varieties as a categorical variable to quantify the relationship between herbicide rates and weed control. Differences in competitive ability among species and varieties influenced herbicide rates necessary for adequate control and the author suggested that his approach could be used to calibrate herbicide rates based on the relative competitive ability of crop and cultivars. Quantifying the relationship between herbicide rates and cultivars might be particularly useful in developing ultra low rate (ULR) practices. The success of ULR weed management depends on the ability of the herbicide used to stunt weed growth enough to give the crop such an advantage in growth that crop yields are unaffected by subsequent weed growth (Alm et al., 2000). This strategy could allow the use of lower rates without compromising control. ULR practices might also reduce selection pressure for herbicide resistance when that resistance is controlled by single alleles. However, if variability in response to herbicide use is controlled by multigene or polygenic inheritance then ultra-low rates may enhance resistance in a weed population by allowing phenotypes with intermediate resistance to survive, produce seed and increase resistance through segregation and recombination (Cousens and Mortimer, 1995). For example, Johnston and Faulkner (1991) applied low rates of paraquat to perennial ryegrass (*Lolium perenne* L.) and allowed survivors to intercross. Screening in this fashion increased the ED_{50} of paraquat of tolerant cultivars 20- to 30-fold. Ultra-low rates are unlikely to be adopted unless non-chemical options can limit weed survival and seed production. Cultivation can improve weed suppression but in cropping systems where cultivation is impractical it may be unwise to adopt ULR weed management unless the crop is strongly competitive. Clearly, the development of more competitive rice cultivars could contribute to the use of ULR weed management in rice.

Liebmann and Gallandt (1997) suggested that information on the competitive ability of currently available cultivars could be used to improve weed management programs. Growers who select weakly competitive cultivars might have to increase herbicide rates or the intensity of mechanical cultivation. Seavers and Wright (1999) grew three crops with five densities of catchweed bedstraw (*Galium aparine* L.). They developed weed competitive values for each crop species based on their relative responses to weed density. The authors suggested that such an approach could be used to determine upper limits above which a crop species or cultivar would require herbicide applications. In addition to providing a rationale for the development of competitive cultivars, quantifying the relationship between varieties and herbicide inputs might allow growers to select currently available cultivars based on expected weed pressure and desired herbicide inputs.

Perhaps the most important weed control tool available to rice farmers and one that is unique to rice is the use of within-season flooding to reduce weed pressure. Relatively few weed species are adapted to flooded conditions and, consequently the weed flora of irrigated continuously flooded rice is less diverse than that of other agronomic crops (Bayer and Hill, 1992). The timing and depth of flooding can play a major role in both the number and abundance of weed species (Williams et al., 1990). Water management is a particularly critical weed control tool for organic rice farmers in California. Organic rice growers rely on a combination of deep water, crop competition and drought stress to suppress the growth of various weed species (A.J. Fischer, pers. obs.). The ability to sequentially tolerate deep water and drought stress would seem to require different crop characteristics than for cultivars selected to compete under conventional rice. Research should be conducted to determine the relationship between water management and cultivar competitive ability in rice.

CONCLUSIONS

The use of competitive crops has been proposed as a tool to maximize herbicide efficiency, address public concerns regarding the use of pesticides, and reduce selection pressure for resistance so that herbicides could be retained as tools in sustainable crop production (Lyon et al., 1996; Fischer et al., 1997, 2000; Gibson et al., 2001). Weed suppression by the crop can decrease weed seed production and thereby limit weed infestations in subsequent years (Jordan, 1993), allow farmers to reduce herbicide applications or cultivation intensity while maintaining current levels of weed control (Christensen, 1994), operate independently of weather conditions that might hinder other management practices (Jannink et al., 2000) and fit readily into existing agronomic practices. There appears to be strong potential for developing competitive high-yielding cultivars in several rice production systems. Competitive rice cultivars could form the backbone of crop-based management strategies that integrate a number of agronomic practices into effective and sustainable weed management systems. However, to optimize the contribution of competitive cultivars, additional research is required on trait heritabilities and genetic correlation to WSA, the relationship between WSA and other desirable agronomic traits, the mechanism and timing of crop:weed competition and the integration of competitive cultivars with existing and novel agronomic practices.

REFERENCES

Akobundu I. O., Ahissou A. Effect of interrow spacing and weeding frequency on the performance of selected rice cultivars on hydromorphic soils of West-Africa. Crop Prot 1985; 4: 71-76.

Alm D.M., Wax L.M., Stoller E.W. Weed suppression for weed management in corn (*Zea mays*) and soybean (*Glycine max*) production systems. Weed Technol 2000; 14: 713-717.

Ampong-Nyarko K., DeDatta S. K. Effects of light and nitrogen and their interaction on the dynamics of rice weed competition. Weed Res 1993; 33: 1-8.

Aspinall D. An analysis of competition between barley and white persicaris. II. Factors determining the course of competition. Ann Appl Biol 1960; 48: 637-654.

Assemat L., Morishima H., Oka H.I. Neighbor effects between rice (*Oryza sativa* L.) and barnyardgrass (*Echinochloa crus-galli* (L.) Beauv.) strains. II. Some experiments on the mechanisms of interaction between plants. Acta Oecol 1981; 2: 63-78.

Ballare C.L., Casal J.J. Light signals perceived by crop and weed plants. Field Crops Res 2000; 67:149-60.

Bayer D.B., Hill JE. "Weeds". In *Integrated Pest Management in Rice*, M. L. Flint, B. P. O. Ohleneger, eds. Oakland, CA: Univ. of California Statewide Integrated Pest Management Project, Univ. of California Division of Agric Nat Resour 1992; pp. 32-55.

Bastiaans L., Kropff M.J., Kempuchetty N., Rajan A., Migo T.R. Can simulation models help design rice cultivars that are more competitive against weeds? Field Crops Res 1997; 51: 101-111.

Becker M., Johnson D.E. "Upland rice yield and resource-base quality as affected by cropping intensity." In 1995 WARDA Annual Report. Bouaké, Cote d'Ivoire: West Africa Rice Development Association, 1995; pp. 52-53.

Boote K.J., Kropff M.J., Bindraban P.S. Physiology and modeling of traits in crop plants: implications for genetic improvement. Agric Syst 2001; 70: 395-420.

Callaway M.B., Forcella F. "Crop tolerance to weeds." In *Crop Improvement for Sustainable Agriculture*, M. B. Callaway, C. A. Francis, eds. Lincoln, NE: University of Nebraska Press, 1992; pp. 101-131.

Carey V. F., Smith R.J., Talbert, R.E. Interference durations of bearded sprangletop (*Leptochloa fascicularis*) in rice (*Oryza sativa*). Weed Sci 1994; 42: 180-83.

Carlson H.L., Hill J.E. Wild oat (*Avena fatua*) competition with spring wheat: effects of nitrogen fertilization. Weed Sci 1985; 34: 29-33.

Caton B.P., Foin T.C., Hill J.E. Mechanisms of competition for light between rice (*Oryza sativa*) and redstem (*Ammannia* spp.). Weed Sci 1997; 45:269-275.

Caton B.P., Foin T.C. and Hill J.E. 1999a. A plant growth model for integrated weed management in direct-seeded rice. I. Development and sensitivity analyses of monoculture growth. Field Crops Res 62, 129-143.

Caton, B.P., Foin T.C., Hill J.E. A plant growth model for integrated weed management in direct-seeded rice. II. Validation testing of water-depth effects and monoculture growth. Field Crops Res 1999b; 62:145-155.

Caton B.P., Foin T.C., Hill J.E. A plant growth model for integrated weed management in direct-seeded rice. III. Interspecific competition for light. Field Crops Res 1999c; 63: 47-61.

Chikoye D., Swanton C.J. Influence of common ragweed (*Ambrosia artemisiifolia*) time of emergence and density on white bean (*Phaseolus vulgaris*). Weed Sci 1995; 43: 375-380.

Christensen S. Crop-weed competition and herbicide performance in cereal species and varieties. Weed Res 1994; 34: 29-36.

Christensen S. Weed suppression ability of spring barley varieties. Weed Res 1995; 35: 241-247.

Cousens R.D., Mortimer M. *Dynamics of Weed Populations*. Cambridge: Cambridge University Press, 1995.

Cousens R.D., Mokhtari S. Seasonal and site variability in the tolerance of wheat cultivars to interference from *Lolium rigidum*. Weed Res 1998; 38: 301-307.

DeDatta S.K., Moomaw J. C., Bantilan, R. T. Effects of varietal type, method of planting, and nitrogen level on competition between rice and weeds. Asian-Pacific Weed Control Interchange 1969; 2: 152-163.

Dingkuhn M. Modelling concepts for the phenotypic plasticity of dry matter and nitrogen partitioning in rice. Agric Syst 1996; 52: 383-397.

Dingkuhn M., Penning de Vries F. W. T., DeDatta S. K., van Laar H. H. "Concepts for a new plant type for direct seeded flooded tropical rice." In *Direct Seeded Rice in the Tropics*. Los Baños, Philippines: International Rice Research Institute, 1991; pp. 17-54.

Dingkuhn M., Jones M.P., Johnson D.E., Sow A. Growth and yield potential of *Oryza sativa* and *O. glaberrima* upland rice cultivars and their interspecific progenies. Field Crops Res 1998; 57: 57-69.

Dingkuhn M., Johnson D.E., Sow A., Audebert A.Y. Relationships between upland rice canopy characteristics and weed competitiveness. Field Crops Res 1999; 61: 79-95.

DiTomaso J.M. Approaches for improving crop competitiveness through the manipulation of fertilization strategies. Weed Sci 1995; 43: 491-497.

Estorninos L.E., Gealy D.R., Talbert R.E. Growth response of rice (*Oryza sativa*) and red rice (*O. sativa*) in a replacement series study. Weed Technol 2002; 16: 401-406.

Evans L.T. *Feeding the Ten Billion: Plants and Population Growth.* Cambridge, UK: Cambridge University Press, 1998.

Exley D.M., Snaydon R.W. Effects of nitrogen fertilizer and emergence date on root and shoot competition between wheat and blackgrass. Weed Res 1992; 32: 175-182.

Fischer A.J. Suppression of junglerice [*Echinochloa colona* (L) Link] by irrigated rice cultivars in Latin America. Agron J 1997; 89: 516-21.

Fischer A.J., Lozano J., Ramirez A., Sanint L.R. Yield loss prediction for integrated weed management in direct- seeded rice. Int J Pest Manage 1993; 39: 175-180.

Fischer A.J., Châtel M., Ramírez H., Lozano J., Guimarães E. Components of early competition between upland rice (*Oryza sativa* L) and *Brachiaria brizantha* (Hochst Ex Arich) Stapf. Int J Pest Manage 1995; 41: 100-103.

Fischer A.J., Ateh C.M., Bayer D.E., Hill J.E. Herbicide-resistant *Echinochloa oryzoides* and *E. phyllopogon* in California *Oryza sativa* fields. Weed Sci 2000; 48: 225-230.

Fischer A.J., Ramírez H.V., Gibson K.D., Da Silveira Pinheiro B. Competitiveness of semidwarf upland rice cultivars against palisadegrass (*Brachiaria brizantha*) and signalgrass (*B. decumbens*). Agron J 2001; 93: 967-973.

Fischer R.A., Miles R. E. The role of spatial pattern in the competition between crop plants and weeds. Math Biosci 1973; 18: 335-350.

Florez J. A., Fischer A. J., Ramírez H., Duque M. C. Predicting rice yield losses caused by multispecies weed competition. Agron J 1999: 91: 87-92.

Fofana B., Rauber R. Weed suppression ability of upland rice under low-input conditions in West Africa. Weed Res 2000; 40: 271-280.

Gamboa W., Vandermeer J. Root and shoot competition in *Phaseolus vulgaris* and *Cyperus rotundus* in Nicaragua. Trop Agric 1988; 29: 21-5.

Garrity D.P., Movillon M., Moody K. Differential weed suppression ability in upland rice cultivars. Agron J 1992: 84: 586-91.

Gibson K.D., Foin T.C., Hill J.E. The relative importance of root and shoot competition between water-seeded rice and *Echinochloa phyllopogon*. Weed Res 1999; 39: 181-190.

Gibson K.D., Fischer A.J., Foin T.C. Shading and the growth and photosynthetic responses of *Ammannia coccinnea*. Weed Res 2001a; 41, 59-67.

Gibson K.D., Hill J. E., Foin T. C., Caton B. P., Fischer A. J. Water-seeded rice cultivars differ in ability to interfere with watergrass. Agron J 2001b; 93: 326-332.

Gibson K.D., Fischer A.J. Relative growth and photosynthetic response of water-seeded rice and *Echinochloa oryzoides* (Ard.) Fritsch to shade. Int J Pest Manage 2001; 47: 305-309.

Gibson K.D., Fischer A.J., Foin T.C., Hill J.E. Implications of delayed *Echinochloa* germination and duration of competition for integrated weed management in water-seeded rice. Weed Res 2002; 42: 351-358.

Gibson K.D., Fischer A. J., Foin T.C., Hill J.E. Crop traits related to weed suppression in water-seeded rice. Weed Sci 2003; 51: 87-93.

Goldberg D.E., Landa K. Competitive effect and response - hierarchies and correlated traits in the early stages of competition. J. Ecol 1991; 79: 1013-1030.

Graf B., Hill J.E. Modeling the competition for light and nitrogen between rice and *Echinochloa crus-galli.* Agric Syst 1992; 40: 345-359.

Graf B., Rakotobe O., Zahner P., Delucchi V., Gutierrez A.P. A simulation-model for the dynamics of rice growth and development. 1. The carbon balance. Agric Syst 1990a; 32: 341-365.

Graf, B., Gutiérrez A.P., Rakotobe O., Zahner P., Delucchi V.A simulation-model for the dynamics of rice growth and development. 2. The competition with weeds for nitrogen and light. Agric Syst 1990b; 32: 367-392.

Holt J.S. Plant responses to light - a potential tool for weed management. Weed Sci 1995; 43: 474-482.

Irons S.M., Burnside O.C. Competitive and allelopathic effects of sunflower (*Helianthus annuus*). Weed Sci 1982; 30: 372-377.

IRRI. *IRRI Rice Almanac.* Manila, Philippines: International Rice Research Institute, 1993.
Jannink J.L., Orf J.H., Jordan N.R., Shaw R.G. 2000. Index selection for weed suppressive ability in soybean. Crop Sci 40: 1087-1094.
Jannink J.L., Jordan N. R., Orf J. H. Feasibility of selection for high weed suppressive ability in soybean: Absence of tradeoffs between rapid initial growth and sustained later growth. Euphytica 2001; 120: 291-300.
Jennings P.R., DeJesus J. Jr. Studies on competition in rice: I. Competition in mixtures of varieties. Evolution 1968; 22: 119-124.
Jennings P.R., Herrera R. M. Studies on competition in rice: II. Competition in segregating populations. Evolution 1968; 22: 332-336.
Jennings P.R., Aquino R.C. Studies on competition in rice: III. The mechanism of competition among phenotypes. Evolution 1968; 22: 529-542.
Johnson D.E., Dingkuhn M., Jones M. P., Mahamane M. C. The influence of rice plant type on the effect of weed competition on *Oryza sativa* and *Oryza glaberrima.* Weed Res 1998; 38: 207-216.
Johnston D.T., Faulkner J.S. "Herbicide resistance in the Graminaceae – a plant breeder's view." In *Herbicide Resistance in Weeds and Crops.* Proceedings of the 19th Long Ashton Symposium. J. C. Caseley, G. W. Cussans, Atkin, R. K. eds. Oxford, UK: Butterworth Heinemann, 1991; pp. 319-330
Jones M. P., Dingkuhn M., Aluko G. K., Semon M. Interspecific *Oryza sativa* L. X *O. glaberrima* Steud progenies in upland rice improvement. Euphytica 1997; 94: 237-246.
Jordan N. Differential interference between soybean (*Glycine max*) varieties and common cocklebur (*Xanthium strumarium*) - a path-analysis. Weed Sci 1992; 40: 614-620.
Jordan N. Prospects for weed control through crop interference. Ecol Applic 1993; 3: 84-91.
Kawano K., González H., Lucena M. Intraspecific competition, competition with weeds and spacing response in rice. Crop Sci 1974; 14: 847-845.
Khush G.S. "Genetic improvement of rice for weed management." In *Herbicides in Asian Rice: Transitions in Weed Management.* R. Naylor, ed. Palo Alto, CA and Manila, Philippines: Institute for International Studies, Stanford University, IRRI, 1996; pp. 201-207.
Kim S.C., Moody K. Effect of plant spacing on the competitive ability of rice growing in association with various weed communities at different nitrogen levels. J Korean Soc Crop Sci 1980; 25: 17-27.
Klepper B., Belford R.K., Rickman R.W. Root and shoot development in winter wheat. Agron J 1984; 76: 117-122.
Kropff M.J., van Laar L.A.P. *Modeling Crop-Weed Interactions.* Wallingford, UK: CAB International, 1993.
Kwon S. L., Smith R. J., Talbert R.E. Interference durations of red rice (*Oryza sativa*) in rice (*Oryza sativa*). Weed Sci 1991a; 39: 363-368.
Kwon S.L., Smith R.J., Talbert R.E. Interference of red rice (*Oryza sativa*) densities in rice (*Oryza sativa*). Weed Sci 1991b; 39:169-174.
Lanning S.P., Talbert L. E., Martin J.M., Blake T.K., Bruckner P.L. Genotype of wheat and barley affects light penetration and wild oat growth. Agron J. 1997; 89: 100-103.
Lemerle D., Verbeek B., Cousens R.D., Coombes N.E. The potential for selecting wheat varieties strongly competitive against weeds. Weed Res 1996; 36: 505-513.
LeStrange M. Competition between rice (*Oryza sativa*) and barnyardgrass (*Echinochloa* spp.): the influence of rice stature, barnyardgrass density and nitrogen fertility. M.S. thesis. University of California, Davis, CA, 1981.
Liebmann M., Gallandt E. "Many little hammers: ecological approaches for management of crop-weed interactions." In *Ecology in Agriculture*, L. Jackson, ed. San Diego, CA: Academic Press, 1997; pp. 291-346.
Lindquist J.L., Kropff M.J. Applications of an ecophysiological model for irrigated rice (*Oryza sativa*) - *Echinochloa* competition. Weed Sci 1996; 44: 52-56.
Lyon D.J., Miller S.D., Wicks G.A. The future of herbicides in weed control systems of the Great Plains. J Prod Agric 1996; 9: 209-15.
Malicki L., Berbeciowa C. Uptake of more important mineral components by common field weeds on loess soils. Acta Agrobotanika 1986; 39: 129-141.
McGregor J.T., Smith R.J., Talbert R.E. Broadleaf signalgrass (*Brachiaria platyphylla*) duration of interference in rice (*Oryza sativa*). Weed Sci 1988; 36: 747-750.
Miller B. C., Foin T. C., Hill J. E. Carice - a rice model for scheduling and evaluating management actions. Agron J 1993; 85: 938-947.
Moncada P., Martinez C. P., Borrero J., Châtel M., Gauch H., Guimarães E., Tohme J., McCouch S.R. Quantitative trait loci for yield and yield components in an *Oryza sativa* x *Oryza rufipogon* BC2F2 population evaluated in an upland environment. Theor App Genet 2001; 102: 41-52.
Moody K. Crop-weed competition. Philippine J Weed Sci 1978; 5: 28-43.

Moody K. "Priorities for Weed Science Research." In *Rice Research in Asia: Progress and Priorities.* R. E. Evenson, R. W. Herdt, M. Hossain, ed. Wallingford, UK: CAB International, 1996; pp. 277-290.

Ni H., Moody K., Robles R. P., Paller E. C. Jr., Lales J. S. *Oryza sativa* plant traits conferring competitive ability against weeds. Weed Sci 2000; 48: 200-204.

Norris R.F. Have ecological and biological studies improved weed control strategies? Proceedings of the 1st International Weed Control Weed Control Congress. Melbourne, Australia, 1992; pp. 7-33.

Ntanos D. A., Koutroubas S.D. Competition of barnyardgrass with rice varieties. J Agron Crop Sci-Z Acker- Pflanzenbau 2000; 184: 241-246.

Olofsdotter M. Getting closer to breeding for competitive ability and the role of allelopathy – an example from rice (*Oryza sativa*). Weed Sci 2001; 5:798-806.

Passioura J. B. Simulation models: science; snake oil, education, or engineering? Agron J 1996; 88: 690-694.

Patterson D.T. Effects of environmental stress on weed/crop interactions. Weed Sci 1995; 43: 483-290.

Peng S., Khush G. S., Cassman K. G. "Evolution of the new plant ideotype for increased yield potential. In *Breaking the Yield Barrier.* K. G. Cassman, ed. Los Baños, Philippines: IRRI, 1994; pp. 5-20.

Perera K. K., Ayres P.G., Gunasena H.P.M. Root growth and the relative importance of root and shoot competition in interaction between rice (*Oryza sativa*) and *Echinochloa crus-galli*. Weed Res 1992; 32: 67-76.

Pester T.A., Burnside O.C., Orff J. H. "Increasing crop competitiveness to weeds through crop breeding." In *Expanding the Context of Weed Management.* D. Buhler, ed. NY: Haworth Press, 1999; pp. 59-76.

Sarkar P.A., Moody K. "Effects of stand establishment techniques on weed population in rice." In: *Weed Control in Rice.* Los Baños, Phillipines: IRRI, 1981; pp. 57-72.

Satorre E.H., Snaydon R.W. A comparison of root and shoot competition between spring cereals and *Avena fatua* L. Weed Res 1992; 32: 45-55.

Seavers G.P., Wright K.J. Crop canopy development and structure influence weed suppression. Weed Res 1999; 39: 319-28.

Slaton N.A., Beyrouty B.R., Wells B.R., Norman R.J., Gbur E. E. Root growth and distribution of two short-season rice genotypes. Plant Soil 1990; 121: 269-78.

Smith R.J. Jr. Weed thresholds in southern United States rice, *Oryza sativa.* Weed Technol 1988; 2: 232-41.

Smith R.J. Jr. Weed competition in rice. Weed Sci 1968; 16: 252-55.

Smith R.J. Jr. Competition of barnyardgrass with rice cultivars. Weed Sci 1974; 22: 423-26.

Suzuki T., Shiraiwa T., Horie T. Competitiveness of four rice cultivars against barnyardgrass, *Echinochloa oryzicola* Vasing, with reference to root and shoot competition. Plant Prod Sci 2002; 5: 77-82.

Tanaka A., Kawano K., Yamaguchi J. *Photosynthesis, respiration and plant type of the tropical rice plant.* Technical Report 7. Los Banos, Philippines: IRRI, 1966.

Teo Y.H., Beyrouty C.A., Norman R. J., Gbur E. E. Nutrient uptake relationship to root characteristics of rice. Plant Soil 1995; 171: 297-302.

Vengris J., Drake W.G., Colby W.G., Bart J. Chemical composition of weeds and accompanying crop plants. Agron J 1953; 45: 213-218.

Vengris J., Colby W.G., Drake M. Plant nutrient competition between weeds and corn. Agron J 1955; 47: 213-16.

Weiner J., Stoll P., Muller-Landau H., Jasentuliyana A. The effects of density, spatial pattern, and competitive symmetry on size variation in simulated plant populations. Am Nat 2001; 158: 438-450.

Williams J., Roberts S., Hill J.E., Scardaci S., Tibbits G. Managing water for weed control in rice. Cal Ag 1990; 44: 7-10.

Yamasue Y., Muruyama H., Inoue H., Matsui T., Kusanagi T. Growth analysis of rice and *Echinochloa oryzicola* Vasing. in mixed stands. J Weed Sci Tech 1997; 42: 365-372.

Yoshida S. *Fundamentals of Rice Crop Science.* Los Banos, Philippines: International Rice Research Institute, 1981.

Zimdahl, R. L. *Fundamentals of Weed Science.* San Diego, CA: Academic Press, 1993.

Author Index

Subject Index

D

F

G

H

I

J

P

R

S

T

V

W

GPSR Compliance

The European Union's (EU) General Product Safety Regulation (GPSR) is a set of rules that requires consumer products to be safe and our obligations to ensure this.

If you have any concerns about our products, you can contact us on ProductSafety@springernature.com

In case Publisher is established outside the EU, the EU authorized representative is:

Springer Nature Customer Service Center GmbH
Europaplatz 3
69115 Heidelberg, Germany

Batch number: 10809436

Printed by Printforce, the Netherlands